HANDBOOK

of

PEST

MANAGEMENT

BOOKS IN SOILS, PLANTS, AND THE ENVIRONMENT

HANDBOOK
of
PEST
MANAGEMENT

edited by
John R. Ruberson

University of Georgia
Tifton, Georgia

CRC Press
Taylor & Francis Group
Boca Raton London New York

CRC Press is an imprint of the
Taylor & Francis Group, an **informa** business

CRC Press
Taylor & Francis Group
6000 Broken Sound Parkway NW, Suite 300
Boca Raton, FL 33487-2742

© 1999 by Taylor & Francis Group, LLC
CRC Press is an imprint of Taylor & Francis Group, an Informa business

No claim to original U.S. Government works

ISBN-13: 978-0-8247-9433-0 (hbk)
ISBN-13: 978-0-367-39961-0 (pbk)

Visit the Taylor & Francis Web site at
http://www.taylorandfrancis.com

and the CRC Press Web site at
http://www.crcpress.com

Foreword

The conceptual basis for pest management has evolved during the latter half of this century, partially in response to the biological and environmental problems caused by the intensive use of chemical pesticides for crop protection. Basic to the concept is the idea that the agroecosystem may be managed in ways to hold pest numbers to tolerable levels. This has been accomplished by integrating the use of chemical, cultural, and biological control methods, including the use of pest-resistant crop varieties and, more recently, transgenic plants. For the plant protection specialist, pest management is seen as the best system for suppressing pests while minimizing the adverse impact of chemical pesticides on the agroecosystem, e.g., the development of pesticide-resistant strains of pests, resurgence of primary pests, secondary pest outbreaks, adverse impact on non-target organisms, and overall contamination of the environment. The farmer implements a pest management system to optimize crop yields and maximize profits. Both the plant protection specialist and the farmer want a pest management system that has minimum risk to crop production and is environmentally friendly and sustainable.

Although very efficient and highly productive systems for producing crops have been developed, pests still cause serious problems. In the United States, pest-inflicted losses to crop yields are estimated to average about 30% annually, despite the fact that most farmers are using the best pest control technology available. In the developing world, pest-inflicted losses are even greater, averaging 50% or more in many areas. Almost complete losses to yield are not uncommon for many subsistence farmers. Because pests have been able to de-

velop defenses against many of the best pest control tactics, including chemical pesticides and resistant varieties, new technology must be developed continually if crop losses are to be held to acceptable levels.

In a global context, pest-inflicted crop losses of this magnitude cannot be
tolerated indefinitely if the human population of the world is to have adequate
food. Approximately 90 million people are likely to be added to the world's
population over the next quarter of a century. Within 25 years the world's population may approach 8 billion people. Providing adequate food for these numbers will not be easy, especially if crop losses to pests cannot be substantially
reduced. This is the challenge to agricultural scientists and farmers.

The authors of the various chapters of the *Handbook of Pest Management*
describe the current status of knowledge of the many facets of pest management. They also provide considerable insight concerning future developments
and trends in crop protection. In doing this, the authors have touched on most
of the important issues related to the development of pest management systems
and the transfer of this technology to farmers. Each chapter includes extensive
references that will be of use to those who might want to conduct a more intensive review of the various topics presented.

The *Handbook of Pest Management* is an excellent compendium of chapters
related to the management of crop pests, including current information on all
classes of crop pests, e.g., invertebrates, pathogens, and vertebrates. The various
chapters broadly cover the field of crop protection and serve as an important
reference to anyone interested in the management of pests. The *Handbook* will
be a valuable addition to the library of any pest management specialist

Perry L. Adkisson
Distinguished Professor of Entomology
Texas A&M University
College Station, Texas

Preface

In 1526–31 the monks of Troyes formally excommunicated the caterpillars
that were plaguing the crops, but added that the interdict would be effective only for lands whose peasants had paid their Church tithes

W. Durant[1]

The human struggle against nature's ravages is a key element in the shaping of
who we are, what we are, what we eat and wear, and, in many cases, where
we live and the language we speak. The ravages of pests have caused extensive starvation, mass migrations, even key military defeats. The need to minimize or avoid the depredations of crop pests has led to a diversity of management schemes, some of questionable value (such as the 5-year program noted
above), and others that stand today as testimonies to the ingenuity and observational skills of our agricultural forebears. Our technology has indeed improved, but we have not eliminated the need to manage pests; on the contrary,
the need is now more acute than ever.

As we prepare to enter a new century and millennium, it is worthwhile to
reflect on the past and internalize its lessons before caroming into an unseen
future. Human population continues to grow at a staggering pace, while the
proportion of that population meeting the food and fiber needs of the remainder continues to dwindle—a greater burden will need to be shouldered by a declining segment of society. Likewise, energy reserves, genetic resources, and
available arable lands are not limitless and must be carefully guarded and utilized. Technological advances of the past century, such as mechanization technology, synthetic pesticides, and genetic engineering, portend greater and more
effective pest management options in the future. Nevertheless, the age-old

methods, such as crop rotation, irrigation, and timing of planting, will continue to play major roles in crop production and pest management. Our focus must be first on the health of the production system and the plants, and second on devising pest management approaches that maintain or enhance that health. The success of producers in the future may depend less on their ability than on their flexibility and adaptability.

Rarely, if ever, does any single practice applied to a crop have a unilateral effect on that crop. For example, the use of fertilizers influences plant health, which in turn alters the susceptibility of the plant to pathogen and arthropod attack. In considering the use of particular pest management tactics, it behooves us to consider each of the tactics as inputs in a functioning, complex ecosystem. In thinking in these terms, we consider the soil, water, nutrients, plants, microbial flora, and invertebrate and vertebrate faunas as part of a highly interactive system where alteration of any component will likely ripple out to other components. Application of such thinking requires greater integration and cooperation among disciplines—a concept with numerous disciples, but few practitioners.

The *Handbook of Pest Management* examines pest management in the context of an ecological production system. The first part (Chapters 1–4) places the agricultural system into its ecological context by examining the crop system as an ecosystem with interfacing biotic and abiotic components, some of which may range over considerable temporal and spatial scales. The second part (Chapters 5–9) considers the plant within the system, its inherent abilities to support beneficial organisms and resist pests, and possibilities for manipulation and improvement of plant defenses. The third part (Chapters 10–24) considers the biologies and ecologies of each of the major groups of plant pests in turn and considers various means for their management. Management methods are presented with the intent to inform the reader of the current status of these methods, rather than to function as a production guide, and also to demonstrate the need for more effort in many of these areas. The final part (Chapters 25–28) examines a set of critical issues that impinge on pest management and will affect future production practices. It is hoped that this volume will stimulate thinking and the growth of more holistic pest management strategies that recognize the intense interconnections within cropping systems, as well as the variable spatial and temporal scales over which pest populations subsist and function. The challenges facing agriculture are considerable and growing. This volume outlines many of these challenges relative to pest management, and points out gaps that must be filled to address these challenges and to move the technology into practical application.

The text is heavily biased toward plant production systems by my own choice. The addition of other systems would have added greatly to an already lengthy tome. Several important issues may be notable by their absence in chapter form (e.g., sustainable agriculture and management of pesticide susceptibility in pests). These topics are addressed in various chapters, and can be located in the index.

I would like to thank the many people who contributed to this effort. The efforts of the numerous authors who gave selflessly of their time and talents is deeply appreciated. I also appreciate the good-natured support of our departmental secretarial staff Jenny Nelms, Carol Ireland, and Elaine Belk. I had the privilege of working with an excellent editorial group at Marcel Dekker, Inc., including Tom Finnegan and Vivian Jao, and especially the pleasant and tenacious Jeanne McFadden. Above all, I appreciate the profound patience, support, and encouragement of a wonderful family, including my parents Jim and Joyce Ruberson, and especially my wife Mary Lu and our children Josh, Christina, Matthew, and Jonathan.

John R. Ruberson

REFERENCE

1. Durant, W. 1957. *The History of Civilization: Part IV. The Reformation*. New York: Simon and Schuster, p. 850.

Contents

PEST MANAGEMENT ISSUES

Contributors

George N. Agrios Professor and Chairman, Department of Plant Pathology, University of Florida, Gainesville, Florida

John N. All Professor, Department of Entomology, University of Georgia, Athens, Georgia

Shawn D. Askew Graduate Research Assistant, Crop Science Department, North Carolina State University, Raleigh, North Carolina

Paul A. Backman* Professor and Director, Department of Plant Pathology and Biological Control Institute, Auburn University, Auburn, Alabama

James L. Baker Professor, Department of Agricultural and Biosystems Engineering, Iowa State University, Ames, Iowa

Alan C. Bartlett Geneticist (Insects), Western Cotton Research Laboratory, Agricultural Research Service, United States Department of Agriculture, Phoenix, Arizona

David C. Bridges Professor, Department of Crop and Soil Sciences, University of Georgia, Griffin, Georgia

---

Current affiliation: Director of Pennsylvania Agricultural Experiment Station and Associate Dean for Research and Graduate Education, College of Agricultural Sciences, Penn State University, University Park, Pennsylvania.

William M. Brown, Jr. Professor, Department of Bioagricultural Sciences and Pest Management, Colorado State University, Fort Collins, Colorado

D. Steven Calhoun Associate Agronomist, Delta Research and Extension Center, Mississippi Agricultural and Forestry Experiment Station, Stoneville, Mississippi

James E. Carpenter Research Entomologist, Insect Biology and Population Management Research Laboratory, Agricultural Research Service, United States Department of Agriculture, Tifton, Georgia

Raghavan Charudattan Professor, Department of Plant Pathology, University of Florida, Gainesville, Florida

Marcel Dicke Uyttenboogaart-Eliasen Professor of Insect–Plant Relationships, Department of Entomology, Wageningen Agricultural University, Wageningen, The Netherlands

Richard A. Dolbeer Wildlife Biologist, APHIS, Wildlife Services, National Wildlife Research Center, United States Department of Agriculture, Sandusky, Ohio

Jerry B. Graves* Professor, Department of Entomology, Louisiana State University Agricultural Center, Baton Rouge, Louisiana

Ronald B. Hammond Associate Professor, Department of Entomology, Ohio Agricultural Research & Development Center, The Ohio State University, Wooster, Ohio

Leon G. Higley Professor, Department of Entomology, University of Nebraska–Lincoln, Lincoln, Nebraska

Dennis L. Isaacson Special Projects Coordinator, Oregon Department of Agriculture, Salem, Oregon

Alan L. Knight Research Entomologist, Agricultural Research Service, United States Department of Agriculture, Yakima, Washington

Marcos Kogan Director and Professor of Entomology, Integrated Plant Protection Center, Oregon State University, Corvallis, Oregon

Wolfram Köller Associate Professor, Department of Plant Pathology, Cornell University, Geneva, New York

---

*Retired.

Joseph A. Kuć* Professor, Department of Plant Pathology, University of Kentucky, Lexington, Kentucky

Douglas A. Landis Associate Professor, Department of Entomology, Michigan State University, East Lansing, Michigan

George W. Langdale* Agricultural Research Service, United States Department of Agriculture, Watkinsville, Georgia

John D. Lattin Professor, Department of Entomology, Oregon State University, Corvallis, Oregon

B. Rogers Leonard Associate Professor, Northeast Research Station, Louisiana State University Agricultural Center, Baton Rouge, Louisiana

Craig M. Liddell Associate Professor, Department of Entomology, Plant Pathology and Weed Science, New Mexico State University, Las Cruces, New Mexico

Paul C. Marino Assistant Professor, Department of Biology, University of Charleston, Charleston, South Carolina

Robert L. Metcalf† Emeritus Professor, Department of Entomology, University of Illinois, Urbana-Champaign, Illinois

James R. Nechols Professor, Department of Entomology, Kansas State University, Manhattan, Kansas

James H. Oard Associate Professor, Department of Agronomy, Louisiana State University Agricultural Center, Baton Rouge, Louisiana

James A. Ottea Associate Professor, Department of Entomology, Louisiana State University Agricultural Center, Baton Rouge, Louisiana

David E. Pedgley* Research Meteorologist, Natural Resources Institute, University of Greenwich, Chatham Maritime, England

---

*Retired.
†Deceased.

Larry P. Pedigo Professor, Department of Entomology, Iowa State University, Ames, Iowa

John R. Ruberson Assistant Professor, Department of Entomology, University of Georgia, Tifton, Georgia

Howard F. Schwartz Professor, Department of Bioagricultural Sciences and Pest Management, Colorado State University, Fort Collins, Colorado

Linnea G. Skoglund Extension Plant Clinic Specialist, Department of Bioagricultural Sciences and Pest Management, Colorado State University, Fort Collins, Colorado

Benjamin R. Stinner Professor, Department of Entomology, Ohio Agricultural Research & Development Center, The Ohio State University, Wooster, Ohio

Norman E. Strobel* Postdoctoral Research Associate, Department of Plant Pathology, University of Kentucky, Lexington, Kentucky

Maurice J. Tauber Professor, Department of Entomology, Cornell University, Ithaca, New York

William K. Vencill Associate Professor of Weed Science, Department of Crop and Soil Sciences, University of Georgia, Athens, Georgia

Thomas J. Weissling[†] Postdoctoral Research Associate, Agricultural Research Service, United States Department of Agriculture, Yakima, Washington

John W. Wilcut Associate Professor, Crop Science Department, North Carolina State University, Raleigh, North Carolina

Mark Wilson Assistant Professor, Department of Plant Pathology, Auburn University, Auburn, Alabama

---

Current affiliations:
*Instructor, Division of Biological Sciences and Nursing, Lexington Community College, Lexington, Kentucky.
[†]Assistant Professor of Entomology, Ft. Lauderdale Research and Education Center, University of Florida, Ft. Lauderdale, Florida.

Richard L. Wilson Research Entomologist, Agricultural Research Service, North Central Regional Plant Introduction Station, United States Department of Agriculture, Iowa State University, Ames, Iowa

Billy R. Wiseman* Research Entomologist, Insect Biology & Population Management Research Laboratory, Agriculture Research Service, United States Department of Agriculture, Tifton, Georgia

Frank G. Zalom Director, Statewide Integrated Pest Management Project, and Department of Entomology, University of California, Davis, California

---

*Retired.

1

Agricultural Systems as Ecosystems

Marcos Kogan and John D. Lattin
Oregon State University, Corvallis, Oregon

I. ECOSYSTEMS: THE EVOLUTION OF A CONCEPT

Sir Arthur Tansley [87] provided the first formal definition of an ecosystem, although the concept's roots emerged many years earlier [55]. Tansley observed that, ". . . the more fundamental conception is . . . the whole system (in the sense of physics), including not only the organism-complex, but also the whole complex of physical factors forming what we call the environment of the biome— the habitat factors in the widest sense." Besides including the environment's physical and biological components, he also suggested the important impact of cultural activities upon ecosystems, noting that, "we cannot confine ourselves to the so-called 'natural' entities and ignore the processes and expressions of vegetation now so abundantly provided us by the activities of man." Thus, Tansley's original consideration of an ecosystem's myriad components provides a backdrop to the topic at hand: agricultural systems as ecosystems.

The degree of conceptual vagueness increases markedly along the hierarchical ecological scale from the population to the ecosystem [67]. There is little disagreement on what constitutes a population, a species, or even a community, but ecosystems have been defined within spatial scales ranging from a single log rotting on the ground to the entire forest, and within temporal scales extending from hours to centuries. Evans [17] argued for an all-inclusive concept of ecosystems—applied to any level of ecological organization—from the individual to the biosphere. This expansion of the concept may have influenced application of the term to minimal scale systems. Yet, ecologists have developed a consensual perception of ecosystems, even while continuing to debate spatial and temporal boundaries, qualitative and quantitative restrictions, and precise definitions.

Agricultural systems exhibit some spatial and temporal parallels with undisturbed ecosystems. However, whether agricultural systems are viewed as ecosystems or not depends, in part, upon individual perspective. Throughout the chapter, we follow the evolutionary path of human foraging: from hunting/gathering through modern industrialized agriculture. Along this path we observe humans gradually taking over regulation of ecosystem functions as well as the direction of ecosystem structure and driving agricultural systems ever so farther away from the natural systems that they displaced, raising the question of the long-term sustainability of modern agroecosystems. As we discuss the issue of sustainability, we introduce the notion of "ecological distance"—a qualitative measure of the difference between the crop community and the natural community that preexisted in the same region. After considering the multiple human impacts on natural ecosystems, we readdress at the end of the chapter the issue of the consideration of agricultural systems as ecosystems.

There is a vast literature on ecosystems as systems including many useful reviews that elaborate on the scientific developments of Tansley's basic ideas. The richness of ecosystems literature is not surprising, since the origins of this area of science are intimately entwined with the development of ecology itself. This literature provides the theoretical background for the ecological approach to agroecosystems analysis and management. Suggested sources for review include Allen and Hoekstra [1], Altieri [2], Bayliss-Smith [5], Costanza et al. [12], Evans [17], Hagen [27], Gliessman [25], Golley [26], Lowrance et al. [52], Major [55], May [56], McIntosh [57], O'Neill et al. [64], Roughgarden et al. [77], and Risser [75] and the references they contain.

A. Agricultural Systems: Ecological and Social/Economic Interactions

For the purpose of this discussion, Tansley's original concept is adopted whereby an ecosystem is: (1) the basic unit of nature on earth, (2) composed of both organisms and environmental physical factors, and (3) an element in a hierarchy of physical systems ranging from the universe as a whole to the atom [26,87]. A brief and yet inclusive definition provides that ecosystems are: "functional systems of complementary relations between living organisms and their environment, delimited by arbitrarily defined boundaries, which in space and time appear to maintain a steady yet dynamic equilibrium" [25]. A clear understanding of the characteristics, processes, and dynamics of natural ecosystems were deemed essential for the scientific analysis of agroecosystems [75], with the major determinant of the degree of differentiation from natural ecosystems being the level of human impact and control [35,61]. Thus, the transition from natural ecosystems to agroecosystems progressively incorporates interactions of three distinct systems: the ecological, the social/economic, and the agricultural.

A representation of the hierarchical relationships among those systems is provided in Figures 1 and 2 [44]. The traditional hierarchy of ecological systems [19] starts with the individual, increases in level of complexity, and expands temporally and spatially to the population, community, ecosystem, and the whole biosphere. By analogy, human social systems similarly increase in level of complexity and expand spatially and temporally from the individual to the household or expanded family, the farm or village, the county or province, and the country or continent. A parallel hierarchy is identified in agricultural systems which encompass crop plants or domestic animals, fields or herds, crop communities, agroecosystems, and regional production systems. We suggest, along with Altieri [2] and other authors writing on agroecology and sustainable agriculture, that agricultural systems are the dynamic end result of the interactions, in ecological time, of ecological and social/economic systems (see Fig. 2). The specific crops that can be grown successfully in a given region are defined primarily by the kinds of crop genotype interactions with the environ-

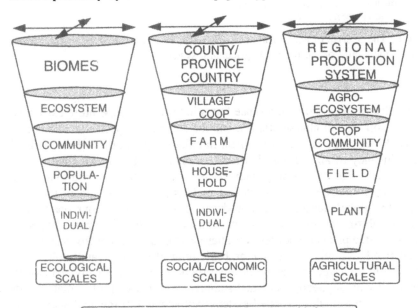

Figure 1 Hierarchical sequence of ecological, social/economic, and agricultural systems. Level of organizational complexity and spatial dimension increases from bottom to top.

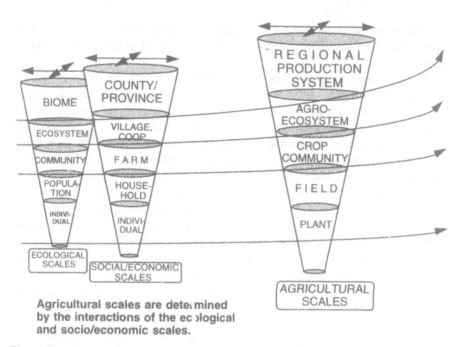

Figure 2 The overlap of ecological and social/economic systems of the various scales determines the nature of the regional agricultural system.

ment that determine a crop's tolerance limits in terms of temperature extremes, water availability, soil characteristics, and the complex of pests that can potentially attack the crop. Success of a plant in colonizing an area will be defined by the stringency of any of the vital requirements that are not satisfied.

For example, coffee is a crop with species and varieties adapted to a range of conditions within the world's tropical regions [48]. Expansion of the crop in Brazil, the globe's largest coffee producer, pushed coffee plantations farther south from subtropical areas to semitemperate zones afflicted by periodical killing frosts. Coffee plants do not recover after years of subsequent frosts and have to be replanted. Having reached the probable limit of its temperature tolerance, the crop could no longer by economically produced and was finally abandoned and replaced with soybean, an annual summer crop that could be rotated with winter cereals in a double cropped system well adapted to the ecological conditions of the region [38]. Introduction of temperate zone crops into the tropics commonly meets with similar problems of hostile physical environmental conditions, often aggravated by intensified pest-induced stresses. Soybean is a good example. A crop that originated in Northeastern China, soybean grows relatively free of serious insect pests in the temperate zones of North and South America, but it suffers severe infestations by a diverse insect pest fauna when planted under the long-season conditions prevailing in subtropical zones [46].

Human societies evolved under the influence of the same environmental pressures which, in the dynamics of human/physical environment interactions, helped shape the social and cultural character of local populations. Acquired food preferences, imposition of dietary laws and requirements, labor force characteristics, land ownership, transportation, and mechanization are some of the components that determine which crops are grown and how they are grown and used. The overlap of those environmental determinants—the scales of the ecological system—and the cultural/sociological character of the local population determine the nature of the agricultural system that prevails in a given region.

Complex interactions evolved in historical times, often through the influence of factors external to the local ecological/social systems [35,54]. Although current globalization of trade has greatly exacerbated the impact of external influences, some basic relationships continue to hold. The industrial, highly mechanized corn/soybean system established in the midwestern United States resulted from the overlap of a physical environment that contains nearly perfect rainfall distribution and temperature regimens during the summer, a fertile soil, and an ideal topography with a farmer population that has the cultural sophistication necessary to operate the complex farm machinery and make the marketing decisions that help maximize a stable income under very competitive world trade conditions. The original grass prairie that covered much of the midwestern United States was largely replaced with a corn/soybean production system that can be sustained only because of an efficient infrastructure for moving crops to the world markets and a pricing system that protects farmers from wild market fluctuations. This model for ecological/social interactions in the shaping of agroecosystems will be revisited in the discussion of integrated pest management (IPM) within the context of the total agroecosystem management.

II. DEVELOPMENT OF THE AGRICULTURAL LANDSCAPE

The world's estimated total land area is about 13.04 billion hectares. About 10% of the total, 1.44 billion ha, is arable land and land planted to permanent crops. Permanent pastures cover another 3.36 billion ha [21]. The global land mass must sustain a human population, estimated at 5,479,758,000 in 1992, growing at a rate of about 2% per year. Population growth exerts enormous pressure on the world's agricultural systems to satisfy food, fiber, and nonfossil fuels needs. The same pressure tends to intensify deforestation of the globe's remaining 3.86 billion ha of forest and woodlands, accelerate desertification of fragile environments, and force into agricultural production semiarid lands and wetlands that are at best marginal for sustainable crop production but are critical for the preservation of biodiversity in these fragile environments.

In spite of increasing reliance on agricultural production, the percentage of the population directly involved in agriculture has declined continuously throughout this century, requiring the remaining farmers to extract the maximum possible yields from their lands (Fig. 3). Combined pressures on these farmers to produce more for an ever-growing human population has significantly accelerated adoption of technologies that are decreasingly labor intensive but increasingly energy and natural resource demanding (see refs. 4, 18, 59, and 88).

The evolution of human societies has been characterized by steadily increasing aggressiveness toward the landscape. A corollary of more drastic disturbance of an area's native ecology seems to be greater challenges to sustainable crop production, including enhanced severity of pest problems. A brief summary of the major trends in cropping systems evolution supplies the backdrop for a discussion on human/crop interactions.

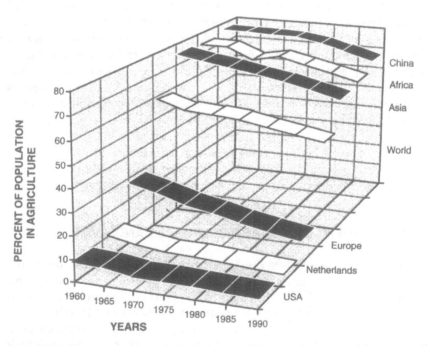

Figure 3 Steady decreases in the percentage of the human population dedicated to agriculture in the world and in representative countries or regions.

A. Lands in Transition

The gradual increase in early human populations altered movement patterns of families and tribes and began to change their feeding habits. Most early peoples were basically nomadic hunter-gatherers. Others tended to locate at sites where resources were concentrated, especially along waterways and coastal regions, where harvests of local foodstuffs—fish, seafood, other game, seeds, fruits, roots, grubs, insect products (honey), and insects themselves [7]—were consistently abundant. These early groups became increasingly sedentary. Those lands could accommodate the migratory habits of low population levels and incipient agriculture, while hunting continued to provide foodstuffs to supplement the diet (see the trilinear theory of the origin of village agriculture, ref. 54).

Fires, often a by-product of human habitation, became a more common practice and helped modify landscapes, except grasslands where fire already was a natural periodic disturbance factor. Concentrated food gathering must have caused depletion of some local foodstuffs, perhaps triggering movement away from established population centers when resource reserves approached critically low levels. The transition from hunting-gathering to sedentary food production seems to have been a gradual process [6]. Throughout history, food availability determined the flux of populations and often caused massive migrations. Bender [6] provides an analysis of hypotheses relating demographic patterns with migration. Evidence is not at all conclusive, but the local build-up of populations beyond the carrying capacity of the region was likely to exert pressure on movements of families and tribes.

These mass migrations temporarily relieved local population pressures, but often led to exploitation of new regions and destruction of additional natural habitats. The same trend appears to have continued and sharply intensified through the centuries and has more recently been greatly aggravated by demand pressure for goods not always connected with essential foodstuffs.

An example is the cycle of sugarcane in Brazil. Although an important high-energy food, sugar could not be considered a staple. After the discovery of Brazil by European navigators, the cycle of sugarcane in Brazil's northeastern states provides a sobering warning of unsustainable agricultural development. Sugarcane production followed a pattern of slash and burn of the original forests of the coastal "zona da mata," a band of tropical forests along the littoral that stretched from the fringes of the Amazon through São Paulo and Paraná states. Sugarcane continued to be planted as a monocrop until the soil nutrient reserves were virtually depleted. Sugarcane plantations still remain, albeit yielding much less than the comparable plantations of São Paulo, but vast acreage was gradually replaced with pasture for cattle and the eventual desertification of large areas reduced agriculture to meager subsistence, causing massive migrations in the last 100 years. Attempts to recover the area involve large irrigation projects that further impact on the ecology of the region [91].

B. Evolution of Cropping Systems

Theories of agriculture's origins are based on archeological findings, particularly seed, pollen, and clay and stone artifacts, and on the ethnographic study of contemporary hunting-gathering cultures [6,30,72]. The theoretical presuppositions advanced by Sauer [80], supported by Carter [9] and others, attempted to locate the cradle of agriculture; that is, one or a few centers of origin whence agriculture diffused to other regions. These hypotheses were reviewed by Harlan [30], who offered the view that the process of cultivating plants for food or fiber likely evolved simultaneously at multiple locations over a period of more than 14,000 years. Harlan [31] argued that the origins of agriculture were diffuse, as opposed to the belief that agriculture—a complex step in the evolution of human societies—was probably the result of a single invention that occurred in one region and diffused to the rest of the world. According to Harlan, then, rather than the diffusion of agriculture from one or a few points to the rest of the world, agriculture's origins were widely distributed in many centers and reinvented by numerous diverse human cultures and societies.

For this discussion, the relevant thrust is not so much how agriculture came to being but rather the dramatic changes that have occurred in the past 500 years on all continents. During these five centuries there were massive introductions of new crops into developing regions of the New World and importation of New World crops into Europe and Asia, regions already possessing well-developed agricultural systems. These crop introductions represented intentional, extensive biological invasions that brought about significant changes in the landscapes of all continents. Transposed crops were exposed to new pests in the new environments and, over time, pests themselves made transcontinental voyages, expanding their geographical range and further compounding agriculture's ecological impacts on natural systems. Along with crops, weeds were intentionally or unintentionally introduced. Many of these weed species were strong competitors that displaced species of the local flora, altering the landscape even beyond the alterations caused by the crops [13].

As mentioned earlier, in the American Midwest, the original prairie grassland gave way to the present broad-scale corn/soybean cropping system: Corn was a New World crop and soybean an import from northeastern China. The commercially successful corn/soybean cropping system continues to evolve even in the immediate era. Both elements of this crop combination are actively traded commodities in world markets, and they thus are agents of drastic social changes. Small farms are gradually being replaced by large farming enterprises, and small rural communities are fading away or coalescing around major urban centers.

Depending on a region's ecological characteristics and the nature of the cropping systems injected into it, resulting outcomes can range from more or less benign to catastrophic. Changes in the tropics have repeatedly resulted in irreversible destruction of natural habitats, gradual decline of subsistence farming, and steadily increased mechanization of farming operations. In turn, these shifts lead to displaced rural populations, declining numbers of small farms, and the rise of farming enterprises predominantly relying on vast monocultures. Evidence suggests that the greater the difference between the characteristics of a cropping system and the natural ecosystem preexisting in the area of introduction, the more severe the ecological disturbance and, consequently, the greater the instability of the cropping system, particularly as it relates to pests. This is the "ecological distance" concept discussed in the next section.

C. Natural Succession, Climax, and Ecological Distance

Ecological communities are dynamic assemblages of organisms that are more or less interconnected through food chains and nutrient cycles that tend to dominate a given environment if undisturbed over relatively long time periods. A disturbed area is colonized by a succession of species, with early successional colonizers being gradually replaced by other species until the community reaches a level, the climax, at which colonization rates and extinction rates are at a minimum. Thus, according to Ricklefs [73], succession represents the local replacement of populations in a regular, albeit not always predictable, progression to a climax stage.

Disturbance of the ecosystem takes many forms. Fire, accidental or manmade, may occur at such an intensity that it removes most vegetation, including trees, leaving behind some enriched materials in the ashes [92], although the soil biota are usually reduced if the upper layers are severely scorched. Such disturbances create open areas of land that permit the planting of desired crops. It seems to be no accident that slash and burn systems are still used in tropical agriculture. Perhaps mimicking the aftermath of a lightning strike or other stand-replacing natural fires, fire-induced disturbances set in motion the process of succession, with its parade of plant species and other life forms, some of them potentially desirable foodstuffs and fiber sources. Ecosystems with relatively short turnover rates might even have provided the stimulus for imitating the results of natural fire, creating an artificial succession to provide fairly predictable foodstuffs rather than waiting for a natural event to occur.

Uncontrolled floods would have enriched riparian zones, resulting in considerable biomass production, including the production of foodstuffs. Predictable flooding regimens, such as that of the Nile River, would have been observed and, if predicted, perhaps would have allowed some degree of planning to take advantage of the soil renewal and water. Controlled flooding may have been the precursor of modern irrigation systems. The nature of the desired food resource would influence any effort to duplicate events that result from naturally occurring disturbances.

Disturbance is the norm in agroecosystems and human intervention has been the most powerful force in shaping agroecosystems, as agriculture has been one of the main sources of natural ecosystems disturbance on a global scale. The evolution from hunting-gathering through early agriculture to present-day native shifting agriculture, permanent subsistence agriculture, and finally industrial-scale agriculture has often led to a drastic departure from the natural climax communities that once existed in the same areas. As suggested earlier, the greater the disparity between the original climactic community and the crop community that replaced it—the *ecological distance* between the natural ecosystem and the introduced agroecosystem—the higher the vulnerability of the agroecosystem to a variety of biotic and abiotic antagonistic factors, including those represented by pest organisms. This concept still awaits rigorous testing to become an operational hypothesis [67]. Indicators of sustainability, both biological and economic [20,28,65], could be used in testing the hypotheses raised by the concept of ecological distance. At the moment, the concept is intuitive, but the following observational evidence may provide preliminary evidence for its operational value.

In South America's humid tropics, shifting agriculture practiced by local populations may exploit forest resources (nuts, roots, fruits), open small patches of land to plant cassava, corn, beans, and possibly vegetables and to raise a few domestic animals. As the forest is cut to make room for field crops, the fragmentation of the original ecosystem is intensified and the new ecosystem remains indefinitely immature, never reaching a new climax, in what becomes a much simplified agroecosystem. The natural forces that contribute to the dynamics of ecological succession, climate, precipitation, the habitat-modifying force of the pioneer vegetation, and grazing [73] continue to operate, placing the agroecosystem under constant pressure of those successional forces. Agroecosystems that are structurally analogous to the original climax tend to be relatively stable, but those that greatly depart from that structure are usually vulnerable to intense abiotic and biotic environmental pressures, including among the latter, weed competition and susceptibility to pathogens and herbivorous insects. These latter systems have differed from the original ecosystems by a "large" ecological distance, conceivably measurable by means of comparative efficiencies of nutrient cycling, energy capture and conversion, biomass productivity, preservation of biodiversity, and the ability to moderate demographic imbalances that result in outbreaks of pest populations.

Examples within both extremes of the ecological distance continuum can be found in both the North American Grand Prairie and the South American Amazon forest. The forest/grassland boundaries in the north-central states of the United States have been variously attributed to the differential tolerance of tree seedlings to fire, to competition for water and nutrients between grass and tree seedlings, and to harsh climatic factors, particularly periodic droughts that kill most young trees growing from seed [47]. In tall-grass prairies, ecosystem function appears to be dominated by nonequilibrium biotic responses to changes in resource limitation [81]. Productivity in these ecosystems is higher during transitional periods when resources are abundant rather than during an equilibrium interval when resources are scarce. These ecosystems are characterized by transient shifts due to unpredictable climate, grazing patterns, and recurrent fires which modify the availability of water, nitrogen, and light.

The prairies occupy a vast area of the middle third of the North American continent. The great prairie of the Midwest is dominated by grasses, both C_3 and C_4 species, legumes, and composites, with the relative dominance of certain species varying with local climatic conditions. Grasses and forbes are hardy and have evolved to tolerate extremes of temperature, periodic droughts, fire, and extensive grazing. For many of the prairie plants, survival depends on extensive subterranean bud and rhizomatous growth [84]. A large portion of the original prairie ecosystem has been replaced by agroecosystems dominated by annual Poaceae and Fabaceae, especially corn, soybean, and small grains. Although fire is not a factor in these agroecosystems, mechanical tillage (by humans) performs a comparable function by removing the crop residue from the surface, disturbing the soil biota—at least in the upper layer of the soil, and promoting germination of dormant weed seeds from the soil seed bank.

In both their basic botanical composition and functional dynamics, the small grain, corn, and soybean ecosystems of the American Midwest are within a short ecological distance of the prairie that preceded them. These systems are remarkably resilient and, if properly managed, can withstand considerable grazing by herbivorous arthropods before productivity is significantly affected. These agroecosystems represent a close match with the original climax, and they are reasonably stable, albeit entirely dependent on human management and high-energy inputs. A model agroecosystem has been proposed for the Midwest [84] that more closely follows the prairie biome structure. It departs radically from the present corn/soybean system and includes perennial grasses, legumes, and composites as seed crops, but none of those currently grown in large-scale monocrops, which include soybean, corn, sorghum, sunflowers, and small grains. The proponents of this system make a strong case for the ecological advantages of the system on a local, regional, and global scale, but they do not provide a socioeconomic impact assessment of their proposal if it is widely adopted.

The Amazon basin, or Amazonia, covers a total are of 756.6 million ha primarily in Brazil and also in eight other South American nations. That land mass is divided into dryland forest (547.3 million ha), wetland forest (65.5

million ha), dry savanna (88.0 million ha), and herbaceous wetland (27.7 million ha). Most of the deforestation is taking place in the dryland forest [16]. For centuries, Amazonia defied human colonization, at least as conceived by Western civilizations, although a considerable population of well-adapted Amerindians lived in harmony with the forest, exploiting its natural resources without significantly or permanently scarring the landscape. The inability to establish a continuing agricultural system in the area was variously ascribed to the incompetence of the population of potential colonizers and the hostile environmental conditions [16]. Most early attempts to establish plantation-type enterprises failed. Fordlandia and Belterra were two colonies established in the late 1920s and early 1930s to produce rubber in plantations of the *Hevea brasiliensis* (rubber tree) established in cleared patches of forest. Commercial production in the colonies was abandoned because of economic, social, and ecological problems. Although the rubber tree is native to the area, when the forest was cleared and the trees planted to a nearly monocrop stand, disease and insect pest problems erupted, seriously reducing the yield potential. Almost 60 years after these experiments, 90% of the rubber produced in the Brazilian Amazonia is extracted from dispersed native stands, just as was done at the turn of the century [3]. The intensive deforestation that proceeded at a rate of 3.6 million ha per year from 1981 through 1990 [94] destroyed about 40 million ha, or 8% of the Brazilian Amazon forest. Cleared land is used mostly for conversion to pastures to raise cattle [40], although in the first few years after clearing the land might be used to produce crops such as rice and cassava. The losses in carbon, nitrogen, and sulfur through the combustion process used for clearing is dramatic, ranging from 50 to 70% of the total aboveground nutrients [39]. These initial losses are compounded over time as the clearing of the forest impacts on rainfall, wind patterns, and average temperatures, and patterns of land use further deplete the residual nutrient reserves. The result is a vastly impoverished ecosystem where before existed some of the organically richest system on earth. The ecological distance between the climactic tropical forest and any agropastoral system is extreme, and the ecological consequences of replacing the forest with field crops and pastures are likely to be disastrous. Invariably, ecological disasters are followed very shortly by equally disastrous economic and social consequences.

D. Role of Scale and Fragmentation on the Agricultural Landscape

An undisturbed landscape presents a mosaic of habitats and ecosystems whose boundaries were determined by a variety of abiotic and biotic factors. At close range, these assemblages of plants and animals often display further recognizable components, themselves forming a mosaic. Viewed from greater distances, the once clearly defined components blur and combine into large units (e.g., forests or grasslands). Thus, the spatial scale at which observations are made, or at which components are recognized or defined, is important [71].

Disturbance, both natural and human caused, is an important element in creating habitat fragmentation that ultimately determines the character of the landscape at different spatial scales. Bark beetles killing several trees in a forest may result in a small gap in the trees visible at close range, but invisible from some distance above the forest canopy. Stand-replacement fires—a major disturbance factor in the Cascade Mountains of the U.S. Pacific Northwest—may alter large portions of the forests and exert an influence on major sectors of the area by initiating succession.

Early agriculturists cleared patches of land or gaps in the vegetation for cultivation; these represented a type of habitat fragmentation paralleling gap formations resulting from natural causes. Multiple small patches in the landscape have a much larger ratio of perimeter (edge) to area than a comparable area within a single large patch. These small cleared areas were vulnerable to reinvasion by the original vegetation and required considerable labor to maintain for agricultural uses. A benefit of a vast edge network is the availability of potential natural enemies that would closely follow the pest colonizers [36]. At greater spatial scales, these early clearings in the landscape would disappear into the clumped vegetative patterns more discernible by their growth form than their species composition. Temporally, such patches would soon be obliterated by the invasion of the surrounding biota if the patches were abandoned.

As agriculture extended its development and influence across the landscape of the world, in many places, the patches of habitats disturbed by agricultural practices became the "landscape" and the original natural landscape remnants reverted to patches. Such landscapes are readily seen from an airplane flying over the intensely cultivated regions of the world. Different impressions of this landscape mosaic are joined at different spatial scales. Historical reconstruction of habitat change will disclose the temporal component of landscape alteration caused by increased agricultural activity. Now, small patches of relatively undisturbed habitats are expected to maintain and provide a variety of beneficial organisms (e.g., insect parasitoids and predators; see Chapter 4) to the agricultural systems. There is considerable advantage in establishing optimal ratios of edge to interior and in ascertaining the role of edges in the dynamics of natural enemy/pest interactions under an IPM system (see Chapter 4). The issue, however, is far from being settled on theoretical grounds [37] and much more needs to be done experimentally. It is unreasonable to expect small patches of undisturbed habitats, often randomly distributed across the landscape, to provide an adequate source of natural control agents to a vast surrounding agricultural landscape. Habitat management, including the manipulation of field edges, is a useful technique for the enhancement of natural enemies [15,34,74].

There is a sharp contrast between the large fields of crops, as commonly found in the United States, which present edges containing very limited habitat diversity and the mosaic of smaller fields usually surrounded by complex hedgerows or adjacent seminatural fields and forests as found in agricultural systems of small holders in Central and South America and some parts of Europe (e.g., Hungary). These latter systems have much greater potential for harboring beneficial organisms—invertebrate and vertebrate alike. The ecological distance between the main agricultural matrix and the surrounding edges of managed, or natural, original vegetation may be a crucial factor in the role that patches play in the total agricultural landscape at every scale.

III. HUMAN ROLES IN MANAGED ECOSYSTEMS

The extent of human intervention in agroecosystem management ranges from gentle practices, such as those of the swidden agriculture based on corn, beans, and squash in the dry tropical areas of Mexico [2], to the total control of all abiotic and biotic factors, as found in hydroponic cultures under greenhouse environments. Along this spectrum, humans have assumed control over all ecosystem functions and structures. While gaining control over those ecosystem components, most humans have grown totally dependent for their survival on ready access to sources of raw or processed food supplies strategically concentrated in all urban areas. Modern technologies in food production, storage, preservation, distribution, and marketing have removed increasingly larger segments of the population from direct food production processes. This simultaneous control over agroecosystem functions and dependency from agroecosystem outputs render humans highly vulnerable to elements that might upset those controls and outputs.

A. Human/Crop Plant or Livestock Mutualism

In the process of domesticating plants and animals for food, fiber, protection, or work, a close mutualistic relationship developed between humans with their domesticates in which neither organism in the relationship probably would survive for long in the absence of the other [41]. Expressions of this mutualistic relationship are the great famines that historically have befallen entire populations following agricultural catastrophes. Perhaps one of the best documented examples, the outbreak of the potato blight in Europe in 1785, caused widespread famine and relocation of entire populations [79]. With a human population growing by about 100 million per year [94], no natural food supplies are capable of meeting the needs of populations that are increasingly urban and incapable of fending for themselves (if supermarkets and grocery stores were suddenly depleted). The result of a global debacle in food supply today would probably spare only remote pockets of rural populations together with members of small tribes that still preserve a close relationship to nature.

Dependency on domesticated plants and animals has led humans to breed cultivars and races that are virtually deprived of any innate ability to out compete other early successional plants, defend themselves against insect pests and diseases, or even to reproduce without human assistance. Humans were thus forced to assume multiple roles in the artificial ecosystems that modern civilization created. Virtually all ecosystem functions are modified and managed to create agroecosystems that streamline energy flow, nutrient cycling, and primary productivity. In this context, humans became direct or indirect agents of nutrient cycling, agents of artificial selection redirecting the evolution of plants and animals, agents of pest population regulation, and moderators of abiotic factors, along with their overwhelming role as primary consumers. Contemporary agroecosystems' extreme dependency upon human assistance is the raison d'être for IPM.

B. Agents of Nutrient Cycling

Under natural conditions, nutrient recycling within ecosystems usually occurs through local processes that decompose organic matter and imports from the atmosphere either through biological nitrogen (N) fixation or the electrochemical results of lightning. Agroecosystems, in turn, depend on nutrient subsidies to maintain adequate productivity levels. A wide variety of fertilizers often replace, or at least augment, locally recycled nutrients in highly managed systems. Under reduced or no-till systems, nutrient cycling may more closely resemble the process that occurs in natural ecosystems; however, human intervention still is essential for optimal yields, because grain harvest removes a major proportion of the accumulated nutrients from the land. For example, harvested soybeans contain about 65% of the total N accumulated by the plant during the season [29]. In the soybean case, the N-fixing capacity of the plant makes up for the deficit that exists by the end of the season and no N subsidies are needed for sustainable economic production. Grain harvest of nonleguminous crops, however, leaves a deficit of nutrient reserves in the soil for the next year's crop even if the stubble is incorporated. In a reduced or no-till system, the primary nutrient sources may be of uniform origin when a single crop species is involved. Such nondiverse plant cover may be deficient in certain nutrients, a situation not found in natural systems where a variety of nutrient sources are associated with diverse plant and perhaps animal species [11,33].

C. Agents of Artificial Selection and Genotype Preservation (Evolutionary Forces)

Natural ecosystems are the contemporary, apparent steady state of complex evolutionary and coevolutionary processes in which competition and selection are major driving forces. In agroecosystems, selection and competition are regulated and modulated; through the techniques of induced mutation and genetic engineering, even the direction and rate of evolution of crops and domestic animals are under human control. These interventions have primarily focused on increasing grain and fiber yields in annual row crops, fruit and vegetable yield, or improving overall cosmetic appearance, color, taste, odor, and texture. They have been only secondarily concerned with increasing crop plants' resistance first to plant pathogens and then to insect pests. Often, however, in the process of selecting for yield, appearance, and taste, plant breeders have eliminated genes for pest resistance and released advanced cultivars that are highly susceptible to insect herbivores and pathogens [43]. The near disaster resulting from the southern corn blight outbreak in 1970 is a frightening example of an unintentional result from releasing hybrid corn selections designed mostly to maximize yields with little information on their susceptibility to potential pests [89].

One highly desirable intervention is the action to protect and preserve genomes through germplasm collections (see Chap. 9). Well-managed collections are dynamic programs directed toward saving endangered genotypes and preserving them under optimal conditions. Scientists travel the world collecting seeds or cuttings of landraces, wild relatives of cultivars, and old varieties. These samples are preserved in central locations and made available to breeders starting new programs, including those aimed at restoring resistance to insect pests and diseases in commercial varieties. These collections, available for most known crops, represent an intelligent investment in the future if a viable agricultural production system is to be maintained [32].

D. Primary Consumers

The primary productivity of most First World agroecosystems has been maximized through agronomical practices based on the use of superior seed, lavish subsidies of fertilizers, irrigation water as needed, and the profligate use of pesticides. The food chain in those agroecosystems has been reduced to a single primary consumer—humans—whose preponderance over potential competing herbivores or predators (of domestic animals) allows for only a small proportion of the crop to be shared. Intolerance for competitors is the main reason for pest control. Thus humans are the main, if not the sole, regulators of the primary producer species in agroecosystems.

These systems involve planting selectively to an optimal density that maximizes light interception and nutrient utilization and harvesting at the time of maximum concentration of the target plant part to be used: root or tuber in potatoes, yam and cassava; stems in sugarcane; foliage in most green salad crops and tobacco; flowers in pyrethrum; fruits in many berry, pome, citrus, and vegetable solanaceous crops; seeds in wheat, rice, beans, and other grains; and seed structures in cotton. An important characteristic of humans as omnivorous primary consumers is the wasteful utilization of food resources, particularly among the more affluent societies. In general, primary consumers in nature are selective feeders and consumption levels are determined by organismal needs, resource availability, and competition. Human consumption rates follow cultural patterns allowed by economic capacity. Thus, food resources are unevenly distributed among the various human populations; consumption in some populations far exceeds organic needs, although in others, it is marginally sufficient to sustain life.

E. Predators—Pest Control as a Density-Dependent Force in Herbivore Population Regulation

Hunting all sorts of vertebrate animals and entomophagy have provided important sources of protein for the diet of many people. Human consumption of insects—entomophagy—is a common practice in Africa, Asia, and among tribes of South American Indians [7]. These, of course, are genuine predaceous roles of humans in nature. However, it is the role of humans as regulators of herbivore populations in agroecosystems that provides the base for the metaphor expounded in this discussion. Modern pesticides, particularly insecticides and rodenticides, are the weapons of choice in human efforts to function as the ultimate predator of invertebrate and vertebrate herbivores in agroecosystems. Economic injury levels trigger a treatment in a direct, density-dependent fashion analogous to the functional response of predator to prey density. This role probably explains the success of insecticides that permit humans to perform equally efficiently at high or low "prey" densities, unlike natural predators that sometimes lag behind exploding prey populations and thus fail to contain the herbivores before they can remove a significant (i.e., economically unacceptable) fraction of the crop.

F. Moderators of Abiotic Forces

Abiotic forces (e.g., climate, atmospheric phenomena, topography, and edaphic factors) exert similar pressures on agroecosystems and natural ecosystems. however, the high biotic diversity characteristic of most natural systems, compared with agroecosystems, may attenuate the local impact of those forces. Natural ecosystems impacted by extremes of temperature, wind, or precipitation, for instance, may have some plant species reduced to dangerously low levels, but most do survive. However, in agroecosystems, especially monocultures with their reduced biological diversity, the dominant crop species may be in great jeopardy under extremely unfavorable climatic conditions. The vulnerability of an individual crop plant to wind, for example, may be quite different from that of an individual plant in a matrix of other plants whose structural characteristics may provide some degree of physical protection. Windbreaks have been used with success to protect both against soil erosion and sandblasting of crop plants. Even the simple arrangement of crops in rows may create problems when oriented to prevailing winds. The configuration of many fields probably has been determined largely by convenience for planting, cultivation, and harvesting procedures, as well as reduction of soil erosion and water capture and retention. In selecting planting practices, however, there is often little consideration given to potential biological impacts. The aerodynamic aspects of field crop systems are seldom considered except in areas normally subjected to high winds. For example, Vomocil and Ramig [90] described measures to attenuate the effect of wind erosion and sandblasting on crops grown in the sandy soils of the Columbia basin of Oregon, an area well known for sustained occurrence of high winds.

Other meteorological factors also may have unexpected influences on pests and natural enemies (see Chapter 3). Changing barometric pressures may modify the biological activities of certain species that result in a heavy "fall out" of components of the aerial "plankton" (e.g., Lepidoptera: Noctuidae) [86]. Prolonged droughts are usually associated with outbreaks of spider mites on annual crops such as cotton and soybean.

Some of these abiotic factors have been brought under control in many highly managed agroecosystems. Irrigation overcomes the negative effects of droughts, windbreaks attenuate the impact of high directional winds, adequate drainage reduces the risk of flood and waterlogging, and heat and wind generators and overhead mist sprayers often protect crops against killing frosts. The ultimate in environment regulation, however, is achieved under greenhouse conditions where production of tropical species is made possible even if the ground all around is covered with snow.

In conclusion, humans perform multiple roles to preserve agroecosystems as functional and efficient agricultural production entities. The cost to the environment and to society of human interference in agroecosystem functions and structure, however, must remain within acceptable bounds. If management requires inordinate amounts of external, nonrenewable inputs, such as those derived from finite reserves of fossil fuels, the system is not sustainable. The ecosystems approach to crop production/protection is likely to provide a path for sustainability; new paradigms are being proposed to achieve this goal (e.g., see ref. 84). Any such scheme must realistically assess the capacity of current production systems to satisfy an exploding human population's mounting demands for food and fiber.

IV. LESSONS FROM NATURAL ECOSYSTEMS

Broadly defined, ecosystems are all-encompassing ecological units. For example, Lewis [51] offers a "utilitarian" classification of terrestrial ecosystems, as presented in Table 1.

Since human presence and interference are no longer limited to highly managed systems, the distinction between natural and managed systems rests basically on the degree of disturbance inflicted by humans within a continuum that starts from a minimum as in the remnants of virgin tropical forests, for instance, through the more accessible temperate forests and savannas and deserts and reaches a maximum in the intensely managed ecosystems that include most agroecosystems. Comparisons between natural and agroecosystems have been offered by many authors. For example, Risser [75] used data from Woodmansee [93] to compare the propensity of natural and agroecosystems to accumulate nutrients and persist through time (Table 2).

Southwood and Way [85] considered seven "agroecosystems" and compared four characteristics likely to impact the effectiveness of biological control in an IPM system. Figure 4 is an adaptation of the qualitative descriptions used by Southwood and Way [85] and Altieri [2] for the characteristics of ecosystems that are most impacted by human interference in the process of creating new agroecosystems. The selected example spans the range from a rain forest, probably a tropical rain forest, to shifting agriculture, dryland, and irrigated polycultures, and finally through dryland monoculture. The characteristics include plant or crop diversity; ecological distance, as defined earlier in this chapter; temporal permanence, as an expression of the transition from a perennial forest to multiyear crops, to annual crops; genetic diversity, ranging from diverse species communities to monocrops that are often reduced to a few basic genotypes; human control; and natural control of pests.

There is ample worldwide interest in modifying agricultural practices to achieve sustainable systems. A composite of environmental, scientific, and political factors [24] is driving these consequential changes, with the concept of the ecosystem being integral to the new emphasis. Instead of considering one crop as a unit, multiple crops are included in agricultural landscapes comprising multiple components with both agricultural and relatively undisturbed patches of original or introduced noncrop vegetation. Mixed crop landscapes represent well-defined spatial units. Since these systems have developed through time,

Table 1 A Utilitarian Classification of Terrestrial
Ecosystems with Focus on Grasslands

Glacial: Ice caps
Forest: Natural forests, including temporary openings
Range
  Natural pastures
    Biomes
      Deserts
      Grasslands
      Shrublands
      Savannas (including open, noncommercial forests)
      Tundra
    Long-enduring pastures of primary succession
      Grazable marshes
      Shrubs and grass successional to forest
  Derived pastures extensively managed
    Derived from forests
    Derived from natural pastures
      Native vegetation
      Introduced vegetation
Cultivated
  Wood lots
  Croplands
  Sown pastures
  Derived pastures intensively managed
Urban-suburban: Residential-industrial areas, airports, etc.

Source: After ref. 51.

there is a temporal component, distinguishing agroecosystems at various levels
of maturity, or the analogue of a "climax."

Achieving sustainability in agroecosystems involves many factors, so the shift
from high-input agriculture to sustainable agriculture is necessarily gradual. The
shift may involve gradual changes of the landscape made up chiefly of single
crops often immediately adjacent to another crop or crops with little, or no,
intervening area. High-input agriculture traditionally relies on intensive pesticide use and minimum utilization of biological control agents. As natural control was not factored into monocrop systems, there was little appreciation for
the preservation of natural enemy reservoirs nor maintenance of noncrop vegetation in and around crop fields. As utilization of natural control agents of all
types becomes more frequent and pesticide usage diminishes, a closer connection between cropping systems and the surrounding landscape emerges. The
matrix in which an agricultural system is imbedded suddenly assumes greater
significance. Patches of nonagricultural vegetation now become an important
element of the agricultural system, as discussed above.

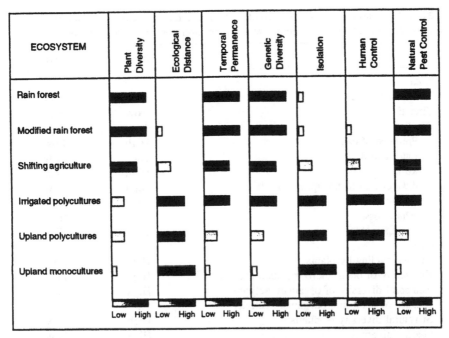

Figure 4 Major natural and agroecosystems characteristics and properties. (Based on refs. 2 and 85.)

V. DIRECTIONS FOR PEST MANAGEMENT AND SUSTAINABILITY OF AGROECOSYSTEMS

Failure of most pest control systems based on single tactics was due to ignorance of, or disregard for, fundamental ecological principles underlying agroecosystems structure and function. Integrated pest management (IPM) evolved from early 1960s efforts to incorporate sound ecological principles into pest population regulation in agroecosystems [10,41,50]. Most IPM definitions (e.g., see refs. 8, 10, 14, 19, 49, 53, 58, 62, 66, 70, 76, 82, and 83) have attempted to capture the notion that the following are key elements of an IPM system:

a. Pests are all biotic components of the crop environment capable of inducing stress and reducing crop yield or quality. The major pest categories are arthropods, microbial pathogens, nematodes, vertebrates, and weeds.

b. Pests coexist with crops and must be controlled if populations or infestations exceed a level—the economic injury level—above which

Table 2 Characteristics that Directly Influence the Propensity of Natural and Cultivated Ecosystems to Accumulate Nutrients and Persist Through Time[a]

	Natural	Cultivated
Abiotic		
Infiltration rates	High	Low
Runoff	Low	High
Erosion	Low	High
Presence of canopy	High	Low
Litter and debris	High	Low
Rocks	High	Low
Soil water loss to transpiration	High	Low
Soil colloids	High	Low
Leaching losses	Low	High
Soil temperature	Low	High
Biotic		
Internal cycling by plants	High	Low
Synchrony of plant-microorganism activity	High	Low
Temporal diversity of organism activity	High	Low
Balance of plant-microorganism activity	1	< 1
Structural diversity of plants	High	Low
Genetic diversity	High	Low
Reproductive potential	High	Low

[a]Data from ref. 93
Source: Based on ref. 75.

      economic losses may be expected to, at least, equal the cost of control measures necessary to prevent those losses. Pests are "managed" to remain below that economic injury level. Thus, IPM systems require adequate methods for assessing population density, infection intensity, and injury levels (see Chap. 27).

c. Management of pests is accomplished through careful consideration for the harmonious use of all available control methods (tactics). These control methods usually fall into four main classes: (a) chemical, (b) biological, (c) host plant resistance, and (d) cultural and physical techniques.

d. IPM systems are designed to minimize adverse environmental impacts, maximize cost/benefit ratios of control measures (particularly chemical controls), and maximize social benefits.

e. Pest and injury level assessments support a decision-making process that takes into account and includes the cost/benefit ratios defined under item 4 (above).

Consequently, IPM is essentially a concatenation of decision-making steps, each of which should consider not only farm gate profitability but also the long-term costs and benefits to society and the environment. Additionally, as sustainable agriculture becomes the agricultural banner of the 1990s, IPM assumes the added role of an operational system for crop protection that contributes to the sustainability of agricultural production by promoting methods that minimize the use of nonrenewable energy inputs (including but not restricted to pesticides) and protects the biodiversity of the crop community [45]. As the component of sustainable agriculture with the most robust ecological foundation, IPM not only contributes to the sustainability of agriculture, it also serves as a model for the practical application of ecological theory [11] and provides a paradigm for the development of the other components of sustainable agricultural systems, for example, such as integrated nutrient management, integrated water management, and integrated livestock management, to culminate in the new synthesis of integrated agroecosystems management (Fig. 5).

The integrated management paradigm represented by IPM suggests the following considerations:

1. Integrated management must rest on solid biological information, and it requires a theoretical base from both the biological and social sciences; consequently, it must involve strong interdisciplinary cooperation in research and development and close public/private sector cooperation in implementation.

2. The integrated management of agricultural systems operates at the interface of the three multidimensional universes: ecological, social/economic, and agricultural, described before as being a hierarchy ordered according to ascending levels of complexity and expanding spatial and temporal scales (see Fig. 2).

3. IPM has been conceived as interactive systems at three possible levels of integration [42]:

Level I: Integration of control methods for single species or species complexes (species/population level integration)

Level II: Integration of the impacts of multiple pest categories (insects, pathogens, and weeds) on the crop and the methods for their control (community level integration)

Level III: Integration of multiple pest impacts and the methods for their control within the context of the entire cropping system (ecosystem level integration)

Similarly, the integrated management of all other agricultural systems components—soil, water, nutrients—also can be conceived at three levels of integration. A fourth level of integration has been considered to account for the

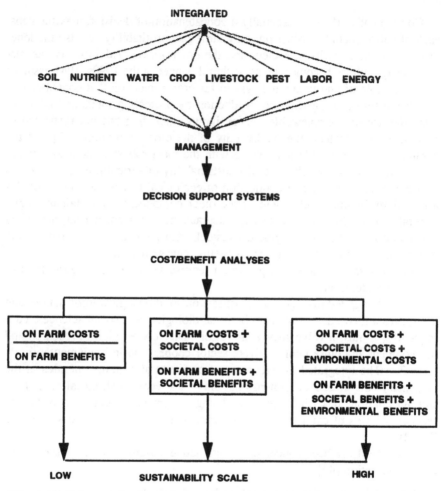

Figure 5 Operational components of a sustainable agroecosystem. Integration takes into account all system components and their interactions. Cost/benefit analyses consider all inputs within the ecological and social scales.

political, legal, social, and psychological constraints impinging on IPM [68,69]. We think that an inclusive agroecosystem concept accounts for all those anthropogenic components as well. Indeed, level III integration represents the convergence of levels I and II management of all other system components including the sociopolitical constraints. At level III, integrated agroecosystem management is achieved. If, at this level, decision support systems have led to adequate evaluations of social and environmental consequences of management actions, then the system should tend toward increased sustainability.

4. Although most IPM and other integrated management programs currently in use are at integration level I, it is advancement to level III that is most likely to yield the results and benefits sought from the IPM approach. These three levels of integration fit within the various scales of the agroecological/socioeconomic universes which, in turn, provide a matrix for programmatic approaches for IPM research and implementation (Fig. 6).

5. Most of the farmers currently following integrative practices are those whose livelihood depends on an understanding of how crop production ecology interacts with the economics of agricultural markets within a sociological context. Farmers, individually and collectively within a region, have accumulated considerable knowledge about time/space relationships in multiple cropping systems, including information about the incidence and impact of pests. This knowledge should be systematically compiled and evaluated as a starting point for proposed programs at any level of integration, and farmers should be part of the ongoing process of sharing information, experimental planning, design and analysis, and feedback as IPM programs are developed.

Sustainable agriculture frequently has been equated with organic farming leading to the conclusion that sustainable agriculture's goal is drastically to reduce, if not totally eliminate, the use of synthetic agrochemicals. However, recognition that human population growth will not soon be curbed has led responsible agroecologists to view agrochemicals as part of the technological arsenal that, when placed within the context of the ecology of agricultural systems, may be used in ways that minimally impact the functions of the natural

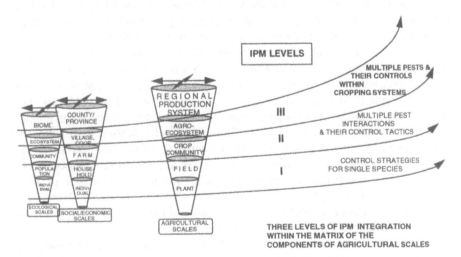

Figure 6 Levels of IPM integration within the context of the ecological, socioeconomic, and agricultural scales.

components of the system, particularly the natural enemies of pests. Thus, a sustainable agroecosystem is one that provides for the long-term maintenance of natural resources and agricultural productivity, has a minimal adverse impact on the environment, guarantees adequate economic returns to farmers, optimizes crop production with minimum necessary chemical inputs, satisfies human needs for food and income, and provides for the social needs of farm families and communities [60]. IPM, a key component of sustainable agriculture, provides a useful paradigm for other agricultural sustainability components, because IPM is rooted in understanding the role of the whole biotic component of agroecosystems in the long-term stability of crop communities.

A. Experimental Designs and Understanding Agroecosystems

If the conceptualization of ecosystems has been the subject of considerable argument, at least some of the controversy may be ascribed to the difficulties in studying these complex systems. Biologists may have the tools to sample, measure, and describe communities of animals and plants, but many of us are less comfortable with the criteria and the tools necessary to measure the physical environmental parameters that are relevant to interpreting and modeling the dynamics of ecosystems [22]. Since Gates wrote the review "Towards Understanding Ecosystems" [22], instrumentation to measure leaf area, temperature, reflectance, photosynthesis, leaf area index, and many other critical plant ecophysiological parameters has been greatly streamlined and made increasingly accessible to laboratories around the world. Miniaturized weather stations with electronic event recorders provide the opportunity to establish networks with reliable data for miniscale or macroscale experiments. Finally, the ubiquitous use of powerful personal computers connected to local area networks provides the means to handle substantial amounts of data in, for the most part, a userfriendly environment. All these advances permit application of theoretical models to real-life situations in the management of agroecosystems. Linear (degree-day driven) and nonlinear models guide management recommendations for important pests in many major crops [78]. These tools and models begin to support a predictive capability in applied biological sciences that has eluded earlier ecologists in their fundamental studies of communities and ecosystems. Modern statistical tools also assist the ecologist in the reduction and interpretation of massive amounts of data generated in community-level studies [23]. However, progress toward understanding agroecosystems and using this understanding to optimize management still faces considerable hurdles.

Progress of IPM to the agroecosystem level of integration hinges on adequate support from ecosystem theory and its application in agroecosystem management. A current impediment for quick progress in agroecosystem research, however, seems to be inadequate experimental designs. As multiple factors must be considered in their singular and interactive effects, the traditional experimental designs and parametric statistical tools are not easily acceptable. Hence, the criticism that a majority of the research concerned with agroecosystems has been either descriptive or simulated in computer models and not validated by experimental data. New experimental approaches are clearly needed to test hypotheses and investigate the influence of interactive multiple factors that determine agroecosystem functions and dynamics.

VI. EPILOGUE: AGROECOSYSTEMS IN THE BIOSPHERE

Agricultural systems have been a significant component of the diverse ecosystems that exist across the Earth's surface. The nature, extent, and character of these agricultural ecosystems vary depending upon location. Some are clearly discernible from great heights (e.g., pivot irrigation systems or contour farming), whereas others meld into a broader landscape matrix where they may be all but invisible from above (e.g., small forest clearings in many parts of the tropics).

Since substantial control of physical and biological parameters may occur in agricultural systems, they may not be accepted by all observers as an ecosystem in the classical sense. It is of interest that Tansley himself [87] felt it appropriate to include anthropogenic impacts on ecosystems as legitimate areas of investigation. Certainly, far greater manipulations of agroecosystems are involved in most agricultural practices. Selected crops are usually the only plants tolerated, generally resulting in the elimination of most, or all, competing plant species. Pest control by means of naturally occurring parasites, predators, and pathogens is usually supplemented by pesticide applications and/or directed releases of biological control agents. In many parts of the world, major crop plants are themselves nonindigenous to the area [63] and many of the pests are also nonnative. This is a major biological difference between most agricultural systems and natural ecosystems, whereas most physical factors are more likely to be similar between agricultural and nonagricultural ecosystems. The creation of vast crop monocultures rarely has a counterpart in natural systems except perhaps early succession after a fire or some other widespread disturbance.

Changing patterns of agriculture, especially those developing low-input or sustainable systems, aspire to approach natural systems. Complete congruence of such systems is unlikely to occur, but the ecological distances between them are likely to narrow. Knowledge of the complexity of unmanaged ecosystems offers intriguing possibilities for reshaping current agricultural systems into more earth-adapted agroecosystems. Concurrently, careful study of agroecosystems can provide valuable insights into the structure and function of natural systems. Agroecosystems have become a permanent, and in certain regions, the dominant element of the biosphere, and it is essential to exert a stewardship of these ecosystems that is maximally in harmony with the rules of nature to the extent that is economically and socially feasible.

ACKNOWLEDGMENTS

We acknowledge the critical reviews provided by our colleagues Peter McEvoy and Dan McGrath. The editorial assistance of Allan Deutsch and the technical assistance of Susan Larson with the typescript are deeply appreciated.

I (J.D.L.) am deeply grateful to my many colleagues of the H.J. Andrews Experimental Forest science team and the greater scientific community of the NSF-supported Long-Term Ecological Research Program. Individually and collectively, they provided a forum where I was exposed to and involved in extensive discussions and research efforts in many aspects of ecosystem science—quite a journey for an insect systematist! Although I did not always understand the words used, the trip was never dull.

REFERENCES

1. Allen, T.F.H., and T.W. Hoekstra. 1992. *Toward a Unified Ecology*. New York: Columbia University Press.
2. Altieri, M.A. 1987. *Agroecology: The Scientific Basis of Alternative Agriculture*. Boulder, Colorado: Westview.
3. Alvim, P. de T. 1982. An appraisal of perennial crops in the Amazon basin. In *Amazonia: Agriculture and Land Use Research*. S.B. Hecht, ed. Cali, Colombia: CIAT, pp 311–328.
4. Avery, D.T. 1991. *Global Food Progress 1991: A Report From Hudson Institute's Center for Global Food Issues*. Indianapolis, Indiana: Hudson Institute.
5. Bayliss-Smith, T.P. 1982. *The Ecology of Agricultural Systems*. Cambridge, U.K: Cambridge University Press.
6. Bender, B. 1975. *Farming in Pre-History: From Hunter-Gatherer to Food-Producer*. New York: St. Martin's Press.
7. Bodenheimer, F.S. 1951. *Insects as Human Food*. The Hague: Junk.
8. Bottrell, D.R. 1979. *Integrated Pest Management*. Washington, D.C.: U.S. Government Printing Office.
9. Carter, G.F. 1977. A hypothesis suggesting a single origin of agriculture. In *Origins of Agriculture*. C.A. Reed, ed. World Anthropology Ser. The Hague: Mouton.

10. Cate, J.R., and M.K. Hinkle. 1994. *Integrated Pest Management: The Path of a Paradigm*. Washington, D.C.: National Audubon Society.

11. Coleman, D.C., E.P. Odum, and D.A. Crossley, Jr. 1992. Soil biology, soil ecology, and global change. *Biol. Fertil. Soils* 14:104–111.

12. Costanza, R., B.G. Norton, and D.B. Haskell, eds. 1992. *Ecosystem Health: New Goals for Environmental Management*. Washington, D.C.: Island Press.

13. Crosby, A.W. 1986. *Ecological Imperialism: The Biological Expansion of Europe, 900–1900*. Cambridge, U.K.: Cambridge University Press.

14. Dover, M.J. 1985. *A Better Mousetrap: Improving Pest Management for Agriculture*. World Resources Institute, Study 4. Washington, D.C.: World Resources Institute.

15. Dowdeswell, W.H. 1987. *Hedgerows and Verges*. London. Allen & Unwin.

16. Eden, M.J. 1990. *Ecology and Land Management in Amazonia*. London: Belhaven Press.

17. Evans, F.C. 1956. Ecosystem as the basic unit in ecology. *Science* 123:1127–1128.

18. Farvar, M.T., and J.P. Milton, eds. 1972. *The Careless Technology: Ecology and International Development*. Garden City, New York, Natural History Press.

19. Flint, M.L., and R. van den Bosch. 1981. *Introduction to Integrated Pest Management*. New York: Plenum Press.

20. Flora, C.B. 1992. Building sustainable agriculture: A new application of farming systems research and extension. *J. Sustainable Agric.* 2:37–47

21. Food and Agriculture Organization (FAO). 1992. *FAO Yearbook: Production*. Statistic Ser. 112, vol. 46.

22. Gates, D.M. 1968. Toward understanding ecosystems. In *Advances in Ecological Research*, J.B. Cragg, ed. New York: Academic Press, pp 5–35.

23. Gauch, H.G., Jr. 1982. *Multivariate Analysis in Community Ecology*. Cambridge U.K.: Cambridge University Press.

24. Ghersa, C.M., M.L. Roush, S.R. Radosevich, and S.M. Coakley. 1994. Coevolution of agroecosystems and weed management. *Bioscience* 44:85–94.

25. Gliessman, S.R. 1990. Agroecology: researching the ecological basis for sustainable agriculture. In *Agroecology: Researching the Ecological Basis for Sustainable Agriculture*, S.R. Gliessman, ed. New York: Springer, pp 3–10.

26. Golley, F.B. 1993. *A History of the Ecosystem Concept in Ecology*. New Haven, Connecticut: Yale University Press.

27. Hagen, J. 1992. *An Entangled Bank: The Origins of Ecosystem Ecology*. New Brunswick, New Jersey: Rutgers University Press.

28. Halloway, J.D., and N.E. Stork. 1991. The dimensions of biodiversity: the use of invertebrates as indicators of human impact. In *Proceedings of the First Workshop on the Ecological Foundations of Sustainable Agriculture*. CASAFA Rep. Ser. 4. Wallingford, U.K.: CAB International, pp 37–63.

29. Hanway, J.J., and C.R. Weber. 1971. Dry matter accumulation in eight soybean (*Glycine max* [L.] Merrill) varieties. *Agron. J.* 63:227–230.

30. Harlan, J.R. 1975. *Crops & Man*. Madison, Wisconsin: American Society of Agronomy Crop Science Society of America.

31. Harlan, J.R. 1986. Plant domestication: diffuse origins and diffusions. In *The Origin and Domestication of Cultivated Plants*. C. Barigozzi, ed. Amsterdam: Elsevier, pp 21–34.

32. Harlan, J.R., and K.J. Starks. 1980. Germplasm resources and needs. In *Breeding Plants Resistant to Insects*. F.G. Maxwell & P.R. Jennings, eds. New York: Wiley, pp 253–273.

33. Hendrix, P.F., D.C. Coleman, and D.A. Crossley, Jr. 1992. Using knowledge of soil nutrient cycling processes to design sustainable agriculture. *J. Sustainable Ag.* 2:63–82.

34. Herzog, D.C., and J.E. Funderburk. 1986. Ecological bases for habitat management and pest cutural control. In *Ecological Theory and Integrated Pest Management Practice*. M. Kogan, ed. New York: Wiley, pp 217–250.

35. Hetch, S.B. 1987. The evolution of agroecological thought. In *Agroecology: The Scientific Basis of Alternative Agriculture*. M.A. Altieri, ed. Boulder, Colorado: Westview, pp 1–20.

36. Hill, D., J. Andrews, N. Sotherton, and J. Hawkins. 1995. Farmland. In *Managing Habitats for Conservation*. W.J. Sutherland & D.A. Hill, eds. Cambridge, U.K.: Cambridge University Press, pp 230–266.

37. Kareiva, P. 1989. Renewing the dialogue between theory and experiments in population ecology. In *Perspectives in Ecological Theory*. J. Roughgarden, R.M. May, & S.A. Levin, eds. Princeton, New Jersey: Princeton University Press, pp. 68–88.

38. Kaster, M., E.F. de Queiroz, and F. Terezawa. 1981. Introdução da soja no Estado do Paraná.. In *A Soja no Brasil. Inst. Tecnol. de Alimentos* (ITAL). S. Miyasaka & J.C. Medina, eds. São Paulo: Campinas, pp 22–24.

39. Kauffman, J.B., D.L. Cummings, D.E. Ward, and R. Babbitt. 1995. Fire in the Brazilian Amazon: Biomass, nutrient pools, and losses in slashed primary forests. *Oecologia* 104:397–408.

40. Kauffman, J.B., R.L. Sanford Jr., D.L. Cummings, I.H. Salcedo, and E.V.S.B. Sampaio. 1993. Biomass and nutrient dynamics associated with slash fires in neotropical dry forests. *Ecology* 74:140–151.

41. Kogan, M., ed. 1986. *Ecological Theory and Integrated Pest Management Practice*. New York: Wiley.

42. Kogan, M. 1988. Integrated pest management theory and practice. *Entomol. Exp. Appl.* 49:59–70.

43. Kogan, M. 1994. Plant resistance in pest management. In *Introduction to Insect Pest Management*. 3rd ed. R.L. Metcalf & W.H. Luckmann, eds. New York: Wiley, pp 73–128.

44. Kogan, M., and D. McGrath. 1993. Integrated pest management: Present dilemmas and future challenges. *An. 14°-Congresso Brasileiro de Entomologia*. São Paulo: Piracicaba, pp 1–16.

45. Kogan, M., and J.D. Lattin. 1993. Insect conservation and pest management. *Biodiversity and Conservation*. 2:242–257.

46. Kogan, M., and S.G. Turnipseed. 1987. Ecology and management of soybean arthropods. *Annu. Rev. Entomol.* 32:507–538.

47. Krebs, C.J. 1972. *Ecology: The Experimental Analysis of Distribution and Abundance*. New York: Harper & Row.
48. Le Pelley, R.H. 1968. *Pests of Coffee*. London: Longmans Gree.
49. Levins, R. 1986. Perspectives in integrated pest management: From an industrial to an ecological model of pest management. In *Ecological Theory and Integrated Pest Management Practice*. M. Kogan, ed. New York: Wiley, pp 1–18.
50. Levins, R., and M. Wilson. 1980. Ecological theory and pest management. *Annu. Rev. Entomol.* 25:287–308.
51. Lewis, J.K. 1969. Range management viewed in the ecosystem framework. In *The Ecosystem Concept in Natural Resource Management*. G.M. Van Dyne, ed. New York: Academic Press, pp 97–187.
52. Lowrance, R., B.R. Stinner, and G. J. House, eds. 1984. *Agricultural Ecosystems: Unifying Concepts*. New York: Wiley.
53. Luckmann, W.H., and R.L. Metcalf. 1982. The pest management concept. In *Introduction to Insect Pest Management*. R.L.Metcalf & W.H.Luckmann, eds. New York: Wiley, pp 1–31.
54. MacNeish, R.S. 1992. *The Origins of Agriculture and Settled Life*. Norman, Oklahoma: University of Oklahoma Press.
55. Major, J. 1969. Historical development of the ecosystem concept. In *The Ecosystem Concept in Natural Resource Management*. G.M. Van Dyne, ed. New York: Academic Press, pp 9–24.
56. May, R.M., ed. 1976. *Theoretical Ecology: Principles and Applications*. Philadelphia: Saunders.
57. McIntosh, R.P. 1985. *The Background of Ecology: Concept and Theory*. Cambridge, U.K.: Cambridge University Press.
58. National Academy of Sciences. 1969. *Insect-Pest Management and Control*. Vol. 3 of *Principles of Plant and Animal Control*. Pub. 1695. Washington, D.C.: National Academy Press.
59. National Research Council. 1989. *Alternative Agriculture*. Washington, D.C.: National Academy Press.
60. National Research Council. 1991. *Toward Sustainability: A Plan for Collaborative Research on Agricultural and Natural Resource Management*. Washington, D.C.: National Academy Press.
61. Odum, E. 1984. Properties of agroecosystems. In *Agricultural Ecosystems*. R. Lowrance, B.R. Stinner, & G.S. House, eds. New York: Wiley, pp 5–11.
62. Office of Technology Assessment. 1979. *Pest Management Strategies in Crop Protection*. Volume I. Washington, D.C.: U.S. Government Printing Office.
63. Office of Technology Assessment. 1993. *Harmful Non-Indigenous Species in the United States*. OTA-F-565. Washington, D.C.: U.S. Government Printing Office.
64. O'Neill, R.V., D.L. DeAngelis, J.B. Waide, and T.F.H. Allen. 1986. *A hierarchical concept of ecosystems*. *Monogr. Pop. Biol.* 23:1–272.
65. Parr, J.F., R.I. Papendick, S.B. Hornick, and R.E. Meyer. 1992. Soil quality: attributes and relationship to alternative and sustainable agriculture. *Am. J. Alternative Ag.* 7:5–10.
66. Pedigo, L.P. 1989. *Entomology and Pest Management*. New York: MacMillan.

67. Peters, R. H. 1991. *A Critique for Ecology*. Cambridge, U.K.: Cambridge University Press.

68. Prokopy, R.J., and B.A. Croft 1994. Apple insect pest management. In *Introduction to Insect Pest Management*. 3rd. ed. R.L. Metcalf & W.H. Luckmann, eds. New York: Wiley, pp 543–558.

69. Prokopy, R.J., M. Christie, S.A.Johnson, and M.T. O'Brien. 1990. Transitional step towards second-stage integrated management of arthropod pests of apple in Massachusetts orchards. *J. Econ. Entomol.*83:2405–2410.

70. Rabb, R.L., and F.E. Guthrie, eds. 1970. *Concepts of Pest Management, Proceedings of a Conference Held at North Carolina State University at Raleigh*, Raleigh, North Carolina: State University.

71. Rabb, R.L. 1978. A sharp focus on insect populations and pest management from a wide-area view. *Bull. Entomol. Soc. Am.* 24:55–61.

72. Reed, C.A., ed. 1977. *Origins of Agriculture*. World Anthropology Ser. The Hague: Mouton.

73. Ricklefs, R.E. 1973. *Ecology*. Newton, Massachusetts: Chiron Press.

74. Risch, S.J., D. Andow, and M.A. Altieri. 1983. Agroecosystem diversity and pest control: data, tentative conclusions and new research directions. *Environ. Entomol.* 12:625–629.

75. Risser, P.G. 1986. Agroecosystems structure, analysis, and modeling. In *Ecological Theory and Integrated Pest Management Practice*. M. Kogan, ed. New York: Wiley, pp 321–343.

76. Rohwer, G.G. 1981. Regulatory plant pest management. In *CRC Handbook of Pest Management in Agriculture*. Vol. 1. D. Pimentel, ed. Boca Raton, Florida: CRC Press, pp 253–267.

77. Roughgarden, J., R.M. May, and S.A. Levin, eds. 1989. *Perspectives in Ecological Theory*. Princeton, New Jersey: Princeton University Press.

78. Ruesink, W.G., and D.W. Onstad. 1994. Systems analysis and modeling in pest management. In *Introduction to Insect Pest Management*. 3rd ed. R.L. Metcalf & W.H. Luckmann, eds. New York: Wiley, pp 393–419.

79. Salaman, R.N. 1949. *The History and Social Influence of the Potato*. Cambridge, U.K.: Cambridge University Press.

80. Sauer, C.O. 1952. *Agricultural Origins and Dispersals*. New York: American Geographic Society.

81. Seastedt, T.R., and A.K. Knapp. 1993. Consequences of nonequilibrium resource availability across multiple timescales: The transient maxima hypothesis. *Am. Naturalist* 141: 621–633.

82. Shepard, M. 1973. Integrated systems of pest management. In *Insect Pest Management: Readings*. M. Shepard, ed. New York: MSS Information Corp., pp 230–266.

83. Smith, R.F., and H.T. Reynolds. 1966. Principles definitions and scope of integrated pest management control. In *Proceedings of the FAO Symposium on Integrated Pest Control*, 11–15 October. Food and Agriculture Organization of the United Nations, Rome, 1965.

84. Soule, J.D., and J.K. Piper. 1992. *Farming in Nature's Image: An Ecological Approach to Agriculture*. Washington, D.C.: Island Press.
85. Southwood, T.R.E., and M.J. Way. 1970. Ecological background to pest management. In *Concepts of Pest Management*. R.L. Rabb & F.E. Guthrie, eds. Raleigh, North Carolina: North Carolina State University, pp 6–29.
86. Szentikirályi, F. 1994. Institute for Plant Protection, Budapest, Hungary. Personal communication.
87. Tansley, A.G. 1935. The use and abuse of vegetation concepts and terms. *Ecology*. 16:284–301.
88. Thomas, W.L., Jr., ed. 1958. *Man's Role in Changing the Face of the Earth*. Chicago: Il: University of Chicago Press.
89. Ullstrup, A.J. 1972. The impacts of the southern corn leaf blight epidemics of 1970–1971. *Annu. Rev. Phytopathol.* 10:37–50.
90. Vomocil, J.A., and R.E. Ramig. 1976. *Wind Erosion Control on Irrigated Columbia Basin Land: A Handbook of Practices*. Corvallis, Oregon: Spec. Rep. Oreg. Agric. Exp. Stn. 466.
91. Webb, K.E. 1974. *The Changing Face of Northeast Brazil*. New York: Columbia University Press.
92. Webster, C.C., and P.N. Wilson. 1980. *Agriculture in the Tropics*. 2nd ed. London: Longman.
93. Woodmansee, R.G. 1984. Comparative nutrient cycles of natural and agricultural ecosystems: A step toward principles. In *Agricultural Ecosystems: Unifying Concepts*. R. Lowrance, B.R. Stinner, & G.J. House, eds. New York: Wiley, pp 145–156.
94. World Resources Institute. 1994. *World Resources 1994–95*. New York: Oxford University Press.

2
Soil Effects on Pest Populations

Craig M. Liddell
New Mexico State University, Las Cruces, New Mexico

I. INTRODUCTION

The soil environment is both highly buffered and heterogeneous in space and time. It consists of a continuum of a largely microscopic matrix of solid particles, air, and water covering virtually the entire land surface of the world. Many factors directly affect the growth, reproduction, and survival of organisms that live in the soil or simply use the soil as a refuge from the aboveground environment. This chapter deals exclusively with soil factors that affect plant pests of various types ranging from bacteria through fungi to insects and weeds. Yet, despite the biological diversity of these pests, the impact of various soil factors on each of them is remarkably similar, facilitating a synthesized, comprehensive view of the effects of the soil on pest populations. All soil-inhabiting organisms must operate within certain limits of temperature and water activity, pH, and salinity and must live in the void space between soil particles. Therefore, it is how the soil mediates variation in these factors and how the soil as a habitat differs from other terrestrial habitats that is the focus of this chapter.

The major effects of the soil environment on pest populations can be classified into three broad categories based on the time and spatial scales over which they vary. Soil factors may vary over (a) short, diurnal, time periods; (b) longer, seasonal, time scales; and (c) pedogenic time scales; and may display a high degree of spatial variation but little temporal variation over decades. The most important soil factor that varies on both a diurnal and seasonal basis is temperature, and this factor probably controls the activity of pests and their hosts more than any other single factor [4,66]. Moisture potential is another important soil factor that varies seasonally but also may vary on short time scales (e.g., after rainstorms). Factors such as pH, texture, cation exchange capacity (CEC), organic matter and free calcium levels, and soil nutritional status tend to vary only slowly over short time scales and are examples of spatially heterogeneous soil properties. Factors that vary diurnally primarily affect pest activity, and factors that vary seasonally and spatially mainly affect pest population dynamics, helping to determine which pests build up and establish.

Soil factors may have either direct or indirect effects on pest populations. Some soil factors, such as temperature, water activity, pH, and free calcium have profound physiological effects on soil organisms. Conversely, soil texture and composition, which result from the parent material and weathering processes that produced the soil, are fundamental soil properties, but they have only an indirect effect on soil pests through the factors listed above.

The three most important soil properties affecting pest populations in soil are temperature, water potential, and microbial activity, and these are worthy of detailed examination. The first two soil properties, soil temperature and water potential, will be considered in this chapter and the third soil property, soil microbiota, is dealt with in other contributions to this volume. Other soil properties that vary spatially, such as nutrient availability and CEC, will be discussed under a general discussion of soil texture and composition.

Temperature and water potential control the activity of virtually all enzyme systems in living cells and therefore fundamentally control the growth, reproduction, and other activities of soilborne pests and their hosts; they also control diffusion rates and the rates of other soil properties and act to provide physical limits for the proper functioning of all living cells. Indeed, the hydraulic and thermal properties of a soil provide the basis for any characterization of soil [32,60,78].

II. PESTS AS SOIL ORGANISMS

Pest organisms are found in every kingdom (with the possible, but unlikely exception of the Archea, a recently characterized "superkingdom") and, therefore, have many adaptations to survive and thrive in the environment. They share one common feature: For some part of their life cycle, they parasitize, compete with, or otherwise interfere with the development of an economically important species. Many pest species survive or reside in soil, because the soil affords protection from the turbulent and extreme aboveground environment, or because their host itself is a soil organism, such as plant roots. The soil generally moderates extremes of atmospheric temperature and water availability, the importance of which can be seen in the evolutionary adaptations of seed plants and insects to survive periods of drought and excessive temperatures. The exoskeleton of insects and the seed coats and cuticle of higher plants provide protection from desiccation, allowing these organisms to survive and grow under dry or otherwise adverse conditions. Insects, nematodes, and other Metazoa, of course, have the advantage of mobility in soil to avoid adverse conditions. In contrast, microbial pests, such as fungi and bacteria, are in direct cellular contact with their environment and must survive and grow in intimate contact with adverse conditions. These differences can be summarized in the following way. All microbial pests must be able to withstand adverse conditions at the cellular or physiological level [11], whereas metazoan organisms are able to withstand adverse conditions through cellular and physiological adaptations and tissue adaptations, such as a cuticle, as well as through behavior. These differences become important as we examine how each type of pest survives and grows in the soil.

All organisms pass through several life stages in the same order, and many of these stages are common among pathogens, arthropod pests, and weeds. The approach taken here will be to address the impact of each soil factor on each of these life stages in their order of occurrence unless clarity dictates otherwise. The basic life stages shared by most organisms include (a) survival during adverse conditions; (b) germination/hatching in response to external stimuli; (c) infection, feeding, growth, and somatic development, which is generally the stage that directly affects the host; (d) reproduction in response to internal or external stimuli; and (e) a return to the quiescent, resting stage. Pests most often come to the attention of human agricultural managers during the growth and development stage where the pest is acquiring nutrition directly from its host. Occasionally, pests affect their hosts at the pest's resting stage, germination/ hatching stage, or at the reproductive stage. In some cases, most notably plant pathogenic fungi and weeds, asexual reproduction greatly magnifies the impact of the growth and development stage by expanding not only the size of the individual pest but also allowing a dramatic increase in the number of individuals at a time when the host may be most susceptible to attack. Soil factors affect all of these life stages in specific and sometimes unexpected ways, yet understanding exactly how the soil affects pest populations ultimately will lead to better pest management practices.

III. TEMPERATURE EFFECTS

The importance of temperature for the activity of pest populations is due to the fundamental relationship between temperature and the kinetics of physical and chemical processes. Temperature strongly influences the rate of chemical reactions and processes such as diffusion, convection, and enzyme kinetics. Temperature, therefore, has a major impact on the growth and activity of all soil organisms [4,33,48,66]. There are thermal cardinal limits and biostatic limits for all organisms [4,35,62,66], meaning that two different sets of high and low temperatures will be lethal and inhibitory to any particular organism. Thermal biostatic limits are nearly always reached well before the cardinal limits, but sometimes biostasis occurs at the cardinal limits. In these cases, biostasis leads to death after a specific length of time that an organism is exposed to the limit temperature. Overall, the thermal properties of soil have a profound impact on the activity and population dynamics of soil pests, and these properties fundamentally depend on the source of heat and the composition, texture, and water content of the soil.

The sun is the source of radiant energy that warms the soil. Consequently, there is both a diurnal and an annual periodicity to the temperature oscillations occurring in the upper layers of the soil. Temperature oscillations at the soil surface are transmitted downward as a phase shifting sine wave of decreasing amplitude with increasing depth [44,48]. Heat also moves upward toward the soil surface from deeper layers at night resulting in complex diurnal oscillations of soil temperature. Nevertheless, seasonal oscillations tend to be quite stable from year to year in soil that is managed the same way each year. Heat transmission in dry soil occurs by conduction through the solid phase, which is generally an inefficient process. Dry soil is, therefore, a poor conductor of heat and has a lower specific heat than wet soil owing to the high specific heat of water. Accordingly, the surface of dry soil rise to much higher temperatures than moist soils, with a given radiation load, but provides far better thermal insulation at depth [37,44]. Overall, soil is generally a poor conductor of heat, and this results in a buffered thermal environment for all soil organisms as compared with the atmosphere [37]. The thermal buffering effect explains why many arthropods seek out refuge in the soil during extreme winter or summer periods [80] and why the upper layers of soil act as an effective seed bank for weeds [82].

A. Temperature and Survival

Survival of all pest species in a dormant state is favored by cool conditions, largely because other antagonistic, pathogenic, or phagotrophic organisms are less active under these conditions. If the temperatures become too low, then some pest species may be killed, although surprisingly few pests adapted to temperate climates are killed by freezing, at least if they are surviving in an appropriate dormant state such as an egg, pupa, seed, or resting spore. Survival tends to decline as the temperature rises, because metabolic rates increase, draining reserves more quickly, and the activity of competing organisms increases. It is also virtually impossible to separate the effects of soil moisture from those of temperature. Soil always dries as it warms and often becomes moister when it cools because of condensation. Unique combinations of moisture and temperature may lead to conditions favoring survival of individual pest species. For example, although most weed seeds occur in the top 15 cm of soil, they survive for far longer at lower depths, probably due to mild temperatures and moderate moisture conditions [63,83]. Many arthropod pests burrow into the soil to avoid surface temperature extremes and in some cases burrow to considerable depths [80]. Most fungi, bacteria, and viruses survive on or in host tissue that affords some protection [49] or produce resistant spores that are able to survive extreme conditions and that survive better at lower soil depths.

Endogenous dormancy or diapause often requires exposure to sustained low temperatures or high temperatures before germination can occur. Many other factors such as a period of stratification, herbicides, fungicides, external chemical triggers such as host plant exudates and exhaustion of specific reserves may trigger germination. These processes are well understood for many seed plants [18,63,66] and for many arthropods [19,77] but not for the majority of soilborne plant pathogens. Some arthropods can pass through several dormant periods ranging from quiescence to true diapause during their life cycle based on adverse environmental conditions [19,77]. In contrast, weed species will usually set seed or exist in a subterranean vegetative state during adverse conditions [63], avoiding the need for a somatic phase wholly able to withstand adverse conditions. Of course, there are several important weeds, such as the nutsedges, that exist primarily in a subterranean vegetative state and reproduce clonally, growing foliage for photosynthate production only during periods suitable for growth [63].

Work on the dormancy of soilborne plant pathogens has concentrated on soil-invading fungi and bacteria rather than the soil-inhabiting pathogens. Soil invaders occur in soil only in a dormant or quiescent state and, after germination, infect aboveground plant parts. Soil-inhabiting fungi and bacteria, on the other hand, can grow and reproduce in soil and infect roots or other subterranean plant parts. Shiekh and Ghaffar [71] showed that specific time-temperature relationships exist for the inactivation of microsclerotia of *Macrophomina phaseolina*. Liddell and Burgess [49] showed that the soil invader *Fusarium verticillioides (F. moniliforme)* could survive for over 2 years under conditions that suppressed the activity of soil microorganisms but survived less than 30 days when temperature and water potentials were conducive to microbial proliferation in the soil. The effect of soil temperature on the survival of soil-inhabiting fungi, such as *Rhizoctonia* and *Pythium ultimum*, is difficult to study, because these organisms grow in native soil and it is difficult to isolate survival of dormant propagules from saprophytic growth in the soil under all conditions. Studies on these fungi tend to concentrate on the saprophytic growth of these organisms, and it is assumed that these organisms will survive a much wider range of conditions than will support saprophytic growth [53]. Soil-inhabiting fungi doubtless can survive a wide range of conditions, and this is due to their ability to increase their biomass in soil under favorable conditions and produce highly resistant resting structures to span adverse periods.

Like the nutsedges and other tuber-forming weed species, soil-inhabiting plant pathogenic fungi are among the greatest soilborne threats to crop health throughout the world for the simple reason that once established in an agricultural field, they are nearly impossible to eradicate.

B. Temperature and Germination/Hatching

Because soils are generally better thermal insulators than conductors, they may develop steep thermal gradients over short distances. This fact is particularly important for organisms that respond to temperature fluctuations for breaking dormancy or molting, because only small zones of soil, at any one time, may be suitable for triggering germination as spring time temperature increases move into the soil. This means that dormant propagules are often found clustered at a single soil depth. Different pests will germinate or hatch in response to a variety of temperatures; however, a cold pretreatment is often required for weed seeds [63] and fungal pathogens. For example, Smith et al. [74] showed that cold temperature pretreatment affected the germinability of sporangiospores of the fungus *Rhizopus stolonifer* but not mitospores of the fungus *Monilinia fructicola*. Temperature fluctuations promote sporulation of the fungus *Psuedocercosporella herpotrichoides*, and knowledge of this relationship allowed wheat stubble management to be applied as a viable control method for this disease [67]. Generally, soil insects hatch and metamorphose under highly specific temperature conditions [80]. Indeed, social insects often cooperate with one another to act as a "superorganism" to maintain appropriate conditions for hatching when needed [82].

C. Temperature and Infection, Somatic Growth, and Reproduction

In the vast majority of cases, any temperature that is conducive or optimal (not usually the same temperature) to germination is also conducive or optimal to infection and somatic growth. It is well known that heat stress can predispose hosts to disease and insect attack [16]. The most severe pest problems often occur when temperatures are optimal for the pest but in an extreme range for the host. These conditions occur most commonly in the tropics, where there exist some of the most difficult conditions for agriculture on earth. High heat loads in the humid tropics lead to the highest known rates of decomposition [17]. High decomposition rates favor high levels of microbial activity, which, in turn, inhibits weak saprophytes, soil-invading fungi, such as *F. verticillioides* (*F. moniliforme*) [49], and favors soil-inhabiting organisms. In the tropics, however, the turnover of biomass is far greater than in temperate zones and soil-inhabiting fungi, weeds, and arthropods grow and reproduce more rapidly at tropical temperatures and readily attack susceptible economic plant species growing at the extreme of their range.

Reproduction of soil pests is often triggered by a combination of cooling soil temperatures and drying soil. Many fungi will reproduce only after temperatures reach an optimal range and sufficient menisci (air/water interfaces) exist in soil to ensure both sufficient water for growth and sufficient aeration for the increased metabolism required for reproduction [33,34,35]. Reproduction by weeds often occurs precociously and in response to management practices, as well as in response to temperature. Often, it is the drying of the soil that occurs as soil temperature increases that triggers reproduction, but reproduction in the fall is predominately triggered by falling night temperatures [83].

Pest control by controlling soil temperature using mulches and irrigation can aid in the control of diseases caused by thermophilic pathogens, such as *Pythium aphanidermatum*, and may delay weed seed germination. These approaches, however, are plagued by adverse effects on the host and the inadvertent favoring of other pests.

IV. MOISTURE EFFECTS

A remarkable quality of soil, which makes it an excellent medium for plant growth, is its capacity to store water and inorganic minerals in a form readily available for plant uptake. This is due to the porous nature of soil and the large specific surface area of soil particles available for physical and chemical interactions. As stated above, the water relations of soil depend largely on the composition and structure of the soil solid matrix [13]. In order to understand the major effect of soils on pest populations, one must clearly understand the basic concepts of soil water relations.

A. Soil Water Content and Water Potential

The simplest descriptions of soil water are the gravimetric and volumetric determinations of the amount of water present in any sample. However, other descriptions are necessary to gain an understanding of the effects of soil water on biological phenomena such as plant growth and the ecology of soil pests. Knowing only water content is insufficient for understanding the dynamic aspects of soil water flow and the availability of water to plants and other soil organisms. It is also necessary to understand the concept of water potential, which describes the energy status of soil water. Water content is important as a measure of capacity, thereby delimiting a boundary on many of the dynamic processes. Detailed information on the water relations of soil, plant, and microbial systems can be found in many sources [11,33,37,44,45,48,57, 60,62,70,72]. Water potential is a valuable concept for describing the movement of water in soil and providing a quantitative basis for describing the water relations of plants and soil organisms. All dynamic aspects of water flow, both saturated and unsaturated flow through soil and across biological membranes, depend on gradients of water potential. Water potential is also one of the critical variables determining the metabolic activity of soil organisms and plants in the soil.

Soil water potential is simply a measure of the chemical potential of water and is related to the free energy of a particular quantity of water [48]. Total water potential is negative in soil and plant systems, since the free energy of water in these systems is lower than pure free water. The units of water potential are energy per unit volume (joules per cubic meter [Jm^{-3}]), which are dimensionally equivalent to units of pressure. Therefore, the preferred SI units (International System of Units) for water potential are the pascal (Pa) or joules per kilogram (Jkg^{-1}). Water potential units in meters or centimeters are also acceptable for some uses, but the use of other units should be discouraged.

Water potential in soil and biological systems is generally considered to comprise four components:

$$\Psi = \pi + \tau + \psi_p + \psi_g \text{ (pascals)} \tag{1}$$

where Ψ is total water potential, π is osmotic potential, τ is matric potential, ψ_p is pressure (or turgor) potential, and ψ_g is gravitational potential. These components are the major parameters relevant to biological systems. It is useful to partition total water potential (Ψ) into these components, since each can be defined and quantified somewhat independently. Moisture potential, on the other hand, is used here as a defined term to describe the contribution to total water potential made by osmotic and matric potential. Osmotic potential is the contribution to the total water potential attributable to the addition of solute molecules to a body of water. The magnitude of the osmotic pressure developed across a semipermeable membrane separating two solutions of different osmotic potentials is equal to the difference in osmotic potential. However, it is important to note that osmotic pressure across a biological membrane also depends on the permeability of the membrane to solute molecules. If solute molecules pass freely through the membrane, it is said to have a reflection coefficient of zero and the osmotic pressure difference across the membrane would be zero. If the membrane is impermeable to a particular solute molecule but remains permeable to water, then the membrane is said to have a reflection coefficient of unity with respect to that molecule. Most membranes have a reflection coefficient of between 0 and 1 for most solute molecules [72]. The osmotic potential of a solution can be calculated easily from the concentration of component solute molecules [48].

Matric potential is the component of total water potential that arises from surface tension and interface effects. Water present in soil spontaneously adopts a distribution of menisci and surface films that lowers its free energy to a minimum. Water in this situation also has a lowered (more negative) matric potential. The effective diameter of the entrance to each soil pore determines the matric potential at which air will enter a water-filled pore. This relationship permits the calculation of the effective pore-size distribution directly from the moisture characteristic of a soil, and it is important in studying the movement of fungal zoospores, nematodes, and bacteria [48].

The water potential of any mass of water depends not only on solute and interface interactions but also upon relative elevation and pressure differences with respect to pure, free water. Although osmotic and matric potential are always negative in sign and may be regarded as negative pressure (or suction), pressure and gravitational potential can be either positive or negative.

Pressure potential becomes important in any confined body of water that is not at atmospheric pressure. In bacterial, plant, and fungal cells which are surrounded by a rigid cell wall, pressure potential is expressed as turgor pressure due to the development of low (more negative) osmotic potential in the cytosol due to solutes in the cell. The cytosol is enclosed by the plasmalemma, which acts as a semipermeable membrane and allows the development of a high turgor pressure within the cell due to migration of water across the plasmalemma into the cytosol, which is the mechanism by which herbaceous plants and fungal mycelia maintain their form. Turgor pressure is also necessary for the growth of plants and fungi. The reader is referred to several excellent sources on the subject of plant and fungal water relations for a fuller description of these phenomena [11,45,57,62,70,72]. For detailed information on how to measure, monitor, and control soil water content and potential in the field and containers, see Liddell [48].

B. The Moisture Characteristic

All porous materials, which are not highly hydrophobic, retain water against gravity or an applied suction, primarily by capillarity and surface tension within the porous body. Water relations of a porous body are thus dominated by matric potential. The soil moisture characteristic (SMC) is therefore defined as the relationship between the water content of a soil and the matric potential of the soil water. It is fundamental to the study of the hydraulics of any soil. The nature of this relationship depends on soil composition, texture, and structure, and it is sensitive to the wetting and drying history of the soil due to hysteresis [48].

The SMC of sandy soils differs markedly from that of loam and clay soils. Consequently, it is important that moisture potential be used in all comparisons between major soil types, at least with respect to their hydraulic properties and effects on soilborne pests. It is easy to determine SMC curves using standard methods [48], and several common reference points are provided as part of any standard soil test that can be used to identify the broad moisture characteristic of any soil. The simplest and most reliable measure is the air-dry moisture content. The soil water potential of air-dry soil is usually in the range −50 to −100 MPa. Very sandy soils have an air-dry moisture content of around 1 % by weight compared with moderate to heavy clay soils with an air-dry moisture content of 7–10%. The determination of the saturated moisture content, where the water potential is approximately 0 MPa, is less reliable, because it is impossible reliably to saturate all soils to the same extent. Field capacity (FC) and permanent wilting point (PWP) are often provided with a soil test where these measures indicate the soil water content at −0.03 MPa and −1.5MPa water potential, respectively. These measures are discussed in more detail below.

Soil organisms such as plants, insects, nematodes, and fungi must be able to reduce the moisture potential of their cells and tissues to a point below the moisture potential of the external environment to obtain water from the soil. This creates a moisture potential gradient along which water flows from the soil to the organism. Plants achieve a lowered tissue potential mainly by evapotranspiration and osmoregulation [45,57,72]. Fungi, however, are able to achieve a significant reduction in tissue moisture potential almost entirely by osmoregulation [11,33,34,54,62,70]. Insects and nematodes obtain their water from their hosts and therefore need not extract water directly from the environment. The lowest tissue moisture potential an organism can achieve is the theoretical lower limit on the "available" water supply for the organism. However, tissue moisture potential is by no means the only determinant of the amount of water available to be taken up from the soil. The availability of water to soil organisms is more complex than this moisture potential gradient concept may indicate.

Cassel and Nielsen [12] define both available water capacity (AWC) and "water availability." These concepts are related but do not represent the same range of available water. AWC is defined by the relation:

$$AWC = FC - PWP \qquad (2)$$

where FC is field capacity and PWP is permanent wilting point. AWC then is the maximum amount of water in a soil profile available to a plant for growth based on the simple potential energy gradient model. Under conditions of low transpiration, it is possible that a plant may extract all the water in the AWC. However, plants often cannot extract all the water from the AWC for a number of reasons. The concept of water availability refers to the water actually available for plant growth in a particular soil at a particular time. For example, plants growing at high transpiration rates require high soil hydraulic conductivity and cannot extract all of the AWC before soil hydraulic conductivity becomes limiting [12].

In addition, many coarse-textured soils (e.g., agricultural loams) drain well before the –1.5 MPa PWP; indeed, many sandy soils drain by –0.5 MPa. This leads to a narrow range of water availability due to the reduction of hydraulic conductivity by several orders of magnitude. Furthermore, as little as 10% of the saturated water content remains at water potentials more negative than –0.5 MPa. Conversely, soils of finer texture, which drain over a much wider range of potentials, can supply water to growing plants at water potentials across the whole range from –0.5 to –1.5 MPa at low transpiration rates.

Permanent wilting point (PWP) has been defined by many authors as the moisture content at a soil matric suction of –1.5 MPa [12]. It is widely regarded to be the lower limit for water availability for plant growth. However, many plants are able to grow at soil water potentials well below PWP if there is any significant amount of water remaining in the soil, as occurs in moderate to heavy clays. In reality, the change in water content from –0.5 to –3.0 MPa is so small in most agricultural soils that it contributes little to plant growth over this range [12]. This suggests that the permanent wilting point is closer to –0.5 MPa than to –1.0 MPa in many agricultural soils and closer to –3.0 MPa in heavy clay soils.

The concept of field capacity has utility in some situations, although it is far from a defined concept. Field capacity is generally regarded as the water content remaining in a field soil 48 h after saturation or after all perceptible drainage has ceased. The widespread use of the –0.03MPa matric suction as the field capacity percentage is only approximate and should be used with care. See Cassel and Nielsen [12] for further information.

Cassel and Nielsen [12] define a simpler concept, known as container capacity, which should not be confused with field capacity. These concepts do not have the same physical basis. Container capacity refers to the moisture content of a containerized soil which has been saturated and allowed to drain until drainage ceases. Container capacity is substantially wetter than field capacity, because water in a field soil drains out of the surface layers at a tension below atmospheric pressure (0 MPa) to deeper layers with a more negative water potential. This means that field capacity in the surface soil depends upon the texture of soil horizons underneath the surface soil. Container capacity, on the other hand, occurs when the soil water has drained from the large pores to atmospheric pressure (0 MPa). Field soils contain few pores that drain near atmospheric pressure, hence these soils barely drain at all when saturated in a container, often leading to a lack of aeration. Hence, properly formulated potting mixes are used for containerized plants. These mixes contain pores large enough to drain at near atmospheric pressure and therefore provide sufficient aeration for root growth. Plants can be grown in fine-textured field soils in containers, but to provide for sufficient aeration, these soils must not be saturated more than transiently [48,52].

C. Water Movement in Soils

Water movement in soils is a complex subject, and readers are referred to other texts for more detailed information [13,25,37,44,48,57,72]. Among soilborne pests, water movement in soils is most important for movement of nematodes, the zoosporic fungi, and bacteria [51]. Water movement also occurs in response to changes in soil temperature, as was discussed earlier, equilibrating not only soil water potential but also soil temperature quickly throughout the soil. Interestingly, larger soil organisms such as plant roots, arthropods, and earthworms may have the greatest impact on soil water flow. It is known that water flows readily along burrows and that arthropod and earthworm burrows dramatically increase soil infiltration [37]. Soil water movement can be monitored in the field by monitoring soil water potential with a number of in situ methods, including tensiometry and psychrometry [48]. However, the use of these methods in the field requires considerable attention to experimental design [56].

D. Impact of Soil Moisture on Pest Populations

The fundamental significance of soil moisture to the survival and growth of soilborne pests depends upon whether there is a deficit or an excess of moisture. The effects of too much or too little moisture may act directly upon the pest or may be mediated through the plant host and other soil microorganisms in many ways.

Low soil moisture conditions occur most commonly in the tropical wet-dry climate, tropical deserts, in the seasonally dry littoral zone, and in marginal temperate soils. Drought itself is the most significant constraint of agricultural production in these regions; however, low soil moisture during any part of a growing season can have profound effects on pest populations and the pest-host interaction in several conflicting ways. Dry soil favors survival of dormant pest propagules such as eggs, seeds, and resting spores and tends to increase the tolerance of these propagules to high temperatures [18,49,65,71,76,83]. Furthermore, in dry or seasonally dry agricultural areas, the effect of low soil moisture predisposing plants to soilborne diseases has been recognized for many years [16].

Overseasoning through a dry period often means survival through hot/dry periods in the absence of a host. Most pests have evolved specialized overseasoning structures such as eggs, seeds, sclerotia, chlamydospores, and various thickened cells to survive during hot, dry conditions. Survival of both *M.*

phaseolina sclerotia and *Bipolaris* spp. conidia is strongly favored by dry conditions. Furthermore, very dry conditions enhance survival at higher temperatures [27,58,71]. Even *F. verticillioides* (*F. moniliforme*), a soil-invading fungus that causes stalk rot of maize and other diseases of tropical crops, lack specialized overseasoning structures but produces resistant hyphae that survive at least 30 months at 35°C and 33% relative humidity (RH) [49]. The survival of other soil-invading or debrisborne fungi such as *Pyrenophora* and *Cephalosporium gramineum* is favored by dry conditions [75,76]. Clearly, unless the temperature reaches the cardinal temperature for a species, a dry period does little to reduce soilborne pests adapted to a climate with periodic drought.

Severe drought stress completely stops germination of pest propagules and somatic growth and reproduction, as well as infection or feeding, whereas favoring survival. Severe drought stress also kills the host plant species, thereby depriving the pest of a food source, which also favors the dormant survival of the pest under these conditions. However, moderate drought stress, which does not kill the host plants, can affect different soil pests by distinct, direct effects on the pest and through changes in host susceptibility.

For example, infection by a pathogen is controlled by the ability of the pathogen to germinate, grow, and/or sporulate to produce primary inoculum and penetrate the host. The effect of low soil moisture on propagule germination and infection of susceptible host tissues varies according to the pathogen. The growth and sporulation of all Oomycete pathogens, such as *Pythium* and *Phytophthora*, is inhibited or stopped by dry conditions so that diseases caused by these fungi can be controlled by draining soil after rainfall or irrigation [8]. The germination of oospores of *Phytophthora* is favored by saturated or very wet soil conditions and is inhibited by moderately dry conditions [10,38]. Diseases caused by pythiaceous pathogens are rarely serious in regions lacking irrigation or seasonal heavy rains that produce localized flooding owing largely to the dependence of these organisms on free water for the dispersal of zoospores [22]. It is highly likely that the infection efficiency of direct germination under drier conditions, when it can occur, is so low that significant disease occurs only in saturated soil, when secondary zoospore release can occur. However, Ristaino and Duniway [64] showed that preinoculation moisture stress led to greater disease levels than postinoculation stress, suggesting that infection efficiency was increased by moisture stress, but tissue susceptibility and therefore colonization of the host was not. Anoxic stress is known to increase both infection rates and disease severity in many diseases caused by Oomycetes [22]. This topic is covered below.

Germination, penetration, and infection by Ascomycetes and Basidiomycetes pathogens have not been as well studied as that of the Oomycetes because of technical challenges of working with soil at below –30kPa water potential with any precision [22]. In one of the few studies specifically designed to study pathogen germination, penetration, and infection under dry soil conditions, several surprising results emerged. Many pathogens such as *F. verticillioides* (*F. moniliforme*), *Fusarium graminearum*, and *M. phaseolina* are thought to be "dryland" pathogens because of their frequent occurrence as root rot and stalk rot fungi in arid and semiarid zones. However, Liddell and Burgess [50] showed conclusively that infection of wheat by *F. graminearum* only occurs in a narrow window of soil matric potential between -0.1 and -1.5 MPa, which is much less than the window of matric potentials conducive for the vegetative growth of the fungus, which extends down to -7.7 Mpa [81]. Disease expression in stalk rot of maize caused by *F. verticillioides* (*F. moniliforme*) [79], crown rot of wheat caused by *F. graminearum* [52], and *Macrophomina* charcoal rot of many crops [9,23,41,59,84] depends on the moisture stress to which the host is exposed. However, it is clear that the relative effects of moisture stress on the pathogen and the host must be evaluated for each pathosystem, because each host and pathogen behaves differently in response to moisture stress. In addition, the soil microflora in different agroecosystems may be sufficiently variable to provide contrasting background antagonism at various moisture potentials, thus resulting in apparently contradictory results on epidemiological studies for the same pathogen [22].

The small number of precise studies on the influence of host water stress on infection and disease development have not yielded a consistent pattern. Host moisture stress after infection of wheat seedlings by *F. graminearum* does not lead to significantly higher infection rates or disease levels over unstressed control plants [7], yet in field trials with mature plants, such an effect has been reported [61]. Conversely, Blanco-Lopez and Jimenez-Diaz [9] showed that infection and disease severity of charcoal rot of sunflower is increased by drought stress. Unexpected and inconsistent results and apparently contradictory results like this may occur as the technical aspects of studies of this kind advance [22,48]. Plants respond to stress through the transduction of stress signals into new gene expression or changes in translation of an existing pool of mRNA, which may lead either to increased susceptibility to pathogens or induced resistance [46]. Hence, each pathosystem must be evaluated independently to differentiate between direct effects of moisture stress on plant growth and indirect effects that induce susceptibility or resistance.

High soil moisture is often a problem, at least seasonally, and may promote Oomycete pathogens by favoring pathogen dispersal [28,43] and disease development [21]. Flooding and heavy rainfall also allow soilborne diseases, such as white mold caused by *Sclerotinia* [69] or web blight of beans caused by *Rhizoctonia* [31] and *Pythium* pod rot [68], to spread to foliar plant parts. On the other hand, periodic flooding may actually help control some diseases and pests, such as *Macrophomina* charcoal rot, by reducing survival of sclerotia in wet soil [1,2,71]. Survival and growth of many soil arthropods and weed seeds in flooded soil is often reduced [63,65]. However, the most important effect of flooding on soilborne diseases is probably hypoxic/anoxic stress of plants, which favors diseases caused by zoosporic pathogens [22,36]. Hypoxia increases host susceptibility and may encourage oospore germination, which often requires light and exogenous nutrients to germinate [3,30], because anoxic plant tissues leak metabolites into the soil [16,21].

Flooding enhances dispersal of zoosporic pathogens and anoxia promotes the development of diseases caused by pythiaceous pathogens. However, flooding may also be beneficial for control of diseases caused by *Fusarium* and *Macrophomina* and for many arthropod and weed pests. Flooding, therefore, should be managed according to the prevailing pest circumstances. Efficient drainage of fields after rainstorms can be used to reduce anoxia and may be combined with irrigation management to control *Phytophthora* diseases [8,21]. Unfortunately, however, many tropical lateritic soils flood readily after rain and cannot be managed easily in this way [43].

V. OTHER SOIL AND EDAPHIC FACTORS

Soil fertility insidiously affects soilborne pests through deficiencies and toxicities that may not be detectable to unaided field personnel. The fact that some soil fertility problems themselves become limiting indicates that the role of poor soil fertility is widespread and likely has a broad and largely underappreciated role in the development of diseases and disorders. Vast areas of acid soils can lead to aluminum and manganese toxicity and the deficiency of phosphorus, calcium, and magnesium [29]. Calcium and manganese are of such profound importance to plant health and the activity of soil pests that further discussion of this topic is beyond the scope of this chapter and recent reviews on this subject should be consulted [26,39]. Soil fertility is perhaps the second greatest overall constraint after drought and may be one of the most manageable. It is generally true that in many regions of the world soil deficiencies and toxicities are of such magnitude as to make pest problems almost irrelevant until the soil problem can be alleviated.

Herbicides are well known to affect nontarget species in several ways. Generally, both plant disease control and increased susceptibility have been reported for different herbicide/disease interactions [6,15,40,55,73]. The effects of herbicides on nonpathogenic components of the soil microbiota also may be important [24]. The effects of herbicides in the development of tropical soilborne disease have not been extensively studied and likely range from little effect to an extreme effect in regions where pesticides are widely applied. Again, the direct effects of herbicides on any pathosystem need to be determined on a case-by-case basis. There have been recent reviews on the subject, including this volume, to which the reader is referred for more detailed information [42,47].

Soilborne pests are affected in numerous ways by edaphic and climatic factors, the most important of which are wind, extreme humidity, intense radiation, soil compaction, air pollution, soil pollution, water pollution, and salinity. Salinity from overfertilization and natural sources can become serious under high evapotranspiration rates [14]. Wind in the intertropical convergence zone also may be severe ad disperse pests over large distances and perhaps increase crop susceptibility through mechanical damage [5].

VI. CONCLUSIONS

The ecology of soil pests and the problems they cause is frequently strongly affected by soil water potential either directly or indirectly. Controlled studies in the laboratory or greenhouse often require the control of soil water potential as an experimental treatment or require soil water potential to be monitored even if it is not an explicit treatment. The methods available to achieve both of these ends may not be ideal but are improving sufficiently that some of the long-standing questions in soilborne plant disease epidemiology and soil pest ecology can be addressed with renewed confidence. Unfortunately, too many studies in soil ecology have failed to monitor soil moisture potential and have reported on properties such as temperature without properly controlling for water potential or even monitoring it, thus confounding results.

The role of the flagellum in zoospore movement in soils and the role of flowing water in the dispersal of zoosporic fungi, bacteria, and nematodes are of profound importance in irrigated soils or seasonally flooded soils [51]. That very small changes in soil matric potential influence the sporulation of zoosporic fungi is beyond doubt [20]. However, evidence is accumulating that small changes in soil matric potential may also influence the behavior of soil pests under much drier conditions, possibly by altering the populations and activity of antagonistic bacteria, fungi, and actinomycetes [22]. The implications here for biological control of soilborne plant diseases are enormous.

Soil pests are an intimate part of the soil biota and their populations and activity are determined by temperature, water availability, nutrition sources, and activity of other soil organisms, as well as many less obvious factors. The study of soil pests can only benefit from the application of a broad range of techniques to broaden our understanding of biological activity in the soil.

REFERENCES

1. Abawi G.S. 1989. Root rots. In *Bean Production Problems in the Tropics,* 2nd ed. H.F. Schwartz & M.A. Pastor-Corales, eds. Cali, Colombia: CIAT, pp 105–157.

2. Abawi, G.S., and M.A. Pastor-Corales. 1990. *Root Rots of Beans in Latin America and Africa: Diagnosis, Research Methodologies and Management Strategies*. Cali, Colombia: CIAT.

3. Ann, P.J., and W.H. Ko. 1988. Induction of oospore germination of *Phytophthora parasitica*. *Phytopathology* 78:335–338.

4. Aragno, M. 1981. Responses of microorganisms to temperature. In *Physiological Plant Ecology I. Responses to the Physical Environment*, P.S. Nobel, C.B. Osmond, O.L. Lange, & M. Ziegler, eds. *Encyclopedia of Plant Physiology, New Series*, Vol. 12A. Berlin: Springer-Verlag, pp 339–369.

5. Aylor, D.E. 1990. The role of intermittent wind in the dispersal of fungal pathogens. *Annu. Rev. Phytopathol.* 28:73–92.

6. Bauske, E.M., and H.W. Kirby. 1992. Effects of dinitroaniline herbicides, carboxin-pentachloronitrobenzene seed treatment, and *Rhizoctonia* disease on soybean. *Plant Dis.* 76:236–239.

7. Beddis, A.L., and L.W. Burgess. 1992. The influence of plant water-stress on infection and colonization of wheat seedlings by *Fusarium graminearum*. *Phytopathology* 82:78–83.

8. Biles, C.L., D.L. Lindsey, and C.M. Liddell. 1992. Control of Phytophthora root rot of chile peppers by irrigation practices and fungicides. *Crop Prot.* 11:225–228.

9. Blanco-Lopez, M.A., and R.M. Jimenez-Diaz. 1983. Effect of irrigation on susceptibility of sunflower to *Macrophomina phaseoli*. *Plant Dis.* 67:1214–1217.

10. Bowers, J.H., and D.J. Mitchell. 1990. Effect of soil-water matric potential and periodic flooding on mortality of pepper caused by *Phytophthora capsici*. *Phytopathology* 80:447–1450.

11. Brown, A.D. 1976. Microbial water stress. *Bacteriol Rev.* 40:803–846.

12. Cassel, D.K., and D.R. Nielsen. Field capacity and available water capacity. In *Methods of Soil Analysis*. Part 1. *Physical and Mineralogical Methods*. 2nd ed. A. Klute, ed. Madison, Wisconsin: ASA and SSSA, pp 901–926.

13. Childs, E.C. 1969. *An Introduction to the Physical Basis of Soil Water Phenomena*. London: Wiley-Interscience.

14. Cloud, G.L., and J.C. Rupe. 1994. Influence of nitrogen, plant-growth stage, and environment on charcoal rot of grain sorghum caused by *Macrophomina phaseolina* (Tassi) Goid. *Plant and Soil* 158:203–210.

15. Cohen, R., B. Blaier and J. Katan. 1992. Chloroacetamide herbicides reduce incidence of *Fusarium* wilt in melons. *Crop Prot.* 11:181–185.

16. Colhoun, J. 1979. Predisposition by the environment. In *Plant Disease*. Vol. 4. J.G. Horsfall and E.B. Cowling, New York: Academic Press, pp 75–92.

17. Cornejo, F.H., A. Varela, and S.J. Wright. 1994. Tropical forest litter decomposition under seasonal drought - nutrient release, fungi and bacteria. *Oikos* 70:183–190.

18. Cousens, R. and M. Mortimer. 1995. *Dynamics of Weed Populations*. Cambridge, UK: Cambridge University Press.

19. Danks, H.V. 1987. *Insect Dormancy: An Ecological Perspective*. Ottawa: Biological Survey of Canada.

20. Duniway, J.M. 1976. Movement of zoospores of *Phytophthora cryptogea* in soils of various textures and matric potentials. *Phytopathology* 66:877–882.

21. Duniway, J.M. 1983. Role of physical factors in the development of *Phytophthora* diseases. In *Phytophthora: Its Biology, Taxonomy, Ecology and Pathology*. D.C. Erwin, S. Bartnicki-Garcia, and P.H. Tsao, eds. St. Paul, Minnesota: APS Press, pp 175–187.

22. Duniway, J.M. and T.R. Gordon. 1986. Water relations and pathogen activity in soil. In *Water, Fungi and Plants*. P.G. Ayres and L. Boddy, eds. Cambridge, UK: Cambridge University Press, pp 119–137.

23. Edmunds, L.K. 1964. Combined relation of plant maturity, temperature, and soil moisture to charcoal stalk rot development in grain sorghum. *Phytopathology* 54:514–517.

24. Elmholt, S. 1991. Side-effects of propiconazole (Tilt 250 EC) on nontarget soil fungi in a field trial compared with natural stress effects. *Microb. Ecol.* 22:99–108.

25. Elzeftawy, A., and M.D. Mifflin. 1983. Vadose moisture migration in semi-arid and arid environments. in *Proceedings of NWWA/U.S. EPA Conference on Characterization and Monitoring of the Vadose (Unsaturated) Zone* (Dec. 8–10, 1983; Las Vegas, Nevada). Worthington, Ohio: National Water Well Association, pp 1–36.

26. Englehard, A.W. 1989. *Management of Diseases with Macro- and Microelements*. St. Paul, Minnesota: APS Press.

27. Filonow, A.B., and D.K. Arora. 1987. Influence of soil matric potential on ^{14}C exudation from fungal propagules. *Can. J. Bot.* 65:2084–2089.

28. Fitt, B.D.L., H. A. McCartney, and P.J. Walklate. 1989. The role of rain in dispersal of pathogen inoculum. *Annu. Rev. Phytopathol.* 27:241–270.

29. Flor, C. and M.T. Thung. 1989. Nutritional Disorders. In *Bean Production Problems in the Tropics*. H.F. Schwartz and M.A. Pastor-Corales, eds. Cali, Colombia: CIAT, pp 571–604.

30. Förster, H., O.K. Ribeiro, and D.C. Erwin. 1983. Factors affecting oospore germination of *Phytophthora megasperma* f. sp. *medicaginis*. *Phytopathology* 73:442–448.

31. Gálvez, G.E., B. Mora, and M.A. Pastor-Corrales. 1989. Web Blight. In *Bean Production Problems in the Tropics*. H.F. Schwartz and M.A. Pastor-Corales, eds. Cali, Colombia: CIAT, pp 195–209.

32. Gardner, W.M. 1986. Water Content. In *Methods of Soil Analysis*. Part 1. *Physical and Mineralogical Methods*. 2nd ed. A. Klute, ed. Madison, Wisconsin: ASA and SSSA, pp 493–544.

33. Griffin, D.M. 1972. *The Ecology of Soil Fungi*. Syracuse, New York: Syracuse University Press.

34. Griffin, D.M. 1977. Water potential and wood decay fungi. *Annu. Rev. Phytopathol.* 15:319–329.

35. Griffin, D.M. 1985. Soil as an environment for the growth of root pathogens. In *Ecology and Management of Soil Borne Plant Pathogens*. C.A. Parker, A.D. Rovira, K.J. Moore, et al. eds. St. Paul, Minnesota: APS Press, pp 187–190.

36. Hancock, J.G., and G.W. Grimes. 1990. Colonization of rootlets of alfalfa by species of *Pythium* in relation to soil-moisture. *Phytopathology* 80:1317–1322.

37. Hillel, D. 1980. *Fundamentals of Soil Physics*. New York: Academic Press.
38. Hord, M.J., and J.B. Ristaino. 1992. Effect of the matric component of soil-water potential on infection of pepper seedlings in soil infested with oospores of *Phytophthora capsici*. *Phytopathology* 82:792–798.
39. Huber, D.M., and R.D. Graham. 1992. Techniques for studying nutrient-disease interactions. In *Methods for Research on Soilborne Phytopathogenic Fungi*. L.L. Singleton, J.D. Milhail, and C.M. Rush, eds. St. Paul, Minnesota: APS Press, pp 204–214.
40. Jeffery, S. and L.W. Burgess. 1990. Growth of *Fusarium graminearum* Schwabe group 1 on media amended with atrazine, chlorsulfuron or glyphosate in relation to temperature and osmotic potential. *Soil Biol. Biochem.* 22:665–670.
41. Jimenez-Diaz, R.M., M.A. Blanco-Lopez, and W.E. Sackstom. 1983. Incidence and distribution of charcoal rot of sunflower caused by *Macrophomina phaseolina* in Spain. *Plant Dis.* 67:1033–1036.
42. Kataria, H.R., and U. Gisi. 1990. Interactions of fungicide herbicide combinations against plant-pathogens and weeds. *Crop Prot.* 9:403–409.
43. Kinal, J., B.L. Shearer, and R.G. Fairman. 1993. Dispersal of *Phytophthora cinnamomi* through lateritic soil by laterally flowing subsurface water. *Plant Dis.* 77:1085–1090.
44. Kirkham, D., and W.L. Powers. 1972. *Advanced Soil Physics*. New York: Wiley-Interscience.
45. Kramer, P.J. 1983. *Water Relations of Plants*. San Diego: Academic Press.
46. Kuhn, D.N. 1987. Plant responses to stress at the molecular level. In T. Kosuge and E.W. Nester, eds. *Plant-Microbe Interactions-Molecular and Genetic Perspectives*. Vol. 2. New York: Macmillan, pp 414–440.
47. Levesque, C.A., and J.E. Rahe. 1992. Herbicide interactions with fungal root pathogens, with special reference to glyphosate. *Annu. Rev. Phytopathol.* 30:579–602.
48. Liddell, C.M. 1992. Measurement and control of soil temperature and water potential. In *Methods for Research on Soilborne Phytopathogenic Fungi*. L.L. Singleton, J.D. Milhail, and C.M. Rush, eds. St. Paul, Minnesota: APS Press, pp 187–203.
49. Liddell, C.M., and L.W. Burgess. 1985. Survival of *Fusarium moniliforme* at controlled temperature and relative humidity. *Trans. Br. Mycol. Soc.* 84:121–130.
50. Liddell, C.M., and L.W. Burgess. 1988. Wax partitioned soil columns to study the influence of soil moisture potential on the infection of wheat by *Fusarium graminearum* Group 1. *Phytopathology* 78:185–189.
51. Liddell, C.M., and J.L. Parke. 1989. Enhanced colonization of pea taproots by a fluorescent pseudomonad biocontrol agent by water infiltration into soil. *Phytopathology* 79:1327–1332.
52. Liddell, C.M., L.W. Burgess, and P.W.J. Taylor. 1986. Reproduction of crown rot of wheat caused by *Fusarium graminearum* Group 1 in the greenhouse. *Plant Dis.* 70:632–635.
53. Lifshitz, R., and J.G. Hancock. 1983. Saprophytic development of *Pythium ultimum*

in soil as a function of water potential and temperature. *Phytopathology* 73:257–261.

54. Luard, E.J. 1982. Accumulation of intracellular solutes by two filamentous fungi in response to growth at low steady-state osmotic potential. *J. Gen. Microb.* 128:2563–2574.

55. Moustafamahmoud, S.M., D.R. Sumner, M.M. Ragab, and M.M. Ragab. 1993. Interaction of fungicides, herbicides, and planting date with seedling disease of cotton caused by *Rhizoctonia solani* AG4. *Plant Dis* 77:79–86.

56. Nielsen, D.R., P.M. Tillotson, and S.R. Viera. 1983. Analyzing field-measured soil-water properties. *Agric. Wat. Manag.* 6:93–109.

57. Nobel, P.S. 1983. *Biophysical Plant Physiology and Ecology*. San Francisco: Freeman.

58. O'Leary, D.J., and J.L. Lockwood. 1988. Debilitation of conidia of *Cochliobolus sativus* at high soil matric potentials. *Soil Biol. Biochem.* 20:239–243.

59. Odvody, G.N., and L.D. Dunkle. 1979. Charcoal stalk rot of sorghum: Effect of environment on host-parasite relations. *Phytopathology* 69:250–254.

60. Papendick, R.I., and G.S. Campbell. 1981. Theory and measurement of water potential. In *Water Potential Relations in Soil Microbiology* (SSSA Special Publication No. 9). J.F. Parr, W.R. Gardner, and L.F. Elliott, eds. Madison, Wisconsin: SSSA, pp 1–22.

61. Papendick, R.I., and R.J. Cook. 1974. Plant water stress and development of Fusarium root rot in wheat subjected to different cultural practices. *Phytopathology* 64:358–363.

62. Papendick, R.I., and D.J. Mulla. 1986. Basic principles of cell and tissue water relations. In *Water, Fungi and Plants* (BMS Symposium No. 11). P.G. Ayers and L. Boddy, eds. Cambridge, UK: Cambridge University Press, pp 1–25.

63. Radosevich, S.R., and J.S. Holt. 1984. *Weed Ecology—Implications for Vegetation Management*. New York: Wiley.

64. Ristaino, J.B., and J.M. Duniway. 1989. Effect of preinoculation and postinoculation water-stress on the severity of Phytophthora root rot in processing tomatoes. *Plant Dis.* 73:349–352.

65. Romoser, W.S., and J.G. Stoffolano. 1994. *The Science of Entomology*. 3rd ed. Dubuque, Iowa: Brown.

66. Rorison, I.H. 1981. Plant growth response to variations in temperature: field and laboratory studies. In *Plants and Their Atmospheric Environment*. J. Grace, E.D. Ford, and P.G. Jarvis, eds. Oxford, UK: Blackwell, pp 313–332.

67. Rowe, R.C., and R.L. Powelson. 1973. Epidemiology of *Cercosporella* root rot of wheat: spore production. *Phytopathology* 63:981–984.

68. Schwartz, H.F. 1989. Additional fungal pathogens. In *Bean Production Problems in the Tropics*. H.F. Schwartz and M.A. Pastor-Corales, eds. Cali, Colombia: CIAT, pp 231–259.

69. Schwartz, H.F., and J.R. Stedman. 1989. White Mold. In *Bean Production Problems in the Tropics*. H.F. Schwartz and M.A. Pastor-Corales, eds. Cali, Colombia: CIAT, pp 211–230.

70. Scott, W.J. 1957. Water relations of food spoilage microorganisms. *Adv. Food Res.* 7:83–127.

71. Shiekh, A.H., and A. Ghaffer. 1987. Time-temperature relationships for the inactivation of sclerotia of *Macrophomina phaseolina*. *Soil Biol. Biochem.* 19:313–315.

72. Slatyer, R.O. 1967. *Plant-Water Relationships*. London: Academic Press.

73. Smiley, R.W., and D.E. Wilkins. 1992. Impact of sulfonylurea herbicides on Rhizoctonia root rot, growth, and yield of winter-wheat. *Plant Dis.* 76:399–404.

74. Smith, W.L., W.H. Miller, and R.D. Bassett. 1965. Effects of temperature and relative humidity on germination of *Rhizopus stolonifer* and *Monilina fructicola* spores. *Phytopahtology* 55:497–602.

75. Specht, L.P., and T.D. Murray. 1989. Sporulation and survival of conidia of *Cephalosporium gramineum* as influenced by soil-pH, soil matric potential, and soil fumigation. *Phytopathology* 79:787–793.

76. Summerell, B.A., and L.W. Burgess. 1989. Factors influencing survival of *Pyrenophora tritici-repentis*—water potential and temperature. *Mycol. Res.* 93:41–45.

77. Tauber, M.J., C.A. Tauber, and S. Masaki. 1986. *Seasonal Adaptations of Insects*. New York: Oxford University Press.

78. Taylor, S.A., and R.D. Jackson. 1986. Temperature. In *Methods of Soil Analysis*. Part 1. *Physical and Mineralogical Methods*. 2nd ed. A. Klute, ed. Madison, Wisconsin: ASA and SSSA, pp 927–940.

79. Trimboli, D.S., and L.W. Burgess. 1983. Reproduction of *Fusarium moniliforme* basal stalk rot and root rot of grain sorghum in the greenhouse. *Plant Dis.* 67:891–894.

80. Villani, M.G., and R.J. Wright. 1990. Environmental influences on soil macroarthropod behavior in agricultural systems. *Annu. Rev. Entomol.* 35:249–269.

81. Wearing, A.H., and L.W. Burgess. 1979. Water Potential and the saprophytic growth of *Fusarium roseum* "graminearum." *Soil Biol. Biochem.* 11:661–667.

82. Wilson, E.O. 1971. *The Social Insects*. Cambridge, Massachusetts: Harvard University Press.

83. Wilson, R.G. 1988. Biology of weed seeds in the soil. In *Weed Management in Agroecosystems: Ecological Approaches*. M.A. Altieri and M. Liebman, eds. Boca Raton, Florida: CRC Press, pp 25–39.

84. Wyllie, T.D., and O.H. Calvert. 1969. Effect of flower removal and podset on the formation of sclerotia and infection of *Glycine max* by *Macrophomina phaseoli*. *Phytopathology* 59:1243–1245.

3
Weather Influences on Pest Movement

David E. Pedgley*
Natural Resources Institute, University of Greenwich, Chatham Maritime, England

I. INTRODUCTION

Strategies for pest management often assume that pest populations are more or less static. Numbers of insects or intensity of disease within a crop may be monitored, sometimes routinely, and emergency action taken when a certain threshold is exceeded. But pest organisms can be highly mobile, with individuals traveling many kilometers before causing harm, such as crop yield reduction or illness in humans and livestock through transmission of disease organisms. Such mobility adds to the difficulties of management, but it also provides opportunities for alternative management strategies: monitoring and controlling pests at their source or during movement (preventive strategies) as well as in the area at risk (defensive strategies). Study of pest mobility is particularly justified if it can lead to a reduction in (a) spread of harm, particularly that caused by an influx of a new genotype, and (b) waste, such as field monitoring or application of management techniques at wrong times or places.

Movement requires energy. Such energy is often beyond the resources of small pest organisms and must be provided by a vector, usually moving animals (including humans) or a fluid environment (water or air). This chapter illustrates the effects of moving air (wind) on long-distance displacement (kilometers or more) of small pest organisms, both fliers (e.g., winged insects) and drifters (e.g., wingless arthropods and microbial pathogens), and on the sub-

---

*Retired.

sequent development of outbreaks. It examines the mechanisms of departure, transit, and arrival, the distances moved and speeds attained, and the consequent problems of monitoring and forecasting movements. This chapter also describes management strategies that have been adopted for some highly mobile pests, and finally it discusses problems in the operational application of these strategies.

II. LONG-DISTANCE MOVEMENTS

This section is an outline of the evidence for long-distance movements of a range of biota. For general references to insect migration and its ecological significance, including the spread of their parasites and predators, see Dingle [7,8], Drake and Gatehouse [11], Gatehouse [13], and Kennedy [19]; and for a review of long-range spore movements, see Davis [5].

A. Insects

Most monocultures have insect pest problems that can be serious because of their intensity and extent. Natural monocultures, such as grasslands and coniferous forests, tend to be perennial, whereas manmade monocultures are usually annual crops. Many insect pests of annual crops are long-distance fliers, with colonizing lifestyles, high fecundity, and short generation times. Rapid pest population growth is adapted to the seasonality of their hosts. Even some natural monocultures have insect pests that fly long distances, such as the African armyworm, *Spodoptera exempta*, on grasslands and the spruce budworm, *Choristoneura fumiferana*, on forests.

Evidence for long-distance flights of insects comes from a variety of field observations (Table 1). Some of these observations involved little more than regular visual monitoring, preferably daily, of population densities at various stages of the life cycle at a fixed site or a network of sites. Other observations utilized equipment and experimental field techniques; for example, traps, trial plots, field surveys, host eradication, or mass rearing and marking. The most elaborate observations require aircraft or radar to measure movement [9,34,43].

The quantity and quality of evidence indicating long-distance flights of insects varies greatly among species. For some, the evidence is strong, but for others, it is still fragmentary and perhaps unconvincing. For most species, there has been little or no attempt to collate evidence for or against long-distance flights, even for species related to those where the evidence is strong. This can be attributed to lack of appreciation of the possible importance of such flights, or to the high cost of field work, particularly at night or at flight levels several hundred meters or more above the ground. Some species, such as the common cutworm, *Agrotis segetum*, and the Egyptian cotton leafworm, *Spodoptera littoralis*, are not usually considered to be long-distance fliers even though there is evidence to the contrary.

Table 1   Evidence for Long-Distance Insect Flight

Direct
1. Mark-and-recapture—markers either artificial (e.g., dye, radioactivity, mutant) or natural (e.g., crop contents, pollen load, phoretic mites, susceptibility to insecticide or disease, color morph).
2. Caught at high altitude (trap on tower, kite, or aircraft).
3. Seen flying at high altitude (radar, searchlights, moon-watching).

Circumstantial
1. Complete absence of species, in all stages of life cycle, during some part(s) of the year.
2. Appearance of immature stages or vectorborne disease far from sources (even though flying adults not seen).
3. Appearance of flying adults before local emergence, especially simultaneously at places far apart, or progressively across country, or with other known migrant species.
4. Appearance of flying adults soon after massive emergence at a distant source.
5. Long flight durations in laboratory tests.
6. Appearance of flying adults but no subsequent immature stages.
7. Appearance of flying adults far from sources (e.g., over seas, deserts, mountains, recently cleared land).
8. Daily fluctuations of adult numbers (in trap or crop) not related to local numbers of immatures.
9. Early arrivals more worn than later ones.
10. Newly emerged adults take off and climb to high altitudes.
11. Newly emerged adults have reproductive diapause (lasting days to months).
12. Last flying females of the season are mostly unmated.
13. Numbers of adults in one year not related to numbers of immatures at end of previous year.
14. Progressive seasonal redistribution of successive generations in the absence of a quiescent phase.
15. Genetic, behavioral, or physiological similarity over wide geographical areas.
16. Change in cropping practice in one area leads to change in insect incidence in another area.

Only a small proportion of an insect population may fly long distances, and the proportion may vary seasonally and geographically across the distribution area. This variability may be a result of genetic differences and under the influence of environmental cues, particularly day length. Laboratory studies of flight duration can provide evidence of potential long-distance flights [4].

Table 2 lists some major insect pests for which there is evidence of long-distance flight. Although the list is intentionally large, it is not exhaustive. A wide range of taxa is represented, but Table 2 emphasizes the well-known importance of long flights by moth species, including leafworms, fruitworms, and stalkworms. Other species are vectors of pathogenic organisms, notably fungi

Table 2 Long-Distance Migrant Insect Pests of Crops (and Some Forest Pests marked *)

Species	Common name	Evidence from[a]					
		1	2	3	4	5	6
Lepidoptera							
Agrotis infusa	Bogong moth	+					+
Agrotis ipsilon	Black or greasy cutworm			+	+	+	+
Agrotis segetum	Common cutworm			+			
Alabama argillacea	American cotton leafworm	+					
Anticarsia gemmatalis	Velvetbean caterpillar	+					
Autographa gamma	Silver-Y moth			+			
*Choristoneura fumiferana**	Spruce budworm	+					
Euxoa auxiliaris	Army cutworm	+					
Chaphalocrocis medinalis	Rice leafroller					+	
Diatraea grandiosella	Southwestern cornborer	+					
Diparopsis castanea	Red bollworm				+		
Earias biplaga	Spiny bollworm				+		
Helicoverpa armigera	Gram podborer			+		+	+
Heliothis virescens	Tobacco budworm	+					
Helicoverpa zea	Corn earworm	+					
Homoeosoma ellectellum	Sunflower moth	+					
Lymantria dispar	Gypsy moth			+		+	
*Malacosoma disstria**	Forest tent caterpillar	+					
Loxostege sticticalis	Beet webworm	+					+
Mythimna convecta	Common armyworm					+	
Mythimna separata	Oriental armyworm			+		+	
Mythimna unipuncta	True armyworm	+					
Ostrinia nubialis	European cornborer	+					
Pectinophora gossypiella	Pink bollworm	+					
Peridroma saucia	Variegated cutworm	+					
Persectania ewingii	Southern armyworm						+
Plathypena scabra	Green cloverworm	+					

Species	Common name											
Plutella xylostella	Diamondback moth	+					+			+		
Pseudoplusia includens	Soybean looper									+		
Spodoptera exempta	African armyworm			+			+			+		
Spodoptera exigua	Beet armyworm			+		+	+			+		
Spodoptera frugiperda	Fall armyworm				+	+	+			+		
Spodoptera littoralis	Egyptian cotton leafworm				+	+						
Torrix viridana*	Green oak tortrix						+					
Trichoplusia ni	Cabbage looper	+				+	+					
Diptera												
Cochliomyia hominivorax	Screwworm fly			+		+	+					
Bactrocera cucurbitae	Melon fly					+						
Drosophila melanogaster	Vinegar fly											
Oscinella frit	Frit fly											
Simulium damnosum	Blackfly											
Homoptera												
Aphis craccivora	Cowpea aphid							+				
Aphis fabae	Black bean aphid						+		+			
Aphis gossypii	Cotton aphid								+			
Sitobion avenae	English grain aphid						+					
Sitobion miscanthi	Grain aphid						+			+		
Myzus persicae	Peach-potato aphid						+		+	+		
Phorodon humuli	Damson-hop aphid						+			+		
Rhopalosiphum fitchii	Apple-grain aphid						+		+		+	
Rhopalosiphum maidis	Corn leaf aphid						+		+		+	
Rhopalosiphum padi	Bird-cherry oat aphid						+					
Schizaphis graminum	Greenbug											+
Cicadulina mbila	Maize leafhopper						+					
Circulifer tenellus	Beet leafhopper						+					
Empoasca fabae	Potato leafhopper						+					
Macrosteles quadrilineatus	6-Spotted leafhopper											

(continued)

Table 2 Continued

Species	Common name	Evidence from[a]					
		1	2	3	4	5	6
Nilaparvata lugens	Brown planthopper					+	
Perkinsiella saccharicida	Sugarcane leafhopper						+
Sogatella furcifera	White-backed planthopper					+	
Bemisia tabaci	Cotton whitefly				+		
Coleoptera							
Anthonomus grandis	Boll weevil	+	+				
*Dendroctonus ponderosae**	Mountain pine beetle	+					
Diabrotica duodecempunctata	Spotted cucumber beetle	+					
Diabrotica virgifera	Western corn rootworm	+					
*Ips typographus**	Spruce bark beetle			+			
Leptinotarsa decemlineata	Colorado potato beetle	+		+			
Heteroptera							
Dysdercus voelkeri	Cotton stainer	+			+		
Murgantia histrionica	Harlequin bug	+					
Nezara viridula	Southern green stink bug	+					
Oncopeltus fasciatus	Milkweed bug	+					
Orthoptera							
Aiolopus simulatrix	Sudan plague locust				+		
Chortoicetes terminifera	Australian plague locust					+	
Locusta migratoria	Migratory locust				+		
Melanoplus sanguinipes	Rocky Mountain locust	+					
Oedaleus senegalensis	Senegalese grasshopper				+		
Schistocerca cancellata	South American locust		+				
Schistocerca gregaria	Desert locust					+	+

[a]1, North America; 2, South America; 3, Europe; 4, Africa; 5, Asia; 6, Australasia.

and viruses, the latter being carried particularly by a variety of species of Homoptera. Much of the evidence for long-distance arthropod movement comes from North America, although it should often be applicable in other continents where the same species occur. However, some of the most extensively studied long-distance fliers (among locusts and moths) are not found in North America. So far, little evidence has come from South America.

B. Microbes

Most monocultures have serious fungal diseases. There is increasing evidence that outbreaks can be caused by long-distance windborne movements of fungal spores. The variety of this evidence (Table 3) is less than for insect pests, and for some species the evidence is only suggestive. As with insects, some observations involve little more than regular visual monitoring (e.g., of disease incidence at a fixed site or a network of sites); others require equipment for field experiments (e.g., rain collectors, traps, trial plots, or field surveys). Monitoring spore-arrival date is clearly more precise than estimating it from subsequent disease development.

Table 4 lists some major fungal crop diseases known to be associated with long-distance spore movements. There are fewer species compared with the insects of Table 2, reflecting the more limited research done on phytopathogen movement. Table 4 emphasizes the importance placed on cereal diseases, particularly those caused by rusts. As with insect pests, much of the evidence about phytopathogen movement comes from North America.

Bacteria are also known to travel long distances. Some may be carried hundreds of kilometers [35], but those in rain splash travel much shorter distances;

Table 3 Evidence for Long-Distance Spore Movements

Direct
1. Appearance of spores of recognizable races (i.e., naturally marked individuals) far from sources and in absence of vectors.
2. Spores trapped at high altitude (by tower, kite, or aircraft).

Circumstantial
1. Absence of the species, in all stages of life cycle, during some part(s) of the year; e.g., the cold or dry season.
2. Appearance of viable spores in rain or sampler or on a sentinel plot or a crop before local production or after massive production at a distant source.
3. Appearance of viable spores or of subsequent disease simultaneously at places far apart.
4. Repeated progressive seasonal redistribution of disease.
5. Random distribution of disease foci across crop and at a uniform plant height (canopy top at arrival date).

Table 4 Fungal Diseases Known To Be, or Possibly, Associated with Long-Distance Movement of Spores

		Evidence from[a]					
		1	2	3	4	5	6
Erysiphe graminis f. sp. *hordei*	Barley powdery mildew	+					
Helminthosporium maydis	Southern corn leaf blight			+			
Hemileia vastatrix	Coffee leaf rust		+				
Melampsora spp.	Poplar rust						+
Mycosphaerella musicola	Banana leaf spot	+	+				
Peronospora tabacina	Tobacco blue mold			+			
Puccinia coronata	Oat crown rust			+			
Puccinia graminis f. sp. *tritici*	Wheat stem rust	+				+	
Puccinia melanocephala	Sugarcane rust	+					
Puccinia polysora	Maize rust				+		
Puccinia recondita	Wheat leaf rust	+					
Puccinia striiformis	Wheat stripe rust						+

[a]1, North America; 2, South America; 3, Europe; 4, Africa; 5, Asia; 6, Australasia.

for example, the *Erwinia* rots (e.g., fireblight on apples and pears [14]) and blackleg (on potatoes [15]).

C. Seeds

The seeds or fruits of some flowering plants drift on the wind either because they are dustlike (e.g., Orchidaceae) and behave similarly to microbes or because they have structures that reduce their fallspeeds. The most well-known structures are the pappose fruits of the Compositae, but others have tufted fruits or seeds (e.g., Onagraceae). Drifting as far as tens of kilometers is indicated by recolonization of areas cleared by fire. Other species have larger, winged fruits (samaras) that fall faster than tufted ones and, therefore, drift shorter distances (e.g., many forest trees, including Aceraceae, Betulaceae, Pinaceae, and Ulmaceae). Some species are weeds because they compete with crops or because they are toxic to livestock (e.g., ragwort, *Senecio jacobaea*).

A few field studies using traps set downwind of sources have demonstrated progressive seed fallout [12,18]. Seedfall density, y seeds per unit area, varies with distance from source, x, in the form:

$$\ln y = a + bx + cx^2, \text{ or } \ln y = a + b(\ln x)$$

These profiles reflect the effects of winnowing (differential settling): they are long-tailed with a very small proportion traveling great distances. Although the studies have been of dispersal of seeds from forest trees, the principles are applicable to smaller plants, but there do not seem to have been any quantitative studies of the long-distance spread of weed seeds. Short-distance spread, quantified by color marking and subsequent trapping of, for example, the pappose fruits of ragwort [23], has been shown to have similar profiles, which vary, as might be expected, with source height and dominance of the species in the natural vegetation.

Similar short-distance profiles have, however, also been found for wingless arthropods, including scale insects and mites; for example, sugar cane scale, *Aulacaspis tegalensis* [16], and the gall mite *Phytopus ribis*, vector of blackcurrant reversion virus [39].

III. WEATHER INFLUENCES ON MOVEMENT

This section deals mainly with those aspects of weather influences that bear upon management strategy. For more extensive discussions, see Pedgley [28], Drake and Farrow [10], Drake and Gatehouse [11], and Burt and Pedgley [2] for insects, and Pedgley [29,31] for spores.

A. Departure

When an insect, whether winged or not, is ready to fly, its departure may be inhibited or enhanced by the weather. There may be lower and upper thresholds of temperature and wind speed below and above which takeoff is more or less impossible because of physiological or aerodynamic constraints. Brief gusts can promote takeoff when most of the time the wind is too weak, or lulls can be briefly favorable when the wind is otherwise too strong. Delay by strong wind can enhance subsequent readiness for takeoff. Rain may or may not have a direct effect, but associated low temperatures or strong winds can be inhibitory. Spores of some fungal species become airborne in the splash from falling drops, whether rain or drip. Some species release their spores when the relative humidity is near saturation, whereas others require warm weather and low humidity.

Weather affects the timing of takeoff—not only from day to day but also from hour to hour. Temperature and wind speed vary from day to day principally through the never-ending influences of large, mobile weather systems and from hour to hour principally through the influence of the solar cycle. For example, in temperate latitudes, the warm, dry weather often found in an anticyclone, favoring insect takeoff and spore release by some species, may be followed the next day by wind, rain, and cool weather of a frontal system, perhaps inhibiting insect takeoff yet favoring spore release in other fungal species. Moreover, the often reduced temperature, high humidity, and weak wind at night, contrasting with the higher temperature, lower humidity, and more gusty wind in the afternoon, lead to the well-known differences in biota that become airborne by night and by day. Decreasing illumination is the main cue for night-flying insects to take off.

B. Transit

Once airborne, the speed of transit is affected by the wind. A winged insect flies through the air, with air speeds mostly < 10 km h^{-1}. Where the air speed is greater than the wind speed (such as occurs typically close to the ground in the flight boundary layer), the insect can progress toward a chosen goal (food, shelter, mate, or egg-laying site), but where the reverse is true (above the layer), then movement is more or less downwind. In wind speeds typical of heights at which nocturnal migrants fly (often 20 km h^{-1} or more), distance traveled is approximately the product of wind speed and flight duration. Insects usually have control over flight duration, but spores and seeds do not. Insects may descend actively to land, whereas spores and seeds are passively deposited (as, indeed, are some wingless insects). An insect may fly actively for seconds, minutes, or hours, varying between species and between individuals within a species.

Winds at altitudes of several hundred meters can be much stronger than at the ground level, especially at night. An individual flying all night in a wind of 50 km h^{-1} is taken 500 km in 10 h, and it has the potential to cross a continent in a sequence of nights. In contrast, a slow flier airborne for 12 min in a wind of 5 km h^{-1} is taken only 1 km.

The evidence that movement is windborne is largely circumstantial but is of several kinds: backtracking from known arrivals to known or likely sources using sequences of windfield maps; repeated association of arrivals with similar wind systems; and seasonal redistribution of species consistent with seasonal changes in wind patterns. Direct evidence is provided by comparing eye or radar measurements of movements of individuals with wind at the time, but distances are limited to, at most, a few kilometers. However, persistent downwind flight at a point may be presumed to extend over many kilometers, and on occasion, this has been demonstrated by more than one radar. The trajectory, or path, taken by a windborne organism, estimated from windfield maps, may be more or less straight or highly curved (even looped) depending on the ever-changing pattern of winds in which the organism is embedded.

Whereas a flying insect can maintain height by wing flapping, a spore or seed cannot prevent fallout under gravity, although fallspeed may be lessened by morphology that may increase drag (by surface hairs or wings) or reduce density (by air-filled cavities). Time spent airborne is increased by atmospheric updrafts, which increase the distance to fall. Updrafts experienced by airborne organisms can be caused by atmospheric turbulence induced by the wind blowing over rough ground. Effects of ground roughness extend upward to heights of some hundreds of meters during the daytime but usually less at night. Sunshine heating the ground induced convective turbulence by day, sometimes to heights of several kilometers in the afternoons of the hottest months. The turbulence from both sources results in a "boundary layer" full of eddies (like those in a river flowing over its rocky bed) mixing airborne organisms both upward and downward except where insect flight actively resists vertical movement. There is, therefore, a tendency for an airborne population to disperse progressively through a greater depth, even to heights where the horizontal pattern of wind speed and direction differs from that near the ground. At night, however, turbulence intensity and depth decrease, and fallout is more likely.

C. Arrival

Landing by a windborne organism may be active, as with an insect descending and ceasing flapping flight, or passive, as with a wingless insect or a spore or seed either impacting on vegetation or settling under gravity. Alternately, transport on the wind into a rain area may cause insects to fly nearer the ground and perhaps enhance landing, or spores may simply be washed out of the air by collision with falling drops. Whether seeds are also prone to such washout does not seem to have been studied. Spores can also be taken aloft by the updrafts that produce rain clouds, and they are deposited as rainout. Because rain is episodic, washout and rainout are patchy.

D. Outbreaks

Sudden appearances of pest insects or disease at intensities causing serious harm may be caused by progressive build-up that has gone unnoticed or by introduction from elsewhere. Outbreaks of windborne pests and diseases typically appear widely at scattered places more or less simultaneously. Organisms have then been transported on a large-scale wind system, but deposition has been particularly intense in some places. Deposition may be a result of active attraction (e.g., flying insects to patchy host plants or sheltered landing sites) or passive concentration (e.g., washout of spores in local heavy rain).

Flying insects can also be concentrated by wind convergence beneath atmospheric updrafts on a range of scales from a few kilometers to hundreds of kilometers [30]. Dense clouds of several species of moths and grasshoppers have been associated with windshift lines—boundaries between airstreams with different directions accompanying the updrafts in rainstorms and weather fronts. It is thought that concentration is brought about by the insects meeting a zone of wind convergence but actively resisting being taken aloft in the updrafts. Any rain may also increase volume density by concentrating flight near the ground. Landing would then lead to localized dense infestations. Some outbreaks of insects can be linked subsequently to particular rainstorms even though there had been no direct observations of insects encountering wind convergence, as is the case for the African armyworm [33]. Satellite images of rain clouds can be used to locate potential zones of wind convergence and hence concentration of flying insects.

For crop diseases, it is possible to estimate the chances of spores reaching a given site at risk from a wide range of more or less distant potential sources. A climatology of endpoints of backtracks can be calculated based on various assumptions about time airborne and behavior of the atmosphere. Correspondingly, the chances of reaching a wide range of sites at risk can be estimated from forwardtracks from a known source. This has been done, for example, in the United States for *Peronospora tabacina*, the pathogen causing tobacco blue mold [6], but the method is applicable in principle to any disease.

In addition to outbreak initiation, the weather also influences outbreak growth, both in size and density. Local spread may not be downwind, and it may be over meters rather than kilometers, often close to the ground and even within the crop. For insects, this is likely to be within winds weak enough to allow flight to a selected goal, such as food, a resting site, or a mate. Outbreak density is affected by the organism's developmental and reproductive rates, which are usually fastest within an optimum condition set (often a temperature and humidity range), but mortality increases as conditions increasingly deviate from optimum values. For example, quantitative relationships have been determined for many insect species (see ref. 41 for discussion), but the results can be conflicting, probably owing in part to the use of laboratory rather than field populations and in part to inadequate control of environmental constraints, such as food availability and quality, and disease incidence.

IV. OPERATIONAL MANAGEMENT STRATEGIES

A. Plume Model

This discussion supports a conceptual model of downwind movement for airborne biota generally, not only microbes and seeds but also insects. Indeed, airborne radar has shown a vast plume of corn earworm moths, *Helicoverpa zea*, advancing more than 400 km downwind in 8 h from a 50-km wide source in the lower Rio Grande valley of Texas and Mexico [43]. Another example, from Africa, is provided by armyworm moths, *Spodoptera exempta*, whose downwind spread between countries over hundreds of kilometers, as a plume from a limited area of caterpillar outbreaks, has been mapped from catches in a network of traps and form the distribution of next-generation outbreaks [33].

An instantaneous source produces a drifting puff that expands as a result of patchy dilution caused by small atmospheric eddies. For insects, this dilution may be aided by the flight of individuals in a variety of directions. A continuous source may be thought of as producing a sequence of overlapping puffs—a plume—that meander downwind under the influence of large atmospheric eddies. An intermittent source produces an intermittent plume. Drifting smoke gives us a picture of plume behavior. But whereas smoke particles are minute and follow closely the air motion, biota have masses that induce significant terminal fallspeeds under gravity. In still air, these are of the order of 1 ms^{-1} for insects, 1 mms^{-1} for microbes, and intermediate values for seeds. Moreover, an actively flying insect may land earlier or stay airborne longer than would happen in simple windborne drift.

The combination of plume movement, dilution, and depletion results in a deposit on vegetation or ground stretching downwind. Deposit density (dose, or biota per unit area) will have a profile decreasing rapidly downwind. Most individuals are deposited near the source, producing a peak density, but a few reach far downwind. The shape and size of this profile will vary greatly both between species and from one occasion to another within a given species. For some, the peak may be at or very close to the source; for others, it will be displaced downwind. With some species, few individuals may reach as far as 1 m (e.g., fungal spores within a crop); for others, the scale may be a millionfold greater, with some insects traveling 1000 km or more. Horizontally, plume shape will be affected not only by dilution and depletion but also by plume meandering in response to varying wind direction and speed. If a plume passes over a wide, unfavorable habitat (e.g., ocean, desert, or mountains), the vast majority of individuals may perish. However, a few colonizers can lead to much harm in subsequent generations, particularly if new genotypes emerge to which hosts are less resistant.

Concentrating mechanisms, notably wind convergence of flying insects and rainout or washout of drifting spores, will greatly increase deposit density, producing "spikes" in the profile corresponding to outbreaks even as early as the next generation. Such concentrated deposition reduces the number of individuals that might otherwise have spread further downwind.

B. Preventative Management

Most pest management strategies are defensive but, where a pest is windborne, preventative management may be possible, at least in principle. To reduce the downwind spread of harm, the manager needs to know (a) source position, size, and timing; (b) plume behavior; and (c) deposit shape, size, and density.

This is easy to state in principle but, in practice, there are likely to be many unknowns. Field surveys may be inadequate to locate the source or to measure its extent sufficiently for an estimate to be made of the time when the biota will become airborne or to estimate the numbers involved. Aerial sampling of the plume may be prohibitively expensive, although a chain of detection radars could monitor widespread insect invasions. Even estimating a plume's ever-changing shape and position from available meteorological data may be inadequate. Routine ground sampling may be insufficiently broad scale. Nevertheless, enough may be known for particular preventive management strategies to be deployed. These strategies include the timely reduction in source size or plume density (spray applications to populations on the ground and in the air) or emergency action (defensive strategy) following the issue of a warning or forecast of windborne movement, particularly for places where outbreaks may occur.

Reduction, or even elimination, of sources has resulted in spectacularly successful decreases in the incidence of harm caused downwind by insects. On the contrary, there do not appear to be any such occasions as yet involving fungal diseases. Three examples of such a preventive management strategy for insect pests will illustrate this principle.

1. *Cochliomyia hominivorax* (screwworm fly). Larvae of this blowfly feed in wounds of live, warm-blooded animals. Formerly, it was a major pest of cattle in the United States, overwintering in the southern states and spreading northward each spring and summer up to several hundred kilometers in each

generation, almost certainly on the wind. Eradication, by release from aircraft of many millions of sterile males, was extended from southeastern to southwestern states [24] and then to Mexico, which was declared free of screwworm flies in 1991 after dispersal of 250 billion sterile flies during 50,000 flying hours [40]. The intention is to eradicate the fly from the whole of Central America. A similarly successful campaign in Libya was started in 1990 following an accidental introduction of the same species [21]. A plan drawn up by two United Nations Agencies, the Food and Agriculture Organization and the International Atomic Energy Agency, was implemented with aircraft releases of sterile males bred in Mexico. No infestations had been reported 9 months after the start of the program, which has prevented a potentially very damaging spread of the fly across Europe and Africa.

2. *Simulium damnosum* (blackfly). This biting fly is the vector of a nematode worm, *Onchocerca volvulus*, that causes the human disease onchocerciasis (river blindness) in Africa. Flies breed in fast-flowing rivers, and in West Africa, they are taken as much as 500 km northeastward on monsoon winds accompanying the seasonal rains [1]. Aerial larviciding of all known and potential breeding sites during 20 years by the Onchocerciasis Control Program of the World Health Organization has resulted in a dramatic reduction of disease transmission. Twenty million people have now been protected from further infection, and seven million children have never been at risk of blindness [20]. The intention is to continue management long enough to allow the pool of adult parasites to die off.

3. *Locusta migratoria manilensis* (Oriental migratory locust). This was a major agricultural pest in China. After 4 years of research on its ecology, an engineering project was established in 1954 to control the waters on the flood lands between two great rivers, the Hwang Ho and the Yangtse Kiang. These lands were the sources of many serious locust plagues. The project progressively extended the cropping area and reduced the locust's habitat; the last outbreak was recorded in 1967 [3].

These three examples illustrate the large time and space scales involved in the regional preventive management of major insect pests. The first two involved the use of may aircraft to attack the pests directly either at their source or downwind. The third involved changing the source to make it an unfavorable breeding habitat. In all three examples, the sources were well defined. Other species are different, as illustrated by the following three insect examples.

1. *Schistocerca gregaria* (desert locust). This species can occur over a vast area of northern and eastern Africa and southwestern Asia, with numbers varying greatly from year to year [27]. During recession years, the population is largely or wholly composed of the solitary phase, with individual adults flying mostly at night, whereas during plague years, there are numerous dense, day-flying swarms composed of the gregarious phase. Both phases can travel

long distances downwind. Controlling swarms and marching bands of the flight-less nymphs ("hoppers"), wherever they may be found, has been the manage-ment strategy for decades, using aircraft as well as ground vehicles because of the great mobility of swarms. But this species breeds, and its bands and swarms form, only in places where erratic desert rains have fallen; there are no per-manent sources. Hence an operational problem is finding the gregarious popu-lations. Understanding the relationship between the insect's population dynam-ics and rainfall permits monitoring to be aided by satellite imagery of both rainstorms and the resulting desert vegetation. This information provides im-portant data on the possible location of incipient locust populations.

2. *Spodoptera exempta* (African armyworm moth). Dense outbreaks of larvae can damage cereal crops and range land in much of sub–Saharan Africa. During the dry season, this species exists at low density in the solitary phase, but the onset of rains can lead to outbreaks of the gregarious phase. The man-agement strategy developed in East Africa has been to locate initial outbreaks of the season, which are in recognized broad areas but with locations varying from year to year [32,33]. Reducing numbers in these locations decreases the risk of next-generation outbreaks downwind. Satellites have again been used to help locate rainstorms associated with concentrations of night-flying parent moths. Although this species travels in huge numbers over wide areas, the pro-portion at densities justifying preventive management while in flight is prob-ably too small to be of concern, unlike the desert locust. Moreover, night-fly-ing moths are more difficult to locate than day-flying locust swarms.

3. *Spodoptera frugiperda* (fall armyworm moth). This is a New World *Spodoptera* species similar to *S. exempta*. Like a number of other moth pests in the United States, it overwinters there only in the extreme southern areas of the country, extending northward in spring and summer, almost certainly on spells of southerly winds. Preventative management has been suggested and, despite the still-limited understanding of its movements, has been successful in southern Florida, where farmers brought forward their corn sowing date so that moths emerging at the end of season were too early to infest spring corn in northern Florida [26].

C. Warnings and Forecasts

It has been stressed for decades that effective pest management should be based on warnings and forecasts of likely population distribution and, for windborne pests, of likely arrival dates in areas at risk. But to be of value, warnings and forecasts must not only reach the user in time but must also be presented in a way that can lead to decisions on necessary action; for example, timing of field surveys or application of management techniques. Precision and accuracy gen-erally decrease with longer forecast periods, but the precision and accuracy needed varies among users. The planning of resources to be used months later requires less precision than rapid field action for an event likely in the next day or two. Our present ability to forecast pest movements in constrained by limited understanding of the mechanisms involved in the movement of most species, as well as their application to a particular situation, and by our inability to forecast the weather more than a few days ahead. To maintain continuous field monitoring of a sporadic pest presents difficulties in sustaining the necessary organizations, particularly during periods of low pest incidence that may last months or years.

For windborne pests, warnings and forecasts of arrivals are based on knowledge of (a) timing of movement out of the source, (b) downwind spread of the resulting plume, and (c) deposition, especially where the deposition is concentrated. Schemes based on such knowledge have existed for decades. One of the first was for the desert locust that was set up for India in 1939 but subsequently expanded and now provided for the whole of the locust's geographical distribution by the Food and Agriculture Organization. It is based on a continuous flow of field records of population sizes and ages and by "dynamic mapping" at regular intervals. Inferred movements are related to past weather, and warnings of imminent movements are based on forecast weather for the following few days. Forecast movements for the subsequent few weeks are expressed as probabilities based on a extensive archive of historical events interpreted in the light of the known current distribution [27]. Forecasts are issued at least monthly to control services in some 60 countries. A similar service is provided for the Australian plague locust, *Chortoicetes terminifera* [17]. Such schemes rely upon prompt reporting and an efficient international communications network.

Monitoring and communication networks have been developed for forecasting services in several European countries. In Britain, emphasis has been placed on aphid monitoring. A network of suction traps was started in 1964 to monitor airborne aphid populations [37,38]. From 1968, records of more than 30 species have been sent weekly to advisory entomologists and have formed the basis for developing methods to forecast size and timing of seasonal movements and the extent of aphidborne viral diseases.

Trap networks also exist for African armyworm moths, complementing field reports of caterpillar infestations. Forecasts are issued weekly to extension services as well as to farmers through the press and radio [25]. In China, regional forecasts of infestation severity by caterpillars of the Oriental armyworm, *Mythimna separata*, are based on monitoring of the overwintering area (south of about 33°N), knowledge of usual intergenerational movements, and statistical relationships between severity and successive generations [44].

In the United States, 2-day forecasts of springtime movements by black cutworm moths, *Agrotis ipsilon*, from Texas to Illinois have been tested using a numerical model that also includes assessment of changes in aerial concentration [22]. Smelser et al. [36] compared trap catches of this species with "moth influx potential"—an assessment of the contributions of moths in a population arriving from various sources resulting from differences in flight durations. They suggest that forecasting could replace labor-intensive field monitoring as a basis for management.

Forecasts of long-distance movements are particularly valuable where damage is immediate (e.g., flying locusts) or is caused by the rapidly developing larval offspring of immigrants (e.g., in moth species whose caterpillars can attain dense outbreak populations). But where only a few individuals arrive, monitoring them and forecasting subsequent population growth is more important than forecasting the arrival. It is unusual for later arrivals to contribute significantly to the growing population. This applies to the rice brown planthopper, *Nilaparvata lugens*, in Japan, where a network of traps is operated annually from April to July to monitor arrivals on strong winds from China [42].

These forecasting schemes have been developed for insects. Although there are several systems for forecasting *development* of disease in crops, relying heavily on meteorological inputs (mostly of recent weather, rarely of forecast weather), there appears to be none that involves windborne *spread* of spores. It is assumed, implicitly or explicitly, that an inoculum is always present on the crop. Where such forecasting systems provide useful results, there may be little incentive to discover whether inclusion of spore mobility might improve them. Persistent inoculum may well often be more significant than windborne arrivals, but that is not the case where year-round persistence is impossible (through cold or drought) and annual onset of disease follows arrival of spores from more or less distant sources. Moreover, a new strain may spread long distances.

V. CONCLUSIONS

The principles of managing windborne insect pests, whether at the source, while airborne, or after arrival, have been known and applied for decades. But they have been applied to only a few species. Their value has not been tested for most species, including some that are major pests and closely related to those where application has been successful. There is a need to collate existing evidence (much of it likely to be unpublished) on the role of long-distance spread of fungal spores. Moreover, there is considerable potential for the development of forecasting schemes with the aim of improving management.

Our present ability to forecast pest movements is constrained by limited understanding of the mechanisms involved in the movement of most species and of their applicability to a particular situation. Research is needed on several fronts (Table 5). Such research is multidisciplinary, possibly international, and sometimes expensive. But what are the costs in relation to the potential reduction in

Table 5 Research Needed on Windborne Pest Movement

1. Standardized, reliable, simple, and inexpensive field sampling methods to quantify population densities.
2. Mapping seasonal changes in distribution area.
3. New techniques to distinguish strains of organisms ("natural markers").
4. Case studies of organisms' windborne movements.
5. Causes of concentration leading to outbreaks.
6. Integration of movement into models of population dynamics.
7. Forecasting schemes for particular species.

losses? Of particular value are case studies of past events using historical records. They provide evidence of the scale of movements and the wind systems involved. But to be convincing, a case study should include the following:

a. precise timing of arrival, preferably directly (e.g., by insect trapping or rain sampling), otherwise indirectly (e.g., from estimated age of a subsequent insect generation or disease focus);
b. calculated backtrack from the arrival site to a known source;
c. evidence of movement from that source to the arrival site (e.g., a natural marker);
d. evidence that individuals are likely to survive travel between source and arrival site.

Many published case studies lack one or more of these constituents and are therefore less convincing.

Until the research has been done and there are prospects of detailed and reliable forecasts, how can a grower apply the existing but limited understanding of windborne pest movements to improve his defensive management and reduce his losses? The grower could:

a. monitor crops for arrival dates of seasonal pests (e.g., flying insects or spores in rain) when television or newspaper weather forecasts show winds will be blowing from known or likely sources, particularly when the crop is at a vulnerable stage;
b. reduce the intensity of infestation by timely application of management techniques after arrival even if only a few individuals are involved, because subsequent development may be rapid;
c. reduce the risk of infecting other crops downwind, including those of other growers, by destroying plant residues on calm days.

The grower must leave preventive management, at distant sources or while the pests are in transit, to organizations with the resources adequate for the likely vast areas involved.

REFERENCES

1. Baker, R.H.A., P. Guillet, A. Seketeli, P. Poudiougo, D. Boakye, M.D. Wilson, and Y. Bissan. 1990. Progress in controlling the reinvasion of windborne vectors into the western area of the Onchocerciasis Control Programme in West Africa. *Phil. Trans. R. Soc. Lond.* B 328:731–750.

2. Burt, P.J.A., and D.E. Pedgley. 1997. Nocturnal insect migration: effects of local winds. *Adv. Ecol. Res.* 27:61–92.

3. Chen, Y., Q. Long, J. Zhu, and Q. Ji. 1981. Studies on the dynamics of the Oriental migratory locust, *Locusta migratoria manilensis*, in the breeding areas of the Hongze Lake. *Act. Ecol. Sin.* 1:37–47.

4. Cooter, R.J. 1993. *The Flight Potential of Insect Pests and its Estimation in the Laboratory: Techniques Limitations and Insights.* Hadleigh, UK: Central Association of Bee-Keepers.

5. Davis, J.M. 1987. Modeling the long-range transport of plant pathogens in the atmosphere. *Annu. Rev. Phytopathol.* 25:169–188.

6. Davis, J.M., and J.F. Monahan. 1991. Climatology of air parcel trajectories related to the atmospheric transport of *Peronospora tabacina*. *Plant Dis.* 75:706–711.

7. Dingle, H. 1985. Migration. In *Comprehensive Insect Physiology, Biochemistry and Pharmacology.* Vol. 9. G.A. Kerkut and L.I. Gilbert, eds. New York: Pergamon, pp 375–415.

8. Dingle, H. 1989. The evolution and significance of migratory flight. In *Insect Flight.* G.J. Goldsworthy and C.H. Wheeler, eds. Boca Raton, Florida: CRC Press, pp 99–114.

9. Drake, V.A. 1993. Insect-monitoring radar: a new source of information for migration research and operational pest forecasting. In *Pest Control and Sustainable Agriculture.* S.A. Corey, D.J. Dall, and W.M. Milne, eds. Melbourne: CSIRO, pp 452–455.

10. Drake, V.A., and R.A. Farrow. 1988. The influence of atmospheric structure and motions on insect migration. *Annu. Rev. Entomol.* 33:183–210.

11. Drake, V.A., and A.G. Gatehouse. 1995. *Insect Migration: Tracking Resources Through Space and Time.* Cambridge, UK: Cambridge University Press.

12. Ford, R.H., T.L. Sharik, and P.P. Feret. 1983. Seed dispersal of the endangered Virginia round-leaf birch (*Betula uber*). *For. Ecol. Manag.* 6:115–128.

13. Gatehouse, A.G. 1997. Behavior and ecological genetics of wind-borne migration by insects. *Annu. Rev. Entomol.* 42:475–502.

14. Glasscock, H.H. 1971. Fireblight epidemic among Kentish apple orchards in 1969. *Ann. Appl. Biol.* 69:137–145.

15. Graham, D.C., and M.D. Harrison. 1975. Potential spread of *Erwinia* species in aerosols. *Phytopathology* 65:739–741.

16. Greathead, D.J. 1972. Dispersal of the sugar cane scale *Aulacaspis tegalensis* (Hem.: Diaspididae) by air currents. *Bull. Entomol. Res.* 61:547–558.

17. Hogstrom, A.W., K.R. Norris, and R.A. Powell. 1982. *Australian Plague Locust Commission: A Review of Functions and Activities.* Canberra, Australia: Department of Primary Industries.

18. Johnson, W.C., D.M. Sharpe, D.L. DeAngelis, D.E. Fields, and R.J. Olson. 1981. Modeling seed dispersal and forest island dynamics. In *Forest Island Dynamics in Man-Dominated Landscapes*. R.L. Burgess and D.M. Sharpe, eds. New York: Springer-Verlag, pp 215–239.

19. Kennedy, J.S. 1985. Migration, behavioral and ecological. In *Migration: Mechanisms and Adaptive Significance*. M.A. Rankin, ed. *Cont. Mar. Sci.* 27(suppl.), pp. 5–26.

20. Le Berre, R., J.F. Walsh, B. Philippon, P. Poudiougo, J.E.E. Henderickx, P. Guillet, A. Sékétéli, D. Quillévéré, J. Grunewald, and R.A. Cheke. 1990. The World Health Organization Onchocerciasis Control Programme: retrospect and prospects. *Phil. Trans. R. Soc. Lond.* B 328:721–729.

21. Lindquist, D.A., and M. Abusowa. 1991. The New World screwworm in North Africa. *World Anim. Rev.* October:2–7.

22. McCorcle, M.D., and J.D. Fast. 1989. Prediction and pest distribution in the Corn Belt: a meteorological analysis. *Proceedings of the 9th Conference of Biometeorology and Aerobiology*. Boston, Massachusetts, pp 298–301.

23. McEvoy, P.B., and C.S. Cox. 1987. Wind dispersal distances in dimorphic achenes of ragwort, *Senecio jacobea*. *Ecology* 68:2006–2015.

24. Novy, J.E. 1991. Screwworm control and eradication in the southern United States of America. *World Anim. Rev.* October:18–27.

25. Odiyo, P.O. 1990. Progress and developments in forecasting outbreaks of the African armyworm, a migrant moth. *Phil. Trans. R. Soc. Lond.* B 328:555–569.

26. Pair, S.D., J.R. Raulston, A.N. Sparks, J.K. Westbrook, and G.K. Douce. 1986. Fall armyworm distribution and population dynamics in southwestern states. *Fla. Entomol.* 69:468–487.

27. Pedgley, D.E. 1981. *Desert Locust Forecasting Manual*. Vols. I & II. London: Centre for Overseas Pest Research.

28. Pedgley, D.E. 1982. *Windborne Pests and Diseases: Meteorology of Airborne Organisms*. Chichester, UK: Ellis Horwood.

29. Pedgley, D.E. 1986. Long distance transport of spores. In *Plant Disease Epidemiology*. K.J. Leonard and W.E. Fry, eds. New York: McGraw-Hill, pp 346–364.

30. Pedgley, D.E. 1990. Concentration of flying insects by the wind. *Phil. Trans. R. Soc. Lond.* B 328:631–653.

31. Pedgley, D.E. 1992. Aerobiology: the atmosphere as a sink and source for microbes. In *Microbial Ecology of Leaves*. J.H. Andrews and S.S. Hirano, eds. New York: Springer-Verlag, pp 43–59.

32. Pedgley, D.E., and D.J.W. Rose. 1982. International workshop on the control of armyworm and other migrant pests in East Africa. *Trop. Pest Manag.* 28:437–440.

33. Pedgley, D.E., W.W. Page, A. Mushi, P. Odiyo, J. Amisi, C.F. Dewhurst, W.R. Dunstan, L.D.C. Fishpool, A.W. Harvey, T. Megenasa, and D.J.W. Rose. 1989. Onset and spread of an African armyworm upsurge. *Ecol. Entomol.* 14:311–333.

34. Reynolds, D.R., and J.R. Riley. 1997. Flight behaviour and migration of insect pests. *National Resources Institute Bulletin 71*. Chatham Maritime, UK: Natural Resources Institute.

35. Rittenberg, S.C. 1939. Investigations on the microbiology of marine air. *J. Mar. Res.* 2:208–217.

36. Smelser, R.B., W.B. Showers, R.H. Shaw, and S.E. Taylor. 1991. Atmospheric trajectory analysis to project longrange migration of black cutworm (Lepidoptera: Noctuidae). *J. Econ. Entomol.* 84:879–885.

37. Tatchell, G.M. 1985. Aphid-control advice to farmers and the use of aphid-monitoring data. *Crop Prot.* 4:39–50.

38. Taylor, L.R., I.P. Woiwod, G.M. Tatchell, M.J. Dupuch, and J. Nicklen. 1982. Synoptic monitoring for migrant insect pests in Great Britain and Western Europe. III. The seasonal distribution of pest aphids and the annual aphid aerofauna over Great Britain. *Rothamsted Experimental Station, Report for 1981*, Part 2:23–121.

39. Thresh, J.M. 1966. Field experiments on the spread of blackcurrant reversion virus and its gall mite vector (*Phytopus ribis*). *Ann. Appl. Biol.* 82:381–406.

40. Vargas-Teran, M. 1991. The New World screwworm in Mexico and Central America. *World Anim. Rev.* October:28–35.

41. Wagner, T.L., H. Wu, P.J.H. Sharpe, R.M. Schoolfield, and R.N. Coulson. 1984. Modeling insect developmental rates: a literature review and application of a biophysical model. *Ann. Entomol. Soc. Am.* 77:208–225.

42. Watanabe, T., K. Sogawa, Y. Hirai, M. Tsurumachi, S. Fukamachi, and Y. Ogawa. 1991. Correlation between migratory flight of rice planthoppers and low-level jet stream in Kyushu, southwestern Japan. *Appl. Entomol. Zool.* 26:215–222.

43. Wolf, W.W., J.K. Westbrook, J. Raulston, S.D. Pair, and S.E. Hobbs. 1990. Recent airborne radar observations of migrant pests in the United States. *Phil. Trans. R.. Soc. Lond.* B 328:619–630.

44. Yan, Y.-L. 1991. Armyworm (*Mythimna separata*) migration and outbreak forecast in China. *Proceedings of the International Seminar on Migration and Dispersal of Agricultural Insects*.Tsukuba, Japan: National Institute of Agro-Environmental Sciences, September, pp 31–39.

4

Landscape Structure and Extra-Field Processes: Impact on Management of Pests and Beneficials

Douglas A. Landis
Michigan State University, East Lansing, Michigan

Paul C. Marino
University of Charleston, Charleston, South Carolina

I. AGRICULTURAL PRODUCTION FROM A LANDSCAPE PERSPECTIVE

Although individual crops or fields can be viewed as fully functional ecosystems in themselves (see Chap. 1), it is clear that crop fields are also influenced by adjacent crop and noncrop habitats. These influences involve the exchange of species (pests and beneficials) and materials (e.g., soil, water, nutrients) and may occur at local or regional spatial scales. From the perspective of pest management, it is important to know how adjacent crop and noncrop habitats influence the population and community dynamics of the pests and beneficials within particular cropping systems. Although integrated pest management (IPM) programs have generally focused on the crop or individual fields as the unit of management [78], and thus tended to ignore the role of extra-field processes, renewed interest in "area-wide" and "preventative pest management" systems necessitates a different approach. Because of the tight coupling of ecological processes occurring within fields and those occurring external to crop fields, it is imperative that pest managers begin to understand this linkage and how to manipulate extra-field habitats to facilitate effective pest management [80].

A useful framework for examining the role of extra-field processes in pest management is found in the field of landscape ecology which seeks to understand the interactions between landscape elements and their impact on the changing structure and function of landscapes [20,21,53,151]. Agricultural landscapes evolve from natural landscapes over time and are unique owing to the nature, frequency, and spatial pattern of their human-induced disturbance regimens. They range from relatively simple environments to complex environments in which crops are embedded in a rich matrix of noncrop habitats. All agricultural landscapes are characterized by abrupt edges (ecotones) as crop fields adjoin neighboring crops, forests, fencerows, riparian zones, old fields, or pastures. The complex array of landscape elements and crop-edge variation results in a diverse set of potential interactions between the species living in crop fields and those living in adjacent and neighboring habitats. This matrix of spatial and temporal associations between landscape elements, in turn, plays a critical role in the distribution and abundance of weeds [98], insects [81], and pathogens [146] in agricultural landscapes.

A. Disturbance and Diversity in Agroecosystems

Agroecosystems are highly disturbed ecosystems. In natural ecosystems, fire, flood, windstorms, insect outbreaks, and so forth, periodically disrupt established communities. The frequency with which disturbances impact natural communities critically affects the structure of those communities [111]. The initial impact of disturbance is to remove individuals from the community with subsequent recolonization of the disturbed area occurring through the process of regrowth and dispersal into the recently disturbed area [119]. As such, highly disturbed natural communities are occupied by species having superior dispersal ability. Such species also tend to be short lived to produce large numbers of offspring, and to be relatively poor competitors [56,99]. These are also the characteristics of species found in early successional communities. In contrast to native ecosystems, many agricultural ecosystems experience extraordinary levels of disturbance. Although a highly disturbed terrestrial ecosystem may have one disturbance event every few years (e.g., fire in grasslands), many agricultural ecosystems experience multiple events per growing season. Thus, many agricultural systems are dominated by weeds, insects, and pathogens highly adapted for rapid colonization and population increase.

Along with high rates of disturbance, agroecosystems are also characterized by low species diversity and by plants with little architectural complexity [85]. The diversity of crop plants in agricultural landscapes is often low by design, whereas the diversity of noncrop plants (weeds) is low by virtue of the disturbance regimen and, in many cases, by the application of herbicides. Crop plants themselves (most of which are annuals) and the early successional weeds found growing in fields have relatively simple architectures. Early successional plant communities with simple architectures have fewer species of insects (pests and beneficials) living on them than later-successional, more architecturally complex plant communities [18,33,60,84,86,100,136,140]. Agroecosystems frequently have an impoverished natural enemy community as a consequence of this low-diversity plant and insect community.

The frequency of disturbance and the successional stage of extra-field communities may differ radically from that found within a crop field. Extra-field communities, unless they are also crop fields, are generally less disturbed, later successional, and architecturally more complex than the crop fields. They, therefore, frequently have higher plant and animal diversity than is found in crop fields. Richer, more complex extra-field communities may provide relatively stable source populations of beneficial arthropods that facilitate pest management, or they may provide relatively stable sources of pests that impede it [35]. Alternatively, they may have little impact on either pest or beneficial populations. The extent to which extra-field habitats will either facilitate or impede pest management will likely depend on the size (local and regional), nature (e.g., early successional versus late successional), and diversity of extra-field communities. Pest managers must, therefore, develop an appreciation for how agricultural landscape structure can influence the interactions of extra-field and within-field processes.

In this chapter, we will explore the interactions of pests and beneficials as influenced by the habitat and landscape structure in which they are embedded. Our focus will be on the exchange of species between agricultural landscape elements, primarily using insects and weeds as examples. Although pathogen biology and epidemiology are also influenced by noncrop habitats, it is beyond the scope of this chapter to address these interactions fully and readers are directed to the following references for more detailed examples [22,59,66,101, 146]. The interactions will be explored at local and landscape scales. The local scale will address problems of field crop/field edge interactions, whereas the landscape scale will address problems involving the larger landscape matrix (i.e., complex vs simple agricultural landscapes). An understanding of the interchange of organisms and materials between landscape elements and the influence of landscape structure on these interchanges is critical for the prediction and management of pest populations within agricultural fields.

II. EXTRA-FIELD HABITATS AND WEEDS

Noncrop habitats have often been assumed to be a major source of weeds invading adjacent crop fields. This invasion occurs mainly through seed dispersal (seed rain). Unfortunately, few studies have examined the seed rain between adjacent habitats in agroecosystems. Seed rain can contribute significantly to the seed bank of fallow fields [5], and there is evidence that seed dispersal can occur over long distances as, for example, has been demonstrated for *Salsola kali* [73] (see also Chap. 3). However, most seeds do not appear to disperse more than a few meters into crop fields from adjacent habitats [63,95,96,98,142]. The highest weed species diversity is found at field edges and in the headlands.

Cavers and Benoit [28] give four principal reasons for the greater diversity of weed seeds along field edges and in headlands: (a) most seeds are dispersed only a short distance from parent plants [27,43,158], (b) many species are only found growing in the uncultivated land along field edges, (c) farm equipment enters fields along the edge or headlands, and (d) higher light intensity at field edges and in headlands than in the field interior enhances the emergence of shade-intolerant species. An additional factor that may encourage the diversity and abundance of seeds and plants at field edges is reduced herbicide usage or dose. The vegetational diversity and structure of extra-field habitats, particularly edge habitats, should therefore have a considerable impact on the diversity and structure of weeds entering crop fields.

Despite our limited knowledge of the movement of weeds into crop fields, we can predict that crop fields bordering recently disturbed land (e.g., highly disturbed fencerows and recently fallowed fields) will have more weed pests entering them than crop fields bordering later successional plant communities (e.g., woodlots and late successional old fields and fencerows). This is because early successional communities are occupied by weedy species that produce large numbers of seeds [138] and are adapted to germinate and establish under conditions of high disturbance, such as those found in crop fields. Although annual weed seeds do not disperse far into crop fields from extra-field habitats, early successional extra-field habitats do provide a steady source population of weed seeds. These can spread throughout crop fields by stepwise establishment and subsequent seed dispersal and through the actions of farm machinery.

A similar prediction also should hold true for biennial and perennial weeds which tend to be more important pests in reduced tillage cropping systems. In these systems, early-successional extra-field habitats with low levels of disturbance will likely provide the greatest source populations of biennial and perennial weeds.

Most extra-field habitats, however, are not composed of early successional plant communities. In Britain, for example, only a small proportion of the field edge flora are significant crop weeds [96,97]. Perennials dominate the edge flora and suppress weedy annuals. In terms of management, Marshall [96] suggests that perennial communities be encouraged along field edges by limiting disturbance of field edges (e.g., use of herbicides). Intact nonweedy perennial communities will inhibit the establishment of annual, biennial, or perennial weeds that could subsequently invade field crops.

We can also predict interactions between the type of edge habitat and field orientation in regard to weed communities. For example, an east-west running fencerow of mature trees should have a different weed community on its shadier, moister, north-facing side than on its sunny, dry, south-facing side. Moreover, the ability of weedy species to germinate and establish along the edge of a crop field (where weed seeds are most likely to be dispersed) should differ between the south- and north-facing sides because of the different microenvironments. Weed seed germination is affected by soil moisture and aeration [31,32,36, 37,108] and by the intensity of solar radiation (see ref. 74 and references therein). Grasses, for example, have a higher germination frequency when soil moisture fluctuates (e.g., south side of wooded fencerows [28]). Also, shade-intolerant species should be uncommon in both the north-facing hedgerow and field edge, whereas shade-tolerant weeds should be relatively more common in these habitats.

A. Extra-Field Habitats and Weed Seed Predation

Although it is clear that the type and the relative position of the edge habitat may influence the number and species of weed seeds dispersing into crop fields, most evidence suggests that weeds growing along field edges are relatively less important as a source of seeds for future field crop weed populations than weeds growing within fields [63,95,96,97,98,122]. However, just as adjacent noncrop habitats provide a source of weed seeds, they also provide a source of potential weed herbivores and weed seed predators that may significantly influence weed populations through intense or selective herbivory and weed seed predation.

Noncrop habitats can provide shelter and food for mammals [113], birds [9,88], and insects [144,145] and thus may provide source populations for predators that selectively forage on weed seeds. Both bird [9,26] and small mammal [26] abundance and diversity are higher in noncrop environments than in adjacent fields. Also, carabid beetles are important invertebrate seed consumers in temperate agroecosystems [10,19,70,75,89,90] and use noncrop habitats as overwintering sites [42,134,135,144,145,153,154]. Noncrop habitats can also provide alternative food resources (host plants) and source populations for predispersal weed seed predators and weed herbivores. These species may in turn greatly decrease seed production (see ref. 34 and references therein). In areas where unmanaged habitats with their associated communities of seed predators and herbivores are adjacent to simple agricultural habitats, predation on weed seeds and herbivory on weeds in the agricultural habitats may be greater.

Despite the well-known influence of postdispersal seed predation on plant populations and community structure in deserts [1,16,17,38,65,120], grasslands [13,112], pastures [118,127], and woodlands [2,62,75,131,132,133], relatively little is known regarding nonmicrobial seed loss in agricultural fields [19,90, 115]. Much of our understanding of the role of extra-crop habitats in influencing postdispersal seed predation is indirect. It focuses on the higher abundance and diversity of potential seed predators in noncrop habitats and not on the consequences of weed seed loss or weed community composition.

The influence of herbivory and predispersal seed predation on plant populations and community structure has also been well studied [34]. However, the focus of these studies has been to examine the use of herbivores and predispersal seed predators (mainly insects) for the biological control of weeds in agroecosystems [8,72,103]. Although there has been some discussion of how alternative plant hosts, for example, may enhance biocontrol efforts (e.g., see ref. 104), as with postdispersal weed seed predators, there has been little direct study of the influence of noncrop habitats on the effectiveness of weed herbivores and predispersal seed predators as biocontrol agents in agroecosystems. The importance of uncultivated land to beneficial insect weed herbivores and predispersal seed predators will likely differ little from the importance of uncultivated land to herbivorous pests [149] (Table 1) that we discuss later in this chapter.

B. Effects of Landscape Structure on Weeds

Scale is of fundamental importance when considering the influence of extra-field habitats on weed seed predation and weed herbivory. Seed predation, for example, can be examined on a within-field scale (i.e., comparing weed seed predation at crop field edges vs crop field interiors). Or it could be examined at the scale of the larger agricultural landscape. This could involve comparing complex agricultural landscapes of small crop fields embedded in a matrix of

Table 1 Biological Relationships Between Noncrop Habitats, Herbivores, and Entomophagous Insects

---

Noncrop habitats may provide herbivorous species:
   Overwintering sites
   Host plants (related or unrelated to crop plant)
   Nectar or pollen resources
   Aggregation sites
   Moderated microclimates
Noncrop habitats may provide entomophagous species:
   Overwintering sites
   Alternative prey (related or unrelated to crop pest)
   Host plants for facultative phytophages
   Overwintering sites
   Nectar or pollen resources
   Aggregation sites
   Moderated microclimates

Source: Modified from ref. 149.

numerous hedgerows and woodlots versus simple agricultural landscapes of large crop fields embedded in a matrix of widely scattered woodlots and hedgerows. At the within-field scale, we would predict greater seed predation near field edges, because noncrop habitats provide food and shelter for potential verte- brate and invertebrate seed predators. At the landscape scale, we would pre- dict greater seed predation in complex agricultural landscapes, because the greater area and density of noncrop habitats should support a more abundant and diverse fauna of potential vertebrate and invertebrate seed predators.

Marino et al. [93] examined within-field patterns of weed seed predation by vertebrate and invertebrate seed predators in Michigan maize fields near and distant from fencerows. The expectation was that overall seed loss would be higher near a fencerow edge than in the field interior, because fencerows pro- vide habitats and resources for source populations of potential seed predators. However, distance from fencerows had no consistent effect on seed loss (Fig. 1). There are at least three possible explanations for this result. First, some seed predators may not respond to fencerows. For example, the presence of ground- cover on crop fields promotes rodent activity [25,26,155,160] and thus possi- bly seed predation. Second, seed predators may respond in different ways to

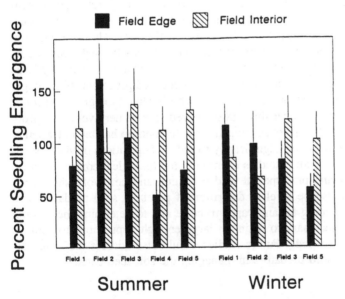

Figure 1   Percentage of seedlings of each species (±1 SE) that emerged, averaged for all weed species. In the summer study, percentage of seedling emergence is relative to emergence in vertebrate + invertebrate exclusion cages. In the winter study, percent- age of seedling emergence is relative to emergence in vertebrate exclusion cages. Lower percentage emergence is equivalent to a higher rate of seed predation.

fencerows with some species foraging within and near fencerows and other species foraging in field interiors with no general "fencerow effect" on seed predation [147]. Third, seed predation may have been examined at an inappropriate spatial scale. The maize fields in Marino et al.'s study were all located in a complex agricultural landscape. Within such a landscape, a field edge versus field interior comparison may not be the appropriate scale with which to examine influences on seed predation. A more appropriate scale may be a comparison of weed seed predation in complex versus simple agricultural landscapes.

Marino and Holder (unpublished data) found that there are large differences for some weed species in the intensity of weed seed predation between a simple and complex agricultural landscape in Mississippi. They compared weed seed loss in three cotton fields located in a complex agricultural landscape (small fields embedded in a mostly forested landscape) with weed seed loss in three cotton fields in a simple agricultural landscape (large fields embedded in a mostly agricultural landscape). The two landscape types were approximately 20 km apart. Four times during the summer of 1994, seeds of seven common Mississippi weeds were placed into the fields near and distant from hedgerow edges at six sites per field. Seeds were left in the fields for 4–5 days at each test period.

Results indicated a significant weed species effect on the percentage of seeds uneaten. For six of the seven weed species, there was more weed seed predation in the complex than in the simple landscape (Fig. 2). These results suggest that effective biological control of weed populations through seed predators likely requires enough suitable noncrop habitat to maintain relatively large source populations of avian, mammalian, and insect seed predators. If the larger landscape scale is the appropriate scale at which effective biological control of weeds may be enhanced, either through seed predation or herbivory, the problem remains as to what constitutes enough suitable noncrop habitat and whether seed predators have the ability significantly to reduce populations of viable seeds in the soil seed bank. Future studies examining the biological control of weed pests through post- and predispersal seed predation and herbivory should address the larger landscape level to determine if potential seed predators and herbivores are responding to landscape structure and to determine the amount of noncrop habitat necessary to maintain large enough populations to achieve substantial control of weeds. Future studies should also address the actual effectiveness of weed seed predators in managing the soil seed bank.

III. EXTRA-FIELD HABITATS AND INSECTS

The importance of uncultivated land to both pest and beneficial insects has been recognized for at least 30 years [149] (see Table 1). These interactions are increasingly being studied in relation to insect behavioral ecology [47], conser-

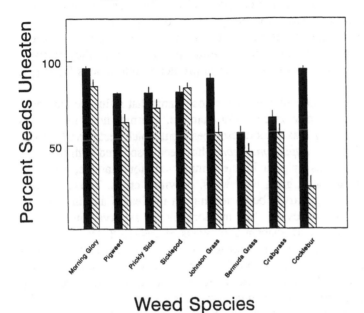

Figure 2 Percentage of seeds remaining uneaten ($\pm$ 1 SE) of seven common Mississippi weeds (common morning glory [*Ipomoea purpurea*], pigweed [*Chenopodium album*], prickly sida [*Sida spinosa*], sicklepod [*Cassia obtusifolia*], Johnson grass [*Sorghum halepense*], crabgrass [*Digitaria ischaemum*], and cocklebur [*Xanthium strumarium*]) in a simple agricultural landscape versus a complex agricultural landscape.

vation of biodiversity [48], and pest management [54,55]. This interchange of species between cultivated and uncultivated habitats can be caused by a variety of factors, including food availability, population density, and abiotic conditions [70]. It also can occur within or between seasons. On a between-season time scale, some pests (e.g., corn rootworms, *Diabrotica* spp.) may overwinter in the field where the future crop is planted [15], whereas others colonize fields each spring from overwintering sites. These may be other crop fields (e.g., European corn borer *Ostrinia nubilalis*) or noncrop habitats (e.g., boll weevil *Anthonomus grandis*) [24,130]. Seasonal movements from extra-field habitats into crop fields may be over short distances. The movements may occur by walking (e.g., Colorado potato beetle, *Leptinotarsa decemlineata* [83]), passive movement on wind (e.g., two-spotted spider mite, *Tetranychus urticae* [14]), or short dispersal flights (e.g., bird-cherry oat aphid, *Rhopalosiphum padi* [45]). Long-distance migration from overwintering areas to crop fields is also well known (e.g., potato leafhopper, *Empoasca fabae* [23]; black cutworm, *Agrotis ipsilon*; and armyworm, *Pseudaletia unipuncta* [61]). Management of the over-wintering site to make it less favorable [107] or rotating crops some distance away from overwintering sites is a common means of managing pest invasion. Monitoring of pest populations in overwintering sites or as they initially move into fields or orchards can also help in predicting infestations [156] (see Chap. 3).

Noncrop habitats can contain alternative host plants that influence the population dynamics of some pest species. For example, alternate hosts in noncrop habitats can provide early-season food resources from which pests invade crops. The common stalk borer, *Papaipema nebris*, has been associated with over 170 host plants [39]. Females oviposit in late summer on, for example, weeds in fencerows, contour terraces, grass waterways, or within crops. The eggs overwinter, and in the spring, larvae begin feeding on the leaves and in stems of the weed hosts. As the weeds die, become dormant, or as larvae outgrow the stem diameter, dispersal onto adjacent corn plants begins [157]. Damage is frequently confined to field edges or to weedy areas of the field [139]. Infestations of the black cutworm, *Agrotis ipsilon*, are also associated with the presence of alternate hosts that provide early-season food resources. Females lay eggs early in the spring on weedy winter annuals in and around crop fields. Larvae switch to crop plants as their weed hosts are removed by tillage or herbicides [128]. Alternative hosts in noncrop areas can also serve to buffer pest populations from cropping practices that adversely affect them. For example, the potato leafhopper, *E. fabae*, utilizes both crop (alfalfa) and noncrop habitats (deciduous trees) following migration into Illinois in the spring [79]. Because cutting and removal of first-harvest alfalfa occurs soon after arrival of migrants, this practice generally prevents the completion of the leafhoppers' first generation on this crop. First-generation *E. fabae* adults produced on noncrop hosts, thus, provide the major source of individuals available to recolonize second-growth alfalfa.

Crop-to-crop movement plays an important role in the life history and population dynamics of many pests. In the case of *E. fabae*, cutting of alfalfa and dispersal of adults from the crop is sometimes associated with increased infestations in other crops such as snap beans, potatoes [51], and soybeans [114]. The corn earworm, *Helicoverpa zea*, is another well-known example of a pest which utilizes a temporal succession of different crop hosts. In the southeastern United States, this may include the use of tobacco, corn, soybeans, and cotton to produce successive generations [102].

Insects also may occur in areas outside of crop fields for reasons unrelated to food. For example, the adults of the European corn borer, *O. nubilalis*, emerge from overwintering sites within corn stubble but they aggregate in stands of tall grasses (*Bromus* or *Setaria* spp.) on the borders of crop fields, in grass waterways, and in ditches [41]. In these so- called "action sites," a favorable microclimate exists and free water from rain or dew collects on the grasses. Female *O. nubilalis* drink at these sites before they produce pheromone and attract males [129]. Following mating, females return to the crop habitat to oviposit.

Natural enemies of crop pests also rely on habitats outside of the particular crop of interest for many of the same reasons that pests do. Many coccinellids [58] and carabids [143] overwinter in sheltered areas outside of crop fields. Many predators utilize prey outside of agricultural fields at some point in their life history. Maredia et al. [91,92] showed that temporal habitat preferences of the seven-spotted lady beetle, *Coccinella septempunctata*, depended on both prey availability and habitat disturbance. Within a localized agricultural landscape, *C. septempunctata* utilized wheat, maize, soybeans, and alfalfa, as well as abandoned fields undergoing succession and *Populus* plantations. Moreover, Perrin [110] emphasized the importance of alternative prey on weeds in noncrop habitats to *C. septempunctata* prior to the appearance of aphids on crop plants. In addition to prey, many predators also utilize plant pollen as a source of nutrition. This may be produced by the crop which also supports the prey [30] or may be blown from surrounding areas [57]. Pollen resources may also be sought out by predators in extra-field habitats. For example, upon emerging from overwintering in the spring, *Coleomegilla maculata* feeds on pollen produced by spring-flowering plants near their overwintering sites, such as trout lily (*Erythronium americanum*) (D. Landis pers. obs.) and dandelion (*Taraxacum officionale*) [91] before moving into alfalfa and wheat habitats where aphid prey are found.

Most parasitoids require alternative food or hosts to complete their life history or to increase their effectiveness as natural enemies [49,67,116,148,149]. Jervis et al. [68] examined the inflorescences of 53 plant species and found hymenopteran parasitoids feeding on 32 species. Other studies have focused on the effects of flower feeding on specific parasitic wasps; for example, *Macrocentrus grandii* [105] and *Diadegma insulare* [64]. Landis and Haas [82] found that parasitism of the European corn borer by the wasp *Eriborus terebrans* was greater along field margins than in field interiors during the first generation of *O. nubilalis*. They hypothesized that increased access to adult food resources (plant nectar, aphid honeydew) near field edges may play a role in the increased efficacy of *E. terebrans* near these sites. In subsequent studies, the longevity of female *E. terebrans* varied significantly when they were confined outdoors in 15-cm diameter screen cages with either water, honey water, flowering plants, or flowering plants plus aphids (Fig. 3). Parasitoid longevity was significantly greater ($F=5.21$, $P>.002$) on flowers versus water alone and varied signifi-

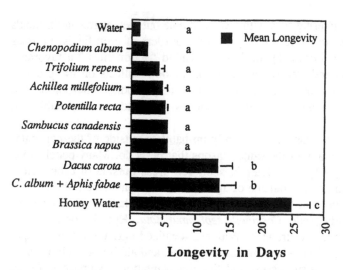

Longevity in Days

Figure 3   Mean longevity (±SEM) of female *Eriborus terebrans* confined in the field in 15-cm diameter screen cages with water, honey water, flowering plants, or flowering plants plus aphids. Longevity recorded to nearest day.

cantly among flowers (*F*=4.98, *P*>.002) with the greatest longevity on *Dacus carota*. Although longevity on *Chenopodium album* flowers alone was poor, longevity on *C. album* with *Aphis fabae* infestations was equal to that on *D. carota*. The presence of aphid honeydew is presumed to be responsible for this effect.

Finally, predators and parasitoids may have microclimate requirements that limit their abundance in particular crop habitats even if prey are present. This may result in seasonal or daily migrations of natural enemies between crop and noncrop habitats which can affect the structure of the natural enemy community at different locations within the fields [11,40,49,50,76,106]. Booij and Noorlander [12] showed that crops with greater early season cover such as winter wheat and peas had greater predator abundance and diversity than late-season crops like carrot and onion. Availability of food, shelter, and a more moderated microclimate is typical of these spring crops.

A.   Effects of Landscape Structure on Insects

1.   Field Edges

Insects move between crop and noncrop habitats for many reasons, and most types of agricultural production are characterized by very abrupt edges; it is not surprising, therefore, that the influence of field edges on pests and beneficials has been frequently studied. Pest movement across field boundaries has been examined for crop-crop borders [14,114], between crop and herbaceous borders [117,124], and between crops and woody borders [29,69,141]. These studies demonstrate that the type of associated vegetation and the physical structure of the border can have a pronounced effect on the movement of insects. For example, the abundance of the carrot fly, *Psila rosae*, in noncrop areas surrounding carrot fields was studied in relation to boundary type and composition. The distribution of the understory plant common nettle, *Urtica dioica*, had the greatest influence on *P. rosae* abundance followed by windbreak density [152].

Modification of microclimate by hedgerow and shelterbelts can also influence the distribution of pests and beneficials (see ref. 109 and references therein). These vertically oriented structures have unique impacts on field microclimates. Relative humidity is altered up to distances into the field equal to eight times the height of the windbreak. Soil moisture, soil temperature, and air temperatures are modified at distances up to 12 times the height, with wind speed altered at distances up to 28 times the height of the windbreak [52]. Combinations of these factors have pronounced impacts on insect distribution. Lewis [87] found many insects accumulate in the calmer air near windbreaks, particularly on the sheltered side. He found 27-fold more predaceous syrphid flies (Diptera: Syrphidae) and 7-fold more predaceous rove beetles (Coleoptera: Staphylinidae) at a distance of 0.9 m (one times the height) on the sheltered side of a windbreak compared with an unsheltered side.

2. Agricultural Landscape

In addition to field border effects, the physical structure of agricultural production systems at larger spatial scales can also influence pest and natural enemy diversity and abundance [137]. The emergence of the science of landscape ecology and associated tools for its study have enhanced investigations of these effects [21,53]. Important components of landscape structural complexity include degree of monoculture, fragmentation and habitat isolation, and the impact of overall landscape structural complexity.

The impact of within-field vegetational diversity (e.g., polyculture, intercropping) on herbivore and natural enemy populations has been extensively studied [6,121,150]. Many of these studies indicate that vegetational diversity within a field can have important effects on herbivore and natural enemy abundance and diversity, although robust theories to accurately predict the direction of response are lacking [4]. Far fewer studies have been directed at the impact of vegetational diversity at the between-field to landscape scales. Andow [3] reviewed the literature on the impact of regional degree of monoculture on insect pest populations. In cotton ecosystems, increases in the regional degree of monoculture led to increases in 13 insect populations and decreases in 2. In wheat, increased monoculture led to increases in three pest populations and no change or decreases in seven. All of the increases in both cotton and wheat were of monophagous pests. Andow concluded that the difference between crop types may be explained by the higher proportion of oligophagous and polyphagous pests in wheat versus cotton, with insect vagility, host-finding ability, and the response of predators and parasitoids as additional potential factors.

Several studies have demonstrated the impact of overall landscape structure on insects. The insect biomass of wheat, barley, maize, sugar beet, and alfalfa in uniform versus mosaic agricultural landscapes has been studied in Poland and Romania [125,126]. Uniform landscapes contained no nonarable habitats, whereas mosaic landscapes contained, for example, shelterbelts, hedges, meadows, small forests, and ponds, comprising about 12% of the study area. A greater biomass and diversity of insects occurred in the mosaic landscapes. All trophic groups (herbivores, saprophytes, predators, and parasites) were generally more abundant in mosaic versus uniform landscapes. This was particularly true for predator and parasitoid groups, which had as much as 70% greater biomass in mosaic landscapes. Ryszkowski et al [126] concluded that natural enemies are more dependent on refuge habitats and the greater abundance of these habitats in the mosaic landscapes resulted in their higher diversity, abundance, and ability to respond to prey numbers. Marino and Landis [94] examined parasitism of the true armyworm, *Pseudaletia unipuncta*, in structurally complex versus simple agricultural landscapes. Overall parasitism in the complex sites was more than three times higher than in the simple sites (13.1 versus 3.4%). Differences were largely attributable to one wasp species, the braconid, *Meterous communis*, which was far more abundant in complex sites. They hypothesized that abundance and proximity of preferred habitats for alternate hosts of *M. communis* may account for the observed differences.

If noncrop habitats play an important role in the effectiveness of biological control in agricultural ecosystems, how much noncrop habitat is sufficient and in what manner should it be distributed? Generalizations regarding the impact of habitat fragmentation on arthropods are difficult to establish, because the impacts are diverse and dependent on the scale of fragmentation [46,123]. In addition, most empirical studies have been done at relatively small spatial scales which may not apply to agricultural landscapes. At the landscape level, theory suggests that increasing fragmentation will result in the loss of species restricted to large undisturbed habitats, increase the abundance of edge species and those requiring two or more elements, and enhance total species coexistence [53]. However, these predictions are dependent on the type of habitat undergoing fragmentation and the habitat(s) that replace it. For example, fragmentation of large monocultural fields by introduction of diverse noncrop habitats may increase species diversity, whereas fragmentation of a landscape composed of diverse noncrop habitats with monocultural fields may have the opposite effect.

Fragmentation of natural ecosystems by agriculture has been the norm, with loss of species and disruption of food webs the general result [44,159]. One reason may be habitat isolation. Kruess and Tscharntke [77] studied the impact of habitat isolation on clover herbivores and parasitoids. They showed that increasing isolation decreased both herbivore and parasitoid abundance. Although all but two of the herbivores colonized distant patches, the species diversity of parasitoids and percentage of parasitism was significantly reduced. They concluded that habitat isolation affects natural enemies more than their herbivore hosts and by releasing pests from biological control potentially contributes to outbreaks. Thomas et al. [144,145] demonstrated this same concept by decreasing habitat isolation through creation of habitats for natural enemies within previously clean-farmed small grain fields. By creation of appropriate overwintering habitats, predator densities and control of aphid populations could be increased. Finally, at the landscape scale, Ryszkowski and Karg [125] detected a similar result due to increased landscape heterogeneity. The correlation between predator and parasitoid biomass and the biomass of their prey was always greater in mosaic versus uniform landscapes, indicating a more responsive trophic structure, which they attributed to the presence of more refuge habitats from which the natural enemies could penetrate crop habitats.

There is evidence that increased fragmentation or near total lack of noncrop habitats in agricultural landscapes may contribute to reduced efficiency of natural enemies, and that by decreasing habitat isolation, the effectiveness of natural control can be restored; however, the question of how much noncrop habitat is enough remains large unknown. One study has addressed this question in regard to conservation of wild pollinators in agricultural landscapes. Banaszak [7] found that although crop habitats (e.g., rape) were important seasonal food sources, refuge habitats including, for example, roadside vegetation, meadows, forests, and shelterbelts were critical for maintaining local populations of wild pollinators. In landscapes with low amounts of refuge habitat (<10% of total area), pollinator populations were reduced; increasing refuge habitat beyond 40% of the total area had proportionally little additional effect. He concluded that in a normally functioning agricultural landscape, the share of cropland should not exceed 75% of the total area.

IV. CONCLUSIONS AND UNIFYING THEMES

Populations of insect and weed pests and their associated natural enemies move between crop and noncrop habitats within agricultural landscapes. Processes which influence the population dynamics of species in noncrop habitats or influence the movement of species between habitats have a direct bearing on the diversity and abundance of pest and beneficial species with crops. Frequent disturbance is characteristic of crop fields, and factors that influence the recolonization process are of primary importance in determining the structure and function of insect and weed communities in agricultural crops. In contrast, natural communities experience much less frequent disturbances and competitive interactions play a much more important role in structuring the community than do colonization events.

The successional stage of the interacting crop and noncrop habitats affect these movement processes owing to influences on the life history characteristics of the species inhabiting habitats in different successional stages. Early-successional habitats are architecturally simple and tend to be inhabited by species with life history characteristics in common with many weed and insect pests (good dispersers, high reproductive capacity, short longevity, low competitive ability). Thus, early-successional noncrop habitats may be a primary source of pest species. On the other hand, predator and parasitoid diversity and abundance increase with increasing successional stage and increasing plant architectural complexity. Thus, later-successional habitats may be relatively important resource habitats and sources of predator and parasitoid species. There are exceptions to these broad generalizations, but we argue that lack of later-successional noncrop habitats in many agroecosystems is likely to be the primary cause of reduced predator and parasitoid diversity and is, therefore, a major factor contributing to pest outbreaks.

The physical structure of the agricultural landscape is the principal template which underlies and directs all of the population and community processes ultimately responsible for determining the diversity of pest and beneficial species present in agricultural fields. At the field level, the presence or absence of different crop and noncrop habitats, their proximity to the crop in question, and the plant and animal diversity found within these habitats can have a critical impact on the management of pest species within associated crops. At the landscape level, the degree of monoculture, habitat fragmentation, habitat isolation, and overall landscape complexity also influence the diversity, distribution, and abundance of pest and beneficial species.

Currently, the array of effective chemical pesticides has allowed pest managers to ignore many of the interactions detailed in this chapter and concentrate on determining and responding to pest abundance at the field level. However, it is widely believed that pest managers need to face the challenge of managing pests with less reliance on the chemical tools so successful in past decades. If these approaches are to succeed and a reduced reliance on chemicals is to be achieved, it is clear that pest managers will need to pay much closer attention to the types of ecological interactions detailed in this chapter.

ACKNOWLEDGMENTS

We wish to thank L. Dyer, E. Grafius, M. Haas, J. Landis, E. J. P. Marshall, D. Orr, K. Renner, A. Roda, and J. Ruberson, who each provided useful comments on earlier versions of this manuscript. Funding from the National Science Foundation, BSR 89-00618 and DEB-92-11771, Rackham Foundation, and USDA Low Input Sustainable Agriculture Program, LWF62-016-02942 has fostered our research examining the impact of agricultural landscape structure on insect natural enemies. We also wish to thank Jan Eschbach for assistance in the final preparation of the manuscript.

REFERENCES

1. Abramsky, Z. 1983. Experiments on seed predation by rodents and ants in the Israeli desert. *Oecologia* 57:328–332.
2. Andersen, A.N. 1987. Effects of seed predation by ants on seedling densities at a woodland site in SE Australia. *Oikos* 48:171–174.
3. Andow, D. 1983. The extent of monoculture and its effects on insect pest populations with particular reference to wheat and cotton, *Agric. Ecosyst. Environ.* 9:25–35.
4. Andow, D. 1991. Vegetational diversity and arthropod population response. *Annu. Rev. Entomol.* 36:561–586.
5. Archibold, O.W., and L. Hume. 1983. A preliminary survey of seed input into fallow fields in Saskatchewan. *Can. J. Bot.* 61:1216–1221.
6. Baliddawa, C.W. 1985. Plant species diversity and crop pest control. *Insect. Sci. Appl.* 6:479–487.
7. Banaszak, J. 1992. Strategy for conservation of wild bees in an agricultural landscape. *Agric. Ecosyst. Environ.* 40:179–192.
8. Bernays, E.A. 1985. Arthropods for weed control in IPM systems. In *Biological Control in Agricultural IPM Systems*. M.A. Hoy and D.C. Herzog, eds. Orlando, Florida: Academic Press, pp 373–391.
9. Best, L.B. 1983. Bird use of fencerows; implications of contemporary fencerow management practices. *Wildlife Soc. Bull.* 11:343–347.
10. Best, R.L., and C.C. Beegle. 1977. Food preferences of five species of carabids commonly found in Iowa cornfields. *Environ. Entomol.* 6:9–12.
11. Boac, J., and J. Pospisil. 1984. Ground (Coleoptera, Carabidae) and rove (Coleoptera, Staphylinidae) beetles of wheat and corn fields and their interaction with surrounding biotopes. *Soviet J. Ecol.* 15:120–130.
12. Booij, C.J.H., and J. Noorlander. 1992. Farming systems and insect predators. *Agric. Ecosyst. Environ.* 40:125–135.
13. Borchert, M.I., and S.K. Jain. 1978. The effect of rodent seed predation on four species of California annual grasses. *Oecologia* 33:101–113.

14. Brandenburg, R.L., and G.G. Kennedy. 1982. Intercrop relationships and spider mite dispersal in a corn/peanut agro-ecosystem. *Entomol. Exp. Appl.* 32:269–276.
15. Branson, T.F. and J.L. Krysan. 1981. Feeding and oviposition behavior and life cycle strategies of *Diabrotica*: an evolutionary view with implications for pest management. *Environ. Entomol.* 10:826–831.
16. Brown, J.H., D.W. Davidson, and O.J. Reichman. 1979. An experimental study of competition between seed-eating desert rodents and ants. *Am. Zool.* 19:1129–1143.
17. Brown, J.H., O.J. Reichman, and D.W. Davidson. 1979. Granivory in desert ecosystems. *Annu. Rev. Ecol. Syst.* 10:201–227.
18. Brown, V.K. 1991. The effects of changes in habitat structure during succession in terrestrial communities. In *Habitat Structure: The Physical Arrangement of Objects in Space.* S. Bell, E.D. McCoy and H. Mushinsky, eds. London: Chapman and Hall, pp 141–168.
19. Brust, G.E., and G.J. House. 1988. Weed seed destruction by arthropods and rodents in low-input soybean agroecosystems. *Am. J. Altern. Agric.* 3:19–25.
20. Bunce, R.G.H., and D.C. Howard, eds. 1990. *Species Dispersal in Agricultural Habitats.* London: Belhaven Press.
21. Bunce, R.G.H., L. Ryszkowski, and M.G. Paoletti. 1993. *Landscape Ecology and Agroecosystems.* Boca Raton, Florida: Lewis.
22. Burdon, J.J., D.R. Marshall, and J.D. Oates. 1992. Interactions between wild and cultivated oats in Australia. *Proceedings of the 4th International Oat Conference.* Adelaide, South Australia: International Oat Conference Committee, Vol. 2, pp 82–87.
23. Carlson, J.D., M.E. Whalon, D.A. Landis, and S.H. Gage. 1992. Springtime weather patterns coincident with long-distance migration of potato leafhopper into Michigan. *Agric. Meterol.* 59:183–206.
24. Carroll, S.C., D.R. Rummel and E. Segarra. 1993. Overwintering by the boll weevil (Coleoptera: Curculionidae) in conservation reserve program grasses on the Texas high plains. *J. Econ. Entomol.* 86:382–393.
25. Castrale, J.S. 1984. Impacts of conservation tillage practices on farmland wildlife in southeastern Indiana. *Federal Aid Annual Report, Proj. No. W-26-R-15. Population Levels and Habitat Use.* Indianapolis: Indiana Department of Natural Resources, Fish and Wildlife Division.
26. Castrale, J.S. 1987. Wildlife use of cultivated fields set aside under the payment of kind (PIK) program. *Indiana Acad. Sci.* 97:173–180.
27. Cavers, P.B. 1983. Seed demography. *Can. J. Bot.* 61:3578–3590.
28. Cavers, P.B. and D.L. Benoit. 1989. Seed banks of arable land. In *Ecology of Soil Seed Banks.* M.A. Leck, V.T. Parker, and R.L. Simpson, eds. San Diego: Academic Press, pp 309–328.
29. Chouinard, G., S.B. Hill, C. Vincent, and N.N. Barthakur. 1992. Border-row sprays for control of the plum curculo in apple orchards: behavioral study. *J. Econ. Entomol.* 85:1307–1317.
30. Coll, M., and D.G. Bottrell. 1991. Microhabitat and resource selection of the European corn borer (Lepidoptera: Pyralidae) and its natural enemies in Maryland field corn. *Environ. Entomol.* 20:526–533.

31. Collis-George, N., and J.E. Sands. 1959. The control of seed germination by moisture as a soil physical property. *Australian J. Agric. Res.* 10:628–636.

32. Collis-George, N., and J.E. Sands. 1962. Comparison of the effects of the physical and chemical components of soil water energy on seed germination. *Australian J. Agric. Res.* 13:575–584.

33. Cornell, H.V. 1986. Oak species attributes and host size influence cynipine wasp species richness. *Ecology* 67:1582–1592.

34. Crawley, M.J. 1989. Insect herbivores and plant population dynamics. *Annu. Rev. Entomol.* 34:531–564.

35. Dambach, C.A. 1948. *A Study of the Ecology and Economic Value of Crop Field Borders*. Columbus: Ohio State University Press.

36. Dasberg, S. 1971. Soil water movement to germinating seeds. *J. Exp. Bot.* 22:999–1008.

37. Dasberg, S. and K. Mendel. 1971. The effect of soil water and aeration on seed germination. *J. Exp. Bot.* 22:992–998.

38. Davidson, D.W., D.A. Samson, and R.S. Inouye. 1985. Granivory in the Chihuahuan Desert: interactions within and between trophic levels. *Ecology* 66:486–502.

39. Decker, G.C. 1931. The biology of the stalk borer, *Papaipema nebris* (Gn.). *Iowa State Agr. Res. Bull.* 143:289–351.

40. Dennis, P. and G.L.A. Fry. 1992. Field margins: can they enhance natural enemy population densities and general arthropod diversity on farmland? *Agric. Ecosyst. Environ.* 40:95–115.

41. DeRozari, M.B., W.B. Showers, and R.H. Shaw. 1977. Environment and the sexual activity of the European corn borer. *Environ. Entomol.* 6:658–665.

42. Desender, K. 1982. Ecological and faunal studies on Coleoptera in agricultural land. II. Hibernation of Carabidae in agro-ecosystems. *Pedobiologia* 23:295–303.

43. Dessaint, F., R. Chadoeuf, and G. Barralis. 1991. Spatial pattern analysis of weed seeds in the cultivated soil seed bank. *J. Appl. Ecol.* 28:721–730.

44. Diamond, J.M., and R.M. May. 1981. Island biogeography and the design of natural reserves. In *Theoretical Ecology: Principles and Applications*. R.M. May, ed. Oxford, UK: Blackwell, pp 238–252.

45. Dixon, A.F.G. 1971. The life-cycle and host preferences of the bird cherry-oat aphid, *Rhopalosiphum padi* L., and their bearing on the theories of host alternation in aphids. *Ann. Appl. Biol.* 68:135–147.

46. Doak, D.F., P.C. Marino, and P.M. Kareiva. 1992. Spatial scale mediates in the influence of habitat fragmentation on dispersal success: implications for conservation. *Theoret. Pop. Biol.* 41:315–336.

47. Duelli, P. 1980. Adaptive dispersal and appetive flight in the green lacewing, *Chrysopa carnea*. *Ecol. Entomol.* 5:213–220.

48. Duelli, P., M. Studer, I. Marchand, and S. Jakob. 1990. Population movements of arthropods between natural and cultivated areas. *Biol. Cons.* 54:193–207.

49. Dyer, L.D. and D.A. Landis. 1996. Effects of habitat, temperature, and sugar availability on longevity on *Eriborus terebrans* (Hymenoptera: Ichneumonidae). *Environ. Entomol.* 25:1192–1201.

50. Dyer, L.D., and D.A. Landis. 1997. Influence of noncrop habitats on the distribution of *Eriborus terebrans* (Hymenoptera: Ichneumonidae) in cornfields. *Environ. Entomol.* 26:924–932.

51. Flanders, K.L., and E.B. Radcliffe. 1989. Origins of potato leafhoppers (Homoptera: Cicadellidae) invading potato and snap bean in Minnesota. *Environ. Entomol.* 18:1015–1024.

52. Forman, R.T.T., and J. Baudry. 1984. Hedgerows and hedgerow networks in landscape ecology. *Environ. Manag.* 8:495–510.

53. Forman, R.T.T., and M Godron. 1986. *Landscape Ecology.* New York: Wiley.

54. Gange, A.C., and M. Llewellyn. 1989. Factors affecting orchard colonization by the black-kneed capsid (*Blepharidopterous angulatus* (Hemiptera: Miridae)) from alder windbreaks. *Ann. Appl. Biol.* 114:221–230.

55. Gravesen, E., and S. Toft. 1987. Grass fields as reservoirs for polyphagous predators (Arthropoda) of aphids (Homoptera: Aphidae). *J. Appl. Entomol.* 104:461–473.

56. Grime, J.P. 1977. Evidence for the existence of three primary strategies in plants and its relevance to ecological and evolutionary theory. *Am. Naturalist* 111:1169–1194.

57. Grout, T.G., and R.I. Richards. 1992. The dietary effect of windbreak pollens on longevity and fecundity of a predacious mite *Euseius addoensis addoensis* (Acari: Phytoseiidae) found in citrus orchards in South Africa. *Bull. Entomol. Res.* 81:317–320.

58. Hagen, K.S. 1962. Biology and ecology of predaceous Coccinellidae. *Annu. Rev. Entomol.* 7:289–326.

59. Hancock, J.F., C.R. Sandoval, D.C. Ramsdell, P.W. Callow, W. Boylen-Pett, K. Hokanson, and T.P. Holtsford. 1993. Disease spread between wild and cultivated blueberries. *Acta Horticulturae* 346:240–245.

60. Hawkins, B.A., and J.H. Lawton. 1987. Species richness for parasitoids of British phytophagous insects. *Nature* 326:788–790.

61. Hendrix, W.H., and W.B. Showers. 1992. Tracing black cutworm and armyworm (Lepidoptera: Noctuidae) northward migration using *Pithecellobium* and *Calliandra* pollen. *Environ. Entomol.* 21:1092–1096.

62. Holmes, P.M. 1990. Dispersal and predation in alien *Acacia. Oecologia* 83:288–290.

63. Hume, L., and O.W. Archibold. 1986. The influence of a weedy habitat on the seed bank of an adjacent cultivated field. *Can. J. Botany* 64:1879–1883.

64. Idris, A.B, and E. Grafius. 1995. Wildflowers as nectar sources for *Diadegma insulare* (Hymenoptera: Ichneumonidae), a parasitoid of diamondback moth (Lepidoptera: Plutellidae). *Environ. Entomol.* 24:1726–1735.

65. Inouye, R.S., G.S. Byers, and J.H. Brown. 1980 Effects of predation and competition on survivorship, fecundity and community structure of desert annuals. *Ecology* 61:1344–1351.

66. Irwin, M.E., and J.M. Thresh. 1990. Epidemiology of barley yellow dwarf: a study in ecological complexity. *Annu. Rev. Phytopathol.* 29:393–424.

67. Jervis, M.A., and N.A.C. Kidd. 1986. Host-feeding strategies in hymenopteran parasitoids. *Biol. Rev.* 61:395–434.

68. Jervis, M.A., N.A.C. Kidd, M.G. Fitton, T. Huddleston, and H.A. Dawah. 1993. Flower-visiting by hymenopteran parasitoids. *J. Nat. Hist.* 27:67–105.

69. Johnson, C.G. 1949. Infestation of a bean field by *Aphis fabae* Scop. in relation to wind direction. *Ann. Appl. Biol.* 33:441–451.

70. Johnson, C.G. 1969. *Migration and Dispersal of Insects by Flight.* London: Methuen.

71. Johnson, H.E., and R.S. Cameron. 1969. Phytophagous ground beetles. *Ann. Entomol. Soc. Am.* 62:909–914.

72. Julien, M.H., J.D. Kerr, and R.R. Chan. 1984. Biological control of weeds: an evaluation. *Prot. Ecol.* 7:3–25.

73. Karpiscak, M.M., and O.M. Grosz. 1979. Dissemination trails of Russian thistle (*Salsola kali* L.) in recently fallowed fields. *J. Arizona Nevada Acad. Sci.* 14:50–52.

74. Karssen, C.M. 1982. Seasonal patterns of dormancy in weed seeds. In *The Physiology and Biochemistry of Seed Development, Dormancy and Germination.* A.A. Khan, ed. Amsterdam: Elsevier, pp 243–269.

75. Kjellsson, G. 1985. Seed fate in a population of *Carex pilulifera* L., II. Seed predation and its consequences for dispersal and the seed bank. *Oecologia* 67:424–429.

76. Kromp, B., and K.H. Steinberger. 1992. Grassy field margins and arthropod diversity: a case study on ground beetles and spiders in eastern Austria (Coleoptera: Carabidae; Arachnida: Aranei, Opiliones). *Agric. Ecosyst. Environ.* 40:71–93.

77. Kruess, A., and T. Tscharntke. 1994. Habitat fragmentation, species loss, and biological control. *Science* 264:1581–1584.

78. Lamp, W.O., and L. Zhao. 1993. Prediction and manipulation of movement by polyphagous, highly mobile pests. *J. Agric. Entomol.* 10:267–281.

79. Lamp, W.O., M.J. Morris, and E.J. Armbrust. 1989. *Empoasca* (Homoptera: Cicadellidae) abundance and species composition in habitats proximate to alfalfa. *Environ. Entomol.* 18:423–428.

80. Landis, D.A. 1994. Integrating biological control into farming systems. *Proceedings of the 1994 Illinois Agricultural Pesticides Conference.* University of Illinois at Champaign-Urbana. College of Agric. Coop. Ext. Ser., pp 61–63.

81. Landis, D.A. 1994. Arthropod sampling in agricultural landscapes: ecological considerations. In *Handbook of Sampling Methods for Arthropods in Agriculture.* L.P. Pedigo and G.D. Buntin, eds. Boca Raton, Florida: CRC Press, pp 16–28.

82. Landis, D.A., and M. Haas. 1992. Influence of landscape structure on abundance and within-field distribution of *Ostrinia nubilalis* Hübner (Lepidoptera: Pyralidae) larval parasitoids in Michigan. *Environ. Entomol.* 21:409–416.

83. Lashomb, J.H., and Y. Ng. 1984. Colonization by Colorado potato beetle, *Leptinotarsa decemlineata* (Say) (Colepotera: Chrysomelidae), in rotated and non-rotated potato fields. *Environ. Entomol.* 13:1352–1356.

84. Lawton, J.H. 1978. Host-plant influences on insect diversity: the effects of space and time. In *Diversity of Insect Faunas* (Symp. R. Entomol. Soc. London 9). L.S. Mound and N. Waloff, eds. Oxford, UK: Blackwell, pp 105–125.

85. Lawton, J.H. 1983. Plant architecture and the diversity of phytophagous insects. *Annu. Rev. Entomol.* 28:23–39.
86. Lawton, J.H., and D. Schröder. 1977. Effects of plant type, size of geographical range and taxonomic isolation on number of insect species associated with British plants. *Nature* 265:137–140.
87. Lewis, T. 1965. The effects of an artificial windbreak on the aerial distribution of flying insects. *Ann. Appl. Biol.* 55:503–512.
88. Lewis, T. 1969. The distribution of insects near a low hedgerow. *J. Appl. Ecol.* 6:443–452.
89. Lund, R.D., and F.T. Turpin. 1977. Carabid damage to weed seeds found in Indiana cornfields. *Environ. Entomol.* 6:695–698.
90. Manley, G.V. 1992. Observations on *Harpalus pennsylvanicus* (Coleoptera: Carabidae) in Michigan seed corn fields. Newsletter of the Michigan Entomological Society, 37:1–2.
91. Maredia, K.M., S.H. Gage, D.A. Landis, and J.M. Scriber. 1992. Habitat use patterns by the seven-spotted lady beetle (Coleoptera: Coccinellidae) in a diverse agricultural landscape. *Biol. Cont.* 2:159–165.
92. Maredia, K.M., S.H. Gage, D.A. Landis, and T.M. Wirth. 1992. Ecological observations on predatory Coccinellidae (Coleoptera) in southwestern Michigan. *Great Lakes Entomol.* 25:265–270.
93. Marino, P.C., K.L. Gross, and D.A. Landis. 1997. Weed seed loss due to predation in Michigan maize fields. *Agric. Ecosyst. Environ.* 66:189–196.
94. Marino, P.C., and D.A. Landis. 1996. Effect of landscape structure on parasitoid diversity and partisitism in agroecosystems. *Ecol. Appl.* 69:276–284.
95. Marshall, E.J.P. 1988. The ecology and management of field margin floras in England. *Outlook Agric.* 17:178–182.
96. Marshall, E.J.P. 1988. The dispersal of plants from field margins. In *Environmental Management in Agriculture*. J.R. Park, ed. London: Belhaven Press, pp 136–143.
97. Marshall, E.J.P. 1989. Distribution patterns of plants associated with arable field edges. *J. Appl. Ecol.* 26:247–257.
98. Marshall, E.J.P. and A. Hopkins. 1990. Plant species composition and dispersal in agricultural land. In *Dispersal in Agricultural Habitats*. R.G.H. Bunce and D.C. Howard, eds. London: Belhaven Press, pp 98–116.
99. MacArthur, R.H., and E.O. Wilson. 1967. *The Theory of Island Biogeography*. Princeton, New Jersey: Princeton University Press.
100. Murdoch, W.W., F.C. Evans, and C.H. Peterson. 1972. Diversity and pattern in plants and insects. *Ecology* 53:819–829.
101. Nagarajan, S., and D.V. Singh. 1990. Long-distance dispersion of rust pathogens. *Annu. Rev. Phytopathol.* 28:139–153.
102. Neunzig, H.H. 1969. *The Biology of the Tobacco Budworm and the Corn Earworm in North Carolina*. Tech. Bull. No. 196, North Carolina Ag. Exp. Sta., Raleigh.
103. Norris, R.F. 1986. Weeds and integrated pest management systems. *Hortscience.* 21:402–410.

104. Oraze, M.J., and A.A. Grigarick. 1992. Biological Control of Ducksalad (*Heteranthera limosa*) by the waterlily aphid (*Rhopalosiphum nymphaeae*) in Rice (*Oryza sativa*). *Weed Sci.* 40:333–336.

105. Orr, D., and J.M. Pleasants. 1996. The potential of native prairie plant species to enhance the effectiveness of the *Ostrinia nubilalis* parasitoid *Macrocentrus grandii*. *J. Kansas Ent. Soc.* 69:133–143.

106. Orr, D.B., D.A. Landis, D.R. Mutch, G.V. Manley, S.A. Stuby, and R.L. King. 1997. Ground cover influence on microclimate and *Trichogramma* (Hymenoptera: Trichogrammatidae) augmentation in seed corn production. *Environ. Entomol.* 26:433–438.

107. Palrang, A.T., A.A. Grigarick, M.J. Oraze, and L.S. Hesler. 1994. Association of levee vegetation to rice water weevil (Coleoptera: Curculionidae) infestation in California rice. *J. Econ. Entomol.* 87:1701–1706.

108. Paraja, M.R., and D.W. Staniforth. 1985. Seed-soil microsite characteristics in relation to weed seed germination. *Weed Sci.* 33:190–195.

109. Pasek, J.E. 1988 Influence of wind and windbreaks on local dispersal of insects. In *Windbreak Technology*. J.R. Brandle, D.L. Hintz, and J.W. Sturrock, eds. New York: Elsevier, pp 539–554.

110. Perrin, R.M. 1975. The role of the perennial stinging nettle, *Urtica dioica*, as a reservoir of beneficial natural enemies. *Ann. Appl. Biol.* 81:289–297.

111. Pickett, S.T.A., and P.S. White, eds. 1985. *The Ecology of Natural Disturbance and Patch Dynamics*. San Diego: Academic Press.

112. Platt, W.J. 1976. The natural history of a fugitive prairie plant (*Mirabilis hirsuta* (Pursh) MacM.). *Oecologia* 22:399–409.

113. Pollard, E. and J. Relton 1970. Hedges. V. A study of small mammals in hedges and uncultivated fields. *J. Appl. Ecol.* 7:549–557.

114. Poston, F.L., and L.P. Pedigo. 1975. Migration of plant bugs and the potato leafhopper in a soybean-alfalfa complex. *Environ. Entomol.* 4:8–10.

115. Povey, F.D., H. Smith, and T.A. Watt. 1993. Predation of annual grass weed seeds in arable field margins. *Ann. Appl. Biol.* 122:323–328.

116. Powell, W. 1986. Enhancing parasitoid activity in crops. In *Insect Parasitoids*. J.Waage and D. Greathead, eds. London: Academic Press, pp 319–340.

117. Price, P.W. 1976. Colonization of crops by arthropods: non-equilibrium communities in soybean fields. *Environ. Entomol.* 5:605–611.

118. Reader, R.J. 1991. Control of seedling emergence by ground cover: a potential mechanism involving seed predation. *Can. J. Bot.* 69:2084–2087.

119. Reice, S.R. 1994. Nonequilibrium determinants of biological community structure. *Am. Scientist.* 82:424–435.

120. Reichman, O.J. 1984. Spatial and temporal variation of seed distributions in Sonoran Desert soils. *J. Biogeogr.* 11:1–11.

121. Risch, S.J., D. Andow, and M. Altieri. 1983. Agroecosystem diversity and pest control: data, tentative conclusions, and new research directions. *Environ. Entomol.* 12:625.

122. Roberts, H.A. 1981. Seed banks in soil. *Adv. Appl. Biol.* 6:1–55.

123. Robinson, G.R., R.D. Holt, M.S. Gaines, S.P. Hamburg, M.L. Johnson, H.S.

Fitch, and E.A. Martinko. 1992. Diverse and contrasting effects of habitat fragmentation. *Science* 257:524–526.

124. Rodenhouse, N.L., G.W. Barrett, D.M. Zimmerman, and J.C. Kemp. 1992. Effects of uncultivated corridors on arthropod abundances and crop yields in soybean agroecosystems. *Agric. Ecosyst. Environ.* 38:179–191.

125. Ryszkowski, L., and J. Karg. 1991. The effect of the structure of agricultural landscape on biomass of insects of the above-ground fauna. *Ekol. Pol.* 39:171–179.

126. Ryszkowski, L., J. Karg, G. Margarit, M.G. Paoletti, and R. Zlotin. 1993. Above-ground insect biomass in agricultural landscapes of Europe. In *Landscape Ecology and Agroecosystems*. R.G.H. Bunce, L. Ryszkowski, and M.G. Paoletti, eds. Boca Raton, Florida: Lewis, pp 71–82.

127. Sarukhan, J. 1974. Studies of the demography: *Ranunculus repens* L. *R. bulbosus* L. and *H. acris* L. II. Reproductive strategies and seed population dynamics. *J. Ecol.* 62:151–177.

128. Sherrod, D.W., J.T. Shaw, and W.H. Luckmann. 1979. Concepts on black cutworm field biology in Illinois. *Environ. Entomol.* 8:191–195.

129. Showers, W.B., G.L. Reed, J.F. Robinson, and M.B. Derozari. 1976. Flight and sexual activity of the European corn borer. *Environ. Entomol.* 5:1099–1104.

130. Showers, W.B., J.F. Witkowski, C.E. Mason, D.D. Calvin, R.A. Higgins, and G.P. Dively. 1989. *European Corn Borer Development and Management*. North Cent. Reg. Publ. 327. Ames: Iowa State University.

131. Smith, T.J., III. 1987. Seed predation in relation to tree dominance and distribution in mangrove forests. *Ecology* 68:266–273.

132. Smith, T.J., III. 1987b. Effects of seed predators and light level on the distribution of *Avicennia marina* (Forsk.) Vierh. in tropical tidal forests. *Estuarine, Coastal Shelf Sci.* 25:43–51.

133. Sork, V.L. 1983. Mammalian seed dispersal of pignut hickory during three fruiting seasons. *Ecology* 64:1049–1056.

134. Southerton, N.W. 1984. The distribution and abundance of predatory arthropods overwintering on farmlands. *Ann. Appl. Biol.* 105:423–429.

135. Southerton, N.W. 1985. The distribution and abundance of predatory Coleoptera overwintering in field boundaries. *Ann. Appl. Biol.* 106:17–21.

136. Southwood, T.R.E., V.K. Brown, and P.M. Reader. 1979. The relationship of plant and insect diversities in succession. *Biol. J. Linnaean Soc.* 12:327–348.

137. Stern, V.M. 1984. Pest and beneficial insects associated with agriculture and riparian systems. In *California Riparian Systems: Ecology, Conservation, and Productive Management*. R.E. Warner and K.M. Hendrix, eds. Berkeley: University of California Press, pp 970–982.

138. Stevens, O.A. 1932. The number and weight of seeds produced by weeds. *Am. J. Bot.* 19:784–794.

139. Stinner, B.R., D.A. McCartney, and W.L. Rubink. 1984. Some observations on ecology of the stalk borer (*Papaipema nebris* (Gn.): Noctuidae) in no-tillage corn agroecosystems. *J. Georgia Entomol. Soc.* 19:229–234.

140. Stinson, C.S.A., and V.K. Brown. 1983. Seasonal changes in the architecture

of natural plant communities and its relevance to insect herbivores. *Oecologia* 56:67–69.

141. Taylor, C.E., and C.G. Johnson. 1954. Wind direction and the infestation of bean fields by *Aphis fabae* Scop. *Ann. Appl. Biol.* 41:107–116.

142. Theaker, A.J., N.D. Boatman, and R.J. Froud-Williams. 1995. Variation in *Bromus sterilis* on farmland: evidence for the origin of field infestations. *J. Appl. Ecol.* 32:47–55.

143. Thiele, H.U. 1977. *Carabid Beetles in Their Environments: A Study on Habitat Selection by Adaptations in Physiology and Behavior.* Berlin: Springer-Verlag.

144. Thomas, M.B., S.D. Wratten, and N.W. Sotherton. 1991. Creation of island habitats in farmland to manipulate populations of beneficial arthropods: predator densities and emigration. *J. Appl. Ecol.* 29:906–917.

145. Thomas, M.B., S.D. Wratten, and N.W. Sotherton. 1992. Creation of island habitats in farmland to manipulate populations of beneficial arthropods: predator densities and species composition. *J. Appl. Ecol.* 29:524–531.

146. Thresh, J.M. 1981. *Pests, Pathogens and Vegetation: The Role of Weeds and Wild Plants in the Ecology of Crop Pests and Diseases.* Boston: Pitman.

147. Tucker, G.M. 1992. Effects of agricultural practices on field use by invertebrate-feeding birds in winter. *J. Appl. Ecol.* 29:779–790.

148. van den Bosch, R., and A.D. Telford. 1964. Environmental modification and biological control. In *Biological Control of Insect Pests and Weeds.* P. DeBach, ed. London: Chapman and Hall, pp 459–488.

149. van Emden, H.F. 1965. The role of uncultivated land in the biology of crop pests and beneficial insects. *Sci. Horticul.* 17: 121.

150. van Emden, H.F. 1990. Plant diversity and natural enemy efficiency in agroecosystems. In *Critical Issues in Biological Control.* M. Mackauer, L.E. Ehler, and J. Roland, eds. Hants, UK: Intercept, pp 63–80.

151. Vos, C.C., and P. Opdam. 1993. *Landscape Ecology of a Stressed Environment.* London: Chapman and Hall.

152. Wainhouse, D., and T.H. Coaker. 1981. The distribution of carrot fly (*Psila rosae*) in relation to the flora of field boundaries. In *Pests, Pathogens and Vegetation: The Role of Weeds and Wild Plants in the Ecology of Crop Pests and Diseases.* J.M. Thresh, ed. Boston: Pitman, pp 263–271.

153. Wallin, H. 1985. Spatial and temporal distribution of some abundant carabid beetles (Coleoptera: Carabidae) in cereal fields and adjacent habitats. *Pedobiologia* 28:19–34.

154. Wallin, H. 1986 Habitat choice of some field inhabiting carabid beetles (Coleoptera: Carabidae) studied by recapture of marked individuals. *Ecol. Entomol.* 11:457–466.

155. Warbuton, D.B., and W.D. Klimstra. 1984. Wildlife use of no-till and conventionally tilled corn fields. *J. Soil Wat. Conser.* 39:327–330.

156. Way, M.J., and M.E. Cammell. 1981. Effects of weeds and weed control on invertebrate pest ecology. In *Pests, Pathogens and Vegetation: The Role of Weeds and Wild Plants in the Ecology of Crop Pests and Diseases.* J.M. Thresh, ed. Boston: Pitman, pp 443–458.

157. Wedberg, J.L., and B.L. Giebink. 1986. Stalk-boring insect pests of corn. UWEX note 5. Madison: University of Wisconsin–Extension.

158. Werner, P.A., I.K. Bradbury, and R.S. Gross. 1980. The biology of Canadian weeds. *Solidago canadensis* L. *Can. J. Plant Sci.* 60:1393–1409.

159. Wilcove, D.S., C.H. McLellan, and A.P. Dobson. 1986. Habitat fragmentation in the temperate zone. In *Conservation Biology: The Science of Scarcity and Diversity*. M. Soule, eds. Sunderland: Sinauer, pp 237–256.

160. Wooley, J.B., Jr., L.B. Best, and W.R. Clark. 1987. Impacts of no-till row cropping on upland wildlife. *Trans. N. Am. Wildlife Nat. Res. Conf.* 50:157–168.

5

Direct and Indirect Effects of Plants on Performance of Beneficial Organisms

Marcel Dicke
Wageningen Agricultural University, Wageningen, The Netherlands

I. INTRODUCTION

The control of herbivorous arthropods by employing their carnivorous enemies is a practice that goes back to the ancient Chinese [37]. Especially in the past 100 years, the application of biological control of arthropod pests through introduction of carnivorous enemies of the herbivorous pest has been practiced extensively [38,78,124,125]. In addition, native species of carnivorous arthropods have most likely suppressed herbivore populations in many agricultural situations and most likely do so even today. First attempts of biological control of plant pathogens through antagonistic action of microorganisms were made in 1921 [9]. However, since the introduction of synthetic pesticides in the 1940s carnivorous arthropods and beneficial fungi and bacteria have usually been eliminated from crop systems where chemical pest control was practiced. The important role of natural enemies in suppressing herbivore populations is strikingly apparent when pesticides induce pest outbreaks resulting from elimination of natural enemies while the herbivores are resistant to the pesticide [37,38].

Another important approach to crop protection is the use of plant cultivars resistant to the pest species. However, just as pest species can become resistant to synthetic pesticides, so too can they develop resistance to plant cultivars. Basically, the stronger the resistance of the host plant, the stronger the selection pressure on the pests to develop resistance. Therefore, plant breeders should ideally develop partially resistant cultivars or cultivars that combine several partial resistance characteristics. Although this approach is intuitively appealing, its success is not guaranteed [74,75]. Pest control strategies often combine resistant cultivars with pesticide applications. However, such pest control programs only consider the crop and the pest organism, while neglecting the pest's natural enemies.

Biological pest control has experienced a revival in the past decades, and its role in pest control is still increasing both in greenhouse crops and in field crops [33,78,125,129,235]. In many instances, the direct financial stimulus for utilization of biological control [124,126,129] has been combined with political measures that stimulate biological control while discouraging the application of synthetic pesticides [130]. This development has been a significant impetus for the use of both biological control and host plant resistance practices. In effect, the two pest control methods have been or will often be integrated. However, it is not a priori obvious that these two strategies are synergistic. Plant characteristics not only affect herbivores but they may affect the plant-inhabiting or visiting natural enemies of herbivores too. This latter aspect has often been neglected, but in the past 15 years, it has become better documented that plant characteristics may decisively affect carnivore performance in many respects [e.g. 20,26,45,50,96,97,127,128,175,229]. Therefore, changes in plant characteristics through breeding for resistance to herbivores is likely to result in changes in the way plants affect the natural enemies. Thus, it is essential to analyze plant-herbivore-natural enemy interactions for crop protection programs in which both host plant resistance and biological control are important components of pest control either intentionally or passively. Otherwise, plant breeding programs may result in a pest control program that is less rather than more effective.

This chapter reviews the current knowledge of the ways in which plants may affect the performance of herbivores' natural enemies, with a focus on carnivorous arthropods, and how the development of integrated pest control programs can benefit from taking such effects into consideration. Thus, this chapter deals with basic aspects as well as the potential or current application in pest control. There is much less knowledge on plants' influences on the effectiveness of antagonists of plant pathogens. Where appropriate, this issue is addressed here, but it necessarily is a minor aspect of this chapter.

II. INTERACTION OF PLANT AND CARNIVORE EFFECTS ON HERBIVORE POPULATION DYNAMICS

To understand the patterns and processes that govern the success or failure of an integration of host plant resistance and biological control, it is necessary to analyze natural systems. It is well known that herbivore population dynamics are affected by both plants and carnivores. However, it is not easy to determine which of the two is the most important factor. Herbivore population dynamics may be affected by plants and carnivores separately, but they also are affected by an interaction of the two [173,175,198]. Thus, when viewed from the plant's perspective, it has two defensive options against herbivores: (a) direct defense, which affects herbivores directly through physical or chemical means, such as thorns, toxins, digestibility reducers; and (b) indirect defense, which influences the effectiveness of carnivores, e.g., through provisioning of alternative food or by attracting carnivores after infestation by herbivores occurs [45,171,175]. In addition to plant characteristics that favor carnivore effectiveness, carnivores can also be hampered by plant characteristics that improve the defense of herbivores against carnivores or that reduce contacts between herbivores and carnivores. Examples are plant toxins that are sequestered by the herbivore or structural refuges in which the herbivore hides from its enemies [45,171] (Fig. 1).

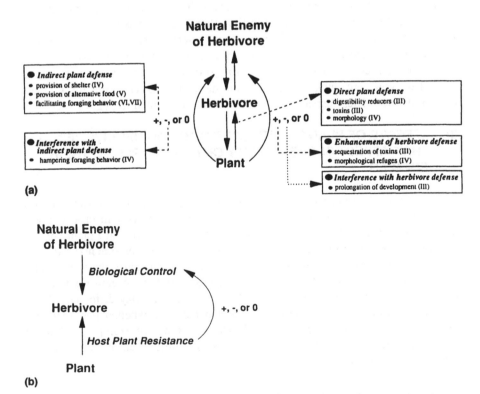

Figure 1 (a) Direct and indirect defense of plants in a tritrophic system of plants, herbivores, and natural enemies of herbivores (modified after ref. 171); numbers in parentheses refer to sections in this chapter; (b) biological control and host plant resistance in a tritrophic context.

The notion that the effects of plants and carnivores on herbivore populations may interact is relatively new [20,175]. The interaction may vary with plant species and carnivore species, and current data suggest that the pattern is rather idiosyncratic. Currently, it is too early to conclude when and why top-down or bottom-up effects are most important [96,173,198]. Yet, the notion itself of interacting effects of plants and carnivores on herbivore population dynamics warrants further investigation and should be taken into consideration in pest control programs. Besides natural systems, considerable knowledge on the interaction of plant and carnivore characteristics has also been gained for agricultural systems. Employing this knowledge and expanding it is of great importance for the development of new, integrated pest control strategies. In subsequent sections, several plant characteristics that affect carnivores either directly or indirectly through the herbivore are presented. On the one hand, plants can affect the attack rate and efficiency of carnivores (see Fig. 1). When attack rate or efficiency are enhanced, this is a form of indirect defense. However, some plant characteristics can interfere with carnivore attack; for example, by hampering foraging behavior. On the other hand, plants can affect the defense of herbivores against their carnivorous enemies (see Fig. 1). Herbivore defense may be enhanced by sequestration of plant toxins or herbivore defense may be hampered; for instance, by prolongation of development which results in the herbivore remaining in a stage that is susceptible to carnivores for a longer period of time. These aspects of the chapter are presented in Figure 1a.

III. EFFECTS OF PLANT CHARACTERISTICS ON HERBIVORES THAT AFFECT SUSCEPTIBILITY OF HERBIVORES TO THEIR ENEMIES

In the context of pest control, classic plant breeding is primarily concerned with the improvement of agronomic properties and effects of certain crop characteristics on herbivores, such as antibiotic or antixenotic secondary plant chemicals that cause nonpreference, mortality, or reduced developmental rates of herbivores. However, increased contents of certain secondary plant chemicals may severely interfere with the effectiveness of beneficial organisms that are used in biological pest control. Thus, different cultivars may differ largely in their susceptibility to pests and pathogens depending on whether pest management is done by chemical or biological means. A range of examples is known where the effectiveness of beneficial organisms is influenced by the impact of plant characteristics on pest organisms.

A. Toxins and Digestibility Reducers

A plant's direct defense may comprise the employment of toxins such as alkaloids, phenolics, and terpenoids. Herbivorous arthropods may overcome this defense in different ways, such as feeding on plant tissue with the lowest concentration of toxins, detoxifying the plant chemicals, or sequestering the compounds (188). If they sequester the plant compounds, they may subsequently employ them actively in defensive actions against their enemies [14,21,142, 164,166]. These options are especially employed by specialist herbivores. Generalist herbivores usually do not thrive well, if at all, on plant species that contain specific toxins; they usually feed on plant species that have more general defenses such as digestibility reducers (see below) [172].

When herbivores sequester plant toxins, this affects interactions with their carnivorous enemies. The carnivores themselves may be seriously harmed by the toxins or alternatively they may avoid feeding on herbivores that sequester plant toxins, they may detoxify the compounds, or they may sequester the compounds themselves and use them third-handedly for their own protection (for reviews, see, e.g., refs. 56, 164, and 188). Usually the more specialized carnivores are better able to cope with herbivore-sequestered plant toxins than are generalist carnivores [172]. This is well documented for tobacco-caterpillar-parasitoid interactions [14,119]. Thus, specific plant toxins result in food chains that are rather simple, consisting predominantly of specialists at the herbivore and carnivore levels. It should be noted here that the classification of specialists and generalists does not necessarily refer to the species level. Generalist species may comprise populations of specialists [65]. This is also reflected in agriculture where generalist pest species may consist of strains that are adapted to a certain host plant [67,68,69,73,163,181,230,231,232].

Parasitoids that are not able to cope with the sequestered plant compounds of their hosts may suffer from severe malformation or mortality [56]. Each parasitoid juvenile feeds on only one host individual, and thus the quality of its food is completely determined by its mother's oviposition choice. Predators, on the other hand, consume several to many prey individuals. They are not completely dependent on the quality of a single prey item. However, feeding on toxic prey may affect their food intake for several days [39,40,138]. Sequestration of secondary plant chemicals is also known to affect the herbivore's susceptibility to pathogens [200]. For instance, tobacco hornworm larvae that feed on diets with low nicotine concentration are very susceptible to the bacterium *Bacillus thuringiensis* (Bt), whereas they are unaffected by the bacterium when feeding on a diet with a high nicotine concentration [120] (Fig. 2). Similar effects were recorded for the effects of tannins on the susceptibility of gypsy moth or corn earworm larvae to viruses or *Bacillus thuringiensis* endotoxin, respectively [95,152].

Tannins and phenolics are better known for their effects on the herbivore developmental rate. They act as digestibility reducers, slowing down herbivore development. Consequently, herbivores remain in a stage that is susceptible to their enemies for a longer period of time, which may increase mortality. Other plant characteristics may also affect the window of the herbivore's vulnerabil-

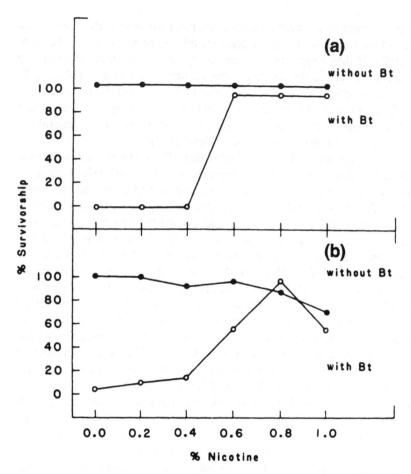

Figure 2 Effect of secondary plant chemicals on herbivore susceptibility to pathogens. Survivorship of *Manduca sexta* larvae reared on synthetic diet containing nicotine (at six concentrations) with *Bacillus thuringiensis* (Bt) (open circles) or without *B. thuringiensis* (closed circles) for two separate experiments (a and b). (Reprinted with permission from ref. 120.)

ity to its enemies. For instance, nutrient availability of the plant may affect the herbivore's developmental rate and subsequently the degree of attack by carnivores [134] as well as the profitability of the herbivore to its parasitoids (e.g., see refs. 79, 202, and 246), predators [107,168], and pathogens (e.g., see ref. 61). However, slow growth does not invariably lead to increased predation or parasitization [36,122]. For instance, those leaves that result in a slow development may provide a much better structural protection against carnivores [36] (see also Section IV below).

B. Herbivore Size Determines Suitability For Carnivores

Herbivore size may affect predators and parasitoids. Predators consume more small prey than large prey. If herbivore size is affected by plant characteristics, such as digestibility reducers, the effect of predators can be more pronounced on resistant varieties than on susceptible varieties [98]. For a parasitoid, each host is a fixed amount of food for one or more offspring. Below a certain size threshold, a host is insufficient for complete offspring development. The effects of plants on herbivore size therefore affect the suitability of the herbivores as hosts for parasitoids. This effect varies with parasitoid species and may thus affect parasitoid competition and local parasitoid species composition and consequently the degree of pest control [135]. This implies that particular cultivars can be unsuitable for the employment of biological control with a certain selected parasitoid species if the host does not reach the size needed by the parasitoid. Parasitoids are often selective as to the sex of the offspring they deposit in the host. In such cases, male offspring will usually be deposited in small hosts and female offspring in large host individuals [72,135]. Because host size depends on plant characteristics, plants can influence parasitoid sex ratio and thereby local population dynamics. The parasitoid sex ratio can also be affected independent of herbivore size by plant characteristics such as total leaf nitrogen [66].

C. Induced Direct Defense or Induced Resistance

It is well documented that herbivory or pathogen infestation changes plant characteristics that affect herbivores [110,216]. For instance, the production of phytoalexins may be induced by pathogen infestation [4,57], or mechanical damage by herbivores can induce the production of proteinase inhibitors that reduce herbivore digestion [190,244]. Induced direct defense appears to be rather non specific. Infestation with one species of herbivore may affect the performance of other herbivore species and even pathogen infestation may induce resistance against herbivorous arthropods and vice versa [111,119,154]. The effects of induced direct defense may last up to several years in natural systems, such as birch-caterpillar interactions [84]. Also, in agricultural systems, induced direct defense can significantly affect herbivore populations [109]. For instance, spider-mite infestation of cotton cotyledons followed by removal of the mites leads to induced resistance: Under field conditions, spider-mite populations on treated plants are lower than on control plants during a substantial period of the growing season [108].

Thus, the phenomena discussed above, whether they relate to specific toxins such as nicotine or digestibility reducers such as tannins can be constitutively effective or after induction [13,123,199]. Although this can affect the time scale during which the effects are present, the effects themselves are not dependent upon the constitutive or induced nature of the phenomenon.

IV. EFFECTS OF PLANT MORPHOLOGY ON CARNIVOROUS ARTHROPODS

Apart from influences of plant characteristics on beneficial organisms via herbivores, plants can also influence the effectiveness of beneficials directly. Plant morphology is one of the characteristics that can strongly determine carnivore effectiveness.

A. Interference with Carnivore Effectiveness

Cuticle thickness, glandular and nonglandular trichomes, and plant architecture are among the many aspects of plant morphology that can affect herbivores [106], and these characteristics have been used in many plant breeding programs to select for cultivars that are less susceptible to herbivores (e.g., see refs 17 and 156). Although characteristics such as cuticle thickness, wax layers, and trichomes may impede herbivore movement and accessibility of plant contents to small herbivores [103], they also have a direct effect on the behavior of carnivores that are of similar size [128,156]. Herbivores move relatively little over a plant once they have selected a suitable feeding site. In contrast, carnivores generally move much more on a plant in search of herbivores, and thus the effects of external morphology on carnivore behavior may exceed the effects on herbivore behavior. The negative effects of many plant morphological characteristics on carnivore behavior have been documented for a range of carnivore species (e.g., see refs. 80, 83, 94, 112, and 114; and see refs. 128 and 156 for reviews). Among the documented effects are (a) impairment of carnivore movements by nonglandular trichomes and thus a reduced rate of encounter with herbivores [94,114,128], (b) entrapment of carnivores by glandular trichomes or hooked trichomes resulting in a carnivore mortality [83,113, 179,180,], and (c) reduced carnivore adhesion on smooth plant surfaces, for example, without any trichomes, resulting in the carnivores falling off the plant [80,112].

Plant morphology is also known to change in response to herbivory. Galling insects, for instance, induce the plant to produce a gall in which the herbivore develops. The morphological characteristics of the gall determine the mortality of the herbivore resulting from its enemies. Thus, the size of the gall affects the vulnerability of the herbivore to parasitoids. The thicker the gall, the lower the chances that a parasitoid ovipositor is able to reach the herbivore

[174,241]. Trichomes and thorns can be induced by herbivory too: After feeding damage, trichome densities on leaves of nettles and alder increased, impeding subsequent herbivory [16,146,178]. Thus, the plant quality for carnivores can change owing to herbivory.

Plant morphological characteristics can provide herbivores with structural refuges such as closed panicles or closed leaf configurations that impede visual herbivore location by their enemies (e.g., see refs. 53 and 167). Herbivores can modify plant structures to construct a refuge; for example, by tying leaves together. This can be so important for the herbivore that they sacrifice a rapid development for it. For instance, larvae of the pyralid moth *Omphalocera munroei* preferentially feed on old, rigid leaves of their host plant rather than on quickly wilting young leaves. On the old leaves, they develop more slowly but acquire better protection against predators than on the young leaves [36].

B. Enhancement of Carnivore Effectiveness

Not all plant morphological characteristics affect carnivores negatively. Plants can also provide carnivores with structures that are used as shelter or as nesting places. The example of *Acacia* trees that have hollow thorns (domatia) in which ants build nests is well known [101,102]. The ants are thus present on the plant and they attack herbivores, both arthropods [101] and vertebrates [137], and even vines from neighboring plants [97,101]. Not only *Acacia* trees provide such domatia. Many plants have structures that can be used as shelter by arthropods. A widespread phenomenon is that of acarodomatia (Greek: "little mite houses"), which have been described from a wide range of plant species [25,99]. These comprise plant structures of various forms, such as invaginations or hair tufts at vein junctions, that are too small for most insects to enter (Figs. 3 and 4). These structures often (more than 50% of the samples) harbor mites that either feed on herbivorous mites or on fungi [157,158,165]; herbivorous mites have seldom been found in these domatia. An explanation for this occupation by predatory and fungivorous mites may be that these mites have a need for a high relative humidity to prevent their eggs from desiccating [165]. For instance, many species of phytoseiid predatory mites are known to lay eggs at sheltered sites [141]. In contrast, many phytophagous mites, such as tetranychid spider mites and eriophyid mites, make their own shelters in the form of webs or galls. Apart from acarodomatia, carnivorous mites were also found to be more abundant on tomentose leaves than on smooth leaves [161,237]. These preferences for hairy substrates and invaginated structures may also be related to the higher relative humidity associated with them [81]. Apart from foliar acarodomatia, growing points such as those found in cassava have also been found to function as acarodomatia and to harbor carnivorous mites [12]. Ex-

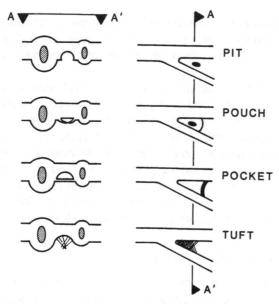

Figure 3   Schematic view of several types of acarodomatia. Left-hand column: cross-section through vein axil; right-hand column: perpendicular view of abaxial surface of leaf showing the domatium in vein axil. A—A′ indicates where cross-section is taken. (Reprinted with permission from ref. 157.)

perimental evidence shows that acarodomatia significantly affect the presence of carnivorous mites on plant leaves as well as their reproduction and consequently the presence of herbivores [12,81,238], which is similar to the case of ants on *Acacia* that benefit the plant by being accommodated on it.

C.   Breeding for Varieties that Enhance the Effectiveness of Carnivores

These examples show that plant morphology can have both negative and positive effects on beneficial organisms. Moreover, a certain morphological trait may differentially affect different species of beneficials [201]. An important question then is whether we can select for plant varieties with morphological characteristics that do not hamper but rather support the activities of beneficial organisms. Studies on natural systems have shown that plant genotypes differ in the expression of morphological characteristics (e.g., see ref. 174). Studies on agricultural systems have shown that breeding for plants with morphological characteristics that favor beneficial organisms is feasible and morphological characteristics often seem to be under simple genetic control by one or two genes [112,128,205]. Pea varieties that differ at only two loci exhibit structural dif-

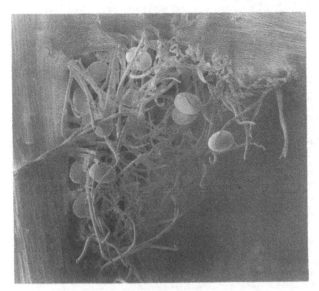

Figure 4 Scanning electron micrograph of eggs of the predator *Typhlodromus occidentalis* laid in a tuft domatium of *Viburnum tinus*. Each egg is approximately 170 μm long, scale bar: 0.6 mm. (Reprinted with permission from P. Grostal and D.J. O'Dowd. 1994. Plants, mites and mutualism: leaf domatia and the abundance and reproduction of mites on *Viburnum tinus* (Caprifoliaceae). *Oecologia* 97:308–315. Copyright of Springer-Verlag, Berlin.)

ferences that significantly alter the effects of coccinellid beetles on aphid population growth [112]. The best example of incorporating plant morphological characteristics into a breeding program with the aim of improving biological control is provided by van Lenteren and de Ponti [128]. Biological control of the greenhouse whitefly with the parasitoid *Encarsia formosa* is successfully applied on a total area of 3700 hectares worldwide in various crops [125]. However, this approach was not successful in all crops. In cucumber, for example, parasitoid movement was hampered by rigid hairs that retained whitefly honeydew. Thus, parasitoid movement was impaired and they also became contaminated with the honeydew, which reduced foraging activity because of the increase in parasitoid preening behavior. This, in turn, resulted in low attack rates. In a breeding program, van Lenteren and de Ponti [128] used a hairless cucumber variety and obtained a half-haired cucumber hybrid by crossing this with commercial, haired varieties (Figs. 5 and 6). Hairiness was determined by one dominant gene with intermediary inheritance. On the hairless variety, the parasitoids moved so fast that they ran over the whitefly larvae without

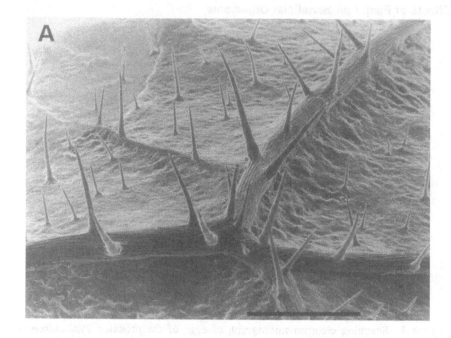

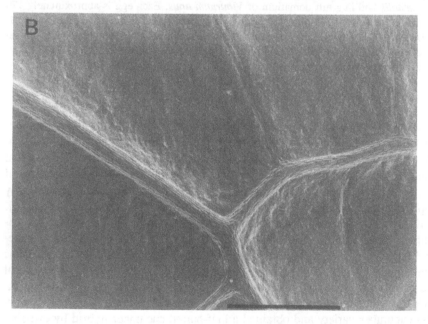

Figure 5 Scanning electron micrograph of the leaf surface of (A) commercially used hairy cucumber variety and (B) hairless cucumber variety. Scale bars: 1 mm. (Courtesy of J.C. van Lenteren, Department of Entomology, Wageningen Agricultural University, Wageningen, The Netherlands.)

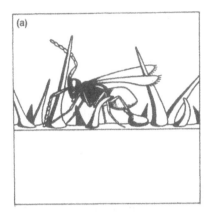

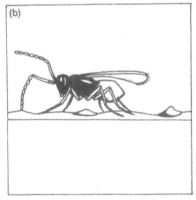

Figure 6 Parasitoid *Encarsia formosa* on (a) hairy and (b) hairless cucumber variety, showing size of hairs relative to parasitoid size. (Courtesy of J.C. van Lenteren, Department of Entomology, Wageningen Agricultural University, Wageningen, The Netherlands.)

noticing them, but on the half-haired variety, they had a walking speed twice that on the haired variety, which resulted in higher encounter rates with host larvae (Fig. 7). In greenhouse experiments, the performance of the parasitoids was better on half-haired hybrids than on a commercial, haired variety, yielding a satisfactory level of control (Fig. 8). The seed of the half-haired hybrid is now available for commercial application by breeding companies. Evaluations under realistic agricultural situations such as that done by van Lenteren and de Ponti [128] is of great importance, as shown by Obrycki [156]. He found that the adverse effects of potato trichomes on predators and parasitoids of aphids were less severe under field conditions than under greenhouse conditions. From his studies, he reached a similar conclusion as van Lenteren and de Ponti [128]: Moderate levels of pubescence can be compatible with biological control.

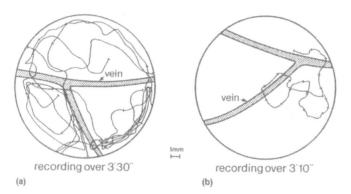

(a) recording over 3'30"

(b) recording over 3'10"

Figure 7 Walking pattern of *Encarsia formosa* on (a) hairless (C–) and (b) hairy (C+) cucumber leaves. The circle is the inner outline of a perspex ring, the veins of the leaves are dotted. The cross lines on the walking pattern indicate the position of the wasp every ten seconds. (After ref. 94; courtesy of J.C. van Lenteren, Department of Entomology, Wageningen Agricultural University, Wageningen, The Netherlands.)

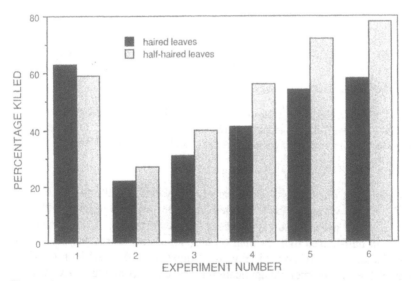

Figure 8 Percentage of whitefly larvae (*Trialeurodes vaporariorum*) killed by *Encarsia formosa* through either parasitism or host feeding in a greenhouse test under realistic agricultural situations. Paired samples *t*-test: reject null-hypothesis of equal parasitism on both cucumber lines ($P < .05$). (After ref. 128; courtesey of J.C. van Lenteren, Department of Entomology, Wageningen Agricultural University, Wageningen, The Netherlands.)

V. NUTRITIONAL ASPECTS OF INTERACTIONS BETWEEN PLANTS AND BENEFICIAL ORGANISMS

Beneficial organisms such as carnivorous arthropods and antagonists of pathogenic microorganisms can use various plant tissues and plant products, such as floral and extrafloral nectar, plant sap, and pollen, as sources of nutrition.

Floral nectar is generally considered as a reward of plants to pollinators. However, floral nectar is also used as food by parasitoids, especially synovigenic species (i.e., species that develop eggs during the adult life and need food to do so), and predatory arthropods. These beneficial organisms are often attracted by flower odors [58,82,203,236] and their presence on the plant may subsequently lead to a reduction of herbivore numbers on the plant [54,209,245]. This attraction of carnivores to flowers is affected by the plant species, which may explain a number of cases where beneficial arthropods are found more on some plant species than on others [209,234]. Some studies indicate that floral nectar may function *primarily* as food for carnivorous bodyguards and not as reward to pollinators; for example, in the wind-pollinated *Croton suberosus* [54].

Many plant species also produce nectar extraflorally in a variety of structures [118] and a range of arthropods have been recorded feeding on them, including predators, such as ants and mites, and parasitoids [3,11,12,19,82, 117,151,218]. Extrafloral nectar is frequently used by carnivorous arthropods as a food source when herbivores are scarce or absent and often the nectar is an inferior food source compared with herbivorous prey, supporting survival and development but not reproduction [12,82]. Yet, an interesting observation is that the production of extrafloral nectar may be induced by herbivory [118]. The effect of extrafloral nectar on carnivore effectiveness is obvious from reports that show that nectar application or the use of nectaried cultivars increases predation or parasitism of herbivores [11,82,201,218]. This effect is often attributable to an increase in longevity during periods of low prey density [10,11,218,236].

Some plants, such as acacia, have special structures called Beltian bodies (named after the discoverer Belt) that contain lipid-rich plant secretions that are collected by ants and transported to their nest in the acacia thorns as food [102]. The production of these food bodies can be induced by carnivore invasion on the plant [184].

Several carnivorous arthropods also thrive well on pollen and may actively forage for pollen in times of prey scarcity [82,160]. Pollen is a much better food source for arthropods than floral or extrafloral nectar. The pollen of at least several plant species contains all nutritional factors required by arthropods for growth and reproduction [82,160].

Several carnivores have been shown to extract plant sap from leaves or employ plant exudates on the leaf surface as a food source in times of prey scarcity [41,151,169,182,189]. This feeding on plant sap may explain why a carnivore species is encountered on one plant species and not on another [34,169].

Plant exudates that diffuse to the leaf surface contain a variety of nutrients [219], but their quantity is usually low, especially on young leaves [23]. In contrast, other phyllosphere deposits such as pollen and aphid honeydew are more abundant, and these generally form a main nutrient source for necrotrophic pathogenic microorganisms. However, the presence of these nutrients also greatly stimulates the colonization of leaves by saprophytic microorganisms [63,64]. Studies where the saprophytes were selectively reduced showed that the outcome of competition between saprophytes and pathogens determines whether disease occurs or whether the antagonism provides protection to the plant [24,52,90,136]. To my knowledge, it is unknown whether the plant affects the competitive ability of the saprophytes analogous to the effects of plants on the effectiveness of carnivorous arthropods. This area needs further exploration. This may be especially important, since there is also an interaction with herbivorous arthropods such as aphids whose honeydew is a food source for both saprophytes and pathogens [52]. Moreover, carnivorous arthropods such as predatory mites may be found abundantly on plants on which no prey is visible while no prey traces are found in the predator gut [12,44,47,191]. In such cases, microorganisms on the phylloplane may have been used as food, since it is known that predatory mites can use fungi as a food source [160]. Extrafloral nectar or honeydew can be colonized by microorganisms, which may provide carnivores with a much better food source than the sugar source by itself, leading to an increase in reproduction through enhancement of processes such as egg maturation [85]. Thus, trophic relationships on the phylloplane can be quite complex, but carbohydrate sources can definitely lead to increased abundance of beneficial organisms.

The fact that plants provide carnivorous arthropods and antagonistic saprophytic microorganisms with food, either directly or indirectly through insect honeydew, may have both quantitative and qualitative effects on beneficial organisms. However, it is impossible to say in general whether this is favorable in terms of pest control. Not only do the beneficials use the plant products but pest organisms may do so as well [82,105,131,132]. For instance, mite predators of the western flower thrips (*Frankliniella occidentalis*) survive on cucumber plants in the presence of pollen, but pollen is also a food source of their prey. Thus, during prey scarcity, pollen supporting the predators favors the outcome of pest control, whereas during a thrips infestation, the pollen enhances the pest organism [183]. In such instances, modeling may help in providing detailed insight into the desirability of a certain plant trait [1,89,227].

VI. CHEMICAL PLANT CUES THAT MEDIATE SEARCHING BEHAVIOR OF BENEFICIAL ORGANISMS

Foraging behavior of carnivorous arthropods is affected to a large extent by chemical cues, both from their herbivorous victims and from the food of the herbivores. This has been extensively investigated and many reviews stress that chemicals of plant origin play an essential role during many stages of host searching by carnivorous arthropods (e.g., see refs. 49, 155, 220, 226, 229, 233, 234, 242, and 243). That carnivores respond to constitutively produced plant volatiles has been known for a long time [155,233,243]. However, constitutive plant odors seem to have a limited information content and may be of particular importance to specialist carnivores that attack specialist herbivores [229]. An exciting discovery was made in the past decade: The production of plant volatiles that attract carnivorous arthropods can be induced by herbivore damage [45,50,222]. In other words, plants have a chemical burglar's alarm. Herbivore-induced plant volatiles appear to play an important role in carnivore foraging behavior (reviewed by Dicke et al. [49], Vet and Dicke [229], Turlings et al. [226], Dicke [43], and Takabayashi et al. [213]) and are likely to play a role in many plant-herbivore-carnivore interactions [43]. This section reviews the knowledge on differences between plant cultivars in attracting carnivores, either constitutively or induced by herbivory, and the relevance of incorporating the effect of plant volatiles on carnivores into plant breeding programs.

A. Constitutive Carnivore Attractants and Plant Breeding

Plant cultivars can differ widely in the amounts of certain secondary chemicals they contain. This is not only true for toxins that affect herbivores after their ingestion but also for volatiles that are emitted by the plant. In a study of cotton, Elzen et al. [59] reported more than a 100-fold difference between different cultivars in the emission of volatile terpenes that attract the parasitoid *Campoletis sonorensis*. That differences in the amount of plant-produced parasitoid attractants can decisively affect pest control was demonstrated in a field study of cabbage-aphid-parasitoid interactions by van Emden [60]. More aphids (*Brevicoryne brassicae*) were found on a "resistant" cultivar than on a "susceptible" cultivar when parasitoids (*Diaeretiella rapae*) were present. He showed that the susceptible cultivar, which produced significantly more (2.4 times) of the parasitoid attractant allyl isothiocyanate, was significantly preferred by the parasitoids and had significantly more mummified aphids than the resistant cultivar. This is one of the few examples where the effect of plant volatiles has been investigated in the field and the laboratory. It shows that the terms *sus-*

ceptible and *resistant* are context specific and that plant breeding practices may be counterproductive if the effect of carnivores is not taken into consideration.

B. Herbivore-Induced Carnivore Attractants: Local and Systemic Effects

It has long been known that carnivores can discriminate from a distance between odors from uninfested plants and odors from plants that are infested by herbivores [229] (Fig. 9). However, it was generally assumed that the volatiles involved were herbivore products. Ten years ago, it was suggested that the distant discriminatory ability of carnivores might well be associated with plant volatiles that are produced in response to herbivore damage [192] (Fig. 10). Evidence that this indeed is the case was first published for a tritrophic system consisting of lima bean plants, herbivorous spider mites (*Tetranychus urticae*),

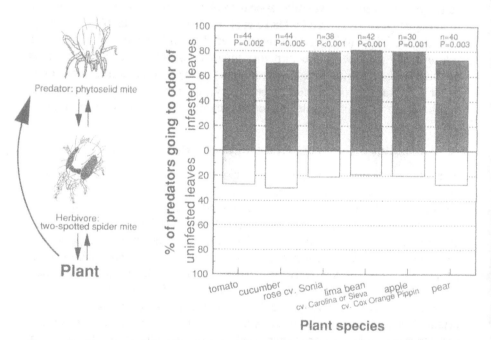

Figure 9   Response of females of the predatory mite *Phytoseiulus persimilis* in a Y-tube olfactometer, when offered the odors from uninfested leaves vs odors of leaves infested by the spider mite *Tetranychus urticae*. The spider mite is a polyphagous herbivore, whereas the predatory mite is a specialist that mainly feeds on spider mites in the genus *Tetranychus*. In all experiments, satiated predators were used, except the experiment with pear, where 24-h starved *P. persimilis* females were used. (Based on data from ref. 45.)

Odor source   % predators choosing this odor source
                when offered vs. appropriate control
                      in a Y-tube olfactometer

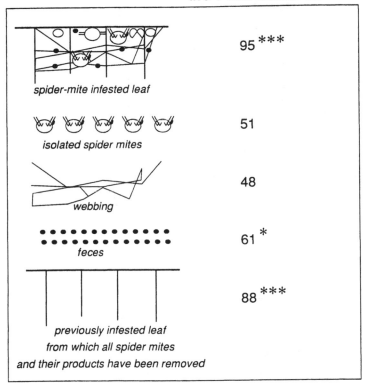

Figure 10 Analysis of components of lima bean–spider mite (*Tetranychus urticae*) complex for their attraction to carnivorous mites (*Phytoseiulus persimilis*). Each odor source was offered against an appropriate control consisting of an uninfested leaf or clean air, in a Y-tube olfactometer. *$P < .05$; ***$P < .001$ (sign test). (Based on data from ref. 194.)

and carnivorous mites (*Phytoseiulus persimilis*) [45,49,50,212] (Fig. 11). Soon afterwards similar data were reported for other tritrophic systems, such as corn–*Spodoptera exigua* caterpillars–*Cotesia marginiventris* parasitoids [221,222, 224,225]; corn–*Pseudaletia separata* caterpillars–*Cotesia kariyai* parasitoids [214]; corn–stemborer caterpillar–*Cotesia flavipes* parasitoids [170]; cabbage–*Pieris* caterpillars–*Cotesia* parasitoids [2,22,70,139 140,207]; and several plant–herbivorous mites–carnivorous mites systems [e.g., see refs. 42,45,47,51,100, 211, and 215]. The induced plant volatiles are released after herbivory and carnivores discriminate between herbivore-damaged and mechanically damaged plants. Moreover, herbivore-induced plant volatiles can even be specific for the

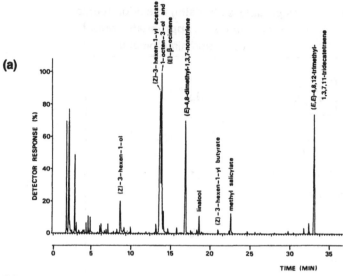

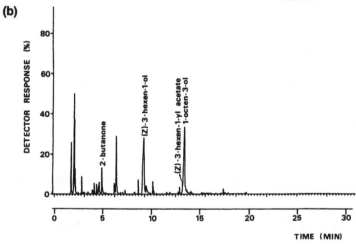

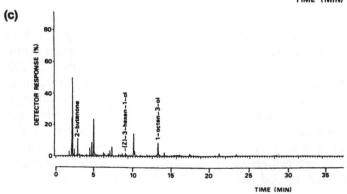

herbivore that inflicts the damage: Carnivores can sometimes discriminate between plants damaged by different herbivore species [49,191,226,229]. Herbivore-induced plant volatiles have been reported for several tritrophic interactions (see above), and their role in extermination of herbivore populations seems to be essential [49,193]. Plant cultivars may differ in the degree of carnivore attraction through herbivore-induced plant volatiles. For instance, the degree of infestation of bean plants by the two-spotted spider mite, *Tetranychus urticae*, that was needed for attraction of the herbivore's predators differed between cultivars [49] (Fig. 12), and chemical analyses showed differences in the composition of the volatile blend emitted by different apple cultivars after infestation by *T. urticae* [211] (Fig. 13).

The production of herbivore-induced plant volatiles is not restricted to the infested leaves. Uninfested leaves of an infested plant also emit them, although in lower amounts [48,49,170,221]. This systemic effect is mediated by elicitors that are transported from the infested to the uninfested leaves [48,212] (Fig. 14). Thus, the odor source related to herbivory is much larger than the herbivore itself: It comprises the whole plant. This adds to the value of herbivore-induced plant volatiles to foraging carnivores. But there is more: The odor source can even be larger than the partly infested plant itself! Downwind neighboring plants that are exposed to the herbivore-induced plant volatiles have been shown to release predator-attracting volatiles [29,30,49; see ref. 46 for review] (Fig. 15). Thus, after herbivore infestation, a relatively (relative to herbivore size) large odor source arises that can be used by carnivores to find their food.

C. Herbivore-Induced Carnivore Attractants and Plant Breeding

So far, more than 15 plants species, more than 10 herbivore species, and more than 10 carnivore species have been used in different combinations in studies on herbivore-induced carnivore attractants [43,45,49,51,213,220,226,229]. Almost all research conducted on herbivore-induced carnivore attractants has been done with agricultural plant species. In all cases, carnivores discriminate between herbivore-infested plants and mechanically damaged plants. The cultivars used in these studies were not selected for their ability to produce herbivore-induced carnivore attractants. It may be expected that plant cultivars currently in use have partly or completely lost much of their ability to attract carnivores through induced volatiles because of the absence of selection for this trait in breeding programs. Therefore, it is of great significance that the

---

Figure 11 Gas chromatograms of volatiles sampled from (a) lima bean leaves infested by the two-spotted spider mite (*Tetranychus urticae*), (b) mechanically damaged lima bean leaves, and (c) undamaged lima bean leaves. Spider-mite infested leaves emit several terpenoids and a phenolic that are not emitted by mechanically damaged or undamaged leaves. (After ref. 50.)

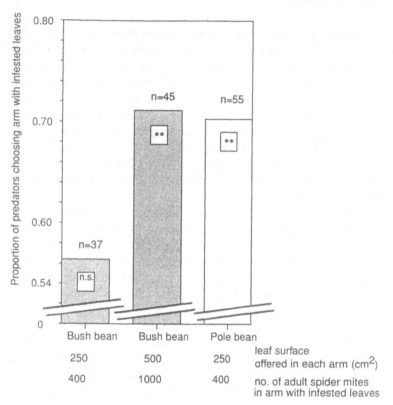

Figure 12 Response of satiated predatory mite (*Phytoseiulus persimilis*) females in a Y-tube olfactometer to bean (*Phaseolus vulgaris*) leaves infested with spider mites or uninfested leaves. The discrimination of the predators between infested and uninfested leaves was investigated for two different bean cultivars: (1) Bush bean "dubbele zonder draad," nr. 697 and (2) Pole bean "Westlandse dubbele" nr. 671, Turkenburg Ltd. The Netherlands. ** $P < .01$; n.s.: $P > .05$, sign test for differences from 50:50 distribution of predators over the two arms. (Reprinted with permission from ref. 49.)

characteristic has been found in all plant species studied so far. It will be interesting to investigate wild relatives of crop plants with respect to the quantity and quality of their herbivore-induced volatiles. If wild relatives produce higher amounts of herbivore-induced volatiles, crossing with wild ancestors may yield cultivars with a higher degree of carnivore attraction.

The use of behavior-modifying chemicals (infochemicals or semiochemicals) in biological control practices with carnivorous arthropods has received much speculation. The spatial distribution of infochemicals relative to that of the pest

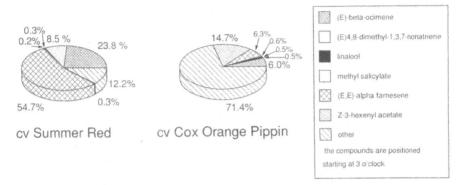

Figure 13 Composition of volatile blend emitted by leaves of two apple cultivars that are infested by two-spotted spider mites (*Tetranychus urticae*). (After data from ref. 211.)

individuals appears to be important in order to improve rather than to impair carnivore foraging behavior (see ref. 49 for discussion). Herbivore-induced plant volatiles combine several advantages over artificial application of infochemicals. First, their spatial distribution is closely linked to that of the herbivores and they are produced in much larger amounts than herbivore volatiles. Second, the application of these infochemicals does not require any action by humans. Basically, it can be seen as the recruitment of biological control agents by the plant after being attacked by herbivores. An important additional aspect for pest control is that these plant-produced volatiles do not need registration under pesticide laws.

Thus, in analogy to the study on constitutive plant volatiles by van Emden [60], the effect of herbivore-induced plant volatiles deserves to be incorporated in plant breeding practices. To do so, one should compare different cultivars for the the production of carnivore attractants in response to herbivore infestation. Recent progress in the basic knowledge of the signal-transduction process may facilitate this comparison of cultivars. It is laborious to compare the induction in response to herbivore infestation for all cultivars. Herbivory can be simulated by application of herbivore oral secretions [139,170,196,222,223]. Several attempts have been made to identify the active components in these herbivore secretions [5,140,223], and recently an elicitor in the oral secretion of caterpillars of the large cabbage white, *Pieris brassicae*, was identified that induces the production of volatiles by cabbage that attract the parasitoid *Cotesia glomerata* [140]. Elicitor identification may facilitate the comparison of cultivars for their ability to produce carnivore attractants.

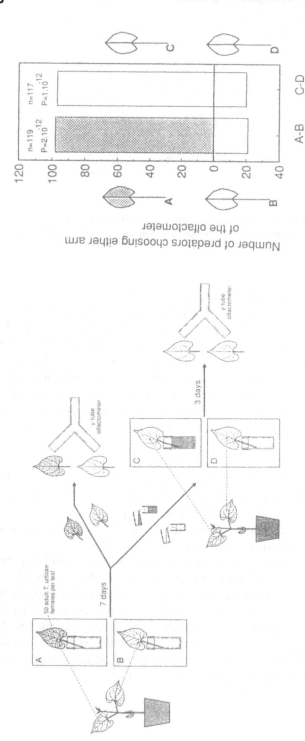

Figure 14 Systemic emission of herbivore-induced carnivore attractants is mediated by internally transported elicitor. Behavioral response of female predatory mites (*Phytoseiulus persimilis*) in a Y-tube olfactometer. Response toward spider-mite (*Tetranychus urticae*) infested lima bean leaves or uninfested lima bean leaves that have been incubated in water in which infested leaves had been standing for the previous 3 days. (Reprinted with permission from ref. 48.)

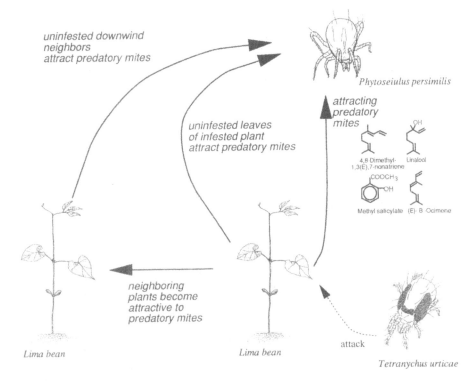

Figure 15 Effects of spider mite (*Tetranychus urticae*) infestation of lima bean plant on behavior of predatory mites (*Phytoseiulus persimilis*) that feed on the spider mites. (After data from refs. 30, 49, and 50.)

VII. VISUAL AND VIBRATIONAL PLANT CHARACTERISTICS THAT MEDIATE SEARCHING BEHAVIOR OF BENEFICIAL ORGANISMS

In contrast to knowledge about the use of chemical stimuli, the role of other stimuli in the foraging behavior of beneficial insects has received much less attention. However, since visual and vibrational stimuli related to plant characteristics are known to affect herbivores [32,147,176], and thus may be considered explicitly or implicitly in plant breeding, it is worthwhile to address the use of these stimuli by carnivorous arthropods during foraging for herbivores.

A. Visual Cues

The use of visual stimuli by beneficial insects has been studied extensively in honeybees. Some aspects have also been studied for predators such as digger

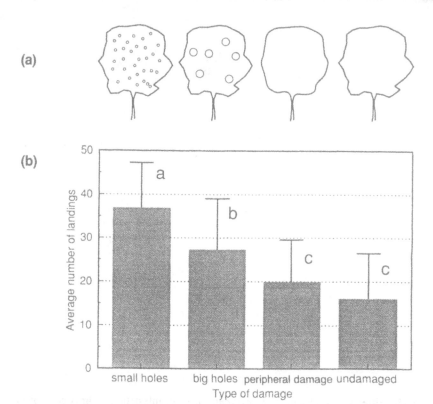

Figure 16 (a) Brussel sprout leaves with different types of mechanical damage (from left): small holes (0.3-cm diameter), visually mimicking feeding damage by larvae of the small cabbage white butterfly, *Pieris rapae*; big holes (1.5-cm diameter), visually mimicking feeding damage by larvae of the large cabbage white butterfly, *Pieris brassicae*; peripheral damage, which does not result in a visual stimulus; and no damage. (b) Effects of different types of mechanical damage on the number of landings of the parasitoid *Cotesia rubecula*, a specialist larval parasitoid of *Pieris rapae*, the small cabbage white butterfly. (Based on data from ref. 236.)

wasps, ants, and coccinellids [77,91,150,187,217] and some parasitoids [71,210, 236,239,240]. The use of visual cues related to feeding damage of herbivores has often been regarded as important for vertebrate predators [86,87,88]. However, invertebrate predators and parasitoids may also use them either alone or in combination with olfactory cues [35,62,149,210,236]. For instance, the parasitoid *Cotesia rubecula* that parasitizes caterpillars of the cryptic herbivore *Pieris rapae* discriminates between feeding holes made by the host and the much larger feeding holes made by the gregariously feeding caterpillars of *Pieris brassicae*

[236] (Fig. 16), which is not a very suitable host for this parasitoid [28]. Plant characteristics such as leaf toughness, trichomes, and the production of volatiles in response to damage may affect patterns of herbivore feeding damage. For instance, when cabbage leaves are tough and thick, caterpillar feeding may result in windows rather than holes, a difference which may significantly affect parasitoid foraging behavior [236].

Apart from herbivore damage itself, visual cues related to herbivore activities may also be due to the response of the plant to herbivore damage, such as the production of galls, the size of which varies with plant genotype [174], or to the herbivore making a hiding place in which it is visually protected [36]. Furthermore, plants may differ in the degree to which herbivores can find hiding places and thus in the degree to which they are visually exposed to their enemies [156].

B. Vibrational Cues

Several species of parasitoids or predators are known to use vibrational cues during host location [15,31,121,144,210,228]. In some cases, the vibrations are known to be transferred through the plant [15,144,210]. For instance, the leafminer parasitoid *Dapsilarthra rufiventris* employs the vibrations of eating host larvae (*Phytomyza ranunculi*) in the mine to discriminate between leaves with an empty mine and leaves in which a larva is present in the mine [210]. The intensity of the vibrations is known to be affected by the herbivore species, some of which move without causing any vibrations. In addition, vibration intensity depends to a large degree on plant characteristics such as thickness and rigidity of the epidermis and cuticle, cell wall thickness, venation structure, and cell turgor [145].

Thus, the degree to which herbivores are detectable through visual and vibrational cues can be influenced by the plant and the plants themselves determine whether they are enemy-free or enemy-rich spaces for herbivores.

VIII. EFFECTS OF PLANT COMMUNITY ON BENEFICIAL ORGANISMS

All effects that have been described above may relate to a crop plant in a monoculture, to several crop plants in a mixed cropping system, or to crop plants and noncrop plants in an agricultural system. Additional crop plants have independent and dependent effects. For instance, they may provide alternative food in the form of extrafloral nectar or pollen which occurs independent of the main crop plant, but additional plants may also affect the detectability of pest damage on the main crop plant as a result of visual or chemical interference [159,197,229]. It has often been reported that monocultures have high population densities of specialist herbivores compared with polycultures [6,7,185]. Two nonexclusive hypotheses have been advanced to explain this effect: the "resource concentration hypothesis" and the "enemies hypothesis" [186]. Basically, the resource concentration hypothesis ascribes the effect to differential influences of monocultures and polycultures on movement and reproduction of monophagous and polyphagous pest insects, whereas the enemies hypothesis seeks an explanation in an enhancement of populations of natural enemies of the pest insects through provision of alternative food and shelter. In addressing the enemies hypothesis, Sheehan [204] correctly urges that we should discriminate between the effects of specialist and generalist carnivorous arthropods. Crop diversification may increase the effectiveness of generalist carnivores through the provision of alternative prey or hosts, but this is not the case for specialist carnivores such as many species of parasitoids [204]. Thus, whether crop diversification reduces pest problems or not may be highly dependent on the characteristics of the additional crop or noncrop plants, the key pest species, and the degree of specialization of the key carnivores that attack them.

This aspect of carnivore specialization has been extended to the degree of specialization in a tritrophic context: Carnivores may be specialists or generalists at both the herbivore level and the plant level, and this may decisively affect their strategies of foraging for herbivores and thus the effects of vegetational diversity [229]. Foraging for herbivores is mediated by chemical and physical stimuli that may be strongly controlled by plants. The incorporation of the effects of plants on carnivore foraging behavior may be essential for obtaining a better understanding of the effect of mixed cropping on carnivore effects on herbivore populations and thus for better ways of manipulating multiple cropping systems. With regard to chemical cues that mediate carnivore foraging behavior, it is expected that multiple cropping will have positive effects on pest reduction because of the more likely disturbance of herbivore foraging behavior than of foraging behavior of their natural enemies (see ref. 229 for discussion).

At present, there is no general pattern known for the effects of vegetational diversity on herbivore population dynamics [8]. At best, we should be aware of potential interference or enhancement of vegetational diversity in crop systems on the effectiveness of beneficial organisms in pest control. In selected cases, knowledge on the effect of other crop plants or noncrop plants on beneficial organisms may be exploited. For instance, one may introduce noncrop plants ("banker plants") with specific herbivores so as to provide a nursery with alternative prey or hosts of carnivores [206,208] or to provide pollen as a food source in a male-sterile crop.

IX. ADAPTATION OF PEST ORGANISMS TO HOST PLANT RESISTANCE

Arthropod pests and pathogens of plants are well known for adapting to plant resistance characteristics [116]. A solution to this adaptation might be to use partially resistant plants that exert less selection pressure on the pest organisms in combination with biological control. These two methods of combatting pests were assumed to be so different that the use of biological control would not affect the rate of adaptation of the pest to the plant's resistance characteristic. However, with increasing knowledge of plants' effects on the outcome of biological control through an interaction with the effectiveness of the biological control agents, it is clear that the original assumption about the two methods exerting independent selection pressures no longer holds. If, for example, plant resistance leads to slower herbivore development and consequently to prolonged exposure to carnivores, biological control is likely to interact with the development of pest adaptation to host plant resistance. This has recently been shown with population genetical models [76,104]. However, the rate of adaptation to partially resistant plants in combination with biological control is slower than that to strongly resistant plant traits in the absence of biological control. Modeling studies have provided insight into what factors can affect adaptation of pest organisms to plant resistance. Among those factors are (a) the mode of inheritance of the pest's adaptation mechanism; (b) the mechanism of interaction with mortality from other sources such as other plant resistance characteristics, biotechnologically introduced *Bacillus thuringiensis* toxins, or carnivorous enemies; and (c) the degree of food choice by the pest organism, within a plant, between individuals of the crop plant, or between crop plants and noncrop plants, and thus its degree of exposure to resistance factors [75]. With the predictions of these models, it is important to see whether they are supported by experimental evidence, but at least they should warn us about potential patterns of adaptation.

Although biological control through carnivorous enemies of herbivores has been used for more than a century, only one case is known where the herbivore has adapted to the attack by one of its enemies. This concerns the larch sawfly, *Pristiphora erichsonii*, for the control of which the parasitoid *Mesoleius tenthredinis* was introduced into Canada [143,177]. Thus, the risk of herbivores becoming resistant to their natural enemies, for example, through encapsulation of parasitoid eggs, should be taken seriously, also because artificial selection on such host characteristics is possible [148]. Yet, it must be realized that the reverse can also be true: Parasitoids may adapt to encapsulation and thus increase their effectiveness as biological control agents [195]. From the past 100 years' experience, it is clear that biological control can be a long-term solution in any pest control program. As a consequence, wherever biological control is feasible, it is desirable to select plant cultivars that are compatible with biological control. In those cases where partial plant resistance is indispensable for attaining the desired level of control, one should realize that the herbivores may adapt to the resistance of the plant unless the resistance is based on morphological traits. However, since this adaptation is likely to occur at a slower rate with a lower level of resistance, the cultivar with the weakest level of resistance will have the longest lifetime in agricultural practice.

In addition to selecting plant cultivars that are compatible with biological control, one may also select natural enemies that are compatible with certain plant characteristics. This may be done either at the species level [124] or within a species of natural enemies by selecting certain strains or genotypes [27,133,162] or manipulating them through genetic engineering [93]. The selection or manipulation of carnivores to make them compatible with pest control has mainly been carried out in the context of pesticide resistance [92,93]. In fact, relative to that of their herbivorous prey, the genetics of carnivorous arthropods has been little studied. So far, no examples are known where natural enemy genotypes that are compatible with certain aspects of host plant resistance have been intentionally selected for. However, there is potential for such selection programs. For instance, parasitoid populations may differ in the degree to which they are affected by secondary plant chemicals such as nicotine. A population of the parasitoid *Cotesia congregata* that was collected from an area where tobacco was reared for over 350 years suffered less from nicotine sequestered by its host, *Manduca sexta*, than a population collected from an area where tobacco has been rarely cultivated for about 250 years [115]. We need to know much more about the genetics of carnivores. When more efforts are made in this field, we may find that new options are available.

X. IMPLICATIONS FOR FUTURE SELECTION PROGRAMS

A wide variety of plant attributes appears to influence the effect of carnivores on herbivore populations. From an applied perspective, this may seen to yield a rather confusing situation: To integrate successfully the breeding of host plants for resistance with the selection of natural enemies for biological control, one must theoretically screen all possible interactions between plants and natural enemies. Although it is valuable to have all this knowledge, it is obviously impractical to obtain all this information before integration of plant breeding and biological control can be started. Thus, priorities must be set. The major lesson to be learned is that although host plant resistance and biological control can be synergistic, it is not self-evident that this is the case [26]. Thus, both plant breeders and entomologists should take into consideration the constraints of each other's practices where integration of host-plant resistance and biological control is to be achieved either intentionally or implicitly. Thus, the following questions must be asked during the procedure of selecting a plant cultivar or a biological control agent.

In a plant breeding project:

a.   Is pest control in the crop under consideration done by chemical means or by biological means? The latter also includes those agricultural practices where biological control is achieved naturally without intentionally introducing carnivores. If chemical control is to be the principal method of managing pests in the crop, further evaluation of the cultivars with respect to carnivorous arthropods is not necessary, but this is an undesirable option from an environmental and medical point of view.

b.   If pest control relies on biological control, what are the main pest organisms in the crop and what are their key natural enemies under the agricultural regime to be practiced? This is information that is gathered by entomologists in a screening of carnivores to select the best species for biological control [124].

c.   What are the main plant traits known to improve or impair the performance of these natural enemies, and is it possible to select for more compatible cultivars without improving herbivore performance or impairing product quality? For instance, for the parasitoid *Encarsia formosa* that is used in biological control of greenhouse whitefly, infochemicals have not been shown decisively to affect their foraging behavior [153], but trichomes are very important [128]. In contrast, for predatory mites that attack spider mites, infochemicals are considered to be of major importance in determining effectiveness of the carnivores [49,100], whereas trichomes may be important on some crop species [83]. This aspect should be investigated in collaboration with entomologists.

In selection of beneficial organisms for biological control:

a.   What is the crop spectrum on which a certain carnivore is to be employed?

b.   What are the main effects of plants on the biology of this carnivore species in general and on those aspects that are most important for the carnivore's effectiveness in particular? See examples on *E. formosa* and predatory mites given above under the third question in the plant breeding project.

c.   What effects on the beneficial species are to be expected on the crop plant for which the carnivore is selected? For instance, some plants

such as cotton and tomato are well known for their glandular trichomes, whereas other crop plants have no glandular trichomes at all
[55,103].

d. Are there alternative carnivore species or strains that are less affected
    by these plant traits and thus potentially more effective as biological
    control agents in this crop?

In order to make a well-based prediction, collaborative studies should be
carried out. In the case of selecting cultivars that are best suitable for biological control, the following aspects should be taken into account: (1) the effect
of different cultivars on the pest organism, (b) the effect of different cultivars
on the beneficial organism, and (c) the effect of different cultivars on the interaction of the beneficial with the pest organism. In the case of selecting natural enemies that are compatible with a certain plant cultivar this concerns: (a)
which species of natural enemies are compatible with the most important plant
trait affecting the carnivores, (b) is it possible to select within a carnivore species for genotypes that are better suited on the cultivar or crop plant under investigation, and (c) is the performance of such selected carnivore species or
genotypes in the interaction with the pest organism satisfactory? Only through
an integrated selection procedure can an optimal integration of host plant resistance and biological control be reached.

The scenario depicted above still involves a large amount of research and
several pitfalls may be encountered when taking this path. For instance, during breeding for resistance against one herbivore species, increased susceptibility against another species may arise [18]. Analogously, there is a risk that
selection for a characteristic that favors one species of beneficial organism interferes with the effectiveness of another beneficial species. Yet, these are the
risks that are inherent to plant breeding. When investing in a program to integrate partial host plant resistance with biological control, this will lead to prolonged use of a resulting cultivar because of the slower rate of herbivore adaptation [76]. Because of potential interference with other herbivore–natural enemy
interactions, it is most profitable to start with projects that deal with crop plants
that have relatively few pest species and where much knowledge exists on the
biological control agents that are used to control these pests. This would favor
starting with greenhouse crops, because biological control in greenhouses is well
established, much knowledge exists on the biological control agents, and the
number of pests is relatively low compared with outdoor crops or perennial
systems such as fruit orchards [129]. The first example that has yielded success actually relates to a greenhouse system: cucumber–greenhouse whitefly–
Encarsia formosa [128] (see Section IV above).

Finally, an approach that yields application in the shortest period of time is
to use the empirical method. If a characteristic is known that improves the effec-

tiveness of a biological control agent, this may be tested under agricultural conditions. If successful for the beneficial organism under consideration while no negative effects are observed for other beneficial species, the characteristic may be incorporated into management programs. Subsequent steps will be to intensify research according to the lines depicted above in order to increase understanding of how, why, and when this aspect is important in the multitrophic system under consideration. In doing so, future programs may be developed more efficiently.

XI. CONCLUSIONS

Host plant resistance and biological control are highly valuable components of environment-friendly pest control. However, these two components are not a priori compatible. The biological control agents are active on the plant and are thus affected by a wide range of plant traits that can influence their effectiveness either directly or through interactions with the pest organism. In order to develop an integrated pest management program that incorporates both methods of pest control, we should consider how the crop affects the beneficial organisms. Ideally, we should incorporate the impact of plants on beneficial organisms both in selection procedures that select for the best plant cultivar for agriculture and in selection procedures that select for beneficial organisms that are most suitable for biological control of pest organisms. It is important to keep in mind that in doing so, we are likely to be forced to set priorities. Not all combinations of traits, either in the plant or in the beneficial organism, may be biologically realistic. For instance, some plant species may have invested considerably in direct defense rather than in indirect defense via natural enemies of herbivores [43]. In crop plants originating from such plants, there may be more possibilities of modifying plant traits that directly affect the herbivore than of traits that indirectly affect the herbivore through its natural enemies. Although the notion that plant traits may decisively influence the effectiveness of beneficial organisms is a recent one, a few studies have already shown that it is very possible to exploit this knowledge and to expand the employment of environmentally sound pest control. These studies show that cooperation of plant breeders and entomologists can lead to innovative new developments in environmentally benign pest control. With respect to plant pathogens, little is known on the influence of plants on the effectiveness of antagonists. The entomological experiences with such influences definitely warrant phytopathological investment in research into this area.

ACKNOWLEDGMENTS

I thank Joop C. van Lenteren and Louise E.M. Vet for constructive comments on an earlier version of the manuscript.

REFERENCES

1. Addicott, J.H. 1981. Stability properties of 2-species models of mutualism: simulations studies. *Oecologia* 49:42–49.
2. Agelopoulos, N.G., and M.A. Keller. 1994. Plant-natural enemy association in the tritrophic system *Cotesia rubecula-Pieris rapae*-Brassicaceae (Cruciferae). III Collection and identification of plant and frass volatiles. *J. Chem. Ecol.* 20:1746–1756.
3. Agnew, C.W., W.L. Sterling, and D.A. Dean. 1982. Influence of cotton nectar on red imported fire ants and other predators. *Environ. Entomol.* 11:629–634.
4. Agrios, G.N. 1988. *Plant Pathology*. 3rd ed. San Diego: Academic Press.
5. Alborn, H.T., T.C.J. Turlings, and J.H. Tumlinson, 1993. An elicitor in caterpillar oral secretions that induces corn seedlings to release parasitoid attractants. In *Plant Signals in Interactions with Other Organisms*. J.C. Schultz and I. Raskin, eds. *Current Topics in Plant Physiology*. Vol. 11. Rockville, Maryland: American Society of Plant Physiologists, pp 256–257.
6. Altieri, M.A., and D.K. Letourneau. 1986. Vegetational diversity and insect pest outbreaks. *CRC Crit. Rev. Plant Sci.* 2:131–169.
7. Andow, D.A. 1983. The extent of monoculture and its effects on insect pest populations with particular reference to wheat and cotton. *Agric. Ecosyst. Environ.* 9:25–35.
8. Andow, D.A. 1991. Vegetational diversity and arthropod population response. *Annu. Rev. Entomol.* 36:561–86.
9. Baker, K.F. 1987. Evolving concepts of biological control of plant pathogens. *Annu. Rev. Phytopathol.* 25:67–85.
10. Bakker, F., and M.E. Klein. 1990. The significance of cassava exudate for predaceous mites. *Symp. Biol. Hung.* 39:437–439.
11. Bakker, F.M., and M.E. Klein. 1992. Transtrophic interactions in cassava. *Exp. Appl. Acarol.* 14:293–311.
12. Bakker, F.M., and M.E. Klein. 1993. Host plant mediated coexistence of predatory mites: the role of extrafloral nectar and extrafoliar domatia on cassava. In *Selecting Phytoseiid Predators for Biological Control, with Emphasis on the Significance of Tri-trophic Interactions*. F. Bakker, ed. Ph.D. dissertation, University of Amsterdam, the Netherlands, pp 33–63.
13. Baldwin, I.T. 1991. Damage-induced alkaloids in wild tobacco. In *Phytochemical Induction by Herbivores*. D.W. Tallamy and M.J. Raupp, eds. New York: Wiley, pp 47–69.
14. Barbosa, P. 1988. Natural enemies and herbivore-plant interactions: influence of

plant allelochemicals and host specificity. In *Novel Aspects of Insect-Plant Interactions*. P. Barbosa and D.K. Letourneau, eds. New York: Wiley, pp 201-229.

15. Barth, F.G., H. Bleckmann, J. Bohnenberger, and E.A. Seyfarth. 1988. Spiders of the genus *Cupiennius* Simon 1891 (Araneae, Ctenidae). II. On the vibratory environment of a wandering spider. *Oecologia* 77:194-201.

16. Baur, R., S. Binder, and G. Benz. 1991. Nonglandular leaf trichomes as short-term inducible defense of the grey alder, *Alnus incana* (L.) against the chrysomelid beetle, *Agelastica alni* L. *Oecologia* 87:219-226.

17. Beck, S.D. 1965. Resistance of plants to insects. *Annu. Rev. Entomol.* 10:207-232.

18. Beck, S.D., and F.G. Maxwell. 1976. Use of plant resistance. In *Theory and Practice of Biological Control*. C.B. Huffaker and P.S. Messenger, ed. New York: Academic Press, pp 615-636.

19. Bentley, B.L. 1977. Extrafloral nectaries and protection by pugnacious body-guards. *Annu. Rev. Ecol. Syst.* 8:407-427.

20. Bergman, J.M., and W.M. Tingey. 1979. Aspects of interaction between plant genotypes and biological control. *Bull. Entomol. Soc. Am.* 25:275-279.

21. Björkman, C., and S. Larsson. 1991. Pine sawfly defence and variation in host plant resin acids: a trade-off with growth. *Ecol. Entomol.* 16:283-289.

22. Blaakmeer, A., J.B.F. Geervliet, J.J.A. van Loon, M.A. Posthumus, T.A. van Beek, and Ae. de Groot. 1995. Comparative headspace analysis of cabbage plants damaged by two species of *Pieris* caterpillars: consequences for in-flight host location by *Cotesia* parasitoids. *Entomol. Exp. Appl.* 73:175-182.

23. Blakeman, J.P. 1972. Effect of plant age on inhibition of *Botrytis cinerea* spores by bacteria on beetroot leaves. *Physiol. Plant Pathol.* 22:143-152.

24. Blakeman, J.P., and N.J. Fokkema. 1982. Potential for biological control of plant diseases on the phylloplane. *Annu. Rev. Phytopathol.* 20:176-192.

25. Bock, J.H., and Y.B. Linhart. 1989. Leaf domatia and mites: a plant protection-mutualism hypothesis. In *The Evolutionary Ecology of Plants*. J.H. Bock and Y.B. Linhart, eds. Boulder, Colorado: Westview Press, pp 341-359.

26. Boethel, D.J., and R.D. Eikenbary. 1986. *Interactions of Plant Resistance and Parasitoids and Predators of Insects*. Chichester, UK: Ellis Horwood.

27. Boulétreau, M. 1986. The genetic and coevolutionary interactions between parasitoids and their hosts. In *Insect Parasitoids*. J.K. Waage and D.J. Greathead, eds. London: Academic Press, pp 169-200.

28. Brodeur, J., and J.B.F. Geervliet. 1992. Host species affecting the performance of the larval parasitoids *Cotesia glomerata* and *Cotesia rubecula* (Hymenoptera: Braconidae). I. Preference for host developmental stage of *Pieris* (Lepidoptera: Pieridae). *Med. Fac. Landbouww. Rijksuniv. Gent* 57/2b:543-545.

29. Bruin, J., and M.W. Sabelis. 1989. Do cotton plants communicate by means of airborne signals?. *Med. Fac. Landbouww. Rijksuniv. Gent* 54/3a:853-859.

30. Bruin, J., M. Dicke, and M.W. Sabelis. 1992. Plants are better protected against spider-mites after exposure to volatiles from infested conspecifics. *Experientia* 48:525-529.

31. Casas, J. 1989. Foraging behaviour of a leafminer parasitoid in the field. *Ecol. Entomol.* 14:257–265.
32. Claridge, M.F., and P.W.F. de Vrijer. 1994. Reproductive behavior: the role of acoustic signal in species recognition and speciation. In *Planthoppers: Their Ecology and Management* R.F. Denno and T.J. Perfect, eds. New York: Academic Press, pp 216–233.
33. Cock, M.J.W. 1985. A review of biological control of pests in the Commonwealth Carribean and Bermuda up to 1982. *C.I.B.C. Tech. Commun.* 9:1–218.
34. Congdon, B.D., and J.A. McMurtry. 1985. Biosystematics of *Euseius* on California citrus and avocado with the description of a new species (Acari: Phytoseiidae). *Int. J. Acarol.* 11:23–30.
35. Cornelius, M.L. 1993. Influence of caterpillar-feeding damage on the foraging behavior of the paper wasp *Mischocyttarus flavitarsis* (Hymenoptera: Vespidae). *J. Insect Behav.* 6:771–781.
36. Damman, H. 1987. Leaf quality and enemy avoidance by the larvae of a pyralid moth. *Ecology* 68:88–97.
37. DeBach, P. 1974. *Biological Control by Natural Enemies*. London: Cambridge University Press.
38. DeBach, P., and D. Rosen. 1991. *Biological Control by Natural Enemies*. 2nd ed. Cambridge: Cambridge University Press.
39. DeMoraes, G.J., and J.A. McMurtry. 1986. Suitability of the spider mite *Tetranychus evansi* as prey for *Phytoseiulus persimilis*. *Entomol. Exp. Appl.* 40:109–115.
40. DeMoraes, G.J., and J.A. McMurtry. 1987. Physiological effect of the host plant on the suitability of *Tetranychus urticae* as prey for *Phytoseiulus persimilis* (Acari: Tetranychidae, Phytoseiidae). *Entomophaga* 32:35–38.
41. Dicke, F.F., and J.L. Jarvis. 1962. The habits and seasonal abundance of *Orius insidiosus* (Say) (Hemiptera-Heteroptera: Anthocoridae) on corn. *J. Kansas Entomol. Soc.* 35:339–344.
42. Dicke, M. 1988. Prey preference of the phytoseiid mite *Typhlodromus pyri*: 1. Response to volatile kairomones. *Exp. Appl. Acarol.* 4:1–13.
43. Dicke, M. 1994. Local and systemic production of volatile herbivore-induced terpenoids: their role in plant-carnivore mutualism. *J. Plant Physiol.* 143:465–472.
44. Dicke, M., and M. de Jong. 1988. Prey preference of the phytoseiid mite *Typhlodromus pyri*: 2. Electrophoretic diet analysis. *Exp. Appl. Acarol.* 4:15–25.
45. Dicke, M., and M.W. Sabelis. 1988. How plants obtain predatory mites as bodyguards. *Neth. J. Zool.* 38:148–165.
46. Dicke, M., J. Bruin, and M.W. Sabelis. 1993. Herbivore-induced plant volatiles mediate plant-carnivore, plant-herbivore and plant-plant interactions: talking plants revisited. In *Plant Signals in Interactions with Other Organisms*. J.C. Schultz and I. Raskin, eds. *Current Topics in Plant Physiology*. Vol. 11. Rockville, Maryland: American Society of Plant Physiologists, pp 87–101.
47. Dicke, M., M.W. Sabelis, and M. de Jong. 1988. Analysis of prey preference of phytoseiid mites as determined with an olfactometer, predation models and electrophoresis. *Exp. Appl. Acarol.* 5:225–241.

48. Dicke, M., P. van Baarlen, R. Wessels, and H. Dijkman. 1993. Herbivory induces systemic production of plant volatiles that attract predators of the herbivore: extraction of endogenous elicitor. *J. Chem. Ecol.* 19:581–599.

49. Dicke, M., M.W. Sabelis, J. Takabayashi, J. Bruin, and M.A. Posthumus. 1990. Plant strategies of manipulating predator-prey interactions through allelochemicals: prospects for application in pest control. *J. Chem. Ecol.* 16:3091–3118.

50. Dicke, M., T.A. van Beek, M.A. Posthumus, N. Ben Dom, H. van Bokhoven, and Ae. de Groot. 1990. Isolation and identification of volatile kairomone that affects acarine predator-prey interactions. Involvement of host plant in its production. *J. Chem. Ecol.* 16:381–396.

51. Dicke, M., M. De Jong, M.P.T. Alers, F.C.T. Stelder, R. Wunderink, and J. Post. 1989. Quality control of mass-reared arthropods: nutritional effects on performance of predatory mites. *J. Appl. Entomol.* 108:462–475.

52. Dik, A.J. 1990. *Population Dynamics of Phyllosphere Yeasts: Influence of Yeasts on Aphid Damage, Diseases and Fungicide Activity in Wheat.* Ph.D. dissertation, University of Utrecht, The Netherlands.

53. Doggett, H. 1964. A note on the incidence of American bollworm *Heliothis armigera* (Hub) (Noctuidae) in sorghum. *East Africa Agric. For. J.* 29:348–349.

54. Domínguez, C.A., R. Dirzo, and S.H. Bullock. 1989. On the function of floral nectar in *Croton suberosus* (Euphorbiaceae). *Oikos* 56:109–114.

55. Duffey, S.S., 1986. Plant glandular trichomes: their partial role in defence against insects. In *Insects and the Plant Surface.* B.E. Juniper and T.R.E. Southwood, eds. London: Edward Arnold, pp 151–172.

56. Duffey, S.S., K.A. Bloem, and B.C. Campbell. 1986. Consequences of sequestration of plant natural products in plant-insect-parasitoid interactions. In *Interactions of Plant Resistance and Parasitoids and Predators of Insects.* D.J. Boethel and R.D. Eikenbary, eds. Chichester, UK: Ellis Horwood, pp 31–60.

57. Ebel, J. 1986. Phytoalexin synthesis: the biochemical analysis of the induction process. *Annu. Rev. Phytopathol.* 24:235–264.

58. Elzen, G.W., H.J. Williams, and S.B. Vinson. 1983. Response by the parasitoid *Campoletis sonorensis* (Hymenoptera: Ichneumonidae) to chemicals (synomones) in plants: implications for host habitat location. *Environ. Entomol.* 12:1873–1877.

59. Elzen, G.W., H.J. Williams, A.A. Bell, R.D. Stipanovic, and S.B. Vinson. 1985. Quantification of volatile terpenes of glanded and glandless *Gossypium hirsutum* L. cultivars and lines by gas chromatography. *J. Agric. Food. Chem.* 33:1079–1082.

60. Emden, H.F. van. 1986. The interaction of plant resistance and natural enemies: effects on populations of sucking insects. In *Interactions of Plant Resistance and Parasitoids and Predators of Insects,* D.J. Boethel and R.D. Eikenbary, eds. Chichester, UK: Ellis Horwood, pp 138–150.

61. Epsky, N.D., and J.L. Capinera. 1994. Influence of herbivore diet on the pathogenesis of *Steinernema carpocapsae* (Nematoda: Steinernematidae). *Environ. Entomol.* 23:487–491.

62. Faeth, S.H., 1990. Structural damage to oak alters natural enemy attack on a leafminer. *Entomol. Exp. Appl.* 57:57–63.

63. Fokkema, N.J. 1971. The effect of pollen in the phyllosphere of rye on colonization by saprophytic fungi and on infection by *Helminthosporium sativum* and other leaf pathogens. *Neth. J. Plant Pathol.* 77(suppl.1):1–60.

64. Fokkema, N.J., I. Riphagen, R.J. Poot, and C. de Jong. 1983. Aphid honeydew, a potential stimulant of *Cochliobolus sativus* and *Septoria notorum* and the competitive role of saprophytic mycoflora. *Trans. British Mycol. Soc.* 81:355–363.

65. Fox, L.R., and P.A. Morrow. 1981. Specialization: species property or local phenomenon?. *Science* 211:887–893.

66. Fox, L.R., D.K. Letourneau, J. Eisenbach, and S. van Nouhuys, 1990. Parasitism rates and sex ratios of a parasitoid wasp: effects of herbivore and plant quality. *Oecologia* 83:414–419.

67. Fry, J.D. 1989. Evolutionary adaptation to host plants in a laboratory population of the phytophagous mite *Tetranychus urticae* Koch. *Oecologia* 81:559–565.

68. Futuyma, D.J., and S.C. Peterson. 1985. Genetic variation in the use of resources by insects. *Annu. Rev. Entomol.* 30:217–238.

69. Futuyma, D.J., and T.E. Philippi. 1987. Genetic variation and covariation in responses to host plants by *Alsophila pometaria* (Lepidoptera: Geometridae). *Evolution* 41:269–279.

70. Geervliet, J.B.F., L.E.M. Vet, and M. Dicke. 1994. Volatiles from damaged plants as major cues in long-range host-searching by the specialist parasitoid *Cotesia rubecula*. *Entomol. Exp. Appl.* 73:289–297.

71. Glas, P.C.G., and L.E.M. Vet. 1983. Host-habitat location and host location by *Diachasma alloeum* Muesebeck (Hym.; Braconidae), a parasitoid of *Rhagoletis pomonella* Walsh (Dipt.; Tephritidae). *Neth. J. Zool.* 33:41–54.

72. Godfray, H.C.J. 1994. *Parasitoids, Behavioral and Evolutionary Ecology*. Princeton, New Jersey: Princeton University Press.

73. Gotoh, T., J. Bruin, M.W. Sabelis, and S.B.J. Menken. 1993. Host race formation in *Tetranychus urticae*: genetic differentiation, host plant preference, and mate choice in a tomato and a cucumber strain. *Entomol. Exp. Appl.* 68:171–178.

74. Gould, F. 1986. Simulation models for predicting durability of insect-resistant germ plasm: A deterministic diploid, two-locus model. *Environ. Entomol.* 15:1–10.

75. Gould, F. 1991. The evolutionary potential of crop pests. *Am. Sci.* 79:496–507.

76. Gould, F., G.G. Kennedy, and M.T. Johnson. 1991. Effects of natural enemies on the rate of herbivore adaptation to resistant host plants. *Entomol. Exp. Appl.* 58:1–14.

77. Gould, J.L., and W.F. Towne, 1988. Honey bee learning. *Adv. Insect Physiol.* 20:55–86.

78. Greathead, D.J., and J.K. Waage. 1983. Opportunities for biological control of agricultural pests in developing countries. *World Bank Tech. Paper* 11:1–44.

79. Greenblatt, J.A., P. Barbosa, and M.E. Montgomery. 1982. Host's diet effects on nitrogen utilization efficiency for two parasitoid species: *Brachymeria intermedia* and *Coccygomimus turionellae*. *Physiol. Entomol.* 7:263–267.

80. Grevstad, F.S., and B.W. Klepetka. 1992. The influence of plant architecture on the foraging efficiencies of a suite of ladybird beetles feeding on aphids. *Oecologia* 92:399–404.

81. Grostal, P., and D.J. O'Dowd. 1994. Plants, mites and mutualism: leaf domatia and the abundance and reproduction of mites on *Viburnum tinus* (Caprifoliaceae). *Oecologia* 97:308–315.

82. Hagen, K.S. 1986. Ecosystem analysis: plant cultivars (HPR), entomophagous species and food supplements. In *Interactions of Plant Resistance and Parasitoids and Predators of Insects*. D.J. Boethel and R.D. Eikenbary, eds. Chichester, UK: Ellis Horwood, pp 151–197.

83. Haren, R.J.F. van, M.M. Steenhuis, M.W. Sabelis, and O.M.B. de Ponti. 1987. Tomato stem trichomes and dispersal success of *Phytoseiulus persimilis* relative to its prey *Tetranychus urticae*. *Exp. Appl. Acarol.* 3:115–121.

84. Haukioja, E. 1990. Induction of defenses in trees. *Annu. Rev. Entomol.* 36:25–42.

85. Heimpel, G., J.A. Rosenheim, and J.M. Adams. 1994. Behavioral ecology of host feeding in *Aphytis* parasitoids. *Norw. J. Agric. Sci.* 16:(suppl.):101–115.

86. Heinrich, B. 1979. Foraging strategies of caterpillars: leaf damage and possible predator avoidance strategies. *Oecologia* 42:325–337.

87. Heinrich, B. 1993. How avian predators constrain caterpillar foraging. In *Caterpillars, Ecological and Evolutionary Constraints on Foraging*. N.E. Stamp and T.M. Casey, eds., New York: Chapman & Hall, pp 224–247.

88. Heinrich, B., and S.L. Collins. 1983. Caterpillar leaf damage, and the game of hide-and-seek with birds. *Ecology* 64:592–602.

89. Heithaus, E.R., D.C. Culver, and A.J. Beattie. 1980. Models of some ant-plant mutualisms. *Am. Nat.* 116:347–361.

90. Hemming, B.C. 1990. Bacteria as antagonists in biological control of plant pathogens. In *New Directions in Biological Control: Alternatives for Suppressing Agricultural Pests and Diseases*. R.R. Baker and P.E. Dunn, eds. New York: Liss, pp 223–242.

91. Hölldobler, B., and E.O. Wilson. 1990. *The Ants*. Berlin: Springer-Verlag.

92. Hoy, M.A. 1985. Recent advances in genetics and genetic improvement of the Phytoseiidae. *Annu. Rev. Entomol.* 30:345–370.

93. Hoy, M.A. 1994. Transgenic arthropod natural enemies for pest management programs. *Norw. J. Agric. Sci.* 16(suppl.):9–39.

94. Hulspas-Jordaan, P.M., and J.C. van Lenteren. 1978. The relationship between host-plant leaf structure and parasitization efficiency of the parasitic wasp *Encarsia formosa* Gahan (Hymenoptera: Aphelinidae). *Med. Fac. Landbouww. Rijksuniv. Gent* 43:431–440.

95. Hunter, M.D., and J.C. Schultz. 1993. Induced plant defenses breached? Phytochemical induction protects an herbivore from disease. *Oecologia* 94:195–203.

96. Hunter, M.D., T. Ohgushi, and P.W. Price. 1992. *Effects of Resource Distribution on Animal-Plant Interactions*. San Diego: Academic Press.

97. Huxley, C.R., and D.F. Cutler. 1991. *Ant-Plant Interactions*. Oxford, UK: Oxford University Press.

98. Isenhour, D.J., B.R. Wiseman, and R.C. Layton. 1989. Enhanced predation by *Orius insidiosus* (Hemiptera: Anthocoridae) on larvae of *Heliothis zea* and *Spodoptera frugiperda* (Lepidoptera: Noctuidae) caused by prey feeding on resistant corn genotypes. *Environ. Entomol.* 18:418–422.

99. Jacobs, M. 1966. On domatia—the viewpoints and some facts. *Proc. Acad. Sci. Amsterdam* 69:275–316.
100. Janssen, A., C.D. Hofker, A.R. Braun, N. Mesa, M.W. Sabelis, and A.C. Bellotti. 1990. Preselecting predatory mites for biological control: the use of an olfactometer. *Bull. Entomol. Res.* 80:177–181.
101. Janzen, D.H. 1967. Interaction of the bull's-horn acacia (*Acacia cornigera* L.) with an ant inhabitant (*Pseudomyrmex ferruginea* F. Smith) in eastern Mexico. *Univ. Kansas Sci. Bull.* 47:315–558.
102. Janzen, D.H. 1973. Evolution of polygynous obligate acacia-ants in Western Mexico. *J. Anim. Ecol.* 42:727–750.
103. Jeffree, C.E. 1986. The cuticle, epicuticular waxes and trichomes of plants, with reference to their structure, functions and evolution. In *Insects and the Plant Surface*. B. Juniper and T.R.E. Southwood, eds. London: Edward Arnold, pp 23–64.
104. Johnson, M.T., and F. Gould. 1992. Interaction of genetically engineered host plant resistance and natural enemies of *Heliothis virescens* (Lepidoptera: Noctuidae) in tobacco. *Environ. Entomol.* 21:586–597.
105. Jolivet, P. 1991. Ants, plants, and beetles: a triangular relationship. In *Ant-Plant Interactions*. C.R. Huxley and D.F. Cutler, eds. Oxford, UK: Oxford University Press, pp 397–406.
106. Juniper, B.E. and Southwood, T.R.E. 1986. *Insects and the Plant Surface*. London: Edward Arnold.
107. Kanda, K. 1987. Effect of fertilization of pasture on larval survival and development of the armyworm, *Pseudaletia separata* Walker, and on its predation by the wolf spider, *Pardosa laura* Karsh. *Jpn. J. Appl. Entomol. Zool.* 31:220–225.
108. Karban, R. 1986. Induced resistance against spider mites in cotton: field verification. *Entomol. Exp. Appl.* 42:239–242.
109. Karban, R. 1991. Inducible resistance in agricultural systems. In *Phytochemical Induction by Herbivores*. D.W. Tallamy and M.J. Raupp, eds. New York: Wiley, pp 403–419.
110. Karban, R., and J.H. Myers. 1989. Induced plant responses to herbivory. *Annu. Rev. Ecol. Syst.* 20:331–348.
111. Karban, R., R. Adamchak, and W.C. Schnathorst. 1987. Induced resistance and interspecific competition between spider mites and a vascular wilt fungus. *Science* 235:678–680.
112. Kareiva, P., and R. Sahakian. 1990. Tritrophic effects of a simple architectural mutation in pea plants. *Nature* 345:433–434.
113. Katanyukul, W., and R. Thurston. 1973. Seasonal parasitism and predation of eggs of the tobacco hornworm on various host plants in Kentucky. *Environ. Entomol.* 2:939–945.
114. Keller, M.A. 1987. Influence of leaf surfaces on movements by the hymenopterous parasitoid *Trichogramma exiguum*. *Entomol. Exp. Appl.* 43:55–59.
115. Kester, K.M., and P. Barbosa. 1991. Behavioral and ecological constraints imposed by plants on insect parasitoids: implications for biological control. *Biol. Control* 1:94–106.

116. Kim, K.C., and B.A. McPheron. 1993. *Evolution of Insect Pests. Patterns of Variation*. New York: Wiley.
117. Knox, R.B., R. Marginson, J. Kenrick, and A.J. Beattie. 1986. The role of extrafloral nectaries in Acacia. In *Insects and the Plant Surface*. B. Juniper and T.R.E. Southwood, eds. London: Edward Arnold, pp 295–307.
118. Koptur, S. 1992. Extrafloral nectary-mediated interactions between insects and plants. In *Insect-Plant Interactions*. Vol. IV. E.A. Bernays, ed. Boca Raton, Florida: CRC Press, pp 81–129.
119. Krischik, V.A. 1991. Specific or generalized plant defense: reciprocal interactions between herbivores and pathogens. In *Microbial Mediation of Plant-Herbivore Interactions*. P. Barbosa, V.A. Krischik, and C.G. Jones, eds. New York: Wiley, pp 309–340.
120. Krischik, V.A., P. Barbosa, and C.F. Reichelderfer. 1988. Three trophic level interactions: allelochemicals, *Manduca sexta* (L.), and *Bacillus thuringiensis* var. *kurstaki* Berliner. *Environ. Entomol.* 17:476–482.
121. Lawrence, P.O. 1981. Host vibration—a cue to host location by the parasite, *Biosteres longicaudatus*. *Oecologia* 48:249–251.
122. Leather, S.R., and P.J. Walsh. 1993. Sub-lethal plant defences: the paradox remains. *Oecologia* 93:153–155.
123. Leather, S.R., A.D. Watt, and G.I. Forrest. 1987. Insect-induced chemical changes in young lodgepole pine (*Pinus contorta*): the effect of previous defoliation on oviposition, growth and survival of the pine beauty moth, *Panolis flammea*. *Ecol. Entomol.* 12:275–281.
124. Lenteren, J.C. van. 1986. Parasitoids in the greenhouse: successes with seasonal inoculative release systems. In *Insect Parasitoids*. J.K. Waage and D.J. Greathead, eds. London: Academic Press, pp 342–374.
125. Lenteren, J.C. van. 1989. World situation of integrated pest management in greenhouses. *ICI Agrochemicals. Proceedings of Symposium: Insect Control Strategies and the Environment*. Fernhurst, United Kingdom: ICI, pp 32–50.
126. Lenteren, J.C. van. 1990. Integrated pest and disease management in protected crops: the inescapable future. *IOBC/WPRS Bull.* 13:91–99.
127. Lenteren, J.C. van. 1991. Biological control in a tritrophic system approach. *Proceedings of Aphid-Plant Interactions: Populations to Molecules*. D.C. Peters, J.A. Webster, and C.S. Chlouber, eds. Stillwater, Oklahoma: pp 3–28.
128. Lenteren, J.C. van, and O.M.B. de Ponti. 1990. Plant-leaf morphology, host-plant resistance and biological control. *Symp. Biol. Hung.* 39:365–386.
129. Lenteren, J.C. van, and J. Woets. 1988. Biological and integrated pest control in greenhouses. *Annu. Rev. Entomol.* 33:239–269.
130. Lenteren, J.C. van, A.K. Minks, and O.M.B. Ponti. 1992. *Biological Control and Integrated Crop Protection: Towards Environmentally Safer Agriculture*. Wageningen: The Netherlands: Pudoc.
131. Letourneau, D.K. 1990. Code of ant-plant mutalism broken by parasite. *Science* 248:215–217.
132. Letourneau, D.K. 1991. Parasitism of ant-plant mutalisms and the novel case of

Piper. In *Ant-Plant Interactions.* C.R. Huxley and D.F. Cutler, eds. Oxford, UK: Oxford University Press, pp 390–396.

133. Lewis, W.J., L.E.M. Vet, J.H. Tumlinson, J.C. van Lenteren, and D.R. Papaj. 1990. Variations in parasitoid foraging behavior: essential element of a sound biological control theory. *Environ. Entomol.* 19:1183–1193.

134. Loader, C., and H. Damman. 1991. Nitrogen content of food plants and vulnerability of *Pieris rapae* to natural enemies. *Ecology* 72:1586–1590.

135. Luck, R.F., and H. Podoler. 1985. Competitive exclusion of *Aphytis lingnanensis* by *A. melinus*: potential role of host size. *Ecology* 66:904–913.

136. Lynch, J.M. 1990. Fungi as antagonists. In *New Directions in Biological Control: Alternatives for Suppressing Agricultural Pests and Diseases.* R.R. Baker and P.E. Dunn, eds. New York: Liss, pp 243–253.

137. Madden, D., and T.P. Young. 1992. Symbiotic ants as an alternative defense against giraffe herbivory in spinescent *Acacia drepanolobium. Oecologia* 91:235–238.

138. Malcolm, S.B. 1989. Disruption of web structure and predatory behavior of a spider by plant-derived chemical defense of an aposematic aphid. *J. Chem. Ecol.* 15:1699–1716.

139. Mattiacci, L., M. Dicke, and M.A. Posthumus. 1994. Induction of parasitoid attracting synomone in Brussels sprouts plants by feeding of *Pieris brassicae* larvae: role of mechanical damage and herbivore elicitor. *J. Chem. Ecol.* 20:2229–2247.

140. Mattiacci, L., M. Dicke, and M.A. Posthumus. 1995. ß-glucosidase: elicitor of herbivore-induced plant odors that attract host-searching parasitic wasps. *Proc. Natl. Acad. Sci. USA* 92:2036–2040.

141. McMurtry, J.A., C.B. Huffaker, and M. van de Vrie. 1970. Ecology of tetranychid mites and their natural enemies: a review. I. Tetranychid enemies: their biological characters and the impact of spray practices. *Hilgardia* 40:331–390.

142. Mendel, Z., D. Blumberg, A. Zehavi, and M. Weissenberg. 1992. Some polyphagous homoptera gain protection from their natural enemies by feeding on the toxic plants *Spartium junceum* and *Erythrina corallodendrum* (Leguminosae). *Chemoecology* 3:118–124.

143. Messenger, P.S., and R. van den Bosch. 1971. The adaptability of introduced biological control agents. In *Biological Control.* C.B. Huffaker, ed. New York: Plenum Press, pp 68–92.

144. Meyhöfer, R., J. Casas, and S. Dorn. 1994. Host location by a parasitoid using leafminer vibrations: characterising the vibrational signals produced by the leafmining host. *Physiol. Entomol.* 19:349–359.

145. Michelsen, A., F. Fink, M. Gogala, and D. Traue. 1982. Plants as transmission channels for insect vibrational songs. *Behav. Ecol. Sociobiol.* 11:269–281.

146. Milewski, A.V., T.P. Young, and D. Madden. 1991. Thorns as induced defenses: experimental evidence. *Oecologia* 86:70–75.

147. Miller, J.R., and M.O. Harris. 1985. Viewing behavior modifying chemicals in the contact of behavior: lessons from the onion fly. In *Semiochemistry Flavors*

and Pheromones. M. Soderlund and E. Acree, eds. Berlin: Walter de Gruyter, pp 3–31.

148. Mollema, C. 1988. *Genetical Aspects of Resistance in a Host-Parasitoid Interaction.* Ph.D dissertation, University of Leiden, The Netherlands, pp 99–107.

149. Montllor, C.B., and E.A. Bernays. 1993. Invertebrate predators and caterpillar foraging. In *Caterpillars, Ecological and Evolutionary Constraints on Foraging.* N.E. Stamp and T.M. Casey, eds. New York: Chapman and Hall, pp 170–202.

150. Nakamuta, K. 1984. Visual orientation of a ladybeetle, *Coccinella septempunctata* L., (Coleoptera: Coccinellidae) towards its prey. *Appl. Entomol. Zool.* 19:82–86.

151. Naranjo, S.E., and R.L. Gibson. 1996. Phytophagy in predaceous Heteroptera: effects on life-history and population dynamics. In *Zoophytophagous Heteroptera: Implications for Life History and Integrated Pest Management.* R. Wiedenmann and O. Alomar, eds. Thomas Say Symposium Proceedings, Entomological Society of America, pp. 57–93.

152. Navon, A., J.D. Hare, and B.A. Federici. 1993. Interactions among *Heliothis virescens* larvae, cotton condensed tannin and the CryIA(c) delta-endotoxin of *Bacillus thuringiensis. J. Chem. Ecol.* 19:2485–2499.

153. Noldus, L.P.J.J., and J.C. van Lenteren. 1990. Host aggregation and parasitoid behaviour: biological control in a closed system. In *Critical Issues in Biological Control.* M. Mackauer, L.E. Ehler, and J. Roland, eds. Andover: Intercept, pp 229–262.

154. Nooij, M.P. de, A. Biere, and E.G.A. Linders. 1992. Interaction of pests and pathogens through host predisposition. In *Pests and Pathogens, Plant Responses to Foliar Attack.* P. G. Ayres, ed. Oxford: Bios Scientific, pp 143–160.

155. Nordlund, D.A., W.J. Lewis, and M.A. Altieri. 1988. Influences of plant-produced allelochemicals on the host/prey selection behavior of entomophagous insects. In *Novel Aspects of Insect-Plant Interactions.* P. Barbosa and D.K. Letourneau, eds. New York: Wiley, pp 65–90.

156. Obrycki, J.J. 1986. The influence of foliar pubescence on entomophagous species. In *Interactions of Plant Resistance and Parasitoids and Predators of Insects.* D.J. Boethel and R.D. Eikenbary, eds. Chichester, UK: Ellis Horwood, pp 61–83.

157. O'Dowd, D.J., and M.F. Willson. 1989. Leaf domatia and mites on Australian plants: ecological and evolutionary implications. *Biol. J. Linn. Soc.* 37:191–236.

158. O'Dowd, D.J., and M.F. Willson. 1991. Associations between mites and leaf domatia. *TREE* 6:170–182.

159. Ohsaki, N., and Y. Sato. 1994. Food plant choice of *Pieris* butterflies as a trade-off between parasitoid avoidance and quality of plants. *Ecology* 74:59–68.

160. Overmeer, W.P.J. 1985. Alternative prey and other food resources. In *Spider Mites, Their Biology, Natural Enemies and Control. World Crop Pests. Vol. 1B.* W. Helle and M.W. Sabelis, eds. Amsterdam: Elsevier, pp 131–139.

161. Overmeer, W.P.J., and A.Q. van Zon. 1984. The preference of *Amblyseius potentillae* (Garman) (Acarina: Phytoseiidae) for certain plant substrates. *Proceedings of the VIth Internation. Congress of Acarology.* Edinburgh: pp 591–596.

162. Pak, G.A. 1988. *Selection of Trichogramma for Inundative Biological Control*. PhD dissertation. Agricultural University, Wageningen, The Netherlands.

163. Pashley, D.P. 1988. Quantitative genetics, development and physiological adaptation in sympatric host strains of fall armyworm. *Evolution* 42:93–102.

164. Pasteels, J.M., M. Rowell-Rahier, and M.J. Raupp. 1988. Plant-derived defense in Chrysomelid beetles. In *Novel Aspects of Insect-Plant Interactions*. P. Barbosa and D.K. Letourneau, eds. New York: Wiley, pp 235–272.

165. Pemberton, R.W., and C.E. Turner. 1989. Occurrence of predatory and fungivorous mites in leaf domatia. *Am. J. Bot.* 76:105–112.

166. Peterson, S.C., N.D. Johnson, and J.L. LeGuyader. 1987. Defensive regurgitation of allelochemicals derived from host cyanogenesis by eastern tent caterpillars. *Ecology* 68:1268–1272.

167. Pimentel, D. 1961. An evaluation of insect resistance in broccoli, brussel sprouts, cabbage, collards, and kale. *J. Econ. Entomol.* 54:156–158.

168. Popov, N.A., and O.A. Khudyakova. 1989. Development of *Phytoseiulus persimilis* (Acarina, Phytoseiidae) fed on *Tetranychus urticae* (Acarina, Tetranychidae) on various food plants. *Acta Entomol. Fenn.* 53:43–46.

169. Porres, M.A., J.A. McMurtry, and R.B. March. 1976. Investigations of leaf sap feeding by three species of phytoseiid mites by labelling with radioactive phosphoric acid (H_3PO_4). *Ann. Entomol. Soc. Am.* 68:871–873.

170. Potting, R.P.J., L.E.M. Vet, and M. Dicke. 1995. Host microhabitat location by the stemborer parasitoid *Cotesia flavipes*: the role of herbivore volatiles and locally and systemically induced plant volatiles. *J. Chem. Ecol.* 21:525–539.

171. Price, P.W. 1986. Ecological aspects of host plant resistance and biological control: interactions among three trophic levels. In *Interactions of Plant Resistance and Parasitoids and Predators of Insects*. D.J. Boethel and R.D. Eikenbary, eds. Chichester, UK: Ellis Horwood, pp 11–30.

172. Price, P.W. 1991. Evolutionary theory of host and parasitoid interactions. *Biol. Contr.* 1:83–93.

173. Price, P.W. 1992. Plant resources as the mechanistic basis for insect herbivore population dynamics. In *Effects of Resource Distribution on Animal-Plant Interactions*. M.D. Hunter, T. Ohgushi, and P.W. Price, eds. New York: Academic Press, pp 139–173.

174. Price, P.W., and K.M. Clancy. 1986. Interactions among three trophic levels: gall size and parasitoid attack. *Ecology* 67:1593–1600.

175. Price, P.W., C.E. Bouton, P. Gross, B.A. McPheron, J.N. Thompson, and A.E. Weis. 1980. Interactions among three trophic levels: influence of plant on interactions between insect herbivores and natural enemies. *Annu. Rev. Ecol. Syst.* 11:41–65.

176. Prokopy, R.J. 1986. Visual and olfactory stimulus interaction in resource finding by insects. In *Mechanisms in Insect Olfaction*. T.L. Payne, M.C. Birch, and C.E.J. Kennedy, eds. Oxford, UK: Oxford University Press, pp 81–89.

177. Pschorn-Walcher, H. 1977. Biological control of forest insects. *Annu. Rev. Entomol.* 22:1–22.

178. Pullin, A.S. and J.E. Gilbert. 1989. The stinging nettle, *Urtica dioica*, increases trichome density after herbivore and mechanical damage. *Oikos* 54:275–280.

179. Putman, W.L. 1955. Bionomics of *Stethorus punctillum* Weise (Coleoptera: Coccinellidae) in Ontario. *Can. Entomol.* 87:9–33.

180. Rabb, R.L., and J.R. Bradley. 1968. The influence of host plants on parasitism of eggs of the tobacco hornworm. *J. Econ. Entomol.* 61:1249–1252.

181. Rausher, M.D. 1984. Tradeoffs in performance on different hosts: evidence from within- and between-site variation in the beetle *Deloyala guttata*. *Evolution* 38:582–595.

182. Ridgway, R.L. and S.L. Jones. 1968. Plant feeding by *Geocoris pallens* and *Nabus americoferus Ann. Entomol. Soc. Am.* 61:232–233.

183. Rijn, P.C.J. van, and M.W. Sabelis. 1990. Pollen availability and its effect on the maintenance of populations of *Amblyseius cucumeris*, a predator of thrips. *SROP/WPRS Bull.* 13:179–184.

184. Risch, S.J. and F.R. Rickson. 1981. Mutualism in which ants must be present before plants produce food bodies. *Nature* 291:149–150.

185. Risch, S.J., D.A. Andow, and M.A. Altieri. 1983. Agroecosystem diversity and pest control: data, tentative conclusions and new research directions. *Environ. Entomol.* 12:625–629.

186. Root, R.B. 1973. Organization of a plant-arthropod association in simple and diverse habitats: the fauna of collards (*Brassica oleracea*). *Ecol. Monogr.* 43:95–124.

187. Rosenheim, J.A. 1987. Host location and exploitation by the cleptoparasitic wasp *Argochrysis armilla*: the role of learning (Hymenoptera: Chrysididae). *Behav. Ecol. Sociobiol.* 21:401–406.

188. Rowell-Rahier, M., and J.M. Pasteels. 1992. Third trophic level influences of plant allelochemics. In *Herbivores: Their Interaction with Secondary Plant Metabolites*. 2nd ed. Vol. 2. G.A. Rosenthal and M.R. Berenbaum, eds. New York: Academic Press, pp 243–277.

189. Ruberson, J.R., M.J. Tauber, and C.A. Tauber. 1986. Plant feeding by *Podisus maculiventris* (Heteroptera: Pentatomidae): effect on survival, development, and preoviposition period. *Environ. Entomol.* 15:894–897.

190. Ryan, C.A., and T.R. Green. 1974. Proteinase inhibitors in natural plant protection. *Rec. Adv. Phytochem.* 8:123–140.

191. Sabelis, M.W. and H.E. van de Baan. 1983. Location of distant spider mite colonies by phytoseiid predators: demonstration of specific kairomones emitted by *Tetranychus urticae* and *Panonychus ulmi*. *Entomol. Exp. Appl.* 33:303–314.

192. Sabelis, M.W., and M. Dicke. 1985. Long-range dispersal and searching behaviour. In *Spider Mites, Their Biology, Natural Enemies and Control. World Crop Pests.* Vol. 1B. W. Helle and M.W. Sabelis, eds. Amsterdam: Elsevier, pp 141–160.

193. Sabelis, M.W., and J. van der Meer. 1986. Local dynamics of the interaction between predatory mites and two-spotted spider mites. In *Dynamics of Physiologically Structured Populations. Springer Lecture Notes in Biomathematics.* Vol. 68. J.A.J. Metz and O. Diekman, eds. Berlin: Springer-Verlag, pp 322–343.

194. Sabelis, M.W., B.P. Afman, and P.J. Slim. 1984. Location of distant spider mite colonies by *Phytoseiulus persimilis*: localization and extraction of a kairomone. *Acarology VI*, 1:431–440.

195. Salt, G., and R. van den Bosch. 1967. The defense reactions of three species of *Hypera* (Coleoptera, Curculionidae) to an ichneumon wasp. *J. Invert. Pathol.* 9:164–177.

196. Sato, Y. 1979. Experimental studies on parasitization by *Apanteles glomeratus.* IV. Factors leading a female to the host. *Physiol. Entomol.* 4:63–70.

197. Sato, Y., and N. Ohsaki. 1987. Host-habitat location by *Apanteles glomeratus* and effect of food-plant exposure on host-parasitism. *Ecol. Entomol.* 12:291–297.

198. Schultz, J.C. 1992. Factoring natural enemies into plant tissue availability to herbivores. In *Effects of Resource Distribution on Animal-Plant Interactions.* M.D. Hunter, T. Ohgushi, and P.W. Price, eds. New York: Academic Press, pp 175–197.

199. Schultz, J.C., and I.T. Baldwin. 1982. Oak leaf quality declines in response to defoliation by gypsy moth larvae. *Science* 217:149–151.

200. Schultz, J.C., and S.T. Keating. 1991. Host-plant-mediated interactions between the gypsy moth and a baculovirus. In *Microbial Mediation of Plant-Herbivore Interactions.* P. Barbosa, V.A. Krischik, and C.G. Jones, eds. New York: Wiley, pp 489–506.

201. Schuster, M.F. and M. Calderon. 1986. Interactions of host plant resistant genotypes and beneficial insects in cotton ecosystems. In *Interactions of Plant Resistance and Parasitoids and Predators of Insects.* D.J. Boethel and R.D. Eikenbary, eds. Chichester, UK: Ellis Horwood, pp 84–97.

202. Senrayan, R., and R.S. Annadurai. 1991. Influence of host's food plant and habitat on *Anastatus ramakrishnae* (Mani) (Hym., Eupelmidae) an egg parasitoid of *Coridius obscurus* (Fab.) (Het., Pentatomidae). *J. Appl. Entomol.* 112:237–243.

203. Shahjahan, M. 1974. Erigeron flowers as a food and attractive odor source for *Peristenus pseudopallipes*, a braconid parasitoid of the tarnished plant bug. *Environ. Entomol.* 3:69–72.

204. Sheehan, W. 1986. Response by specialist and generalist natural enemies to agroecosystem diversification: a selective review. *Environ. Entomol.* 15:456–461.

205. Southwood, T.R.E. 1986. Plant surfaces and insects - an overview. In *Insects and the Plant Surface.* B. Juniper and T.R.E. Southwood, eds. London: Edward Arnold, pp 1–22.

206. Stacey, D.L. 1977. 'Banker' plant production of *Encarsia formosa* Gahan and its use in the control of glasshouse whitefly on tomatoes. *Plant Pathol.* 26:63–66.

207. Steinberg, S., M. Dicke, and L.E.M. Vet. 1993. Relative importance of infochemicals from first and second trophic level in long-range host location by the larval parasitoid *Cotesia glomerata. J. Chem. Ecol.* 19:47–59.

208. Stengard-Hansen, L. 1983. Introduction of *Aphidoletes aphidimyza* (Rond.) (Diptera: Cecidomyiidae) from an open rearing unit for the control of aphids in glasshouses. *SROP/WPRS Bull.* 6:146–150.

209. Streams, F.A., M. Shahjahan, and H.G. LeMasurier. 1968. Influence of plants on the parasitization of the tarnished plant by *Leiophron pallipes. J. Econ. Entomol.* 61:996–999.

210. Sugimoto, T., Y. Shimono, Y. Hata, A. Nakai, and M. Yahara. 1988. Foraging for patchily-distributed leaf-miners by the parasitoid, *Dapsilarthra rufiventris*

(Hymenoptera: Braconidae). III. Visual and acoustic cues to a close range patch-location. *Jpn. J. Appl. Entomol. Zool.* 23:113-121.

211. Takabayashi, J., M. Dicke, and M.A. Posthumus. 1991. Variation in composition of predator-attracting allelochemicals emitted by herbivore-infested plants: relative influence of plant and herbivore. *Chemoecology* 2:1-6.

212. Takabayashi, J., M. Dicke, and M.A. Posthumus. 1991. Induction of indirect defence against spider-mites in uninfested lima bean leaves. *Phytochemistry* 30:1459-1462.

213. Takabayashi, J., M. Dicke, and M.A. Posthumus. 1994a. Volatile herbivore-induced terpenoids in plant-mite interactions: variation caused by biotic and abiotic factors. *J. Chem. Ecol.* 20:1329-1354.

214. Takabayashi, J., S. Takahashi, M. Dicke, and M.A. Posthumus. 1995. Effect of the developmental stage of the herbivore *Pseudaletia separata* on the production of herbivore-induced synomone by corn plants. *J. Chem. Ecol.* 21:273-287.

215. Takabayashi, J., M. Dicke, S. Takahashi, M.A. Posthumus, and T.A. van Beek. 1994b. Leaf age affects composition of herbivore-induced synomones and attraction of predatory mites. *J. Chem. Ecol.* 20:373-386.

216. Tallamy, D.W., and M.J. Raupp. 1991. *Phytochemical Induction by Herbivores*. New York: Wiley.

217. Tinbergen, N., and W. Kruyt. 1938. Über die Orientierung des Bienenwolfes (*Philanthus triangulum* Fabr.). III. Die Bevorzugung bestimmter Wegmarken. *Zeitschr. Vergl. Physiol.* 25:292-334.

218. Treacy, M.F., J.H. Benedict, M.H. Walmsley, J.D. Lopez, and R.K. Morrison. 1987. Parasitism of bollworm (Lepidoptera: Noctuidae) eggs on nectaried and nectariless cotton. *Environ. Entomol.* 16:420-423.

219. Tukey, H.B. 1971. Leaching substrates from plants. In *Ecology of Leaf Surface Microorganisms*. T.F. Preece and C.H. Dickinson. London: Academic Press, pp 143-150.

220. Tumlinson, J.H., T.C.J. Turlings, and W.J. Lewis, 1992. The semiochemical complexes that mediate insect parasitoid foraging. *Agric. Zool. Rev.* 5:221-252.

221. Turlings, T.C.J., and J.H. Tumlinson. 1992. Systemic release of chemical signals by herbivore-injured corn. *Proc. Natl. Acad. Sci. USA* 89:8399-8402.

222. Turlings, T.C.J., J.H. Tumlinson, and W.J. Lewis, 1990. Exploitation of herbivore-induced plant odors by host-seeking parasitic wasps. *Science* 250:1251-1253.

223. Turlings, T.C.J., P. McCall, H.T. Alborn, and J.H. Tumlinson. 1993. An elicitor in caterpillar oral secretions that induces corn seedlings to emit chemical signals attractive to parasitic wasps. *J. Chem. Ecol.* 19:411-425.

224. Turlings, T.C.J., J.H. Tumlinson, F.J. Eller, and W.J. Lewis. 1991. Larval-damaged plants: source of volatile synomones that guide the parasitoid *Cotesia marginiventris* to the micro-habitat of its hosts. *Entomol. Exp. Appl.* 58:75-82.

225. Turlings, T.C.J., J.H. Tumlinson, R.R. Heath, A.T. Proveaux, and R.E. Doolittle. 1991. Isolation and identification of allelochemicals that attract the larval parasitoid, *Cotesia marginiventris* (Cresson), to the microhabitat of one of its hosts. *J. Chem. Ecol.* 17:2235-2251.

226. Turlings, T.C.J., F.L. Wäckers, L.E.M. Vet, W.J. Lewis, and J.H. Tumlinson.

1993. Learning of host-finding cues by hymenopterous parasitoids. In *Insect Learning*. D.R. Papaj and A.C. Lewis, eds. New York: Chapman & Hall, pp 51–78.

227. Vandermeer, J.H., and D.H. Boucher. 1978. Varieties of mutualisitic interaction in population models. *J. Theoret. Biol.* 74:549–558.

228. Vet, L.E.M., and J.J.M. van Alphen. 1985. A comparative functional approach to the host detection behaviour of parasitic wasps. I. A qualitative study on Eucoilidae and Alysiinae. *Oikos* 44:478–486.

229. Vet, L.E.M., and M. Dicke. 1992. Ecology of infochemical use by natural enemies in a tritrophic context. *Annu. Rev. Entomol.* 37:141–172.

230. Via, S. 1984. The quantitative genetics of polyphagy in an insect herbivore. I. Genotype-environment interaction in larval performance on different host plant species. *Evolution* 38:881–895.

231. Via, S. 1984. The quantitative genetics of polyphagy in an insect herbivore. II. Genetic correlations in larval performance within and among host plants. *Evolution* 38:896–905.

232. Via, S. 1990. Ecological genetics and host adaptation in herbivorous insects: the experimental study of evolution in natural and agricultural systems. *Annu. Rev. Entomol.* 35:421–446.

233. Vinson, S.B. 1976. Host selection by insect parasitoids. *Annu. Rev. Entomol.* 21:109–134.

234. Vinson, S.B. 1981. Habitat location. In *Semiochemicals: Their Role in Pest Control*. D.A. Nordlund, R.L. Jones, and W.J. Lewis, eds. New York: Wiley, pp 51–77.

235. Voegelé, J., J.K. Waage, and J.C. van Lenteren. 1986. *Trichogramma and Other Egg Parasites*. Paris: INRA.

236. Wäckers, F.L. 1994. *Multisensory Foraging by Hymenopterous Parasitoids*. Ph.D. dissertation, Wageningen Agricultural University, The Netherlands.

237. Walter, D.E. 1992. Leaf surface structure and the distribution of *Phytoseius* mites (Acarina: Phytoseiidae) in south-eastern Australian forests. *Aust. J. Zool.* 40:593–603.

238. Walter, D.E., and D.J. O'Dowd. 1992. Leaf morphology and predators: effect of leaf domatia on the abundance of predatory mites (Acari:Phytoseiidae). *Environ. Entomol.* 21:478–484.

239. Wardle, A.R. 1990. Learning of host microhabitat colour by *Exeristes roborator* (F.) (Hymenoptera: Ichneumonidae). *Anim. Behav.* 39:914–923.

240. Wardle, A.R., and J.H. Borden. 1990. Learning of host microhabitat form by *Exeristes roborator* (F.) (Hymenoptera: Ichneumonidae). *J. Insect Behav.* 3:251–263.

241. Weis, A.E. 1983. Patterns of parasitism by *Torymus capite* on hosts distributed in small patches. *J. Anim. Ecol.* 52:867–877.

242. Weseloh, R.M. 1981. Host location by parasitoids. In *Semiochemicals: Their Role in Pest Control*. D.A. Nordlund, R.L. Jones, and W.J. Lewis, eds. New York: Wiley, pp 79–95.

243. Williams, H.J., G.W. Elzen and S.B. Vinson. 1988. Parasitoid-host-plant inter-

actions, emphasizing cotton (*Gossypium*). In *Novel Aspects of Insect-Plant Interactions*. P. Barbosa and D.K. Letourneau, eds. New York: Wiley, pp 171–199.

244. Wolfson, J.L. 1991. The effects of induced plant proteinase inhibitors on herbivorous insects. In *Phytochemical Induction by Herbivores*. D.W. Tallamy and M.J. Raupp, eds. New York: Wiley, pp 223–243.

245. Yano, S. 1994. Flower nectar of an autogamous perennial *Rorippa indica* as an indirect defense mechanism against herbivorous insects. *Res. Popul. Ecol.* 36:63–71.

246. Zohdy, N.Z.M. 1976. On the effect of the food of *Myzus persicae* Sulz. on the hymenopterous parasite *Aphelinus asychis* Walker. *Oecologia* 26:185–191.

6

Mechanisms of Plant Resistance Against Arthropod Pests

Billy R. Wiseman*
United States Department of Agriculture, Tifton, Georgia†

I. INTRODUCTION

A knowledge of pest-crop interactions in the management of arthropod pests is of paramount importance. The development and use of a particular crop plant species and appropriate cultivars of that plant species are the base from which all management strategies must arise. If a crop cultivar is susceptible to pests (i.e., it is readily attacked and damaged by the pest), then chemical control is more likely to be the primary method used to reduce arthropod damage. However, if a crop cultivar is resistant (e.g., it is inherently less damaged or less infested than comparison cultivars [34]), pest management options are substantially broadened.

II. VALUE OF PLANT RESISTANCE

Luginbill [26] indicated that the most effective method of combating arthropods that attack plants is to grow arthropod-resistant varieties. He further showed that resistant plants returned about $300 for each $1 invested in research and development. McMillian and Wiseman [28] also estimated that $20 was returned

---

*Retired.
†All programs and services of the United States Department of Agriculture are offered on a non-discriminatory basis without regard to race, color, national origin, religion, sex, age, marital status, or handicap.

to growers for each $1 invested by the United States Department of Agriculture (USDA) between 1950 and 1970 for research on resistance in corn to the corn earworm, *Helicoverpa zea.*

The Purdue-USDA small grains improvement program has estimated that a $3.4 billion increase in farm income is attributable to improved cultivars of wheat with resistance to the Hessian fly, *Mayetiola destructor* [40]. The annual return exceeded $4.6 million per scientific year invested and calculated over a 64-year period of the program. Buntin and Raymer [3] reported that the economic benefit of using resistant wheat cultivars in Georgia to control the Hessian fly averages $104/hectares (ha).

Resistant plants also can permit economical production of a crop where it was previously precluded because of extreme pest pressure. For example, before 1951, production of sweet corn was unprofitable in the southeastern United States even with the use of pesticides. However, in 1951 "Ioana" sweet corn, with low to intermediate levels of resistance to the corn earworm, was released to growers. This single release allowed growers to produce sweet corn with pesticides. Today, higher levels of resistance in sweet corn are available to growers, as evidenced by the present use of Ioana in many studies as a susceptible check. Another example of a crop that can now be grown more profitably because of resistance to an arthropod is hybrid sorghum with resistance to the greenbug, *Schizaphis graminum* (Rondani). Dharmaratne et al. [9] reported that insecticide use on sorghum increased dramatically after the greenbug became a pest in 1968. However, they found that cultivation of resistant sorghums is more profitable than growing nonresistant sorghum with or without insecticides. Net returns were largest ($82.27/acre) for the greenbug-resistant sorghum with insecticides, followed by greenbug-resistant sorghums without insecticides ($81.04/acre), nonresistant sorghums with insecticides ($80.50/acre), and lastly nonresistant sorghums without insecticides ($66.76/acre).

III. CLASSIFICATIONS OF RESISTANCE

Conventional classifications of resistance have been reported by Painter [34,35], Horber [17], and Gallun and Khush [12] as:

  a. *Immunity:* An immune plant is one which a specific arthropod will not damage or use under any known condition. Thus, an immune plant is a nonhost.
  b. *High resistance:* A cultivar with high resistance is one which possesses attributes that result in little or minor damage by a specific arthropod under a given set of conditions.
  c. *Moderate resistance:* A moderate or intermediate level of resistance

in a cultivar results from any one of at least three situations: (a) a mixture of phenotypically high- and low-resistance plants; (b) plants homozygous for genes which under a given environmental condition produce an intermediate level of injury; and (c) a single clone which is heterozygous for incomplete dominance for high resistance.

d. *Low resistance:* A low level of plant resistance results in less damage or infestation by an insect than the average for a susceptible cultivar.

e. *Susceptibility:* A susceptible cultivar is heavily damaged by an arthropod pest species.

Painter [34] attributed resistance to heritable qualities of the plant. A resistant plant is always resistant to a specific pest species under given environmental conditions; if, however, the environment changes, the level of resistance may vary. Mutations in a resistant plant genotype may or may not result in a change in the level of resistance, whereas its predecessor remains resistant to an arthropod pest. An arthropod pest may form new biotypes, whereas the original biotype remains susceptible to a resistant plant genotype. Resistance in plants, however, appears more stable and persistent than does the ability of arthropod pests to overcome it, as evidence by the number of biotypes in certain crop-arthropod relationships; for example, the Hessian fly and wheat. There are only about 14 arthropod species out of more than 1000 attacking domestic crops that have formed biotypes in seven crop-arthropod relationships [42].

IV. MECHANISMS OF RESISTANCE

Painter [37] proposed three mechanisms of resistance to arthropod pests: *nonpreference, antibiosis*, and *tolerance*. Within each of these mechanisms, the level of resistance may vary from high to low. The bases of the resistance mechanisms may be morphological or chemical factors. Resistant cultivars may possess varying combinations of one or more of these mechanisms of resistance.

A. Nonpreference

The resistance mechanism of nonpreference involves a group of plant characters (and arthropod responses) that causes an arthropod to avoid a plant or plant part for oviposition, food, shelter, or any combination of the three. Painter [37] delineated nonpreference among cultivars into two distinct actions of choice by arthropods: (a) a choice to oviposit, establish, or feed when only one cultivar or plant part is available; and (b) a choice to oviposit, establish, or feed when more than one cultivar is available. Owens [33] described these two types of nonpreference as relative and absolute. The term *antixenosis* was proposed by Kogan and Ortman [24] as a substitute for nonpreference, because it parallels antibiosis more closely than does nonpreference, which indicates a response by the insect rather than a property of the plant. However, most researchers in plant resistance to arthropods continue to use the term *nonpreference*, because a majority of the scientists at a biennial plant resistance workshop (Gainesville, FL, 1978) adopted its continued use.

B. Antibiosis

The term *antibiosis* is used for the mechanism of resistance that results in adverse effects on the arthropod's life history when a resistant plant is used for food [34]. The effects on an arthropod feeding on a plant with this type of resistance may be death of the neonate (larva or nymph), reduced food consumption resulting in a lower weight, increased developmental time, low food reserves, death in the prepupal or pupal stage, reduced weight of pupae, and/or reduced fecundity.

C. Tolerance

The resistance mechanism of tolerance allows the plant to grow and reproduce or repair injury despite supporting a density of arthropods approximately equal to what would be damaging to a susceptible cultivar [34]. This mechanism has been difficult to define in several crop-arthropod relationships because in many cases equal or increased yield of the tolerant cultivar has been difficult to determine. Tolerance is influenced more by environmental factors than is nonpreference or antibiosis, adding to the difficulty of characterizing this mechanism in cultivars.

Resistant cultivars often possess combinations of these resistance mechanisms, especially with regard to nonpreference and antibiosis. With a combination of resistance mechanisms, a cultivar that is nonpreferred does not require the same level of antibiosis or tolerance that a more preferred cultivar must possess to attain the same level of resistance. Thus, different cultivars may express an equal level of resistance with varying amounts of the different mechanisms of resistance.

The development of crop cultivars with multiple mechanisms of resistance, multiple genes for resistance, and, more specifically, multiple pest resistance continues to be a challenge (see Chapter 8). Both biological and genetic bases for resistance must be identified successfully to incorporate multiple resistance characters into a cultivar. Sometimes the agronomic and resistance factors are antagonistic. However, with a sound biological basis and restriction fragment length polymorphism (RFLP) analysis of the plant, resistance gene transfer may now be accomplished more easily and more rapidly than before [46] (see Chapter 8). This may be especially true for gene transfer to adapted plant germplasm when more exotic plant material is found to possess resistance to insects.

Many times, insufficient plant material has been evaluated to find adequate levels of resistance to transfer to more adapted plant materials that can be used commercially. In some cases, being able to rear an arthropod pest artificially can enhance a search for higher levels of resistance by applying higher levels of arthropod pressure. When an arthropod pest can be reared, artificial infestations of eggs, larvae, or nymphs can enhance the search for resistance by testing in locations where the arthropod does not occur or by testing earlier in the season. Studies on the mechanisms or bases of resistance can be readily completed when adequate supplies of the insect are available. But when high levels of plant resistance have been found and incorporated into adapted cultivars, their use in the management of the pest can be demonstrated, as described below.

V. RESISTANCE MECHANISMS AND INTEGRATED PEST MANAGEMENT

As mentioned earlier, knowledge of arthropod-plant interactions, or mechanisms of resistance, is important for the correct use of resistant cultivars in an integrated pest management (IPM) system. Arthropod-plant interactions may occur at several levels: between or among crop species, among varieties within a crop species, within a single variety in a field, and finally on an individual plant [49] (see Chap. 14). Interactions also occur between an arthropod and mechanisms of cultivar resistance [34]. A few specific examples of successes of plant resistance in IPM will be given. Even though there are many examples of successes of plant resistance, a few examples exist that fall short of expectations. Some may say that when an arthropod biotype develops and overcomes a formerly resistant cultivar, plant resistance is unsuccessful in managing the pest. However, it must be pointed out that a resistant cultivar remains resistant to the original arthropod population. There also are examples where a cultivar was developed for resistance to a given arthropod species and when released it was found to be highly susceptible to another arthropod species.

Resistant cultivars have traditionally been used in one of two ways: (a) as a primary method for controlling pests, or (b) as an adjunct to other control components of IPM programs [36]. The following will illustrate how each mechanism of resistance may operate independently as a primary method in controlling arthropod populations and/or limiting damage.

A. Use Of Resistant Cultivars As A Primary Method Of Control

The result of using resistant cultivars on the arthropod has often been specific, cumulative, and persistent control of pests [36]. Adkisson and Dyck [1] stated that reduction in pest populations achieved through the use of resistant plants is constant, cumulative, and practically without cost to the grower. However, higher cost for seed of a resistant cultivar may occur with the advent of biotechnologically transformed cultivars.

There are many examples of the use of resistant cultivars as the primary method of suppressing arthropod density and/or reducing damage. Plant resistance was historically sought for crops where it was the only mode of plant protection available [16]. For example, resistant American grape rootstocks were first used in 1870 to control the grape phylloxera, *Phylloxera* spp., in France to save the French wine industry. The resistant rootstocks were employed on a worldwide basis, resulting in effective control of *Phylloxera* spp. for more than 100 years [43]. Wheat cultivars resistant to the Hessian fly and the wheat stem sawfly, *Cephus cinctus*, are also present-day examples of resistant cultivars used as a primary method of pest control; the planting of some 8.6 million hectares of corn hybrids resistant to the European corn borer, *Ostrinia nubilalis*, is another example [41].

B. Tolerance

A crop or plant may be tolerant if it can yield well despite infestations that seriously damage more susceptible plants [34]. Arthropod populations on tolerant cultivars are not reduced nor affected adversely. Antagonistic interactions between pests and the tolerant cultivar are not believed to occur, since the phenomenon with this resistance mechanism is entirely a plant response. Of the 2000 corn plant introductions evaluated for resistance to rootworms (*Diabrotica* spp.) by Wilson and Peters [48], only 6 were found to be tolerant to infestations.

Another example of tolerance is the many corn hybrids that have long, tight husks with copious silks (more than 10 g) and no antibiosis, but which limit damage by the corn earworm while reducing arthropod numbers [56]. Also, a sweet corn hybrid, "471-U6 × 81-1," is tolerant to corn earworm damage, because larvae feed in the silk channel and complete their development without any adverse effects and without inflicting significant damage to the ear [53]. A tolerant variety also offers a farmer an opportunity to use several alternative methods of pest control: (a) pesticides at a reduced rate [29,51] (Fig. 1); (b) parasites or predators [44,45] (Fig. 2); (c) cultural control (e.g., early plantings); and (d) insect pathogens [11].

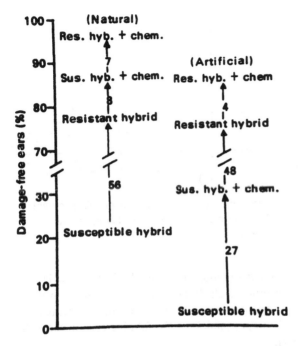

Figure 1 Percentage of ears that are damage-free on resistant or susceptible sweet corn hybrids combined with insecticides and natural or artificial infestation of corn earworm. (From ref. 29.)

C. Nonpreference

A crop or variety may not be preferred when plants possess characters that stimulate arthropod responses in which pests use them less for oviposition, shelter, food, or any combination of the three [34]. Two examples will illustrate the value of nonpreference resistance in the control of arthropod populations. Dahms [8] showed in a simple mathematical model (Table 1) differences in the development of an aphid population over 50 days on a nonpreferred and susceptible cultivar. Assuming that one aphid colonizes a nonpreferred cultivar, whereas two aphids colonize a susceptible cultivar, at the end of 50 days, 15,502 aphids will have developed on the resistant cultivar compared with 31,004 aphids on a susceptible cultivar. In a quite different example, resistance in sorghum to the sorghum midge, *Contarinia sorghicola*, is composed of two nonpreference factors: reduced attractiveness to the adult midge and a contact oviposition deterrent within the sorghum spikelet that increases the probing time by oviposit-

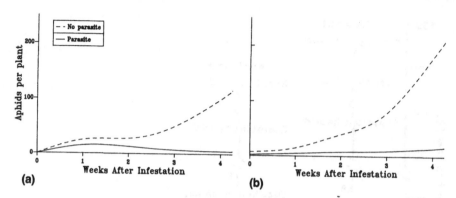

Figure 2 Increase of greenbugs in the absence and presence of one female parasite caged on greenbug-resistant (Will, a) and susceptible (Rogers, b) barley infested at the rate of three aphids/plant. (From ref. 44.)

ing females and ultimately contributes to reduce fecundity [47]. Teetes [47] also reported that the economic threshold levels (ETLs) for the resistant sorghum hybrids range from 1.0 to 6.0 midges/panicle, whereas susceptible hybrids have an ETL ranging from 0.2 to 1.2 midges/panicle; a fivefold increase in the ETL for nonpreferred-resistant hybrids.

Table 1 Effect of Nonpreference on an Arthropod Population[a]

Days after first reproduction	Total living aphids with indicated number of viviparous females on plant cultivar	
	Nonpreferred	Preferred
5	6	12
10	11	22
20	76	152
30	450	900
40	2620	5240
50	15,502	31,004

[a]Beginning with one aphid on the nonpreferred cultivar and two on the preferred cultivar; aphids reproduce one per day for 20 days; nymphs mature in 10 days; no nymphal mortality. Assumption was made that two aphids initiated the colony on the preferred cultivar compared to one on the nonpreferred.
Source: Modified from ref. 8.

D. Antibiosis

Antibiosis is the mechanism of resistance that produces adverse effects on an arthropod's life history when a resistant plant is used for food [34]. Several examples illustrate the antibiotic effects of resistant plants. Beginning as early as 1792 [15], a high level of antibiosis against the Hessian fly in resistant wheats has been the major control component for reducing damaging populations of this pest. Dahms [8] illustrated the differential rate of aphid development on population growth when nymphs matured in 5 (susceptible variety), 10 (intermediate antibiosis), and 20 days (high antibiosis) with adults reproducing at a rate of one offspring/day for 20 days (Table 2). At the end of 5 days, there were no differences in the aphid populations on the various plants; however, by 25 days, there were 10 times more aphids on susceptible plants than on plants exhibiting intermediate resistance. By 50 days, there were more than 125 times more aphids on susceptible plants than on plants with intermediate resistance, and over 3300 times more aphids on susceptible plants than on plants exhibiting a high level of antibiosis.

Mortality of immature arthropods is one of the most important factors limiting the increase of arthropod populations. The effects of 0, 50, and 90% mortality on an aphid population over 50 days is illustrated in Table 3. There is almost a 15-fold increase in the aphid population at 50 days for 0 (15,502) versus 50% (1178) mortality, but an over 100-fold increase in the aphid population at 50 days for 50% (1178) versus 90% (9) mortality, and more than 1000-

Table 2 Effect of a Cultivar Resistance and Resultant Differential Developmental Rates on Rate of Increase in an Aphid Population[a]

Days after first reproduction	Total living aphids on cultivars		
	Resistant	Low resistant	Susceptible
5	6	6	6
25	35	175	1,751
50	585	15,502	1,955,056

[a]Starting with one viviparous female; adults produce one nymph per day for 20 days; assumes no nymphal mortality. Each level of resistance began with one viviparous female. Nymphs matured in 5, 10, and 20 days on the susceptible, low-resistant, and high-resistant cultivars, respectively, before additional reproduction occurred.
Source: Modified from ref. 8.

Table 3 Effect of Resistance-Related Nymphal Mortality on Growth of an Aphid Population[a]

Days after first reproduction	Total living aphids with indicated percentage of nymphal mortality and level of resistance		
	0%	10%	90%
5	6	5	2
25	175	141	4
50	15,502	10,431	9

[a]Starting with one viviparous female; nymphs mature in 10 days; adults produce one nymph per day for 20 days. Cultivar with 0% mortality is considered susceptible; 10% mortality exhibits low resistance, and 90% mortality is highly resistant.
Source: Modified from ref. 8.

fold increase in the aphid population at 50 days for 0 (15,502) versus 90% (9) mortality. At 90% mortality, there was essentially no increase in the arthropod population.

In a more dramatic illustration (Table 4), Dahms [8] showed the population effects of a resistant alfalfa, cv. "Lahontan," and a susceptible cv. "Chilean" alfalfa on a population of the spotted alfalfa aphid, *Therioaphis maculata* (Buckton). Using four factors of antibiosis in an actual cumulative model, there

Table 4 Cumulative Effect of Four Antibiosis Factors[a] on an Aphid Population on a Resistant and Susceptible Cultivar

Days after first reproduction	Total living aphids with indicated antibiosis factors	
	Lahontan (R) Reproduce 2.5/d for 13 d; nymphs mature in 9 d with 90% mortality	Chilean (S) Reproduce 4/d for 13 d; nymphs mature in 6 d with 10% mortality
5	2	19
25	12	58,489
50	87	1,216,252,841

[a]Antibiotic factors were: days from nymph to maturity; adult longevity (days); number of nymphs produced per day; and nymphal mortality (percent).
Source: Modified from ref. 8.

were more than 9 times as many aphids on Chilean than on Lahontan alfalfa after 5 days; more than 4800 times as many aphids on Chilean than on Lahontan alfalfa after 25 days; and almost 14 million times as many aphids on Chilean than Lahontan alfalfa after 50 days.

In the studies in our laboratory, we have found that resistant corn lines reduced populations of the fall armyworm, *Spodoptera frugiperda*, by 50% [55]. Also, Wiseman and Isenhour [52] found that resistant corn silks slowed the growth of corn earworm larvae, thereby extending their life cycle by about 20 days, and reduced egg production by almost 65% per generation.

Transgenic maize plants expressing an insecticidal or antibiotic protein derived from the entomopathogenic bacterium *Bacillus thuringiensis* killed from 95 to 100% of the European corn borer larvae in field tests [25]. It was also noted that damage to the leaves by larvae was reduced by more than 50% and tunneling in the stems was reduced by 97% in the transgenic plant compared with that of the control [25]. This level of resistance could preclude the use of most IPM components other than cultural control for the control of the European corn borer.

VI. RESISTANT CULTIVARS AND INTEGRATED PEST MANAGEMENT

Resistant cultivars often are used effectively in combination with other control components of IPM. When a resistant cultivar is developed, it may possess resistance to more than one arthropod species. Hence, we can improve pest management of several arthropod species with a single cultivar. For example, Overman [32] reported that several of Dekalb's experimental hybrids are resistant to several leaf-feeding lepidopterous species. Therefore, the development and use of resistant cultivars should discourage the release and use of susceptible ones for any IPM system. Adkisson and Dyck [1] stated that IPM systems are designed to suppress pest populations below crop-damaging levels, not to replace chemical pesticides. They further stated that a resistant variety can provide a foundation on which to build an integrated control system. Therefore, the following is presented to illustrate that resistant cultivars can be used in combination with other components of IPM.

A. Use of Resistant Cultivars as a Component of Integrated Pest Management

With a low degree of plant resistance to pests, other components must be used to achieve adequate pest control that permit growers to attain the desired margin or profit. Highly resistant crop cultivars, in most cases, dramatically reduce or eliminate the need for most other control components for the target pest. But this does not necessarily have to be the practice. For example, with a highly tolerant cultivar, other control tactics are needed to keep pest populations in check. In addition, even with highly antibiotic or nonpreferred resistance, other control components should be used to achieve greater overall success in reducing populations of a pest.

B. Plant Resistance and Insecticides

Resistant cultivars generally are compatible with insecticidal control of pests. For many years, corn, especially sweet corn, could not be grown economically in the southern United States until moderate resistance to the corn earworm was introduced into Ioana sweet corn hybrids [27]. Sweet corn could then be produced with insecticidal control of corn earworm larvae. Wiseman et al. [51] demonstrated that using a resistant (tolerant) sweet corn hybrid in combination with supplemental insecticide treatment reduced losses from the corn earworm. By using a resistant cultivar, a reduction of 7.5 kg active ingredient/ha of insecticides was realized through a reduction in both the rate and the number of insecticide applications. In another example, the use of sorghum hybrids resistant to the greenbug, *Schizaphis graminum*, biotype C, permitted the use of extremely low rates of insecticides [7]. Using low rates of insecticide preserved natural biological control agents that prevented resurgence of greenbug populations. Conversely, insecticide treatments induced resurgence of the brown planthopper, *Nilaparvata lugens*, on resistant varieties [39]. In the latter case, compatibility between a resistant cultivar and the insecticide used was lacking. However, in this instance, the antagonisms between the brown planthopper and the insecticide may be an isolated case and may not apply to all insecticides, the planthopper, or the use of resistant varieties of rice. Kennedy et al. [23] also reported antagonisms between resistance and insecticide where they found neonate *Helicoverpa zea* on foliage of *Lycopersicon hirsutum* were tolerant to the carbamate insecticide carbaryl.

C. Plant Resistance and Biocontrol Agents

Plant resistance and biocontrol agents should be compatible in reducing populations of a pest. Gould et al. [13] developed a model that suggested that selection for adaptation to a resistant plant is lower at a given level of pest population suppression when that suppression is achieved by the combined action of plant resistance and natural enemies than by strong resistance alone. Isenhour and Wiseman [18] found no change in pupation, weight of pupae, or time to adult eclosion of the parasitoid, *Campoletis sonorensis*, when host fall armyworm larvae had fed on resistant corn foliage. In fact, they found that the combined effects of the parasite and plant resistance were additive and beneficial in reducing consumption of foliage, weight of larvae and pupae, and number of fall armyworms [19]. Pair et al. [38] found that fewer fall armyworms established on "Pioneer X304C" and "Antigua 2D-118," but the rate of their parasitism was higher on these two resistant lines than on susceptible lines. In another study, Isenhour and Wiseman [19] found the effects of combining *C. sonorensis* and plant resistance was additive in reducing the growth of fall armyworm larvae. The life cycle of the parasite was prolonged but pest larval development was no longer than the developmental period of unparasitized fall armyworm larvae feeding on resistant plants. Thus, in this system, the life cycle of larvae surviving on a resistant cultivar, although prolonged, would continue to be synchronized with development of their parasitoids when resistant cultivars are used. Boethel and Eikenbary [2], however, reported that certain resistance features of plants, especially glandular trichomes [30], may adversely affect natural enemies. For example, Campbell and Duffey [4] showed that resistant tomatoes with tomatine adversely affects a parasitoid of the tomato fruitworm, *Hyposoter exiguae*. Kennedy et al. [22] reported that 2-tridecanone/glandular trichome–mediated resistance in a wild tomato adversely affected several species of parasitoids and predators of the tomato fruitworm, *H. zea*. The rates of parasitism or predation, and parasitoid survival, were lower on "PI134417" foliage than on susceptible tomato foliage. Farrar and Kennedy [10] found differences in parasitism between two tachinid flies, *Archytas marmoratus* and *Eucelatoria bryani*. *A. marmoratus*, which larviposits on its host's food plant, deposited significantly fewer larvae on a resistant host than on a susceptible host, whereas *E. bryani*, which larviposits directly into its host, was not affected by the methyl ketone–mediated resistance [10]. Also, a high level of resistance in soybean to the soybean looper, *Pseudoplusia includens*, is detrimental to the parasitoid, *Microplitis demolitor* [57].

It may be important to note that in some cases where adverse interactions between resistant plants and the biocontrol agents occur, pest reduction capacities of resistant plants may be sacrificed in favor of attempts to promote biological control. It has been stated that many of the International Rice Research Institute's (IRRI) rices have such high levels of resistance that biocontrol agents are finding it difficult to increase their population; however, insect losses are substantially reduced. If a pest insect can readily adjust to resistant plants, then can we assume that biocontrol agents also may adjust to the changes in their host? This may be the case with the rice pests in question. But, what about other pests and parasitoid interactions in the same system? Other pests and parasitoids within the same system affected by one pest are not likely to be affected adversely, because most resistant cultivars do not confer multiple pest resistance.

The use of resistant cultivars combined with predators and parasitoids also has proven to be beneficial in reducing insect numbers. Wiseman et al. [54]

found that populations of *Orius insidiosus*, a predator of corn earworm larvae, were higher on a tolerant sweet corn hybrid (471-U6 × 81-1) than on susceptible ones, indicating compatibility of the effects of varietal resistance and the predator on *H. zea* larvae. Isenhour et al. [20] found that although fall armyworm and corn earworm larvae were feeding on a resistant cultivar, the efficacy of *O. insidiosus* improved. They demonstrated that predation increases on larvae on the resistant cultivar relative to predation on the susceptible cultivar, particularly at a high prey density. The resistant host plant slowed larval growth, thus allowing the predator to prey on older larvae, whereas larvae of similar age feeding on susceptible cultivars were too large for the predator to attack.

Obrycki et al. [31] found through mechanical exclusion experiments that naturally occurring aphid predators (primarily Coccinellidae and Chrysopidae) and hymenopterous parasitoids (primarily Aphidiidae) reduced aphid density >65% on potato hybrids with varying densities of glandular trichomes. A similar relationship was observed when predators and parasitoids were present, but aphid densities were lower on all potato clones. Their results indicated that biological control and plant resistance were compatible and complementary tactics in integrated aphid management on potatoes.

Hamm and Wiseman [14] showed that the susceptibility of fall armyworm larvae to a nuclear polyhedrosis virus was inversely related to the growth and vigor of the larvae, which was, in turn, directly related to the level of resistance of the host plant. They concluded that the virus was more effective in controlling fall armyworm larvae when used on a resistant cultivar than on a susceptible one, thus demonstrating that the resistant cultivar could enhance performance by the virus in an IPM system. More recently, Wiseman and Hamm [50] found that Elcar (a nuclear polyhedrosis virus; Crops Genetics International, Columbia, MD) and corn silk resistance to larvae of the corn earworm could be combined effectively to enhance the control of this pest. They found that when corn earworm larvae fed on the resistant corn silks, their growth was severely stunted and Elcar could more effectively kill neonates (Fig. 3), and 4-day-old to 8-day-old larvae (Table 5). Elcar caused 98 and 87% mortality of 4- and 8-day-old corn earworm larvae that had fed on resistant silks compared with 69 and 3% mortality, respectively, of those that had fed on control diets for 4 and 8 days (Table 5).

D. Plant Resistance and Inherited Sterility

Carpenter and Wiseman [5,6] found that plant resistance and inherited sterility were compatible control strategies for managing larvae of both the fall armyworm and corn earworm. They found that differences between the resistant and susceptible cultivars were similar when normal or substerile fall armyworm or corn earworm larvae were used.

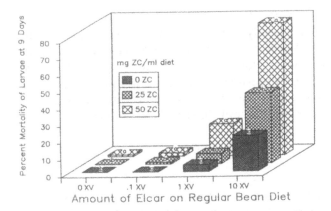

Figure 3 Control of corn earworm neonates with Elcar after the larvae were fed on resistant Zapalote Chico (ZC) corn silks at 25 or 50 mg silk/ml diet. (From ref. 50.)

E. Plant Resistance and Cultural Control

Advantages gained by farmers using cultural methods combined with growing resistant crop cultivars to control arthropod pests would certainly be greater than using cultural controls with a susceptible cultivar. Adkisson and Dyck [1] stated that resistant varieties, including those that can be manipulated to evade pest attack, are highly desirable in a cultural control system that maintains pest populations below the economic threshold while preserving the natural enemies. The use of "early plantings" in combination with resistant cultivars has been recommended for many years to avoid arthropod attacks on wheat, cotton, sorghum, and corn. Crop rotation as a form of cultural control to avoid damaging

Table 5 Control of 4- and 8-Day-Old Corn Earworm Larvae with Elcar and Resistant Corn Silks

Larval age (days)	Larval weight on		Larval mortality on			
	NS	RS	NS		RS	
4	24.1	3.6	0	*	98	
8	557.6 *	20.2	0	*	87	

Weights of larvae or percentage of mortality separated by * are significantly different $P<.05$. NS, no resistant silk; RS, resistant silk.
Source: From ref. 50.

populations of the corn rootworm, *Diabrotica* spp., has been used in the U.S. Corn Belt for many years. Recently, early mechanical harvesting and drying of field-infested corn was demonstrated to reduce the abundance of the maize weevil, *Sitophilus zeamais*, as the grain was placed in storage [21]. Also, fewer weevils were found on the weevil-resistant corn than in susceptible corn. Thus, the use of weevil-resistant corns, early harvesting, and adequate drying of the grain would reduce the number of weevils in corn during storage.

VII. CONCLUSIONS

Plant resistance, in the classic sense or as transgenic plants, used alone can be effective in reducing arthropod damage to crops and/or populations. It also has shown that plant resistance can be a principal cornerstone of IPM to further enhance the control of arthropod pests. But why is there only limited use of resistant cultivars in IPM systems? Teetes [47] identified a number of reasons for the limited impact that resistant cultivars have had in crop production: (a) failure of entomologists and plant breeders to complete their work after locating or developing an arthropod resistant germplasm; (b) failure of growers to accept and use arthropod-resistant cultivars; (c) overreliance on insecticides; (d) separation of crop production and crop protection; and (e) failure to produce and distribute adequate information about the pest and the resistant cultivar. These reasons or similar ones may explain the limited application of IPM of arthropod pests, nematodes, diseases, and weeds in practice today. However, one idea remains sound: The use of a resistant cultivar provides crop protection that is biologically, ecologically, economically, and socially acceptable [47]. This holds true whether the cultivar is resistant to an arthropod pest(s), a nematode, a disease, or a weed. This method of crop protection extends further to the use of resistant cultivars when used alone or as a multiple resistant cultivar (arthropod, nematode, disease, and weed) and whether used in combination with other IPM systems. But, most importantly, the resistant cultivar is the base from which management decisions can be made and will become even more important in the future with an increased emphasis on reduced pesticide usage and sustainable agriculture systems.

REFERENCES

1. Adkisson, P.L., and V.A. Dyck. 1980. Resistant varieties in pest management systems. In *Breeding Plants Resistant to Insects*. F.G. Maxwell and P.R. Jennings, eds. New York: Wiley, pp 233–251.

2. Boethel, D.J., and R.D. Eikenbary. 1986. *Interactions of Plant Resistance and Parasitoids and Predators of Insects*. West Sussex: Ellis Horwood.

3. Buntin, G.D., and P.L. Raymer. 1989. Hessian fly (Diptera: Cecidomyiidae) damage and forage production of winter wheat. *J. Econ. Entomol.* 82:301–306.

4. Campbell, B.C., and S.S. Duffey. 1979. Tomatine and parasitic wasps: Potential incompatibility of plant antibiosis with biological control. *Science* 205:700–702.

5. Carpenter, J.E., and B.R. Wiseman. 1992. *Spodoptera frugiperda* (Lepidoptera: Noctuidae) development and damage potential as affected by inherited sterility and host plant resistance. *Environ. Entomol.* 21:57–60.

6. Carpenter, J.E., and B.R. Wiseman. 1992. Effects of inherited sterility and insect resistant dent-corn silks on *Helicoverpa zea* (Lepidoptera: Noctuidae) development. *J. Entomol. Sci.* 27:413–420.

7. Cate, J.R., D.G. Bottrell, and G.L. Teetes. 1973. Management of the greenbug on grain sorghums. I. Testing foliar treatments of insecticides against greenbug and corn leaf aphid. *J. Econ. Entomol.* 66:945–951.

8. Dahms, R.G. 1972. The role of host plant resistance in integrated insect control. In *The Control of Sorghum Shoot Fly*. M.G. Jotwani and W.R. Young, eds. New Delhi: Oxford and IBH, pp 152–167.

9. Dharmaratne, G.S., R.D. Lacewell, J.R. Stoll, and G.L. Teetes. 1986. Economic impact of greenbug resistant grain sorghum varieties: Texas Blacklands. *Misc. Pub. Texas Agric. Exp. Sta.* 1585.

10. Farrar, R.R., and G. Kennedy. 1993. Field cage performance of two tachinid parasitoids of the tomato fruitworm on insect resistant and susceptible tomato lines. *Entomol. Exp. Appl.* 67:73–78.

11. Fernandez, A.T., H.M. Graham, M.J. Lukefahr, H.R. Bullock, and N.S. Hernandez, Jr. 1969. A field test comparing resistant varieties plus applications of polyhedral virus with insecticides for control of *Heliothis* spp. and other pests. *J. Econ. Entomol.* 62:173–177.

12. Gallun, R.L., and G.S. Khush. 1980. Genetic factors affecting expression and stability of resistance. In *Breeding Plants for Resistance to Insects*. F.G. Maxwell and P.R. Jennings, eds. New York: Wiley, pp 63–85.

13. Gould, F., G.G. Kennedy, and M.T. Johnson. 1991. Effects of natural enemies on the rate of herbivore adaptation to resistant host plants. *Entomol. Exp. Appl.* 58:1–14.

14. Hamm, J.J., and B.R. Wiseman. 1986. Plant resistance and nuclear polyhedrosis virus for suppression of the fall armyworm (Lepidoptera: Noctuidae). *Florida Entomol.* 69:541–549.

15. Havens, J.N. 1792. Observations on the Hessian Fly. *N.Y. Society of Agric. Trans., Arts and Manufact.* 1:89–107.

16. Horber, E. 1972. Plant resistance to insects. *USDA Agric. Sci. Rev.* 10:1–18.

17. Horber, E. 1980. Types and classification of resistance. In *Breeding Plants for Resistance to Insects*. F.G. Maxwell and P.R. Jennings, eds. New York: Wiley, pp 63–85.

18. Isenhour, D.J., and B.R. Wiseman. 1987. Foliage consumption and development of the fall armyworm (Lepidoptera: Noctuidae) as affected by the interactions of

a parasitoid, *Campoletis sonorensis* (Hymenoptera: Ichneumonidae), and resistant corn genotypes. *Environ. Entomol.* 16:1181–1184.

19. Isenhour, D.J., and B.R. Wiseman. 1989. Parasitism of the fall armyworm (Lepidoptera: Noctuidae) by *Campoletis sonorensis* (Hymenoptera: Ichneumonidae) as affected by host feeding on silks of *Zea mays* L. cv. Zapalote Chico. *Environ. Entomol.* 18:394–397.

20. Isenhour, D.J., B.R. Wiseman, and R.C. Layton. 1989. Enhanced predation by *Orius insidiosus* (Hemiptera: Anthocoridae) on larvae of *Heliothis zea* and *Spodoptera frugiperda* (Lepidoptera: Noctuidae) caused by prey feeding on resistant corn genotypes. *Environ. Entomol.* 18:418–422.

21. Keever, D.W., B.R. Wiseman, and N.W. Widstrom. 1988. Effects of threshing and drying on maize weevil populations in field-infested corn. *J. Econ. Entomol.* 81:727–730.

22. Kennedy, G.G., R.R. Farrar, Jr., and R.K. Kashyap. 1991. 2-Tridecanone-glandular trichome-mediated insect resistance in tomato: effect on parasitoids and predators of *Heliothis zea*. *Naturally Occurring Pest Bioregulators*. P.A. Hedin, ed. ACS Symposium Ser. 449. Washington, D.C.: American Chemical Society, pp 150–165.

23. Kennedy, G.G., R.R. Farrar, and M.R. Riskallah. 1987. Induced tolerance of neonate *Heliothis zea* to host plant allelochemicals and carbaryl following incubation of eggs on foliage of *Lycopersicon hirsutum* f. *glabratum*. *Oecologia* 73:615–620.

24. Kogan, M., and E.E. Ortman. 1978. Antixenosis--a new term proposed to replace Painter's "nonpreference" modality of resistance. *Bull. Entomol. Soc. Am.* 24:175–176.

25. Koziel, M.G., G.L. Beland, C. Bowman, N.B. Carozzi, R. Crenshaw, L. Crossland, J. Dawson, N. Desai, M. Hill, S. Kadwell, K. Launis, K. Lewis, D. Maddox, K. McPherson, M.R. Meghji, E. Merlin, R. Rhodes, G.W. Warren, M. Wright, and S.V. Evola. 1993. Field performance of elite transgenic maize plants expressing an insecticidal protein derived from *Bacillus thuringiensis*. *Bio/Technology* 11:194–200.

26. Luginbill, P., Jr. 1969. Developing resistant plants - the ideal method of controlling insects. *USDA. ARS Prod. Res. Rep.* 111.

27. Maxwell, F.G. 1972. Host plant resistance to insects--nutritional and pest management relationships. In *Insect and Mite Nutrition*. J.G. Rodriguez, ed. Amsterdam: North Holland, pp 599–609.

28. McMillian, W.W., and B.R. Wiseman. 1972. Host plant resistance: a twentieth century look at the relationship between *Zea mays* L. and *Heliothis zea* (Boddie). *Florida Agric. Exp. Sta. Monograph Ser.* 2.

29. McMillian, W.W., B.R. Wiseman, N.W. Widstrom, and E.A. Harrell. 1972. Resistant sweet corn hybrid plus insecticide to reduce losses from corn earworms. *J. Econ. Entomol.* 65:229–231.

30. Obrycki, J.J. 1986. The influence of foliar pubescence on entomophagous species. In *Interactions of Plant Resistance and Parasitoids and Predators of Insects*. D.J. Boethel and R.D. Eikenbary, eds. Chichester, England: Ellis Horwood, pp 61–83.

31. Obrycki, J.J., M.J. Tauber, and W.M. Tingey. 1983. Predator and parasitoid interaction with aphid-resistant potatoes to reduce aphid densities: A two-year field study. *J. Econ. Entomol.* 76:456–462.

32. Overman, J.L. 1989. A maize breeding program for development of hybrids with resistance to multiple species of leaf-feeding and stalk-boring lepidoptera. In *Toward Insect Resistant Maize for the Third World: Proc. of the International Symposium on Methodologies for Developing Host Plant Resistance to Maize Insects.* J.A. Mihm, ed. El Batan, Mexico, D.F.: International Maize and Wheat Improvement Center, pp 235–243.

33. Owens, J. 1974. An explanation of terms used in insect resistance to plants. *Iowa State J. Res.* 49:513–517.

34. Painter, R.H. 1951. *Insect Resistance in Crop Plants*. New York: Macmillan.

35. Painter, R.H. 1958. Resistance to insects. *Annu. Rev. Entomol.* 3:267–290.

36. Painter, R.H. 1966. Plant resistance as a means of controlling insects and reducing their damage. In *Pest Control by Chemical, Biological, Genetic, and Physical Means--A Symposium*. E.F. Knipling, ed. USDA-ARS-33-110, Washington, D.C.: U.S. Department of Agriculture, pp 138–148.

37. Painter, R.H. 1968. Crops that resist insects provide a way to increase world food supply. *Kansas Agric. Exp. Sta. Bull.* 520.

38. Pair, S.D., B.R. Wiseman, and A.N. Sparks. 1986. Influence of four corn cultivars on fall armyworm (Lepidoptera: Noctuidae) establishment and parasitism. *Florida Entomol.* 69:566–570.

39. Reissig, W.H., E.A. Heinrichs, and S.L. Valencia. 1982. Insecticide-induced resurgence of the brown planthopper, *Nilaparvata lugens*, on rice varieties with different levels of resistance. *Environ. Entomol.* 11:164–168.

40. Roberts, J.J., J.E. Foster, and F.L. Patterson. 1988. The Purdue-USDA small grain improvement program - a model of research productivity. *J. Prod. Agric.* 1:239–241.

41. Schalk, J.M., and R.H. Ratcliffe. 1976. Evaluation of ARS programs on alternative methods of insect control: host plant resistance to insects. *Bull. Entomol. Soc. Am.* 22:7–10.

42. Smith, C.M. 1989. *Plant Resistance to Insects: A Fundamental Approach*. New York: Wiley.

43. Smith, E.H. 1992. The grape phylloxera—a celebration of its own. *Am. Entomol.* 38:212–221.

44. Starks, K.J., R. Muniappan, and R.D. Eikenbary. 1972. Interaction between plant resistance and parasitism against the greenbug on barley and sorghum. *Ann. Entomol. Soc. Am.* 65:650–655.

45. Starks, K.J., E.A. Wood, Jr., and R.L. Burton. 1974. Relationships of plant resistance and *Lysiphlebus testaceipes* to population levels of the greenbug on grain sorghum. *Environ. Entomol.* 3:950–952.

46. Tanksley, S.D., N.D. Young, A.H. Paterson, and M.W. Bonierbale. 1989. RFLP mapping in plant breeding: new tools for an old science. *Bio/Technology* 7:257–264.

47. Teetes, G.L. 1985. Insect resistant sorghums in pest management. *Insect Sci. Appl.* 6:443–451.

48. Wilson, R.L., and D.C. Peters. 1973. Plant introductions of *Zea mays* as sources of corn rootworm tolerance. *J. Econ. Entomol.* 66:101–104.

49. Wiseman, B.R. 1982. The importance of *Heliothis*—crop interactions in the management of the pest. International Crops Research Institute for the Semi-Arid Tropics. *Proceedings of an International Workshop in Heliothis Management*. Nov. 15–20, 1981. Patancheru, A.P. India, pp 209–222.

50. Wiseman, B.R., and J.J. Hamm. 1993. Elcar and resistant corn silks enhance mortality of corn earworm (Lepidoptera: Noctuidae) larvae. *Biol. Control* 3:337–342.

51. Wiseman, B.R., E.A. Harrell, and W.W. McMillian. 1973. Continuation of tests of resistant sweet corn hybrid plus insecticides to reduce losses from corn earworm. *Environ. Entomol.* 2:919–920.

52. Wiseman, B.R., and D.J. Isenhour. 1990. Effects of resistant maize silks on corn earworm (Lep.: Noct.) biology: A laboratory study. *J. Econ. Entomol.* 83:614–617.

53. Wiseman, B.R., W.W. McMillian, and N.W. Widstrom. 1972. Tolerance as a mechanism of resistance in corn to the corn earworm. *J. Econ. Entomol.* 65:835–837.

54. Wiseman, B.R., W.W. McMillian, and N.W. Widstrom. 1976. Feeding of corn earworm in the laboratory on excised silks of selected corn entries with notes on *Orius insidiosus*. *Flor. Entomol.* 59:305–308.

55. Wiseman, B.R., N.W. Widstrom, and W.W. McMillian. 1981. Effects of 'Antigua 2D-118' resistant corn on fall armyworm feeding and survival. *Flor. Entomol.* 64:515–519.

56. Wiseman, B.R., N.W. Widstrom, and W.W. McMillian. 1984. Increased seasonal losses in field corn to corn earworm. *J. Georgia Entomol. Soc.* 19:41–43.

57. Yanes, J., Jr., and D.J. Boethel. 1983. Effect of a resistant soybean genotype on the development of the soybean looper (Lep.: Noct.) and an introduced parasitoid, *Microplitis demolitor* Wilkinson (Hymenop.: Brac.). *Environ. Entomol.* 12:1270–1274.

7
Plant Resistance to Pathogens

Norman E. Strobel* and Joseph A. Kuć†
University of Kentucky, Lexington, Kentucky

I. INTRODUCTION

Plant diseases are caused by many different infectious agents including fungi, bacteria, viruses, and nematodes (see Chap. 10). Unlike their relatively minor role in causing disease in animals, fungi are a major cause of plant disease. Plant diseases cause an estimated annual loss of at least 9.1 billion dollars in the United States alone [2] owing to reduced yield and quality and postharvest losses. In some developing countries, postharvest disease causes the loss of more than 30% of the harvest [2]. The production of substances toxic to humans and other animals by fungi growing on diseased food crops (mycotoxins) can result in severe health problems and further effectively reduce the crop available for consumption. The development of disease is conditioned by the dynamic interactions of plants, pathogens, and the environment. The extent of crop loss depends upon the severity of disease and on its effects on marketable plant parts. Disease can affect yield indirectly through declines in photosynthesis due to loss of healthy leaf area (foliar diseases [128]) and reduced uptake of water and nutrients (root diseases [21]). Disease can also affect yield directly through infection of the marketable portions of plants, which may reduce the value of food or fiber or in some cases render it unmarketable.

Plant diseases have contributed to the development of famines in recent historical times and continue to threaten our food supplies. The Irish potato famine during the 1840s [42,63,97] and the Bengali rice famine of 1943 [80] resulted in the deaths of millions of human beings. Both famines were the result

---

**Current affiliation*: Lexington Community College, Lexington, Kentucky.
†Retired.

of complex interacting factors which today remain pertinent in many parts of the world. These include (a) heavy dependence on a single staple crop, (b) the widespread planting of disease-susceptible cultivars, (c) environmental conditions highly conducive to disease development, and (d) social, political, and economic factors which prevented the average person from obtaining alternate foodstuffs. Additional factors which could contribute to disease-induced food shortfalls in the modern world include (a) the narrow genetic base of staple crops, (b) continuous culture of monocrops, (c) soil depletion, (d) loss of effective pesticides (due to concern about their adverse effects on nontarget organisms and the environment, government regulations, and the development by pathogens of pesticide resistance), (e) development and rapid dissemination of new pathogen variants with greater potential to cause plant diseases, and (f) a burgeoning human population that challenges the ability of farmers to produce sufficient food to sustain it and the societal will to distribute that food.

Diverse strategies implemented in an integrated manner can reduce crop losses due to plant disease. These include cultural practices (e.g., crop rotation, sanitation, planting of pathogen-free seed, maintenance of adequate soil fertility, structure and water relations; see Chapter 11), application of pesticides (see Chapter 13), and the use of crop varieties able to resist pathogens. The latter strategy offers several advantages to growers, including the possibility of maintaining or increasing crop yields while permitting reductions in pesticide use leading to environmental, social, and economic benefits.

In this chapter, we discuss the nature of those forms of plant disease resistance that have been developed through conventional plant breeding practices, the mechanisms by which they act to restrict development of a pathogen, the influence of environmental factors on the expression of disease resistance, novel strategies for enhancing resistance, and the implementation of conventional and novel types of disease resistance in the context of integrated pest management (IPM). A considerable literature is available documenting the nature and practical application of conventionally derived disease resistance, and we cannot provide a comprehensive treatment of these topics. However, we have attempted to present basic principles, gleaned from decades of experience with conventionally derived resistance, which can guide the deployment of novel strategies for enhancement of plant disease resistance now under development.

II.   RESISTANCE DEVELOPED BY CONVENTIONAL PLANT BREEDING

Resistance can be defined as the ability of a plant to restrict the growth and reproduction of a potential pathogen relative to that which would occur in a fully susceptible cultivar [44]. Resistance can be viewed from three distinct perspectives: the *magnitude* of its effect, its *genetic basis*, and whether or not it is *differential* with respect to pathogen isolates [33]. The effects (or magnitude) of resistance on disease development may be large (e.g., complete) or relatively small (e.g., partial). The inheritance of resistance may be controlled by single (monogenic) or multiple (polygenic) plant genes. Resistance is differential if it is expressed against some but not all isolates of a pathogen, or if some isolates are less affected than others by the resistance. The availability of a series of host genotypes that differentiate among pathogen isolates allows the identification of pathogen races. Differential resistance is also called race-specific resistance, whereas nondifferential resistance is also called race-nonspecific resistance. Although all possible combinations of the three aspects of resistance (small or large magnitude, single or multiple gene inheritance, race specificity or nonspecificity) can and do occur, it is commonly observed that race-specific resistance is monogenically inherited and has a large effect (i.e., complete resistance), whereas race-nonspecific resistance is commonly observed to be polygenically inherited by the host and to have a smaller effect [33,44] (i.e., partial resistance). Race-specific resistance has a large effect in those many cases in which host cells that come into direct contact with an incompatible pathogenic race undergo rapid necrosis, a phenomenon termed the hypersensitive response (HR) [37]. Although some host cells die in such plant-pathogen interactions, the HR results in prevention of further development and reproduction of the pathogen.

Race-specific resistance is often governed by single complementary genes in the host and the pathogen [29,51]. It has been broadly theorized that the host gene for resistance (R gene) plays a role in the race-specific recognition of the pathogen, whereas the corresponding pathogenic gene (avirulence, or avr, gene) is involved in the production of a race-specific elicitor, a molecule which is recognized by the host and serves to trigger the cascade of molecular and biochemical events which culminate in the HR. Recent molecular biological studies have provided evidence that supports this theory [9,12,24,59,71,119].

In race-nonspecific resistance, a particular line or cultivar of a plant species is resistant to all members of a pathogen population, and this resistance is commonly partial. Although the pathogen is able to complete its life cycle on a partially resistant host, its growth and reproduction are impaired, to a greater or lesser extent, relative to that on a fully susceptible cultivar. Race-nonspecific resistance is often polygenically inherited in the host, and in such cases it is difficult, even with modern biological approaches, to dissect out the individual host genes involved in order to characterize their functional contributions. Because cultivars with partial resistance to one pathogen species can be fully susceptible to a second pathogenic species, it is possible that some of the genes may be involved in race-nonspecific recognition of individual pathogenic species. However, the molecular mechanisms by which expression of partial resistance can be triggered in a race-nonspecific, yet pathogen-specific manner remain to be determined.

Race-specific and race-nonspecific resistance also differ in their utility to agriculture [33]. In theory, race-specific resistance would appear to be superior to race-nonspecific resistance, because it often has a greater capacity to restrict the pathogen. In practice, however, race-specific resistance has often been found to be of limited utility for disease control [33]. The high degree of genetic variability and rapid reproductive rates in many pathogen populations render race-specific resistance vulnerable to the selection of new pathogenic races of the pathogen not recognized by the R gene(s) present in a particular cultivar. If the cultivar lacks race-nonspecific resistance, it may be highly susceptible to these new pathogen races [33]. Race-specific resistance is particularly short-lived for fungal pathogens which colonize living host cells during all or part of their development, such as rusts and powdery mildews, and the potato late blight fungus, *Phytophthora infestans*, respectively. In these cases, breakdown of plant resistance may result from the intense selection pressure race-specific resistance exerts on such pathogens. Race-specific resistance can be usefully deployed against those relatively fewer pathogens with limited genetic variability and slow rates of reproduction. For such pathogens, race-specific resistance can provide relatively durable disease control.

Race-nonspecific, partial resistance reduces to a greater or lesser extent the rate at which pathogens grow and reproduce in plant tissues and hence delays development of disease epidemics. But such resistance often needs to be supplemented with other measures to suppress disease development to levels which do not cause economic losses [33]. However, the complex polygenic basis of race-nonspecific resistance, in most instances, renders it far less vulnerable to the development of pathogenic races which might overcome it. Also, partial resistance confers less selective advantage on strains with increased virulence, because the less virulent strains are also able to reproduce and contribute genetic material to the pool. Because of this durability, race-nonspecific resistance is an important tool for the management of diseases caused by pathogens that readily overcome race-specific resistance [33].

III. MECHANISMS FOR PLANT DISEASE RESISTANCE

A. Hypersensitive Response

Two salient features of the HR are rapid host cell death and severe restriction of pathogen development. The relationship between these aspects of the HR, and the mechanisms by which they are mediated, are the subjects of intense research and debate. The generation of reactive oxygen species (ROS) by the host with resultant peroxidation of host membrane lipids is associated with, and may have a causal role in, the HR [1,19,23,26,27,50,52,84]. Because ROS and products of lipid peroxidation are toxic to microorganisms as well as plant cells, these materials may also contribute to pathogen restriction by the HR. Evidence from studies using metabolic inhibitors indicates that hypersensitive cell death is a metabolic process which requires signal transduction, gene expression, de novo synthesis of proteins, and energy production and utilization. The HR has been suggested to be a form of programmed cell death, or apoptosis, which regulates cell survival in animal systems [38,94]. Most pathogens against which the HR is typically expressed require living host cells during at least the early portion, if not the entirety, of their interaction with plants. It is easy to understand how host cell death might prevent the further development of these pathogens. However, a wide variety of other putative defense mechanisms are expressed during HR development and in plant-pathogen interactions which result in a lesser degree of pathogen restriction, as in many cases of race-nonspecific resistance.

B. Cell Wall Modifications and Papillae

Plant cell walls can be modified in response to pathogens in ways that retard colonization by pathogens [91,118]. Various shikimate-derived phenolic compounds, including flavanols [110] and acetate malonate–derived fatty acid derivatives, can be incorporated into cell walls (lignin and lignin-like materials [54,91,118]) or deposited as a coating upon the surface of cell walls (suberin [78,86]). These chemical modifications may limit development of pathogens by increasing wall resistance to hydrolysis by the pathogen's enzymes [92,111] and possibly by restricting diffusion of water and nutrients to pathogens. Peroxidases participate in lignin formation and cross linking of phenols to plant walls by oxidatively polymerizing phenolics in the presence of plant-generated hydrogen peroxide. Oxidative cross linking of plant cell wall constituents can occur within minutes of the recognition of incompatible pathogens [13]. Phenolic radicals produced as intermediates during the oxidative polymerization of phenolics have direct antimicrobial activity in vitro [117], and they may inhibit pathogen development in vivo. Modification of cell walls by enrichment with hydroxyproline-rich glycoproteins (HRGPs) during plant-pathogen interactions has also been suggested to contribute to pathogen restriction [30]. The HRGPs are resistant to hydrolysis by proteases and carbohydrases. Papillae formed in host cells by the deposition of glucan polymers (e.g., callose) can also serve as physical barriers to fungal penetration and colonization of plant tissues [15].

C. Antimicrobial Proteins

A variety of proteins with antimicrobial activity are produced by plants in response to infection. Some of these "pathogenesis-related" (PR) proteins [107] are enzymes, such as chitinases and beta-1,3-glucanases, which can inhibit growth of fungal pathogens by degrading their cell walls [107]. Further, the fungal wall fragments released by the action of these enzymes may function as elicitors (nonspecific) of other plant defense mechanisms [134]. Certain chitinases degrade the walls of bacterial cells [107]. Other of the PR proteins have no known enzymatic activity but are associated with resistance to disease and function by mechanisms which remain to be identified. Among these are osmotin [67] and a protein designed as PR-1a [3]. The potential contribution of the PR proteins to plant disease resistance is suggested by studies in which their constitutive expression in transgenic plants has sometimes resulted in enhanced resistance to pathogens, usually seen as a delay in symptom development [3,8,14,46].

D. Low Molecular Weight Antimicrobial Compounds

Higher plants also produce many different low molecular weight compounds that have direct antimicrobial activity and may participate in disease resistance. The term *phytoanticipins* has been proposed to designate those low molecular weight antimicrobial compounds that are present in plants prior to infection (they are constitutive [124]), whereas the term *phytoalexins* designates low molecular weight antimicrobial compounds produced in response to infection (they are induced [83,124]). Synthesis and accumulation of phytoalexins, however, can also be elicited by numerous abiotic stresses. Some phytoanticipins resemble phytoalexins in that their content in plants is markedly increased owing to de novo synthesis following infection [124]. The chemical structures of phytoanticipins and phytoalexins vary between plant species, and their modes of action against pathogens also vary [57,83].

The contribution of phytoanticipins and phytoalexins to disease resistance is supported by an extensive literature [57]. Recently, molecular biological studies lend additional support for the role of these compounds in resistance: mutant plants, impaired in their ability to synthesize the compounds, exhibit increased susceptibility to pathogens [36,70]. Some pathogens produce enzymes that degrade and detoxify the phytoanticipins and phytoalexins produced by the host plant species to which the pathogen is adapted [34,123]. However, some nonpathogens of a host also produce such enzymes and some pathogens retain their ability to cause disease without producing such enzymes [122].

IV. RELATIONSHIP BETWEEN RESISTANCE MECHANISMS AND DISEASE RESISTANCE

Resistant and susceptible cultivars of plants are distinguished not by the presence or absence of genes that encode resistance mechanisms (for these are present in all cultivars of a species) but by the rapidity and magnitude with which resistance mechanisms are expressed following infection. The diverse resistance mechanisms inherent to each plant species are often expressed in a coordinated manner in response to infection. Individual defense mechanisms may exert only a modest inhibitory effect on a pathogen. It is, however, possible that the combined expression of multiple mechanisms, acting additively or synergistically, can be sufficient to restrict pathogen growth and reproduction significantly if these are expressed early enough in the plant-pathogen interaction.

The timing of expression of resistance mechanisms is likely a key determinant of their efficacy in limiting pathogens. For example, race-specific resistance is often characterized by the rapid recognition of pathogens and the expression, at early stages of the plant-pathogen interaction, of a series of defense responses, including those that result in the death of those host cells which are interacting directly with the pathogen (e.g., due to generation of ROS) and the development in neighboring uninfected cells of physiochemical barriers to further pathogen colonization (e.g., lignification and accumulation of low and high molecular weight antimicrobial agents such as the phytoalexins and PR proteins). In many cases of race-nonspecific resistance, host recognition of and response to pathogens appear to occur later in the host-pathogen interaction than in most race-specific interactions. This delay results in a lower degree of pathogen restriction (i.e., partial resistance), although a similar or identical array of defensive mechanisms are deployed. Although resistance mechanisms are also expressed during the interaction of susceptible plants with pathogens, they are expressed at levels too low to inhibit the pathogen until the disease is well developed. However, it is likely that in the absence of this delayed expression of resistance, plant disease would be even more severe in susceptible plants. The delayed response may be part of an evolutionary pattern for survival and coexistence of plant and pathogen.

V. ENVIRONMENTAL EFFECTS ON THE EXPRESSION OF DISEASE RESISTANCE

Plant disease results from the interaction of a susceptible host with a virulent pathogen in a conducive environment (see Chapter 10). The influence of environmental factors on disease development has been studied most intensively with respect to their effects on the initiation of infections by pathogens [20,21]. For example, it is well known that infection of plants by many fungal and bacterial pathogens is favored by the presence of free water on the leaf surface (due to precipitation or dew) and by temperatures conducive to pathogen development.

Environmental variables can also influence the expression of disease resistance by plants. This is probably of greater significance for partial resistance in which defensive mechanisms have a more moderate effect on pathogen development than does the HR. In partial resistance, the balance between host and pathogen may be more sensitive to environmental variables that alter the host's ability to mobilize energy-requiring resistance mechanisms. For example, because green plants obtain energy for disease resistance and other metabolic processes via photosynthesis, light intensity can have a marked effect on the outcome of plant-pathogen interactions. In experiments controlled for other variables, shading increased the susceptibility of moderately (and race-nonspecifically) resistant potato cultivars to the late blight fungus, *Phytophthora infestans*, but had no effect on disease development in a cultivar that was already highly susceptible under normal light levels (i.e., lacked partial resistance [98,125]). Interestingly, potato cultivars differed in the degree to which their race-nonspecific resistance was diminished by low light levels. Pennypacker et al. [87] found that an alfalfa clone's partial, race-nonspecific resistance to the wilt fungus, *Verticillium albo-atrum*, was lost when light intensity was reduced to 40% of controls in a growth chamber study. An alfalfa clone lacking resistance to the pathogen did not become more susceptible to the pathogen at the lower light level. It was suggested that resistance loss resulted from observed reduction in carbon assimilation that may have diminished the capacity of plants to express carbon- and energy-intensive resistance mechanisms [87], including perhaps the coating of vessel cell walls with suberin [78,86] and the accumulation of phytoalexins [64]. We have observed that development of the anthracnose disease caused by the fungus *Colletotrichum lagenarium* is more severe on cucumbers with varying degrees of race-nonspecific resistance grown in the greenhouse under seasonally low light levels compared with high light levels (our unpublished observations). Besides low light levels, a variety of other stressful conditions may diminish plant resistance to disease, probably by reducing the plant's vigor and ability to mobilize energy-requiring defenses against pathogen attack. Examples include drought [5,21] and air pollutants [5]. It is important to note that stresses may differentially affect development of different diseases. In general, such stresses may enhance the development of diseases caused by nectrotrophic pathogens, which do not require living host cells, and suppress the development of diseases caused by biotrophic pathogens, which thrive on otherwise healthy plant tissues [5].

VI. NOVEL STRATEGIES FOR ENHANCEMENT OF PLANT DISEASE RESISTANCE

A. Systemic Induced Resistance to Disease

The infection of plants by necrotizing pathogens can induce a systemic enhancement of disease resistance that persists for weeks or months and is expressed against a broad spectrum of pathogens, including fungi, bacteria, and necrotizing viruses [56,58,69,93]. This systemic induced resistance (SIR) is expressed as a reduction in the number and/or size of lesions resulting from challenge inoculations with pathogens. Disease symptoms are commonly reduced by 50–90% relative to noninduced controls. SIR has been demonstrated in over 20 crop species and under both greenhouse and field conditions [18,58]. Besides pathogens, SIR can be triggered by select nonpathogens and chemicals [53,73,76,130]. In cucumber plants, the anthracnose fungus, *Colletotrichum lagenarium*, induces resistance not only to itself but to at least 12 other pathogens [56]. This broad-spectrum activity of SIR stands in sharp contrast to the previously described types of resistance, which are only active against particular races or species of pathogens. This contrast may be of immense practical value, because crop plants may be infected by several pathogens in a typical agricultural setting. Thus, the physiological and molecular bases of SIR are under intense investigation. SIR triggered by necrotizing pathogens is characterized by systemic increases in peroxidase activity and the systemic accumulation of PR proteins (see Section III.C above) prior to challenge inoculation with pathogens [41,81,129,132,133]. Following challenge inoculation, peroxidase activity and PR protein content increase still further, and a number of other defense mechanisms (which vary with the host species) are expressed with greater rapidity and magnitude than in noninduced plants responding to the same pathogen. For example, induced-resistant cucumber tissues respond to pathogen challenge or elicitor treatment with rapid and intense expression of phenolic cell wall modifications (involving lignin and lignin-like materials), papilla formation, increases in chitinase activity, and production of reactive oxygen species [25,40,55,104,112,131]. Rapid expression of resistance to diverse pathogens in SIR suggests that the phenomenon entails broad changes (at the regulatory level?) in the plant's ability to perceive and respond to pathogens.

In addition to mechanisms which may directly limit pathogen development in plants, SIR may also be mediated by physiological changes that protect plants against necrosis per se. Cucumber and tobacco plants treated with a variety of SIR inducers exhibit heightened resistance to damage by the herbicide paraquat [109] and by cupric chloride (our unpublished observations). These compounds harm plant tissues by promoting generation of reactive oxygen species and peroxidation of membrane lipids. Oxidative damage appears to have a causal role in the necrotization of tissues by pathogens [1,23,27,84]. Thus, enhanced resistance to paraquat damage may indicate the enhancement in SIR of defense mechanisms that participate in protection from the damage caused by necrotizing pathogens but do not necessarily inhibit pathogen development directly. The mechanisms by which SIR protects plants from paraquat and copper damage are unknown, but enhancement of the activity of enzymes which protect against oxidative damage (superoxide dismutase, catalase, ascorbate peroxidase, and glutathione reductase) was not detected in induced-resistant tissues. The systemic increases in cytokinin content that occur in SIR [95,114] may function to protect plants from necrosis-causing chemicals and pathogens. Exogenous cytokinins can increase plant resistance to necrotizing pathogens [75,79] and the prooxidant air pollutant ozone [82,116] and delay senescence of plant tissues [65].

Although SIR appears to be a promising form of resistance, we have encountered significant variability in the efficacy of treatments for triggering SIR, particularly under field conditions. The sources and mechanisms of this variability are unknown, but environmental stresses (e.g., low light intensity, drought) may contribute to the problem. Methods for reducing this variability or for precise quantification of SIR's relationship to environmental factors may be essential to the practical application of SIR. The molecular basis of SIR is being actively investigated because of its possible direct applications for disease management, and because it can guide the development of other novel forms of disease resistance through genetic engineering.

B. Genetic Engineering of Plant Disease Resistance

The advent of transgenic techniques has made possible numerous novel approaches for the enhancement of plant disease resistance [22,62,89,108]. Typically, these approaches involve constitutive or pathogen-inducible expression in transgenic plants of plant-, pathogen-, or animal-derived genes encoding proteins or polypeptides which may have direct antimicrobial activity or which may catalyze the biosynthesis of other antimicrobial substances. We present here a brief sketch of some of the approaches that have been taken together with comments on their possible advantages and drawbacks and suggested directions for future research.

1. Plant-Derived Genes

Pathogenesis-Related (PR) Proteins. Systemic accumulation of PR proteins is consistently associated with SIR, and it is thought that these proteins may have a role in the expression of the broad-spectrum resistance conferred by SIR. The various PR proteins can inhibit pathogen growth in vitro by digestion of the pathogen's cell walls (as in the case of some isoforms of the enzymes chitinase and beta-1,3-glucanase) or by as yet unknown mechanisms (e.g., osmotin and PR-1a). Expression of a soybean-derived glucanase in tobacco confers foliar resistance to the black shank fungus, *Phytophthora parasitica* var. *nicotianae* [134]. This pathogen contains beta-1,3-glucan but no chitin in its walls. Yoshikawa and coworkers [134] presented evidence that the genetically engineered resistance resulted not from direct inhibition of the pathogen by the enzyme, but from the elicitation of other plant defenses by fungal wall fragments (nonspecific elicitors) released by the enzyme's action. In contrast, constitutive expression of various beta-1,3-glucanase isozymes derived from tobacco did not protect tobacco plants from *P. parasitica* var. *nicotianae* or from a second oomycetous pathogen, *Peronospora tabacina* [3], although both fungi have beta-1,3-glucans in their cell walls. For pathogens whose cell walls contain both chitin and glucan, expression of either chitinase or glucanase alone has often provided only modest or no protection [8,14,46,66,77], whereas coexpression of chitinase and beta-1,3-glucanase has provided considerable protection from these fungi [136]. These findings are consistent with observations that chitinases and beta-1,3-glucanases individually have little inhibitory effect on in vitro development of fungi whose walls contain both polymers, but when combined have synergistic inhibitory activity toward these fungi [72].

Expression of the PR protein osmotin delayed the development of disease symptoms caused by the late blight fungus, *Phytophthora infestans*, in transgenic potato, but did not affect symptom development in transgenic tobacco inoculated with the black shank fungus [67]. Purified osmotin was more inhibitory to *P. infestans* than to the black shank fungus in vitro [67]. Expression of the PR protein-1a in transgenic tobacco resulted in reductions and delays in symptom development caused by two oomycetous pathogens, *P. parasitica* var. *nicotianae* and *P. tabacina* [3].

Because pathogens appear to vary with respect to their in vitro and in vivo sensitivity to the various PR proteins and their isozymes [3,66,99], it seems unlikely that constitutive expression of one or a few PR proteins will confer the broad-spectrum disease resistance associated with SIR. However, results obtained in some host–PR protein–pathogen combinations have been sufficiently encouraging to suggest that constitutive expression of PR proteins, particularly in combinations, may have utility as an aid for plant disease management.

Thionins. Thionins are cysteine-rich polypeptides of plant origin that have antimicrobial activity in vitro. Constitutive expression of a thionin gene from barley enhanced resistance of transgenic tobacco plants to two bacterial pathogens, *Pseudomonas syringae* pv. *tabaci* and *P. syringae* pv. *syringae* [17]. The spectrum of pathogens against which thionins may be active in transgenic plants and the durability of the resistance they confer remain to be determined.

Peroxidases. Peroxidases are enzymes which catalyze the oxidative polymerization of phenolics forming products such as lignin and antimicrobial radicals. Peroxidases can also generate hydrogen peroxide, an antimicrobial compound, from NADH and NADPH [85]. Because increases in the activity of peroxidases are associated with the expression of conventionally derived and induced resistance, it was considered possible that overexpression of peroxidases in transgenic plants would heighten their resistance to pathogens. However, this approach has not led to increases in disease resistance [61] (J.A. Kuć et al., unpublished observations), and undesirable side effects on plant growth and development have been observed [60]. The failure of overexpressed peroxidases to enhance disease resistance may be due to a number of factors, including an insufficient supply of phenolics and hydrogen peroxide at critical times in plant-pathogen interactions. It may also indicate that peroxidase-mediated defense mechanisms are alone insufficient to confer resistance, or that peroxidases do not have a major role in disease resistance.

Phytoalexin Biosynthetic Enzymes. Because phytoalexins can be toxic not only to pathogens but also to plant cells (and, potentially, to the consumers of plant products), constitutive expression of phytoalexins is probably not a viable option. Many phytoalexins require multiple enzymes for their synthesis [57,83], and the transfer and appropriate expression of such complex biosynthetic pathways is likely to be difficult to achieve. However, the transfer to tobacco of the gene for stilbene synthase from grape enabled tobacco to synthesize the phytoalexin resveratrol and conferred enhanced (partial) resistance to *Botrytis cinerea*, a fungal pathogen [39]. Success in this case was made possible by several factors. Resveratrol is synthesized from precursors which are commonly found in plants, including those which do not normally produce this phytoalexin, and resveratrol synthesis was accomplished in tobacco by one enzyme, stilbene synthase, which is encoded by a single gene. Further, the gene for this enzyme was not constitutively expressed but was under control of its native pathogen-inducible promoter, with the consequence that the phytoalexin was produced only at infection sites. Also, resveratrol was produced early in the transgenic plant-pathogen interaction, a totally unexpected result that may be the key to resistance in this case, particularly if other defense mechanisms also show a pattern of unusually early expression in these transgenic plants. Whether resveratrol can protect plants from a broad spectrum of pathogens and whether pathogens can readily develop mechanisms for detoxification of the phytoalexin remain to be determined.

Antiviral Proteins. Constitutive expression of a gene encoding an antiviral protein from pokeweed protected tobacco and potato plants from three unrelated viruses, potato X (potexvirus), potato Y (potyvirus), and cucumber mosaic (cucumovirus) [68]. If resistance achieved by this or related means proves to be durable, it may be of major benefit for the management of viral diseases, for which no effective pesticides are currently available. Other antiviral proteins have also been identified in plants (e.g., see ref. 28) and may find similar utility.

2. Pathogen-Derived Genes

Plant pathogens themselves are being considered sources of genes encoding products that may enhance disease resistance. Transformation of plants to constitutively express genes derived from viral pathogens (e.g., those encoding viral coat proteins or replicases) often results in a substantial enhancement of resistance to those viruses from which the genes were cloned and sometimes to related viruses [47,96,113,135]. Whether or not such pathogen-derived viral resistance will be durable or will be overcome by selection of viral variants unaffected by it remains to be determined. Transgenic manipulation of pathogen genes encoding elicitors of the HR (avr genes) holds potential for broadening the utility of the HR for disease management by directing it to be expressed in a manner that is nonspecific with respect to pathogenic race [22,126]. As previously noted, the HR is a highly effective mechanism for restricting pathogen development, but only for those races which trigger its expression. Broader utility of the HR could theoretically be achieved by transforming plants with chimeric genes consisting of structural portions of pathogen avr genes placed under the regulation of promoters from other plant genes, the expression of which is elicited by pathogens in a race-nonspecific manner [126]. In plants which also contain R genes encoding factors for recognition of these heretofore race-specific elicitors, pathogens which do not ordinarily trigger the HR would elicit this powerful defense mechanism. Provided that the scheme will work in practice, it is questionable whether it will confer resistance to pathogens that do not require living host cells. Also, because many putative disease resistance mechanisms can be elicited by wounding, chemical treatments, and other stresses, there may be some difficulty in finding promoter regions that respond only to pathogens.

Some pathogens produce toxins that play an important role in disease development by damaging plant cells. Often the pathogens which produce these toxins are unaffected by them. A bacterial pathogen of bean plants, *Pseudomonas syringae* pv. *phaseolicola*, produces phaseolotoxin, which inhibits the bean enzyme ornithyl transcarbamylase, but does not inhibit the same enzyme in the bacterium [35]. Transgenic bean plants expressing the bacterial enzyme are insensitive to the toxin and respond to the pathogen by forming small necrotic lesions resembling the HR [35]. This approach can apparently provide highly effective disease resistance, but only to specific pathogens.

3. Animal-Derived Genes

Antimicrobial polypeptide and protein components of animal immune systems are also being investigated for their potential to increase disease resistance of transgenic plants in which they are expressed. Constitutive expression of an insect-derived gene encoding the polypeptide cercopin increases the resistance of transgenic tobacco plants to the wilt bacterium, *Pseudomonas solanacearum* [48]. However, cercopin expression did not increase resistance of tobacco to a second bacterial pathogen, *Pseudomonas syringae* pv. *tabaci* [45]. Antibodies encoded by animal-derived genes are also being studied, but these can be expected to confer resistance in a highly pathogen-specific manner, as they do in animal systems.

4. Other Approaches Suggested by Systemic Induced Resistance

As noted previously, plants with SIR respond more quickly and effectively to the presence of diverse pathogens and are less damaged by the prooxidant herbicide paraquat and metal oxidants such as Cu^{2+} as compared with noninduced plants. Identification of the molecular bases of these phenomena and the subsequent development of transgenic plants which mimic induced-resistant plants may result in crops with high levels of resistance to a broad spectrum of pathogens. Enhanced resistance to environmental stresses is another possible benefit of this approach. A paraquat-resistant tobacco line, selected by in vitro exposure of calli to the herbicide, is cross resistant to necrotizing pathogens, a pathogen-produced toxin (fusaric acid), heat and cold [7], and sulfur dioxide [115]. This startling finding is consistent with the emerging view that harm caused to plants by necrotizing pathogens and many environmental stresses is mediated in part by oxidative damage [7,101]. The multiple resistances of the aforementioned tobacco line are apparently due to the enhanced activities of enzymes which protect against oxidative damage [115]. Increasing the activities of such enzymes, notably superoxide dismutase and glutathione reductase, by transgenic means can heighten plant resistance to damage caused by paraquat and environmental stresses [4,100]. Logic suggests that these transgenic plants would also exhibit increased resistance to necrotizing pathogens. If so, these enzymes may serve as useful complements to other agents which enhance disease resistance through direct antimicrobial activity. Increased expression of antioxidant-related enzymes and other cellular protectants could potentially ameliorate the suppression of antimicrobial defenses by environmental stresses.

C. Perspectives on Novel Resistance Strategies

The ideal resistance strategy would be one that affords durable plant protection of sufficient magnitude and reliability that it substantially reduces or eliminates the need for pesticides to manage the multiple pathogens which may cause disease in a plant. It appears unlikely that any single novel resistance strategy discussed above can meet these criteria. Systemic induced resistance typically affords a high level of protection from diverse pathogens and is likely to be durable with respect to the selection of pathogen isolates that can overcome it. However, SIR is ineffective or marginally effective against systemic viral infections and can vary in its performance, especially under field conditions. Most of the remaining strategies discussed are likely to exhibit a greater degree of pathogen specificity than SIR, and some would be highly pathogen specific. The efficacy of most strategies involving transgenics would also likely be subject to environmental variability. Consider the case of transgenic tobacco expressing a soybean beta-1,3-glucanase and resultant resistance to the black shank fungus [134]. Other plant defenses (such as phytoalexin accumulation) triggered by hyphal wall fragments (elicitors) released by the enzyme are thought likely to mediate this resistance [134]. The expression of these other mechanisms and thus the resistance of the transgenic plants is likely to be conditioned by environmental factors. It is probable that resistance mechanisms inherent to parental plants contribute substantially to pathogen restriction in other transgenic plants, the resistance of which would also be conditioned by environmental factors. The highest degree and broadest spectrum of resistance obtainable will likely result from combinations of resistance factors; for example, SIR or transgenic mimics of SIR (perhaps plants expressing multiple PR proteins and antioxidant-related enzymes) together with antiviral gene products. However, no degree of intellectualizing about the possible additive or synergistic properties of resistance mechanisms can substitute for the testing of induced-resistant and transgenic plants under field conditions to evaluate fully their potential to express resistance and other beneficial agronomic qualities. Present evidence suggests that the most durable types of novel resistance will be partial but may be sufficient to reduce the amount of pesticides required for practical disease management (as is the case for some conventionally derived durable resistance).

VII. EPIDEMIOLOGICAL PRINCIPLES AND THE APPLICATION OF RESISTANCE FOR INTEGRATED MANAGEMENT OF PLANT DISEASE

Modern integrated management of plant diseases depends largely on the understanding and application of epidemiology. This branch of plant pathology relates the interactions of plants, pathogens, and environmental factors through time to the degree of disease development and yield losses, and can thereby guide the making of decisions regarding deployment of disease management strategies. An in-depth discussion of the relation of epidemiology to plant disease management is beyond the scope of this chapter, and the two following paragraphs are a summary of information and ideas derived from other more detailed treatments of the topic, to which the interested reader is referred [10,16,33,120,121].

For epidemiological purposes, pathogens and the diseases they cause can be placed into two categories based on the number of reproductive or disease cycles that may potentially occur within each cropping season. Monocyclic pathogens infect and reproduce upon a host just once per season, and so diseases caused by such pathogens also occur with this frequency. Inoculum produced on a host plant by a monocyclic pathogen does not disperse to and infect other plants within the same season, but survives to infect plants in a subsequent season. In contrast, polycyclic pathogens are those which can undergo multiple cycles of infection and reproduction on their host plants in each growing season. Polycyclic pathogens typically produce large amounts of inoculum in relatively short periods of time, which is dispersed to other plants (or other parts of the same plant) that can support another cycle of disease development. Under conditions conducive to disease development (i.e., presence of a susceptible host and a favorable environment), polycyclic pathogens and the diseases they cause can increase exponentially, resulting in severe crop devastation. Both the Irish potato famine and the Bengali rice famine resulted from crop losses caused by polycyclic pathogens in years when environmental factors were highly conducive for, and hosts highly susceptible to, disease development. We note that distinctions between monocyclic and polycylic habits are general rather than absolute. For example, certain nematodes are polycyclic in southern latitudes and monocyclic in northern latitudes. Furthermore, the number of disease and reproductive cycles of these nematodes is dependent on crop susceptibility as well as environmental conditions.

The infective and reproductive patterns of monocyclic and polycyclic pathogens determine which strategies are appropriate for their management. Diseases caused by monocyclic pathogens are most effectively suppressed by means which reduce the amount and/or efficacy of initial inoculum (that present at the beginning of the cropping season). In contrast, reducing the amount or efficacy of initial inoculum may not substantially control a pathogen that undergoes many infection cycles. Thus, diseases caused by polycyclic pathogens are most effectively managed by measures which reduce the rate of increase of the pathogen and disease over the season. For example, diseases caused by monocyclic pathogens such as some soilborne fungi and nematodes are best managed by crop rotation and soil treatments (often with chemicals), which reduce the amount of initial inoculum, and by host resistance, which limits the efficacy of that inoculum. The goal is to reduce both the number of infected plants and the extent of damage (yield loss) occurring in those plants which do become infected. Race-specific resistance can often be useful for the management of diseases caused by such monocyclic pathogens owing to their slow generation times and low genetic variability relative to polycyclic pathogens. Because pathogens which cause polycyclic diseases have high rates of reproduction and are typically genetically diverse, race-specific resistance is often of less value for their management. Although the effects of partial resistance on a single cycle of pathogen and disease development may be modest, these partial effects are amplified over the course of multiple disease cycles such that they may have a major impact on the rate of disease increase over the cropping season and thus on crop yields. However, the effects of partial resistance alone are often insufficient to suppress disease development to acceptable levels, particularly under environmental conditions highly favorable for disease development. In these cases, partial resistance is typically supplemented with cultural practices as well as pesticides (e.g., fungicidal and bactericidal sprays applied to plant foliage). Spraying pesticides on a routine schedule to ensure crop quality and yield has been a commonplace practice for many crops.

However, less intensive pesticide use can in many cases suppress disease development to levels that do not cause economically important losses. Computer models of the precise influences of environmental factors such as moisture and temperature on disease development can guide the grower to use pesticides as they are needed rather than as a routine insurance policy. The flexibility and utility of some disease models are increased in some cases by the availability of systemic pesticides that can be effective when applied up to several days after infections have been initiated, in contrast to protectant materials which are effective only if present on aerial plant surfaces before infection occurs. Computer models of value to integrated pest management (IPM) have been developed for a number of plant diseases, including fire blight [105,106] and scab [11,49] of apple, late blight of potato [31,32,102,103], early blight of tomato [88], early leaf spot of peanut [6], and *Botrytis* blight of onion [127]. The effects of host resistance on disease development have been incorporated into some of these models, enabling still further reductions in pesticide use [31,32,102,103]. The Campbell Soup Company's IPM programs for contract growers of tomatoes, celery, and carrots have enabled growers to reduce pesticide use by an average of 59% [43]. Much of this reduction has resulted from the planting of disease-resistant varieties and from the use of computer models and weather monitoring to guide the timely application of pesticides for disease control [43].

Several problems limit the widespread application of models for integrated management of plant diseases. Among the most important are economic in nature: models are expensive to develop and hence are available for relatively few, high-value crops. Models have not been developed for numerous crops of lesser economic value, for which disease management is also rendered more difficult by limited availability of pesticides and of disease-resistant cultivars. Other problems are biological in nature: crops (major and minor) are typically affected by more than one major pathogen in a given agricultural setting, and models for one disease may not be used to full benefit if other major diseases not encompassed by the model must be controlled by routine spraying [74]. Some progress has been made in the development of models that encompass more than one disease [102,103]. Cultivars which combine desirable agronomic/horticultural characteristics with multiple disease resistance of sufficient degree to permit reductions in pesticide usage are of limited availability for many crops. Novel strategies for enhancing plant disease resistance, such as SIR, may provide higher levels of resistance to multiple pathogens in existing cultivars preferred for their agronomic/horticultural characteristics. This would provide expanded opportunities for integrated management of plant disease.

Although novel strategies for enhancing plant disease resistance appear to hold promise for improving integrated management of plant diseases, much of what we know about these strategies is based upon results of one-time challenges with pathogens (single disease cycles) conducted in controlled environments (greenhouses or growth chambers). Field testing has been conducted on a relatively small scale and in a limited number of locations. For novel resistance strategies to find practical application in an IPM context, they must be evaluated, in controlled and field environments, with respect to (a) magnitude of effect, (b) specificity, (c) reliability in diverse environments, (d) durability, (e) crop productivity and profitability, and (f) safety for the environment [47,90] and human health. Such evaluations will require a renewed commitment of public support for applied agricultural research.

REFERENCES

1. Adam, A., T. Farkas, G. Somlyai, M. Hevesi, and Z. Kiraly. 1989. Consequence of O_2^- generation during a bacterially induced hypersensitive reaction in tobacco: deterioration of membrane lipids. *Physiol. Mol. Plant Pathol.* 34:13–26.
2. Agrios, G.N. 1988. *Plant Pathology*. 3rd ed. San Diego: Academic Press.
3. Alexander, D., R.M. Goodman, M. Gut-Rella, C. Glascock, K. Weymann, L. Friedrich, D. Maddox, P. Ahl-Goy, T. Luntz, E. Ward, and J. Ryals. 1993. Increased tolerance to two oomycete pathogens in transgenic tobacco expressing pathogenesis-related protein 1a. *Proc. Natl. Acad. Sci. USA* 90:7327–7331.
4. Aono, M., A. Kubo, H. Sagi, K. Tanaka, and N. Kondo. 1993. Enhanced tolerance to photooxidative stress of transgenic *Nicotiana tabacum* with high chloroplastic glutathione reductase activity. *Plant Cell Physiol.* 34:129–135.
5. Ayres, P.G. 1984. The interaction between environmental stress injury and biotic disease physiology. *Annu. Rev. Phytopathol.* 22:53–75.
6. Bailey, J.E., G.L. Johnson, and S.J. Toth, Jr. 1994. Evolution of a weather-based peanut leaf spray advisory in North Carolina. *Plant Dis.* 78:530–535.
7. Barna, B., A.L. Adam, and Z. Kiraly. 1993. Juvenility and resistance of a superoxide-tolerant plant to diseases and other stresses. *Naturwissenschaften* 80:420–422.

8. Benhamou, N., J. Broglie, I. Chet, and R. Broglie. 1993. Cytology of infection of 35S-bean chitinase transgenic canola plants by *Rhizoctonia solani*: cytochemical aspects of chitin breakdown *in vivo*. *Plant J.* 4:295–305.

9. Bent, A.F., B.N. Kunkel, D. Dahlbeck, K.L. Brown, R. Schmidt, J. Giraudat, J. Leung, and B.J. Staskawicz. 1994. *RSP2* of *Arabidopsis thaliana*: a leucine-rich repeat class of plant disease resistance genes. *Science* 265:1856–1860.

10. Berger, R.D. 1977. Application of epidemiological principles to achieve plant disease control. *Annu. Rev. Phytopathol.* 15:165–183.

11. Blaise, Ph., P.A. Arneson, and C. Gessler, 1987. APPLESCAB. A teaching aid on microcomputers. *Plant Dis.* 71:574–578.

12. Bonas, U., R.E. Stall, and B.J. Staskawicz. 1989. Genetic and structural characterization of the avirulence gene *avrBs3* from *Xanthomonas vesicatoria* pv. *vesicatoria*. *Mol. Gen. Genet.* 218:127–136.

13. Bradley, D.J., P. Kjellbom, and C.J. Lamb. 1992. Elicitor- and wound-induced oxidative cross-linking of a proline-rich plant cell wall protein: a novel, rapid defense response. *Cell* 70:21–30.

14. Broglie, K., I. Chet, M. Holliday, R. Cressman, P. Biddle, S. Knowlton, J. Mauvais, and R. Biddle. 1991. Transgenic plants with enhanced resistance to the fungal pathogen *Rhizoctonia solani*. *Science* 254:1194–1197.

15. Cadena-Gomez, G., and R.L. Nicholson. 1987. Papilla formation and associated peroxidase activity: a non-specific response to attempted fungal penetration of maize. *Physiol. Mol. Plant Pathol.* 31:51–67.

16. Campbell, C.L., and L.V. Madden. 1990. *Introduction to Plant Disease Epidemiology*. New York: Wiley.

17. Carmona, M.J., A. Molina, J.A. Fernandez, J.J. Lopez-Fando, and F. Garcia-Olmedo. 1993. Expression of the alpha-thionin gene from barley in tobacco confers enhanced resistance to bacterial pathogens. *Plant J.* 3:457–462.

18. Caruso, F.L., and J. Kuć. 1977. Field protection of cucumber, watermelon, and muskmelon against *Colletotrichum lagenarium* by *Colletotrichum lagenarium*. *Phytopathology* 67:1290–1292.

19. Chai, H.B., and N. Doke. 1987. Superoxide anion generation: a response of potato leaves to infection with *Phytophthora infestans*. *Phytopathology* 77:645–649.

20. Colhoun, J. 1973. Effects of environmental factors on plant disease. *Annu. Rev. Phytopathol.* 11:343–364.

21. Cook, R.J., and R.I. Papendick. 1972. Influence of water potential of soils and plants on root disease. *Annu. Rev. Phytopathol.* 10:349–374.

22. Cornelissen, B.J.C., and L.S. Melcher. 1993. Strategies for control of fungal diseases with transgenic plants. *Plant Physiol.* 101:709–712.

23. Croft, K.P.C., C.R. Vorsey, and A.J. Slusarenko. 1990. Mechanism of hypersensitive cell collapse: Correlation of increased lipoxygenase activity with membrane damage in leaves of *Phaseolus vulgaris* (L.) inoculated with an avirulent isolate of *Pseudomonas syringae* pv. *phaseolicola*. *Physiol. Mol. Plant Pathol.* 36:49–62.

24. Dangl, J.L., C. Ritter, M.J. Gibbon, L.A. Mur, R.J. Wood, S. Goss, J. Mansfield, J.D. Taylor, and A. Vivian. 1992. Functional homologs of the

Arabidopsis RPM1 disease resistance gene in bean and pea. *Plant Cell* 4:1359–1369.

25. Dean, R.A., and J. Kuć. 1988. Rapid lignification in response to wounding and infection as a mechanism for induced systemic protection in cucumber. *Physiol. Mol. Plant Pathol.* 31:69–81.

26. Doke, N. 1983. Involvement of superoxide anion generation in the hypersensitive response of potato tuber tissues to infection with an incompatible race of *Phytophthora infestans* and to the hyphal wall components. *Physiol. Mol. Plant Pathol.* 23:345–357.

27. Doke, N., and Y. Ohashi. 1988. Involvement of an O_2^- generating system in the induction of necrotic lesions on tobacco leaves infected with tobacco mosaic virus. *Physiol. Mol. Plant Pathol.* 32:163–175.

28. Edelman, O., N. Ilan, G. Grafi, N. Sher, Y. Stram, D. Novic, N. Tal, I. Sela, and M. Rubenstein. 1990. Two antiviral proteins from tobacco: purification and characterization with monoclonal antibodies to human b-interferon. *Proc. Natl. Acad. Sci. USA* 87:588–592.

29. Ellingboe, A.H. 1976. Genetics of host-parasite interactions. In *Physiological Plant Pathology*. ed. R. Heietefuss and P.H. Williams, eds. Berlin: Springer-Verlag, pp 761–778.

30. Esquerre-Tugaye, M.T., C. Lafitte, D. Mazau, A. Toppan, and A. Touze. 1979. Cell surfaces in plant-microorganism interactions. II. Evidence for the accumulation of hydroxyproline-rich glycoproteins in the cell wall of diseased plants as a defense mechanism. *Plant Physiol.* 64:320–326.

31. Fry, W.E. 1977. Integrated control of potato late blight: Effects of polygenic resistance and techniques of timing fungicide applications. *Phytopathology* 67:415–420.

32. Fry, W.E. 1978. Quantification of general resistance of potato cultivars and fungicide effects for integrated control of potato late blight. *Phytopathology* 68:1650–1655.

33. Fry, W.E. 1982. *Principles of Plant Disease Management.* New York: Academic Press.

34. Fry. W.E., and P.H. Evans. 1977. Association of formamide hydrolase with fungal pathogenicity to cyanogenic plants. *Phytopathology* 67:494–500.

35. Fuente-Martinez, J.M. de la, G. Mosqueda-Cano, A. Alvarez-Morales, and L. Herrera-Estrella. 1992. Expression of a bacterial phaseolotoxin-resistant ornithyl transcarbamylase in transgenic tobacco confers resistance to *Pseudomonas syringae* pv. *phaseolicola*. *Bio/Technology* 10:905–909.

36. Glazebrook, J., and F.M. Ausubel. 1994. Isolation of phytoalexin-deficient mutants of *Arabidopsis thaliana* and characterization of their interactions with bacterial pathogens. *Proc. Natl. Acad. Sci. USA* 91:8955–8959.

37. Goodman, R.N., and A.J. Novacky. 1994. *The Hypersensitive Reaction in Plants to Pathogens: A Resistance Phenomenon.* St. Paul, Minnesota: APS Press.

38. Greenberg, J.T., A. Guo, D.F. Klessig, and F.M. Ausubel. 1994. Programmed cell death in plants: a pathogen-triggered response activated coordinately with multiple defense mechanisms. *Cell* 77:551–563.

39. Hain, R., H.J. Reif, E. Krause, R. Langebartels, H. Kindl, B. Vornam, W. Wiese, S. Schmelzer, P.H. Schreier, D.H. Stocker, and K. Stenzel. 1993. Disease resistance results from foreign phytoalexin expression in a novel plant. *Nature* 361:153–156.

40. Hammerschmidt, R., and J. Kuć. 1982. Lignification as a mechanism for induced systemic resistance in cucumber. *Physiol. Mol. Plant Pathol.* 20:61–71.

41. Hammerschmidt, R., E.M. Nuckles, and J. Kuć. 1982. Association of enhanced peroxidase activity with induced resistance of cucumber to *Colletotrichum lagenarium. Physiol. Mol. Plant Pathol.* 20:73–82.

42. Hampson, M.C. 1992. Some thoughts on demography of the great potato famine. *Plant Dis.* 76:1284–1286.

43. Hasan, A.B., and W.R. Reinert. 1994. Developing and implementing IPM strategies to assist farmers: an industry approach. *Plant Dis.* 78:545–550.

44. Heath, M.C. 1981. A generalized concept of host-parasite specificity. *Phytopathology* 71:1121–1123.

45. Hightower, R., C. Baden, E. Penzes, and P. Dunsmuir. 1994. The expression of cercopin peptide in transgenic tobacco does not confer resistance to *Pseudomonas syringae* pv. *tabaci. Plant Cell Rep.* 13:295–299.

46. Howie, W., L. Joe, E. Newbigin, T. Suslow, and P. Dunsmuir. 1994. Transgenic tobacco plants which express the *chiA* gene from *Serratia marcescens* have enhanced tolerance to *Rhizoctonia solani. Transgen. Res.* 3:90–98.

47. Hull, R. 1994. Resistance to plant viruses: obtaining genes by non-conventional means. *Euphytica* 75:195–205.

48. Jaynes, J.M., P. Nagpala, L. Destefano-Beltran, J.H. Huang, J.H. Kim, T. Denny, and S. Cetiner. 1993. Expression of a cercopin B lytic peptide analog in transgenic tobacco confers enhanced resistance to bacterial wilt caused by *Pseudomonas solanacearum. Plant Sci.* 89:43–53.

49. Jones, A.L., S.L. Lillevik, P.D. Fisher, and T.C. Stebbins. 1980. A microcomputer-based instrument to predict primary apple scab infection periods. *Plant Dis.* 64:69–72.

50. Kato, S., and T. Misawa. 1976. Lipid peroxidation during the appearance of hypersensitive reaction in cowpea leaves infected with cucumber mosaic virus. *Ann. Phytopathol. Soc. Jpn.* 42:472–480.

51. Keen, N.T. 1990. Gene-for-gene complementarity in plant-pathogen interactions. *Annu. Rev. Genet.* 24:447–463.

52. Keppler, L.D., and A. Novacky. 1987. The initiation of membrane lipid peroxidation during bacteria-induced hypersensitive reaction. *Physiol. Mol. Plant Pathol.* 30:233–245.

53. Kessman, H., T. Staub, C. Hoffman, T. Maetzke, J. Herzog, E. Ward, S. Uknes, and J. Ryals. 1994. Induction of systemic acquired disease resistance in plants by chemicals. *Annu. Rev. Phytopathol.* 32:439–459.

54. Kneusel, R.E., U. Matern and K. Nicolay. 1989. Formation of *trans*-caffeoyl CoA from *trans*-4-coumaryl CoA by ZN^{++} dependent enzymes in cultured plant cells and its activation by an elicitor-induced pH shift. *Arch. Biochem. Biophys.* 269:455–462.

55. Kovats, K., A. Binder, and H.R. Hohl. 1991. Cytology of induced systemic resistance of cucumber to *Colletotrichum lagenarium*. *Mycopathologia* 183:484–490.
56. Kuć, J. 1987. Plant immunization and its applicability for disease control. In *Innovative Approaches to Plant Disease Control*. I. Chet, ed. New York: Wiley, pp. 257–274.
57. Kuć, J. 1995. Phytoalexins, stress metabolism and disease resistance. *Annu. Rev. Phytopathol.* 33:275–297.
58. Kuć, J., and N.E. Strobel. 1992. Induced resistance using pathogens and nonpathogens. In *Biological Control of Plant Diseases*. E.C. Tjamos, G.C. Papavizas, and R.J. Cook, eds. New York: Plenum Press, pp 295–303.
59. Kunkel, B.N., A.F. Bent, D. Dahlbeck, R.W. Innes, and B.J. Staskawicz. 1993. *RSP2*, an *Arabidopsis* disease resistance locus specifying recognition of *Pseudomonas syringae* strains expressing the avirulence gene *avrRpt2*. *Plant Cell* 5:865–875.
60. Lagrimini, L.M., S. Bradford, and S. Rothstein. 1990. Peroxidase-induced wilting in transgenic tobacco plants. *Plant Cell* 2:7–18.
61. Lagrimini, L.M., J. Vaughn, W.A. Erb, and S.A. Miller. 1993. Peroxidase overproduction in tomato: wound-induced polyphenol deposition and disease resistance. *HortScience* 28:218–221.
62. Lamb, C.J., J.A. Ryals, E.R. Ward, and R.A. Dixon. 1992. Emerging strategies for enhancing crop resistance to microbial pathogens. *Bio/Technology* 10:1436–1445.
63. Large, E.C. 1940. *The Advance of the Fungi*. New York: Henry Holt.
64. Latunde-Dada, A.O., R.A. Dixon, and J.A. Lucas. 1987. Induction of phytoalexin biosynthesis in resistant and susceptible lucerne callus lines infected with *Verticillium albo-atrum*. *Physiol. Mol. Plant Pathol.* 31:179–186.
65. Leshem, Y.Y. 1984. Interactions of cytokinins with lipid-associated oxy free radicals during senescence: a prospective mode of cytokinin action. *Can. J. Bot.* 62:2943–2949.
66. Linthorst, H.J.M., R.L.J. Meuwissen, S. Kauffman, and J.F. Bol. 1989. Constitutive expression of pathogenesis-related proteins PR-1, GRP, and PR-S in tobacco has no effect on virus infection. *Plant Cell* 1:285–291.
67. Liu, D., K.G. Raghotama, P.M. Hasegawa, and R.A. Bressan. 1994. Osmotin overexpression in potato delays development of disease symptoms. *Proc. Natl. Acad. Sci. USA* 91:1888–1892.
68. Lodge, J.K., W.K. Kaniewski, and N.E. Tumer. 1993. Broad-spectrum virus resistance in transgenic plants expressing pokeweed antiviral protein. *Proc. Natl. Acad. Sci. USA* 90:7089–7093.
69. Madamanchi, N.R., and J. Kuć. 1991. Induced systemic resistance in plants. In *The Fungal Spore and Disease Initiation in Plants and Animals*. G.T. Cole and H.C. Hoch, eds. New York: Plenum Press, pp 347–363.
70. Maher, E.A., N.J. Bate, W. Ni, Y. Elkind, R.A. Dixon, and C.J. Lamb. 1994. Increased disease susceptibility of transgenic tobacco plants with suppressed levels of preformed phenylpropanoid products. *Proc. Natl. Acad. Sci. USA* 91:7802–7806.

71. Martin, G.B., S.H. Brommonschenkel, J. Chunwongse, A. Frary, M.W. Ganal, R. Spivey, T. Wu, E.D. Earle, and S.D. Tanksley. 1993. Map-based cloning of a protein kinase gene conferring disease resistance in tomato. *Science* 262:1432–1436.

72. Mauch, F., B. Mauch-Mani, and T. Boller. 1988. Antifungal hydrolases in pea tissue. II. Inhibition of fungal growth by combinations of chitinase and beta-1,3-glucanases. *Plant Physiol.* 88:936–942.

73. Maurhofer, M., C. Hase, P. Meuwly, J.P. Metraux, and G. Defago. 1994. Induction of systemic resistance of tobacco to tobacco necrosis virus by the root-colonizing *Pseudomonas fluorescens* strain CHA0: Influence of the *gacA* gene and of pyoveridine production. *Phytopathology* 84:139–146.

74. Merwin, I., S.K. Brown, D.A. Rosenberger, D.R. Cooley, and L.P. Berkett. 1994. Scabresistant apples for the northeastern United States: new prospects and old problems. *Plant Dis.* 78:4–10.

75. Mills, P.R., E.J. Guissin, and R.K.S. Wood. 1986. Induction of resistance in cucumber to *Colletotrichum lagenarium* by 6-benzylaminopurine. *J. Phytopathol.* 116:11–17.

76. Mucharromah, E., and J. Kuć. 1991. Oxalate and phosphates induce systemic resistance against diseases caused by fungi, bacteria and viruses in cucumber. *Crop Prot.* 10:265–270.

77. Neuhaus, J.M., P. Ahl-Goy, U. Hinz, S. Flores, and F. Meins. 1991. High-level expression of a tobacco chitinase gene in *Nicotiana sylvestris*. Susceptibility of transgenic plants to *Cercospora nicotinae* infection. *Plant Mol. Biol.* 16:141–151.

78. Newcombe, G., and J. Robb. 1988. The function and relative importance of the vascular coating response in highly resistant, moderately resistant and susceptible alfalfa infected by *Verticillium albo-atrum*. *Physiol. Mol. Plant Pathol.* 33:47–58.

79. Novacky, A. 1972. Suppression of the bacterially induced hypersensitive reaction by cytokinins. *Physiol. Mol. Plant Pathol.* 2:101–104.

80. Padmanabhan, S.Y. 1973. The great Bengal famine. *Annu. Rev. Phytopathol.* 11:11–26.

81. Pan, S.Q., X.S. Ye, and J. Kuć. 1991. Association of B-1,3-glucanase activity and isoform pattern with systemic resistance to blue mold induced by stem injection with *Peronospora tabacina* or leaf inoculation with tobacco mosaic virus. *Physiol. Mol. Plant Pathol.* 39:25–39.

82. Pauls, K.P., and J.E. Thompson. 1982. Effects of cytokinins and antioxidants on the susceptibility of membranes to ozone damage. *Plant Cell Physiol.* 23:821–832.

83. Paxton, J.D. 1982. Phytoalexins. In *Active Defense Mechanisms in Plants*. R.K.S. Wood, ed. New York: Plenum Press, pp 344–346.

84. Peever, T.L., and V.J. Higgins. 1989. Electrolyte leakage, lipoxygenase, and lipid peroxidation induced in tomato leaf tissue by specific and nonspecific elicitors from *Cladosporium fulvum*. *Plant Physiol.* 90:867–875.

85. Peng, M., and J. Kuć. 1992. Peroxidase-generated hydrogen peroxide as a source

of antifungal activity in vitro and on tobacco leaf disks. *Phytopathology* 82:696–699.

86. Pennypacker, B.W., and K.T. Leath. 1993. Anatomical response of resistant alfalfa infected with *Verticillium albo-atrum*. *Phytopathology* 83:80–85.

87. Pennypacker, B.W., D.P. Knievel, M.L. Risius, and K.T. Leath. 1994. Photosynthetic photon flux density X pathogen interaction in growth of alfalfa infected with *Verticillium albo-atrum*. *Phytopathology* 84:1350–1358.

88. Pitblado, R.E. 1992. The development and implementation of TOM-CAST: A weather-timed fungicide spray program for field tomatoes. *Ministry of Agriculture and Food*, Ontario, Canada.

89. Raman, K.V., and D.W. Altman. 1994. Biotechnology initiative to achieve plant pest and disease resistance. *Crop Prot.* 13:591–596.

90. Raybould, A., and A.J. Gray. 1994. Will hybrids of genetically modified crops invade natural communities? *Trends Ecol. Evol.* 9:85–89.

91. Ride, J.P. 1978. The role of cell wall alterations in resistance to fungi. *Ann. Appl. Biol.* 89:302–306.

92. Ride, J.P. 1980. The effect of induced lignification on the resistance of wheat cell walls to fungal degradation. *Physiol. Mol. Plant Pathol.* 16:187–196.

93. Ryals, J., Uknes, and E. Ward. 1994. Systemic acquired resistance. *Plant Physiol.* 104:1109–1112.

94. Sandstrom, P.A., M.D. Mannie, and T.M. Buttke. 1994. Inhibition of activation-induced death in T cell hybridomas by thiol antioxidants: oxidative stress as a mediator of apoptosis. *J. Leukoc. Biol.* 55:221–226.

95. Sarhan, A.R.T., Z. Kiraly, I. Sziraki, and V. Smedegaard-Petersen. 1991. Increased levels of cytokinins in barley leaves having the systemic acquired resistance to *Bipolaris sorokiniana* (Sacc.) Shoemaker. *J. Phytopathol.* 131:101–108.

96. Scholtof, K.B.G., H.B. Scholtof, and A.O. Jackson. 1993. Control of plant virus diseases by pathogen-derived resistance in transgenic plants. *Plant Physiol.* 102:7–12.

97. Schumann, G.L. 1991. *Plant Diseases: Their Biology and Social Impact*. St. Paul, Minnesota: APS Press.

98. Schumann, G.L., and H.D. Thurston. 1977. Light intensity as a factor in field evaluations of general resistance of potatoes to *Phytophthora infestans*. *Phytopathology* 67:1400–1402.

99. Sela-Buurlage, M.B., A.S. Ponstein, S.A. Bres-Vloemans, L.S. Melchers, P.J.M. van den Elzen, B.J.C. Cornelissen, and P.J.M. van den Elzen. 1993. Only specific tobacco (*Nicotiana tabacum*) chitinases and beta-1,3-glucanases exhibit antifungal activity. *Plant Physiol.* 101:857–863.

100. Sen Gupta, A., J.L. Heinen, A.S. Holaday, J.J. Burke, and R.D. Allen. 1993. Increased resistance to oxidative stress in transgenic plants that overexpress chloroplastic Cu/Zn superoxide dismutase. *Proc. Natl. Acad. Sci. USA* 90:1629–1633.

101. Shaaltiel, Y., A. Glazer, P.F. Bocion, and J. Gressel. 1988. Cross tolerance to herbicidal and environmental oxidants of plant types tolerant to paraquat, sulfur dioxide, and ozone. *Pestic. Biochem. Physiol.* 31:13–23.

102. Shtienberg, D., and W.E Fry. 1990. Quantitive analysis of host resistance, fun-

gicide and weather effects on potato early and late blight using computer simulation models. *Am. Pot. J.* 67:277–286.

103. Shtienberg, D., R. Raposo, S.N. Bergeron, D.E. Legard, A.T. Dyer, and W.E. Fry. 1994. Incorporation of cultivar resistance in a reduced-sprays strategy to suppress early and late blights on potato. *Plant Dis.* 78:23–26.

104. Siegrist, J., W. Jeblick, and H. Kauss. 1994. Defense responses in infected and elicited cucumber (*Cucumis sativus* L.) hypocotyl segments exhibiting acquired resistance. *Plant Physiol.* 105:1365–1374.

105. Steiner, P.W. 1990. Predicting apple blossom infections by *Erwinia amylovora* using the MARYBLYT model. *Acta Hortic.* 273:139–148.

106. Steiner, P.W. 1990. Predicting canker, shoot and trauma blight phases of apple fire blight epidemics using the MARYBLYT model. *Acta Hortic.* 273:149–158.

107. Stintzi, A.T. Heitz, V. Prasad, S. Wiedemann-Merdinoglu, S. Kauffmann, P. Geoffrey, M. Legrand, and B. Fritig. 1993. Plant "pathogenesis-related" proteins and their role in defense against pathogens. *Biochimie* 75:687–706.

108. Strittmater, G., and D. Wegener. 1993. Genetic engineering of disease and pest resistance in plants: present state of the art. *Naturforsch. Z.* 48c:673–688.

109. Strobel, N.E., and J.A. Kuć. 1994. Systemic induced cross-resistance by plant pathogens and the prooxidant herbicide paraquat (abstr.). *Phytopathology* 84:1100.

110. Strobel, N.E., and W.A. Sinclair. 1991. Role of flavanolic wall infusions in the resistance induced by *Laccaria bicolor* to *Fusarium oxysporum* in primary roots of Douglas-fir. *Phytopathology* 81:420–425.

111. Strobel, N.E., and W.A. Sinclair. 1992. Role of mycorrhizal fungi in tree defense against fungal pathogens of roots. In *Defense Mechanisms of Woody Plants Against Fungi*. R.A. Blanchette and A.R. Biggs, eds. Berlin: Springer-Verlag, pp 321–353.

112. Stumm, D., and C. Gessler, 1986. Role of papillae in the induced systemic resistance of cucumbers against *Colletotrichum lagenarium*. *Physiol. Mol. Plant Pathol.* 29:405–410.

113. Sturtevant, A.P., and R.N. Beachy. 1993. Virus resistance in transgenic plants: coat protein-mediated resistance. In *Transgenic Plants: Fundamentals and Applications*. A. Hiatt, ed. New York: Marcel Dekker, pp 93–112.

114. Sziraki, I., E. Balazs, and Z. Kiraly. 1980. Role of different stresses in inducing systemic acquired resistance to TMV and increasing cytokinin level in tobacco. *Physiol. Mol. Plant Pathol.* 16:277–284.

115. Tanaka, K., I. Furusawa, N. Kondo, and K. Tanaka. 1988. SO2 tolerance of tobacco plants regenerated from paraquat-tolerant cultures. *Plant Cell Physiol.* 29:743–746.

116. Tomlinson, H., and S. Rich. 1973. Anti-senescent compounds reduce injury and steroid changes in ozonated leaves and their chloroplasts. *Phytopathology* 63:903–906.

117. Urs, R.R., and J.M. Dunleavy. 1975. Enhancement of the bactericidal activity of a peroxidase system by phenolic compounds. *Phytopathology* 65:686–690.

118. Vance, C.P.T.K. Kirk, and R.T. Sherwood 1980. Lignification as a mechanism of disease resistance. *Annu. Rev. Phytopathol.* 18:259–288.

119. Van den Ackerveken, G.F.J.M., J.A.L. VanKan, and P.J.G.M. De Wit. 1992. Molecular analysis of the avirulence gene *avr9* of the fungal tomato pathogen *Cladosporium fulvum* fully supports the gene-for-gene hypothesis. *Plant J.* 2:359–366.

120. Van der Plank, J.E. 1963. *Plant Diseases: Epidemics and Control.* New York: Academic Press.

121. Van der Plank, J.E. 1968. *Disease Resistance in Plants.* New York: Academic Press.

122. VanEtten, H., D. Funnell-Baerg, C. Wasmann, and K. McCluskey. 1994. Location of pathogenicity genes on dispensable chromosomes of *Nectria haematococca* MPVI. *Antonie Van Leeuwenhoek* 65:263–267.

123. VanEtten, H.D., D.E. Matthews, and P.S. Matthews. 1989. Phytoalexin detoxification: importance for pathogenicity and practical implications. *Annu. Rev. Phytopathol.* 27:143–164.

124. VanEtten, H.D., J.W. Mansfield, J.A. Bailey, and E.E. Farmer. 1994. Two classes of plant antibiotics: phytoalexins versus "Phytoanticipins". *Plant Cell* 6:1191–1192.

125. Victoria, J.I., and H.D. Thurston. 1974. Light intensity effects on lesion size caused by *Phytophthora infestans* on potato leaves. *Phytopathology* 64:753–754.

126. Vidhyasekaran, P. 1993. Avirulence gene for crop disease management. In *Genetic Engineering, Molecular Biology and Tissue Culture for Crop Pest and Disease Management.* P. Vidhyasekaran, ed. Delhi: Daya, pp 65–74.

127. Vincelli, P.C., and J.W. Lorbeer. 1989. BLIGHT-ALERT: A weather-based predictive system for timing fungicide applications on onion before infections of *Botrytis squamosa. Phytopathology* 79:493–498.

128. Waggoner, P.E., and R.D. Berger. 1987. Defoliation, disease and growth. *Phytopathology* 77:393–398.

129. Ward, E.R., S.J. Uknes, S.G. William, S.S. Dincher, D.L. Wiederhold, D.C. Alexander, P. Ahl-Goy, J.P. Metraux, and J.A. Ryals. 1991. Coordinate gene activity in response to agents that induce systemic acquired resistance. *Plant Cell* 3:1085–1094.

130. Wei, G., J.W. Kloepper, and S. Tuzun. 1991. Induction of systemic resistance of cucumber to *Colletotrichum orbiculare* by select strains of plant growth-promoting rhizobacteria. *Phytopathology* 81:1508–1512.

131. Xuei, X.L., U. Jarlfors, and J. Kuć. 1988. Ultrastructural changes associated with induced systemic resistance of cucumber to disease: host response and development of *Colletotrichum lagenarium* in systemically protected leaves. *Can. J. Bot.* 66:1028–1038.

132. Ye, X.S., S.Q. Pan, and J. Kuć. 1989. Pathogenesis-related proteins and systemic resistance to blue mold and tobacco mosaic virus induced by tobacco mosaic virus, *Peronospora tabacina* and aspirin. *Physiol. Mol. Plant Pathol.* 35:161–175.

133. Ye, X.S., S.Q. Pan, and J. Kuć. 1990. Association of pathogenesis-related proteins and activities of peroxidase, beta-1,3-glucanase and chitinase with systemic induced resistance to blue mold of tobacco but not to systemic tobacco mosaic virus. *Physiol. Mol. Plant Pathol.* 36:523–531.

134. Yoshikawa, M., M. Tsuda, and Y. Takeuchi. 1993. Resistance to fungal diseases in transgenic tobacco plants expressing the phytoalexin elicitor-releasing factor, beta-1,3-endoglucanase from soybean. *Naturwissenschaften* 80:417–420.

135. Zaitlin, M.J.M. Anderson, K.L. Perry, L. Zhang, and P. Palukaitis. 1994. Specificity of replicase-mediated resistance to cucumber mosaic virus. *Virology* 201:200–205.

136. Zhu, Q., E.A Maher, S. Masoud, R.A. Dixon, and C.J. Lamb. 1994. Enhanced protection against fungal attack by constitutive co-expression of chitinase and glucanase genes in transgenic tobacco. *Bio/Technology* 12:807–812.

8
Approaches to Plant Breeding and Genetic Engineering

D. Steven Calhoun
Mississippi Agricultural and Forestry Experiment Station,
Stoneville, Mississippi

James H. Oard
Louisiana State University Agricultural Center, Baton Rouge, Louisiana

I. INTRODUCTION

> *The plant breeder is an applied evolutionist working towards defined objectives by tolerably well understood methods.* [131].

Plant breeding, like evolution, requires genetic variation, selection of desired genotypes, and a degree or form of isolation to preserve the desired characteristics. In addition, modern plant breeding is pursued with specific objectives in mind and with at least a basic understanding of the genetic principles involved. This chapter addresses plant breeding as it relates to developing plants with resistance to pests, including diseases, insects, and nematodes. The term *resistance* is used here in the broadest sense, including any reduction in plant damage and/or pest fitness. The first section is a brief discussion of basic plant breeding principles, including consideration of breeding objectives, elementary genetic principles, and various selection schemes. The second section focuses on breeding for pest resistance using traditional methods. The final selection deals with the use of emerging biotechnologies in breeding for pest resistance.

II. PLANT BREEDING BASICS

A. Objectives and Priorities

Successful plant breeding efforts have specific and well-defined objectives. In most applied breeding programs (programs which develop new varieties or cultivars that will be used directly by farmers), objectives are determined primarily by economic considerations, with emphasis placed on plant characteristics most likely to produce added income to the user of the new cultivar. The primary objective in these programs is usually increased yield, because yield generally has the most direct influence on producer income and thus grower acceptance and economic success of a new cultivar. Depending on the crop, quality of the product (e.g., nutritional content, shelf life, suitability for industrial processing) may also be a major breeding objective. Pest resistance, in applied breeding programs, is pursued as a high-priority objective only if resistance enhances yield or quality, or if resistance can significantly reduce the cost of production by replacing crop-protection chemicals without sacrificing yield or quality. It should be noted that breeding for high yielding ability per se often confers a level of tolerance-type resistance even when breeding is conducted under nonchallenged conditions [21,92]. In many situations, particularly with relatively low-value crops, appropriate pesticides are not available or are prohibitively expensive. These situations make genetic resistance essential, as in the case of disease resistance in wheat and soybeans. In such cases, a pest's impact on yield often makes pest resistance the primary breeding objective. Cultivars that are only partially resistant can increase net crop value by responding more efficiently to pesticides [114,145,150]. Even in applied breeding programs, resistance can, in some circumstance, stand with yield and quality as a top breeding priority.

More basic breeding programs, including many state or federally funded programs, focus on developing germplasm lines that possess various useful traits that will ultimately be used as breeding stocks by applied breeders to develop new cultivars. The objective of germplasm enhancement is to make new and useful traits readily accessible to applied breeders with a minimum of genetic obstacles. Germplasm enhancement programs can relax their emphasis on yield and quality somewhat with the assumption that resistance factors or other useful traits will be incorporated into high-yielding, high-quality cultivars at a later time. These more basic programs must still focus their efforts on traits with the greatest potential for significant economic, social, and/or environmental impact. A disease, insect, or nematode must be a major pest of the crop species in order for resistance to qualify as a worthwhile objective even in a less applied breeding program.

Finite resources require that priorities be established and that available assets be directed accordingly. Both applied and basic breeding programs require concerted effort sustained over many years for the development of improved cultivars or germplasm lines. Therefore, efforts at breeding for resistance in both types of programs should be directed toward pests of key importance. A pest that consistently causes a major yield or quality reduction and that is expensive (monetarily or environmentally) to control with chemicals or cultural practices is a good target for a genetic resistance program.

B. Genetic Principles

Although some successful crop improvement efforts have, in the past, been carried out with little or no knowledge of genetic principles, a basic understanding of these principles is valuable in developing an efficient and effective breeding program. The following is not intended as a comprehensive discussion of our current knowledge of genetics but rather a brief listing of essential concepts. These concepts are necessary for understanding and evaluating different approaches to breeding for pest resistance.

1. Heritable Traits

To be useful to breeders, a plant characteristic must be heritable; that is, the characteristic must be passed with reasonable reliability from parent to offspring. Heritable traits fall into two general categories: (a) qualitatively inherited or monogenic traits, and (b) quantitatively inherited or polygenic traits [40].

Qualitative traits are controlled by one or a few genes and different genotypes can be readily assigned to distinct classes—resistant or susceptible, for example. Qualitative traits include such things as the color of cotyledons in peas (studied in Mendel's classic research [99]), flower color in soybeans, presence or absence of awns in wheat, leaf shape in cotton, and many others. When working with a qualitative trait, it is useful to know the number of loci and alleles affecting the trait. Using qualitative traits in a breeding program is relatively easy, because the trait is either present or absent and expression of the trait is affected little, if at all, by the environment.

Quantitative traits are controlled by many genes (thus the term *polygenic*) interacting with each other and often with the environment. Many different genotypes are therefore possible, and expression of polygenic traits displays continuous variation among genotypes. Phenotypes could range, for example, from highly susceptible to highly resistant with continuous gradation of intermediates. Many agronomically important traits such as yield and maturity are quantitatively inherited. When working with quantitative traits, it is useful to know something about the degree to which a trait, such as yielding ability, is passed from parents to offspring. The degree of similarity for a given trait between parents and offspring (or other relatives) is known as heritability. Traits with high heritability (i.e., offspring closely resemble their parents) are easier to breed for than traits with a low heritability. For a more complete discussion of heritability and its measurement, see Falconer [39]. Even highly heritable polygenic traits are generally more difficult to breed for than simply inherited traits, because (a) breeding for polygenic traits requires identifying plants with desirable alleles at a large proportion of the numerous loci affecting the trait in question, and (b) identifying those individuals is complicated by the effect of the environment on the expression of the trait.

The preceding are generalizations regarding quantitative and qualitative traits; it should be noted that there are exceptions. In a few cases, traits controlled by a single gene are expressed in a quantitative manner with more or less continuous variation among genotypes.

2. Recombination and Segregation

The ideal cultivar or crop variety does not exist. All currently used cultivars, regardless of species, have certain flaws or deficiencies that plant breeders try to remedy through genetics or that other agriculturalists try to remedy through crop management. In many cases, the deficiencies in one genotype are different than those in another genotype. For example, many commercial genotypes are high yielding but are susceptible to important pests. Certain other genotypes are low yielding but resistant to one or more pests. The challenge for the plant breeder is to combine high yielding ability from the former with pest resistance from the latter. This recombination is accomplished in sexually reproducing plants through hybridization. For a complete discussion of methodology for hybridizing major crop species, see Fehr and Hadley [41].

If the parental plants are homozygous, or "true breeding," the first generation after hybridization will be genetically uniform. This generation is termed the F_1, or first filial, generation. After self-pollination, the following (F_2) and subsequent generations will "segregate," or produce many different types of plants, some resembling one parent or the other and most plants expressing some traits from both parents (the symbol S, rather than F, is sometimes used to denote generations of self-pollination; see ref. 40 for a description of this system).

The mechanics and expected outcomes of segregation at one or more loci are beyond the scope of this chapter, and they may be found in standard genetics text books (e.g., ref. 81). However, the main points that need to be understood with regard to recombination and segregation are as follows:

    a. Thousands of different genotypes are produced through recombination of dissimilar parents.

    b. Significant segregation may continue for several generations after hybridization (e.g., an individual F_2 plant will produce some F_3 plants that may or may not resemble the F_2 plant). When parents differ for only a few genes, true breeding lines can be isolated as early as the F_2 generation. When parents are dissimilar for polygenic traits or for several monogenic traits, noticeable segregation continues for many generations. Continuous self-pollination will decrease heterozygosity and eventually result in homozygous plants whose progeny will not segregate.

c. It is primarily through recombination that new genetic variation is obtained (mutation provides infrequent, although sometimes important, new variation), from which genetic improvement is gained.

3. Correlated Traits

In many cases, two or more traits may be positively or negatively correlated [39]. For example, genotypes possessing certain pest resistance gene(s) in a population may tend to be lower yielding than plants in the population without the gene(s). Phenotypic correlations of this type may be due to genetic linkage (to be discussed below) of the two traits. If this is the case, the correlation will be transitory and will not occur in all populations. Genotypic correlations that are consistently observed in all populations are usually due to pleiotropy or one set of genes affecting more than one trait. Genetic correlations can be an advantage when improvement of one trait causes an improvement in other trait(s) as well. Likewise, negative genetic correlation among traits can be a disadvantage when trying to simultaneously improve traits such as yield and pest resistance.

C. Breeding Methods

Selection in plant breeding is the identification and isolation of desirable individuals or populations of individuals from an original heterogeneous population. In some cases, the original population may be a naturally occurring land race, but generally the population under selection is made up of recombinants resulting from planned hybridization.

Several schemes of selection are used by various breeders, depending on the mode of reproduction of the crop (open pollinated or self-pollinated), the nature of the germplasm, the objectives of the program, and the personal preferences of the breeder. Fehr [40] gives an in-depth discussion of various selection schemes; a brief description of the primary schemes is given here.

1. *Pedigree method:* Pedigree selection, sometimes called "ear-to-row" or "plant-to-row" selection, usually begins with the F_2 generation. Individual F_2 plants are selected, based on phenotype, and harvested separately. Seed from each selected plant is sown in a separate row or plot the next season. From

within desirable F_2-derived F_3 (symbolized, $F_{2:3}$) populations, individual F_3 plants are selected, harvested, and resown individually as before. The process continues until no noticeable segregation occurs in the progeny of the selected plants. Selections from among the resulting homogenous populations are each then bulk harvested and later evaluated in replicated tests.

2. *Bulk method:* Bulk selection differs from pedigree selection in that desirable plants in early generations are harvested en masse rather than individually. In an advanced generation (usually F_4 or greater), individual plants are selected and harvested separately and planted in individual rows. Selection is then made among rows for uniformity and desirability, with selected rows harvested individually to begin replicated testing.

3. *Single-seed descent method:* In a single-seed descent system, all selection is delayed until an advanced generation, when progeny will be expected to be near homozygous (true breeding). Generations are advanced as quickly as possible, often in systems involving multiple generations per year and usually with large numbers of individuals included in each generation. The classic procedure of single-seed descent is to harvest one seed from each plant. However, the concept of single-seed descent is also served by harvesting several seeds from each plant in the population and planting a sample of the bulked seed the next generation. At about the F_5 generation, individual plants are selected and planted the next generation for evaluation and seed increase prior to replicated testing.

4. *Recurrent selection:* The preceding three methods are used primarily with self-pollinated species, although they can be used in open pollinated species if manual self-pollination is practiced. Recurrent selection, a method of population improvement, is best suited to open pollinated species. Desirable individuals are selected from a population and allowed to intercross to form a new population, from which desirable individuals are again selected for intercrossing. The cycle of selection and intercrossing is repeated with the objective of continually improving the population by accumulating and increasing the frequency of desirable alleles. This method can also be used in self-pollinated species if a means is available to facilitate extensive intercrossing. When used in self-pollinated species, the objective is usually ultimately to select from within the improved population and develop pure lines.

5. *Hybrids:* In some crops (e.g., corn and grain sorghum) commercial cultivars are primarily hybrids. Seed of hybrid cultivars is produced by using pollen from one genotype to fertilize ovules of another genotype. Hybrid seed production thus requires an economical means (either genetic or physical) of preventing self-pollination by the seed parent and of transferring pollen from the pollen parent to the seed parent. The purpose of this rather complicated seed production system is to take advantage of the heterosis or "hybrid vigor" often

observed in the offspring of dissimilar individuals. The parents used to produce hybrid seed are inbred lines developed using breeding methods described above. Various combinations of inbred lines are evaluated to identify the most productive hybrids. In addition to heterosis, the use of hybrid cultivars offers the potential to combine traits (e.g., high yield and pest resistance) from two inbred lines without the long selfing and selection process required to produce inbred cultivars.

III. BREEDING FOR HOST PLANT RESISTANCE

A. Breeding Strategies

Pest resistance can be a quantitative or qualitative trait. Resistance that is conditioned by one or a few major genes is sometimes referred to as specific, monogenic, or vertical resistance and generally is inherited in a qualitative manner. Resistance conditioned by many minor genes is referred to as general, polygenic, or horizontal resistance and generally is inherited in a quantitative manner. The terms used above to describe types of resistance have different meanings to and implications for the various disciplines involved in plant resistance [107]. In any case, they represent extremes in a continuum of specificity and mode of inheritance [109,115].

Specific resistance is expressed only against some biotypes (or pathotypes or races or strains or isolates) of a pest, whereas general resistance is expressed equally against all biotypes of a pest [151]. Specific resistance is the easiest type of resistance to manipulate in a breeding program, because it is controlled by one or a few genes, and the resistance genes can be readily identified and transferred from one genotype to another. However, specific resistance is often prone to "breaking down" as new or rare, virulent biotypes of the pest become predominant. A few of the many possible examples of major genes losing effectiveness include major genes for potato blight resistance [129], *Sr27* gene for stem rust resistance in triticale [94], various genes for yellow rust resistance in wheat [141], and greenbug resistance in "Amigo" wheat [113].

The genetic relationship between a host, possessing various resistance gene(s), and its pest, possessing various virulence gene(s), has been described as the gene-for-gene relationship [43]. Gene-for-gene relationships have been demonstrated for both diseases [97] and insects [48,70].

General resistance, although less vulnerable to genetic changes in the pest population [48], is more difficult to utilize in a breeding program owing to the problem of transferring multiple genes from one genotype to another. Different breeding strategies are used for different types of resistance.

1. Breeding Strategies for Specific Resistance

One strategy for utilizing specific resistance is to develop cultivars with a single major resistance gene that acts against the predominant biotype of the pest. This is the most vulnerable system, because the genetic composition of pest populations can change rapidly. Despite its inherent vulnerability, this is currently the most common strategy among breeders because (a) the monogenic resistance factors are convenient to transfer to new cultivars, and (b) the level of resistance is often quite high, at least for as long as avirulent biotypes predominate in the pest population. When using a single major gene for resistance, the efficacy of the resistance may be prolonged by strategies that promote the continued predominance of avirulent pest biotypes [93].

Another strategy for utilizing specific resistance is to develop multiline cultivars. Multilines are mixtures of seed from different genotypes, each possessing a major gene that acts against a different pest biotype. Multiline cultivars are, in theory, buffered against shifts in the pest population, because some genotypes are likely to be resistant to whatever biotype occurs and will reduce the spread of the pest to susceptible genotypes [40]. Multilines also tend to mitigate pest population shifts by allowing reproduction of various biotypes on certain of the component lines. The primary disadvantages of the multiline approach are that (a) considerable effort is required to transfer a number of genes to agronomically superior genotypes, and (b) the component lines must be not only agronomically superior but also similar in morphology and phenology to allow for mechanized crop management.

A third strategy is to incorporate several resistance genes that are effective against different biotypes into a single cultivar. This strategy is known as pyramiding [110,123]. To overcome the resistance, a new biotype must possess virulence genes for each of the resistance genes in the host. It has been suggested that the ability of a pest organism to survive in nature is reduced as the number of virulence genes increases [151]. The primary disadvantages of pyramiding are the difficulty of incorporating several major genes into a single superior cultivar, and the need for extensive screening against various individual biotypes to ensure that each of the desirable alleles is present.

Before leaving the topic of specific resistance (in this case, with emphasis on the monogenic nature of the resistance), it is important to note that this type of resistance is not necessarily short lived. Examples can be found of qualitatively inherited resistance that has been at least partially effective for many years and continues to be so [133, 135; see also reference to *Sr2* gene for stem rust and *Lr13/Lr34* genes for leaf rust resistance in ref. 128]. Russell [119] cites numerous examples. Resistance that is conditioned by one or a few genes should not be entirely ignored simply because the resistance *may* not be durable.

2. Breeding Strategies for General Resistance

General resistance is considered more desirable than specific resistance, because it provides some level of protection against many biotypes and is less likely to cause or suffer from changes in pest populations. The fact that relatively few parasitic organisms are key pests is evidence that general or horizontal resistance for the majority of potential pests has been achieved and maintained without direct selection [129]. However, developing general/polygenic resistance to key pests is a difficult objective to achieve. Mayor [90] has outlined many of the breeding-related obstacles to developing this type of resistance.

Although general resistance to key pests is more difficult than specific resistance to detect and to manipulate due to its polygenic nature, it appears to be fairly ubiquitous and in many cases highly heritable [129]. Selecting genotypes with moderate, rather than very high, levels of resistance is the primary approach that has been proposed for developing cultivars with horizontal resistance [69]. In this paradigm, avoiding immune or highly resistant genotypes is necessary to eliminate the possibility of selecting for transient specific resistance [115]. The development of "slow rusting" wheat genotypes, selected in the adult plant stage, is an example of this approach [28].

Selecting for pest tolerance (discussed in Chapter 6) is another means of developing general resistance [4]. Although this type of resistance may not maximize production during a given year or pest outbreak, as is possible with highly resistant cultivars, the long-term performance may be more stable. Because there is no selection pressure on the pest populations, shifts to more virulent biotypes are unlikely. Tolerance is also compatible with other biocontrol factors (especially in the case of insect pests), because pest populations are not reduced to the point that predators and parasites are eliminated [54].

Breeding approaches that may be useful in developing cultivars with general resistance include delaying selection until an advanced generation (using single-seed descent or a similar method) to ensure that selected plants will breed true and screening very large populations. Recurrent selection is also useful in accumulating desirable alleles, and it has been successful in developing insect-resistant alfalfa cultivars [53,103]. Because of the difficulty in combining multiple alleles for resistance with multiple desirable alleles for yield and other agronomic traits, it is advisable to use as resistance sources genotypes that are already relatively well adapted rather than using exotic germplasm. Simmonds [129] has written an extensive review on the genetics of horizontal disease resistance, including some strategies for identifying genotypes with horizontal resistance and citing examples from breeding programs aimed at developing horizontal resistance.

3. Breeding for Morphologically Based Resistance

In most cases, monogenic resistance is based on a biochemical or physiological incompatibility of the pest with the host. Allelochemicals or physiological reactions (such as hypersensitivity) in the resistant plant prevent avirulent biotypes from utilizing it as a host. When the pest evolves mechanisms for metabolizing the offending alleochemical or overcoming the physiological reaction, a new and virulent biotype emerges. In some cases, monogenic resistance is based on a physical obstacle or nutritional deficiency for the pest, and can have a more general and durable effect on the pest as a species. Examples include:

a. Dense trichomes on various plant species which restrict certain insect pests' ability to move about or to reach feeding sites on the plant [77,131,158,159,160]

b. Absence of trichomes, which in cotton [85] and soybean [77] reduces *Heliothis* spp. oviposition by depriving females of their preferred oviposition site against or attached to a trichome

c. Awns on wheat, which increase the likelihood of dislodgement of the grain aphid *Sitobion avenae* [1]

d. Absence of extrafloral nectaries in cotton, which deprives *Heliothis* spp. [86] and *Lygus* spp. [13] of a key carbonhydrate source

Resistance may also derive from a plant character that minimizes damage caused by serving as a host. One example is the solid stem trait in wheat which minimizes lodging caused by the wheat sawfly [95].

Some morphological traits confer resistance to one pest while increasing susceptibility to others. As noted above, sparse pubescence in cotton confers resistance to *Heliothis* spp. but increases sensitivity to plant bugs. The frego bract trait in cotton provides resistance (antixenosis) to boll weevils but increases sensitivity to plant bugs [137]. A single morphological trait may provide resistance to one life stage of a pest, whereas conferring susceptibility to another. Leaf pubescence encourages oviposition by lepidopteran adults but decreases feeding by larvae [77].

In many (if not most) cases, morphological traits are simply inherited, making them convenient traits to manipulate in a breeding program. Another advantage of morphologically based resistance, in addition to its durability, is that it can be selected for based on the appearance of the plant and in the absence of the pest [33]. As will be seen in the next section, manipulating pest populations and the plant's exposure to the pest is one of the greatest challenges in breeding for pest resistance.

B. Evaluating Genotypes for Pest Resistance

Numerous books are devoted wholly or partially to describing specific techniques for evaluating plants and plant populations for pest resistance [24,33,55,

89,119,136,137,139]. This section will address general principles that should be observed rather than in-depth specifics about alternative techniques, and it will emphasize approaches appropriate to large-scale screening designed to answer the question Is this genotype resistant, in the broadest sense? rather than Why and how is this genotype resistant? Although questions of how and why are important, a breeder must be able to evaluate a large number of genotypes quickly; once the more promising genotypes are identified, more detailed evaluations can be designed.

Regardless of the crop or pest species, evaluation/screening techniques should meet certain criteria. The methodology should be as simple as possible in terms of producing the desired conditions for evaluation and in terms of the type of data that are collected, so that large numbers of genotypes can be evaluated quickly. The method should make efficient use of time and physical resources. It should be capable of discerning various levels of resistance, since durable, horizontal resistance is often partial resistance. The method should produce repeatable results from year to year and from different breeding programs. Finally, the results should reflect the outcome that will be experienced under large scale field conditions—resistance detected in a petri disk is of little value if it is not expressed in a producer's field [118].

1. Evaluating Plants or Pests

Pest resistance can be viewed and measured in terms of how different plants respond to pest pressure or in terms of how pests respond to different host plants. Screening methods focusing on the plant can be designed to detect differences among genotypes in, for example, defoliation rates [5,117,140], characteristic symptom expression [22,38,44], plant growth rate [18,20], root galling [61,167], resistance to uprooting [107], chlorosis [44], necrosis [165], mortality [140], proportion of plants affected [26], and yield loss [18,23,165] under uniform pest pressure. Aids such as soft x-ray photography [49,155] may be used to make damage symptoms more readily visible. Screening for resistance evaluated as the effect of plant genotype on pest fitness would look for differences in, for example, growth rates of pest individuals [154], size of lesions [122], disease progress curves [6,72,132], fecundity [102], survival [118], and life cycle duration [168]. Antixenosis (an entomological term discussed in Chapter 7) is another pest-based measure of resistance.

Plant-based measures of resistance are, for screening purposes, usually based on visual rating scales (discussed below) and are thus faster than most pest-based evaluation methods which often require counting and/or weighing the pest organism. In the case of plant diseases, visual rating scales are often used to estimate success of a pathogen in utilizing a host [37] and may combine plant reaction and pathogen success [10]. In some cases, such as reniform nematodes in cotton, acute symptoms of pest attack are not evident, and pest reproductive rate is the only suitable measure of resistance [56].

2. Managing Pest Pressure

One of the greatest challenges in evaluating genotypes for pest resistance is achieving an appropriate degree of pest pressure. Pest pressure may be understood to be the number of pest organisms actually attacking each host, or the number of pest organisms that have an opportunity to attack each host plant, depending on the nature of the breeding objectives and the type of resistance sought. Pressure that is too high may mask valuable partial resistance; pressure that is too low may permit a high proportion of escapes that will be misclassified as resistant. The most useful pest pressure is one that maximizes differences in the plant material evaluated. Pest pressure must also be relatively uniform across a screening test, or mechanisms must be in place to account for nonuniformity. Multiple replications, when adequate seed and other resources are available, or extensive use of susceptible and resistant checks in single replication evaluations will aid in accounting for variability in pest pressure in a test. Screening evaluations may be conducted in the field, in cage situations, or in a controlled environment such as a greenhouse or laboratory. Each has advantages and disadvantages in terms of the ease and precision of pest pressure manipulation, as well as in terms of resource allocation and applicability to actual commercial situations.

Field Evaluation. Field evaluations most closely resemble commercial situations, but caution should be applied even here. Resistance factors that derive primarily from antixenosis, for example, may be evident in small plots but may not be expressed in large commercial plantings. Interplot interference, the influence neighboring plots may exert by generating too much or too little inoculum, is another potential danger of small plot field trials [129]. Provided that land and normal agricultural equipment are available, field evaluations generally require little in the way of additional specialized equipment and facilities. They can, therefore, be among the least expensive evaluations to conduct, but they are generally also the least precise and are used primarily in the early stages of cultivar development.

The greatest disadvantage of field evaluation is the difficulty in obtaining appropriate pest pressure. Total reliance on natural infestation frequently results in pressure that is too low or occurs at an improper stage of crop development for discriminating evaluations. Various methods have been used to manipulate pressure from diverse pests. If selective insecticides are available, insect pest populations can be increased by eliminating beneficial insects [11]. In some cases, crop management practices such as excessive fertilization [47] or irrigation timing [143] can enhance pest pressure. Trap crops or spreader rows of susceptible genotypes may be planted around and/or between test plots to build up inoculum [67] or insect pest populations [8]. Trap crops may need to be cut to force insect pests into test plots [78]. Crop rotations with susceptible cultivars or species can enhance populations of soilborne pests such as nematodes [14].

Artificial infestation (or inoculation) under favorable conditions for the pest is the most reliable, but also the most expensive means of obtaining desirable levels of pest pressure. Artificial infestation requires a means to mass culture the pest organism or economically to collect organisms in the field. Breeding for disease resistance has progressed more rapidly than for insect resistance [29], which is due in large part to the availability of technologies to economically mass culture pest pathogens on artificial media or plant products [119]. Propagules of plant pathogens such as cereal rusts can also be economically collected from infected plants in the field or greenhouse [119]. Mass rearing and processing techniques for many insect pests on artificial media [64] or on living plants [12,116] are also available.

Care must be taken when culturing pests for artificial infestation/inoculation to ensure that shifts in the pest population are minimized. When rearing insects, especially, it is common for a colony to become adapted to laboratory conditions and lose adaptation to field conditions.

Once large quantities of insects or disease propagules have been collected, they must be distributed in a test. Critical to this phase of artificial infestation are (a) uniform and economical distribution of inoculum, (b) maintenance of inoculum viability during distribution, and (c) high rate of pest establishment on susceptible genotypes. Disease inoculum has been distributed by spraying a water- or oil-based suspension of spores [28], mixing spores with talc and distributing with a dust blower [119], spreading infected crop residue, applying infected oat seeds or other culture medium, injecting plants with a syringe charged with inoculum suspensions [28], piercing test plants with a wooden toothpick bearing inoculum [31], and many other means. It is important to be aware that certain inoculation methods, particularly those causing unnatural tissue injury, may bypass possible resistance mechanisms. One of the most commonly used means of distributing insects in tests is the "bazooka" insect applicator [100]. Techniques have also been developed to spray insect eggs in a xanthene gum suspension [147].

Cage Evaluations. Cage evaluations are applicable primarily to entomological studies. Cages used can vary in size from small clip-on cages for a single leaf [102] to large field cages or screenhouses [75] of a hectare or more in size. These studies have the advantages of a high degree of control over pest pressure and precise evaluation of pest-based resistance effects. Owing to the high labor and material requirements of these studies, they are not generally used in preliminary screening of large numbers of genotypes.

Greenhouse and Laboratory Evaluations. Indoor evaluations have the advantages of better controlled environmental conditions and more easily controlled pest pressure (including selection of pest biotype) than field evaluations. However, it is important to structure and test indoor evaluations to ensure that results are indicative of field performance [91]. In most cases, it is preferable to use resistant and susceptible genotypes characterized in the field as guides in developing indoor screening systems—indoor classification should agree with field performance.

Indoor screening can use seedlings or adult plants [119,137]. The use of seedlings allows for screening of a greater number of genotypes in a limited amount of time and space, and this can be a very economical means of evaluation. Adult plant screening may be needed to identify horizontal resistance; in some cases, this is even referred to as "adult plant resistance" [126]. Effective insect resistance screening has also been done with excised plant parts [134].

Environmental conditions can be manipulated as a tool to manage pest pressure in indoor systems, but it is important to understand the implications of such manipulations. Varying temperature or other factors can enhance or mask expression of resistance [121,126]. Again, the rule of thumb should be: Do not rely on results from indoor screening unless they reflect field performance.

3. Rating Systems

Visual rating scales, rather than direct measurements, are frequently used in preliminary screening tests. Scales range from very simple (e.g., 1=resistant, 2=intermediate, 3=susceptible) to very sophisticated. Scales may be either plant based, indicating degree of damage, or pest-based, indicating success in host utilization. The danger in using visual rating scales is that they are inherently subjective. The following steps may be taken to enhance precision and repeatability of results:

a. Allow a wide enough range in the scale to accommodate various intermediate levels of resistance.
b. Clearly describe appearance of each level in the scale. The use of photographs or drawings depicting levels is also useful (e.g., see ref. 37).
c. Establish a relationship between the rating scale and a measurable indicator of plant damage (such as percentage of defoliation) or pest success (such as insect numbers or percentage of diseased leaf area). Measurements can then be taken periodically to ensure the scale is being properly used by all evaluators.

    d.   Separate assessment of plant reaction (such as infection type) and pest success (such as disease severity) using one or more scales for each.

    e.   Use a numeric scale, avoiding zero, to permit statistical analysis.

    f.   Use frequent susceptible and resistant checks.

Examples of rating scales are shown in Tables 1 and 2. Statistical analysis of rating scale data may require transformation or other mathematical manipulation [80]. It needs to be emphasized that subjective, visual rating scales are valuable in preliminary screening of large numbers of genotypes. Visual rating should be followed up with replicated tests involving direct and objective measurements.

IV. USE OF EMERGING TECHNOLOGIES

Recent technological developments promise to improve the efficiency of selecting for pest resistance and to expand the range of resistance traits available to breeders. Various types of molecular markers are being used increasingly as aids in selecting for resistance genes without the need for laborious and relatively imprecise bioassays. Advances in genetic engineering promise the potential of removing genetic barriers to open up the genomes of both plant and animal kingdoms as sources of resistance genes.

Table 1 Rating Scales Used to Measure Insect Resistance in Sorghum

Score	Greenbug damage	Sorghum Midge (% damage)
0	–	0
1	No red spots on leaves	1–10
2	Red spots on leaves	11–20
3	Part of one leaf dead	21–30
4	One leaf dead	31–40
5	Two leaves dead	41–50
6	Four leaves dead	51–60
7	Six leaves dead	61–70
8	Eight leaves dead	71–80
9	Entire plant dead	81–90
10	–	91–100

Source: From ref. 64.

Table 2   Rating Scale for Reaction of Barley to Russian Wheat Aphid

Score	Damage symptoms	Chlorotic area (%)
1	Little or no symptom expression	0–5
2	Minimal striping on a few tillers	6–10
3	Distinct striping on several tillers	11–20
4	As above, plus leaf rolling	21–30
5	As above, plus leaf tip necrosis	31–50
6	As above, but extensive leaf necrosis	>50

Source: From ref. 21.

A.   Use of Molecular Markers in Selecting for Pest Resistance

When the genes for two or more traits are located close to each other on a chromosome, they are said to be linked. If the genes are closely linked, they will usually occur together in the same individual in segregating populations. With the use of new DNA analysis techniques, it is possible to identify thousands of DNA segments (analogous to genes) whether or not they have a perceptible influence on whole-plant phenotype. With such a large number of identifiable "genes," or molecular markers, it is often possible to identify a marker that is closely linked to a gene conferring pest resistance. Using these markers, breeders can identify plants that carry resistance genes without the need for bioassays.

Numerous methods and modifications of methods are currently available for fingerprinting DNA in the search for molecular markers. The most common systems include restriction fragment length polymorphism (RFLP) [15], random amplified polymorphism detection (RAPD) [163], and polymerase chain reaction (PCR) [101]. It is beyond the scope of this chapter to describe these in detail, as has been done by several investigators [52,153]. Basically, the methods involve extracting and processing plant DNA, then separating DNA fragments via differential migration rates on an agarose gel, which is processed to make DNA fragments visible, and finally identifying diagnostic banding patterns on the gels.

In order for molecular markers to be useful in this context, they must be closely linked to genes conferring resistance. This involves unequivocally identifying resistant and susceptible plants and then finding molecular markers or polymorphisms that are always associated with resistance and never with susceptibility. In Table 3 are but a few of the recently identified molecular markers for resistance to various pests.

In order to be economical, molecular analysis must be less expensive (and time consuming) than bioassays, or provide improved accuracy to justify the

Table 3 Molecular Markers Linked to Pest Resistance

Crop	Pest	Marker type	Reference
Barley	ym4 virus	RFLP	*Theoret. Appl. Genet.* 86:689–693
Barley	Russian wheat aphid	PCR	*Crop Sci.* 34:655–569
Barley	Powdery mildew	RFLP	*Genome* 35:1019–1025
Common bean	Mosaic virus	RAPD	*Crop Sci.* 34:1061–1066
Common bean	Rust	RAPD	*Theoret. Appl. Genet.* 85:745–749
Lettuce	Downy mildew	RAPD	*Genome* 34:1021–1027
Maize	Northern corn leaf blight	RFLP	*Theoret. Appl. Genet.* 87:537–544
Maize	European corn borer	RFLP	*Heredity* 70:648–659
Mungbean	Bruchid	RFLP	*Theoret. Appl. Genet.* 84:839–844
Oats	Crown rust	RAPD	*Genome* 36:818–820
Oats	Stem rust	RAPD	*Theoret. Appl. Genet.* 85:702–705
Oats	Crown rust	RFLP	*Crop Sci.* 34:940–944
Pea	Mosaic virus	RFLP	*Theoret. Appl. Genet.* 85:609–615
Soybean	Phytophthora	RAPD	*Soyb. Genet. News* 20:112–117
Tomato	Various insects	RFLP	*Crop Sci.* 27:797–803
Tomato	Pseudomonas	RAPD	*Proc. Natl Acad. Sci. USA* 88:2336–2340
Tomato	Root-knot nematode	RFLP	*Plant J.* 2:971–982
Tomato	Root-knot nematode	RFLP	*Theoret. Appl. Genet.* 82:529–536
Tomato	Root-knot nematode	PCR	*Plant Mol. Int. J. Fund. Res. Genet. Eng.* 16:647–611
Wheat	Powdery mildew	RFLP	*Theoret. Appl. Genet.* 86:959–963
Tomato	Root-knot nematode	RFLP	*Theoret. Appl. Genet.* 81:661–667

RFLP, restriction fragment length polymorphism; PCR, polymerase chain reaction; RAPD, random amplified polymorphism detection.

added expense. In most cases at present, molecular techniques are expensive and require fairly elaborate equipment. As the technology develops and costs decline, the use of marker-assisted selection may play an important role in applied breeding programs.

B. Gene Transfer and Production of Engineered Plants

Despite many successes, there are certain constraints and problems that remain unsolved by traditional breeding methods. One serious drawback is that new sources of resistance may not be readily available to the breeder when needed or sources may only be found in unrelated species that are incompatible for crossing. Resistance is sometimes found in unadapted land races or weedy relatives that exhibit unacceptable agronomic characters. Introgression of even single-gene resistance characters via conventional interspecific crosses into elite cultivars or populations usually requires a minimum of 5–10 years. Owing to this extended time period and the inevitable linkage of one or more undesirable traits, few plant breeders willingly engage in extensive backcross programs to enhance host resistance levels.

To overcome this problem in crop variety development, some private and public plant breeders have developed techniques to transfer individual cloned genes into cultured cells of crop plants for enhanced levels of pest resistance. Successful transfer and expression of genes in over 47 crop plant species have been reported [45]. The principal advantage of this approach is that pest resistance can, in theory, be rapidly introduced into elite varieties without compromising other desirable agronomic traits. In practice, this technology provides another tool for plant breeders rapidly to enhance germplasm for pest resistance and certain other characters.

1. Methods for Gene Transfer and Production of Engineered Crops

Bacteria-Mediated Transfer. The most efficient methods for delivery of cloned genes to dicotyledonous crop plants exploits the natural infection process of the soil bacterium *Agrobacterium tumefaciens* [156,169]. This bacterium completes its life cycle by inserting its own DNA into a chromosome of the host plant. Under natural conditions, virulent strains of *A. tumefaciens* infect a wide array of dicotyledonous hosts in 144 genera and 60 families [3] to cause formation of galls or tumors on roots and shoots. To avoid the deleterious effects of galls and tumors on the host plant, certain genetic material from *A. tumefaciens* has been removed (disarmed). This modification can be further exploited, because genes of interest that are inserted between the flanking borders are also transferred to the host. Recently, *A. tumefaciens* has been engineered for successful gene transfer to cereal crops such as rice [58] and maize [62].

Electroporation or Chemical Treatment of Cultured Protoplasts. Modifications of *A. tumefaciens* have allowed the efficient transfer of cloned genes to such crops as tomato, tobacco, potato, and others in the nightshade family (Solanaceae). Transformation efficiency by this method, unfortunately, is low for certain crops that are not members of this family. Cotton is one exception of economic importance, although only certain genotypes can be transformed easily. To overcome these problems, electroporation (opening small pores in cell membranes using electrical current in order to permit entry of desired DNA fragments) or chemical treatment of cultured protoplasts (individual plant cells with cell walls enzymatically removed) have been used for gene transfer [127]. These approaches have generally proven to be successful, but high transformation rates are strain specific and sterility of regenerated plants from in vitro culture is common in many crops.

Particle Bombardment. An alternative to bacterial and protoplast methods of gene transfer is the particle bombardment, or "biolistic," procedure [105,125]. In this process, micron-sized particles of gold or tungsten are coated with DNA and accelerated to high speeds sufficient for nonlethal penetration of cell walls and membranes. Because transfer of genes in this system is a physical process, particle bombardment presumably is host-range independent, unlike techniques utilizing *A. tumefaciens.* Indeed, many of the major crops in the United States have taken up genes via particle bombardment [71], and field trials to evaluate the transferred genes have been conducted in cotton, soybean, maize, and other crops. Commercial availability of insect and herbicide-tolerant cotton and maize developed in this manner is tentatively scheduled for the 1990s.

Although successful in several cases, there are certain limitations of the particle bombardment process that make it less than the ideal gene transfer system. A primary example is the high initial purchase price ($10,000 or greater) of a commercially available biolistic device. For some public laboratories or small companies, this cost can be prohibitive. Shop-built devices can be constructed at minimal cost, but their efficiency and reliability will be unknown. Another drawback is the relatively low transformation rates of particle bombardment which can be 10- to 100-fold less than those of *A. tumefaciens.* Finally, particle bombardment can give rise to plants where only a portion of a plant is transformed (chimerism). Solutions to these and other related problems are urgently needed if particle bombardment is to serve as a general and reliable tool for plant transformation.

2. Use of Cloned Genes in Engineered Plants for Pest Management

Genetic engineering has been used to introduce into crop plants foreign genes for the control of insect pests and viral and fungal pathogens.

Engineered Genes for Insect Resistance.   The primary arthropod targets of genetic engineering have been lepidopteran and coleopteran pests.

Bacillus Thuringiensis.   Successful control of certain insects in the orders Lepidoptera, Coleoptera, and Diptera has resulted from insect exposure to toxic proteins, sometimes referred to as delta endotoxins, produced by *Bacillus thuringiensis* (Bt), a gram-positive, spore-forming soil bacterium [35,60]. Approximately 20 Bt genes that encode delta endotoxins specific against lepidopteran insects have been isolated and characterized in detail [60]. All genes have been placed into the *cry*IA(a), *cry*IA(b), *cry*IA(c), *cry*IB, *cry*IC, or *cry*ID class. Additional genes effective against the coleopteran Colorado potato beetle, *Leptinotarsa decemlineata*, have been placed in the *cry*IIIA category [96]. The majority of the Bt genes have been isolated from extrachromosomal molecules (plasmids) that apparently can be transferred among different strains of the bacterium.

In recent years, various Bt genes have been cloned, sequenced, and characterized at the biochemical level [60]. In early work, certain genes were transferred and expressed in tobacco, *Nicotianum tabacum* [149]; tomato, *Lycopersicum esculentum* [42]; and cotton [148]. These unmodified genes conferred a low level of protection against certain lepidopteran pests, but the Bt proteins were generally less than 0.01% of the total soluble leaf protein. To overcome these problems for effective control of the cotton bollworm, *Helicoverpa zea*, native *cry*IA(b) and *cry*IA(c) genes from Bt var. *kurstaki* were extensively modified by the Monsanto Company to increase plant protection levels in the greenhouse 5- and 70-fold over the sensitive untransformed controls and increase the Bt protein content to 0.1% of the total cell protein content [112]. The *cry*IA(c) gene provided a twofold increase in control over the *cry*IA(b) gene, whereas a hybrid gene composed of *cry*IA(c) and *cry*IA(b) sequences provided a fivefold advantage over the *cry*IA(b) gene. In a greenhouse study [65], growth and survival of the tobacco budworm, *Heliothis virescens*, larvae on cotton lines carrying the *cryIA*(c) and *cry*IA(b) genes were significantly less than for larvae on the sensitive "Coker 312" control plants. Field trials in Mississippi with the same transgenic cotton lines [65] sustained 2.5-fold less damaged squares, 4-fold fewer larvae, and 1.5- to 10-fold less damaged bolls than the control plants. Under tobacco budworm infestation, lint yields were significantly higher in the transgenic material as compared with the Coker 312 control. No significant differences in yield were detected between the transgenics and Coker 312 under insect-free conditions. Additional greenhouse and field studies in Texas [16] with the Monsanto engineered cotton lines resulted in control levels similar to the Mississippi trials for tobacco budworm. Damage to bolls and flower buds by the bollworm was reduced 9- and 25-fold, respectively, on transgenic versus control plants in the greenhouse. Field trials with the Monsanto lines in Arizona [164] showed effective control of the pink bollworm, *Pectinophora gossypiella*; the cotton leaf perforator, *Bucculatrix thurberiella*; beet armyworm,

Spodoptera exigua; and the saltmarsh caterpillar, *Estigmene acrea*. All of the previously mentioned studies showed good pest control levels, but boll size and seed weight were smaller for certain transgenic lines versus control plants in the Mississippi trials [66]. Variation in lint percentage and fiber properties for some of these transgenic lines has also been reported [148,166]. Through extensive breeding efforts, agronomically acceptable varieties of transgenic Bt cotton have been developed, and in 1996, were planted on more than one million hectares in the United States.

In addition to the cotton studies previously mentioned, the potential of cloned Bt genes for control of the European corn borer, *Ostrinia nubilalis* in maize, *Zea mays*, was recently evaluated under field conditions [73]. A synthetic *cry*IA(b) gene was transferred by particle bombardment to maize, where the Bt insecticidal toxin was produced and expressed at high levels. Field-grown transgenic plants were infested with neonate larvae and with up to 96 times the economic threshold of second-generation egg masses. Foliar damage from first generation insects was two- to threefold greater in the sensitive controls versus the transgenic material that expressed the modified *cry*IA(b) gene. After the second insect generation, internal stalk damage was reduced by an average of 14-fold in five transgenic populations as compared with the untransformed controls.

The unmodified Bt gene was also evaluated in early experiments with transgenic tomato and shown to provide adequate control for infestations of tobacco hornworm, *Manduca sexta*, but acceptable protection was not observed against the tomato fruitworm, *Helicoverpa zea*, or tomato pinworm, *Keiferia lycopersicella* [32]. Subsequent experiments [152] showed that modified *cry*IA(b) and *cry*IC genes in transgenic tomato and tobacco provided increased resistance against *M. sexta, H. virescens,* and *S. exigua*.

Several lepidopteran pests cause significant economic damage to rice, *Oryza sativa*, both in tropical and temperate environments [57]. The *cry*IA(b) gene was recently modified to reflect rice sequences and introduced into a line extensively used in laboratory studies [46]. Significant protection was observed in transformed lines against the striped stem borer, *Chilo suppressalis*, but only moderate levels of protection were shown against the leaf folder, *Cnaphalocrosis medinalis*.

*Cry*IIIA insecticidal proteins from Bt var. *tenebrionis* exhibit high activity levels against the Colorado potato beetle [74]. The *cry*IIIA gene was recently modified and transferred to potato for protection against this economically important pest [111]. High levels of *cry*IIIA protein were expressed in the transgenic potato that displayed excellent resistance in multiple field trials to all growth phases of the beetle. Similar results were obtained with modified *cry*IIIA genes expressed in potato [2] and tobacco plants [144]. Transgenic potato lines carrying a modified *cry*IA(b) gene showed good resistance against the potato tuberworm, *Phthorimea operculella*, in the leaf and in tubers stored up to 7 months [63].

Trypsin Inhibitors.   Certain plant compounds that inhibit serine protease activity have been shown to provide protection against insect attack [140]. The gene for the Bowman-Birk type trypsin inhibitor from cowpea, *Vigna unguiculata*, has been cloned [59] and transferred to tobacco to enhance resistance levels against the tobacco budworm.

Durability of Transgene Resistance.   The insect resistance genes described above are essentially monogenic in nature, and they have the potential to lose their effectiveness in the same way as other specific resistance. Considerable worry and attention has been given to the question of how durable the resistance will be and how to prolong its effectiveness.

Jenkins [65] has recently outlined a general approach to prolong the usefulness of specific insect resistance genes in cotton that could be applied to other crop situations. The principal idea behind this strategy is that modified resistance genes should form part of, but never replace, an overall integrated pest management scheme. With multiple pest problems for cotton and most other crops, a single approach cannot simultaneously address all problems. In practice, this could mean combining multiple engineered genes with different properties and sites of action with natural plant resistance traits such as pubescence, thick leaf cuticle layers, and canopy type. Proper cultural practices would complement the genetic components and frequent monitoring of pest populations for breakdown of resistance would be required. Others [50,51,79] have proposed expressing modified genes in specific organs of crops or planting mixtures of transgenic and nontransgenic material to limit the development of insect resistance. One recent approach to mitigate selection pressure used a chemically responsive promoter to control expression of a modified Bt gene in tobacco [162]; infested crop plants would be treated with a chemical to trigger onset of resistance. A potential disadvantage to this strategy is that the crop plants must be adequately scouted and sprayed with the chemical inducer in a timely manner for proper expression of the resistance gene.

Engineered Resistance to Virus Diseases.   Extensive research in the last 5 years has produced numerous engineered crop plants that show resistance to various viral groups [142]. The different modes of entry, replication, and transmission of distinct pathogenic viruses have led to at least six different strategies to develop virus-free crops [108]. The first and most common approach has been to clone and transfer genes that encode proteins, known as coat proteins, that envelope the viral particles [142]. These engineered proteins are thought to prevent unraveling of the capsid proteins or to bind to cellular receptors that allow viral entry. One recent example of coat-protein engineering involved the yellow leaf curl geminivirus that is transmitted by the sweetpotato whitefly, *Bemisia tabaci* [76]. Under proper field conditions, this virus can cause complete tomato crop loss in tropical and temperature regions [25]. The capsid coat protein gene V1 from the yellow leaf curl virus was cloned and transferred to tomato by *Agrobacterium* [76]. Transformed and susceptible control plants were inoculated with the virus by the whitefly vector. Detection of the virus and onset of disease symptoms were delayed 30 days as compared with the controls. Neither the virus nor the symptoms were detected 4 months after initial inoculation of the transformed material. A second inoculation of the transformants led to milder symptoms and disappearance of the virus in 4.5 months. This study is the first to demonstrate coat protein protection in transformed plants against a geminivirus. Other examples of coat-protein genes include resistance against lettuce mosaic polyvirus [34] and tobacco mosaic virus [124]. Details of the coat-protein strategy for virus resistance were recently published [142].

A second and related method uses coding sequences of untranslatable coat-protein genes [108]. High levels of resistance were observed in transgenic tobacco that carried the untranslated "N" gene coding sequence of the tomato spotted wilt virus. Yet another approach [104] used viral sequences placed in the opposite orientation to produce an "antisense" vector that inhibited tobacco mosaic virus infection in tobacco 100-fold over inoculated controls. A defective, 54K replicase gene from cucumber mosaic virus was transferred in another scheme [7] to induce high resistance levels in tobacco. High levels of resistance to tobacco mosaic virus were detected when the 54K gene, a so called "nonstructural" gene, was expressed in transgenic tobacco [84]. A similar study transferred genes for viral nonstructural proteins that conferred high resistance levels against the tobacco vein mottling virus in transformed tobacco plants [87].

The aforementioned studies showed engineered resistance to a specific or related viral group. Distantly related viruses were almost never controlled by these different methods. In contrast, a "ribosome-inhibiting protein" has been isolated from pokeweed, *Phytolacca americana*, and found to act as a general inhibitor of infection for several unrelated viruses [82]. A gene for this protein was introduced into tobacco where it conferred high resistance levels when challenged with different viral strains. The inhibitory factor is thought to act by interfering with initial synthesis of viral proteins and thus may provide a general method to develop broad spectrum resistance. Experiments with additional viral and host strains will be needed to assess the practical significance of these initial experiments.

Engineered Genes Against Fungi and Bacteria. The best characterized antifungal proteins to date are chitinases and ß-glucanases [161] that break up cell wall components and inhibit growth of many fungi. The chitinases have been placed into four classes of which Class I is localized primarily in plant vacuoles and Class II proteins are generally found in extracellular spaces. A new chitinase Class V has recently been described [106]. It is active under laboratory conditions against *Trichoderma viride* and *Alternaria radicina*. A Class I endochitinase gene isolated from the common bean, *Phaseolus vulgaris*, provided enhanced levels of resistance in tobacco against the fungus *Rhizoctonia solani* [19]. High resistance levels were also obtained against the brown-spot pathogen, *Alternaria longipes*, when a chitinase gene from the bacterium *Serratia marcescens* was transferred to and expressed in tobacco [36].

The ß-glucanases fall into three classes where Class I proteins are antifungal [27] and Classes II and III are primarily antiviral in nature [157]. Modified Class I chitinase and ß-1,3-glucanase genes produced high resistance levels in tobacco when introduced together [98]. Other antifungal agents include the "ribosome-inhibiting protein" that confers increase resistance in tobacco to *Rhizoctonia solani* [83] and a small, cysteine-rich seed protein from *Raphanus sativus* that displays antifungal properties [146].

The *Pto* gene in tomato confers resistance to specific races of *Pseudomonas syringae* pv. *tomato*, the causal agent of bacterial speck. This gene has been cloned, sequenced, and transferred to susceptible varieties of tomato [88]. Although the susceptible plants showed typical symptoms of bacterial speck, transformed lines carrying the *Pto* gene in greenhouse studies showed high levels of resistance against a purified strain of *P. syringae*. Inheritance studies showed resistance in the transgenic plants was controlled by a single, dominant gene. Sequence analysis indicated that the cloned *Pto* gene is a protein kinase and suggested that the *Pto* locus is involved in the induction of host defenses. In a similar case [138], the cloned rice *Xa21* gene was transferred to rice and provided resistance against the bacterial pathogen *Xanthomonas oryzae* pv. *oryzae*, race 6. The *XA21* locus also encodes for a protein kinase that may be involved in perception and activation of an intracellular defense response. Finally, the *RPS2* gene from *Arabidopsis* was cloned and transferred to tomato for resistance to *Pseudomonas syringae* [17]. The structure of *RPS2* is different from the *Pto* and *Xa21* defense genes, but it is also postulated to modulate defense signals within the host.

In addition to identifying the transferring resistance genes, scientists involved with engineered genes for long-term control of diseases have generally concentrated on the goal of stabilizing the composition of pathogen populations. One idea has been to design vectors with promoters that are tissue or organ-specific and inducible by a wide range of pathogen strains [30]. A modification of this approach would involve the transfer of a specific resistance gene with a corresponding avirulence fungal gene into the host plant to create nonspecific pathogen resistance. Owing to the nature of this problem, several years of field and laboratory work will be required to select the best management schemes that provide long-term benefits.

3. Limitations and Future Prospects of Engineering for Pest Management

There are at least four potential limitations to the broader use of engineering plant varieties for pest management:

1. *Identification of new resistance genes*: In spite of the successes with Bt and certain viruses and fungi, little is known about the genetics and biochemistry of resistance against many economically important plant pests. Clearly, more basic research that can be transferred to applied situations is urgently needed. In addition, it is desirable that new resistance genes do not promote the selection of new pest strains that will overcome the engineered resistance. One approach would be to develop resistance levels that maintain pest populations just below the economic threshold to prolong the usefulness of the introduced gene(s).

2. *Efficient delivery and expression of cloned genes*: Agrobacterium-mediated gene transfer has been touted as the preferred method for gene delivery, but as previously mentioned, the efficiency and ease of transformation can be substantially reduced for certain dicot crops. Moreover, only certain varieties within a crop are routinely amenable to efficient gene transfer by this method. Particle bombardment was developed to deliver genes to any variety, but even this method can be hampered by problems of in vitro culture (e.g., male sterility) in certain crop varieties.

3. *Regulatory policies and statutes*: The United States Food and Drug Administration has recently approved Calgene's FLAVR SAVR tomatoes for marketing [9], which should make it easier for other engineered crops to obtain governmental approval. In other rulings, the U.S. Patent and Trademark Office has recently awarded patents to three private companies for genetically engineered cotton and all plants containing modified chitinase or Bt genes [9]. Scientists may need to obtain a license from these companies to conduct transgenic research in these areas. This legal requirement may constrain future use of gene transfer to develop new pest-resistant varieties. Clearly, a uniform statute under a single governmental agency that applies to all transgenic plants is needed to resolve this and related regulatory issues.

4. *Public Acceptance of Engineered Food and Fiber*: Several of the advances in gene transfer technology and isolation of pest resistance genes have been market driven. Future success in this area will also depend upon public willingness to consider engineered plants as safe and reliable as crops developed by traditional methods.

V. CONCLUSIONS

Past success in breeding for pest resistance has helped sustain the productivity and (for the most part) profitability of production agriculture, and it has helped to relive the potential environmental hazards associated with sole reliance on chemical pest control. The growing, and at times irrational, public concern for the environment, and world economic pressures command continued and renewed efforts to enhance genetic crop defenses through breeding. Advances in biotechnology, both marker-assisted selection and gene transfer technologies, have opened new frontiers in breeding for pest resistance. However, they are not a panacea. The work of incorporating new genes, even using new selection tools, into new, agronomically superior cultivars with continually improved yield and quality, and evaluating those new cultivars under "real world" pest pressure will require the continued and strengthened *team* effort of specialists in applied breeding, pathology, entomology, nematology, and biotechnology. Basic research in mechanisms of host and pest interactions should continue together with field evaluations to develop the best long-term approach for combining different resistance genes and agronomic performance.

REFERENCES

1. Acreman, T.M., and A.F.G. Dixon. 1986. The role of awns in the resistance of cereals to the grain aphid, *Sitobion avenae. Ann. Appl. Biol.* 61:289–294.
2. Adang, M.J., M.S. Brody, G. Cardineau, N. Eagan, R.T. Roush, C.K. Shoemaker, A. Jones, J.V. Oakes, and K.E. McBride. 1993. The reconstruction and expression of a *Bacillus thuringiensis cry*IIIA gene in protoplasts and potato plants. *Plant Mol. Biol.* 21:1131–1145.
3. Agrios, G.N. 1978. *Plant Pathology.* 2nd ed. New York: Academic Press.
4. Alexander, H.M., and P.J. Bramel-Cox. 1991. Sustainability of genetic resistance. In *Plant Breeding and Sustainable Agriculture: Considerations for Objective and Methods, CSSA Spec. Publ. 18.* D.A. Sleper, T.C. Barker, and P.J. Bramel-Cox, eds. Madison, Wisconsin: Crop Science Society of America, pp 11–27.
5. All, J.N., H.R. Boerman, and J.W. Todd. 1989. Screening soybean genotypes in the greenhouse for resistance to insects. *Crop Sci.* 29:1156–1159.
6. Amorim, L., A. Bergamin-Filho, and B. Hau. 1993. Analysis of progress curves of sugarcane smut on different cultivars using functions of double sigmoid pattern. *Phytopathology.* 83:933–936.
7. Anderson, J.M., P. Palukaitis, and M. Zaitlin. 1992. A defective replicase gene induces resistance to cucumber mosaic virus in transgenic tobacco plants. *Proc. Natl. Acad. Sci. USA* 89:8759–8763.
8. Anderson, M.D., and G.J. Brewer. 1991. Mechanisms of hybrid sunflower re-

sistance to the sunflower midge (Diptera: Cecidomyiidae). *J. Econ. Entomol.* 84:1060–1065.

9. Anonymous. 1994. Broad patenting trend continues. *NBIAP News Report*, p 3.

10. Anonymous. *Rust Scoring Guide*. El Batan, Mexico: International Center for Maize and Wheat Improvement.

11. Armstrong, A.M. 1991. Field evaluations of pigeon pea genotypes for resistance against pod borers. *J. Agric. Univ. P.R.* 75:73–78.

12. Bailey, J.C., A.L. Scales, and W.R. Meredith, Jr. 1984. Tarnished plant bug (Heteroptera: Miridae) nymph numbers decreased on caged nectariless cottons. *J. Econ. Entomol.* 77:68–69.

13. Bailey, J.C. 1986. Infesting cotton with tarnished plant bug (Heteroptera: Miridae) nymphs reared by improved laboratory rearing methods. *J. Econ. Entomol.* 79:1410–1412.

14. Barfield, M.E., D.S. Calhoun, C. Overstreet, and W. D. Caldwell. 1993. Field performance of selected root-knot nematode resistant cotton genotypes. *Proceedings of the Beltwide Cotton Conference.* Vol. 2. Memphis, Tennessee: National Cotton Council, pp 621–623.

15. Beckmann, J.S., and M. Soller. 1983. Restriction fragment length polymorphisms in genetic improvement: methodologies, mapping and costs. *Theoret. Apply. Genet.* 67:35–43.

16. Benedict, J.H., E.S. Sachs, D.W. Altman, W.R. Deaton, R.J. Kohel, D.R. Ring, and S.A. Beberich. 1996. Field performance of cotton expressing transgenic cryIA insecticidal proteins for resistance to *Heliothis virescens* and *Helicoverpa zea* (Lepidoptera: Noctuidae). *J. Econ. Entomol.* 89:230–238.

17. Bent, A.F., B.N. Kunkel, D. Dahlbeck, K.L. Brown, R. Schmidt, J. Giraudat, J. Leung, and B. Staskawicz. 1994. RPS2 of *Arabidopsis thaliana*: a leucine-rich repeat class of plant disease resistance genes. *Science* 265:1856–1860.

18. Bowman, D.T., and C.C. Green. 1991. Screening cotton germplasm for Columbia lance and reniform nematode resistance. *Proceedings of Beltwide Cotton Conference.* Vol. 1. Memphis, Tennessee: National Cotton Council, pp 551–552.

19. Broglie, K., I. Chet, M., Holiday, R. Cressman, P. Biddle, S. Knowlton, C.J. Mauvais, and R. Broglie. 1991. Transgenic plants with enhanced resistance to the fungal pathogen *Rhizoctonia solani*. *Science* 254:1194–1197.

20. Busey, P., R.M. Giglin-Davis, and B.J. Center. 1993. Resistance in *Stenotaphrum* to the sting nematode. *Crop Sci.* 33:1066–1070.

21. Calhoun, D.S., G. Gebeyehu, A. Miranda, S. Rajaram, and M. van Ginkel. 1994. Choosing evaluation environments to increase wheat yield under drought stress. *Crop Sci.* 34:673–678.

22. Calhoun, D.S., P.A. Burnett, J. Robinson, and H.E. Vivar. 1991. Field resistance to Russian wheat aphid: I. Symptom expression. *Crop Sci.* 31:1464–1467.

23. Calhoun, D.S., P.A. Burnett, J. Robinson, H.E. Vivar, and L. Gilchrist. 1991. Field resistance to Russian wheat aphid: II. Yield assessment. *Crop Sci.* 31:1468–1472.

24. Chesnokov, P.G. 1962. *Methods of Investigating Plant Resistance to Pests.* Published for the National Science Foundation and U.S. Department of Agriculture by Israel Program for Scientific Translation.

25. Cohen, S., and F.E. Nitzany. 1966. Transmission and host range of the tomato yellow leaf curl virus. *Phytopathology*. 78:127–1131.

26. Comstock, J.C., J.D. Miller, and D.F. Farr. 1994. First report of dry top rot of sugarcane in Florida: symptomology, cultivar reactions, and effect on stalk water flow rate. *Plant Dis*. 78:428–431.

27. Cornelissen, B.J.C., and L.D. Melchers. 1993. Strategies for control of fungal diseases with transgenic plants. *Plant Physiol*. 101:709–712.

28. Das, M.K., S. Rajaram, C.C. Mundt, and W.R. Kronstad. 1992. Inheritance of slowrusting resistance to leaf rust in wheat. *Crop Sci*. 32:1452–1456.

29. de Ponti, O.M.B., and C. Mollema. 1992. Emerging breeding strategies for insect resistance. *Proc. Symposium on Plant Breeding in the 1990's*. H.T. Stalker and J.P. Murphy, eds. Wallingford, UK: CAB International, pp 323–346.

30. de Wit, P.J.G.M. 1992. Molecular characterization of gene-for -gene systems in plant-fungus interactions and the application of avirulence genes in control of plant pathogens. *Annu. Rev. Phytopathol*. 30:391–418.

31. De Leon, C., and S. Pandey. 1989. Improvement of resistance to ear and stalk rots and agronomic traits in tropical maize gene pools. *Crop Sci*. 29:12–17.

32. Delannay, X., B.J. LaVallee, R.K. Proksch, R.L. Fuchs, S.R. Sims, J.T. Greenplate, P.G. Marrone, R.B. Dodson, J.J. Augustine, J.G. Layton, and D.A. Fischhoff. 1989. Field performance of transgenic tomato plants expressing the *Bacillus thuringiensis* var. *kurstaki* insect control protein. *Bio/Technology* 7:1265–1269.

33. Dent, D. 1991. *Insect Pest Management*. Wallingford, UK: CAB International.

34. Dianant, S., F. Blaise, C. Kusiak, S. Astier-Manifacier, and J. Albouy. 1993. Heterologous resistance to potato virus Y in transgenic tobacco plants expressing the coat protein gene of lettuce mosaic potyvirus. *Phytopathology*. 83:818–824.

35. Dulmage, H.T. 1981. Insecticidal activity of isolates of *Bacillus thuringiensis* and their potential for pest control. In *Microbial Control of Pests and Plant Diseases 1970-1980*. H.D. Burges, ed. New York: Academic Press, pp 193–222.

36. Dunsmuir, P. and T. Suslow. 1989. Structure and regulation of organ-and tissue-specific genes: chitinase genes in plants. In *Cell Culture and Somatic Genetics of Plants*, Vol. 6. J. Schell and I. Vasil, eds. New York: Academic Press, pp 215–227.

37. Duveiller, E. 1994. A pictorial series of disease assessment keys for bacterial leaf streak of cereals. *Plant Dis*. 78:137–141.

38. Ehlenfeldt, M.K., A.W. Stretch, and A.D. Draper. 1993. Sources of genetic resistance to red ringspot virus in a breeding blueberry population. *HortSci*. 28:207–208.

39. Falconer, D.S. 1981. *Introduction to Quantitative Genetics*. New York: Longman.

40. Fehr, W.R. 1987. *Principles of Cultivar Development*. Vol. 1. New York: Macmillan.

41. Fehr, W.R., and H.H. Hadley. 1980. *Hybridization of Crop Plants*. Madison, Wisconsin: Crop Science Society of America.

42. Fischhoff, D.A., K.S. Bowdish, F.J. Perlak, P.G. Marrone, S.M. McCormick,

J.G. Niedermeyer, D.A. Dean, K.K. Kretzmer, E.J. Mayer, D.E. Rochester, S.G. Rogers, and R.T. Fraley. 1987. Insect tolerant transgenic tomato plants. *Bio/Technology* 5:807–813.

43. Flor, H.H. 1956. The complementary genic systems in flax and flax rust. *Adv. Genet.* 8:29–54.

44. Formusoh, E.S., G.E. Wilde, J.H. Hatchett and R.D. Collins. 1994. Resistance to the Russian wheat aphid (Homoptera: Aphididae) in wheat and wheat-related hybrids. *J. Econ. Entomol.* 87:241–244.

45. Fraley, R.T. 1992. Sustaining the food supply. *Bio/Technology* 10:40–43.

46. Fujimoto, H., K. Itoh, M. Yamamoto, J. Kyozuka, and K. Shimamoto. 1993. Insect resistant rice generated by introduction of a modified δ-endotoxin gene by *Bacillus thuringiensis. Bio/Technology* 11:1151–1155.

47. Gahukar, R.T. 1990. Reaction of locally improved pearl millets to three insect pests and two diseases in Senegal. *J. Econ. Entomol.* 83:2102–2106.

48. Gallun, R.L., and G.S. Khush. 1980. Genetic factors affecting the expression and stability of resistance. In *Breeding Plants Resistant to Insects.* F.G. Maxwell and P.R. Jennings, eds. New York: Wiley, pp 63–86.

49. George, B.W., F.D. Wilson, and R.L. Wilson. 1983. Methods of evaluating cotton for resistance to pink bollworm, cotton leaf perforator, & lygus bugs. In *Host Plant Resistance Research Methods for Insects, Diseases, Nematodes and Spider Mites in Cotton.* So. Coop. Ser. Bul. 280. Starkville, Mississippi: Miss. Agric. Exp. Stn., pp 41–45.

50. Gould, F. 1988. Evolutionary biology and genetically engineered crops. *Bioscience* 38:26–33.

51. Gould, F. 1988. Genetic engineering, integrated pest management and the evolution of pests. *TIBTECH.* 6:515–518.

52. Gustofson, J.F., and R. Appels, eds. 1988. *Chromosome Structure and Function.* New York: Plenum Press.

53. Hanson, C.H., T.H. Busbice, R.R. Hill, O.J. Hunt, and A.J. Oaks. 1972. Directed mass selection for developing pest resistance and conserving germplasm in alfalfa. *J. Environ. Qual.* 1:106–111.

54. Hare, J.D. 1983. Manipulation of host suitability for herbivore pest management. In *Variable Plants and Herbivores in Natural and Managed Systems.* R.F. Denno and M.S. McClure, eds. New York: Academic Press, pp 655–580.

55. Harris, M.K., ed. 1979. *Biology and Breeding for Resistance to Arthropods and Pathogens in Agricultural Plants. Proc. International Short Course in Host Plant Resistance.* College Station, Texas: Texas A& M University.

56. Heald, C.M., and A.F. Robinson. 1990. Screening for resistance to *Rotylenchulus* species. In. *Methods for Evaluating Plant Species for Resistance to Plant-parasitic Nematodes.* J.L Star, ed. Hyattsville, Maryland: Society of Nematologists, pp 42–50.

57. Herdt, R.W. 1991. Research priorities for rice biotechnology. In *Rice Biotechnology.* G.S. Kush and G.H. Toenniessen, eds. Wallingford, UK: CAB International, pp 19–54.

58. Hiei, Y., S. Ohta, T. Komari, and T. Kumashiro. 1994. Efficient transforma-

tion of rice (*Oryza sativa* L.) mediated by *Agrobacterium tumefaciens* and sequence analysis of the boundaries of the T-DNA. *Plant J.* 6:271–282.

59. Hilder, V.A., R.F. Barker, R.A. Samour, A.M.R. Gatehouse, J.A. Gatehouse, and D. Boulter. 1989. Protein and cDNA sequences of Bowman-Birk protease inhibitors from the cowpea (*Vigna unguiculata* Walp.). *Plant Mol. Biol.* 13:701–710.

60. Hofte, H., and H.R. Whitely. 1989. Insecticidal crystal proteins of *Bacillus thuringiensis. Microbiol. Rev.* 53:249–255.

61. Holbrook, C.C., and J.P. Noe. 1992. Resistance to the peanut root-knot nematode (*Meloidogyne arenaria*) in *Arachis hypogaea. Peanut Sci.* 19:35–37.

62. Ishida, Y., H. Saito, S. Ohta, Y. Hiei, T. Komari, and T. Kumashiro. 1996. High efficiency transformation of maize (*Zea mays* L.) mediated by *Agrobacterium tumefaciens. Nature Biotech.* 14:745–750.

63. Jansens, S., M. Cornelissen, R. de Clercq, A. Reynaerts, and M. Peferoen. 1995. *Phthorimaea operculella* (Lepidoptera: Gelechiidae) resistance in potato by expression of the *Bacillus thuringiensis cry*IA(b) insecticidal crystal protein. *J. Econ. Entomol.* 88:1469–1476.

64. Jenkins, J.N., and W.L. Parrott. 1992. Equipment for mechanically harvesting eggs of *Heliothis virescens* (Lepidoptera: Noctuidae). *J. Econ. Entomol.* 85:2496–2499.

65. Jenkins, J.N. 1993. Use of *Bacillus thuringiensis* genes in transgenic cotton to control Lepidopterous insects. *Am. Chem. Soc. Symp.* 524:267–280.

66. Jenkins, J.N., W.L. Parrott, and J.C. McCarty. 1991. Field performance of transgenic cotton containing the B.t. gene. *Proceedings of the Beltwide Cotton Conference.* Vol. 1. Memphis, Tennessee: National Cotton Council, p 576.

67. Johnson, D.A., and E.C. Gilmore. 1979. Breeding for resistance to pathogens in wheat. *Biology and Breeding for Resistance to Arthropods and Pathogens in Agricultural Plants, Proc. International Short Course in Host Plant Resistance.* College Station, Texas: Texas A&M University, pp 261–302.

68. Johnson, J.W., and G.L. Teetes. 1979. Breeding for arthropod resistance in sorghum. *Biology and Breeding for Resistance to Arthropods and Pathogens in Agricultural Plants, Proc. Int'l Short Course in Host Plant Resistance.* College Station, Texas: Texas A&M University, pp 168–180.

69. Johnson, D.A., and Gilmore. 1980. Breeding for resistance to pathogens in wheat. *Biology and Breeding for Resistance to Arthropods and Pathogens in Agricultural Plants, Proceedings of the International Short Course in Host Plant Resistance.* College Station, Texas: Texas A&M University, pp 263–275.

70. Kindler, S.D., and S.M. Spomer. 1986. Biotypic status of xi greenbug (Homoptera: Aphididae) isolates. *Environ. Entomol.* 15:567–572.

71. Klein, T.M., R. Arentzen, P.A. Lewis, and S. Fitzpatrick-McElligott. 1992. Transformation of microbes, plants, and animals by particle bombardment. *Bio/Technology* 10:286–291.

72. Knott, D.R., and B. Yadav. 1993. The mechanism and inheritance of adult plant leaf rust resistance in 12 wheat lines. *Genome* 36:872–883.

73. Koziel, M.G., G.L. Beland, C. Bowman, N.B. Carozzi, R. Crenshaw, L.

Crossland, J. Dawson, N. Desai, M. Hill, S. Kadwell, K. Launis, K. Lewis, D. Maddox, K. McPherson, M.R. Meghji, E. Merlin, R. Rhodes, G.W. Warren, M. Wright, and S.V. Evola. 1993. Field performance of elite transgene maize plants expressing an insecticidal protein derived from *Bacillus thuringiensis*. *Bio/Technology* 11:194–200.

74. Kreig, A, A.M. Huger, G.A. Langenbruch, and W. Schnetter. 1983. *Bacillus thuringiensis* var. *tenebrionis*: ein neuer gegenüber Larven von coleopteren wirksamer Pathotyp. *Z. Angew. Entomol.* 96:500–508.

75. Kumar, H., E.M.O. Nyangiri, and G.O. Asino. 1993. Colonization responses and damage by *Chilo partellus* (Lepidoptera: Pyralidae) to four variably resistant cultivars of maize. *J. Econ. Entomol.* 86:739–746.

76. Kunik, T., R. Salomon, D. Zamir, N. Navot, M. Zeidan, I. Michelson, Y. Gafni and II. Czosnek. 1994. Transgenic tomato plants expressing the tomato yellow leaf curl virus capsid protein are resistant to the virus. *Bio/Technology* 12:500–504.

77. Lambert, L., R.M. Beach, T.C. Lilen, and J.W. Todd. 1992. Soybean pubescence and its influence on larval development and oviposition preference of Lepidopterous insects. *Crop Sci.* 32:463–466.

78. Laster, M.L., and W.R. Meredith, Jr. 1974. Evaluating the response of cotton cultivars to tarnished plant bug injury. *J. Econ. Entomol.* 67:686–688.

79. Leeper, J.R., R.T. Roush, and H.T. Reynolds. 1986. Preventing or managing resistance in arthropods. In *Pesticide Resistance: Strategies and Tactics for Management*. National Research Council, Washington, D.C.: National Academy Press, pp 335–346.

80. Lefkovitch, L.P. 1991. Analysis of rating scale data. *Can. J. Plant Sci.* 71:571–573.

81. Lewin, B. 1987. *Genes*. 3rd ed. New York: Wiley.

82. Lodge, J.K., W.K. Kaniewski, and N.E. Tumer. 1993. Broad-spectrum virus resistance in transgenic plants expressing pokeweed antiviral protein. *Proc. Natl. Acad. Sci. USA* 90:7089–7093.

83. Logemann, J., G. Jach, H. Tommerup, J. Mundy, and J. Schell. 1992. Expression of a barley ribosome-inactivating protein leads to increased fungal protection in transgenic tobacco plants. *Biol/Technology* 10:305–308.

84. Lomonosseff, G.P. 1993. Virus Resistance mediated by a nonstructural viral gene sequence. In *Transgenic Plants*. A. Hiatt, ed. New York: Marcel Dekker, pp 79–91.

85. Lukefahr, M.J., J.E. Houghtaling, and H.M Graham. 1971. Suppression of *Heliothis* populations with glabrous cotton strains. *J. Econ. Entomol.* 64:486–488.

86. Lukefahr, M.J., J.E. Houghtaling, and D.G. Cruhm. 1975. Suppression of *Heliothis* sap. with combinations of resistant characters. *J. Econ. Entomol.* 68:743–746.

87. Maiti, I.B., J.F. Murphy, J.G. Shaw, and A.G. Hunt, 1993. Plants that express a polyvirus proteinase gene are resistant to virus infection. *Proc. Natl. Acad. Sci. USA* 90:6110–6114.

88. Martin, G.B., S.H. Brommonschenkel, J. Chunwongse, A. Frary, M.W. Ganal,

R. Spivey, T. Wu, E. Earle, and S.D. Tanksley. 1993. Map-based cloning of a protein kinase gene conferring disease resistance in tomato. *Science* 262:1432–1436.

89. Maxwell, F.G., and P.R. Jennings, eds. 1980. *Breeding Plants Resistant to Insects.* New York: Wiley.

90. Mayo, O. 1987. *The Theory of Plant Breeding.* Oxford, UK: Clarendon Press.

91. McBlain, B.A., J.K. Hacker, M.M Zimmerly, and A.F. Schmitthenner. 1991. Tolerance to phytophthora rot in soybean: II. Evaluation of three tolerance screening methods. *Crop Sci.* 31:1412–1217.

92. McCarty, J.C., Jr., and J.N. Jenkins. Cotton cultivar performance under two levels of tobacco budworm—1988-1992. Miss. Agric. Forestry Exp. Sta. Tech. Bull. 197.

93. McGuaghey, W.M., and M.E. Whalon. 1992. Managing insect resistance to *Bacillus thuringiensis* toxins. *Science* 258:1451–1455.

94. McIntosh, R.A. 1988. The role of specific genes in breeding for durable stem rust resistance in wheat and triticale. In *Breeding Strategies for Resistance to the Rusts of Wheat.* N.W. Simmonds and S. Rajaram, ed. Mexico, D.F.: International Center for Maize and Wheat Improvement, pp. 1–9.

95. McKinzie, H. 1965. Inheritance of sawfly reaction and stem solidness in spring wheat crosses. *Can. J. Plant Sci.* 45:583–589.

96. McPherson, S.A., F.J. Perlak, R.L. Fuchs, P.G. Marrone, P.B. Lavrik, and D.A. Fischhoff. 1988. Characterization of the coleopteran specific protein gene of *Bacillus thuringiensis* var. *tenebrionis. Bio/Technology* 6:61–66.

97. McVey, D.V., and D.L. Long. 1993. Genes for leaf rust resistance in hard red winter wheat cultivars and parental lines. *Crop Sci.* 33:1373–1381.

98. Melchers, L.S., M.B. Sela-Buurlage, S.A. Vloemans, C.P. Woloshuk, J.S.C. van Roekel, J. Pen, P.J.M. van den Elzen, and B.J.C. Cornelissen. 1993. Extracellular targeting of the vacuolar tobacco proteins AP24, chitinase and ß-1,3 glucanase in transgenic plants. *Plant Mol Biol.* 21:583–593.

99. Mendel, G. 1865. *Experiments in Plant Hybridization.* (Translated from German by the Royal Horticultural Society of London and reproduced in *Classic Papers in Genetics.* J.A. Peters, ed. Englewood Cliffs, New Jersey: Prentice-Hall.)

100. Mihm, J.A. 1982. *Techniques for Efficient Mass Rearing and Infestation in Screening for Host Plant Resistance to Corn Earworm, Heliothis zea.* El Batan, Mexico: International Center for Maize and Wheat Improvement.

101. Mullis, K.B., and F. Faloona. 1987. Specific synthesis of DNA in vitro via a polymerase chain reaction. *Methods Enzymol.* 155:335–350.

102. Munthali, D.C. 1992. Effect of cassava variety on the biology of *Bemisia afer* (Preisner & Hosny, Hemiptera: Aleyrodidae). *Insect Sci. Appl.* 13:459–465.

103. Neilson, M.W., and W.F. Lehman. 1980. Breeding approaches in alfalfa. In *Breeding Plants Resistant to Insects.* F.G. Maxwell and P.R. Jennings, eds. New York: Wiley, pp 277–311.

104. Nelson, A., D.A. Roth, and J.D. Johnson. 1993. Tobacco mosaic virus infection of transgenic *Nicotiana tabacum* plants is inhibited by antisense constructs directed at the 5′ region of viral RNA. *Gene* 127:227–232.

105. Oard, J.H. 1991. Physical methods for the transformation of plant cells. *Biotech. Adv.* 9:1–11.

106. Ohl, S., M. Apotheker-de Groot, J.A. van der Knapp, A.A. Ponstein, M.B. Sela-Burrlage, J.F. Bol, B.J.C. Cornelissen, H.J.M. Linthorst, and L.S. Melchers. 1994. A new class of tobacco chitinases homologous to bacterial exo-chitinases is active against fungi *in vitro*. *J. Cell. Biochem.* 18A(suppl):90.

107. Ortman, E.E., D.C. Peters, and P.J. Fitzgerald. 1968. Vertical pull technique for evaluating tolerance of corn rootworm systems to northern and western corn rootworm. *J. Econ. Entomol.* 61:373–375.

108. Pang, S.Z., J.L. Slightom, and D. Gonsalves. 1993. Different mechanisms protect transgenic tobacco against tomato spotted wilt and impatiens necrotic spot Tospoviruses. *Bio/Technology* 11:819–824.

109. Parlevliet, J.E., and J.C. Zadoks. 1977. The integrated concept of disease resistance: a new view including horizontal and vertical resistance in plants. *Euphytica* 26:5–21.

110. Pederson, W.L., and S. Leath. 1988. Pyramiding major genes for resistance to maintain residual effects. *Annu. Rev. Phytopathol.* 26:369–378.

111. Perlak, F.J., T.B. Stone, Y.M. Musskkopf, L.J. Peterson, G.B. Parker, S.A. McPherson, J. Wyman, S. Love, G. Reed, D. Biever, and D.A. Fischhoff. 1993. Genetically improved potatoes: protection from damage by Colorado potato beetles. *Plant Mol. Biol.* 22:313–321.

112. Perlak., F.J.R.W. Deaton, T.A. Armstrong, R.I. Fuchs, S.R. Sims, J.T. Greenplate, and D.A. Fischhoff. 1990. Insect resistant cotton plants. *Bio/Technology* 8:939–943.

113. Porter, K.B., G.L. Peterson, and O. Vise. 1982. A new greenbug biotype. *Crop Sci.* 22:847–850.

114. Robertson, C.A., III, D.S. Calhoun, B.R. Leonard, and S.H. Moore. 1993. Bollworm/tobacco budworm management in insect resistant cotton genotypes. *Proceedings of the Beltwide Cotton Conference.* Memphis, Tennessee: National Cotton Council, pp 593–597.

115. Robinson, R.A. 1980. New concepts in breeding for disease resistance. *Annu. Rev. Phytopathol.* 18:189–210.

116. Robinson, J., and P.A. Burnett. 1992. Greenhouse rearing and field infestation of Russian wheat aphid using triticale as an example. *Southwest Entomol.* 17:17–22.

117. Rowan, G.B., H.R. Boerma, J.N. All, and J.W. Todd. 1993. Soybean maturity effect on expression of resistance to Lepidopterous insects. *Crop Sci.* 33:433–436.

118. Rowan, G.B., H.R. Boerma, J.N. All, and J.W. Todd. 1991. Soybean cultivar resistance to defoliating insects. *Crop Sci.* 31:678–682.

119. Russell, G.E. 1978. *Plant Breeding for Pest and Disease Resistance.* Boston: Butterworth.

120. Ryan, C.A. 1981. Proteinase inhibitors. In *The Biochemistry of Plants.* Vol. 6. A Marcus, ed. New York: Academic Press, pp 351–370.

121. Salim, M., and R.C. Saxena. 1991. Temperature stress and varietal resistance in rice: Effects on whitebacked planthopper. *Crop Sci.* 31:1620–1625.

122. Salter, R., J.E. Miller-Garvin, and D.R. Viands. 1994. Breeding for resistance to alfalfa root rot caused by *Fusarium* species. *Crop Sci.* 34:1213–1217.

123. Samborski, D.J., and P.L. Dyck. 1982. Enhancement of resistance to *Puccinia recondita* by interactions of resistance genes in wheat. *Can. J. Plant Pathol.* 4:152–156.

124. Sanders, P.R., B. Sammons, W. Kaniewski, L. Haley, J. Layton, B.J. LaVallee, X. Dellannay, and N. Tumer. 1992. Field resistance of transgenic tomato expressing the tobacco mosaic virus or tomato mosaic virus coat protein genes. *Phytopathology.* 82:683–690.

125. Sanford, J.C., F.D. Smith, and J.A. Russell. 1993. Optimizing the biolistic process for different biological applications. *Methods Enzym.* 217:438–509.

126. Schultz, T.R., and R.F. Line. 1992. High-temperature, adult-plane resistance to wheat stripe rust and effects on yield components. *Agron. J.* 84:170–175.

127. Shillito, R., M. Saul, J. Paszkowski, M. Muller, and I. Potrikus. 1985. High efficiency direct gene transfer to plants. *Bio/Technology* 3:1099–1103.

128. Simmonds, N.W., and S. Rajaram, eds. 1988. *Breeding Strategies for Resistance to the Rusts of Wheat.* El Batan, Mexico: International Center for Maize and Wheat Improvement.

129. Simmonds, N.W. 1991. Genetics of horizontal resistance to diseases of crops. *Biol. Rev.* 66:189–241.

130. Simmonds. N.W. 1979. *Principles of Crop Improvement.* New York: Longman.

131. Singh, B.B.H.H. Hadley, and R.L. Bernard. 1971. Morphology of pubescence in soybean and its relationship to plant vigor. *Crop Sci.* 11:13–16.

132. Singh, R.P. 1993. Resistance to leaf rust in 26 Mexican wheat cultivars. *Crop. Sci.* 33:633–637.

133. Singh, R.P. 1992. Association between gene Lr34 for leaf rust resistance and leaf tip necrosis in wheat. *Crop Sci.* 32:874–878.

134. Sinha, N.K., and D.G. McLaren. 1989. Screening for resistance to tomato fruitworm and cabbage looper among tomato accessions. *Crop Sci.* 29:861–868.

135. Skovmand, B., R.D. Wilcoxson, B.L. Shearer, and R.E. Stucher. 1978. Inheritance of slow rusting to stem rust in wheat. *Euphytica* 27:95–107.

136. Smith, C.M., Z.R. Khan, and M.D. Pathak. 1994. *Techniques for Evaluating Insect Resistance in Crop Plants.* Boca Raton, Florida: CRC Press.

137. Smith, C.M. 1989. *Plant Resistance to Insects: A Fundamental Approach.* New York: Wiley.

138. Song, W.Y., S.L. Wang, L.L. Chen, H.S. Kim, L.Y. Pi, T. Holsten, J. Gardner, B. Wang, W.X. Zhai, L.H. Zhu, C. Fauquet, and P. Ronald. 1995. A receptor kinase-like protein encoded by the rice disease resistance gene Xa21. *Science* 270:1804–1806.

139. Star, J.L., ed. 1990. *Methods for Evaluating Plant Species for Resistance to Plant-Parasitic Nematodes.* Hyattsville, Maryland: Society of Nematologists.

140. Stephens, P.A., C.D. Nickell, and S.M. Lim. 1993. Sudden death syndrome development in soybean cultivars differing in resistance to *Fusarium solani. Crop Sci.* 33:63–66.

141. Stubbs, R.W. 1988. Pathogenicity and lysis of yellow (stripe) rust of wheat and

its significance in a global context. In *Breeding Strategies for Resistance to the Rusts of Wheat*. N.W. Simmonds and S. Rajaram, eds. El Batan, Mexico: International Center for Maize and Wheat Improvement, pp 23–38.

142. Sturtevant, A.P., and R.N. Beachy. 1993. Virus resistance in transgenic plants: coat protein-mediated resistance. In *Transgenic Plants*. A. Hiatt, ed. New York: Marcel Dekker, pp 93–112.

143. Subbarao, K.V., J.P. Snow, G.T. Berggren, J.P. Damicone, and G.B. Padgett. 1992. Analysis of stem canker epidemics in irrigated and nonirrigated conditions on differentially susceptible soybean cultivars. *Phytopathology* 82:1251–1256.

144. Sutton, D.W., P.K. Havstad, and J.D. Kemp. 1992. Synthetic *cry*IIIA gene from *Bacillus thuringiensis* improved for high expression in plants. *Transgenic Res.* 1:228–236.

145. Teetes, G.L., M.I. Beccrra, and G.C. Peterson. 1986. Sorghum midge (Diptera: Cecidomyiidae) management with resistant sorghum and insecticide. *J. Econ. Entomol.* 79:1091–1095.

146. Terras, F.R.G., S. Torrekens, F. van Leuven, R.W. Osborn, J. Vanderleyden, B.P.A. Cammue, and W.F. Broekaert. 1993. A new family of basic cysteine-rich antifungal proteins from Brassicaceae species. *FEBS Lett.* 316:233–240.

147. Thompson, G.D., D.L. Paroonagian, P.K. Leonard, L.A. Pavan, and S.P. Nolting. 1994. A rapid method for artificially infesting research plots with Heliothis eggs. *Proceedings of the Beltwide Cotton Conference*. Memphis, Tennessee: National Cotton Council, pp 1139–1140.

148. Umbeck, P., G. Johnson, K. Barton, and W. Swain. 1987. Genetically transformed cotton (*Gossypium hirsutum* L.) plants. *Bio/Technology* 5:263–266.

149. Vaeck, M., A. Reynaerts, H. Hofte, S. Jansens, M. de Beuckeleer, C. Dean, M. Zabeau, M. van Montagu, and J. Leemans. 1987. Transgenic plants protected from insect attack. *Nature* 328:33–37.

150. Van Emden, H.F. 1990. The interaction of host plant resistance with other control measures. *Proceedings of the Brighton Crop Protection Conference*. National Council, Vol. 3., pp 939–949.

151. Van der Plank, J.E. 1963. *Plant Disease: Epidemics and Control*. New York: Academic Press.

152. van der Salm, T., D. Bosch, G. Honee, L. Feng, E. Munsterman, P. Bakker, W.J. Stiekema, and B. Visser. 1994. Insect resistance of transgenic plants that express modified *Bacillus thuringiensis cry*IA(b) and *cry*IC genes: a resistance strategy. *Plant Mol. Biol.* 26:51–59.

153. Vasil, I.K., ed. 1987. *Advances in Cellular and Molecular Biology of Plants*. Vol. 1. Dordrecht, The Netherlands: Kluwer.

154. Videla, G.W., F.M. Davis, W.P. Williams, and S. Seong-Ng. 1992. Fall armyworm (Lepidoptera: Noctuidae) larval growth and survivorship on susceptible and resistant corn at different vegetative growth stages. *J. Econ. Entomol.* 85:2486–2491.

155. Villani, M.G., and F. Gould. 1986. Use of radiographs for movement analysis of the corn wireworm, *Melanotus communis* (Coleoptera: Elateridae). *Environ. Entomol.* 15:462–464.

156. Walden, R. 1988. *Genetic Transformation in Plants*. London: Open University Press.
157. Ward, E.R., G.B. Payne, M.B. Moyer, S.C. Williams, S.S. Dincher, K.C. Sharkey, J.H. Beck, H.T. Taylor, P. Ahl-Goy, F. Meins, and J.A. Ryals. 1991. Differential regulation of ß-1,3-glucanase messenger RNAs in response to pathogen infection. *Plant Physiol.* 96:390–397.
158. Webster, J.A., D.H. Smith, H. Rathke, and C.E. Cress. 1975. Resistance to cereal leaf beetle in wheat: density and length of leaf-surface pubescence in four wheat lines. *Crop Sci.* 15:199–202.
159. Webster, J.A. 1975. *Association of Plant Hairs and Insect Resistance: An Annotated Bibliography*. USDA-ARS Misc. Publ. no. 1297. U.S. Government Printing Office, Washington, D.C.
160. Webster, J.A., C. Inayatullah, M. Hamissou, and K.A. Mirkes. 1994. Leaf pubescence effects in wheat on yellow sugarcane aphids and greenbugs (Homoptera: Aphididae). *J. Econ. Entomol.* 87:231–240.
161. Wessels, J.G.H., and J.H. Sietsma. 1981. Fungal cell walls: a survey. In *Encyclopedia of Plant Pathology*. New Series, Vol. 13B. W. Tanner and F.A. Lewis, ed. Berlin: Springer-Verlag, pp 352–394.
162. Williams, S., L. Friedrich, S. Dincher, N. Carozzi, H. Kessman, E. Ward, and J. Ryals. 1992. Chemical regulation of *Bacillus thuringiensis* δ-endotoxin expression in transgenic plants. *Bio/Technology* 10:540–543.
163. Williams, J.G.K., A.R.K. Kubelick, J.L. Livak, J.A. Rafalsoa, and S.V. Tingey. 1990. DNA polymorphisms amplified by arbitrary primers are useful genetic markers. *Nucleic Acids Res.* 18:6531–6535.
164. Wilson F.D., and H.M. Flint. 1991. Field performance of cotton genetically modified to express insecticidal protein from *Bacillus thuringiensis*. *Proceedings of the Beltwide Cotton Conference*. Vol. 1. Memphis, Tennessee: National Cotton Council, p. 579.
165. Wilson, J.P., and R.N. Gates. 1993. Forage yield losses in hybrid pearl millet due to leaf blight caused primarily by *Pyricularia grisae*. *Phytopathology* 83:739–743.
166. Wilson, F.D., H.M. Flint, W.R. Deaton, D.A. Fischhoff, F.A. Perlak, T.A. Armstrong, R.L. Fuchs, S.A. Berberich, N.A. Parks, and B.R. Stapp. 1992. Resistance of cotton lines containing a *Bacillus thuringiensis* toxin to pink bollworm (Lepidoptera: Gelechiidae) and other insects. *J. Econ. Entomol.* 85:1516–1521.
167. Windham, G.L., and G.A. Pederson. 1991. Reaction of *Trifolium repens* cultivars and germplasms to *Meloidogyne incognita*. *J. Nematol.* 23:593–597.
168. Zeng, F., G. Pederson, M. Ellsbury, and F. Davis. 1993. Demographic statistics for the pea aphid (Homoptera: Aphididae) on resistant and susceptible red clovers. *J. Econ. Entomol.* 86:1852–1856.
169. Zuban, J., V. Citovsky, D. Warnick, and P. Zambryski. 1994. Travels of the *Agrobacterium* T-DNA complex: tunneling through biological membranes (abstr.). *J. Biol. Biochem.* 18A(suppl.):79.

9
Acquisition and Maintenance of Resistant Germplasm*

Richard L. Wilson
United States Department of Agriculture, Iowa State University, Ames, Iowa

I. INTRODUCTION

The use of appropriate crop species and genotypes is the foundation upon which agricultural production is built; therefore, availability of germplasm for selecting appropriate crops and varieties is critical to the future of agriculture. Germplasm has been variously defined as ". . . the source of the genetic potential of living organisms" [70], or ". . . the array of plant materials, assembled or not, that serve as a basis for crop improvement, or related research" [12].

Nature has provided extensive genetic resources for agricultural use, but the current, narrow genetic base deployed in extensive monocultures is a tenuous situation, described as "thin ice" by Wilkes [71]. For many years agriculturists have been aware of the genetic vulnerability upon which world agriculture is built. The great potato famine of Ireland and the outbreaks of wheat stem rust in 1954 and southern leaf blight in 1970 underscore the need to strengthen the genetic bases of major crops [72]. This need was further emphasized in 1972 when the U.S. National Research Council examined the genetic diversity of corn, wheat, sorghum, pearl millet, rice, potato, sugar beet, sweet potato, soybean, peanut, dry bean, snap bean, pea, cotton, and several vegetable crops. The Council concluded that these crops are vulnerable to pests because their genetic bases are too narrow [45].

*Journal Paper No. J-16033 of the Iowa Agriculture and Home Economics Experiment Station, Ames, Iowa. Project No. 1018.

At the same time as the need for expanding available genetic resources grows, the availability of those resources is declining. Germplasm diversity is a natural resource which is being eroded by a variety of forces, such as human destruction of the environment. Once that germplasm is lost, it can never be recovered. Nabhan [42], for example, has reviewed the irretrievable loss of such germplasm in Native American culture.

Wild germplasm will be of vital importance for the future of agriculture, as it has been historically [63]. Harlan and Starks [25] pointed out that wild races and wild species related to domesticated crops could provide important sources of insect resistance, as they have provided sources of disease resistance. They further observed that plant collections tend to be weakest in these wild races and related species, although these materials have repeatedly yielded valuable resistance traits. Maunder [38] also cited examples to support the use of exotic germplasm in U.S. breeding programs.

Including exotic germplasm in breeding programs offers the best approach to reducing genetic vulnerability while simultaneously broadening the genetic variability available for breeding programs [38]. Plant introduction (the process of taking plants from their natural habitat and placing them in a new, different habitat) is hardly a new idea. The historical records indicate that the Sumerians were importing plants from Asia Minor around 2500 BC, and 1000 years later, the return of an expedition searching for plants in 1500 BC was recorded on a temple wall in Thebes [58].

There is thus a great need to broaden the diversity of agriculture's genetic underpinnings to promote agricultural stability and to increase productivity potential; and conserving and collecting and preserving wild germplasm will be a key to this diversification. Even though plant breeders prefer using advanced breeding stocks, circumstances arise in which desired genes are not present in available material. In such cases, the availability of exotic germplasm is vital, and germplasm collections can help fill this gap. One outstanding example of such a collection is the U.S. National Germplasm System (NPGS) [16]. This system will be used as the framework for this chapter's discussion of the acquisition and maintenance of resistant germplasm.

II. U.S. NATIONAL PLANT GERMPLASM SYSTEM

As the growing threat to worldwide germplasm has been recognized, the NPGS has greatly increased its efforts in acquiring and maintaining plant germplasm, and its role in conserving germplasm has become even more important.

Historical overviews of the NPGS have been presented by various investigators [1,5,20,32,35,44,58,60,62,64,69]. The NPGS is a federal, state, and industry partnership. The germplasm within the system is available without cost to qualified users (e.g., plant scientists) throughout the world [59]. The four components comprising the NPGS were described by Jones and Gillette [35]: (a) plant introduction facilities and activities, (b) maintenance and evaluation facilities for long-term and medium-term storage of germplasm, (c) an effective information system, and (d) crop germplasm committees to advise NPGS personnel.

A. Plant Introduction Stations

The four regional plant introduction stations, their locations, and their dates of inception are listed in Table 1. The varied germplasm collections held at these stations are considered "active collections": Plant scientists throughout the world can request seed or vegetatively propagated material from these stations for use in their research.

The regional plant introduction stations were authorized as part of the 1946 Research and Marketing Act passed by the U.S. Congress. Each station has its own history, some of which is documented; for example, the North Central Regional Plant Introduction Station [74] and the National Research Support Program-6 (formerly the Inter-Regional Potato Program) [3].

The basic mission of the plant introduction stations is (a) to cooperate and participate in a coordinated program of foreign and domestic plant exploration and the introduction of germplasm that is potentially valuable for agricultural and industrial uses; (b) to multiply, evaluate, and maintain introduced materials and provide back-up accessions to the National Seed Storage Laboratory (NSSL); (c) to distribute materials, maintain records of use and potential value;

Table 1 Location of the Major Facilities for Seed-Propagated Germplasm of the National Plant Germplasm System

Name	Location	Established
National Seed Storage Laboratory	Ft. Collins, CO	1958
Regional Plant Introduction Stations	Northeast: Geneva, NY	1953
	Southern: Griffin, GA	1949
	North Central: Ames, IA	1948
	Western: Pullman, WA	1952
National Germplasm Resources Laboratory	Beltsville, MD	1990
Plant Germplasm Quarantine Office	Glenn Dale, MD	1919
National Small Grains Collection	Aberdeen, ID	1948[a]
National Research Support Program-6[b]	Sturgeon Bay, WI	1950
U.S. National Arboretum	Washington, D.C.	1927

[a]Moved to Aberdeen, ID in 1988.
[b]Formerly Inter-Regional Potato Program (IR-1); name changed in 1993.

and (d) to publish research results and to produce and distribute seed lists [62].
The basic mission has changed little since 1975, although there is more emphasis
on worldwide cooperation among the NPGS and international germplasm centers.

Currently, the NPGS holds 9995 species representing 1476 genera and 181
families and encompassing 431,337 accessions. A list of species, by common
name, and the locations where they are stored is available through the U.S. Department of Agriculture's (USDA) Agricultural Research Service Information
Service [2]. A compilation of the larger collections (more than 1000 accessions)
maintained within the NPGS is shown in Table 2.

B. National Seed Storage Laboratory

The National Seed Storage Laboratory (NSSL), located in Ft. Collins, Colorado, was established in 1958. A large addition to the facility was constructed
in 1992 that provides sufficient space to store 1 million germplasm samples. It
was designed to withstand earthquakes, floods, tornadoes, and vandalism. Our
national germplasm treasure is well protected. The plant germplasm (called the
base collection) is held for long-term storage at this facility. The basic mission
of the NSSL is to preserve the NPGS base collection and to conduct research
on aspects of long-term seed viability and storage [1].

C. Clonal Repositories

Much of the germplasm within the NPGS cannot be propagated easily by seed,
so germplasm must be maintained vegetatively. Clonal repositories provide locations throughout the NPGS to store and propagate these types of germplasm.
Table 3 lists the national clonal germplasm repositories, the years of their establishment, and the common names of their major holdings.

D. Special Collections

The NPGS includes special germplasm collection sites, either genetic stocks
(maintaining specific genes within a crop) or crop specific (e.g., cotton, flax).
These collections were established separately from the regional plant introduction stations for one of two reasons: (a) because other locations might have better
climatic conditions for seed increase, or (b) the location of the curator, based
on where the person was living at the time, dictated the collection's location.
Table 4 lists the genetic stock collections, their locations, and the number of
their holdings within the NPGS. Table 5 lists the crop-specific collections, their
locations, and the number of holdings within the NPGS.

Table 2 Genera Comprising More than 1000 Accessions in Plant Germplasm Collections and Their Locations Within the National Plant Germplasm System (United States)[a]

Genus	Number of accessions	Location of holdings
Abelmoschus	1926	Griffin, GA
Aegilops	3598	Aberdeen, ID
Amaranthus	3153	Ames, IA
Andropogon	1067	Ft. Collins, CO
Arachis	9250	Griffin, GA
Avena	21244	Aberdeen, ID
Beta	1474	Pullman, WA
Brassica (oilseed)	3091	Ames, IA
Brassica (vegetable)	2087	Geneva, NY
Bromus	1033	Pullman, WA
Cajanus	4278	Griffin, GA
Capsicum	3562	Griffin, GA
Carthamus	2314	Pullman, WA
Cicer	4348	Pullman, WA
Citrullus	1503	Griffin, GA
Cucumis	4728	Ames, IA
Cucurbita	1319	Griffin, GA
Dactylis	1195	Pullman, WA
Elymus	1637	Pullman, WA
Eragrostis	1280	Pullman, WA
Festuca	1673	Pullman, WA
Glycine	13,737	Urbana, IL
Gossypium	6007	College Station, TX
Helianthus	3565	Ames, IA
Hordeum	27,126	Aberdeen, ID
Lactuca	1256	Pullman, WA
Lens	2839	Pullman, WA
Linum	2659	Fargo, ND
Lolium	1101	Pullman, WA
Lycopersicon	5489	Geneva, NY
Malus	3548	Geneva, NY
Medicago	6762	Pullman, WA
Nicotiana	2077	Oxford, NC
Oryza	16,488	Aberdeen, ID
Oryza	1790	Ft. Collins, CO
Panicum	2094	Ames, IA
Paspalum	1495	Griffin, GA

(*continued*)

Table 2 Continued

Genus	Number of accessions	Location of holdings
Phaseolus	12,704	Pullman, WA
Pisum	3754	Pullman, WA
Prunus	1727	Davis, CA
Pyrus	1618	Corvallis, OR
Rubus	1417	Corvallis, OR
Saccharum	2089	Miami, FL
Secale	1912	Aberdeen, ID
Sesamum	1075	Griffin, GA
Setaria	1340	Ames, IA
Solanum	1912	Glenn Dale, MD
Solanum	5239	Sturgeon Bay, WI
Solanum	3268	Ft. Collins, CO
Sorghum	12,619	Ft. Collins, CO
Sorghum	29,254	Griffin, GA
Trifolium	2954	Pullman, WA
Trifolium	1827	Griffin, GA
Triticum	46,754	Aberdeen, ID
Triticum	1151	Ft. Collins, CO
Vicia	1730	Pullman, WA
Vigna	12,254	Griffin, GA
Vitis	2417	Davis, CA
Vitis	1426	Geneva, NY
Zea	12,915	Ames, IA
Zea	19,756	Ft. Collins, CO

[a]Data from the GRIN database.

E. Germplasm Resources Information Network

As the germplasm collections grew, so did the volume of associated data. A survey conducted in 1976–1977 to determine the adequacy of the current information system indicated that more organization (i.e., a better way to keep track of the germplasm) was needed to satisfy the germplasm community. The U.S. Department of Agriculture, Agricultural Research Service (ARS) began using a new system in 1983 called the Germplasm Resources Information Network (GRIN).

Located in Beltsville, Maryland, GRIN gave the NPGS a centralized means of managing the vast amount of data associated with the over 431,000 accessions maintained in the system [38]. The GRIN database contains three basic types of data: (a) passport data comprising plant identification numbers, common names and taxonomic classification, collector name, place of collection,

Table 3 Location, Year of Establishment, and Principal Holdings of the National Germplasm Repositories in the United States

Location	Year established	Principal holdings
Corvallis, OR	1980	Blackberry, blueberry, cranberry, currant, filbert, gooseberry, hops, mint, pear, raspberry, strawberry
Davis, CA	1981	Almond, cherry, fig, grape, kiwifruit, mulberry, olive, peach, persimmon, pistachio, pomegranate, plum, walnut
Miami, FL	1984	Avocado, banana, carambola, Chinese date, cocoa, coffee, jujube, lychee, palm, sugarcane, tropical citrus
Mayaguez, PR	1984	Bamboo, banana, cacao, cassava, cocoyam, monkeypod nut, plantain, tropical yam
Geneva, NY	1985	Apple, grape
Brownwood, TX	1984	Chestnut, hickory, pecan
Hilo, HI	1987	Barbados cherry, breadfruit, guava, jackfruit, lychee, macadamia, papaya, passion fruit, peach palm, pili nut, pineapple, pulasan, rambutan
Riverside, CA	1987	Citrus, date

Table 4 Location of Genetic Stock Collections and Number of Holdings in the United States

Crop	Location	No. of holdings
Barley (*Hordeum*)	Ft. Collins, Colorado	225
Clover (*Trifolium*)	Lexington, Kentucky	40
Cotton (*Gossypium*)	College Station, Texas	600
Crucifers (*Brassica*)	Madison, Wisconsin	68
Lettuce (*Lactuca*)	Salinas, California	2873
Maize (*Zea*)	Urbana, Illinois	2000
Pea (*Pisum*)	Pullman, Washington	450
Pearl millet (*Pennisetum*)	Tifton, Georgia	2629
Pepper (*Capsicum*)	Las Cruces, New Mexico	135
Potato (*Solanum*)	Sturgeon Bay, Wisconsin	500
Sorghum (*Sorghum*)	College Station, Texas	300
Soybean (*Glycine*)	Urbana, Illinois	645
Soybean (*Glycine*)[a]	Ames, Iowa	3000
Tobacco (*Nicotiana*)	Oxford, North Carolina	30
Tomato (*Lycopersicon*)	Davis, California	3000
Wheat (*Triticum*)	Columbia, Missouri	588

[a]This is a cytogenetic stock collection.

Table 5 Crop-Specific Collections Within the U.S. National Plant Germplasm System

Crop	Location	Number of holdings
Apples (*Malus*) and grapes (*Vitis*)	Geneva, NY	4657[a]
Citrus (*Citrus*) and date (*Phoenix*)	Riverside, CA	970[b]
Clover (*Trifolium*)	Lexington, KY	1440
Cotton (*Gossypium*)	College Station, TX	5944
Flax (*Linum*)	Fargo, ND	2659
Lettuce (*Lactuca*)	Salinas, CA	2683
Pearl Millets (*Pennisetum*)	Tifton, CA	5703
Pecan (*Carya*), hickory (*Carya*), chestnut (*Castanea*)	Brownwood, TX	756
Potato (*Solanum*)	Sturgeon Bay, WI	4305
Small grains[c] and rice (*Oryza*)	Aberdeen, ID	118,229
Sorghum (*Sorghum*)	Mayaguez, PR	773
Soybean (*Glycine*)	Urbana, IL	14,382
Tobacco (*Nicotiana*)	Oxford, NC	2300
Various tropical species	Hilo, HI	835

[a]3207 apple accessions and 1450 grape accessions.
[b]Seventy of the accessions are date (*Phoenix*).
[c]Collections of wheat (*Triticum*), oats (*Avena*), barley (*Hordeum*), and rye (*Secale*).

crop name, and the form in which the germplasm was received (e.g., seed, live plant); (b) evaluation or observation data, including information on growth characteristics, disease and insect resistance (or susceptibility), fruit or grain quality, and time to maturity; and (c) inventory comprising availability for germplasm distribution, germination records, and maintenance techniques [40,41].

The effective use of germplasm depends on the availability and quality of information associated with the germplasm [22]. There are two chief uses for genebank germplasm: (a) to provide genetic diversity for plant breeders, and (b) to meet the needs of breeders in the future [30]. This list, however, is too narrow, because, in addition to plant breeders, the germplasm is used by many other plant scientists, entomologists, and educators in general. If genetic resources are to be used, the seeds must germinate or clones must be viable. Further, there must be information on characteristics of each accession to help breeders make informed decisions, and this information must be readily available in a computerized data base [30].

There are two methods for obtaining germplasm from the NPGS for research. First, workers may request germplasm directly from the curator by writing or phoning. Second, one can submit a computer request through the GRIN system. The GRIN system is available worldwide, free of cost, to any scientist who has a personal computer, and access to the world wide web. To examine germplasm data or request germplasm, the scientist must log-on to the world wide web site: www.ars-grin.gov.

F. Advisory Committees

National Crop Germplasm Committees (CGCs) are composed of interdisciplinary peer groups of federal, state, and industry scientists who advise the states and USDA of on all aspects of germplasm relating to a particular commodity. They prioritize the needs of the commodity in the areas of germplasm acquisition, maintenance, characterization, evaluation, and enhancement [59].

In addition, the germplasm centers are guided by Technical Advisory Committees (TACs). The TACs are composed of representatives appointed by the cooperating states' experiment station directors. The TACs usually meet annually to review programs and budget.

III. ACQUISITION OF GERMPLASM

The objective of the NPGS is ". . . to provide the maximum amount of genetic diversity within a minimum number of accessions of each species of interest to the United States . . ." [21].

Where does our germplasm originate? Creech and Reitz [12] listed three sources of germplasm: (a) wild species and primitive forms of crops in primary centers of diversity, (b) plant migrants to secondary centers of culture where their diversity may be augmented by natural selection, and (c) the unique products of plant breeding and genetic engineering. Breese [4] and Hawkes [28] described crop diversity as including obsolete cultivar and genetic stocks, land races and primitive cultivars, and genetically related wild and weedy species. Wilkes [72] added to the above lists varieties in current use, special genetic stocks, and organisms of no apparent immediate use but which are part of the ecosystem. The acquisition policy for the NPGS lists five categories of germplasm: (a) cultivars, (b) germplasm releases and genetic stock collections, (c) landraces and primitive cultivars, (d) wild relatives of crops, and (e) other species [21].

Plucknett et al. [49] noted three principles that generally guide the collection, conservation, and exchange of germplasm. First, when an accession is collected, a sample should be left in the country of origin for their use. If there are no suitable storage facilities, then duplicate material is usually stored elsewhere until the country of origin can manage the material. Second, germplasm is to be made available at no cost to all researchers who can effectively use it. And third, all long-term collections must be duplicated and maintained in other locations for safety reasons.

Ninety-nine percent of the crops planted in the United States are not native to this country. Without plant introduction, the United States would be limited to growing relatively few crops such as sunflower, strawberry, blueberry, cranberry, and pecans [59].

Germplasm collected in the past and saved may help solve some of our plant breeding problems now and in the future [36]. Chang [7] noted that crop germplasm is vital to our scientific efforts to increase crop production and enhance nutrition for humans.

From a plant breeder's point of view, to develop a program for breeding insect-resistant varieties, germplasm collections must be screened for resistant plant types. The search may focus on adapted varieties, introduced or exotic materials, or related species [57].

One of the National Germplasm Resources Laboratory's functions is cataloguing incoming germplasm accessions. They also assign plant introduction (PI) numbers and distribute the new accessions to the curators. These duties were previously handled by the Plant Introduction Office and its predecessors since 1898 when the USDA Section of Seed and Plant Introduction was established [46].

New plant germplasm is continually being added to the NPGS through exploration, exchange, cooperation with U.S. agencies (e.g., the Soil Conservation Service assembles collections and exchanges them with other countries), agreements (e.g., *Crop Science* Registration program or bilateral agreements between countries), special projects (e.g., Latin American Maize Project), breeding programs and other significant sources (e.g., travelers or hobbyists) [66].

A. Plant Exploration

Worldwide, the International Board for Plant Genetic Resources (IBPGR) has identified acquisition priorities for world collections [33]. In the United States, the need for additional germplasm for a particular crop is often identified by the crop germplasm committees. Perdue and Christenson [48] described in detail the operation of the NPGS plant exploration system.

Most plant explorers prefer to collect new accessions from identified plant centers of origin and diversity located throughout the world. The Russian botanist N.I. Vavilov identified where he thought the centers of plant origin and diversity were located throughout the world. He concluded that there were eight centers with three subcenters for plant origin and diversity [66]. Later, Darlington [13] expanded the number of centers to 16. More recently, Harlan determined that a crop-by-crop analysis showed the crop origin and diversity situation to be much more complex than the simplistic view conceived by Vavilov. [24].

In searching for plants with new sources of insect resistance, Harris [26] suggested that resistance obtained from plants that evolved in the absence of the insect pest may have an advantage in that the resistance may be polygenic, which could result in disrupting an insect's growth and development in several ways.

If the CGCs feel there is a lack of representative germplasm (e.g., material from a particular part of the world, related species), they will recommend that the NPGS acquire additional germplasm to fill the shortage. In some cases, exchange of material with other countries will suffice, but sometimes a collecting trip is required.

Plant exploration is an age-old activity that continues to the present. Singh et al. [61] reviewed areas of the world that have plant resources needing to be collected. IBPGR also has assembled a priority list of plant species needing to be collected around the world [33].

During the 6-year period from 1989 to 1994, there were 74 NPGS-sponsored plant exploration trips to 34 countries around the world. Germplasm was collected for 24 crop species and for several general areas categorized as forage grasses, forage and food legumes, fruits and nuts, Asian vegetables, oilseed crops, woody landscape plants, and endangered species. The 74 collection trips involved the services of 189 principal collectors and many other cooperating scientists, graduate students, and interpreters. The collecting efforts by these participants have added large amounts of new genetic material to the NPGS as well as to the germplasm collections of the host countries.

B. Germplasm Exchange

Exchange of germplasm with other germplasm centers is the leading source of new germplasm for the NPGS (68). There are eight International Agricultural Research Centers (IARCs) within the Consultative Group on International Agricultural Research (CGIAR) which exchange germplasm with the United States. A list of the eight IARCs and their principal holdings is given in Table 6. Many other countries also have national germplasm collections and will exchange germplasm with the United States.

With an almost continual flow of germplasm into the United States, it becomes necessary to monitor it for imported pests. The U.S. National Plant

Table 6 International Agricultural Research Centers Within the Consultative Group on International Agricultural Research[a]

Center	Acronym	Year established	Location country	Research crops
International Rice Research Institute	IRRI	1960	Philippines	Rice
International Maize and Wheat Improvement Center	CIMMYT	1964	Mexico	Maize, wheat, barley, triticale
International Institute of Tropical Agriculture	IITA	1965	Nigeria	Maize, rice, cowpea, cassava, yams, sweet potaoes
International Center for Agriculture in the Tropics	CIAT	1968	Colombia	Beans, cassava, pastures, rice
West Africa Rice Development Association	WARDA	1971	Liberia	Rice
International Potato Center	CIP	1972	Peru	Potato
International Crops Research Institute for the Semi-Arid Tropics	ICRISAT	1972	India	Chickpea, groundnut, millet, pigeonpea, pear, sorghum
International Center for Agricultural Research in Dry Areas	ICARDA	1976	Syria	Barley, chickpea, faba beans, forages, triticale, wheat, lentil

[a]*Source:* Information from ref. 47.

Germplasm Quarantine Center in Beltsville, Maryland, is responsible for examining introduced germplasm for pests. The center, managed by the USDA's Agricultural Research Service (ARS), was designed, constructed, and is maintained by the Animal and Plant Health Inspection Service (APHIS). APHIS personnel determine the genera to be quarantined, tests to be performed, and when to release the germplasm for general use. Disease monitoring is the center's main function, but detecting introduced insect pests is also an important and necessary function [67].

C. Donations

Most plant breeders accumulate and improve germplasm during their careers. When these breeders retire or are deceased, their valuable germplasm collections may be lost, or, if not properly maintained, they may lose their viability. Ideally, plant breeders nearing retirement should make plans to dispose of or transfer their collections to an appropriate agency such as the NPGS. If individuals are aware of collections that are in danger of being lost, they should contact the NPGS.

The NPGS has received many germplasm collections from plant breeders in the past. For example, a small collection of 35 maize accessions was donated to the NPGS by J. C. Eldredge in 1960; in 1990 the NPGS received 1600 maize accessions from W. C. Galinat, who had inherited the bulk of the accessions from P. C. Mangelsdorf; and in 1992, 200 teosinte accessions were donated by H. Iltis, who originally obtained many of them from G. Beadle (M. Millard, personal communication). The maize obtained from J. C. Eldredge has been evaluated and sources of resistance to the corn earworm, *Helicoverpa zea*, and European corn borer, *Ostrinia nubilalis* have been identified [75].

IV. GERMPLASM PRESERVATION AND CONSERVATION

A. Need for Conservation

Thousands of species (plants and animals) are becoming extinct as a result of human activities [50]. The massive disturbances to our environment caused by humanity are continually driving species to the point of extinction. Humanity's impact will continue to increase in the future as our population grows. World population has more than doubled in only 43 years—from 2.5 billion in 1950 to 5.5 billion in 1993. At the present rate of increase, the earth's population may reach 11 billion by 2045 and 14 billion by 2100 [39]. The high population produces a major threat to earth's biological diversity, mainly by destroying habitat via climatic change and human activity [30].

Hoyt [31] noted that by the middle of the next century as many as 60,000 plant species may become extinct if our destruction of nature continues at the present rate. Accordingly, "the conservation of crop genetic resources—the plants that feed us and their wild relatives—is one of the most important issues for mankind today" [31]. It is, therefore, critical that means be available for storing plant germplasm and retaining its viability for subsequent regeneration.

B. Storage

Winters [76] discussed the high percentage of the early seed collections in the NPGS (e.g., 70% of the soybeans, 98% of the clovers, and 66% of the oats) that have been lost since 1898, because no provisions were made to preserve them. Safeguarding genetic diversity after collecting is vital, because genetic erosion is a constant threat inside as well as outside genebanks [11,30].

1. Ex Situ Conservation

Ex situ conservation—maintaining germplasm outside its natural environment— is the predominant method of conserving crop germplasm. It is generally conducted in genebanks, botanical gardens, and arboreta [22]. Good storage conditions are necessary, because they lengthen the regeneration cycle, reduce costs of storage, and help maintain the genetic integrity of the samples [27]. There is some disagreement about the utility of ex situ methods for conserving germplasm. Hamilton [23] believes that field sampling may exclude important genes and that important genotype x environment interactions are missing from genebanks. He also believes that germplasm viability declines in relation to the length of storage, which results in a loss of genes. Natural selection in genebanks via standard regeneration techniques will remove certain alleles from the population. Namkoong [43] showed that plant collections made at any one time, and maintained statically, cannot capture all the useful alleles in a given natural population, because the population is undergoing continual evolution. Ex situ conservation disrupts the process of evolution, and seed banks may mislabel germplasm or fail to regenerate them in time when germinations are low [31].

Procedures for handling desiccation-tolerant seeds are well established [19]. Roos [56] discussed long-term storage, seed longevity, seed deterioration, conventional storage, and cryopreservation of such seed. Roberts and Ellis [54] discussed the deterioration of desiccation-tolerant seed in storage. They presented seed-survival curves for several seed types and noted that the IBPGR recommended that desiccation-tolerant seeds be stored at a relative humidity of 5 ± 1% and a temperature of –18°C or less. Roberts [53] stated that, for most species, reducing moisture and temperature will increase a seed's period of viability. A decrease in oxygen pressure also prolongs seed viability. Viability can be affected by pre- and postharvest conditions, as well as the occurrence of intraspecific genetic polymorphisms for different viability characteristics.

Each germplasm storage facility within the NPGS has developed an operations manual that documents germplasm management procedures. These include information on seed/plant storage conditions, germination procedures, and regeneration techniques. In general, desiccation-tolerant seed is stored from 0 to 4°C at a relative humidity of 25–40% [9].

Desiccation-intolerant seeds present different problems for handling germplasm. Some species that produce desiccation-intolerant seeds grow in aquatic habitats, whereas others are large-seeded woody perennials (e.g., rubber, cocoa, coconut, most tropical fruits, and some timber species like oak, chestnut, and horse chestnut). These seeds are killed when their moisture content is reduced below a relatively high level [55]. Therefore, seeds of these types cannot be preserved by the same methods such as those used for desiccation-tolerant seed. Most of these plant types are preserved by vegetative cuttings. New techniques (some mentioned below) are being researched to handle desiccation-intolerant seeds.

Towill and Roos [65] thoroughly discuss the nontraditional techniques used to preserve germplasm ex situ. These preservation techniques are classified as: a) *in vitro*, b) meristem culture, and c) clonal propagation.

In Vitro Techniques.   In vitro techniques, such as cryopreservation (storing germplasm in liquid nitrogen at –196°C) provide for the long-term conservation of germplasm (100+ years). Routine methods are now available for the in vitro preservation of all types of crop tissue cultures [78]. In addition to cryopreservation, in vitro methods also include another culture, excised embryos, pollen storage, and use of bud/shoot tips. These in vitro techniques hold great promise for the preservation of desiccation-intolerant and vegetatively propagated plant species [51]. Germplasm that may be damaged or very old might be rescued in vitro with embryo culture techniques, but these procedures are costly and there are risks of genetic change and contamination, and cultures can be lost by human error [77].

Meristem Culture.   Meristem/shoot tip cultures provide another method for germplasm preservation. The use of meristem/shoot tip cultures has five distinct advantages for germplasm preservation: (a) exact genotypes can be conserved indefinitely, free from viruses or pathogens, and without loss of genetic integrity; (b) it is advantageous for root crops like potatoes, yams, cassava, and sweet potatoes, because their seed production is poor; (c) the loss of vegetatively propagated materials resulting from natural disasters or pathogens in the field can be eliminated, (d) a long regeneration cycle is possible; and (e) it is good for forest and fruit tree germplasm, because they have 10- to 20-year cycles for seed production [27].

Clonal Propagation.   Clonal propagation of germplasm (vegetative propagation by means of root or stem cuttings or from pieces of rhizome) is also an important preservation method. De Langhe [14] explained the general usage of clonal propagation and argued that this method can preserve a valuable combination of quality factors such as vigor, pest resistance, fruit size, fruit color, or shape. It is also beneficial for crops that do not produce seed easily and for crops with a long juvenile stage before seed production.

2. In Situ Conservation

In situ conservation—maintaining germplasm in natural environments—may offer distinct advantages for many plant species, because it preserves both genetic and ecological information [23]. It is useful mainly with wild species, wild crop relatives, forest and pasture species [22], and land races. Plants from agroecosystems of village farming and natural vegetation where near relatives of crop plants and weedy forms exist would also benefit from this method [72]. Conservation in situ demands establishment of nature or biosphere reserves, national parks, or special laws to protect endangered or threatened species [34].

In situ and ex situ methods are complementary [10,31], but both methods have their weaknesses. For example, protected areas remain vulnerable to loss (e.g., severe weather) or destruction (e.g., human encroachment on the land). Likewise, seed banks are vulnerable to losses by natural disasters, human mistakes, and technical problems such as power failures, fires, and floods [31].

When new germplasm is added to the NPGS, there is a risk of introducing both insect and disease pests into the collections. Thus, plant material acquired from foreign countries must be inspected at the National Plant Quarantine Center for foreign pests. Likewise, plant material grown within the NPGS has the potential of being infested or infected with native pests. Crop-specific curators have the responsibility of examining their collections for pests and are generally successful in controlling pests. Occasionally, a stored grain pest will be found in storage, but this is rare with current inspection methods.

C. Regeneration of Germplasm

Crop-specific curators regenerate seed supplies when incoming samples contain too few seeds to meet needs, when germination drops below a certain level, or when their existing seed supply is reduced [30]. Hawkes [27] recommended testing for germination every 5–10 years. Breese [4] believed seed stocks should be regenerated frequently, because when the viability drops below 85%, mutational events significantly increase. The IBPGR and others recommend regeneration when germination falls below 85% [30,49].

Much of the germplasm within the NPGS is cross pollinated and traditionally was regenerated by hand pollination. In the late 1970s, at the North Central Regional Plant Introduction Station, Ames, IA (NCRPIS), field cages containing a nucleus hive of honeybees were developed to control pollinate several crop species [18]. At present, the NCRPIS averages 800–1000 field cages per growing season for controlled pollination of regenerated seed stocks. Recently, the NCRPIS has begun to experiment with other insect pollinators (e.g., alfalfa leafcutting bees, bumblebees, and the mason bee, *Osmia cornifrons*) to determine if they are superior to honeybees for pollinating *Brassica* spp., *Cucumis* spp., *Cuphea* spp., and *Helianthus* spp. Other NPGS sites are currently considering the use of honeybees for controlled pollinations of their crops.

V. EVALUATION

Chapman noted that, "Until a collection has been evaluated and something is known about the material it contains, it has little practical value" [6]. He also observed that the main reason for establishing and maintaining crop collections is for their potential use in breeding [6]. Evaluation is a key step in this process and deserves more attention. When discussing genetic diversity, Duvick [15] noted that it is widely advocated but poorly understood: " We don't know where to get [useful traits], because our germplasm banks as yet have done very little to describe or catalogue their collections in regard to useful traits."

Within the NPGS, there are two research entomologists and five plant pathologists who evaluate the germplasm collections for pest resistance. They encounter several roadblocks to their evaluation research: (a) there are too many crops for them to evaluate, (b) the total number of accessions within each crop is high, (c) each crop may have multiple pests to consider, and (d) there are insufficient resources to support highly aggressive programs. There are cooperating state, private, and other ARS scientists who are actively involved with evaluating germplasm for new sources of pest resistance. Nevertheless, much germplasm remains to be examined.

What determines the procedure for the entomologist or pathologist to follow when evaluating germplasm for pest resistance? There are several areas to consider: (a) which crop is to be evaluated when there are so many from which to choose, b) which insect or pathogen should be tested, and c) which germplasm from a specific crop should be evaluated?

Frequently, the choice of crop for evaluation depends on the predominant crop grown in a particular area. Commodity groups tend to promote work on their crops. Funding for evaluation is needed but it is not generally available for many crops. The CGCs have access to limited funding for evaluation of their crops. Entomologists and plant pathologists interested in evaluation should submit requests for these funds to the chairperson of the CGC.

The choice of which pest to consider in germplasm evaluation may depend on the priority list developed by the CGCs or may be determined by local or regional needs. A source of test insects or pathogens is necessary to effectively evaluate for resistance. This usually requires rearing or culturing the pest to ensure adequate supplies of test pests when they are needed. If laboratory rearing or culturing is not available (i.e., rearing or culturing techniques are not established), then evaluation will have to be performed using natural populations of pests.

The selection of a germplasm to test can be accomplished in several ways. One way is to examine all the available germplasm in a collection. This can be rather time consuming for the large collections. A particular evaluation technique and the amount of available help will influence the practical number of accessions that can be evaluated in a given time period. Another way to select germplasm for evaluation is to look at specific agronomic traits that interest the breeder (e.g., maturity, plant height, seed color). Germplasm selected for evaluation may be the result of passport data (e.g., country of origin, altitude or latitude of collection). Selection of specific traits will narrow the germplasm base to be evaluated.

There is considerable interest, especially when evaluating the larger germplasm collections, to develop core (or evaluation subsets) collections. These smaller collections (perhaps 10% of the collection) would contain a cross section representative of the larger collection. Thus an entomologist looking for resistance in maize to a particular pest could initially evaluate the smaller subset rather than having to begin evaluation of the nearly 13,000 accessions in the total collection.

The data generated by evaluation testing are entered into the GRIN system so that it is available to scientists throughout the United States and the world.

VI. FUTURE OUTLOOK

Holden [29] questioned, "Do we really need to keep collecting when our genebanks are full already?" He proposed that we concentrate on our current collections and develop newer techniques of in vitro culture to complement or replace conventional methods. More international cooperation is needed to determine duplication of material and handling methods and to detect gaps in our collections. Cohen et al. [10] believe that the future of resource conservation lies in closer connections with breeders, a partnership of conservation and biotechnology, and efficient networks for information exchange. The Panel on Biodiversity Research Priorities [47] recommended that national biological inventories be organized, funded, and strengthened for each country, and that screening of plants and other living organisms for features potentially beneficial to humanity should be systematized and accelerated through strengthened programs.

Williams [73] stated, "An enormous amount of crop genetic resources has been collected in the past two decades. Emphasis is now shifting to the postcollecting phases and to problems of storing material not easily conserved as seed." Chang et al. [8] believe that sound management is the key to having an effective germplasm system. Both Williams and Chang et al. believe that biotechnology is expected to be complementary with conservation.

Matthews and Saunders [37] discussed a method of germplasm storage in DNA libraries where genes would be maintained in test tubes. They also said that the DNA libraries could be used to increase genetic diversity through gene transfer.

More work is needed in the future in the area of germplasm enhancement: the transfer of genes for desirable traits from wild, unadapted germplasm to adapted germplasm with utility in advanced breeding programs [17]. Roath [52] reviewed the evaluation and enhancement efforts within the NPGS.

Using the tools of biotechnology, the transfer of insect "resistant" genes from exotic germplasm or from unrelated genera will become more commonplace (see Chap. 9). Perhaps new technology will enable plant scientists to "build" genes in the laboratory that, when placed into a crop plant, will express resistance to pests.

There is much interest, both nationally and internationally, in increasing the use and understanding of sustainable agriculture which involves, in part, reducing chemical inputs into our agricultural systems. With less reliance on chemicals, pest-resistant crops will be more important in the years to come. A reduction in chemical use not only implies lower costs of production but also reduced detrimental effects on the environment. Maintaining pest-resistant germplasm is of great importance to the future of agriculture. As broad a sampling of germplasm as possible must be acquired and saved, because we cannot predict which genes will be needed in the future and germplasm losses are irretrievable.

REFERENCES

1. Anonymous. 1977. *National Seed Storage Laboratory*. U.S. Department of Agriculture, Agricultural Research Service, Western Region. U.S. Government Printing Office, Washington, DC, pp. 793–035.
2. ARS Information Service. 1990. *Seeds for Our Future: The U.S. National Plant Germplasm System*. Program Aid 1470. Washington DC: U.S. Department of Agriculture.

3. Bamberg, J., ed. 1993. *IR-1, The Inter-Regional Potato Program*, The United States Potato Genebank. Antigo, Wisconsin: Palmer Publications.

4. Breese, L. 1989. Multiplication and regeneration of germplasm. In *IBPGR Training Courses: Lecture Series 2, Scientific Management of Germplasm: Characterization, Evaluation, and Enhancement*. H.T. Stalker & C. Chapman, eds. Rome: International Board for Plant Genetic Resources, pp 17–21.

5. Burgess S., ed. 1971. *The National Program for Conservation of Crop Germ Plasm*. State Agricultural Experimental Station Directors Association for the North Central, Northeastern, Southern, and Western Regions of the United States, and the Agricultural Res. Servey, U.S. Department of Agriculture, Athens, Georgia: University of Georgia.

6. Chapman, C. 1989. Principles of germplasm evaluation. In *IBPGR Training Courses: Lecture Series. 2. Scientific Management of Germplasm: Characterization, Evaluation and Enhancement*. H.T. Stalker & C. Chapman, eds. Rome: International Board for Plant Genetic Resources, pp 55–63.

7. Chang, T.T. 1987. Saving crop germplasm. *Prog. in Agric.* 30:62–63.

8. Chang, T., S.M. Dietz, & M.N. Westwood. 1980. Management and use of plant germplasm collections. In *Biotic Diversity and Germplasm Preservation, Global Imperatives*. L. Knutson & A.K. Stoner, The Netherlands: Kluwer, pp 127–159.

9. Clark, R.L. 1980. Seed maintenance and storage. In *Plant Breeding Reviews: The National Plant Germplasm System of the United States*. Vol. 7. J. Janick, ed. Portland, Oregon: Timber Press, pp 95–110.

10. Cohen, J.I., J.T. Williams, D.L. Plucknett, & H. Shands. 1991. *Ex situ* conservation of plant genetic resources: global development and environmental concerns. *Science* 253:866-872.

11. Committee on Germplasm Resources. 1978. *Conservation of Germplasm Resources: An Imperative*. Washington, DC: National Academy of Science.

12. Creech, J.L., & L.P. Reitz. 1971. Plant germplasm: new and for tomorrow. *Adv. Agron.* 23:1–49.

13. Darlington, C.D. 1973. *Chromosome Botany and the Origins of Cultivated Plants*. 3d ed. London: Allen & Unwin.

14. De Langhe, E.A.L. 1984. The role of in vitro techniques in germplasm conservation. In *Crop Genetic Resources: Conservation & Evaluation*. J.H.W. Holden & J.T. Williams, eds. London: Allen and Unwin, pp 131–137.

15. Duvick, D.N. 1982. Improved conventional strategies and methods for selection and utilization of germplasm. In *Chemistry and World Food Supplies: The New Frontiers*. L.W. Shemilt, ed. Oxford: Pergamon Press, pp 577–584.

16. Duvick, D.N. 1986. Expectations for the future. *Econ. Bot.* 40:289–297.

17. Elgin, J.H., Jr., & P.A. Miller. 1989. Enhancement of plant germplasm. In *Biotic Diversity and Germplasm Preservation, Global Imperatives*. L. Knutson & A.K. Stoner, eds. The Netherlands: Kluwer, pp 311–322.

18. Ellis, M.D., G.S. Jackson, W.H. Skrdla, & H.C. Spencer. 1981. Use of honey bees for controlled interpollination of plant germplasm collections. *HortScience* 16:488–491.

19. Ellis, R.H., T.D. Hong, & E.H. Roberts. 1985. *Handbook of Seed Technology for Genebanks*. Rome: International Board for Plant Genetic Resources.

20. Fisher, H.H. 1969. Plant introduction in the United States. *Plant Introduction Newsletter*. 22:13–22.
21. Fitzgerald, P.J., K. Hummer, J. McFerson, R. Nelson, C. Sperling, G. White, & T. Williams. 1989. Acquisition policy for the U.S. National Plant Germplasm System. *Diversity* 5:31.
22. Ford-Lloyd, B., & M. Jackson. 1986. *Plant Genetic Resources: An Introduction to Their Conservation and Use*. London: Edward Arnold.
23. Hamilton, M.B. 1994. *Ex situ* conservation of wild plant species: Time to reassess the genetic assumptions and implications of seed banks. *Conserv. Biol.* 8:39–49.
24. Harlan, J.R. 1992. *Crops & Man*. Madison, Wisconsin: American Society of Agronomy, Inc., and Crop Science Society of America, Inc.
25. Harlan, J.R., & K.J. Starks. 1980. Germplasm resources and needs. In *Breeding Plants Resistant to Insects*. F.G. Maxwell & P.R. Jennings, New York: Wiley, pp 253–273.
26. Harris, M.K. 1975. Allopatric resistance: searching for sources of insect resistance for use in agriculture. *Environ. Entomol.* 4:661–669.
27. Hawkes, J.G. 1981. Germplasm collection, preservation, and use. In *Plant Breeding II*. K.J. Frey, ed. Ames, Iowa: The Iowa State University Press, pp 57–83.
28. Hawkes, J.G. 1983. *The Diversity of Crop Plants*. Cambridge, Massachusetts: Harvard University Press.
29. Holden, J.H.W. 1984. The second ten years. In *Crop Genetic Resources: Conservation & Evaluation*. J.H.W. Holden & J.T. Williams, pp 277–285. London: Allen and Unwin.
30. Holden, J., J. Peacock, & T. Williams. 1993. *Genes, Crops and the Environment*. Cambridge, Massachusettts: Cambridge University Press.
31. Hoyt, E. 1988. *Conserving the Wild Relatives of Crops*. IBPGR-IUCN-WWF. Rome: IBPGR.
32. Hyland, H.L. 1984. History of plant introduction in the United States. In *Plant Genetic Resources: A Conservative Imperative*. C.W. Yeatman, D. Kafton, & G. Wilkes, Boulder, Colorado: Westview Press, pp 5–14.
33. IBPGR Secretariat. 1981. *Revised Priorities Among Crops and Regions*. AGP: IBPGR/81/34. Rome: IBPGR.
34. Jain, S.K. 1975. Population structure and the effects of breeding system. In *Crop Genetic Resources for Today and Tomorrow*. O.H. Frankel J.G. Hawkes, ed. Cambridge, Massachusetts: Cambridge University Press, pp 15–36.
35. Jones, Q., & S. Gillette. 1982. *The NPGS: An Overview*. Special Rept. 1, Laboratory for Information Science in Agriculture, Colorado State University, Ft. Collins, Colorado: Diversity.
36. Lindstrom, L. 1970. They don't throw anything away. *The Furrow* 75:34–35.
37. Matthews, B.F., & J.A. Saunders. 1989. Gene transfer in plants. In *Biotic Diversity and Germplasm Preservation, Global Imperatives*. L. Knutson & A.K. Stoner, eds. The Netherlands: Kluwer, pp 275–291.
38. Maunder, A.B. 1992. Identification of useful germplasm for practical plant breeding programs. In *Plant Breeding in the 1990's*. H.T. Stalker & J.P. Murphy, eds. Wallingford, UK: C.A.B. International, pp 147–169.

39. Miller, G.T., Jr. 1995. *Environmental Science: Working with the Earth*. 5th ed. Belmont, California: Wadsworth.

40. Mowder, J.D., & A.K. Stoner. 1989a. Plant germplasm information systms. In *Biotic Diversity and Germplasm Preservation, Global Imperatives*. L. Knutson & A.K. Stoner. The Netherlands: Kluwer, pp 419–426.

41. Mowder, J.D., & A.K. Stoner. 1989b. Information systems. In *Plant Breeding Reviews: The National Plant Germplasm System of the United States*. Vol. 7. J. Janick, ed. Portland, Oregon: Timber Press, pp 57–66.

42. Nabhan, G.P. 1989. *Enduring Seeds: Native American Agriculture and Wild Plant Conservation*. San Francisco: North Point Press.

43. Namkoong, G. 1989. Population genetics and the dynamics of conservation. In *Biotic Diversity and Germplasm Preservation, Global Imperatives*. L. Knutson & A.K. Stoner, eds. The Netherlands: Kluwer, pp 161–181.

44. National Plant Germplasm Committee. 1978. *The National Plant Germplasm System*. U.S. Department of Agriculture, Science and Education Administration, Program Aid No. 1188.

45. National Research Council. 1972. *Genetic Vulnerability of Major Crops*. Washington, DC: National Academy of Science Press.

46. National Research Council. 1991. *Managing Global Genetic Resources: The U.S. National Plant Germplasm System*. Washington, DC: National Academy of Science Press.

47. Panel of Biodiversity Research Priorities of the Board on Science and Technology for International Development. 1992. *Conserving Biodiversity: A Research Agenda for Development Agencies*. Washington, DC: National Academy Press.

48. Perdue, R.E., & G.M. Christenson. 1980. Plant exploration. In *Plant Breeding Reviews: The National Plant Germplasm System of the United States*. Vol. 7. J. Janick, ed. Portland, Oregon: Timber Press, pp 67–94.

49. Plucknett, D.L., N.J.H. Smith, J.T. Williams, & N.M. Anishetty. 1987. *Gene Banks and the World's Food*. Princeton, New Jersey: Princeton University Press.

50. Primack, R.B. 1993. *Essentials of Conservation Biology*. Sunderland, Massachusetts: Sinauer Associates.

51. Reed, S.M. 1989. *In vitro* conservation of germplasm. In *IBPGR Training Courses: Lecture Series 2, Scientific Management of Germplasm: Characterization, Evaluation and Enhancement*. ed. H.T. Stalker & C. Chapman, eds. Rome: International Board for Plant Genetic Resources, pp 23–30.

52. Roath, W.W. 1989. Evaluation and enhancement. In *Plant Breeding Reviews: The National Plant Germplasm System of the United States*. Vol. 7. J. Janick, ed. Portland, Oregon: Timber Press, pp 183–212.

53. Roberts, E.H. 1973. Predicting the storage life of seeds. *Seed Sci. & Technol.*, 1:499–514.

54. Roberts, E.H., & R.H. Ellis. 1984. The implications of the deterioration of orthodox seeds during storage for genetic resources conservation. In *Crop Genetic Resources: Conservation & Evaluation*. J.H.W. Holden & J.T. Williams, eds. London: Allen & Unwin.

55. Roberts, E.H., M.W. King, & R.H. Ellis. 1984. In *Crop Genetic Resources:*

Conservation & Evaluation. J.H.W. Holden & J.T. Williams, eds. London: Allen & Unwin, pp 38–52.

56. Roos, E.E. 1989. Long-term seed storage. In *Plant Breeding Reviews: The National Plant Germplasm System of the United States*. Vol. 7. J. Janick, ed. Portland, Oregon: Timber Press, pp 129–158.

57. Russell, W.A. 1975. Breeding and genetics in the control of insect pests. *Iowa State J. Res.* 49:527–551.

58. Ryerson, K.A. 1967. The history of plant exploration and introduction in the United States Department of Agriculture. In *Proceedings of the International Symposium on Plant Introduction*. Tegucigalpa, Honduras: Escuela Agricola Panamericana, pp 1–19.

59. Shands H.L., & V.A. Sisson. 1989. Plant germplasm maintenance: An example of a national program. In *IBPGR Training Courses: Lecture Series. 2, Scientific Management of Germplasm: Characterization, Evaluation and Enhancement*. H.T. Stalker & C. Chapman, eds. Rome: International Board for Plant Genetic Resources, pp 7–15.

60. Shands, H.L., P.J. Fitzgerald, & S.A. Eberhart. 1989. Program for plant germplasm preservation in the United States. In *Biotic Diversity and Germplasm Preservation, Global Imperatives*. L. Knutson & A.K. Stoner, eds. The Netherlands: Kluwer, pp 97–115.

61. Singh, H.B., R.K. Arora, & M.W. Hardas. 1975. Untapped plant resources. *Proc. Indian Natl. Sci. Acad.* 41:194–203.

62. Skrdla, W.H. 1975. The U.S. plant introduction system. *HortScience* 10:570–574.

63. Stalker, H.T. 1989. Utilizing wild species for crop improvement. In *IBPGR Training Courses: Lecture Series 2, Scientific Management of Germplasm: Characterization, Evaluation and Enhancement*. H.T. Stalker & C. Chapman, eds. Rome: International Board for Plant Genetic Resources, pp 139–154.

64. Steyn, R. 1976. *Plant Genes for the Future: Research for a Better Iowa*. Bulletin AR-5, Iowa Agriculture and Home Economics Experimental Station, Ames.

65. Towill, L.E., & E.E. Roos. 1989. Techniques for preserving of plant germplasm. In *Biotic Diversity and Germplasm Preservation, Global Imperatives*. L. Knutson & A.K. Stoner, eds. The Netherlands: Kluwer, pp 379–403.

66. Vavilov, N.I. 1951. The origin, variation, immunity and breeding of cultivated plants. *Chronica Botanica* 13:1–366. (Translated from Russian by K.S. Chester.)

67. Waterworth, H. 1993. Processing foreign plant germ plasm at the national plant germplasm quarantine center. *Plant Dis.* 77:854–860.

68. White, G.A., & J.A. Briggs. 1989. Plant germplasm acquisition and exchange. In *Biotic Diversity and Germplasm Preservation, Global Imperatives*. L. Knutson & A.K. Stoner, eds. The Netherlands: Kluwer, pp 405–417.

69. White, G.A., H.L. Shands, & G.R. Lovell. 1989. History and operation of the national plant germplasm system. In *Plant Breeding Reviews: The National Plant Germplasm System of the United States*. Vol. 7. J. Janick, ed. Portland, Oregon: Timber Press. pp 5–56.

70. Wilkes, G. 1983. Current status of crop plant germplasm. *CRC Crit. Rev. Plant Sci.* 1:133–181.

71. Wilkes, G. 1984. Germplasm conservation toward the year 2000: Potential for new crops and enhancement of present crops. In *Plant Genetic Resources: A Conservative Imperative*. C.W. Yeatman, D. Kafton, & G. Wilkes, Boulder, Colorado: Westview Press, pp 5–14.

72. Wilkes, G. 1989. Germplasm preservation: objectives and needs. In *Biotic Diversity and Germplasm Preservation, Global Imperatives*. L. Knutson & A.K. Stoner, The Netherlands: Kluwer, pp 13–41.

73. Williams, J.T. 1989. Plant germplasm preservation: a global perspective. In *Biotic Diversity and Germplasm Preservation, Global Imperatives*, ed. L. Knutson & A.K. Stoner, eds. The Netherlands: Kluwer, pp 81–96.

74. Wilson, R.L., R.L. Clark, & M.P. Widrlechner. 1985. A brief history of the North Central Regional Plant Introduction Station and a list of a genera maintained. *Proc. Iowa Acad. Sci.* 92:63–66.

75. Wilson, R.L., B.R. Wiseman, & G.L. Reed. 1991. Evaluation of J.C. Eldredge popcorn collection for resistance to corn earworm, fall armyworm (Lepidoptera: Noctuidae), and European corn borer (Lepidoptera: Pyralidae). *J. Econ. Entomol.* 84:693–698.

76. Winters, H.F. 1967. Handling and storage of plant materials and seed. In *Proceedings of the International Symposium on Plant Introduction*. Tegucigalpa, Honduras: Escuela Agricola Panamericana, pp 151–156.

77. Withers, L.A. 1989. In vitro conservation and germplasm utilization. In *The Use of Plant Genetic Resources*. A.H.D. Brown, O.H. Frankel, D.R. Marshall, & J.T. Williams, Cambridge, UK: Cambridge University Press, pp 309–334.

78. Withers, L.A., & P.J. King. 1980. A simple freezing unit and cryopreservation method for plant cell suspensions. *Cryoletters* 1:213–320.

10
General Overview of Plant Pathogenic Organisms

George N. Agrios
University of Florida, Gainesville, Florida

I. INTRODUCTION

It is commonly observed that wherever plants are grown for food or fiber they are fed upon and damaged by insects and mites, and that they compete with other plants (i.e., weeds) for water, nutrients, and sometimes sunlight. In addition, however, plants must fend off and survive numerous diseases. Some diseases are caused by adverse environmental conditions, such as low temperatures, droughts, nutrient deficiencies or toxicities, air pollutants, and the like. Most importantly, however, plants must avoid or survive diseases caused by organisms that live in or on the plant, obtain nutrients from the plant, and through their presence and through substances they produce, affect the normal functions of the plants to the point where plants become diseased, produce less, and frequently die from the disease.

II. KINDS OF PLANT PATHOGENIC ORGANISMS

The plant pathogenic organisms that cause disease in plants are basically different members of the same groups of microorganisms that cause disease in humans and animals. Such microorganisms include fungi, bacteria, and the closely related mycoplasma-like organisms (mollicutes), viruses, nematodes, and protozoa [1]. In addition, certain higher plants grow parasitically on other plants and cause disease on these plants.

A. Fungi

Fungi are usually filamentous microorganisms whose body, the mycelium, consists of branching cylindrical strands, known as hyphae. Hyphae are 1–5 µm in diameter, and the whole mycelium may be a few to several millimeters or centimeters long. Fungi lack chlorophyll; they survive by feeding either on dead organic matter (saprophytic fungi) or on living plants or animals (parasitic fungi), thereby causing the host to become diseased. Fungi multiply by producing spores both asexually and sexually following a fertilization process. The types of spores and the structures on or in which the spores are produced are used to distinguish and identify the various plant pathogenic fungi. Fungi cause the majority of all plant diseases, including, for example, many leaf spots, fruit spots and blights, stem and root rots, wilts, cankers, downy mildews, powdery mildews, rusts, smuts, and galls [1,9].

B. Bacteria and Mycoplasma-Like Organisms

Bacteria and mycoplasma-like organisms (mollicutes) are single-celled microorganisms whose genetic material (DNA) is not bound by a membrane; therefore, they are prokaryotes [7]. Bacteria have a cell membrane and a rigid cell wall; the latter gives them their typical rod or spherical shape. Most plant pathogenic bacteria are rod shaped, about 1×3 µm in size, and may have one or more flagella which help them move for short distances in liquid media. Mollicutes have a membrane but no cell wall, and their shape, therefore, is quite variable. Mollicutes have no flagella, but some of them are helical and seem capable of moving. Bacteria multiply by binary fission, each of them splitting into two bacteria every 20–80 min, provided conditions (food, moisture, and temperature) are favorable. Bacteria cause symptoms similar to those caused by fungi. Mollicutes cause the diseases known as yellows, shoot and root proliferations, and some tree declines. Plant pathogenic bacteria survive in or on plant tissues; some of them can also survive in the soil or in their insect vectors. Mollicutes survive only in living host plants and in their insect vectors.

C. Viruses and Viroids

Viruses and viroids are self-replicating nucleic acid molecules that can cause disease in their host [10]. In viruses, the nucleic acid may be either DNA or RNA and is always surrounded by a coat made of one or a few kinds of protein molecules. In viroids, the nucleic acid is a small RNA molecule and is naked. Viruses may be spherical (about 25–70 nm in diameter), rod shaped (15 $\times$ 300 nm), or threadlike (10 $\times$ 500 to 10 $\times$ 2000 nm in size). Some viruses are surrounded by a membrane. Viroids have only about a tenth or less as much

RNA as is present in viruses. Viruses and viroids mutiply (i.e., replicate) by taking over the host cell machinery and directing it to produce viral and viroid components, which the host cell assembles into viruses and viroids. The viral nucleic acid contains genes that are translated into proteins of known function (e.g., viral coat protein, virus replication protein [replicase], viral movement protein, viral transmission protein). Viroid RNA does not appear to code for any specific protein, but viroids can do all the things viruses do, such as multiply, make plants diseased, and spread. Viruses and viroids survive almost entirely in living plant cells. All viruses and viroids are spread by vegetative propagating organs (e.g., grafts, tubers). In addition, several are spread by sap transfer through contact or handling, by seed, through pollen, and through bridging of plants by the parasitic plant known as dodder (*Cuscuta* sp.). Some viruses are also transmitted by specific vectors such as insects (especially aphids, leafhoppers, whiteflies, and thrips) and a few by mites, nematodes, and fungi. Viruses and viroids cause numerous diseases that appear primarily as mottling of leaves and fruit (mosaic), ring spots, yellowing, plant stunting, leaf and fruit malformations, flower sterility, and reduced yield quantity and quality, and in some cases, these pathogens cause death of shoots and of entire plants.

D. Nematodes

Nematodes are wormlike animals that feed mostly on roots and other below-ground parts of plants, although a few of them feed on stems, leaves, or fruit [4]. Most nematodes are transparent and wormlike, about 30×1000 μm or more in size, but the females of some species are spherical and pear shaped and may have an opaque outer coat. All plant pathogenic nematodes have a spear (stylet) with which they pierce plant cells and suck out the cell contents. Some nematodes are sedentary (i.e., once they find and pierce a host, they remain at that location for life); others are migratory, feeding and moving on to other roots and other plants. Most sedentary nematodes are endoparasitic (feed and remain inside the plants), whereas most migratory nematodes are ectoparasitic (feed from outside the plants). Nematodes multiply by laying eggs, which may overseason or hatch immediately and produce four successive juvenile stages prior to the adult stage. Plant pathogenic nematodes can cause root lesions or galls, kill root tips to produce excessive root branching, cause stem or leaf lesions and distortions and seed galls, and generally drastically reduce plant growth and yield.

E. Protozoa

Protozoa are small 1×5 μm trypanosomatid flagellate animals that feed and multiply in the phloem of some tropical trees, such as palms and coffee, and cause disease [1]. More frequently, flagellate protozoa are found in the laticiferous cells of plants in the Euphorbiaceae family and in several other families, but it is not certain whether they cause disease in these plants. Plant-infecting protozoa are believed to spread by insect vectors of the genera *Lincus* and *Ochlerus* of the family Pentatomidae.

F. Parasitic Higher Plants

Parasitic higher plants, such as the dodders (*Cuscuta*), broomrapes (*Orobanche*), mistletoe (*Viscum*), dwarf mistletoe (*Arceuthobium*), and witchweed (*Striga*), are higher plants that live parasitically on other plants and cause disease [1]. Parasitic plants produce flowers and seeds, but all of them lack true roots and some have little or no chlorophyll and, therefore, must obtain all the water and mineral nutrients and all or much of the carbohydrates from the plants they parasitize. Parasitized plants are stunted, produce little or no yield, and, in some cases, are killed.

III. EFFECTS OF PATHOGENS ON PLANTS: SYMPTOMS

Depending on the kind of pathogen that affects a plant and on the part of the plant affected, the effects of the pathogen on the plant, that is, the symptoms exhibited by the affected plant, may vary greatly [1]. Viruses, viroids, mollicutes, protozoa, and some bacteria and fungi invade the plant systemically. These pathogens either reach almost all cells (most viruses and viroids) or are confined mainly to the vascular system of the plant; that is, the phloem (some viruses, all mollicutes, protozoa, some bacteria) or the xylem (vascular fungi and bacteria). Systemically infected plants may become stunted, chlorotic (yellowish), and decline; their leaves and shoots may die back; and they may produce few flowers and fruit, or they may wilt and die. In many viral diseases, leaves, flowers, fruit, and stems may develop various discolorations (mosaics, mottles) and malformations. In some systemic diseases, affected plants produce galls and excessive numbers of shoots and roots.

Most plant diseases are caused by fungi which, along with most plant pathogenic bacteria, cause more or less localized infections. Such infections appear as small or large necrotic spots on leaves, flowers, fruit, stem, and roots. Sometimes the spots enlarge or coalesce to produce large necrotic areas on these organs, or the spots continue to enlarge killing entire leaves, shoots, and stems, thereby causing the so-called blights. Similar spread of localized infections on fruits, roots, and other underground food storage organs, such as bulbs and tubers, results in fruit rots, root rots, or rots of the various storage organs.

Most plant diseases occur on growing plants in the field, but many of them continue to develop and cause rots on various plant structures even after these are harvested. However, even when such organs are healthy when harvested, they frequently become infected with fungi and bacteria after harvest and may become partially or totally rotten within a few days.

IV. KINDS AND AMOUNTS OF LOSSES CAUSED BY PLANT DISEASES

The most obvious losses caused by plant diseases are the result of reduced quantities of produce from diseased plants or loss of the diseased plants. Additional losses are caused by the reduced quality of produce from diseased plants or when the produce itself is infected and shows symptoms such as spots or rotting. Infected produce is not only less usable, but it imposes additional costs for handling and processing (e.g., sorting, peeling). In many cases, infected food and feed products also contain mycotoxins, produced by some plant-infecting fungi, which make infected or contaminated produce poisonous to humans and animals and, therefore, unfit for consumption. Lower quantity and quality of produce caused by plant diseases results in lower income for the affected producers and higher prices for all consumers.

In addition to the above, plant diseases cause several other kinds of losses. The principal additional loss is the cost of controlling or managing the various diseases in each cultivated crop. Such losses include labor and equipment costs for a number of cultural practices (such as soil fumigation, solarization, certain kinds of plowing, water draining, bed formation, the necessary scouting for pathogens, purchase of more expensive certified pathogen-free seed and transplants) carried out primarily for controlling plant diseases,. Furthermore, the costs for materials, labor, and equipment for application of required pesticides often represent a significant portion of the production costs for many crops, particularly fruits and vegetables, and these costs are actually losses imposed on producers by the various plant diseases. The application of pesticides to control plant diseases may also result in hidden additional costs from chemical contamination of water, foodstuffs, and the environment in general.

Several other losses or costs are also caused by plant diseases. For example, one of the main reasons for crop rotation is to avoid build-up of pathogens which occur when the same crop is planted in a field year after year. Crop rotation, however, requires additional land which must be left fallow or planted to a different, less profitable crop. Similarly, producers must often give up susceptible, productive crops or cultivars of a crop for less desirable or less productive ones which are resistant to one or more severe diseases. Finally, producers, ship-pers, wholesalers, retailers, and consumers often must provide refrigeration for produce from harvest until consumption to protect produce from being attacked and rendered unusable by plant pathogens.

The losses caused by different plant diseases vary with the particular host-pathogen combination, the location, the environmental conditions prevailing each year, and the control measures practiced [13]. The quantity of produce lost may range from slight to 100% and may take place in the field or after harvest, in the absence of, or in spite of, control measures. In general, in each location, only a few diseases affect a crop each year and cause measurable losses. Some of these diseases occur in a location year after year and may cause losses of 1–10% or, if not controlled, may destroy the majority or the whole crop. It has been estimated that plant diseases before and after harvest destroy about 12% of all produce [3,11]. This is the same percentage of produce destroyed by insects and slightly higher than the percentage of produce (about 10%) lost to competition by weeds [3,11].

V. ECOLOGY AND BIOLOGY OF PLANT PATHOGENS

Plant pathogens may consist of single molecules (viroids), or two or a few molecules (viruses), or they may be more complex microorganisms requiring a moist organismal environment for their survival and multiplication (bacteria, mollicutes, protozoa, fungi, and nematodes). As a result, most plant pathogens complete their life cycle from year to year being entirely associated with, and usually within, their hosts. For example, all viroids and viruses, all mollicutes and protozoa, most fungi, and most bacteria complete their life cycles entirely within their host plants. This is particularly true for pathogens affecting perennial plants systemically and, to some extent, for pathogens affecting stems and roots of perennial plants. Most fungi, bacteria, and nematodes affecting annual plants or leaves and fruits of perennial plants complete their life cycles partly on or in their hosts and partly in plant debris or in the soil. As long as the host is alive, pathogens survive in it in their vegetative form, although some of them, such as fungi and nematodes, may also produce reproductive structures (fungal spores and nematode eggs). When the host dies, most types of pathogens within it (viruses, mollicutes, protozoa, most bacteria, and many fungi) also die. Such pathogens, however, may survive in some of the host seeds (viruses, bacteria, fungi), in vegetative propagative organs (all pathogens), in alternate hosts (viruses, mollicutes, protozoa, bacteria, some fungi), or in plant debris and in the soil (some fungi, bacteria, nematodes). Fungi and nematodes surviving in plant debris and in the soil generally exist there in the form of hardy spores (fungi) or eggs (nematodes). In a few cases, plant pathogens are carried from infected to healthy plants by vectors such as insects, nematodes, and fungi, and, in these cases, the pathogens may survive part of their life cycle in their vectors.

The biology of the various types of pathogens varies greatly. Viroids consist of a small RNA molecule that is not known to code for even one protein, yet these pathogens are capable of inducing host cells to reproduce the RNA molecule and to make the host exhibit disease symptoms. Viroids seem to have no vectors but are spread from plant to plant in vegetative propagative organs and on tools used in various cultural practices.

Viruses consist of relatively large RNA or DNA molecules, each surrounded by one or more protein molecules and, in some viruses, by a membrane. Some viruses also carry one or a few molecules of a replicase, an enzyme needed for the multiplication of these viruses. Viruses, like viroids, also induce the host cell to reproduce the viral components (i.e., nucleic acid and proteins) as they are coded by genes on the viral nucleic acid(s). Viruses, in most cases, also induce disease symptoms in the host. Viruses are spread from plant to plant on vegetatively propagative materials, some by seed, pollen, or sap, and some by vectors such as insects, nematodes, and fungi.

Bacteria, mollicutes, and protozoa are microscopic single-celled microbes that live either within plant cells (some bacteria and all mollicutes and protozoa) or between cells (most bacteria). All of these organisms absorb nutrients from the plant cells while they release, within or around cells, biologically active metabolites (e.g., enzymes, toxins). Some of these metabolites cause, or add to, the disease symptoms of the host. These pathogens multiply in the host via fission and, particularly the bacteria, produce large numbers of progeny in the host in a relatively short time. Although most bacteria spread from host to host primarily via contact or windblown rain, some bacteria and all mollicutes and probably protozoa spread via infected propagative materials and by specific insect vectors, especially planthoppers and leafhoppers.

Nematodes are semimicroscopic wormlike animals that parasitize primarily the roots of host plants, although some also parasitize leaves, stems, and seeds of their hosts. Nematodes feed on plant tissues either from the outside or by partially or totally entering these organs. Nematodes lay eggs inside or outside their hosts, and the eggs hatch to produce juveniles. The latter can move around for short distances and infect the same or other host plants. Most nematodes survive part of the year in the soil as eggs, juveniles, or adults. Nematodes are carried over long distances on transplants or in soil carried on human or animal feet or farm equipment.

Fungi are the most varied, most numerous, most complex, and most important plant pathogens. All fungal plant pathogens enter plant organs, grow into or between plant cells, and absorb nutrients from them. Many, and probably all, fungi secrete biologically active substances (e.g., enzymes, toxins, growth regulators) in the tissues where they grow, and these substances cause, or contribute to, the production of disease symptoms. A few fungi, such as those causing the powdery mildews, remain mostly outside their hosts, sending only feeding organs (haustoria) into them. Most fungi grow through plant tissues before or after the tissues are killed by the fungus. Some fungi enter only the water-conducting vascular tissues (xylem) of plants and cause plants to wilt. Almost all fungi produce numerous spores within the host or at the host's surface. The spores may be carried, for example, by air, water, or animals, to other plants, which they infect. Some kinds of spores can survive adverse temperatures and moisture conditions until favorable weather returns. The spores can survive unfavorable weather on the host, in plant debris, on or in seeds, or in the soil. Some fungi, however, can also survive in the same areas as mycelium.

In many cases, possibly in the majority of cases, plants in a field are infected by more than one pathogen. In most such infections, the effects caused by the various pathogens are additive, equaling the sum of those caused by each pathogen separately. In some cases, however, infection of a plant by one pathogen, for example, a virus, a vascular wilt-causing fungus, or a nematode, makes the plant more susceptible to the next pathogen. In such disease complexes, the combined effect of multiple pathogens on the plants may be greater than the sum of the damage caused by each pathogen separately.

VI. DETECTION AND IDENTIFICATION OF PLANT PATHOGENS

Of all plant pathogens, only the parasitic higher plants and a few fungi that produce characteristic, relatively large, fruiting structures such as mushrooms can be detected and identified with the naked eye. All others require either a dissecting microscope (nematodes, some fungi), a compound microscope (most fungi and all bacteria, mollicutes, and protozoa), or an electron microscope (viruses, viroids, mollicutes) to be visualized. Usually, however, most of these also require additional tests in order to be properly identified.

The presence of plant pathogens on a diseased plant is, of course, deduced by the fact that such a plant exhibits certain symptoms, such as leaf spots or mottling, stem or root galls or lesions, or wilting of the plant. Sometimes, the symptoms on a specific plant are characteristic of and diagnostic for a particular disease caused by a specific pathogen, such as apple scab symptoms caused by the fungus *Venturia inaequalis* (Fig. 1). In such cases, someone with expertise in plant disease detection and identification can quickly identify the pathogens of such diseases with no additional examinations or testing. Usually, however, similar symptoms, such as leaf spots, fruit spots, and root lesions, can be caused by several different pathogens. In these cases, portions of the infected tissue are examined under a dissecting microscope, compound microscope, or

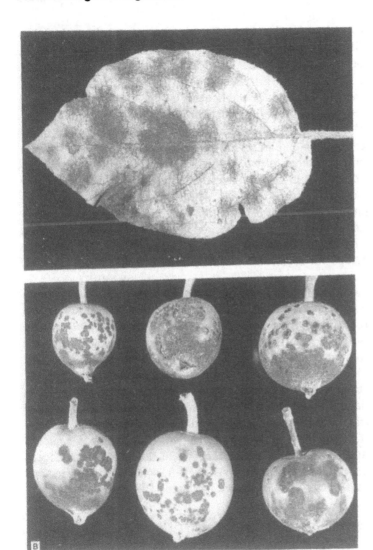

Figure 1 Symptoms of apple scab disease on leaves (A) and fruit (B) caused by the fungus *Venturia inaequalis*. (From Ref. 1.)

electron microscope with the expectation of finding a pathogen that fits the description of one of the pathogens known to cause such a disease on the particular host. This search, however, is not always fruitful, because pathogens are often hard to find in infected tissues and because they are often admixed with other, usually saprophytic, microorganisms that make it difficult or impossible to identify the pathogen conclusively.

With pathogens such as fungi and bacteria, attempts are usually made to isolate them from diseased tissue and grow them in pure culture in test tubes or culture plates. There the pathogens can be observed more easily and over time can be allowed to form their characteristic spores (fungi) or growth forms (bacteria), and they also can be used for additional tests; for example, to inoculate host plants and test for pathogenicity or to test their physiological properties. In the last several decades, selective media have been developed which allow only fungi or bacteria belonging to certain genera or species to grow, whereas they inhibit the growth of other related or unrelated microorganisms. For pathogens that cannot be grown in culture (viruses, viroids) or can be grown only with difficulty (mollicutes, some bacteria), detection and identification are attempted by grafting infected plant parts onto healthy indicator plants or by sap transmission (viruses, viroids). Virus detection and approximate identification can also be made by examining specially stained young infected cells under the compound microscope for crystalline or amorphous inclusion bodies induced by the virus.

For several years, serological tests, especially the enzyme-linked immunosorbent assay, usually referred to as ELISA, have been used extensively and effectively to detect and identify viruses and other pathogens in plant sap or after purification [8]. Serological tests depend on the ability of antibodies, produced in a warm-blooded animal against one or more proteins of the pathogen injected into the animal, to react with the same proteins upon contact and under appropriate conditions to give a visible manifestation of such a reaction (Fig. 2). In another type of diagnostic test, bacteria are identified by automated analysis of their fatty acids and comparison of the fatty acid profile of the unknown bacterium with those of known bacteria. In more recently developed tests, the nucleic acids of viruses and of ribosomes or mitochondria of the unknown pathogens are restricted (cut) with specific enzymes and their fragments are analyzed and compared with fragment profiles of known pathogens [5]. Alternatively, spe-

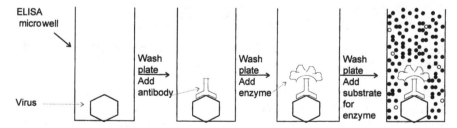

Figure 2 Schematic representation of a simple ELISA test. The colorless substrate changes color in the presence of virus, whereas it remains colorless in wells where virus (and, therefore, enzyme) are absent.

cific nucleic acid fragments are labeled with radioactive isotopes or with color-generating compounds, and these fragments are used as probes to hybridize with potentially complementary nucleic acids of unknown pathogens, which are thereby identified. For more precise identification and distinction of closely related pathogens or pathogen races and strains, the nucleic acid of the pathogens, or of their ribosomes or mitochondria, is isolated and the sequence of its bases is determined and compared to those of known races or strains for identification.

VII. MECHANISMS BY WHICH PATHOGENS CAUSE DISEASE IN PLANTS

Pathogens attack plants because, through evolution, the groups of the microorganisms to which they belong developed the ability to penetrate and feed on (or parasitize) plants. Some pathogens, known as obligate parasites, became so well adapted to this source of food that they lost the ability to survive without parasitizing living host plants. Others, the so-called facultative parasites, can feed on or in living host plants but also can survive by feeding on dead organic matter.

The most obvious way that pathogens affect plants is through removal of foodstuffs that would normally be used by the plant. In many plant diseases, however, infected plants show effects that are much greater than can be explained only by the loss of the plant's foodstuffs to the pathogen. For example, in some diseases, a few spots on a leaf may cause the whole leaf to turn yellow and fall off. In leaf, fruit, stem, and root spots, infected tissues often become soft before they are killed, and in rots of fleshy tissues, softening and disintegration of tissues is the main effect observed. In some diseases, entire plants wilt and die, although a relatively small population of the pathogen is present in the vascular system of the stem. In still other diseases, infected leaves, stems, or roots become malformed and develop galls or proliferate excessively in the presence of the pathogen. In still other diseases, entire plants remain stunted, their leaves become chlorotic, flowers may abort, or parts of or entire plants are killed. The ways pathogens cause these effects on plants are many and complex, but basically they affect plants by disturbing or interfering with one or more of the plant's basic functions. These functions include water and nutrient absorption, translocation, and transpiration; photosynthesis; photosynthate translocation and assimilation; energy generation through respiration; cell division, enlargement and differentiation; structural integrity; and reproduction.

The mechanisms by which pathogens disturb the basic functions of the plant are sometimes a result of the physical presence of the pathogen; for example, reduction of photosynthesis by the presence on leaves of slime mold, sooty mold, and powdery mildew fungi. In most cases, however, the disturbances on the plants are caused by chemical substances, particularly enzymes, toxins, growth regulators, and polysaccharides secreted by the pathogens. Some of the pathogens (viruses, viroids, and certain bacteria) disturb normal plant functions by taking over the plant cell nucleic acid replication and redirecting it to functions that, collectively, are detrimental to the plants. Different types of pathogen enzymes affect different functions of the plant. For example, pathogen pectinases, cellulases, and ligninases alter or destroy the structural integrity of cell walls; proteinases destroy cell membranes or affect other cellular structural proteins or enzymes; amylases break down starch, and so on. Several types of pathogenic fungi and bacteria affect their host plants by producing toxins. These compounds act directly on the plant protoplast by reacting and interfering with the function of a structural protein (membrane), an enzyme, or a nucleic acid, and thus resulting in the malfunction or death of the cell. Many plant pathogens also produce in their hosts hormone-like compounds such as auxins, cytokinins, gibberellins, abscisic acid, and ethylene. These compounds have growth-regulatory activity and may stimulate or suppress senescence, or they may affect division and enlargement of host cells in which they occur. Finally, many pathogens secrete in their environment viscous polysaccharides which, in some cases, may clog cell pores or xylem vessels and may help in the production of symptoms in diseases such as wilts.

VIII. DEFENSE MECHANISMS OF PLANTS AGAINST PATHOGENS

Plants have survived in nature over the millennia and continue to survive in the presence of pathogens, because they have developed defense mechanisms against them (see Chap. 7). Such mechanisms inhibit or reduce the entrance, invasion by, or multiplication of pathogens in plants. The defense mechanisms may consist of defense structures that are present regardless of whether or not the pathogen is in contact with the host (preexisting or constitutive defenses) [1]. Such structures may consist of a thick, hard wax and cuticle over the epidermal cells; thick, hard cell walls; small, raised stomata and lenticels; a thick mat of hairs on the plant surface; and so on. Several defense structures, however, develop only after the plant has been attacked by the pathogen (inducible defenses). Such defense structures, formed in response to infection by pathogens, include the formation of cork layers around the point of infection, formation of abscission layers around infection points in leaves, formation of tyloses in xylem vessels ahead of invading pathogens, deposition of gums around lesions caused by pathogens, formation of callose deposits or sheaths around invading pathogens, and necrosis of invaded and adjacent cells. All of these responses restrict the pathogen to one or a few cells and protect the rest of the plant from becoming invaded.

Much more important than the structural defenses, however, are defenses based on the presence or production in the host of chemical substances that are toxic to the invading pathogen. The preexisting toxic substances are often various types of phenolic compounds or hydrolytic enzymes and may be present in cells or may be exuded to the plant surface. On the other hand, pathogens may fail to infect plants because the latter lack certain recognition structures or molecules, lack receptor molecules or sites for pathogen toxins, or lack essential nutrients required by the pathogen.

Plants more commonly defend themselves against pathogens through a number of metabolic changes which occur in plant cells in response to the attacking pathogen. The defensive metabolic changes are numerous. Some of them occur in sequence, whereas several occur in combination and their effects are probably additive. Such an aggregation of defensive metabolic changes occurs in the "hypersensitive reaction" which is expressed in incompatible combinations of plant cultivars and pathogens. In such combinations, attacked cells and cells surrounding them show loss of selective permeability of membranes, increased respiration, accumulation and oxidation of phenolic compounds, production of phytoalexins, and finally cell death and collapse. Pathogens surrounded by such defenses are usually immobilized and die, and they consequently fail to cause further infection and disease. However, even in cases in which infection becomes established and the pathogen invades the host to a lesser or greater extent, a number of defensive metabolic changes occur. Depending on the rapidity of development, the magnitude, and possibly the number of different defensive metabolic changes, the pathogen may proceed unchecked and may invade and kill the entire plant, or it may be stopped and become isolated within a small leaf spot or root lesion thereby sparing the rest of the plant from the pathogen's destructive potential.

Some of the defensive metabolic changes induced in the host plant by infecting pathogens include the production of several toxic phenolic compounds at various degrees of oxidation; the production of phytoalexins, which are specific toxic substances produced in appreciable amounts in plants only after infection or injury; the induced synthesis of enzymes involved in the synthesis of phenolic compounds; the production of fungitoxic phenolics from nontoxic complexes of phenolics with sugar and other molecules; the activation of phenol-oxidizing enzymes of the host; the inactivation of pathogen pectinolytic and other enzymes; the formation of substances that resist the enzymes of the pathogens; the release of fungitoxic cyanides from nontoxic substances; and the detoxification of pathogen toxins.

Although the above defensive metabolic changes are produced in resistant hosts in response to an infecting pathogen, it is also possible to induce resistance in a susceptible host by preinoculating such host plants with various microorganisms or by pretreatment with various chemical or physical agents. Such induced resistance is at first localized around the infection sites, but a few weeks later, it can be detected even in uninoculated leaves. In induced resistance as well as in other resistances, such as an hypersensitive reaction, a number of new proteins appear which are called pathogenesis-related proteins (see Chapter 7).

IX. GENETICS OF PLANT DISEASE INDUCTION AND DEVELOPMENT

Plant diseases are the result of interactions between compatible host plants and pathogens. In simplified terms, what makes some host plants compatible and others incompatible with their pathogens is the presence in the compatible plants of genes that determine susceptibility to the pathogen and in the incompatible ones of genes that determine resistance to the pathogen. At the same time, a pathogen strain may be compatible with its host because it contains one or more pathogenicity (virulence) genes toward the host, whereas an incompatible pathogen strain contains one or more avirulence genes towards the host [1] (see Chapter 7). However, it is not the presence of a particular gene in the host or the pathogen, but the appropriate combination of host resistance or susceptibility genes on the one hand and pathogen virulence or avirulence genes on the other hand that determines the plant reaction as compatible (i.e., susceptible and, therefore, diseased) or incompatible (i.e., resistant). Thus, if two plants, one of which has a resistance gene, R, and the other a susceptibility (lack of resistance) gene, r, react with two pathogens, one of which has a virulence gene, V, and the other an avirulence gene, v, their reactions are as depicted in Figure 3.

The quadratic check (Fig. 3) shows that the only plants that remain resistant [vR(-)] are those that have dominant genes for resistance (R) and react with pathogens that have recessive genes for virulence (v); that is, dominant genes for avirulence. Plants that have no resistance (r) or are attacked by virulence genes (V) that attack the specific resistance gene (R) of the host are susceptible and become diseased.

Each plant, of course, has many different genes of resistance which enable it to survive in the presence of many pathogens and many races of each pathogen. Each resistance gene codes for a protein which either by itself makes the host resistant to a particular pathogen (an unlikely or rare occurrence) or, usually, it initiates a series ("cascade") of biochemical reactions that leads to production of one or more of the previously mentioned structural or biochemical defense mechanisms against the pathogen(s).

Resistance genes are generally triggered into action by substances, called elicitors, which are produced by specific avirulence genes in the pathogen [14]

Pathogen Genes	Host Genes	
	R	r
V	VR(+)	Vr(+)
v	vR(-)	vr(+)

Figure 3 Disease reaction types in a host-pathogen system in which resistance or susceptibility is controlled by one gene. Plus signs indicate compatible (susceptible) reaction in which infection (disease) develops. Minus signs indicate incompatible reaction between host and pathogen (resistance) in which no infection (and, therefore, no disease) develops.

(see Chapter 7). The defense mechanisms set into action by some plant resistance genes are sufficient to stop further infection by the pathogen. Such resistance genes are often referred to as major genes for resistance [15]. The resistance provided by major genes is known as vertical resistance, because the presence of major genes in the host makes the host completely resistant to one or more pathogen races, whereas their absence from the host makes the host totally susceptible to the same pathogen.

Many, probably most, genes for resistance, however, are not quite as powerful and effective against pathogens as are the major genes. Rather, each such resistance gene, known as minor gene for resistance, triggers the production of a structural or biochemical defense mechanism which by itself provides only minor protection against a pathogen, but together with several other such genes, contributes to a significant level of resistance in the host plant. Characteristic of minor genes for resistance is that they are usually present in fairly large numbers in all plants. The plant resistance provided by minor genes is known as horizontal resistance, because it is effective against many pathogens and because, since it is the result of reactions of several minor genes in each host plant, even if some of the minor genes are absent, overcome, or bypassed by a pathogen, the remaining minor genes continue to provide a level of resistance to the pathogen. Horizontal resistance alone, however, seldom protects plants completely or sufficiently, and, besides, it may be affected significantly by prevailing weather conditions. Moreover, genes for horizontal resistance are more plentiful in hosts growing in the wild, whereas they tend to be lost from intensively bred cultivated crop plants.

Horizontal and vertical resistance usually coexist in host plants. However, because vertical resistance is easier to detect and very effective when present, vertical resistance has been much more sought after by plant breeders. Therefore, wherever available and possible, vertical resistance has been incorporated into most popular cultivars of the various crop plants. Unfortunately, because vertical resistance depends on one, two, or, rarely, three or four major genes, the pathogens against which the products of these genes are aimed develop, through intrinsic natural genetic variability, new individuals that possess mechanisms by which they can overcome or bypass the products of these genes for resistance. Once such pathogenic individuals appear, they can infect and multiply unchecked in the previously resistant host variety. As a result of this change in the pathogen, the resistance of this variety is said to have "broken down." The previously resistant variety is now susceptible to the progeny of the new pathogenic individuals, which collectively comprise a new race of the pathogen.

Depending on the particular host-pathogen combination, and even on the specific major gene for resistance, vertical resistance provided by a single major gene for resistance usually lasts no more than 2–5 years [15]. Usually within that time, a new pathogenic race appears that can overcome the resistance and may cause more or less severe losses to the crop. As a result, the now susceptible host variety must be replaced by another variety containing new and different genes for resistance. How long a variety with vertical resistance remains resistant also depends greatly on the size of the pathogen population that exists on adjacent susceptible varieties. Large populations of pathogens provide large amounts of inoculum that lands and establishes contact with the resistant variety and, in turn, increases the probability for appearance of mutant individuals that can survive in the "hostile" environment of the resistant variety. To avoid sudden and frequent breakdowns of disease resistance, efforts are being made to incorporate as many major and minor genes of resistance as possible in the same cultivars, to use different sets of genes in different host varieties or geographical areas, and to keep the amount of inoculum reaching the resistant varieties as low as possible.

X. FACTORS AFFECTING DEVELOPMENT OF PLANT DISEASE EPIDEMICS

It is obvious that for a plant disease to be initiated and to develop, a pathogen must first come in contact with a plant. Such a contact between a plant and a pathogen is insufficient, however, to guarantee initiation and development of disease. First, the plant must be of a genus, species, and variety that is susceptible to that pathogen. Second, the plant must be at the appropriate growth stage, such as seed, seedling, or blooming stages, with immature or mature leaves or fruit, or the plant must be in a senescent stage; pathogens often specialize in attacking one or more of these organs at the appropriate stage of their development. Contact of the pathogen with plant parts that it cannot attack, or at a time when susceptible parts such as flowers, leaves, or fruit are absent or too young or too senescent for infection, will not result in disease.

Similarly, the pathogen that comes in contact with the plant must, of course, be pathogenic to that kind of plant, but it must also be of a race that can attack the particular plant variety by overcoming its specific genetic resistance. Furthermore, the pathogen must arrive at the host plant as, or soon change to, its pathogenic state. For example, fungal pathogens must arrive as active mycelium, or as spores that will soon germinate to produce mycelium, bacteria must be out of their dormancy, nematodes must arrive as infective juveniles or adults, and so on.

Even if all of the above conditions are satisfied, however, little or no disease may be initiated and develop unless certain environmental conditions of temperature and, often, moisture are also satisfied. Plants and their pathogens are active and grow and multiply only within certain ranges of temperatures with minimum, optimum, and maximum temperatures for each of them. Most pathogens are most active, multiply faster, and cause the most severe disease when the temperature is at or near the optimum for growth of the pathogen, which is also usually close to the optimum of the susceptible host stage. In many cases, however, pathogens invade their hosts and cause more severe disease at temperatures which are less than optimum for the pathogen as long as the same temperatures are, comparatively, even less favorable for the development of the host plant's defenses. Generally, however, disease and pathogen growth and multiplication occur at varying rates within a fairly broad range of temperatures, and they accelerate or slow down as the prevailing temperature approaches or departs from the optimum.

The presence of certain levels of moisture in the environment of the host and of the pathogen is absolutely essential for the initiation of disease and the subsequent spread of some pathogens, such as fungi, bacteria, and nematodes. Fungi require moisture (free water or high relative humidity) for their spores to germinate and for the germ tube (mycelium) to penetrate and infect the host plant. Subsequently, fungi require moisture for spore production, release, spread, and again germination and infection. Bacteria require moisture for activation out of dormancy, movement, and penetration and infection, as well as for release from the host and subsequent spread to other hosts. Nematodes require moisture for egg hatching, breaking of dormancy in juveniles or adults, and survival and movement of juveniles and adults in the soil or on host surfaces. However, moisture is not required by these pathogens once they have infected their hosts, nor is it required for infection by and spread of the other pathogens (viruses, mollicutes, and protozoa), since these multiply inside the host cells and are transmitted to other hosts by various vectors. Moisture levels, however, do affect the survival and, particularly, the movement of the vectors and, thereby, indirectly affect the survival and movement of these pathogens.

Although the coexistence of the three components of disease—susceptible host, virulent pathogen, favorable environmental conditions—may lead to initiation and development of disease, it still may not lead to the development of an epidemic. An epidemic is defined as any increase of disease in a population. In more practical terms, an epidemic is the spread of a pathogen to and development of disease in many individuals of a population over a relatively large area and within a relatively short time. The more susceptible the host plants, the more closely they are planted, the more succulent or more stressed the plants are, and the larger the areas planted with genetically identical susceptible plants, the greater will be the possibility for and size of an epidemic. Similarly, the more virulent the pathogen, the more abundant the initial inoculum, the greater its rate of, for example, reproduction and spread, the greater will be the possibility for and size of the epidemic. Finally, the less favorable are factors such as temperature, moisture, and nutrition for the host plants, while these and other conditions, such as wind rate and direction, are favorable are for the pathogen and its vectors, the greater will be the possibility for and size of the epidemic. The interrelationship of these factors is shown in the so-called disease triangle (Fig. 4). The size of the epidemic or the amount of disease is proportional to the sum total of these factors, as long as none of the factors are zero.

The amount of the three components and, therefore, the amount of disease produced, are affected by another component: time. The specific point in time

Figure 4 The disease triangle. The amount of disease is proportional to the quantities of favorable host, pathogen, and environmental conditions converging at a given time and space. (From Ref. 1.)

when a particular event in disease development occurs determines the developmental stage of the host, the type and amount of inoculum, and the environmental conditions (climate) likely to prevail. The length of time during which certain environmental conditions prevail (e.g., temperature, rain, dew, wind) also affects the host and the pathogen and determines, for example, whether and how much spore movement and germination, infection, spore release and spread, and vector movement will occur. Time (timing and duration of events) is often weighted equally with the other three components and is represented as one of the four sides of the so-called disease tetrahedron or disease pyramid.

Finally, all of the above factors (host, pathogen, environment, and time) can be greatly influenced by a fifth component: humans. Humans affect the kinds (e.g., annuals, perennials, cereals, vegetables), varieties, and resistance of plants planted, as well as time of planting. By the resistance of the kinds and varieties of plants planted, humans influence the kinds (e.g., fungi, viruses, nematodes), races, and population size of pathogens present. Also, by the cultural practices and the biological and chemical controls they apply, humans influence the amounts of primary and secondary inoculum available to attack the plants. Humans may even modify the environments in which diseases develop by delaying or speeding up planting or harvesting, adding windbreaks, planting in raised beds or in more widely spaced rows, regulating the humidity in storage areas, and so forth.

When the combinations and successive progression of the right sets of environmental conditions—moisture, temperature, and wind or insect vector—coincide with the susceptible stage or stages of large expanses of genetically uniform plants, and with the production, spread, inoculation, penetration, infection, and reproduction of the pathogen, a major epidemic is likely to develop unless humans intervene with sufficient and effective control measures aimed against the pathogen. The size and rate of development of epidemics vary greatly with the particular host-pathogen combination, but the size of the initial inoculum and the frequent repetition of favorable environmental conditions (mostly temperature, rainfall, and wind) in successive locations of disease development are considered as the most important factors determining the severity of most epidemics.

XI. PATHOGEN TRAITS THAT CAN BE TARGETED FOR PLANT DISEASE CONTROL

The diverse biology of each of the plant pathogen types (fungi, bacteria, viruses, nematodes, protozoa, parasitic higher plants) precludes generalizations on their traits that can be targeted for plant disease control. Taken separately, however, each pathogen type has certain traits that provide opportunities for targeting a variety of management or control measures to reduce the threat of catastrophic disease outbreaks by these pathogens.

A. Fungal Traits

Fungi, the pathogens that cause the majority of plant diseases and the most common and most severe epidemics, have the most complex life cycles and therefore provide the most points at which they are vulnerable to plant disease control measures. For example, depending on geographical area, fungi affecting annual crops or leaves, blossoms, and fruit of perennial plants must overwinter or oversummer in the absence of the host. Such fungi overseason mostly as spores or mycelium in plant debris on or in the soil, on or in seeds of their hosts, and as spores or mycelium in the soil. Fungi affecting other parts of perennial hosts overwinter on stem lesions (cankers), in the vascular system, or in root lesions. Removal and/or destruction, or at least significant reduction of the overseasoning (primary) inoculum of fungi (and of all other pathogens), is often one of the most desirable and most effective targets of plant disease control. A number of cultural practices (e.g., plowing under of debris, crop rotation, pruning out deceased stems), physical treatments (e.g., steam sterilization of soil, hot water or hot air treatment of seeds, budwood; physical separation of seeds from fungal sclerotia), biological control treatments (e.g., addition of hyperparasitic or antagonistic microorganisms, addition of soil amendments), and chemical controls (soil fumigation, soil drenching, seed treatments) are aimed at reducing or eliminating the overseasoning inoculum (see Chapters 11 and 12). Some fungi require two alternate hosts to complete their life cycles; elimination of one of them stops the disease from developing in the other.

The surviving primary inoculum must be transported to the susceptible part of the host plant on which it must adhere, germinate, penetrate the host surface, establish infection within or between the host cells, invade and colonize the host tissues, sporulate (reproduce), and spread to other parts of the plant, other plants, other fields, and so forth. Each of these stages provides an opportunity for intervention by humans and for the development of appropriate controls targeting those stages. Thus, spore germination and penetration require free moisture or high relative humidity as well as a favorable temperature range, which we can attempt to modify by, for example, changing planting date, irrigation timing, or plant density. More commonly, we use these requirements of the fungus to develop an effective disease-forecasting program that allows timely application of chemical or biological controls on the plant surfaces. The need for attachment and penetration by the fungus can be targeted for selection of genetic modification of host plants so their surfaces will produce more hair; a thicker waxy layer, cuticle and cell walls that are harder for the pathogen to penetrate; or stomata that are fewer, raised, or closed at the time of penetration. Fungal penetration may also be minimized by reducing wounding of plants caused by cultural practices or by insects.

Initiation and establishment of infection by a fungus (and other pathogens), that is, establishment of the parasitic relationship of the pathogen in its host, is the most critical stage in disease initiation and development. Establishment of infection sets in motion all the weapons of attack possessed by the plant pathogen and all the structural and biochemical defense mechanisms at the disposal of the host plant, whether preexisting or triggered by the pathogen. It is this stage of disease development at which most of the plant breeding for disease resistance is aimed. Genes for resistance, both major and minor, have been found in cultivated or wild host plants against most pathogens of the host, and their incorporation in desirable cultivated varieties is an ongoing process in all plant breeding centers. Establishment of infection, however, as well as invasion and colonization of host tissues, and reproduction (sporulation) of the pathogen, can be influenced considerably by the nutritional state of the host and by prevailing temperature and moisture. They can also be reduced or eliminated by the use of chemical controls (see Chap. 13). Sporulation may also be interfered with by appropriate biological controls (see Chap. 12). Finally, the spread of fungal pathogens by means of spores, sclerotia, and, in some cases, of mycelium, depends on their transport by wind, splashing rain, runoff water, contact by humans, animals, and equipment and on transport, for example, by seeds, transplants, or insects. Each of these means of pathogen spread can be targeted for improving disease control. Planting of trees as windbreaks, changing irrigation methods, reducing handling or eliminating plant contact with contaminated hands and equipment, using disease-free or treated seeds and transplants, and controlling insect vectors of the pathogen can help reduce the amount of disease (see Chap. 11).

B. Bacterial Traits

Traits of plant pathogenic bacteria that can be targeted for plant disease control are, in general, quite similar to those of plant pathogenic fungi, and so are the measures that can be taken to control them. Again, overseasoning of bacteria is the most critical stage for their survival and for perpetuation of the disease from one season to the next and is, therefore, the most common stage at which control measures are aimed. Actually, most plant pathogenic bacteria survive primarily in contact with their host (e.g., seeds, tubers, transplants, buds). Thus, using pathogen-free or treated propagating organs helps immensely toward controlling (avoiding) these diseases. Adherence, penetration, establishment of infection, invasion and colonization of host tissues, and multiplication of bacteria can also be targeted by plant breeding programs for resistance to bacterial diseases. Spread of bacteria to other plants is by the same agents as is that of fungi (except bacteria are seldom windblown) and can be targeted for the same types of control measures.

C. Viral Traits

Viruses almost always overseason in the host plant or its propagative organs such as seeds, tubers, and bulbs or in alternate hosts. Therefore, elimination of virus-infected hosts and alternate hosts (cultivated or wild) reduces or eliminates the virus inoculum that could infect the desirable host plant. Similarly, use of, for example, virus-free seed, budwood, and tubers, is indispensable for control of viral diseases. Viruses are spread from plant to plant mostly by insects, but a few of them are spread by nematodes or fungi, and some are spread through infected pollen or plant sap. Control or management of these vectors is necessary, therefore, for control of such viruses. Viruses are not, at least in practice, susceptible to control by any kind of chemicals. However, viruses survive and replicate (multiply) only within living plant cells. As a result, the diseases they cause can often be controlled by incorporating genes for virus resistance in the cultivated plants. Such genes are usually derived from resistant plants. However, because of the peculiarities of replication of viruses in plant cells, it has been shown recently that plant resistance to viruses can be obtained by inoculating plants with mild strains of the virus [6] (see Chap. 7). Moreover, plant resistance to viruses can be obtained by incorporating into the plant genome, through genetic engineering techniques, genetic material derived from the virus itself [2,12]. Such genetic material, while on the virus codes for various proteins, such as the viral coat protein, or for portions of proteins. How these viral genes, when incorporated in the plant genome, make the plant resistant to the virus is not yet known.

D. Nematode Traits

Nematodes overseason as eggs, juveniles, or adults in the soil or in association with their host plants. Survival and movement of nematodes depend on favorable soil temperature and moisture. Therefore, excessive soil heating by solarization, drying by summer cultivation, or excessive moisture by flooding of the soil reduces or eliminates nematodes from the upper soil layers where plant roots develop. Nematodes must also compete with and survive in the presence of numerous other soil microorganisms, some of which directly attack nematodes (e.g., nematophagous fungi) or are antagonistic to them. Therefore, promoting the increase of such antagonistic microorganisms helps reduce nematode populations and the amount of disease they cause. Some plants (e.g., marigolds) produce substances toxic to nematodes, whereas others (trap plants) support nematode growth up to a certain stage but do not permit them to develop into adults so they can reproduce. Obviously, planting such plants as rotation or interrow crops can reduce the number of nematodes that may be available to attack the desirable crop. Finally, nematodes are sensitive to several chemicals which, when applied as gases, liquids, or granules, disperse and kill any nematodes with which they come in contact.

XII. GENERAL APPROACHES FOR PLANT DISEASE CONTROL

In Section XI, a number of methods were listed that can provide considerable levels of control of plant disease. Each method is suitable for control of diseases caused by some, but not other, pathogens and for some, but not all, the diseases caused by the same pathogen. Here only a brief listing of the general approaches for plant disease control will be given [1]. Further detail is provided in the appropriate chapters in this volume.

A. Exclusion of the Pathogen from the Host Plant

Wherever possible, by far the most effective approaches for plant disease control are those that exclude the pathogen from the host plant. They include regulatory controls such as pathogen quarantines and inspections at fields, warehouses, or national borders; the use of pathogen-free seed and other propagative organs; seed, tuber, transplant, and other propagative organ certification of freedom from certain pathogens; and avoidance of diseases by growing crops where certain pathogens do not yet exist.

B. Eradication or Reduction of the Pathogen Inoculum

a. *Cultural Controls*: such as host eradication; crop rotation; sanitation of fields, warehouses, equipment, and personnel; creating moisture or temperature conditions unfavorable to the pathogen; the use of polyethylene traps and mulches to reduce insect vectors in the crop; and soil solarization with clear plastic (see Chapter 11).

b. *Biological Controls*: such as planting in pathogen-suppressive soils; adding antagonistic microorganisms to the soil, roots, or plant surfaces; planting trap plants to attract the pathogen or its vectors away from crop; or planting antagonistic plants that release substances toxic to certain pathogens (see Chapter 12).

c. *Physical Controls*: such as soil sterilization with heat; hot water or hot air treatment of propagative organs; drying stored seeds and fruit; refrigeration of fleshy fruits and vegetables; radiation of fruits and vegetables with ultraviolet or x-rays (see Chapter 11).

d. *Chemical Controls*: such as soil treatment with chemical (fumigation, chemigation, etc.); disinfestation of warehouse; control of insect vectors with insecticides (see Chapter 13).

C. Immunization or Improvement of Host Resistance

Approaches that immunize or improve the resistance of the host include cross protection of plants from certain viruses by previous inoculation of the plants with mild strains of the same viruses; induced resistance in host plants to some diseases by previous inoculation of still resistant stages of the plants with the same or other pathogens, or treatment with certain chemicals; and improving the growing conditions of host plants so they can better defend themselves when attacked by pathogens (see Chapter 7).

This group of control approaches also includes the most important current means of controlling plant diseases and is likely to become even more important in the future. This type of control involves the development and use of plant varieties resistant to each of the serious diseases of a particular crop. Resistant plant varieties have so far been produced by appropriate selection and conventional breeding techniques, but in the future, their production is expected to be expedited by a variety of very promising genetic engineering techniques (see Chapter 8).

D. Direct Protection of Plants from Pathogens

a. *Biological Controls*: such as fungal and bacterial antagonists of fungi in soil, roots, foliage, and on harvested fruit; hypovirulent strains; bacterial antagonists of other bacteria and of fungi; bacteriophages; nematophagous fungi (see Chapter 12). So far, very few biological controls of plant pathogens have proven to be effective in practice, and this is not expected to change drastically for several more years.
b. *Chemical Controls*: by the use of fungicides, bactericides, (and nematicides) as foliar sprays and dusts; as seed treatments; as soil treatments; for treatment of tree wounds; and for control of postharvest diseases of plants and plant produce (see Chapter 13).

XIII. SOCIAL, ECONOMIC, AND ENVIRONMENTAL IMPLICATIONS OF PLANT DISEASE CONTROL

Controlling plant diseases prevents them from destroying or reducing the quantity and quality of foodstuffs and beverages, animal feeds, fibers, forest products, and ornamental plants so that these products can be available to humans. For agrarian societies, control of plant diseases reduces crop losses from diseases, thereby preventing the hunger, malnutrition, and sometimes death that result from such losses. For any society, control of plant diseases prevents the loss of produce and the higher prices, reduced availability, and narrowed choice of products resulting from plant diseases. Plant disease control reduces the production of plant produce that may be poisonous and unfit for consumption. Plant disease control, on the other hand, even when done by knowledgeable persons, leads to increased work and/or increased costs, the need for additional land, and the necessity to plant only resistant crops and varieties. At the same time, the need for plant disease control is forcing studies on their causes and control, the development of diagnostic and monitoring techniques and equipment, and the development of materials, equipment, and techniques to control plant diseases. In general, in developing countries, plant diseases and their control affect, for example, the amount of food and fiber available for families of agrarian communities and the whole countries, and this may cause local or widespread hunger and famine. In developed, industrialized countries, plant diseases and their control cause primarily economic losses to some (those affected), but not all, growers, usually result in higher prices, and, for the poorer people, may limit the availability of produce or may lower the quality of available produce.

The economic impacts of plant diseases and their control can be direct, for example, the loss of income, cost of control measures (materials, labor, equipment); or indirect, for example, the use of more expensive seed or transplants, abandoning cultivation of profitable but susceptible crops or varieties, changing and often increasing cultural practices to reduce disease, and increasing sanitation measures in the field and in the warehouse.

The environmental impacts of plant diseases and their control are also multifaceted. Plant diseases damage or kill cultivated and wild plants and as such bring about changes in the natural ecosystem of plant communities. Some plant species may be annihilated by a plant disease. To control plant diseases, growers sometimes, for example, burn plant debris or plow it under, or leave the land fallow—in all cases increasing soil erosion by wind and rain. The need for control of plant diseases has resulted in the production of resistant crops and varieties which are composed of plants that would probably never develop by themselves but which now cover huge expanses of our planet. The need for biological control of plant diseases is yet to be met, but introduction or selective promotion of biocontrol agents (microorganisms) in soils or on the surface of plants and plant produce is likely to have far-reaching consequences in our environment and the natural microfloras of these habitats.

Finally, by far the greatest impacts on our environment have been caused, and will continue to be caused, by the diverse toxic chemical substances used to control plant diseases. Aside from any accidental release or misuse of such chemicals from the time of their production in the manufacturing plant until they are applied by the grower, many of these substances have varying levels of toxicity to plants, animals, and microorganisms. Some chemicals persist for a long time, some react with other chemicals or with soil or plant components, and some find their way, although probably only in small amounts, into the food chain and into surface and underground water supplies through which they can reach animals and humans. The complete effects of these occurrences are not yet known and may be minimal, but continued annual broadcasting of billions of pounds of toxic chemicals to our environment for the purpose of controlling plant diseases is likely to affect our environment adversely and also the animals and humans that live in it. It is also likely that societies and their governments will not allow such pollution to continue and, therefore, future control of plant diseases may have to depend primarily or exclusively on nonchemical control measures.

XIV. CONCLUSIONS

Plant diseases are caused primarily by certain plant pathogenic fungi, viruses, bacteria, and nematodes. These microorganisms attack plants in the field or in storage and destroy more than 12% of the annual world production of food and other plant products. Plant diseases frequently develop in well-prescribed patterns that are the result of prevailing environmental conditions, especially temperature and moisture. Plant diseases can usually be avoided or controlled by practices that keep the pathogen from entering the area where host plants grow; by cultural practices that make it difficult for the pathogen to survive, move, infect, or multiply; by breeding plants that contain genes for resistance to the pathogen; by using antagonistic microorganisms to control the pathogen biologically, and by applying toxic chemicals on plants, seeds, or soil to kill the pathogen or inhibit it from growing and causing infection. Chemical controls, although still widely used, are likely to become more limited in the future because of their potential damage to the environment.

REFERENCES

1. Agrios, G.N. 1997. *Plant Pathology*. 4th ed. San Diego: Academic Press.
2. Beachy, R.N., S. Loesch-Fries, & N.E. Tumer. 1990. Coat protein-mediated resistance against virus infection. *Annu. Rev. Phytopathol.* 28: 341–363.
3. Cramer, H.H. 1967. Plant Protection and Crop Production. (Translated from German by J.H. Edwards.) *Pflanzenschutz-Nachr.* Vol. 20. Leverkusen: Farbenfabriken Bayer.
4. Dropkin, V.H. 1989. *Nematology*. 2nd ed. New York: Wiley.

5. Duncan, J.M., & L. Torrance, eds. 1992. *Techniques for the Rapid Detection of Plant Pathogens*. Oxford, UK:Blackwell.
6. Fulton, R.W. 1986. Practices and precautions in the use of cross protection for plant virus disease control. *Annu. Rev. Phytopathol.* 24:67–81.
7. Goto, M. 1992. *Fundamentals of Bacterial Plant Pathology*. San Diego: Academic Press.
8. Hampton, R., E. Ball, & S. DeBoers. 1990. *Serological Methods for the Detection and Identification of Viral and Bacterial Plant Pathogens*. St. Paul, Minnesota: APS Press.
9. Horsfall, J.G., & E.B. Cowling, eds. 1977–1980. *Plant Disease*, Vols. 1–5. New York: Academic Press.
10. Matthews, R.E.F. 1991. *Plant Virology*. 3rd ed. San Diego: Academic Press.
11. Oerke, E.C., H.-W. Dehne, F. Schönbeck, & A. Weber. 1994. *Crop Production and Crop Protection: Estimated Losses in Major Food and Cash Crops*. Amsterdam: Elsevier.
12. Prins, M., P. Haan, R. Luyten, M. Van Veller, M.Q.J.M. Van Grinsven, & R. Goldbach. 1995. Broad resistance to tospoviruses in transgenic tobacco plants expressing three tospoviral nucleoprotein gene sequences. *Mol. Plant Microbe Interact.* 8:85–91.
13. Singh, U.S., A.N. Mukhopadhyay, J. Kumar, & H. S. Chaube. 1992. *Plant Diseases of International Importance*. Englewood Cliffs, New Jersey: Prentice Hall.
14. Staskawitz, B.J., F.M. Ausubel, B.J. Baker, J.G. Ellis, & J.D.G. Jones. 1995. Molecular genetics of plant disease resistance. *Science* 268:661–667.
15. Vanderplank, J.E. 1984. *Disease Resistance in Plants*. 2nd ed. Orlando, Florida: Academic Press.

11
Cultural Approaches to Managing Plant Pathogens

Linnea G. Skoglund, Howard F. Schwartz, and William M. Brown, Jr.
Colorado State University, Fort Collins, Colorado

I. INTRODUCTION

Ever since neolithic farmers first tilled the soils and planted crops, people have used cultural practices to manage plant health. Although diseases and their causes as we know them were not recognized until the mid 1800s, these practices impacted plant pathogens. The ancient Egyptians recognized the benefits of rotation and fallow. The Incas of Peru instituted farming systems that incorporated complex rotations of potatoes, grains, and vegetables with burning, manuring, fallowing, and other practices. Eradication of barberry to control stem rust of wheat is said to have been first instigated by farmers in 1660 in Rouen, France. Laws were enacted in Connecticut in 1729, Massachusetts in 1755, and Rhode Island in 1766 for barberry eradication [36,50].

More recently, farmers have used numerous cultural practices in various combinations in a science-based effort to reduce the impact of plant pathogens on crops. For example, the amount of primary inoculum, rate of inoculum build-up, and length of time during which a susceptible host is exposed to pathogens are affected by cultural practices such as sanitation, crop sequence, tillage, irrigation, planting date, and proximity to infection source [32]. Cultural practices such as crop rotation, alteration of soil pH, sanitation, and adjustment of planting time and harvest to avoid peak outbreaks of pathogen activity also are used to complement genetic resistance in crop plants [22]. Many cultural practices also contribute to enhancing the soil biodiversity and stimulating competition for plant pathogen survival. Hall and Nasser [19] discuss a range of cul-

tural practices used to manage diseases of white beans in Ontario. These practices are presented in Table 1.

II. CROPPING SYSTEMS

Cropping systems influence pathogen survival, disease development [46], and
management success. Examples of different cropping systems include:

Monocropping: repetitive growing of one crop on a field each year.

Multiple cropping: the intensification of cropping in time and space, such
as growing two or more crops simultaneously on the same field in a
year.

Sequential cropping: intensification in time such as growing two or more
crops sequentially on the same field per year where the succeeding crop
is planted after the preceding crop is harvested (double, triple, qua-
druple, ratoon cropping).

Intercropping: intensification in time and space such as growing two or
more crops intermixed in the same field (mixed, row, strip, relay) [16].

Monoculture systems tend to experience increased incidence of root diseases,
possibly due to occupation of old root channels by successive susceptible crop

Table 1   Cultural Practices Used to Manage Diseases of White Beans [19]

Cultural practice	Bean diseases managed
Rotation	Fungal and bacterial diseases
Pathogen-free seed	Anthracnose, bacterial diseases
Weed control	White mold, bacterial diseases
Tillage (hilling, chisel plowing, plowing-in)	Root rots, anthracnose, white mold, bacterial diseases
Enhanced drainage	Root rots
Wider plant spacing	White mold
Crop separation	Bacterial diseases
Soil fertility	Root rots
High soil organic matter	Root rots
Control of volunteer bean plants	Bacterial diseases
Restricted movement in fields	Bacterial diseases
Soil temperature at planting	Seed decay, root rots
Timely harvest	White mold
Upright plant architecture	White mold
Planting rows parallel to the wind	White mold
Roguing	Bacterial diseases
Appropriate seed depth	Seed decay, root rots

root systems. Continuous monoculture appears to reduce the diversity of root-associated fungi and other beneficial soil organisms, with concomitant reductions in net root growth. Therefore, the greater yields associated with crop rotations, often referred to as the "rotation effect," can be partly or largely explained by the lower incidence of root diseases as roots of different plant species explore different soil strata [2,43,46].

In the low-input agriculture of developing countries, the use of nonhost crops in interplantings can significantly reduce the rate of viral spread in the field [48]. Some crop associations result in increased relative humidity and shade, which may favor the incidence of diseases owing to microclimate modifications imposed by the predominant crop. In such cases, it is necessary to modify the spatial arrangement of the associations to alter the microclimate and thus minimize the benefits of shade and humidity to pathogens. In general, the shielding effect of the companion crops against airborne pathogens should more than offset the microclimatic advantage pathogens may derive from the dense and/or layered foilage of mixed crops. In summary, mixtures of different crop species can buffer against disease losses by reducing the rate of disease development, reducing spore dissemination, or modifying microenvironmental conditions such as humidity, light, temperature, and air movement [3].

III. SANITATION AND DEBRIS MANAGEMENT

Sanitation systems that remove or prevent the introduction of pathogens are commonly used in agriculture. Sanitation consists of all activities that eliminate or reduce the amount of inoculum present in a plant, field, or storage facility and thereby reduce spread of pathogens to healthy plants and plant products. Thus, plowing under or removing and properly disposing of infected leaves, pruning infected or dead branches, and removing or destroying any plant or plant debris that may harbor the pathogen all serve to delay the initial outbreak and reduce the amount of disease that will develop later [1].

Plowing down infested crop residues decreases potential inoculum. Surface residues may affect plant diseases in several ways by providing habitats for overwintering, survival, growth, and multiplication of plant pathogens [7]. Some pathogens within diseased tissues continue to grow and multiply on crop residue after the crop is harvested. In other cases, pathogens from soil, weeds, and other hosts colonize residues from disease-free crops.

Surface residues often alter soil moisture, temperature, aeration, density, organic content, nutrition, and types and population levels of microorganisms. Such changes may, in turn, affect the growth pattern and configuration of roots. Leaving plant debris on the surface or partially buried in the soil may allow pathogens to overwinter or to survive until the next crop is planted, but conditions favorable for biological control of plant pathogens may also be increased [46].

Survival of fungal pathogens in residues also can be affected by nutrients released from decomposing residues [7]. These nutrients may stimulate germination of fungal propagules and subsequent mycelial growth and sporulation, thereby enhancing pathogen survival. On the other hand, nutrients also can stimulate competitive organisms that suppress disease. Soilborne fungal pathogens associated with surface tillage also may be decreased by the type and the amount and timing of fertilizer application [7].

Roguing (removal of infected plants) performed during the initial stages of an epidemic acts as a disease-removal mechanism to reduce the spread of pathogens to healthy plants [6]. This practice decreases the total reservoir of secondary inoculum and decreases the rate of disease increase. Roguing of infected plants is used frequently to reduce losses from insect-transmitted diseases [1,17,50].

Composting methods can affect suppression of some disease agents. Control of pathogens through composting may directly kill plant pathogens present in compost during and after the high-temperature composting process. Some pathogens that produce sclerotia and other protected survival structures are not as easily controlled by composting. Specific physical, chemical, and biological properties of composts may have a major effect on compost suppressiveness. For example, the activity of antagonists involved in biological control is enhanced by nutrients released in compost [21]. Mechanisms by which these antagonistic microorganisms affect pathogen populations are generally considered to be (a) direct parasitism and death of the pathogen, (b) competition with the pathogen for food, (c) direct toxic effects on the pathogen by antibiotic substances released by the antagonist, and (d) indirect toxic effects on the pathogen by volatile substances, such as ethylene, released by the metabolic activities of the antagonist [1,21,35].

IV. FIRE AND HEAT

Burning is possibly the oldest form of disease and pest management known. Heat used to burn or flame crop residue has been applied to cereals, especially in minimum tillage systems [32]. Until recently, fire was routinely used in California rice fields to control *Sclerotium oryzae* and facilitate the elimination of straw so the fields could be worked (W. Brown, personal observations). In 1983, an outbreak of cephalosporium stripe (*Cephalosporium gramineum*) of wheat in the U.S. Pacific Northwest was so severe that extensive areas had to be burned. Doupnik et al. [15] have shown that burning is the most effective means of controlling cephalosporium stripe in Nebraska. Grass seed fields in the Willamette Valley of Oregon are still routinely burned to control blind seed disease caused by *Gloeotinia temulenta* and other pathogens [20]. However, this practice is now criticized because of its contribution to air pollution.

Flaming with propane burners has been used by Oregon mint growers to control mint rust caused by *Puccinia menthae* and verticillium wilt caused by *Verticillium dahliae*, both of which overwinter on mint stubble [13,24,25]. This technique is very effective and causes minimal air pollution from smoke. For a more complete review of the use of fire for disease control, see a review of the subject by Hardison [20].

Heat, in the form of steam, is used commonly to control soilborne pathogens in greenhouse production. Aerated steam is piped either into or below growing beds by free flow or under pressure. Temperatures of 72–82°C are sufficient to kill plant pathogens as well as most insects and weed seeds in soil. Steam can be applied in mobile steam boxes, in boxes with electric heating elements, for dry baking or burning, or with continuous belt mechanical treaters. To date, the use of steam has proven to be the most effective method for management of soilborne pathogens in greenhouse systems [5].

Solarization is a more recent and increasingly popular form of heat treatment for soil. The process utilizes clear plastic and allows solar radiation to enter the moist soil; the generated heat is then trapped by the plastic [12]. It has been used successfully to decrease cabbage yellows (*Fusarium oxysporum* f. sp. *conglutinans*) inoculum [49] and to suppress other pathovars of *Fusarium oxysporum* [40]. It is used in California to treat established plantings of pistachio trees for verticillium wilt [4], and it is widely used in Israel and other areas with high solar energy and high-value crops [29]. The effect on the pathogen is not directly attributable to heat, because the temperature levels attained are not lethal; rather the temperature stimulates an increase in soil-inhabiting, beneficial bacteria, which in turn suppress the pathogens. Katan and DeVay [29] have thoroughly reviewed the solarization process and the mechanisms by which it functions.

V. VOLUNTEER AND WEED CONTROL

Control of crop volunteers (plants resulting from unharvested crop seed or vegetative regrowth) and weeds is important in several disease situations where the disease causal agent and/or its vector can survive and reproduce on volunteers or weeds. Wheat streak mosaic virus and its wheat curl mite vector (*Aceria tosichella*) both survive on volunteer wheat, thus providing a "green bridge" for disease development [52]. Many fungi and bacteria also survive in weed hosts or, like the bacterial blight pathogens (*Pseudomonas syringae* pv. *syringae*, *Pseudomonas syringae* pv. *phaseolicola*, and *Xanthomonas campestris* pv.

phaseoli) and rust pathogen (*Uromyces appendiculatus*) of beans, build-up on crop volunteers early in the season and supply a ready source of inoculum for the newly developing bean crop [30,37].

VI. REMOVAL OF ALTERNATE HOSTS AND REFUGIA

Removal or management of alternate hosts has long been known to be an effective management tool for control of wheat stem rust (*Puccinia graminis* f.sp. *tritici*), which uses the barberry as an alternate host and white pine blister rust (*Cronartium ribicola*), which uses a *Ribes* sp. as an alternate host [1,6,50].

If there is a refugia (an alternative host for overwintering) for the vector of a pathogen, elimination of the vector while on this alternative host becomes the focus of the management technique. Such is the case with control of the green peach aphid, a vector of potato leaf roll virus. Aphid control targets population on their overwintering *Prunus* host before they can move into virus-infected herbaceous hosts in the spring where the aphids acquire the virus and transmit it to potato plants [10].

VII. ROTATION

Rotation is the basis of effective root disease control. Many soilborne pathogens are limited to certain classes of crops, such as legumes or grasses. Therefore, rotation of nonhosts with hosts serves to reduce inoculum for subsequent susceptible crops. Diseases such as fusarium wilt (*Fusarium oxysporum*), cephalosporium stripe (*Cephalosporium gramineum*), nematodes, and even some soilborne vector-transmitted viruses (e.g., rhizomania of sugar beets, soilborne mosaic of wheat, and potato stem mottle) are effectively controlled by rotations. It is essential to maintain economically viable rotations that alternate between major crop families such as wheat and small grains, corn, tomato/potato/egg plant, legumes, cole crops, and cucurbits [1,17].

Crop rotations inhibit the development of large pathogen populations, reduce pathogen populations, and reduce selection pressure for specific races of a pathogen. This is especially true for soil-*invading* fungi—those that survive only on living plants or on host residue. Such fungi do not form long-lived survival structures and so do not exist freely in the soil. In the absence of the host, acceptable substrate is depleted by the saprophytic phase and inoculum levels begin to diminish. Crop rotation is less effective for soil-*inhabiting* fungi—those that form persistent survival structures or live as saprophytes for long periods.

VIII. PATHOGEN-FREE SEED AND PROPAGATING MATERIAL

Exclusion of plant pathogens by pathogen-free seed or propagating material has long been the basis of quarantine and phytosanitary programs [28]. The dissemination of potato pathogens in seed potatoes is a well-known example of movement of plant pathogens over long distances. The use of pathogen-free certified seed then becomes a form of exclusion and possibly one of the most critical bases of all integrated plant disease management approaches [8]. Bacterial pathogens disseminated on or in contaminated seed are common in small grains, vegetables, and legume crops [39]. Many grape viruses have been successfully controlled through the use of certified planting stock programs [33].

Another form of exclusion is reliance on isolation to reduce initial inoculum by locating transplant beds at a distance from field plantings, the use of barrier crops, interplanting of unrelated crops, and avoiding movement of personnel or equipment through fields when foliage is wet [6].

IX. TILLAGE

Tillage practices directly influence physical and chemical properties of the soil, soil moisture and temperature, root growth and nutrient uptake, soil biodiversity, and populations of pathogens and vectors of plant pathogens. In turn, these factors may influence the viability and variability of plant pathogens and the susceptibility or resistance of the host [14,41]. Different tillage practices require variations in (a) the kind, rate, and time of fertilizer application; (b) pesticide use; (c) plant spacing; (d) irrigation; and (e) other cultural practices that may indirectly influence plant diseases. Any change in the soil or plant canopy environment may influence the development of epidemics.

Common tillage practices include:

Conventional tillage: a traditional tillage system, which typically begins with a primary deep tillage operation followed by some secondary tillage for seedbed preparation. This has been the common way to dispose of infested plant residue; burying debris provides an opportunity for antagonistic organisms to break down infested tissue and effectively depress pathogen populations [41].

Minimum tillage: that minimum amount of tillage required to create the proper soil condition for seed germination and plant establishment. This frequently provides the opportunity for plant pathogens to survive and infect the following crops. In cases where this system is used, effective rotation with nonhosts is critical [46].

Deep tillage: chiseling or ripping while leaving debris on the soil surface. This provides a better environment for root growth and development as well as a more complete filling of the soil profile with water. In many instances, this is sufficient to minimize the potential impact of disease, because a more vigorous and less stressed plant develops [14].

Soil compaction is a problem commonly associated with tillage. It can be produced with wheel traffic, tillage and planting implements, and any equipment used for ground spraying, cultivating, harvesting, or transport. Compaction may also occur in no-till (direct drilled) systems, in which there is no tillage after harvest and the only soil disturbance is during planting [2]. The management of soil compaction requires an awareness of when and where compaction is produced, when it becomes excessive and harmful, how long it lasts, and how it affects root health and soilborne pathogens [14].

Excessive soil compaction can decrease air-filled porosity, degrade soil aggregation, and increase soil resistance (bulk density) sufficiently to impede root growth and impede drainage. Increased root disease can result from each of these factors associated with excessive compaction. The form and rate of root growth, location of pathogen inoculum, and ecology of both the root and pathogen are all influenced by soil structure as characterized by pore shape and continuity, aggregate size and density, and planes of weakness and fracture.

Tillage can have an effect on disease potential. Generally, any system that maintains high residues on the soil surface will enhance development of diseases such as take-all (*Gaeumannomyces graminis* var. *tritici*), tan spot (*Pyrenophora tritici-repentis*), and cephalosporium stripe (*Cephalosporium gramineum*) of wheat [46,52]. In cases where high residues are required for moisture and/ or erosion control, alternative chemical control, biological control, and/or disease resistance become critical [7].

Residues fully incorporated (plowing) or partially incorporated (surface tillage or subtillage) in soil decompose faster than residues left undisturbed (notillage) on the soil surface. However, with plowing and surface tillage, there is greater opportunity for decomposing residues to come in contact with roots and predispose them to root pathogens [7,51]. At the same time, decomposing residues can enhance soil competitors or biological agents that would compete with or slow down pathogens and their diseases.

Diseased roots, colonized dead roots, and debris from diseased parts of the shoot and fruit are the most common sources of inoculum for the next crop. Inoculum and crop residue after moldboard plowing may be concentrated at lower depths in the plow layer, but other forms of primary tillage and associated secondary tillage may mix inoculum and residue in the top 10 cm and facilitate pathogen exposure and disease development in subsequent crops.

The ecofallow system differs from most surface-tillage practices in that one crop is planted directly into the residue of a different crop rather than into the residue of the same crop. Stalk rot (*Fusarium moniliforme*) severity in sorghum is less after a wheat crop because more moisture is retained in the ecofallow system, thereby reducing water stress. The stalk rot pathogen does the most damage when host plants are under stress. Additionally, the reduction of diseases associated with reduced tillage is achieved through tillage system rotations [7].

Smucker [43] points out that roots from the same or different plants often compete for similar spaces within the soil, thereby resulting in the clustering or grouping of roots in the root zone. Factors that cause roots to congregate within certain regions of the root zone include (a) preferential horizontal growth resulting from root geotropism within each soil horizon, (b) frequencies of branching along primary and secondary parent roots, (c) root accumulations at planes of weakness within the bulk soil matrix, (d) duration of plant growth, and (e) above- and belowground environmental conditions.

Root clustering spatially limits root access to nutrients and water stored in adjacent soil volumes and increases exposure to soilborne plant pathogens through greater concentration of root exudates that support growth of host-specific root pathogens. Therefore, root clustering increases the potential for greater abiotic and biotic stresses. Root clustering in soils could be reduced by deep soil tillage, crop rotations, and the planting of bridge or other crops with different root-system geometry [43].

The physical environment of untilled soil differs from that of plowed soil and can influence pathogen activity and plant stress. Untilled soils are generally wetter, cooler, and more compact than plowed soils. Tillage operations incorporate residues into the soil, and this not only increases soil porosity but also creates greater potential for water loss through surface evaporation. Another characteristic of untilled but not compacted soil is that it generally has less resistance to rainwater infiltration than tilled soil.

Other studies have shown untilled soils have a substantially lower temperature near the ground surface than tilled soils [7]. Lower soil temperatures in the spring and early summer with reduced tillage may retard seed germination and seedling development of many crops, thus providing more opportunity for invasion by damping-off fungi. Therefore, the importance of removal of residues from the area of emerging plants, as accomplished with surface tillage, is obvious.

X. PLANT SPACING AND OTHER FACTORS

Plant spacing can influence the rate of disease increase. Airborne pathogens can spread faster among plants that are closely spaced. Viruses spread more readily to adjacent plants than those at a distance. Dense plantings also maintain higher humidity and leaf wetness that can enhance development of foliar diseases and some soilborne pathogens, such as white mold [17]. Finally, dense plantings enhance root contact among plants which can also help spread disease.

Conversely, severity of some diseases is reduced by increasing plant densities. This is true when inoculum levels are low and the pathogen is systemic and monocyclic. Denser plantings of several crops will experience lower losses from fusarium wilt than sparser plantings [17].

Row orientation (compass direction) is frequently overlooked in management systems, especially where temperatures are limiting for a particular crop. Orientation can be used to reduce high soil temperature by shading. Orientation also can be used to increase soil temperature for early season or marginal crops. Row orientation parallel to prevailing winds was observed to decrease white mold of beans, probably owing to increased air circulation and light penetration which reduced periods of tissue wetness [18].

Finally, reducing the time of exposure to pathogens by planting larger and more vigorous transplants, using short-season cultivars, and maintaining adequate soil fertility and moisture to avoid any slowdown of crop growth can be very important factors in disease reduction [6].

XI. PLANTING DATE MODIFICATION

Planting date modification is used frequently to escape periods of pathogen-spread by vectors or other agents. One of the easiest ways to decrease losses during an epidemic is to sow or plant a crop at a time of year that is less favorable for disease development [6] or is more favorable to the crop. This technique is very effectively used to manage wheat streak mosaic and barley yellow dwarf viruses in the plains states of the United States [52]. Barley stripe rust (*Puccinia striiformis*) management in the San Luis Valley of Colorado depends on early planting to prevent infection of young plants by urediospore showers originating from Texas and Mexico [9].

XII. MULCHING

The use of mulches to maintain moisture and minimize plant stress can contribute to enhanced plant vigor and escape from infection [7,21]. Any practice that minimizes host plant stress contributes to reduction in infection and disease expression. Mulches can also act as barriers between developing fruit and vegetables by reducing and preventing contact with pathogens in the soil [53]. Mulching is becoming more widely used as a means to repel virus-bearing aphids. Reflective foil mulches provide background color that deters the aphid's ability to recognize and land on host plants [1].

XIII. WATER MANAGEMENT

Irrigation timing, frequency, and type (sprinkler vs surface, subsurface, or drip), as well as the efficiency of excess water drainage contribute to disease development or suppression. Irrigation practices influence the rhizosphere and phyllosphere climates, duration of the growing season, relief of or contribution to crop stress, survival and management of inoculum, and dispersal of inoculum in the crop [32]. Sprinkler irrigation, in some instances, can enhance spread and development of foliar pathogens, whereas flood systems may enhance dissemination and development of soilborne pathogens.

The time and scope of disease development in irrigated crops is affected by (a) macroclimatic weather factors (rain, atmospheric humidity, temperature, solar radiation) through their effects on the climate of the plant canopy, formation of dew, and viability of airborne spores deposited on susceptible plant tissue; (b) pathogen and host characteristics, foremost among which are the time and method of spore dispersal, drought resistance of spores, and speed of the infection process; and (c) the technique, frequency, and time of day chosen for water application. All of these operate within the background of crop rotation, crop density, and other cultural factors.

Dispersal of root- and stem-infecting pathogens by irrigation water has been demonstrated in relation to several diseases either by propagules carried in water or by active spread of mycelia in the irrigated soil. For example, Rotem and Palti [34] showed that high levels of moisture favor pathogens dependent upon water for dissemination of zoospores. Nematodes also are easily moved in surface water [44].

The higher the air temperature and the stronger the solar radiation, the greater the effect of sprinkle irrigation on leaf temperature reduction. This is of considerable significance in arid to semiarid climates, where leaf temperature may often be marginal for infection, and the temperature decrease due to irrigation may determine the success or failure of the infection process on a moderately hot day.

In areas of low rainfall, persistence of various pathogens in plant debris is inversely related to the amount of irrigation water applied; persistence may be longest in nonirrigated soil. Irrigation in these cases may play an important role in diminishing the amount of inoculum surviving from one season to the next. In addition to irrigation technique (furrow, flooding, subsoil, trickle, overhead sprinkling), pronounced effects are exerted on pathogens and the potential host by the amount and rate of water given at each application, the interval between successive applications, and the time of day when water is applied.

XIV. FERTILITY MANAGEMENT

Fertility management is critical to healthy plant development. Nitrogen over-abundance results in the production of young, succulent growth and may pro-long the vegetative period and delay maturity of the plant. These effects may prolong the plant's susceptibility to pathogens that normally attack such tissues, such as the fungus responsible for white mold of dry beans. Conversely, plants suffering from a lack of nitrogen may be weaker, slower growing, faster ag-ing, and more susceptible to less aggressive pathogens, such as the fungus re-sponsible for fusarium root rot of beans [1,17].

It is possible, however, that the form of nitrogen (ammonium vs nitrate) rather than the amount of nitrogen that is available to the host or pathogen may affect disease severity or resistance [17]. Take-all (*Gaeumannomyces graminis*) of turf and wheat is frequently managed by changing from nitrate to ammonified forms of nitrogen [42]. Huber and Watson [26] present a list of diseases affected by inorganic forms of nitrogen.

Although nitrogen, because of it's profound effects on plant growth, has been the most extensively studied nutrient in relation to disease development, stud-ies with other elements such as phosphorus, potassium, calcium, and micronu-trients have revealed similar relationships between nutrient level and suscepti-bility or resistance to certain pathogens and their diseases. Chloride has been shown to have a suppressive effect on take-all of wheat [47]. In general, plants receiving balanced nutrition in which all required elements are supplied in ap-propriate amounts are more capable of resisting, tolerating, or escaping new infections and of limiting existing infections than when one or more nutrients are supplied in excessive or deficient amounts.

Survival of fungal pathogens in residues may be affected by nutrients released from decomposing residues [7]. These nutrients may stimulate germination of fungal propagules and subsequent mycelial growth and sporulation and, thereby, extend pathogen survival. These nutrients also can stimulate competitive organ-isms that will suppress disease.

Palti [32] lists several nutrient effects on the host, including (a) vigor of growth, which in turn may affect the crop climate, and the ability of crops to recover from or become more prone to pathogen attack; (b) anatomical and histological characteristics, such as thickness of cuticle and epidermis, or lig-nification of tissues, and biochemical and physiological reactions, including the formation of corky layers and phytoalexins; (c) rate of growth, making it pos-sible to shorten stages at which plants are most susceptible to pathogens (seedlings) or to regulate the period in which such stages occur, and influence the time of maturation and senescence; and (d) water economy–proper nutrition may enable plants, under certain conditions, to make fuller use of the soil's water potential. Soil fertility also may have direct effects on pathogens [32]: (a) the pathogen's rate of penetration, colonization and reproduction; (b) toxic effects of some fertilizers, such as urea, on pathogens; and (c) relative rate of development of pathogens and their competitors and antagonists, especially in the soil.

XV. SOIL pH MODIFICATION

Modification of soil pH has been used to control take-all in turf [42] and wheat [52] by utilizing ammonified nitrogen fertilizers to lower the soil pH. Club root (*Plasmodiophora brassicae*) in crucifers [11] and scab (*Streptomyces* sp.) of potatoes [23] are controlled by the addition of lime. Although such procedures may be cost effective on a farm using ammonified fertilizers, in general, heavy liming in agriculture is cost prohibitive.

XVI. HARVEST PRACTICES

Harvest practices can have an impact on the development of diseases in the harvested crop. Timely harvest shortens the period during which the plant is exposed to inoculum and reduces the probability of exposure to environmental conditions favorable for disease development [19].

Appropriate timing of harvest means essentially an attempt to escape disease; the intention is to harvest either before environmental conditions become favorable to the pathogen or before the crop becomes highly susceptible to age-related pathogens. Harvest practices should minimize injury to the crop that might promote pathogen spread by physically carrying inoculum from plant to plant, and by simultaneously wounding plants and facilitating entry of parasites into the organs harvested and into the plants left in the field [32].

Early cutting of alfalfa is a commonly used technique to minimize the impact of foliar diseases. By removing infected foliage before it falls to the ground and becomes a source of inoculum, disease incidence in subsequent cuttings is reduced [45].

XVII. STORAGE PRACTICES

Handling, curing, and storage management practices are extremely important to prevent the development and spread of pathogens during the storage process.

For example, low temperature ($<5°C$) and good ventilation to prevent CO_2 accumulation and water films are essential to minimize storage losses in potatoes [23]. Rapid cooling before shipping and/or storage is important for high-moisture fruits and vegetables coming from the field. Postharvest losses of peaches due to monilinia brown rot (*Monilinia fructicola*) are greatly reduced by rapid cooling of fruit immediately after picking [31].

Once in storage, an optimum balance of temperature and relative humidity (RH) for each crop must be maintained. Rots of onion caused by *Penicillium* spp., *Aspergillus* spp., and *Rhizopus* spp. are controlled best at temperatures below 15°C and RH of 50% or less [38]. In some situations, atmospheric composition is manipulated in storage to provide atmospheres that are suppressive to microorganism growth. Controlled atmosphere (low oxygen 1%, high nitrogen 99%) is commonly used for control of fungal rots caused by *Penicillium* spp., *Alternaria* spp., and *Mucor* spp. of apples and pears [27].

XVIII. CONCLUSIONS

Cultural practices are vital to managing pathogens and plant diseases. Along with resistance and pesticides, they are integral parts of any disease management program. Cultural practices are often the simplest and most cost-effective disease control measures to implement. Although among the oldest forms of management, cultural practices have effects on many host-pathogen systems that are not always well understood. Combining an understanding of how cultural practices affect the physical, chemical, and microbiological environment and an understanding of the host-pathogen relationship will lead to better management systems and reduced losses due to plant diseases.

REFERENCES

1. Agrios, G.N. 1988. *Plant Pathology*. 3rd ed. New York: Academic Press.
2. Allmaras, R.R., J.M. Kraft, & D.E. Miller. 1988. Effects of soil compaction and incorporated crop residue on root health. *Annu. Rev. Phytopathol.* 26:219–243.
3. Altieri, M.A., & M. Liebman. 1986. Insect, weed, and plant disease management in multiple cropping systems. In *Multiple Cropping Systems*. C.A. Francis, ed. New York: Macmillan, pp 183–218.
4. Ashworth, L.J., & S.A. Ganoa. 1982. Use of clear polyethylene mulch for control of Verticillium wilt in established pistachio nutgroves. *Phytopathology* 72:243–246.
5. Baker, K.F. 1957. *The U.C. System for Producing Healthy Container-grown Plants*. Davis, California: University of California Press.

6. Berger, R.D. 1977. Application of epidemiological principles to achieve plant disease control. *Annu. Rev. Phytopathol.* 15:165–183.

7. Boosalis, M.G., B.L. Doupnik, & J.E. Watkins. 1986. Effect of surface tillage on plant diseases. In *No-Tillage and Surface-Tillage Agriculture: The Tillage Revolution*. M.A. Sprague & G.B. Triplett, eds. New York: Wiley, pp 389–408.

8. Brown, W.M., Jr 1992. The role of IPM in contemporary agriculture and environmental issues. In *Successful Implementation of Integrated Pest Management for Agricultural Crops*. A.A. Leslie & G.W. Cuperus, Boca Raton, Florida: CRC Press, pp 171–179.

9. Brown, W.M., Jr., V.R. Velasco, & J.P. Hill. 1995. Integrated management of barley stripe rust. *Phytopathology* 85:1037.

10. Brown, W.M., W.M. Cranshaw, R.D. Davidson, D.G. Holm, R. Klein, K.W. Knutson, C.H. Livingston, M.D. Harrison, & G.A. McIntyre. 1986. Current Status of Potato Leafroll Virus Disease in the San Luis Valley. *Colorado State University Extension Bulletin 536A.*

11. Chupp, C., & A.A. MacNab. 1986. *Vegetable Diseases and Their Control*. New York: Wiley.

12. Conway, K.E., & L.S. Pickett. undated. Solar heating (solarization) of soil in garden plots for control of soilborne plant diseases. *Oklahoma State University Extension Facts 7640.*

13. Coykendall, I.J. 1968. Application of LP-gas flaming for disease control in mint oil and grass seed production. *Proc. 5th Annu. Symp. Thermal Agric.* 75 pp.

14. Croissant, R.L., H.F. Schwartz, & P.D. Ayers. 1991. Soil compaction and tillage effects on dry bean yields. *J. Prod. Agric.* 4:461–464.

15. Doupnik, B., Jr., M.G. Boosalis, G. Wicks, & D. Smika. 1975. Ecofallow reduces stalk rot in grain sorghum. *Phytopathology* 65:1021–1022.

16. Francis, C.A. 1986. Introduction: distribution and importance of multiple cropping. In *Multiple Cropping Systems*. C.A. Francis, ed. New York: Macmillan, pp 183–218.

17. Fry, W.E. 1982. *Principles of Plant Disease Management*. New York: Academic Press.

18. Haas, J.H., & B. Bolwyn. 1972. Ecology and epidemiology of Sclerotinia wilt of white beans in Ontario. *Can. J. Plant Sci.* 52:525–533.

19. Hall, R., & L.C.B. Nasser. 1996. Practice and precept in cultural management of bean diseases. *Can. J. Plant Pathol.* 18:176–185.

20. Hardison, J.R. 1976. Fire and flame for plant disease control. *Annu. Rev. Phytopathol.* 14:355–379.

21. Hoitink, H.A.J., & P.C. Fahy. 1986. Basis for the control of soilborne plant pathogens with composts. *Annu. Rev. Phytopathol.* 24:93–114.

22. Holtzer, T.O., R.L. Anderson, M.P. McMullen, & F.B. Peairs. 1996. Integrated pest management of insects, plant pathogens, and weeds in dryland cropping systems of the Great Plains. *J. Prod. Agric.* 9:200–208.

23. Hooker, W.J. 1981. *Compendium of Potato Diseases*. St Paul, Minnesota: APS Press.

24. Horner, C.E. 1965. Control of mint rust by propane gas flaming and contact herbicide. *Plant Dis. Rep.* 49:393–395.
25. Horner, C.E., & H.L. Dooley. 1965. Propane flaming kills *Verticillium dahliae* in peppermint stubble. *Plant Dis. Rep.* 49:581–582.
26. Huber, D.M., & R.D. Watson. 1974. Nitrogen form and plant disease. *Annu. Rev. Phytopathol.* 12:139–165.
27. Jones, A.L., & H.S. Aldwinckle. 1995. *Compendium of Apple and Pear Diseases.* St. Paul, Minnesota: APS Press.
28. Kahn, R.P. 1991. Exclusion as a plant disease control strategy. *Annu. Rev. Phytopathology* 29:219–246.
29. Katan, J., & J.E. DeVay. 1991. *Soil Solarization.* Boca Raton, Florida: CRC Press.
30. Legard, D.E., & H.F. Schwartz. 1987. Sources and management of *Pseudomonas syringae* pv. *phaseolicola* and *Pseudomonas syringae* pv. *syringae* epiphytes on dry beans in Colorado. *Phytopathology* 77:1503–1509.
31. Ogawa, J.M., E.I. Zehr, G.W. Bird, D.F. Ritchie, L. Uriu, & J.E. Uyemoto. 1995. *Compendium of Stone Fruit Diseases.* St. Paul, Minnesota: APS Press.
32. Palti, J. 1981. *Cultural Practices and Infectious Crop Diseases.* New York: Springer-Verlag.
33. Pearson, R.C., & A.C. Goheen. 1988. *Compendium of Grape Diseases.* St. Paul, Minnesota: APS Press.
34. Rotem, J., & J. Palti. 1969. Irrigation and plant diseases. *Annu. Rev. Phytopathol.* 7:267–288.
35. Schippers, B., A.W. Bakker, & P.A.H.M. Bakker. 1987. Interactions of deleterious and beneficial rhizosphere microorganisms and the effect of cropping practices. *Annu. Rev. Phytopathol.* 25:339–358.
36. Schumann, G.L. 1991. *Plant Diseases: Their Biology and Social Impact.* St. Paul, Minnesota: APS Press.
37. Schwartz, H.F., M.S. McMillan, and G.F. Lancaster. 1994. Volunteer beans with rust and white mold inocula in winter wheat. *Plant Dis.* 78:1216.
38. Schwartz, H.F. & S.K. Mohan. 1995. *Compendium of Onion and Garlic Diseases.* St. Paul, Minnesota: APS Press.
39. Schwartz, H.F., & F.J. Morales. 1989. Seed Pathology. *In Bean Production Problems in the Tropics.* H.F. Schwartz & M.A. Pastor-Corrales, eds. Cali, Colombia: CIAT, pp 413–431.
40. Skoglund, L.G., C. Rasmussen-Dykes, & W.M. Brown, Jr. 1986. Effects of combined chemical and solar treatment on population densities of *Fusarium* spp. *Phytopathology* 76:846.
41. Skoglund, L.G., & W.M. Brown, Jr. 1988. Effects of tillage regimes and herbicides on *Fusarium* species associated with corn stalk rot. *Can. J. Plant Pathol.* 10:332–338.
42. Smiley, R.W., P.H. Dernoeden, & B.B. Clarke. 1992. *Compendium of Turfgrass Diseases.* 2nd ed. St. Paul, Minnesota: APS Press.
43. Smucker, A.J.M. 1993. Soil environmental modifications of root dynamics and measurement. *Annu. Rev. Phytopathol.* 31:191–216.
44. Steadman, J.R., C.R., Maier, H.F. Schwartz, & E.D. Kerr. 1975. Pollution of

surface irrigation waters by plant pathogenic organisms. *Water Resources Bull.* 11:796–804.

45. Stuteville, D.L. & D.C. Erwin, 1990. *Compendium of Alfalfa Diseases.* 2nd ed. St. Paul, Minnesota: APS Press.

46. Sumner, D.R., B. Doupnik Jr., & M.G. Boosalis. 1981. Effects of reduced tillage and multiple cropping on plant diseases. *Annu. Rev. Phytopathol.* 19:167–187.

47. Taylor, R.G., T.L. Jackson, R.I. Powelson, & N.W. Christensen. 1983. Chloride, nitrogen form, lime, and planting date effects on take-all root rot of winter wheat. *Plant Dis.* 67:1116–1120.

48. Thurston, H.D. 1990. Plant disease management practices of traditional farmers. *Plant Dis.* 74:96–102.

49. Villapudua, J.R., & D.E. Munnecke. 1986. Solar heating and amendments control cabbage yellows. *Calif. Agric.* 40:11–13.

50. Walker, J.C. 1969. *Plant Pathology.* New York: McGraw-Hill.

51. Watkins, J.E., & M.G. Boosalis. 1994. Plant disease incidence as influenced by conservation tillage systems. In *Managing Agricultural Residues.* P.W. Unger, ed. Boca Raton, Florida: Lewis, pp 261–283.

52. Wiese, M.V. 1987. *Compendium of Wheat Diseases.* 2nd ed. St. Paul, Minnesota: APS Press.

53. Zitter, T. A., D.L. Hopkins, & C.E. Thomas. 1996. *Compendium of Cucurbit Diseases.* St. Paul, Minnesota: APS Press.

12
Biological Control of Plant Pathogens

Mark Wilson and Paul A. Backman*
Auburn University, Auburn, Alabama

I. INTRODUCTION

Biological control of plant pathogens can no longer be considered to be of just academic interest. In 1995, there were reported to be at least 30 commercial biocontrol agents for the control of soilborne plant pathogens on the world market [95]. Further, there were at least 14 commercial biocontrol agents of plant pathogens available in the United States (Table 1). Hence, at least for some diseases, biological control represents a viable alternative approach to chemical pesticides, host resistance, or crop rotations for disease management.

The current preponderance of products for biological control of soilborne pathogens is reflective of the fact that traditional seed treatment fungicides have been relatively ineffective for the control of these pathogens; hence, greater effort has been devoted to the development of alternative control measures. In contrast, fungicides for control of foliar fungal pathogens and copper bactericides or antibiotics for control of foliar bacterial pathogens have previously been very effective. Recent impetus for the development of biological control products for foliar, floral, and postharvest pathogens has derived from: (a) the occurrence of fungicide, copper, and antibiotic resistance in foliar pathogen populations, (b) concern about pesticide residues on fruit and vegetable produce, and (c) the revocation, or threatened revocation, of registrations for several widely used foliar fungicides. In addition, there have been mandates by several national governments to increase the use of pesticide alternatives; an example is the U.S. policy to increase the proportion of agricultural lands utilizing integrated pest management. Such programs will no doubt hasten the adoption of biological approaches for the management of foliar, soilborne, and postharvest plant pathogens.

Current affiliation: College of Agricultural Sciences, Penn State University, University Park, Pennsylvania.

Table 1 Commercially Available Biocontrol Agents of Plant Pathogens[a]

Product	Biocontrol Agent	Target Pathogen or Disease	Producer or Distributor
Actinovate	*Streptomyces lydicus*	*Fusarium, Pythium, Rhizoctonia*	San Jacinto Environmental Supplies
AQ10	*Ampelomyces quisqualis*	Powdery mildew	Ecogen
Aspire	*Candida oleophila*	Postharvest *Botrytis, Penicillium*	Ecogen
BioSave 10 and 11	*Pseudomonas syringae*	Postharvest pathogens	Ecoscience
BlightBan A506	*Pseudomonas fluorescens*	*Erwinia amylovora*, Ice + *Pseudomonas syringae*	Plant Health Technologies
Deny	*Burkholderia cepacia*	*Fusarium, Pythium, Rhizoctonia*	CCT Corp.
Galltrol A	*Agrobacterium radiobacter* K84	*Agrobacterium tumefaciens*	AgBioChem
Kodiak	*Bacillus subtilis*	*Fusarium, Rhizoctonia*	Gustafson
Epic	*Bacillus subtilis*	*Fusarium, Rhizoctonia*	Gustafson
Mycostop	*Streptomyces griseoviridis*	*Alternaria, Fusarium, Phomopsis*	AgBio Development
Promot	*Trichoderma*	Soilborne pathogens	J.H. Biotech
SoilGard 12G	*Gliocladium virens*	*Pythium; Rhizoctonia*	Grace Biopesticides
Bio-Trek 22G	*Trichoderma harzianum*	Soilborne pathogens	Bioworks
Trichodex 25WP (not yet EPA registered)	*Trichoderma harzianum*	*Botrytis cinerea*	Makhteshim Chemical Works

EPA, U.S. Environmental Protection Agency.
[a]Data for the United States in 1995.

This chapter discusses biological control of foliar, soilborne, and postharvest plant pathogens, concentrating on approaches utilizing individual cultured organisms or defined mixtures of cultured organisms rather than on undefined mixtures of indigenous organisms present in composts, compost extracts, or suppressive soils. The chapter concentrates on commercial or near-commercial approaches, but it also introduces organisms or approaches which either are several years from commercial development or currently are not considered commercially viable. The latter are included because, in the authors' opinion, the area of commercial biological control of plant pathogens is still in its infancy. An organism or approach not considered commercially viable in 1995 may be exploited very successfully in a niche market of the future, in which agricultural sustainability and reduced environmental impact are of a higher priority.

II. BIOLOGICAL CONTROL OF PLANT PATHOGENS ON AERIAL PLANT SURFACES

Biological approaches for the control of pathogens on aerial plant surfaces have been reviewed extensively over the past 20 years [6–8,13,42,43,53, 54,123,124,130]. During this period, most approaches employed for the biological control of diseases of aerial plant surfaces have consisted of the use of a single, empirically selected biocontrol agent to antagonize a single pathogen. Recently, however, some novel approaches have been developed, including (a) the use of a single biocontrol agent to inhibit several pathogens on the same host, (b) mixtures of biocontrol agents, (c) foliar "immunization," (d) rhizobacterial-induced systemic resistance; and (e) integration of biological and chemical agents. In this section, both traditional and novel approaches to the biological control of fungal, bacterial, and viral pathogens of aerial plant surfaces are discussed.

A. Preharvest Fungal Pathogens

1. Necrotrophic Fungal Pathogens

Biological approaches to the control of the necrotroph (pathogen which grows on dead and dying plant tissues) *Botrytis cinerea* have been directed toward the inhibition of infection, or alternatively the suppression of sporulation and dissemination. Conidia of *B. cinerea* typically require exogenous nutrients during germination and germ tube elongation; hence, these pathogens are subject to competition for these nutrients with the indigenous saprophytic microbial community on foliar surfaces [19,20]. Foliar applications of both saprophytic bacteria and yeasts have been reported to have some effect in reducing infection by *B. cinerea* [47,118]. Suppression of sporulation of *B. cinerea* has been effectively achieved through foliar applications of the saprophytic fungi *Trichoderma* spp., *Penicillium* spp., and *Gliocladium roseum* in various hosts, including strawberry [131–134], grape [62], and cucumber [48]. Field use of these biocontrol agents is not limited to spray application, since conidia of *G. roseum* also can be disseminated by bees [113,133]. In some cases, these biocontrol agents have been integrated with chemical fungicides to provide superior disease suppression [45,46,48]. The apparent success, in particular with *Trichoderma* spp. and *G. roseum*, suggests that there is potential for commercial development of these biocontrol agents for the control of necrotrophic pathogens. The biocontrol agent Trichodex 25WP, based on *Trichoderma harzianum* isolate T39, is commercially available in several countries for control of *B. cinerea*, but it has not yet been registered by the U.S. Environmental Protection Agency (EPA) (see Table 1).

Some reduction in the severity of disease caused by the necrotrophic pathogen *Sclerotinia sclerotiorum* can be achieved through the use of antagonistic bacteria [126,165,166] and fungi [14,15,167,168] to inhibit ascospore germination on petals. Disease control under field conditions, however, is limited by the colonization and survival of the applied organisms (G.Y. Yuen, personal communication). An alternative approach to the use of antagonistic bacteria and fungi involves the use of the mycoparasite *Coniothyrium minitans*. Foliar applications of *C. minitans* have not been successful in reducing disease [139,155], but it is possible that applications to crop residues may reduce pathogen survival [153]. *C. minitans* also may be used in a soil application to reduce sclerotial survival; however, even the few remaining sclerotia which germinate carpogenically will produce sufficient ascospores to initiate substantial infection [153]. Hence, Whipps and Gerlagh [153] concluded that it is unlikely that *C. minitans* will be commercialized for control of *S. sclerotiorum*. Nevertheless, the possibility remains that the combined use of *C. minitans* to reduce survival of sclerotia, particularly in minimum-till systems, and foliar application of antagonistic bacteria or fungi to reduce ascospore germination on petals could provide successful control of *S. sclerotiorum* under field conditions.

2. Biotrophic Fungal Pathogens

Spores of biotrophic fungi (those that produce haustoria to obtain nutrients from living plant tissue), powdery mildews, downy mildews, and rusts typically do not require exogenous nutrients during germination, precluding the use of nutrient competition as above. Further, host penetration occurs within a short period following germination, limiting the use of antibiotic-producing antagonists to suppress germination. For these reasons, biological approaches for the control of biotrophic fungal pathogens to date have been directed primarily toward the suppression of pathogen sporulation and dissemination using mycoparasites.

Biological control of powdery mildews (Erysiphales) on various plant hosts has been achieved using the mycoparasites *Ampelomyces quisqualis* [57,70,129], *Stephanoascus flocculosus* [71], and *Verticillium lecanii* [145]. The mycoparasite *A. quisqualis* isolate M-10 was recently released in the United States as the commercial product AQ10 for control of powdery mildew (see Table 1). Sporulation and dissemination of rusts also may be suppressed using the mycoparasite *V. lecanii* [152]. Additionally, some success has been achieved using foliar applications of competitive or antibiotic-producing bacilli [11,13] or pseudomonads [85] to reduce spore germination. There have been few reports on the use of foliar applications of either antagonistic bacteria or mycoparasites for the control of downy mildews (Peronosporales), although *V. lecanii* has been reported to parasitize *Peronospora parasitica* [65].

In the future, induced systemic resistance may prove to be one of the most effective biological approaches to the control of downy mildews and other biotrophic pathogens. The effectiveness of foliar applications of antagonistic bacteria or mycoparasitic fungi are both limited by the development of biotrophic pathogens in the interior of the leaf tissue. In contrast, induced systemic resistance is not limited by this constraint. Stem injections of sporangiospores of *Peronospora tabacina* provide significant protection against subsequent infection of tobacco by *P. tabacina* [142]. Although stem injections may be impractical on a commercial scale, systemic resistance also may be induced in tobacco by seed treatment with select strains of rhizobacteria (S. Tuzun, personal communication).

Some level of biological control of hemibiotrophic fungal pathogens (those that are facultatively saprotrophic) can be achieved through inhibition of spore germination, suppression of sporulation, or by induction of systemic resistance. Infection by the early leaf spot pathogen of peanut, *Cercospora arachidicola*, was reduced through foliar applications of chitinolytic strains of *Bacillus cereus* in combination with colloidal chitin [78]. Alternaria leaf spot of tobacco, caused by *Alternaria alternata*, can be reduced by foliar application of either antagonistic bacteria or nonpathogenic *A. alternata* isolates [123,124]. However, unless registration of fungicides conventionally used for control of these pathogens is revoked, it is unlikely that a commercial biocontrol agent for foliar application will be developed. On the other hand, the development of seed treatments with rhizobacteria which induce systemic resistance is a possibility. Select strains of plant growth-promoting rhizobacteria (PGPR) were effective in reducing the severity of anthracnose, caused by *Colletotrichum orbiculare*, in cucumber when applied to the seed [77,148].

B. Postharvest Fungal Pathogens

More biocontrol agents of postharvest fungal pathogens have been commercialized than biocontrol agents for fungal pathogens of aerial plant surfaces. In 1995, three postharvest biocontrol agents were commercialized: Aspire, BioSave 10, and BioSave 11™ (see Table 1). The successful commercialization of postharvest biocontrol agents probably results from several factors: (a) few fungicides are available for use in postharvest situations, (b) ease of application of biocontrol agents to the potential infection court, (c) a controlled environment which enhances biocontrol agent persistence and activity, and (d) a public desire for pesticide-free fruit. Biological approaches to the control of postharvest fungal pathogens will be illustrated with reference to biological control of brown rot in peaches, green mold of citrus, and storage rots of apples and pears. For greater detail, readers are referred to recent reviews on this subject [156,157].

Biological control of brown rot of peach, caused by *Monilinia fructicola*, can be achieved by incorporation of *Bacillus subtilis* strain B-3 into the fruit wax coating [114,115,116]. Pilot tests demonstrated that this approach could be effective on a commercial scale [116]. However, because the primary mode of action of this biocontrol agent is the production of antifungal iturin peptide antibiotics [61,98], this agent is unlikely to be developed commercially because of concerns over human toxicity. Nevertheless, iturin-producing *Bacillus* spp. are registered and sold commercially as seed and root inoculants [10,140], since the iturins produced in the rhizosphere are unlikely to present any significant risk to human health.

Biological control of green mold of citrus, caused by *Penicillium digitatum*, has been demonstrated with the yeasts *Debaryomyces hansenii* [41] and *Pichia (Candida) guillermondii* [23] and with the bacterium *Pseudomonas (Burkholderia) cepacia* [66,121]. In pilot scale studies, incorporation of *P. guillermondii* isolate US-7 into the fruit wax coating with a reduced concentration of the fungicide thiabendazole provided the same level of control of green mold as the regular commercial concentration of the fungicide [23].

Apples and pears are subject to postharvest decays caused by the pathogens *Penicillium expansum*, *Botrytis cinerea*, and *Mucor* spp. Biological control of these pathogens has been reported in several studies using the bacterium: *P. syringae* [68] and the yeasts *Candida oleophila* [102]; *Kloeckera apiculata* and *P. guillermondii* [100], and *Sporobolomyces roseus* [69]. Two *P. syringae* strains were commercialized in 1995 as the products BioSave 10 and BioSave 11 for control of blue mold and gray mold on apple and pear [72] (W. Janisiewicz, personal communication) and the yeast *C. oleophila* was released as the product Aspire for the control of postharvest pathogens of apple.

An interesting development in the control of postharvest diseases is the use of hormetic doses of ultraviolet C (UVC) radiation (doses insufficient to cause surface sterilization) for treatment of fruit, which elicits or induces some degree of protection against fungal pathogens. Such hormetic doses of UVC have recently been combined with the yeast biocontrol agent *D. hansenii* for control of *M. fructicola* brown rot in peaches, *P. digitatum* blue mold in tangerines, and *Rhizopus stolonifer* in tomatoes, where disease control was comparable to that achieved by fungicide dips [125].

C. Bacterial Pathogens

Biological approaches for control of bacterial pathogens of aerial plant surfaces are generally less well developed even than biological control of fungal pathogens of aerial surfaces, perhaps because these diseases were historically controlled effectively with bactericides or host plant resistance. An exception to this generality is the biocontrol agent *Pseudomonas fluorescens* A506, marketed as the product BlightBan A506 in 1996, for the control of fire blight of apple and pear, caused by *E. amylovora* [74,87,99]. The biocontrol agent *P. fluorescens* A506 appears to prevent blossom colonization by *E. amylovora* by prior utilization of nutrients or other resources associated with the blossom [158]. This means that the biocontrol agent must be applied to blossoms prior to the arrival of immigrant *E. amylovora*. This can be achieved either by spray application [74,87,99] or by dissemination with honeybees [137]. Certain *Pantoea agglomerans* (synonym *Erwinia herbicola*) strains are also effective against *E. amylovora* [49,74,160], and a recent interesting approach is the use of mixtures of *P. fluorescens* A506 with *P. agglomerans* (*E. herbicola*) strain C9-1 [127]. Strain mixtures may be selected empirically through field observations of superior performance or strategically through predicted complementarity of mechanisms. Both *P. fluorescens* A506 and a streptomycin-resistant derivative of *P. agglomerans* (*E. herbicola*) C9-1 can be integrated with the antibiotic streptomycin in an orchard spray program [128].

Although research on biological control of other foliar bacterial pathogens has lagged behind the development of biocontrol agents for fire blight, the increasing prevalence of resistance to copper bactericides (currently the primary chemical control agents) among pathovars of *Pseudomonas syringae* and *Xanthomonas campestris* probably will prompt greater activity in this area. Some effort has addressed the development of biocontrol of bacterial speck of tomato, caused by *Pseudomonas syringae* pv. tomato, using either naturally occurring saprophytic bacteria [30] or engineered nonpathogenic strains of the pathogen [32,86]. Under field conditions, with pathogens not yet resistant to copper, it may be possible to integrate copper-tolerant biocontrol agents with a copper bactericide to achieve superior disease control [32].

Induced systemic resistance may also be valuable for control of foliar bacterial diseases. In this case, the inducing agent may be applied as a seed treatment or as a foliar "immunization" (Tuzun and Backman, unpublished). Biological control through induced systemic resistance, achieved by seed bacterization with PGPR, has been observed with *P. syringae* pv. lachrymans in cucumber [90], *P. syringae* pv. phaseolicola in bean [5], and *P. syringae* pv. tomato in tomato (M. Wilson and Ji, unpublished data). Foliar "immunization" has been achieved by the introduction of nonpathogenic or incompatible *X. campestris* strains into the leaf tissue through hydathodes or stomates, using a polysilicone (Silwet) surfactant, to provide protection against black rot of cabbage, caused by *X. campestris* pv. campestris [73]. Whether this technology will be adopted in the future remains unclear.

D. Viral Pathogens

Viral infection can be reduced or prevented by cross protection, in which a mild strain of the target virus is introduced into the host plant prior to infection by the severe strain [58,59]. Cross protection has been reported to provide some level of control of several viral pathogens, including tobacco mosaic virus [119], cocoa swollen shoot virus [67], cucumber mosaic virus [38], zucchini yellow mosaic virus [146], and soybean mosaic virus [79]. Cross protection, however, has been used most widely for control of two viruses, citrus tristeza virus (CTV) and papaya ringspot virus (PRV). CTV is a semipersistent aphid-transmitted closterovirus which causes stem pitting of lime, grapefruit, and sweet orange, and it also causes a decline of sweet orange on sour orange rootstocks. Cross protection with naturally occurring mild strains of CTV is somewhat effective in reducing the symptoms of stem pitting, but protection against tristeza decline under severe disease pressure, as found in Brazil and South Africa, generally is not economically viable [59]. Cross protection for controlling PRV in papaya using mild nitrous acid–induced PRV strains has been tested on a large scale in Taiwan with some success [101,147,163]. There are, however, a number of problems with classic cross protection: (a) the mild strain of the virus may multiply in the inoculated plants, providing inoculum for the infection of nearby hosts upon which the "mild" strain is more virulent; (b) a second virus may interact synergistically to cause more severe disease symptoms; (c) mild strains have not been identified for many severe strains; and (d) the approach is excessively costly for an annual crop [58]. More recently, the use of transgenic plants expressing viral coat–protein (CP) genes has superseded classic cross protection in some applications [52, 60]. Although successful, this approach is not be discussed further in this chapter dealing primarily with introduced organisms as biocontrol agents (see Chapter 7).

An alternative approach to either cross protection or CP-mediated virus resistance is induced systemic resistance. PGPR-mediated induced systemic resistance has been reported against tobacco necrosis virus in tobacco [97] and cucumber mosaic virus in cucumber [88]. One advantage of induced systemic resistance is that resistance usually is induced against a range of pathogens, whereas in cross-protection resistance typically is effective only against a single virus (see Chapter 7).

III. BIOLOGICAL CONTROL OF SOILBORNE PLANT PATHOGENS

Biological control of soilborne pathogens has received more attention than biological control of foliar or floral pathogens. This effort has resulted in the development of upward of 30 commercial products for biological control of soilborne plant pathogens [94], approximately 10 of which are available in the United States (see Table 1). Some of the factors responsible for the substantial effort directed toward biocontrol of soilborne pathogens and the relatively high level of success (judged by the number of commercialized products) include (a) the lack of effective host plant resistance for many soilborne pathogens, (b) absence of bactericides and relative lack of fungicides for use against soilborne pathogens, (c) cost of treating large fields with a chemical agent, (d) difficulty of protecting all belowground infection courts with a chemical agent, (e) relative ease of protecting those infection courts using a seed-inoculated biocontrol agent; and (f) greater abundance of information on the ecology of soilborne pathogens compared with foliar pathogens.

In this section, approaches for biological control of several important fungal pathogens (*Fusarium* spp.; *Rhizoctonia solani*; *Pythium* spp.; *Gaeumannomyces graminis*) and bacterial pathogens (*Agrobacterium tumefaciens*; *Erwinia carotovora*; *Ralstonia solanacearum*) will be discussed. For greater detail on biological control of soilborne pathogens, readers are referred to other reviews on the subject [12,22,36,63,94,95,149].

A. Fungal Pathogens

1. *Fusarium* spp.

Fusarium spp. cause wilts and crown rots in several economically important crops throughout the world, hence there has been extensive interest in the biological control of diseases caused by this genus [3,109]. Several different organisms have been investigated for their potential as biocontrol agents of *Fusarium* spp., including nonpathogenic *F. oxysporum* [3,83,84], *Bacillus subtilis* [10,17], *Streptomyces* spp. [104,105], *Trichoderma harzianum* [35,120], *Pseudomonas* spp. [64,80–82], and vesicular-arbuscular mycorrhizae (VAM) [35].

Soils of the Chateaurenard region of France are suppressive to *Fusarium* wilts caused by *Fusarium oxysporum* [3]. Studies of this suppressive soil have led to a greater understanding of the interactions between nonpathogenic *F. oxysporum* strains and pseudomonads which contribute to this phenomenon. The nonpathogenic *F. oxysporum* isolate Fo47 provided control of fusarium wilt of flax (caused by *F. oxysporum* f.sp. *lini*), *Fusarium* wilt of tomato (caused by *F. oxysporum* f.sp. *lycopersici*), and fusarium wilt of carnation (caused by *F. oxysporum* f.sp. *dianthi*) [3]. In association with a fluorescent *Pseudomonas* sp., *F. oxysporum* Fo47 also gave control of crown and root rot of tomato (caused by *F. oxysporum* f.sp. *radicis-lycopersici*) [3]. Apparently, the nonpathogenic *F. oxysporum* Fo47 competes with the pathogen for carbon in the rhizosphere [34], and the intensity of competition with the pathogen is enhanced by the siderophore pseudobactin, produced by the *Pseudomonas* spp.[83,84]. A nonpathogenic *F. oxysporum* is available commercially in France under the name Fusaclean for the control of fusarium wilt of tomato and carnation [94].

Bacillus subtilis strain A-13 was originally isolated by Broadbent in Australia [18]. During industrial investigation of this strain as a potential biocontrol agent, a derivative of A-13, designated GB03, was selected by host passage through cotton. *B. subtilis* strain GB03 is now marketed as Kodiak (see Table 1) for control of several soilborne pathogens, including *Fusarium* wilt of cotton, caused by *F. oxysporum* f.sp. *vasinfectum* [10,17]. Cotton seed treatments with *B. subtilis* GB03 combined with a fungicide (carboxin-PCNB-metalaxyl) provide control of *Fusarium* wilt of cotton superior to the use of fungicides alone [17]. *B. subtilis* GB03 also stimulates cotton root mass, which contributes to improved yields [10]. *B. subtilis* strain GB07, marketed as Epic (see Table 1), is used in conjunction with GB03 in treated cotton seed (P.M. Brannen, personal communication).

Three additional biological products are available for control of *Fusarium* wilts (see Table 1): (a) Mycostop, based on *Streptomyces griseoviridis*, is used for biological control of *Fusarium* wilt in vegetables and ornamentals under greenhouse cultivation; (b) F-stop, based on *Trichoderma harzianum* (Table 1), is registered in the United States for control of *Fusarium* spp. in row crops [94] (Table 1); and (c) Deny liquid biological fungicide, is based upon a *Burkholderia cepacia* strain which has previously been marketed under several names including Intercept and Blue Circle.

In the future, induced systemic resistance may prove to be one of the most effective ways to control *Fusarium* wilts. PGPR-mediated induced systemic resistance has been reported against *F. oxysporum* f.sp. *dianthi* [144], *F. oxysporum* f.sp. *raphani* [81], and *F. oxysporum* f.sp. *cucumerinum* [89]. Systemic resistance against *F. oxysporum* f.sp. *raphani* also may be induced by certain root-colonizing fungi [82]. In the future, seed treatments with combinations of bacterial or fungal inducing agents could provide optimal induction of systemic resistance under differing environmental and edaphic conditions.

2. *Rhizoctonia solani*

Rhizoctonia solani causes pre- and postemergence damping-off of seedlings in many crops, as well as stem rots, brown patch of turf, and diseases of aboveground plant parts (e.g., sheath blight of rice); hence, successful biological control of *R. solani* would be of great economic benefit to agriculture. Fungal antagonists that have been used to control diseases caused by *R. solani* include *Trichoderma hamatum* [26], *T. harzianum* [29,44,159], *T. viride* [29], binucleate nonpathogenic *Rhizoctonia* spp. [166], *Verticillium biguttatum* [107], and *Gliocladium virens* [91–93]. *G. virens* strain GL-21, isolated from soil in Beltsville, Maryland, has provided effective control of damping-off due to *R. solani* [91–93]. This strain has been commercialized as the product SoilGard 12G (previously called GlioGard) (see Table 1) for control of damping-off in glasshouse production of ornamentals and vegetables [94].

Commercial bacterial biocontrol agents of *R. solani* damping-off and root rot include Deny, based on *B. cepacia*; Mycostop, based on *Streptomyces griseoviridis* [105]; and Kodiak and Epic™, based on *B. subtilis* strains GB03 and GB07, respectively [10] (see Table 1). The biocontrol agent Dagger G, based on *Pseudomonas fluorescens* [94], apparently is no longer available commercially.

B. subtilis strain GB03 provides significant control of late-season *Rhizoctonia* root rot in peanut and cotton, when used in combination with seed-treatment fungicides [10]. In the 1994 growing season, *B. subtilis* GB03, as Kodiak, was used to treat more than 2 million hectares of cotton and peanuts in the United States for control of *Fusarium* and *Rhizoctonia* [10]. This represents the first large-scale utilization of a bacterial biocontrol agent of soilborne plant pathogens in U.S. agriculture.

Sheath blight of rice, caused by *R. solani*, is one of the most important diseases of rice in Asia [103,135]. The coordinated efforts of several Asian countries, organized by the International Rice Research Institute (IRRI), have targeted the development of biocontrol agents of this disease [103]. In China, yield-increasing bacteria (YIB), the majority of which are bacilli, have been used in agriculture for several years [25,135]. *B. subtilis* strain B-908, which was isolated from rice rhizosphere soil in China, provides significant control of sheath blight when applied either as a soil treatment or a foliar spray [135]. The control provided by B-908 may result from induced systemic resistance [135].

3. *Pythium* spp.

The pre- and postemergence damping-off diseases and root rots caused by *Pythium* spp. are as numerous as those caused by *R. solani*; hence, pathogens of this genus have also been the focus of considerable attention. The various approaches and antagonists used for the biological control of *Pythium* spp. have been discussed in a comprehensive review by Whipps and Lumsden [154]. Several of the commercial biological products used for the control of *R. solani* are also effective against *Pythium* spp. These include, *G. virens* GL-21 (SoilGard 12G), *T. harzianum* (F-stop), and *B. cepacia* (Deny™) [94] (Table 1).

An interesting alternative to the use of pseudomonads, bacilli, *Trichoderma* spp., and *Gliocladium* spp., is the use of mycoparasitic *Pythium oligandrum* strains. Mycoparasitic strains of *P. oligandrum* can be incorporated into seed coats to provide protection against damping-off caused by *P. ultimum* [4,96] and *P. splendens* [136]. A commercial product based on *P. oligandrum*, Polygandron, has been produced by a company in Czechoslovakia and has been tested on sugarbeet and other field crops in Europe [154].

Although not yet commercialized, *Enterobacter cloacae* strain EcCT-501 is an effective biocontrol agent of cotton seed and seedling rots caused by *P. ultimum* [110]. This strain is particularly interesting, because its mode of action may complement other biocontrol agents with mechanisms such as antibiosis or mycoparasitism. Apparently, *E. cloacae* EcCT-501 catabolizes cotton seed exudates, particularly long chain fatty acids such as linoleic acid, which stimulate sporangial germination in *P. ultimum* [143]. Successful and, more importantly, consistent biological control of *Pythium* diseases might be achieved through the combined use of agents such as *E. cloacae* which reduce sporangial germination, pseudomonads and bacilli to produce a zone of antibiosis around the seed, and mycoparasitic *P. oligandrum* to parasitize hyphae of *Pythium* in the spermosphere or rhizosphere.

4.   *Gaeumannomyces graminis* var. *tritici*

Unlike the other fungal pathogens discussed here, biological control approaches for take-all of wheat, caused by *Gaeumannomyces graminis* var. *tritici*, have been based almost exclusively upon seed treatments with fluorescent *Pseudomonas* spp., originally isolated from suppressive soils [31,150,151]. Two of the most effective isolates, *P. fluorescens* 2-79 and *Pseudomonas aureofaciens* 30-84, protect wheat roots against *G. graminis* var. *tritici* primarily by the production of phenazine antibiotics [31]. *P. aureofaciens* Qc69, which has performed most consistently to date in a variety of soil types, does not produce any antibiotics in vitro but does produce hydrogen cyanide (HCN) [31]. Whether HCN production is the primary mode of action with this strain is unclear, since *G. graminis* var. *tritici* appears to be able to detoxify HCN in vitro (L. S. Pierson, personal communication). Further improvement of non–antibiotic-producing strains such as *P. aureofaciens* Qc69 that show consistent efficacy, may be possibly through the cloning and transfer of genes encoding phenotypes involved in biocontrol of take-all. Transfer of phenazine or phloroglucinol biosynthetic functions from other *Pseudomonas* strains to *P. aureofaciens* Qc69 has provided enhanced control of take-all, at least under controlled conditions [31].

5. *Sclerotinia* spp.

Two very different approaches are available for the biological control of diseases caused by *Sclerotinia* spp. The first approach employs *Sporidesmium sclerotivorum*, an obligate mycoparasite of sclerotia, particularly of the Sclerotiniaceae [1,55]. Biological control of lettuce drop, caused by *Sclerotinia minor* using *S. sclerotivorum*, is facilitated by the monocyclic nature of the disease and also the aggregated distribution of pathogen sclerotia in soil. By application of the mycoparasite to crop debris before it was disked into the soil, control of lettuce drop under field conditions was achieved with biocontrol agent inoculum levels as low as 0.2 kg/ha [1,2,55].

The second approach involves the use of hypovirulent isolates of the pathogen. Hypovirulent isolates of *Sclerotinia* spp. exhibit a reduced ability to cause disease and are usually infected with double-stranded RNA (dsRNA) viruses [16]. These dsRNA viruses are transmissible during hyphal anastomosis (hyphal fusion) causing hypovirulence in the recipient fungal isolate; hence, hypovirulent *Sclerotinia* spp. have potential for the biological control of *Sclerotinia* diseases, such as lettuce drop (caused by *S. minor*) and dollar spot of turf (caused by *S. homeocarpa*) [16]. In field experiments, hypovirulent *S. homeocarpa* isolate 12B significantly reduced the severity of dollar spot disease [16]. One of the limiting factors to the use of hypovirulent isolates, however, is that the approach is only feasible in pathogenic species or geographical areas with relatively few vegetative compatibility groups, since the dsRNA virus cannot be transmitted in the absence of hyphal anastomosis.

B. Bacterial Pathogens

1. *Agrobacterium tumefaciens*

Biological control of crown gall, caused by the pathogen *Agrobacterium tumefaciens*, with the biocontrol agent *A. radiobacter* strain K84 has become one of the success stories in biological control of bacterial diseases [28,51,106]. Biological control of crown gall using *A. radiobacter* strain K84 has been practiced commercially since 1973 [28]. Currently, there are at least three commercial products on the world market, based upon *A. radiobacter* strain K84, for control of crown gall in stone fruits and ornamentals, variously called Galltrol (see Table 1), Norbac 84-C, and Diegall [94]. Control of *A. tumefaciens* by *A. radiobacter* strain K84 is due, in part, to the production of agrocins 84 [106] and 434 [40]. Production of agrocin 84 is encoded by the plasmid pAgK84; this plasmid also carries genes encoding resistance to agrocin 84 and conjugal transfer functions (Tra). In order to prevent conjugal transfer of agrocin resistance to agrocin 84-sensitive *A. tumefaciens* strains, a Tra-minus derivative of K84 was generated and designated K1026 [75]. The engineered *A. radiobacter* strain K1026 was registered by the Australian government in 1991 and is now marketed as the product Nogall [28]. Unfortunately, neither strain K84 nor K1026 has any effect against crown gall of grape, caused by *A. vitis*. Nontumorigenic strains of *A. vitis* may provide some biological control [21], but as yet no such strains have been commercialized.

2. *Erwinia carotovora*

The pathogen *Erwinia carotovora* subsp. *carotovora* causes potato seed piece decay and black leg, and it also causes aerial softrots in several crops. Inhibition of potato seed piece decay can be achieved through the use of antagonistic pseudomonads [161,162]. Strains of the sugarbeet wound pathogen *Erwinia carotovora* subsp. *betavasculorum* inhibit colonization of potato tissues by *E. carotovora* susbsp. *atroseptica* by antibiosis [9]. *E. carotovora* susbsp. *betavasculorum* cannot be used as a biocontrol agent, however, because it also macerates potato tissue through production of pectolytic enzymes [33]. Recently, mutant derivatives of *E. carotovora* subsp. *betavasculorum* strain 168, which do not secrete pectate lyases (Out-minus mutants) and are reduced in virulence, have been produced [33]. Through further mutagenesis, it may be possible to create avirulent mutants of *E. carotovora* subsp. *betavasculorum* which can be used in biological control of *E. carotovora* subsp. *carotovora* soft rot of potato [33].

3. *Ralstonia solanacearum*

Ralstonia solanacearum causes bacterial wilt of several economically important crop plants in subtropical and tropical areas. Reduced wilt severity has been effected through the use of various rhizobacteria, including saprophytic *Pseudomonas* spp. [76,112], incompatible strains of *R. solanacearum* [76], naturally occurring avirulent strains of *R. solanacearum* [27,76], and Tn5-induced avirulent mutants of *R. solanacearum* [56,138]. Unfortunately, extensive genetic diversity within and between races of *R. solanacearum* may mean that a single avirulent mutant may not be effective against all strains of the pathogen, and mixtures of biocontrol strains may be necessary [122]. Nevertheless, a nonpathogenic strain of *R. solanacearum* is available in France as the commercial product PSSOL [94], although no information is available concerning its geographical range of effectiveness.

IV. PROMISING AREAS IN BIOLOGICAL CONTROL OF PLANT PATHOGENS

Biologically induced systemic resistance will probably be investigated for the control of both foliar and soilborne plant pathogens in an increasing number of cropping systems. Systemic resistance induced by seed or soil treatments with PGPR has been reported to provide protection against a wide range of plant pathogens. These include: (a) viruses, such as cucumber mosaic virus [88] and tobacco necrosis virus [97]; (b) foliar bacterial pathogens, such as *P. syringae* pv. lachrymans [90] and *P. syringae* pv. phaseolicola [5]; (c) vascular bacterial pathogens, such as *Erwinia tracheiphila* [141]; (d) foliar fungal pathogens, such as *Colletotrichum orbiculare* [141]; and (e) soilborne fungal pathogens, such as *Fusarium oxysporum* f.sp. *dianthi* [144] and *F. oxysporum* f.sp. *cucumerinum* [89]. Future demonstration of rhizobacterial-induced systemic resistance in a wider range of crops and pathosystems will probably increase the acceptance and utilization of this biological approach to pathogen control.

Although endophytic bacteria, such as *Clavibacter xyli* subsp. *cynodontis*, have received a great deal of attention as a possible delivery system for insecticidal toxins [37,50], the use of endophytic bacteria for biocontrol of plant pathogens has received much less attention. Endophytic bacteria and fungi may occur more widely in crop plants than previously recognized and may provide multiple beneficial effects upon the host plant, including suppression of plant pathogens. In preliminary studies, endophytic bacteria from cotton provided control of *Fusarium* wilt, caused by *F. oxysporum* f.sp. *vasinfectum*, under controlled conditions [24]. In addition to selection methods for beneficial endophytic strains, the development and implementation of endophytes for biological control of plant pathogens will also require modification of currently available delivery systems for seed bacterization [108]. Like endophytic bacteria and fungi, vesicular-arbuscular mycorrhizae (VAM) may also be manipulated in the future to provide biological control of soilborne pathogens [35, 111,117].

V. CONCLUSIONS

Hopefully, this chapter has demonstrated that for each pathosystem there usually are several different biological approaches available that can be employed, either alone or in combination, to achieve significant reductions in disease severity under agricultural conditions. The fact that several of these approaches and organisms are now commercialized and are being employed in production agriculture appears to be a vindication of the potential of the biological approach, and suggests that numerous opportunities lie ahead.

ACKNOWLEDGMENTS

We wish to acknowledge Phil Brannen, Joe Kloepper, Bob Lumsden, Sandy Pierson, John Sutton, and Gary Yuen for their careful reviews of the manuscript.

Disclaimer

Reference to any commercial materials or products in this chapter does not constitute a recommendation or endorsement of their use.

REFERENCES

1. Adams, P.B., & D.R. Fravel. 1990. Economical biological control of *Sclerotinia* lettuce drop by *Sporidesmium sclerotivorum*. *Phytopathology* 80:1120–1124.
2. Adams, P.B., & D.R. Fravel. 1993. Dynamics of *Sporidesmium*, a naturally-occurring fungal mycoparasite. In *Pest Management: Biologically Based Technologies*. R.D. Lumsden & J.L. Vaughn, eds. Washington, DC: American Chemical Society, pp 189–195.
3. Alabouvette, C., P. Lemanceau, & C. Steinberg. 1993. Recent advances in the biological control of Fusarium wilts. *Pestic. Sci.* 37:365–373.
4. Al-Hamndani, A.M., R.S. Lutchmeah, & Cooke, R.C. 1983. Biological control of *Pythium ultimum*-induced damping-off by treating cress seed with the mycoparasite *Pythium oligandrum*. *Plant Pathol.* 32:449–454.
5. Alstrom, S. 1991. Induction of disease resistance in common bean susceptible to halo blight bacterial pathogen after seed bacterization with rhizosphere pseudomonads. *J. Gen. Appl. Microbiol.* 37:495–501.
6. Andrews, J.H. 1985. Strategies for selecting antagonistic microorganisms from the phylloplane. In *Biological Control on the Phylloplane*. C.E. Windels & S.E. Lindow, eds. Minneapolis, Minnesota: American Phytopathological Society, pp 31–44.
7. Andrews, J.H. 1990. Biological control in the phyllosphere: realistic goal or false hope? *Can. J. Plant Pathol.* 12:300–307.
8. Andrews, J.H. 1992. Biological control in the phyllosphere. *Annu. Rev. Phytopathol.* 30:603–635.
9. Axelrood, P.E., M. Rella, & M.N. Schroth. 1988. Role of antibiosis in competition of *Erwinia* strains in potato infection courts. *Appl. Environ. Microbiol.* 54:1222–1229.

10. Backman, P.A., P.M. Brannen, & W.F. Mahaffee. 1994. Plant response and disease control following seed inoculation with *Bacillus subtilis*. In *Improving Plant Productivity with Rhizosphere Bacteria*. M.H. Ryder, P.M. Stephens, & G.D. Bowen, eds. Adelaide, Australia: CSIRO, pp 3–8.

11. Baker, C.J., J.R. Stavely, & N. Mock. 1985. Biocontrol of bean rust by *Bacillus subtilis* under field conditions. *Plant Dis.* 69:770–772.

12. Becker, O.J., & F.J. Schwinn. 1993. Control of soil-borne pathogens with living bacteria and fungi: status and outlook. *Pest. Sci.* 37:355–363.

13. Blakeman, J.P., & N.J. Fokkema. 1982. Potential for biocontrol of plant diseases on the phylloplane. *Annu. Rev. Phytopathol.* 20:167–192.

14. Boland, G.J., & J.E. Hunter. 1988. Influence of *Alternaria alternata* and *Cladosporium cladosporioides* on white mold of bean caused by *Sclerotinia sclerotiorum*. *Can. J. Plant Pathol.* 10:172–177.

15. Boland, G.J., & G.D. Inglis. 1980. Antagonism of white mold (*Sclerotinia sclerotiorum*) of bean by fungi from bean and rapeseed flowers. *Can. J. Bot.* 67:1775–1781.

16. Boland, G.J., M.S. Melzer, & T. Zhou. 1995. Hypovirulence in *Sclerotinia* spp. In *Proceedings of the 6th International Symposium on Microbiology of Aerial Plant Surfaces*. C.E. Morris, P. Nicot, eds. Marseille: Plenum Press, p. 48.

17. Brannen, P.M., & P.A. Backman. 1994. Suppression of Fusarium wilt of cotton with *Bacillus subtilis* hopper box formulations. In *Improving Plant Productivity with Rhizosphere Bacteria*. M.H. Ryder, P.M. Stephens, & G.D. Bowen, eds. Adelaide, Australia: CSIRO, pp 83–85.

18. Broadbent, P., K.F. Baker, N. Franks, & J. Holland. 1977. Effect of *Bacillus* spp. on increased growth of seedlings in steamed and nontreated soil. *Phytopathology* 67:1027–1034.

19. Brodie, I.D.S., & J.P. Blakeman. 1975. Competition for carbon compounds by a leaf surface bacterium and conidia of *Botrytis cinerea*. *Physiol Plant Pathol.* 6:125–135.

20. Brodie, I.D.S., & J.P. Blakeman. 1976. Competition for exogenous substrates *in vitro* by leaf surface microorganisms and germination of conidia of *Botrytis cinerea*. *Physiol Plant Pathol.* 9:227–239.

21. Burr, T.J., & C.L. Reid. 1994. Biological control of grape crown gall with nontumorigenic *Agrobacterium vitis* strain F2/5. *Am. J. Enol. Vitic.* 45:213–219.

22. Campbell, R. 1994. Biological control of soil-borne diseases: some present problems and different approaches. *Crop Protection* 13:4–12.

23. Chalutz, E., S. Droby, B. Hofstein, B. Fridlender, C. Wilson, M. Wisniewski, & D. Timar. 1993. Pilot testing of *Pichia guillermondii*—a biocontrol agent of postharvest diseases of citrus fruit. *JOBC Bull.* 16:119–122.

24. Chen, C., E.M. Bauske, G. Musson, R. Rodriguez-Kabana, & J.W. Kloepper. 1995. Biological control of Fusarium wilt on cotton by use of endophytic bacteria. *Biol. Control* 5:83–91.

25. Chen, Y., R. Mei, S. Lu, L. Liu, & J.W. Kloepper. 1995. The use of yield increasing bacteria (YIB) as plant growth-promoting rhizobacteria in Chinese ag-

riculture. In *Management of Soilborne Diseases*. R. Utkhede & V.K. Gupta, eds. New Delhi: Kalyani, pp

26. Chet, I., & R. Baker. 1981. Isolation and biocontrol potential of *Trichoderma hamatum* from soil naturally suppressive to *Rhizoctonia solani*. *Phytopathology* 71:286–290.

27. Ciampi-Panno, L., C. Fernandez, P. Bustamante, N. Andrade, S. Ojeda, & A. Contreras. 1989. Biological control of bacterial wilt of potatoes caused by *Pseudomonas solanacearum*. *Am. Pot. J.* 66:315–332.

28. Clare, B.G. 1993. *Agrobacterium*: Biological plant disease control. In *Advanced Engineered Pesticides*. L. Kim, ed. New York: Marcel Dekker, pp 129–146.

29. Coley-Smith, J.R., C.J. Ridout, & C.M. Mitchell. 1991. Control of bottom rot disease of lettuce (*Rhizoctonia solani*) using preparations of *Trichoderma viride*, *T. harzianum* or tolclofos methyl. *Plant Pathol.* 40:359–366.

30. Colin, J.E., & Z. Chafik. 1986. Comparison of biological and chemical treatments for control of bacterial speck of tomato under field conditions in Morocco. *Plant Dis.* 70:1048–1050.

31. Cook, R.J. 1994. Problems and progress in the biological control of wheat take-all. *Plant Pathol.* 43:429–437.

32. Cooksey, D.A. 1988. Reduction of infection by *Pseudomonas syringae* pv. *tomato* using a non-pathogenic, copper-resistant strain combined with a copper bactericide. *Phytopathology*, 78:601–603.

33. Costa, J.M., & J.E. Loper. 1994. Derivation of mutants of *Erwinia carotovora* subsp. *betavasculorum* deficient in export of pectolytic enzymes with potential for biological control of potato soft rot. *Appl. Environ. Microbiol.* 60:2278–2285.

34. Couteaudier, Y., & C. Alabouvette. 1990. Quantitative comparison of *Fusarium oxysporum* competitiveness in relation to carbon utilization. *FEMS Microbiol. Ecol.* 74:261–268.

35. Datnoff, L.E., S. Nemec, & K. Pernezny. 1995. Biological control of Fusarium crown and root rot of tomato in Florida using *Trichoderma harzianum* and *Glomus intraradices*. *Biol. Control* 5:427–431.

36. Deacon, J.W., & L.A. Berry. 1993. Biocontrol of soil-borne plant pathogens: concepts and their application. *Pestic. Sci.* 37:417–426.

37. Dimock, M., J. Turner, & J. Lampel. 1994. Endophytic microorganisms for delivery of genetically engineered microbial pesticides in plants. In *Advanced Engineered Pesticides*. L. Kim, ed. New York: Marcel Dekker, pp 85–97.

38. Dodds, J.A., S.Q. Lee, & M. Tiffany. 1985. Cross protection between strains of cucumber mosaic virus: effect of host and type of inoculum on accumulation of virions and double-stranded RNA of the challenge strain. *Virology* 144:301–309.

39. Doherty, M.A., & T.F. Preece. 1978. *Bacillus cereus* prevents germination of uredospores of *Puccinia allii* and the development of rust disease of leek, *Allium parium*, in controlled environments. *Physiol. Plant Pathol.* 12:123–132.

40. Donner, S.C., D.A. Jones, N.C. McClure, G.M. Rosewarne, M.E. Tate, A. Kerr, N.N. Fajardo, & B.G. Clare. 1993. Agrocin 434, a new plasmid encoded agrocin from the biocontrol *Agrobacterium* strains K84 and K1026, which inhibits biovar 2 agrobacteria. *Physiol. Mol. Plant Pathol.* 42:185–194.

41. Droby, S., E. Chalutz, C.L. Wilson, & M. Wisniewski. 1989. Characterization of the biocontrol activity of *Debaryomyces hansenii* in the control of *Penicillium digitatum* on grapefruit. *Can. J. Microbiol.* 35:794–800.
42. Dubos, B. 1987. Fungal antagonism in aerial agrobiocenoses. In *Innovative Approaches to Plant Disease Control*. I. Chet, ed. New York: Wiley, pp 107–135.
43. Elad, Y. 1993. Microbial suppression of infection by foliar plant pathogens. *IOBC Bull.* 16:3–7.
44. Elad, Y., & Y. Hadar. 1981. Biological control of *Rhizoctonia solani* by *Trichoderma harzianum* in carnation. *Plant Dis.* 65:675–677.
45. Elad, Y., & G. Zimand. 1991. Experience in integrated chemical-biological control of grey mould (*Botrytis cinerea*). *Working Group: Integrated Control in Protected Crops Under Mediterranean Climate*, Alassio, Italy: IOBC, pp 195–199.
46. Elad, Y., & G. Zimand. 1992. Integration of biological and chemical control for grey mould. In *Recent Advances* in Botrytis *Research*. K. Verhoeff, N.E. Malathrakis, & B. Williamson, eds. Wageningen, Germany: Pudoc Scientific, pp. 272–276.
47. Elad, Y., J. Kohl, & N.J. Fokkema. 1994. Control of infection and sporulation of *Botrytis cinerea* on bean and tomato by saprophytic yeasts. *Phytopathology* 84:1193–1200.
48. Elad, Y., G. Zimand, Y. Zaqs, S. Zuriel, & I. Chet. 1993. Use of *Trichoderma harzianum* in combination or alternation with fungicides to control cucumber grey mould (*Botrytis cinerea*) under commercial greenhouse conditions. *Plant Pathol.* 42:324–332.
49. Epton, H.A.S., M. Wilson, S. Nicholson, & D.C. Sigee. 1994. Biological control of *Erwinia amylovora* with *Erwinia herbicola*. In *Ecology of Plant Pathogens*. J.P. Blakeman & B. Williamson, eds. Surrey, UK: CAB International, pp 335–352.
50. Fahey, J.W., M.B. Dimock, S.F. Tomasino, J.M. Taylor, & P. Carlson. 1992. Genetically engineered endophytes as biocontrol agents: a case study from industry. In *Microbial Ecology of Leaves*. S.S. Hirano & J.H. Andrews, eds. New York: Springer-Verlag, pp 401–411.
51. Farrand, S.K. 1990. *Agrobacterium radiobacter* strain K84: a model biocontrol system. In *New Directions in Biological Control: Alternatives for Suppressing Agricultural Pests and Diseases*. R. Baker & P.E. Dunn, eds. New York: Liss, pp 679–691.
52. Fitchen, J.H., & R.N. Beachy, 1993. Genetically engineered protection against viruses in transgenic plants. *Annu. Rev. Phytopathol.* 47:739–763.
53. Fokkema, N.J. 1976. Antagonism between fungal saprophytes and pathogens on aerial plant surfaces. In *Microbiology of Aerial Plant Surfaces*. C.H. Dickinson & T.F. Preece, eds. New York: Academic Press, pp 487–506.
54. Fokkema, N.J. 1993. Opportunities and problems of control of foliar pathogens with micro-organisms. *Pestic. Sci.* 37:411–416.
55. Fravel, D.R., P.B. Adams, & W.E. Potts. 1992. Use of disease progress curves to study the effects of the biochemical agent *Sporidesmium sclerotivorum* on lettuce drop. *Biocontrol Sci. Technol.* 2:341–348.

56. Frey, P., P. Prior, C. Marie, A. Kotoujansky, D. Trigalet-Demery, & A. Trigalet. 1994. Hrp⁻ mutants of *Pseudomonoas solanacearum* as potential biocontrol agents of tomato bacterial wilt. *Appl. Environ. Microbiol.* 60:3175–3181.

57. Fridlender, B., M. Keren-Zur, A. Bercowitz, D. Nessim, C. Katz, D. Beit-Din, & R. Hofstein. 1994. Control of powdery mildew by *Ampelomyces quisqualis*: an example for the development of a commercial biofungicide. *Proceedings of the 4th Siconbiol Symposium on Biological Control.* Granada, RS, Brazil. EMBRAPA/CPAC: Documents, 6, p 167.

58. Fulton, R.W. 1986. Practices and precautions in the use of cross protection for plant virus disease control. *Annu. Rev. Phytopathol.* 24:67–81.

59. Gonsalves, D., & S.M. Garnsey, 1989. Cross-protection techniques for control of plant virus diseases in the tropics. *Plant Dis.* 73:592–597.

60. Gonsalves, D., & J.L. Slighton. 1993. Coat protein-mediated protection: analysis of transgenic plants for resistance in a variety of crops. *Virology* 4:397–405.

61. Gueldner, R.C., C.C. Reilly, P.L. Pusey, C.E. Costello, R.F. Arrendale, R.H. Cox, D.S. Himmelsbach, F.G. Crumley, & H.G. Cutler. 1988. Isolation and identification of iturins as antifungal peptides in biological control of peach brown rot with *Bacillus subtilis*. *J. Agric. Food Chem.* 36:366–370.

62. Gullino, M.L., & A. Garibaldi. 1988. Biological and integrated control of grey mould of grapevine: results in Italy. *EPPO Bull.* 18:9–12.

63. Harman, G.E. 1991. Control of soil-borne plant pathogens with biological control agents. *Biol. Cult. Tests* 6:5–8.

64. Hebbar, K.P., D. Atkinson, W. Tucker, & P.J. Dart. 1992. Suppression of *Fusarium moniliforme* by maize root-associated *Pseudomonas cepacia*. *Soil Biol. Biochem.* 24:1009–1020.

65. Hijwegen, T. & J.A.A.M. Dirven 1993. Mycoparasitism of powdery and downy mildews. *IOBC Bull.* 16:76–77.

66. Huang, Y., B.J. Deverall, & S.C. Morris, 1993. Effect of *Pseudomonas cepacia* on postharvest biocontrol of infection by *Penicillium digitatum* and on wound responses of citrus fruit. *Australasian Plant Pathol.* 22:84–93.

67. Hughes, J.d'A., & L.A.A. Ollennu. 1994. Mild strain protection of cocoa in Ghana against cocoa swollen shoot virus - a review. *Plant Pathol.* 43:442–457.

68. Janisiewicz, W.J., & A. Marchi. 1992. Control of storage rots on various pear cultivars with saprophytic strain of *Pseudomonas syringae*. *Plant Dis.* 76:555–560.

69. Janisiewicz, W.J., D.L. Peterson, & R. Bors. 1994. Control of storage decay of apples with *Sporobolomyces roseus*. *Plant Dis.* 78:466–470.

70. Jarvis, W.R., & K. Slingsby. 1977. The control of powdery mildew of greenhouse cucumber by water sprays and *Ampelomyces quisqualis*. *Plant Dis. Rep.* 61:728–730.

71. Jarvis, W.R., L.A. Shaw, & J.A. Traquair. 1989. Factors affecting antagonism of cucumber powdery mildew by *Stephanoascus flocculosus* and *S. rugulosus*. *Mycol. Res.* 92:162–165.

72. Jeffers, S.N., & T.S. Wright. 1994. Comparison of four promising biological

control agents for managing postharvest diseases of apples and pears. *Phytopathology* 84:1082.

73. Jetiyanon, K., P.A. Backman, N.K. Zidack, S. Tuzun, & J. Shaw. 1993. Use of a polysilicone surfactant for delivery of resistance-inducing strains of *Xanthomonas campestris* pvs. for control of black rot of cabbage. *Proc. 6th Int. Congr. Plant Pathol.* 6:65.

74. Johnson, K.B., V.O. Stockwell, R.J. McLaughlin, D. Sugar, J.E. Loper, & R.G. Roberts. 1993. Effect of antagonistic bacteria on establishment of honey bee-dispersed *Erwinia amylovora* in pear blossoms and on fire blight control. *Phytopathology* 83:995–1002.

75. Jones, D.A., & A. Kerr. 1989. *Agrobacterium radiobacter* strain K1026, a genetically engineered derivative of strain K84, for biological control of crown gall. *Plant Dis.* 73:15–18.

76. Kempe, J., & L. Sequeira. 1983. Biological control of bacterial wilt of potatoes: attempts to induce resistance by treating tubers with bacteria. *Plant Dis.* 67:499–503.

77. Kloepper, J.W., G. Wei, & S. Tuzun. 1992. Rhizosphere population dynamics and internal colonization of cucumber by plant growth-promoting rhizobacteria which induce systemic resistance to *Colletotrichum orbiculare*. In *Biological Control of Plant Diseases: Progress and Challenges for the Future*. E.C. Tjamos, G.C. Papavizas, & R.J. Cook, eds. New York: Plenum Press, pp 185–191.

78. Kokalis-Burelle, N., P.A. Backman, R. Rodriguez-Kábana, & L.D. Ploper. 1992. Potential for biological control of early leafspot of peanut using *Bacillus cereus* and chitin as foliar amendments. *Biol. Control* 2:321–328.

79. Kosaka, Y., & T. Fukunishi. 1994. Application of cross-protection to the control of black soybean mosaic disease. *Plant Dis.* 78:339–341.

80. Leeman, M., F.M. Den Ouden, J.A. Van Pelt, C. Cornelissen, A. Matamala-Garros, P.A.H.M. Bakker, & B. Schippers. 1995. Suppression of Fusarium wilt of radish by co-inoculation of fluorescent *Pseudomonas* spp. and root colonizing fungi. *Eur. J. Plant Pathol.* 102: 21–31.

81. Leeman, M., J.A. Van Pelt, F.M. Den Ouden, M. Heinsbroek, P.A.H.M. Bakker, & B. Schippers. 1995. Induction of systemic resistance by *Pseudomonas fluorescens* in radish cultivars differing in susceptibility to Fusarium wilt, using a novel bioassay. *Eur. J. Plant Pathol.* (in press).

82. Leeman, M., Van Pelt, J.A. Hendrickx, M.J., Scheffer, R.J., Bakker, P.A.H.M., & Schippers, B. 1995c. Biocontrol of Fusarium wilt of radish in commercial greenhouse trials by seed treatment with *Pseudomonas fluorescens* WCS374. *Phytopathology* 85:1301–1305.

83. Lemanceau, P., P.A.H.M. Bakker, W.J de Kogel, C. Alabouvette, & B. Schippers. 1992. Effect of pseudobactin 358 production by *Pseudomonas putida* WCS358 on suppression of Fusarium wilt of carnations by nonpathogenic *Fusarium oxysporum* Fo47. *Appl. Environ. Microbiol.* 58:2978–2982.

84. Lemanceau, P., P.A.H.M. Bakker, W.J. DeKogel, C. Alabouvette, & B. Schippers. 1993. Antagonistic effect of nonpathogenic *Fusarium oxysporum* Fo47

and pseudobactin 358 upon pathogenic *Fusarium oxysporum* f. sp. *dianthi*. *Appl. Environ. Microbiol.* 59:74–82.

85. Levy, E., Z. Eyal, & I. Chet. 1988. Suppression of *Septoria tritici* leaf blotch and leaf rust on wheat seedling leaves by pseudomonads. *Plant Pathol.* 37:551–557.

86. Lindemann, J. 1985. Genetic manipulation of microorganisms for biological control. In *Biological Control on the Phylloplane*. ed. C.E. Windels & S.E. Lindow. Minneapolis, Minnesota: American Phytopathological Society, pp 116–130.

87. Lindow, S.E. 1992. Integrated control of frost injury, fire blight, and fruit russet of pear with a blossom application of an antagonistic bacterium. *Phytopathology* 82:1129.

88. Liu, L., J.W. Kloepper, & S. Tuzun. 1992. Induction of systemic resistance against cucumber mosaic virus by seed inoculation with select rhizobacteria strains. *Phytophathology* 82:1108.

89. Liu, L., J.W. Kloepper, & S. Tuzun. 1995. Induction of systemic resistance in cucumber against Fusarium wilt by plant growth-promoting rhizobacteria. *Phytopathology* 85:695–698.

90. Liu, L., J.W. Kloepper, & S. Tuzun. 1995. Induction of systemic resistance in cucumber against bacterial angular leaf spot by plant growth-promoting rhizobacteria. *Phytopathology* 85:843–847.

91. Lumsden, R.D., & J.C. Locke. 1989. Biological control of damping-off caused by *Pythium ultimum* and *Rhizoctonia solani* with *Gliocladium virens* in soilless mix. *Phytopathology* 79:361–366.

92. Lumsden, R.D., & J.F. Walter. 1995. Development of the biocontrol fungus *Gliocladium virens*: risk assessment and approval for horticultural use. In *Biological Control: Benefits and Risks*. H.M. T. Hokkanen & J.M. Lynch, eds. Cambridge, UK: Cambridge University Press, pp 263–269.

93. Lumsden, R.D., J.A. Lewis, & J.C. Locke. 1993. Managing soilborne fungal pathogens with fungal antagonists. *Pest Management: Biologically Based Technologies*. R.D. Lumsden & J.L. Vaughn, eds. Washington, DC: American Chemical Society, pp 196–203.

94. Lumsden, R.D., J.A. Lewis, & D.R. Fravel. 1995. Formulation and delivery of biocontrol agents for use against soilborne pathogens. *Biological Pest Control Agents: Formulation and Delivery*. F.R. Hall & J.W. Barry, eds. Washington, DC: American Chemical Society.

95. Mahaffee, W.F., & J.W. Kloepper. 1994. Applications of plant growth-promoting rhizobacteria in sustainable agriculture. In *Soil Biota: Management in Sustainable Farming Systems*. C.E. Pankhurst, B.M. Doube, V.V.S.R. Gupta, P.R. Grace, eds. Adelaide, Australia: CSIRO.

96. Martin, F.N., & J.G. Hancock. 1987. The use of *Pythium oligandrum* for biological control of preemergence damping-off caused by *P. ultimum*. *Phytopathology* 77:1013–1020.

97. Maurhofer, M., C. Hase, P. Meuwly, J.-P. Metraux, & G. Defago, 1994. Induction of systemic resistance of tobacco to tobacco necrosis virus by the root-colonizing *Pseudomonas fluorescens* strain CHAO: Influence of the *gacA* gene and of pyoverdine production. *Phtyopathology* 84:139–146.

98. McKeen, C.D., C.C. Reilly, & P.L. Pusey. 1986. Production and partial characterization of antifungal substances antagonistic to *Monilinia fructicola* from *Bacillus subtilis*. *Phytopathology* 76:136–139.

99. McLaughlin, R.J., & R.G. Roberts. 1992. Biological control of fire blight with *Pseudomonas fluorescens* strain A506 and two strains of *Erwinia herbicola*. *Phytopathology* 82:1129.

100. McLaughlin, R.J., C.L. Wilson, S. Droby, R. Ben-Arie, & E. Chalutz. 1992. Biological control of postharvest diseases of grape, peach, and apple with the yeasts *Kloekera apiculata* and *Candida guilliermondii*. *Plant Dis.* 76:470–473.

101. McMillan, R.T., & D. Gonsalves. 1987. Effectiveness of cross-protection by a mild mutant of papaya ringspot virus for control of ringspot disease of papaya in Florida. *Proc. Fla. State Hortic. Soc.* 100:294–296.

102. Mercier, J., & C.L. Wilson. 1994. Colonization of apple wounds by naturally occurring microflora and introduced *Candida oleophila* and their effect on infection by *Botrytis cinerea* during storage. *Biol. Control* 4:138–144.

103. Mew, T.W., A.M. Rosales, & G.V. Maningas. 1994. Biological control of Rhizoctonia sheath blight and blast of rice. In *Improving Plant Productivity with Rhizosphere Bacteria*. M.H. Ryder, P.M. Stephens, & G.D. Bowen, eds. Adelaide, Australia: CSIRO, pp 9–13.

104. Mohammadi, O. 1994. Commercial development of Mycostop fungicide. In *Improving Plant Productivity with Rhizobacteria*. M.H. Ryder, P.M. Stephens & G.D. Bowen, eds. Adelaide, Australia: CSIRO, pp 282–284.

105. Mohammadi, O., & M.L. Lahdenpera. 1994. Impact of application method on efficacy of Mycostop biofungicide. In *Improving Plant Productivity with Rhizosphere Bacteria*. M.H. Ryder, P.M. Stephens, G.D. Bowen, eds. Adelaide, Australia: CSIRO, pp 279–281.

106. Moore, L.W., & G. Warren. 1979. *Agrobacterium radiobacter* strain 84 and biological control of crown gall. *Annu. Rev. Phytopathol.* 17:163–179.

107. Morris, A.C., J.R. Coley-Smith, & J.M. Whipps. 1995. The ability of the mycoparasite *Verticillium biguttatum* to infect *Rhizoctonia solani* and other pathogenic fungi. *Mycol. Res.* 99:997–1003.

108. Musson, G., & J.W. Kloepper. 1994. Development of delivery systems for introducing endophytic bacteria into cotton. *Phytopathology* 84:1112–1113.

109. Naik, M.K., & B. Sen. 1992. Biocontrol of plant disease caused by *Fusarium* species. In *Recent Developments in Biocontrol of Plant Disease*. K.G. Mukerji, J.P. Tewari, D.K. Arora, & G. Saxena, eds. New Delhi: Aditya Books, pp 38–51.

110. Nelson, E.B. 1988. Biological control of *Pythium* seed rot and premergence damping-off of cotton with *Enterobacter cloacae* and *Erwinia herbicola* applied as seed treatments. *Plant Dis.* 72:140–142.

111. Nemec, S., & L. Datnoff. 1993. Pepper and tomato cultivar responses to inoculation with *Glomus intraradices*. *Adv. Hortic. Sci.* 7:161–164.

112. Peixoto, A.R., R.L.R. Mariano, & S.J. Michereff. 1994. Screening of fluorescent *Pseudomonas* spp. for the control of *Pseudomonas solanacearum* on tomato. In *Improving Plant Productivity with Rhizosphere Bacteria*. M.H. Ryder, P.M. Stephens, & G.D. Bowen, eds. Adelaide, Australia: CSIRO, pp 51–53.

113. Peng, G., J.C. Sutton, & P.G. Kevan. 1992. Effectiveness of honeybees for applying the biocontrol agent *Gliocladium roseum* to strawberry flowers to suppress *Botrytis cinerea*. *Can. J. Plant Pathol.* 14:117–129.

114. Pusey, P.L., & C.L. Wilson. 1984. Postharvest biological control of stone fruit brown rot by *Bacillus subtilis*. *Plant Dis.* 68:753–756.

115. Pusey, P.L., C.L. Wilson, M.W. Hotchkiss, & J.D. Franklin. 1986. Compatibility of *Bacillus subtilis* for postharvest control of peach brown rot with commercial fruit waxes, dicloran, and cold-storage conditions. *Plant Dis.* 70:587–590.

116. Pusey, P.L., M.W. Hotchkiss, H.T. Dulmage, R.A. Baumgardner, E.I. Zehr, C.C. Reilly, & C.L. Wilson. 1988. Pilot tests for commercial production and application of *Bacillus subtilis* (B-3) for postharvest control of peach brown rot. *Plant Dis.* 72:622–626.

117. Schenck, N.C. 1987. Vesicular-arbuscular mycorrhizal fungi and the control of fungal root diseases. In *Innovative Approaches to Plant Disease Control*. I. Chet, ed. New York: Wiley, pp. 179–191.

118. Seddon, B., & S.G. Edwards. 1993. Analysis of and strategies for the biocontrol of *Botrytis cinerea* by *Bacillus brevis* on protected chinese cabbage. *IOBC Bull.* 16:38–41.

119. Sherwood, J.L., & R.W. Fulton. 1982. The specific involvement of coat protein in tobacco mosaic virus cross protection. *Virology* 119:150–158.

120. Sivan, A., & I. Chet. 1986. Biological control of *Fusarium* spp. in cotton, wheat and muskmelon by *Trichoderma harzianum*. *J. Phytopathol.* 116:39–47.

121. Smilanick, J.L., & R. Denis-Arrue. 1992. Control of green mold of lemons with *Pseudomonas* species. *Plant Dis.* 76:481–485.

122. Smith, J. 1994. Genetic diversity within *Burkholderia solanacearum* race 3 and the development of a biological control agent. *Mol. Ecol.* 3:607.

123. Spurr, H.W. 1981. Experiments on foliar disease control using bacterial antagonists. In *Microbial Ecology of the Phylloplane*. J.P. Blakeman, ed. London: Academic Press, pp 369–381.

124. Spurr, H.W. & G.R. Knudsen. 1985. Biological control of leaf diseases with bacteria. In *Biological Control on the Phylloplane*. C.E. Windels & S.E. Lindow, eds. Minneapolis, Minnesota: American Phytopathological Society Press, pp 45–62.

125. Stevens, C., V.A. Khan, J.Y. Lu, C.L. Wilson, P.L. Pusey, M.K. Kahwe, Y. Mafolo, J. Liu, E. Chalutz, & S. Droby. 1994. Reduction of storage rots of fruits and vegetable by combining UV-C application and biocontrol strategies. *Phytopathology* 84:1152–1153.

126. Stevenson, W.R., R.V. James, G. Yuen, J.L. Parke, & A.E. Joy. 1994. Evaluation of biological control of white mold on snap bean. *Biol. Cult. Tests* 9:46.

127. Stockwell, V.O., K.B. Johnson, & J.E. Loper. 1992. Establishment of bacterial antagonists on blossoms of pear. *Phytopathology* 82:1128.

128. Stockwell, V.O., K.B. Johnson, & J.E. Loper. 1993. Compatibility of bacterial antagonists with antibiotics used for management of fire blight. *Phytopathology* 83:1383.

129. Stzjenberg, A., S. Galper, S. Mazar, & N. Lisker. 1989. *Ampelomyces quisqualis*

for biological and integrated control of powdery mildews in Israel. *J. Phytopathology* 124:285–295.

130. Sutton, J.C. 1994. Biocontrol of aerial plant diseases; perspectives and application of epidemiology and microbiol ecology. In *Proceedings of the 4th Siconbiol Symposium on Biological Control.* Granado, RS, Brasil, EMBRAPA/CPACT: Documents, 6, pp 140–150.

131. Sutton, J.C. 1995. Evaluation of micro-organisms for biocontrol: *Botrytis cinerea* and strawberry, a case study. In *Advances in Plant Pathology.* Arizona Press, pp 171–188.

132. Sutton, J.C., & G. Peng. 1993. Biocontrol of *Botrytis cinerea* in strawberry leaves. *Phytopathology* 83:615–621.

133. Sutton, J.C., & G. Peng. 1993. Manipulation and vectoring of biocontrol organisms to manage foliage and fruit diseases in cropping systems. *Annu. Rev. Phytopathol.* 31:473–493.

134. Sutton, J.C. & G. Peng. 1993. Biosuppression of inoculum production by *Botrytis cinerea* in strawberry leaves. *IOBC Bull.* 16:47–52.

135. Tang, W.H. 1994. Yield-increasing bacteria (YIB) and biocontrol of sheath blight of rice. In *Improving Plant Productivity with Rhizosphere Bacteria.* M.H. Ryder, P.M. Stephens, G.D. Bowen, eds. Adelaide, Australia: CSIRO, pp 267–273.

136. Thingaard, K., H. Larsen, & J. Hockenhull. 1988. Antagonistic *Pythium* against pathogenic *Pythium* on cucumber roots. *EPPO Bull.* 18:91–94.

137. Thomson, S.V., D.R. Hansen, K.M. Flint, & J.D. Vandenberg. 1992. Dissemination of bacteria antagonistic to *Erwinia amylovora* by honey bees. *Plant Dis.* 76:1052–1056.

138. Trigalet, A., & D. Trigalet-Demery. 1990. Use of avirulent mutants of *Pseudomonas solanacearum* for the biocontrol of bacterial wilt of tomato plants. *Physiol. Mol. Plant Pathol.* 36:27–38.

139. Trutman, P., P.J. Keane, & P.R. Merriman. 1982. Biological control of *Sclerotinia sclerotiorium* on aerial parts of plants by the hyperparasite *Coniothyrium minitans. Trans. Br. Mycol. Soc.* 78:521–529.

140. Turner, J.T., & P.A. Backman. 1991. Factors relating to peanut yield increases after seed treatment with *Bacillus subtilis. Plant Dis.* 75:347–353.

141. Tuzun, S., & J.W. Kloepper. 1995. Practical application and implementation of induced resistance. In *Induced Resistance to Disease in Plants.* R. Hammerschmidt & J. Kuć, eds. The Netherlands: Kluwer, pp 152–168.

142. Tuzun, S., J. Juarez, W.C. Nesmith, & J. Kuć. 1992. Induction of systemic resistance in tobacco against metalaxyl-tolerant strains of *Peronospora tabacina* and the natural occurrence of the phenomenon in Mexico. *Phytopathology* 82:425–429.

143. Van Dijk, K.V., & E.B. Nelson. 1994. Loss of biological control ability in an *Enterobacter cloacae* mutant unable to catabolize linoleic acid. *Phytopathology* 84:1091.

144. Van Peer, R., G.J. Niemann, & B. Schippers. 1991. Induced resistance and phytoalexin accumulation in biological control of Fusarium wilt of carnation by *Pseudomonas* sp. strain WCS417r. *Phytopathology* 81:728–734.

145. Verhaar, M.A., P.A.C. Van Strien, & T. Hijwegen. 1993. Biological control of cucumber powdery mildew (*Sphaerotheca fuliginea*) by *Verticillium lecanii* and *Sporothrix cf. flocculosa*. *IOBC Bull*. 16:79–81.

146. Walkley, D.G.A., H. Lecoq, R. Collier, & S. Dobson. 1992. Studies on the control of zucchini yellow mosaic virus in courgettes by mild strain protection. *Plant Pathol*. 41:762–771.

147. Wang, H.-L., S.-D. Yeh, R.-J. Chiu, & D. Gonsalves. 1987. Effectiveness of cross-protection by mild mutants of papaya ringspot virus for control of ringspot disease of papaya in Taiwan. *Plant Dis*. 71:491–497.

148. Wei, G., J.W. Kloepper, & S. Tuzun. 1991. Induction of systemic resistance of cucumber to *Colletotrichum orbiculare* by select strains of plant growth-promoting rhizobacteria. *Phytopathology* 81:1508–1512.

149. Weller, D.M. 1988. Biological control of soilborne plant pathogens in the rhizosphere with bacteria. *Annu. Rev. Phytopathol*. 26:379–407.

150. Weller, D.M., & R.J. Cook. 1983. Suppression of take-all of wheat by seed treatments with fluorescent pseudomonads. *Phytopathology* 73:463–469.

151. Weller, D.M., B.X. Zhang, & R.J. Cook. 1985. Application of a rapid screening test for selection of bacteria suppressive to take-all of wheat. *Plant Dis*. 69:710–713.

152. Whipps, J.M. 1993. A review of white rust (*Puccinia horiana* Henn.) disease on chrysanthemum and the potential for its biological control with *Verticillium lecanii* (Zimm.) Viegas. *Ann. Appl. Biol*. 122:173–187.

153. Whipps, J.M., & M. Gerlagh. 1992. Biology of *Coniothyrium minitans* and its potential for use in disease biocontrol. *Mycol. Res*. 96:897–907.

154. Whipps, J.M., & R.D. Lumsden. 1991. Biological control of *Pythium* species. *Biocontrol Sci. Technol*. 1:75–90.

155. Whipps, J.M., S.P. Budge, & S.J. Mitchell. 1993. Observations on sclerotial mycoparasites of *Sclerotinia sclerotiorum*. *Mycol. Res*. 97:697–700.

156. Wilson, C.L., & P.L. Pusey. 1985. Potential for biological control of postharvest plant diseases. *Plant Dis*. 69:375–378.

157. Wilson, C.L., & M.E. Wisniewski. 1989. Biological control of postharvest diseases of fruits and vegetables: an emerging technology. *Annu. Rev. Phytopathol*. 27:425–441.

158. Wilson, M., & S.E. Lindow. 1993. Interactions between the biological control agent *Pseudomonas fluorescens* A506 and *Erwinia amylovora* in pear blossoms. *Phytopathology* 83:117–123.

159. Wilson, M., E.K. Crawford, & R. Campbell. 1987. Biological control by *Trichoderma harzianum* of damping-off of lettuce caused by *Rhizoctonia solani*. *EPPO Bull*. 18:83–89.

160. Wilson, M., H.A.S. Epton, & D.C. Sigee. 1990. Biological control of fire blight of hawthorn (*Crataegus monogyna*) with *Erwinia herbicola* under protected conditions. *Plant Pathol*. 39:301–308.

161. Xu, G.W., & D.C. Gross. 1986. Field evaluation of the interactions among fluorescent pseudomonads, *Erwinia carotovora*, and potato yields. *Phytopathology* 76:423–430.

162. Xu, G.W., & D.C. Gross, 1986. Selection of fluorescent pseudomonads antago-
      nistic to *Erwinia carotovora* and suppressive of potato seedpiece decay. *Phyto-
      pathology* 76:414–422.
163. Yeh, S.-D., H.-L. Wang, R.-J. Chiu, & D. Gonsalves. 1986. Control of papaya
      ringspot virus by seedling inoculation with mild virus strains. *FFTC Extension
      Bull. No. 662.*
164. Yuen, G.Y., M.L. Craig, E.D. Kerr, & J.R. Steadman. 1994. Influences of
      antagonist population levels, blossom development stage, and canopy tempera-
      ture on the inhibition of *Sclerotinia sclerotiorum* on dry edible bean by *Erwinia
      herbicola*. *Phytopathology* 84:495–501.
165. Yuen, G.Y., G. Godoy, J.R. Steadman, E.D. Kerr, & M.L. Craig. 1991. Epi-
      phytic colonization of dry edible bean by bacteria antagonistic to *Sclerotinia
      sclerotiorum* and potential for biological control of white mold disease. *Biol.
      Control* 1:293–301.
166. Yuen, G.Y., M.L. Craig, & L.J. Giesler. 1994. Biological control of *Rhizocto-
      nia solani* on tall fescue using fungal antagonists. *Plant Dis.* 78:118–123.
167. Zhou, T., & R.D. Reeleder. 1989. Application of *Epicoccum purpurascens* spores
      to control white mold of snap bean. *Plant Dis.* 73:639–642.
168. Zhou, T., R.D. Reeleder, & S.A. Sparace. 1991. Interactions between *Sclerotinia
      sclerotiorum* and *Epicoccum purpurascens*. *Can. J. Bot.* 69:2503–2510.

13
Chemical Approaches to Managing Plant Pathogens

Wolfram Köller
Cornell University, Geneva, New York

I. INTRODUCTION

A. History of Chemical Disease Control

Plant diseases, particularly the smuts and rusts, have been described throughout history as serious problems in food and fiber production. However, it was not until the late 1900s that microorganisms were discovered as the causal agents of plant diseases. Microorganisms causing plant as well as human diseases include fungi, bacteria, and viruses, although the ranking in importance is different, with fungal organisms causing the vast majority of plant diseases. Although fungi comprise the most important group of plant pathogens, the number of infectious species is relatively small. Among the over 100,000 fungal organisms described, only a small fraction (approximately 200) is known to colonize living plants.

The long-recognized importance of fungi as the major cause of plant diseases makes it understandable that chemical countermeasures were and still are largely restricted to fungal pathogens. Very few bactericides are commercially available for control of bacterial plant diseases, and similar to human diseases caused by viruses, chemicals interfering specifically with viral reproduction are in their infancy of discovery. The first remedy for fungal plant diseases was the dusting of plants with sulfur long before fungi were discovered as causal organisms. The dawn of a more systematic development of chemical fungicides was the 1885 introduction of the Bordeaux mixture, a mixture of copper sulfate with calcium hydroxide (lime). Similar to sulfur, copper fungicides are still in wide use. The transition from these early inorganic compounds to organic fungicides was triggered by advances in synthetic organic chemistry and biological screening methodologies. The origin of fully synthetic fungicides can be dated to the 1940s when the dithiocarbamates were discovered.

In spite of their different chemistries and routes of discovery, the first generation of organic fungicides share one important feature with their inorganic predecessors: Their activity is not specific to fungal organisms, and any noticeable penetration into plant tissue would cause severe phytotoxic symptoms. Therefore, they must be confined to plant surfaces, and prophylaxis rather than therapy of fungal diseases is inherent to all nonspecific fungicides.

Fungicides useful in the therapy of diseases already established in host tissues must meet two criteria. First, they must penetrate through the plant cuticle, the hydrophobic surface layer protecting plant tissues, and then they must be translocated within the tissue in order to reach the growing pathogen. These properties are referred to as systemicity. Secondly, they must have a fungus-specific mode of action to avoid any interference with steps in plant growth and development. Cuticular uptake and translocation as the first requirement of systemicity is dependent on the physicochemical properties of a compound, whereas fungal specificity is mandated by biochemical parameters. It is thus not surprising that some of our modem fungicides are fungus-specific but not systemic.

The first specific and systemic fungicides—the carboxamides, the hydroxy-pyrimidines, and the benzimidazoles—were discovered and introduced in the late 1960s, and systemic fungicides remain the major focus of fungicide discovery and development efforts. The most serious drawback of specific fungicides was the sometimes rapid development of resistance. Although resistance already was commonplace among insect pests, resistance to all of our early and strictly protective fungicides with nonspecific modes of action never became a serious problem. In contrast, fungicide resistance has severely restricted the usefulness of almost all of our fungus-specific fungicides, and many of the older protective compounds remain indispensable in the current management of resistance to modern fungicides.

More recently, a fourth class of fungicides has gained in importance and attention. This new generation of chemicals is not targeted toward metabolic steps in fungal organisms and is neither fungicidal nor fungistatic. Rather, they interfere with specific steps in host-pathogen interactions either by preventing penetration of pathogens into host plants or by activating the defense system of plants against invading pathogens. They are, therefore, referred to as disease control agents with indirect modes of action. The major advantages of this fourth generation of chemicals are their long residual activities requiring less frequent applications and the current lack of resistance development.

B. Volume and Value of Chemical Disease Control

The early use of fungicides was largely restricted to the dressing of seeds for control of seed- and soilborne diseases; foliar applications were only common in horticultural crops such as grapes, banana, coffee, fruit and nut trees, and vegetables, with potatoes representing an early example of a field crop treated with fungicides. Although seed dressing remains a general practice, control of foliar diseases has grown in importance over the past 25 years and now includes numerous field crops such as rice and cereals.

In 1991, end users spent approximately 5.5 billion U.S. dollars worldwide for fungicides, a proportion representing approximately 20% of the entire pesticide sales [49]. Systemic fungicides represented the largest share of fungicides sold, followed by nonspecific organics and the inorganics sulfur and copper (Fig. 1A). The use of fungicides is surprisingly disproportionate on a global scale, with Western Europe consuming almost half of all fungicides sold (Fig 1B). This disproportion is also reflected in the spectrum of crops routinely treated with fungicides. In 1991, almost 40% of all fungicides were used in cereal (particularly wheat) and rice production [49] (Fig. 1C). It must also be noted that within the segment of horticultural crops, almost half of the fungicides were used for control of diseases caused by the Oomycetes, with over 80% of the oomycete fungicides applied to grapes, potatoes, and vegetables [67].

In summary, almost half of all fungicides used in 1991 were applied to wheat, rice, and grapes. This emphasis on only a few crops is explained by the globally disproportionate use pattern. For example, 78% of all fungicides applied to wheat were used in Western Europe, which contributes 13% to world wheat production [57]. In comparison, only 3% of the wheat fungicides were used in North America, which produces 17% of the world wheat supply. Very similarly, 66% of rice fungicides were used in Japan, which contributes less than

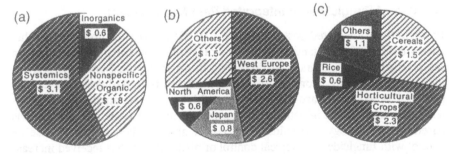

Figure 1   Volume of chemical disease control in 1991. (a) Distribution among different classes of fungicides. (b) Global use of fungicides. (c) Crops treated with fungicides.

10% to global rice production [57]. This disproportionate use pattern is largely explained by differences in the intensity of agricultural production, which in turn is influenced by peculiarities of agricultural price support systems implemented in different countries.

A list of the economically most important diseases is provided in Table 1. Fungal plant pathogens belong, like all fungi, to the kingdom Myceta, divided into the lower and higher fungi. Among the lower fungi, the Oomycetes are by far the most important plant pathogens. As summarized by Griffith et al. [23], the phylogenetic relationship between Oomycetes and the higher fungi is more distant than anticipated in the past, and this relative unrelatedness makes it understandable that many of the modern fungicides used to control Oomycetes are not active against diseases caused by the higher fungi (and vice versa).

The higher fungi include three groups: the Ascomycetes, the Deuteromycetes (imperfect fungi), and the Basidiomycetes, with important plant pathogens included in each group. Many pathogens belong to the Deuteromycetes, because their sexual stages have not been identified. The majority of pathogens listed in this group of imperfect fungi share similarities with the Ascomycetes and, in most cases, they have been classified as Ascomycetes once the perfect stage was identified. Sometimes confusing to practitioners engaged in plant disease control, new classifications are accompanied by new names assigned to the sexual stages of such species. According to general convention, the names of sexual stages are supposed to be used. To mention only a few examples, *Fusarium solani* is now referred to as *Nectria hematococca*, *Pyricularia oryzae* became *Magnaporthe grisea*, and *Botrytis cinerea* became *Botryotinia fuckeliana*. With the exception of *Rhizoctonia solani*, now named *Tanatephorus cucumeris*, the Basidiomycetes have been spared from major name changes. In Table 1, an attempt has been made to find a compromise between scientific convention and the terminology most familiar to practitioners.

C. Use of Fungicides in Integrated Pest Management

By definition, integrated pest management (IPM) is a pest control strategy that uses a combination of tactics without relying solely on chemical pesticides in the production of safe and nutritious food in an economical and environmentally sustainable way. Control of plant diseases has been traditionally a combination of breeding disease-resistant crop cultivars, of sanitation defined as interference with disease initiation and propagation, of forecasting infection periods based on the biology and epidemiology of a given pathogen, and of disease control with fungicides. Biological control of plant diseases has received increasing attention, although economically viable and sufficiently effective biocontrol agents have not been developed for important crops (see Chap. 12). Although the definition of IPM calls for the integration of all available options of pest

Table 1 Important Pathogens and the Diseases They Cause in Crops

Class	Subdivisions/species	Diseases
Oomycetes	Phytophthora spp.	Potato late blight; root diseases
	Phythium spp.	Damping-off (seedlings), root diseases
Peronosporaceae		Downy mildews
	Plasmopara viticola	grape
	Perenospora nicotianae	tobacco
	Bremia lactuccae	lettuce
	Pseudoperonospora cubensis	Cucurbits
Ascomycetes, Deuteromycetes (= Imperfect Fungi)	Erysiphales	Powdery mildews
	Erysiphe spp.	cereals, cucurbits
	Podosphaera leucotricha	apple
	Sphaerotheca spp.	cucurbits, roses, peaches
	Uncinula necator	grapes
	Colletotrichum (Glomerella) spp.	Anthracnose
	Fusarium (Gibberella, Nectria) spp.	Foot and root rots, wilts, wheat scab
	Helminthosporium (Pyrenophora, Cochliobolus) spp.	Leaf spots (cereals, and vegetables), net blotch barley, leaf blight maize
	Septoria spp.	Leaf and glume blotch (wheat), leaf spots
	Venturia spp.	Scab (apple, pear, pecan)
	Monilinia spp.	Rots of stone fruits and nuts
	Sclerotinia spp.	Soft rot (vegetables)
	Alternaria spp.	Leaf spots and blights
	Cercospora (Mycosphaerella) spp.	Leaf spots and blights, sigatoka
	Botrytis cinerea	Gray mold
	Pyricularia oryzae	Rice blast
Basidiomycetes	Puccinia spp.	Rusts (cereals, maize)
	Uromyces spp.	Bean rusts
	Hemileia vastatrix	Coffee rust
	Ustilago spp.	Smuts
	Tilletia spp.	Bunts
	Rhizoctonia solani	Stem and root diseases, leaf blights

control, the concept remains primarily focused on the mere reduction of pesticide use. This goal can be achieved by reducing the number of applications of a given pesticide per season, supplemented with nonchemical tactics, or by employing more recently introduced low-rate pesticides for long-lasting control of a given pest. Both approaches have been discussed as being consistent with the IPM goal of pesticide use reductions.

A quantitative approach to the suitability of pesticides in IPM programs was attempted by Kovach et al. [46] by introducing a quantitative seasonal environmental impact quotient (EIQ) for particular pesticides, including fungicides. This first approach to a quantitative assessment of pesticide qualities in IPM programs has been criticized for several reasons: The rationale of the weighted formula used in the calculation of EIQs was questioned, the data used for EIQ calculations were not always of highest quality, and strategies for coping with pesticide resistance were not acknowledged. It was also questioned, whether static EIQ values assigned to a given pesticide could objectively reflect the risks and benefits of the pesticide in the specific context of a particular crop. The discussion demonstrates that objective data retrieval remains a major challenge.

A different approach to the assessment of fungicides in IPM programs might be to reemphasize the relatively old tradition of not applying a fungicide if it is not needed. Soil fungicides are not needed if sizes of pathogenic populations are too small to cause damage. Foliar fungicides are not needed if the environmental conditions are unfavorable for plant infections. The latter parameter is determined primarily by the duration of leaf wetness caused by rain and sometimes dew, by the temperatures recorded at leaf wetness episodes, and the amount of inoculum present at a particular site. The prediction of relevant infection periods mandating application of fungicides has been accomplished for many diseases. In spite of improvements made, it remains more practical to apply a therapeutic fungicide after an infection has occurred rather than to apply a prophylactic fungicide in anticipation of such an event.

The discussion of relative merits of prophylactic and therapeutic fungicides in IPM programs might become less important with the current discovery and development of fungicides that combine specific and systemic properties with lasting residual activities. A single application of such long-lasting fungicides would control diseases over long parts of the growing season. They could be applied prophylactically in anticipation of recurring infection periods and still would reduce the number of fungicide applications required for season-long control. In view of current progress being made in the quality of chemical fungicides, the engineering of disease-resistant crops, and the potential interplay of both technologies, the definition of IPM relating to disease control might experience some major adjustments in the future.

D. Purpose of This Chapter

The purpose of this chapter is to present a systematic overview of the fungicides currently available. It is organized according to chemical groups with similar modes of action and, consequently, similar restrictions and benefits in the control of plant diseases. A treatise of bactericides will not be provided, because chemical control of bacterial diseases is restricted to very few compounds not necessarily available in all countries.

Several comprehensive books on fungicides have been published. The two-volume set *Fungicides*, edited by Torgeson and published in 1967 [74], is still an invaluable source of information on older protective fungicides which remain in wide use. A second two-volume set *Antifungal Compounds*, edited by Siegel and Sisler [69], also presents detailed information on early fungicides, but it also covers several more general subject areas and comprehensive information on the first groups of systemic fungicides. The most current comprehensive account on fungicides, edited by Lyr, is entitled *Modern Selective Fungicides* [48]. A comprehensive review article dedicated to fungicides was provided by Buchenauer [8]. The book *Target Sites of Fungicide Action*, edited by Köller is focused on the modes of action of modern antifungals [42]. Specific references given in this chapter will only guide to more comprehensive information provided in the past; only new information not yet included in these sources will be referenced in greater detail. No specific references will be given if information is easily obtained from the sources listed above. The individual fungicide groups will be treated according to their benefits and restrictions inherent to their modes of action; brief comments also will be made regarding problems with resistance.

II. FUNGICIDES WITH NONSPECIFIC MODES OF ACTION

A. Restrictions and Benefits Inherent to a Nonspecific Mode of Action

With the exception of fluazinam, all fungicides with nonspecific modes of action were introduced prior to the late 1960s. Nonspecific action implies that numerous metabolic steps in fungal organisms are affected. All of our nonspecific fungicides bind to or irreversibly modify reactive groups of numerous enzymes and other biologically important compounds. The chemically most reactive group in biological systems is the thiol residue of cysteine, and indiscriminate blockage of essential cysteine residues is the major cause for the fungitoxic action of nonspecific fungicides. In some cases, older mode of action studies had indicated a more specific inhibition of particular enzymes. However, the significance of these more specific modes of action was rarely clarified.

The indiscriminate action of nonspecific fungicides makes it likely that metabolism in plant cells would also be inhibited if these fungicides were to penetrate plant tissue. Potentially detrimental uptake by plants is prevented by the plant cuticle, a hydrophobic layer covering all aboveground parts of herbaceous plants and leaves of woody plants [29]. The major function of cuticles is to protect plants from uncontrolled water loss; however, penetration of potentially phytotoxic fungicides also is prevented.

Confinement of nonspecific fungicides to the plant surface limits their usefulness in therapeutic applications, because the development of a pathogen will only be controlled until infections become established beneath the cuticle. Although the time span of pathogen establishment within the host tissue varies, after-infection activities of nonspecific fungicides are very restricted. Consequently, this class of fungicides is largely used in prophylactic programs by providing a protective deposit on cuticular surfaces. This mode of disease control demands that new plant growth must be protected by repeated applications at relatively narrow intervals. Although redistribution of protective fungicides by rain to new and unprotected plant surfaces has been documented, the contribution of this redistribution depends on the specifics of the crop to be protected. As a general rule, nonspecific fungicides must be applied in advance of infection periods to be protectively effective, and treatments must be repeated at relatively narrow intervals to protect new plant growth.

The benefit of a nonspecific mode of action has been the positive experience with fungicide resistance [41]. Although a detoxification mechanism leading to resistance to nonspecific fungicides is feasible and has indeed been documented in some laboratory studies [41], no protective fungicide has experienced serious problems with practical resistance. In fact, the continued wide use of these fungicides is partly explained by their value in the management of fungicide resistance to modern fungicides. The lack of resistance development can also be considered as evidence that more specific modes of action suggested for some of the protective fungicides are of low practical relevance. A second reason for the continued use of purely protective fungicides is the generally broad spectrum of diseases they control and the lack of registered alternatives, in particular for disease control in "minor crops."

B. Sulfur, Copper Fungicides, and Organometals

As mentioned above, elemental sulfur was the first chemical fungicide used on a broad scale. Sulfur was used as dust or sprayed as a suspension, primarily for control of powdery mildews but also for some other diseases. Sulfur is still widely used in agriculture, with more recent improvement of formulations providing more uniform suspensions and, thus, more uniform plant coverage.

Although several explanations exist for the antifungal mode of sulfur action, the most plausible theory implicates a slow air oxidation of sulfur to sulfur dioxide. In contact with moisture, sulfurous acid is formed, and the highly acidic environment on leaf surfaces prevents fungal spores from germinating. The speed of sulfur oxidation is temperature dependent and explains why sulfur efficacies are relatively low at cool temperatures. The acidic environment created by sulfur can cause severe phytotoxicity. It is well known that some plants (e.g., raspberries, many cucurbits) are "sulfur shy" and that "sulfur burns" on other crops can occur at high temperatures. The prolonged use of sulfur at a location may also require soil management to prevent soil acidification. The continued use of sulfur is mandated by low costs and by the broad certification of sulfur as a "natural" fungicide allowable for organic farming.

Copper fungicides are intimately connected with the discovery and broad introduction of the bordeaux mixture by Millardet in 1885, a fungicide originally used for the control of grape downy mildew in France. The Bordeaux mixture is a compound of soluble copper sulfate and calcium hydroxide (lime), forming water-insoluble copper hydroxide. Soluble copper ions are known to bind tightly to sulfhydryl groups such as cysteine residues and are thus biocidal. Unfortunately, free copper ions also penetrate through plant cuticles and cause severe phytotoxicity. The solution to this problem was to cover plant surfaces with a water-insoluble copper salt. Fungal spore germination is inhibited by these insoluble copper deposits, because active growth of germ tubes is accompanied by an acidification of the area immediately surrounding the spore. In these microscopically small acidified areas, sufficient amounts of copper ions are dissolved from the water-insoluble copper deposits, and further germ tube development is inhibited.

To minimize potential problems with phytotoxicity, numerous insoluble copper salts ("fixed coppers") have been developed and are still broadly used as economical fungicides for the control of many diseases. Control levels achieved are not always high, but they vary among diseases and are often sufficient. Copper fungicides are also used for the control of some bacterial diseases. In contrast to fungal diseases, resistance of bacteria to copper has become a recent problem [72]. Copper ions are not degraded in soil and can accumulate to high levels at locations with long copper fungicide histories. Like sulfur, the coppers are broadly certified as "natural" fungicides in organic farming.

Copper fungicides were not only developed and marketed as inorganic salts and oxides, but also as organometals, for example, as salts of fatty acids (copper linoleate) and as complexes with hydroxyquinolate (oxime copper). More important members of this intermediate class of organometals were the mercury fungicides (e.g., phenyl mercuric acetate), introduced and widely used in the early part of the century as seed dressing fungicides and as foliar sprays. Their high acute toxicity and accumulation in the environment led to a ban in all developed countries. Substantial use restrictions mandated by a relatively high applicator toxicity were also imposed on the "phenyltins" (fentin hydroxide and acetate).

Despite risks inherent to the class of organometals, they have positive attributes shared with some other early fungicides withdrawn in many developed countries. Their production is economical and relatively simple, patent protection has long expired, and levels of disease control are sufficient in many cases. It is, therefore, understandable that their production and use continues in several countries around the globe.

C. Carbamates and Ethylenebisdithiocarbamate Fungicides

The carbamates and ethylenebisdithiocarbamates (EBDCs), discovered in the 1930s and 1940s, were the first fully synthetic fungicides. The carbamates ferbam, ziram, and thiram still are used for control of a broad spectrum of diseases (Fig. 2). The second generation of carbamate fungicides were the EBDCs, of which maneb, zineb, mancozeb, metiram and propineb remain as widely used fungicides.

With the exception of thiram, both the carbamates and EBDCs were initially produced as sodium salts of the weakly acidic dithiocarbamic acid group. The discovery that salts of iron and, more importantly, manganese and zinc greatly improved the stability of these fungicides led to their final success. It is interesting to note that the mode of action of the carbamates might be different from

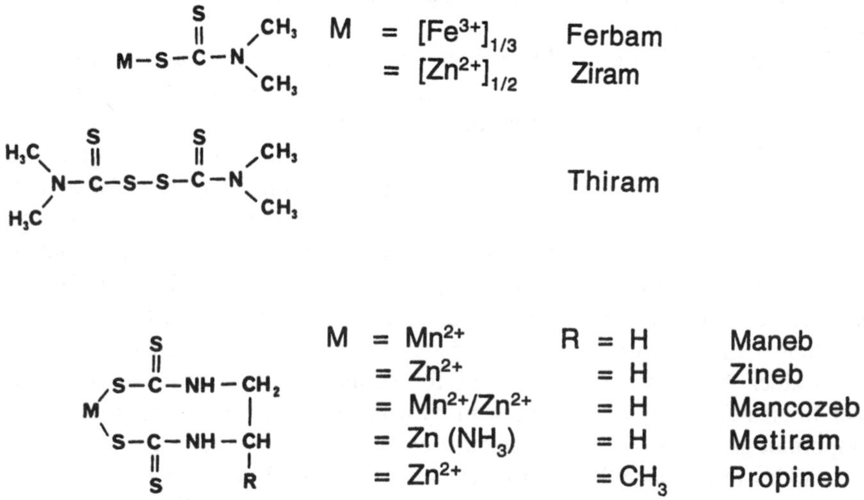

Figure 2 Carbamate fungicides.

the EBDCs. Complexation of copper and, thus, a function as a "copper carrier" was postulated to be important for the carbamates, whereas the slow generation of highly reactive ethylene diisothiocarbamate was generally assumed to be important for the EBDC mode of action [68]. Regardless of some unresolved questions, both groups of fungicides act as nonspecific inhibitors and are used for prophylactic control of fungal diseases.

The EBDCs are generally more effective than the carbamates, but both groups are broad-spectrum fungicides primarily used to control diseases caused by Oomycetes and Ascomycetes/Deuteromycetes. For unknown reasons, both groups are virtually inactive in the control of powdery mildews. Activity against Basidiomycetes is useful in some cases but is generally less pronounced than against other pathogens. The potential conversion of the EBDCs (with the exception of propineb) into ethylenethiourea (ETU), a potential carcinogen, has imposed use restrictions on EBDCs in several countries. Some of these restrictions relate to differences in national safety standards, and because these standards are periodically revised, restrictions on EBDC use are not static. Side effects of EBDCs on predatory mites have been reported [31, 32] and are of concern in some IPM programs.

D. Trichloromethylthiocarboximides

The fungicidal property of this class was discovered in the early 1950s, with captan being introduced as the first representative. Captafol, folpet, and the structurally different, but mechanistically related, dichlofluanid were introduced later (Fig. 3). Captan, captafol, and folpet registrations have been withdrawn in several, but not all, European countries because of some evidence for carcinogenic activities. Reregistration of captan was recently granted in the United

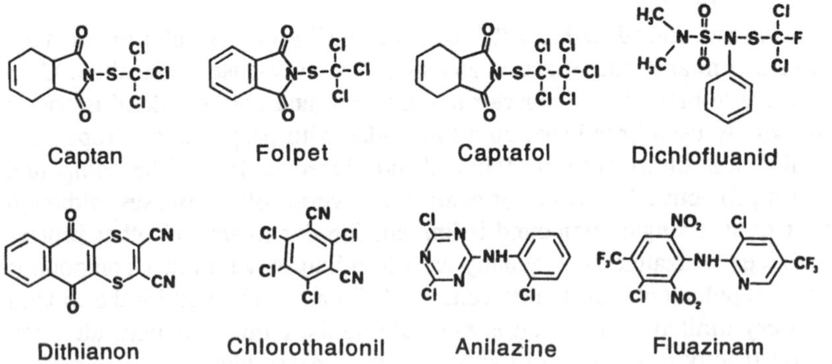

Figure 3 Trichloromethylthiocarboximides and miscellaneous fungicides with nonspecific modes of action.

States, with several use restrictions being imposed. In spite of several regulatory actions, these fungicides remain in wide use in many countries around the globe.

The fungitoxic activity of this class of fungicides is inherent to the $-NSCCl_3$ group in captan and captafol and to related groups in folpet and dichlofluanid (Fig. 3). The group reacts with bases such as thiol groups, and highly reactive thiophosgene is formed during this reaction. Thiophosgene can also be formed directly during water hydrolysis, particularly under alkaline conditions. Accessibility of the toxiphore to water hydrolysis makes these fungicides prone to slow degeneration in aqueous solutions. Penetration of the cuticle leading to phytotoxicity has been reported under certain conditions, especially when used in combination with mineral oils.

The trichloromethylthiocarboximide fungicides are used as seed dressing materials and as protective foliar fungicides. The target spectrum is broad and includes all fungal classes, with best control levels achieved for diseases caused by Ascomycetes/Deuteromycetes. As with the carbamates, powdery mildew control represents an exception.

E. Miscellaneous Nonspecific Fungicides

Some of the important nonspecific fungicides are chemically unrelated; structures of these compounds are given in Figure 3. Dithianon was introduced in 1963 as a protective fungicide primarily against a broad spectrum of foliar diseases caused by Ascomycetes/Deuteromycetes, although activity against some Oomycetes and rusts has been reported. As with many other protective compounds, powdery mildew activities are not sufficient. Dithianon reacts nonspecifically with thiol groups in biological systems and decomposes in alkaline spray solutions. Similar to captan, phytotoxicity has been reported in applications with mineral oils.

Chlorothalonil, introduced in 1964, contains nitril groups similar to dithianon and shares a similar mode of action and spectrum of diseases controlled. Control levels are in many cases superior to dithianon, and chlorothalonil is one of the most widely used broad-spectrum fungicides with prophylactic properties.

Anizaline was discovered in 1955 and introduced in 1966. The compound is used as a protective fungicide for control of several foliar diseases, although practical use as a single compound is limited. Similar to several other protective fungicides, anizaline is frequently used in mixtures with other compounds in order to supplement insufficient control of diseases achieved by the mixing partner. Very limited information is available on the mode of action, although nonspecific reactivity toward sulfhydryl groups appears likely.

Fluazinam, introduced in 1987, is the latest addition to the arsenal of nonspecific fungicides. The spectrum of diseases controlled is broad, but full commercial development has not been achieved as yet. Fluazinam has good prophylactic but no therapeutic activities and inhibits, as surface deposit, spore germination and appressoria formation [44,45]. The mode of fluazinam action remains somewhat controversial. Originally, a simple uncoupling of mitochondrial electron transport observed with isolated mitrochondria was reported as an explanation of activity [5,73]. Later, additional reactivity toward sulfhydryl groups was suggested [2]. The most recent account considers both activities as important, dependent on the pathogen [3]. It must be emphasized that neither suggested mode of action is fungus specific.

III. FUNGICIDES WITH SPECIFIC MODES OF ACTION

A. Introduction

A distinction between nonspecific fungicides inhibiting numerous sites in fungal metabolism and fungicides that bind to or inhibit a single fungus-specific target site is not always easy to make, and a specific mode of action is not necessarily related to the year of introduction of a particular compound. However, two criteria met by all fungicides listed below suggest fungal specificity, even if their respective target sites remain elusive. Either systemic activity not accompanied with phytotoxicity was evident or problems with resistance have been encountered. Both criteria are strong evidence for a specific mode of action, although in many cases, detailed molecular information is not available.

B. Aromatic Hydrocarbons and Dicarboximides

1. Properties and Mode of Action

Aromatic hydrocarbons are a relatively old and chemically diverse group of fungicides and include diphenyl, o-phenylphenol, pentachlorophenol (PCP), dinocab, and chloroneb, which are of restricted current importance either because of resistance or because more efficacious alternatives became available. Dicloran (Fig. 4), introduced in 1959, remains in relatively wide use in many countries for pre- or postharvest control of a relatively narrow yet important spectrum of pathogens such as species of *Botrytis, Monilinia, Sclerotinia*, and *Sclerotia*. Quintozene (Fig. 4) better known as PCNB and etridiazole (Fig. 4) are both used as broad-spectrum soil and seed-dressing fungicides, with acceptable control of Oomycetes only achieved with etridiazole. Some quintozene is still used as a foliar fungicide with a disease spectrum similar to dicloran. The most recent addition to the group of aromatic hydrocarbons is tolcofos-methyl (Fig. 4), a phosphoorganic fungicide introduced in 1973 with high *Rhizoctonia*

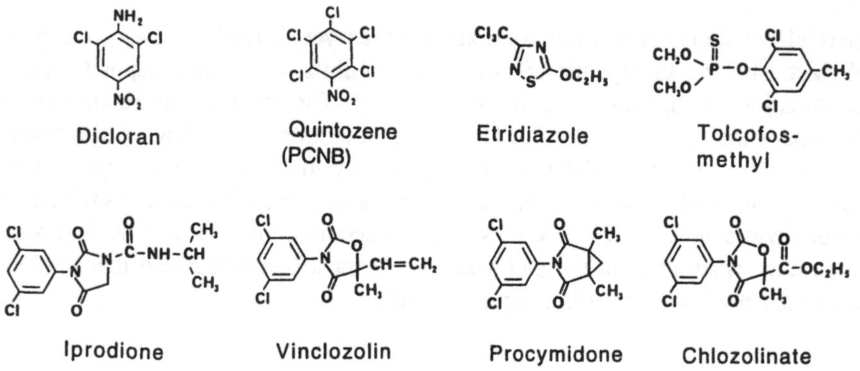

Figure 4 Aromatic hydrocarbons and dicarboximides.

specificity and in use for the control of several diseases caused by *Rhizocto-nia*, primarily in soil applications.

The structures of the aromatic hydrocarbons (see Fig. 4) described above are relatively unrelated, and they have been grouped together only because they share mechanistic similarities and relatively consistent cross-resistance patterns. These similarities extend to the group of dicarboximides introduced in the early 1970s, with iprodione, vinclozolin, procymidone, and chlozolinate in current use as commercial products (see Fig. 4). The spectrum of diseases controlled with dicarboximides is very similar to dicloran, although *Alternaria* and *Rhizoc-tonia* spp. must be added to the list of target pathogens. The dicarboximides are fungus-specific, but they are not systemic and are used essentially as pro-phylactic fungicides. Although spore germination is less sensitive than myce-lial growth when tested in vitro, full germ tube development required for in-fection is strongly affected when in contact with leaf surface deposits.

The mode of action of aromatic hydrocarbons and dicarboximides remains under discussion. One hypothesis, derived primarily from work with *Botrytis cinerea* and *Mucor mucedo*, is that active oxygen species generated under the influence of the fungicides leads to the peroxidation of unsaturated fatty acids thereby disturbing membrane functions. Several NADPH-dependent flavin en-zymes were shown to be inhibited in vitro, and this inhibition was considered to be responsible for the origin of active oxygen. The lipid peroxidation mecha-nism was not confirmed for *Ustilago maydis* [61]. A gene conferring dicarbo-ximide resistance in a laboratory mutant of *U. maydis* was cloned recently, and the protein product of the gene was identified as a putative protein kinase [60]. It remains to be clarified whether this protein kinase is involved in the mode of action or only in the mechanism of resistance.

In contrast to the results with *U. maydis*, recent evidence reported for *B. cinerea* and *Alternaria alternata* has related dicarboximide resistance at least partly to an enhanced level of oxidative protective enzymes such as catalase [70,71]. These findings are in support of the lipid peroxidation mechanism proposed originally. In summary, neither the mode of action nor the mechanism of resistance of the aromatic hydrocarbons and dicarboximides are fully understood at present.

2. Resistance

In general, dicarboximide-resistant isolates are also resistant to the aromatic hydrocarbons, although exceptions exist. Development of resistance to dicarboximides was first observed with *Botrytis cinerea* after several years of intensive dicarboximide use in European vineyards [41]. Practical problems remain largely restricted to this pathogen, although resistance of *Alternaria brassicicola* has been reported recently [30]. The development of resistance in *B. cinerea* was accompanied by the selection of isolates with relatively low resistance factors. Although levels of disease control with dicarboximides were impaired under these conditions, they remained satisfactory in many cases [47].

The frequency of isolates with relatively low levels of resistance decreased after dicarboximide use was discontinued, suggesting that resistance was accompanied by a fitness penalty. To allow resensitization of *Botrytis* populations between dicarboximide applications and, thus, to prolong efficacy of dicarboximides against gray mold, the number of applications allowed per season was restricted. The strategy was later modified by the recommendation to mix dicarboximides with a nonspecific protective fungicide active against the disease [41].

Although this strategy was successful for several years, it did not prevent the selection of a more stable resistant population. The decline of resistance between dicarboximide applications became less pronounced, and a greenhouse study suggested that the relative contribution of dicarboximides to the overall control achieved with fungicide mixtures can become minimal [75]. The current resistance situation is relatively stable in many European vineyards with long histories of dicarboximide use. One or two dicarboximide applications per season still provide noticeable gray mold control, but frequencies of resistant isolates increase rapidly to prohibitively high levels. Dicarboximide resistance has not yet been reported for *Sclerotinia*, *Sclerotia*, *Monilinia*, and *Rhizoctonia* spp. despite a long use history with these pathogens.

It is well known that under laboratory conditions, spontaneous mutations toward resistance to aromatic hydrocarbons and dicarboximides occur frequently in mycelial cultures of *Botrytis cinerea* and other fungi. However, the relevance of such observations to field populations remains largely unexplored. The question has gained in importance since laboratory mutants of *B. cinerea* were found to be cross resistant to the entirely unrelated new class of phenylpyrrole fungi-

cides to be described below. For the laboratory mutants, the trait of cross resistance segregated as a single or closely linked gene [27]. However, a large
number of dicarboximide-resistant field isolates tested was not cross resistant
to the phenylpyrrole fludioxonil [17,27]. This result implies that dicarboximide-
resistant mutants commonly generated in laboratory experiments express dissimilar mechanisms of resistance to those important in field populations.

C. Carboxamides and Hydroxypyrimidines

1. Properties and Mode of Action

The carboxamides and the hydroxypyrimidines are completely independent
classes of fungicides, and in the past, the two classes were treated separately
and in great detail. This emphasis is hardly justified in view of the current
importance of these fungicides; it simply reflected the fact that these two classes
were the first fully systemic and fungus-specific fungicides.

The first carboxamide representatives, carboxin and oxycarboxin (Fig. 5),
were introduced in 1966. Both compounds are still in use as commercial fungicides. Although several other carboxamides such as benodanil, mepronil,
fenfuram, and metsulfovax were introduced later, their current commercial
availability is restricted. Flutolanil (Fig. 5), introduced in 1981, and currently
used on a broader scale, and thifluzamide (Fig. 5), introduced in 1990 [59],
and under current commercial development, should be mentioned as late additions to this class of fungicides. All carboxamides are highly systemic and, thus,
are of therapeutic value.

The carboxamides are highly specific toward Basidiomycetes. Carboxin is
used as a seed dressing material against *Ustilago* and *Rhizoctonia* spp., whereas
oxycarboxin remains in limited use as a fungicide against rusts in greenhouse

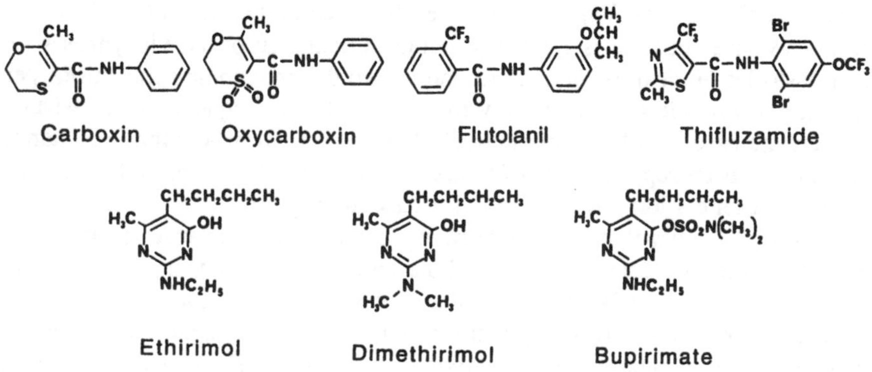

Figure 5 Carboxamides and hydroxypyrimidines.

production of ornamentals and, potentially, against coffee rust in the field. Flutolanil is used primarily as a foliar fungicide for control of rice sheath blight caused by *Rhizoctonia solani*, with good activity against several other diseases caused by Basidiomycetes. Thifluzamide shares this basidiomycete spectrum of diseases, with additional, although weak, effects on some diseases caused by Ascomycetes [59].

The carboxamide mode of action has been clarified in great detail as inhibition of mitochondrial succinate oxidation as part of the respiratory electron chain. The enzyme system affected is the succinate-ubiquinone oxidoreductase, also referred to as the mitochondrial complex II. Carboxamides bind to one of the complex II subunits and disrupt the electron transfer to ubiquinone. It was shown recently that the exchange of one amino acid in the carboxamide target protein is responsible for resistance of *Ustilago maydis* [6]. Carboxamides acted also as respiration inhibitors when tested with mitochondria isolated from plants and mammals, although doses higher than for fungal mitochondria were required to achieve this effect. Increased metabolism of the inhibitors might contribute to their high degree of fungal specificity. Specificity toward Basidiomycetes was attributed to uptake barriers present in other fungi rather than biochemical reasons; carboxamides with activity against Ascomycetes have been synthesized but not yet introduced.

The hydroxypyrimidines were introduced in 1968, with ethirimol, dimethirimol, and bupirimate (see Fig. 5) still in limited use as commercial products. All members of this class are highly specific toward powdery mildews and lack activity against other pathogens. This high specificity combined with free systemic movement made ethirimol a versatile tool for control of barley powdery mildew, and it was one of the first fungicides used in the control of foliar diseases on cereals. Seed dressing with ethirimol provided mildew control on barley seedlings for 6–10 weeks, and mildew control in later parts of the season was provided by a foliar application. Unfortunately, the control of wheat powdery mildew was much weaker. Ethirimol became less important as other cereal fungicides introduced in the late 1970s controlled powdery mildews on both barley and wheat and as problems with resistance emerged. Dimethirimol remains in limited use as powdery mildew fungicide in cucurbit production; bupirimate is still used against powdery mildew on some perennial crops such as apples.

The mode of hydroxypyrimidine action has been clarified to a certain degree. These fungicides inhibit adenosine deaminase, a ubiquitous enzyme involved in the biosynthesis of nucleic acids. How this mode of action relates to the pronounced specificity of hydroxypyrimidines toward powdery mildews remains largely unexplored. Moreover, the mechanism of powdery mildew resistance to the hydroxypyrimidines is not known.

2. Resistance

Both the carboxamides and the hydroxypyrimidines are fungus-specific inhibitors. This specificity contributed to the development of resistance, although the impact of practical resistance was not equally severe. For example, carboxin was primarily used as seed dressing against *Ustilago* spp., and although a target mutation conferring carboxin resistance in *U. maydis* [6] would normally signal a high risk of resistance, first evidence of the practical resistance of *U. nuda* was only reported after more than 20 years of continued use [41]. In general, carboxamide resistance of *Ustilago* spp. has not yet caused great concern.

This slow emergence of dicarboximide resistance is most likely explained by the slow generation time inherent to the biology of *Ustilago* spp. and by the fact that carboxins were increasingly replaced by other seed dressing fungicides. In contrast, rusts controlled with oxycarboxin developed resistance rapidly [41]. The lack of widespread problems with resistance in rusts is probably due to the very limited use of these fungicides in rust control. It must be mentioned, however, that resistance has not become a recognized problem in the control of foliar diseases caused by Rhizoctonia with flutolanil.

Resistance to hydroxypyrimidines was first reported for dimethirimol used against powdery mildew in greenhouse cucumber production [41]. Shortly after, a gradual shift of barley powdery mildew populations toward ethirimol resistance caused a substantial decline in efficacy. Problems with resistance were overcome by the introduction of the more efficacious class of azole fungicides. Reports indicating the decline of hydroxypyrimidine resistance in powdery mildew populations exist, although the practical impact of this decline remains largely unexplored. Resistance of apple powdery mildew to bupirimate has not become a problem.

D. Benzimidazoles

1. Properties and Mode of Action

The class of benzimidazole fungicides, first commercially introduced in the early 1970s, is considered as the breakthrough of fungus-specific and systemic fungicides valuable for after-infection therapy of plant diseases. In contrast to the highly disease-specific carboxamides and hydroxypyrimidines described above, the benzimidazoles combined systemic activities with an amazingly broad and yet unsurpassed spectrum of activities against diseases caused by Ascomycetes/Deuteromycetes. Activity against Basidiomyctes was less pronounced, and it was completely absent against Oomycetes. In spite of these limitations, the benzimidazoles with their systemic properties and therapeutic values became a rapidly accepted class of fungicides during the 1970s and 1980s. As described below, their value has decreased because of widespread resistance.

The first benzimidazoles discovered in the early 1960s were thiabendazole (TBZ) and fuberidazole. The use of these fungicides was largely restricted to the postharvest control of fruit diseases and to the seed dressing of cereals. Benzimidazoles broadly used as foliar fungicides were benomyl, thiophanate-methyl, and carbendazime, all introduced in the late 1960s (Fig. 6). As discovered later, these three compounds are closely related; both benomyl and thiophanate-methyl rapidly convert to carbendazime in aqueous solutions.

The benzimidazole mode of action was identified as the specific binding to fungal (but not plant and mammalian) β-tubulin. Binding to fungal β-tubulin prevents the polymerization of microtubules, which are primarily responsible for physical separation of dividing nuclei; consequently, benzimidazoles inhibit cell division. Negative side effects of benzimidazoles on plants were not observed; a hormonal cytokinin effect was often considered as a beneficial effect contributing to yield increases. The reason for benomyl phytotoxicity reported in the early 1990s in the United States is still not fully explained.

2. Resistance

It was discovered early that benzimidazole-resistant mutants of fungi could be easily generated in laboratory mutation experiments, and such mutants were readily recovered from field populations of pathogens shortly after exposure to benzimidazoles [41]. Typically, these mutants were entirely immune to any feasible inhibitor dose, and this high level of resistance was related to the absence of inhibitor binding to β-tubulin isolated from the mutants. The early warning signs for selection of highly resistant pathogenic populations became reality soon after the benzimidazoles were used on a large scale [41]. Resistance has by now affected the majority of diseases effectively controlled with benzimidazoles. The immediate response to this resistance situation was the recommendation to mix the benzimidazoles with older prophylactic fungicides in order to control the resistant subpopulations [41].

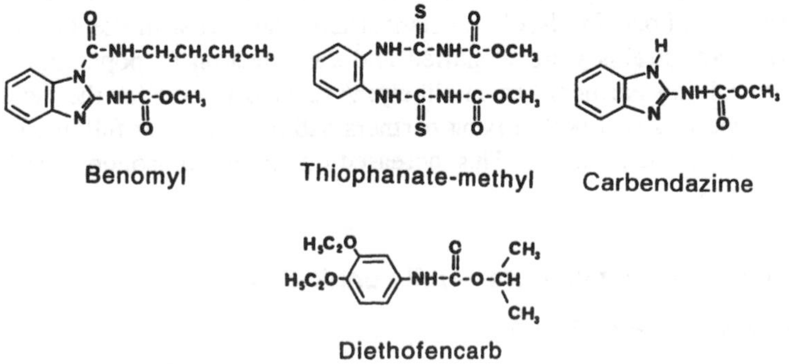

Benomyl Thiophanate-methyl Carbendazime

Diethofencarb

Figure 6 Benzimidazoles.

This antiresistance strategy relied heavily on relative contributions of both mixing partners to disease control. In many cases, contributions of benzimidazoles eroded and sometimes became insignificant. However, regional differences soon became apparent and remain important. Benzimidazoles are still used to control numerous plant diseases, usually in mixture with older protective fungicides. Relative benzimidazole contributions to overall disease control are rarely known in particular situations and might range from substantial to insignificant. Despite these limitations, benzimidazoles continue to be useful for the control of numerous diseases. Their original efficacies, however, have eroded substantially owing to the development of resistance.

A novel strategy to cope with fungicide resistance was also connected to the group of benzimidazoles. Certain phenylcarbamates with good antifungal activities against benzimidazole-resistant fungal mutants but without any activity against sensitive wild-type strains were discovered in 1980. These antagonists bound tightly to β-tubulins of resistant mutants but lacked affinity to the same target present in sensitive genotypes. The strong antagonism opened the opportunity to introduce a mixture of β-tubulin inhibitors with activities toward both benzimidazole-sensitive and benzimidazole-resistant isolates. A mixture containing the phenylcarbamate diethofencarb (see Fig. 6) was introduced for control of gray mold. Soon after, however, the first cases of *Botrytis cinerea* strains resistant to both β-tubulin inhibitors were reported [41,63]. The reason for the unexpected development of double resistance was recently clarified by Koenraadt et al. [39,40]. In an analysis of β-tubulin sequences of *Venturia inaequalis* field isolates with differing levels of benomyl resistance and diethofencarb sensitivities, four different mutational changes of amino acids in β-tubulin all conferred benzimidazole resistance. However, only two of these options were sensitive to diethofencarb.

A benzimidazole-diethofencarb mixture is currently used in the control of gray mold, primarily in Europe. In general, however, the novel strategy of mixing a benzimidazole with an antagonist such as diethofencarb to control both benzimidazole-sensitive and benzimidazole-resistant strains was met with skepticism after double-resistant strains were identified and the selection of populations resistant to both mixing partners became likely. A second reason for the lack of broad acceptance was that both mixing partners had to be used at full application rates in order to be effective. This increased rate imposed economic and regulatory handicaps.

E. Sterol Biosynthesis Inhibitors—DMI Fungicides

1. Properties and Mode of Action

The first antifungals acting as sterol demethylation inhibitors (DMIs)—miconazole, clotrimazol, and triarimol—were independently discovered and patented in 1968. None of these original DMIs were developed into agricultural fungicides, but they paved the way for an unprecedented number of commercial products used in agriculture and medicine. All aspects of DMI fungicides have been summarized in great detail elsewhere [8,42,48].

DMIs have become a widely used class of systemic fungicides in the control of a broad spectrum of plant diseases. Unlike other fungicides, however, this broad spectrum of activity is not necessarily inherent to a single DMI representative; rather, it is achieved by the development of over 30 DMI fungicides available worldwide (Figs. 7 and 8), all with slightly different properties and introduced over the past 20 years, with first commercial use in the mid 1970s. This long history of DMI development makes it understandable that commercial interest in early DMI representatives such as triadimefon or nuarimol has declined either because of resistance development or because of a relatively narrow spectrum of diseases controlled. The current importance of DMI fungicides lies in the control of cereal diseases in Europe. As an entire class, however, the DMI fungicides are widely used worldwide to control many diseases in a great number of crops.

DMIs are active against Ascomycetes/Deuteromycetes and Basidiomycetes but inactive against Oomycetes. The first generation of DMIs introduced in the late 1970s and early 1980s had excellent powdery mildew and rust activities on a variety of crops when used as foliar fungicides, and some had very potent activity against *Ustilago* and *Tilletia* spp. when used as seed dressing on cereals. Several additional seedborne cereal diseases were controlled in addition, and systemic activity protected barley and wheat seedlings from early powdery mildew infections.

The second generation of DMIs provided control of a broader spectrum of foliar diseases, not only on cereals but also on many other crops. At present,

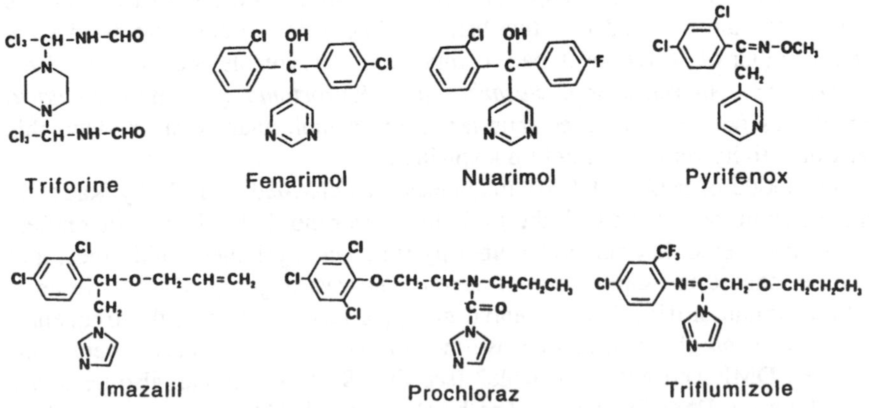

Figure 7 DMI fungicides containing an essential group other than a triazole.

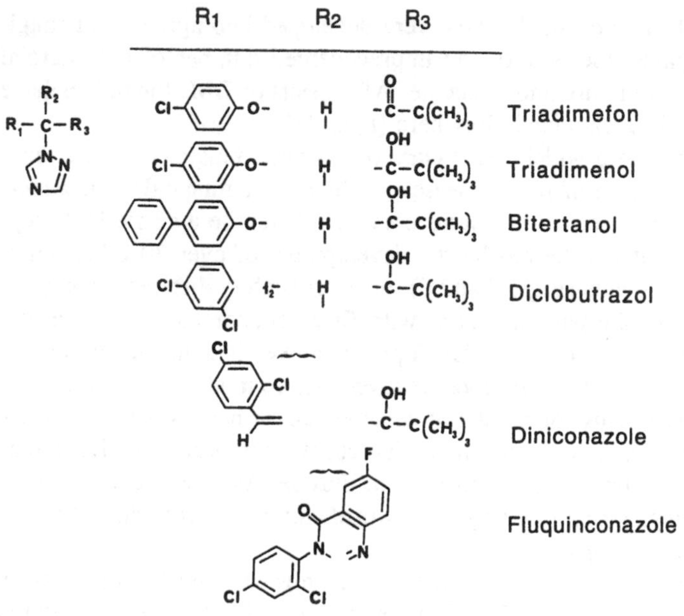

Figure 8 DMI fungicides containing a triazole.

DMIs are used as foliar fungicides against powdery mildews and rusts on many field and horticultural crops and ornamentals, against most of the important cereal diseases and several turf diseases, against *Venturia* and *Monilinia* spp. on fruit and nut trees, against *Cercospora* diseases of peanut and sugar beet, and against *Mycospherella* spp. on banana. Many related diseases on various other crops must be added to this list. It is interesting to note that despite the great number of DMIs developed, commercially acceptable control of *Colleto-trichum* spp., *Alternaria* spp., *Fusarium* spp., *Rhizoctonia* spp., and *Pyricularia oryzae* has not been achieved thus far. From a molecular point of view, this lack of activity remains largely unexplained.

The mode of action of DMI fungicides was clarified in 1972 by Ragsdale and Sisler in their work with the pyrimidine triarimol [62]. Shortly thereafter, it was discovered that numerous other pyrimidines, pyridines, imidazoles, and triazoles shared the same mode of action. The vast majority of DMIs in agriculture contain a triazole as essential group, a fact reflected in the frequently used term *azole fungicides*. Because the common names of more recently introduced DMIs end with "conazole" (see Fig. 8), this term can also be found describing the DMI fungicides. The biochemisty of DMI action in sterol biosynthesis is exceptionally well understood.

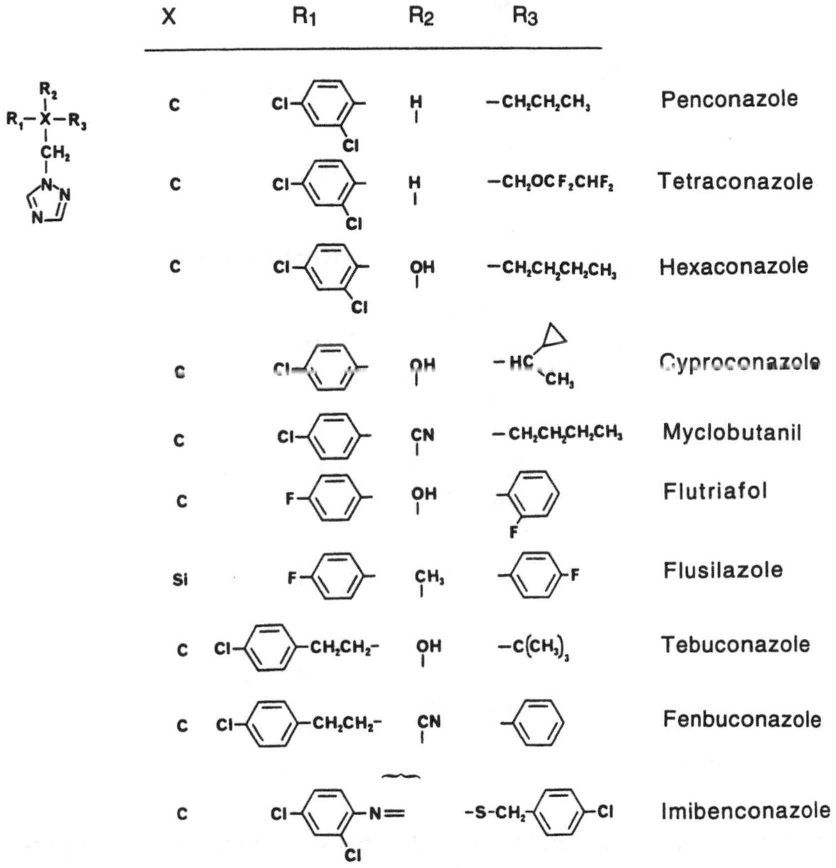

	X	R₁	R₂	R₃	

The general structure is:

R₁—X—R₃ with R₂ and CH₂ attached to X, and CH₂ connected to a triazole ring (N-N=N).

X	R₁	R₂	R₃	Name
C	Cl,Cl-phenyl	H	—CH₂CH₂CH₃	Penconazole
C	Cl,Cl-phenyl	H	—CH₂OCF₂CHF₂	Tetraconazole
C	Cl,Cl-phenyl	OH	—CH₂CH₂CH₂CH₃	Hexaconazole
C	Cl-phenyl	OH	—HC(cyclopropyl)CH₃	Cyproconazole
C	Cl-phenyl	CN	—CH₂CH₂CH₂CH₃	Myclobutanil
C	F-phenyl	OH	F-phenyl	Flutriafol
Si	F-phenyl	CH₃	phenyl-F	Flusilazole
C	Cl-phenyl-CH₂CH₂—	OH	—C(CH₃)₃	Tebuconazole
C	Cl-phenyl-CH₂CH₂—	CN	phenyl	Fenbuconazole
C	Cl,Cl-phenyl-N=		—S-CH₂-phenyl-Cl	Imibenconazole

Figure 8 Continued.

Sterols are vital components of biological membranes in all higher organisms. In most higher fungi, ergosterol is the predominant sterol present in membranes. The initial sterol precursor to be converted to the endproduct ergosterol is normally 24-methylenedihydrolanosterol. The earliest biosynthetic step accomplished in the structural rearrangement of this precursor is the removal of a methyl group in the 14 position, and the sterol demethylase responsible for this conversion was identified as a specific cytochrome P450 system. All DMIs contain a nitrogen in an aromatic ring, and the sterol-specific demethylation in the 14 position is inhibited by binding of this nitrogen to an essential porphyrin iron atom of cytochrome P450. In fact, the aromatic ring nitrogen present in pyridines, pyrimidines, imidazoles, and triazoles is the only essential feature of all but one (triforine) DMI fungicides. The additional DMI substituents direct the exact positioning of this essential nitrogen by binding into the active

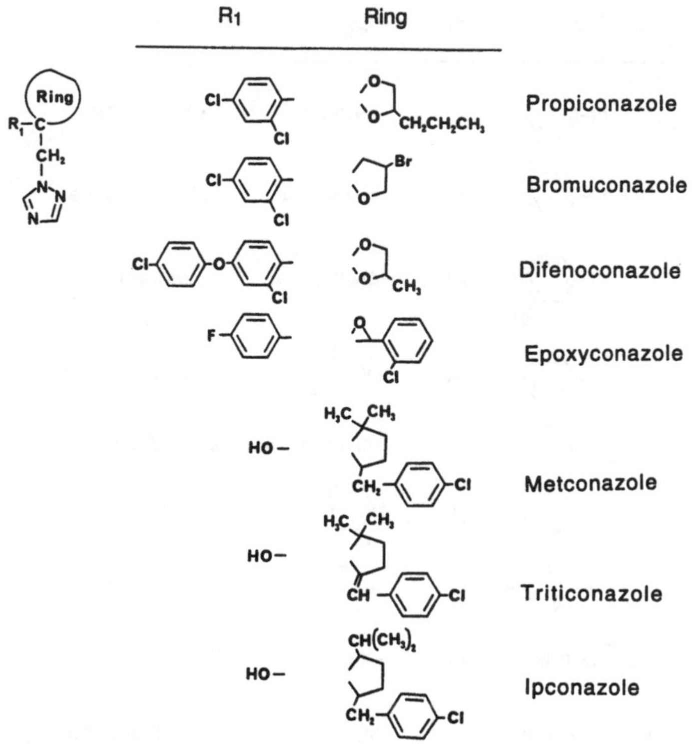

Figure 8 Continued.

enzyme site normally occupied by the methylated sterol precursor. In the presence of a DMI, sterol precursors with methyl groups in the 14 position will accumulate and will be incorporated into membranes, where they disturb normal membrane functions. Additional hormonal and regulatory side effects of DMI action caused by the rapid depletion of ergosterol from inhibited cells have been discussed, but they are not fully understood.

Side effects of the synthesis of plant sterols and gibberellic acid [42] have been observed for several DMIs. The severity of these plant growth–regulator effects, predominantly expressed as stunting of new growth, is dependent on the plant species and the structure and systemic property of a particular DMI. Although side effects on plants are not noticed under recommended conditions, they have led to restrictions in the development and usefulness of several DMIs.

2. Resistance

The first cases of DMI resistance were reported for triadimefon in the control of powdery mildew of barley and wheat [41]. Resistance has since been reported for powdery mildew on cucurbits and grapes, the postharvest control of *Penicillium italicum*, apple scab, some turf diseases, *Rhynchosporium secalis* on barley, and black sigatoka of banana [24,28,41,43,65]. Control achieved for many other diseases, including rusts, remains stable.

Considering the over 20 years of DMI use and the number of diseases controlled with DMIs, it is evident that the development of resistance was qualitatively different from the benzimidazoles. As described above, benzimidazole resistance developed rapidly in many cases, whereas DMI resistance developed gradually and proceeded more slowly. This difference between DMIs and benzimidazoles is related to a different response of pathogen populations to the selection pressure of a given fungicide class and, ultimately, to the molecular mechanisms of resistance functioning in fungal populations.

The benzimidazole mechanism of resistance is based on the mutation of the inhibitor binding site and generates resistant mutants virtually immune to any inhibitor dose. A very small, yet finite, population of such mutants existed in wild-type populations of pathogens prior to the use of benzimidazoles. This highly resistant subpopulation was not controlled to any degree at recommended field rates, and respective isolates increased rapidly in frequencies. Once the benzimidazole-insensitive and uncontrolled population had reached a size that compromised overall disease control, the fact of practical resistance was established [41]. Under such conditions, two separate populations either sensitive or immune to benzimidazoles were established, and the resistant subpopulations had to be controlled by other means, most often with protective fungicides present in mixtures.

It was soon recognized that DMI sensitivities among isolates in wild-type pathogenic populations varied greatly, with rare extremes separated by a factor of several hundred [41,43,65]. Upon exposure to DMIs, the vast majority of isolates was fully controlled, but some isolates with reduced sensitivity were less effectively controlled than their fully sensitive counterparts [43,65]. This lower level of control achieved for initially infrequent isolates led to the selection of a population insufficiently controlled and, thus, DMI resistant. The slower and more gradual emergence of practical DMI resistance relative to the benzimidazoles is likely due to the fact that benzimidazole-resistant strains were not controlled to any degree, whereas DMI-resistant strains remained suppressed at a variable, yet appreciable, level [43].

The degree of control achieved against DMI-resistant subpopulations appears to be important for future prospects of DMI resistance management. It is widely accepted that DMI sensitivities of phenotypes with gradually increasing levels of resistance are determined by several genes with additive effects, and evidence suggests that the number of these genes is finite [28]. If indeed finite, a fully stable and sustainable situation could be reached once all genotypes of a pathogen population would contain all possible DMI resistance genes. The long-term usefulness of DMIs under such an extreme yet stable situation would depend on the level of control still achieved with a particular DMI, which depends on the intrinsic potency of the DMI in relation to the recommended field rate [43]. The two parameters are not equal for all DMIs, and highly efficacious DMIs applied at high rates could provide appreciable control at fully resistant sites, where other DMIs would fail to control the disease. Unfortunately, information pertinent to control levels achieved for DMI-resistant populations is lacking at present.

DMI resistance in cereal powdery mildew was managed successfully by introducing DMI mixtures with morpholines, a class of sterol biosynthesis inhibitors described below. The management of resistance of grape powdery mildew appears to be less clear at present, with some DMIs remaining effective in vineyards resistant to other DMIs. The strategy for managing resistance in apple scab control is to apply mixtures of DMIs and older protective fungicides, similar to the management of bezimidazole resistance.

The mechanisms of DMI resistance are poorly understood. Various models, primarily based on laboratory mutants, have been proposed [42]. Increased ejection of DMIs from exposed cells appears to be one of the resistance mechanisms relevant to field populations [12,34]. A mutational change of the DMI target site lanosterol demethylase, which would be similar to the mechanism of benzimidazole resistance, has not been identified in any field isolate. Decreased inhibitor binding to the target site has only been reported for laboratory mutants [42], and such mutants were not necessarily DMI-resistant [36]. In summary, the molecular mechanisms operative in pathogen field populations remain poorly understood.

F.   Sterol Biosynthesis Inhibitors—Morpholines

1.   Properties and Mode of Action

The three morpholine fungicides widely used are tridemorph, fenpropimorph, and fenpropidine (Fig. 9). Although fenpropidine is chemically not a morpholine, many properties are similar to those of fenpropimorph. The same applies to spiroxamine (KWG 4168), a fungicide at a late stage of development [14]. The major use of the morpholines is in the control of cereal powdery mildews. Tridemorph is also used to control black sigatoka on banana; fenpropimorph, fenpropidine, and spiroxamine have additional activities against some other cereal diseases.

The mode of action of morpholines is the inhibition of fungal sterol biosynthesis with target enzymes different from the DMIs. Two separate steps in the biosynthesis of ergosterol are affected, although to different degrees dependent on the inhibitor. The first step of inhibition is immediately adjacent to the DMI target site. The demethylation of lanosterol leads to a double bond in the 14

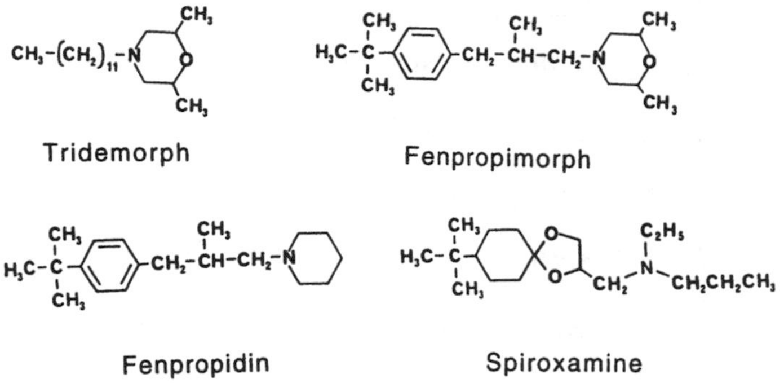

Tridemorph   Fenpropimorph

Fenpropidin   Spiroxamine

Figure 9 Morpholines.

position, which then is reduced by a sterol 14-reductase. This reductase is inhibited by the morpholines. The second target enzyme inhibited by morpholines is a C-8 sterol isomerase involved in the rearrangement of a double bond in later steps of sterol biosynthesis.

2. Resistance

In spite of over 25 years of use in the control of cereal powdery mildew and in spite of a specific mode of action, resistance has not become a limiting factor. Although shifts in isolate sensitivities toward reduced morpholine sensitivities were reported [16], respective isolates appear to be sufficiently controlled under field conditions. Reduced morpholine sensitivities in barley powdery mildews were inherited via one or several genes, with differences observed for fenpropimorph/fenpropidine and tridemorph [7].

G. Phenylamides and Other Oomycete-Specific Fungicides

1. Properties and Mode of Action

Foliar diseases caused by Oomycetes were primarily controlled with copper or EBDC fungicides prior to the introduction of the first representatives of the phenylamides, furalaxyl and metalaxyl, in the late 1970s. These new fungicides combined an oomycete-specific mode of action with systemic properties and became widely used fungicides with therapeutic value. Although metalaxyl experienced the broadest acceptance, current phenylamides include furalaxyl, benalaxyl, ofurace, and oxadicyl (Fig. 10). Although furalaxyl is used only as a soil fungicide, all other representatives are also applied as foliar fungicides for the control of potato late blight and downy mildews. Only one of the two isomers of phenylamides is biologically active, and the pure active isomer metalaxyl M was introduced recently as a commercial fungicide [56].

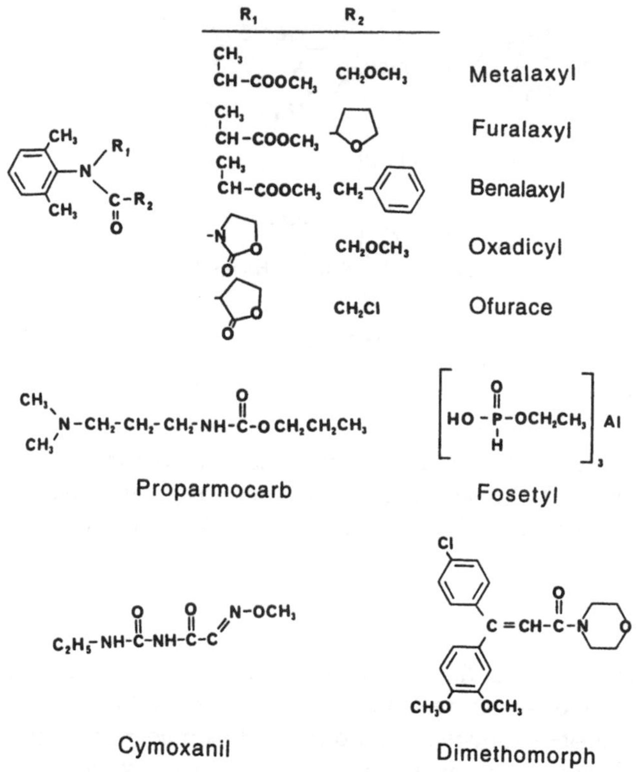

Figure 10 Phenylamides and other oomycete-specific fungicides.

The mode of action of phenylamides is inhibition of the RNA polymerase responsible for synthesis of ribosomal RNA, but no explanation for the high oomycete specificity has been forthcoming. It was also observed that the mechanism of pathogen resistance was prevention of inhibitor binding to the target site. Like the mode of action, no recent progress has been made in understanding the mechanism of resistance in greater detail.

While phenylamides were introduced in the late 1970s and early 1980s, three additional oomycete-specific fungicides, propamocarb, cymoxanil, and fosetyl (Fig. 10), were discovered and developed. Their market acceptance remained moderate until resistance to phenylamides affected the control of many diseases in the late 1980s. Dimethomorph (see Fig. 10) was introduced as an additional oomycete fungicide in 1986. All of the above compounds are used in foliar applications against various downy mildews and, with the exception of fosetyl, late blight of potato and tomato. Fosetyl is the only current fungicide transported into roots after foliar applications, and this property is exploited in the control of several root diseases. It must also be mentioned that fosetyl shows activity against some diseases caused by pathogens other than Oomycetes, including some bacterial diseases.

Fosetyl is taken up readily by plants and is useful in therapeutic applications. In contrast, proparmocarb and cymoxanil display restricted systemicity and are primarily used in prophylactic programs. The same applies to dimethomorph, although systemic translocation within plants can be pronounced [10] and uptake can be improved by the addition of certain adjuvants [22]. The modes of action of the above fungicides are not well understood. Proparmocarb appears to act on phospholipid synthesis [64], whereas dimethomorph has activity on the synthesis of cell walls [1]. Two modes of action have been proposed for fosetyl: either acting directly through phosphonic acid generated in plants or by the stimulation of plant defense reactions [23].

2. Resistance

Resistance of *Phytophthora infestans* and downy mildews to the phenylamides developed rapidly [41]. Resistant genotypes were virtually immune to the inhibitors and were not controlled at any practical dose. The rapid development of resistance combined with a target site mutation responsible for abolished inhibitor binding was similar to the benzimidazole case described above. The resistance management strategy implemented was also very similar and relied on the employment of mixtures of phenylamides, primarily with the protective EBDC mancozeb [41]. At present, the relative contribution of phenylamides in such mixtures is rarely known. Under high levels of pathogen resistance, such mixtures may lose their therapeutic value, because disease control would rely heavily on mancozeb. A different resistance management approach was developed and partly implemented for oxadicyl used as a triple mixture with mancozeb and cymoxanil, with synergistic effects being reported for the control of phenylamide-resistant isolates [48]. Thus far, control of diseases caused by soilborne Oomycetes has been unaffected by phenylamide resistance.

No resistance has been reported for proparmocarb, cymoxanil, fosetyl, or dimethomorph. However, mutants of *Phytophthora parasitica* resistant to dimethomorph were easily obtained in the laboratory, and although the resistance levels were smaller than those for metalaxyl, the risk of resistance development was considered high [9].

H. Anilinopyrimidine, Phenylpyrrole, and Strobilurin Fungicides

1. Properties and Modes of Action

Several classes of specific fungicides have been introduced since 1990 and have become commercially available. The class of anilinopyrimidines is currently represented by pyrimethanil, mepanipyrim, and cyprodinil (Fig. 11). These fungicides are specific against Ascomycetes/Deuteromycetes, with current em-

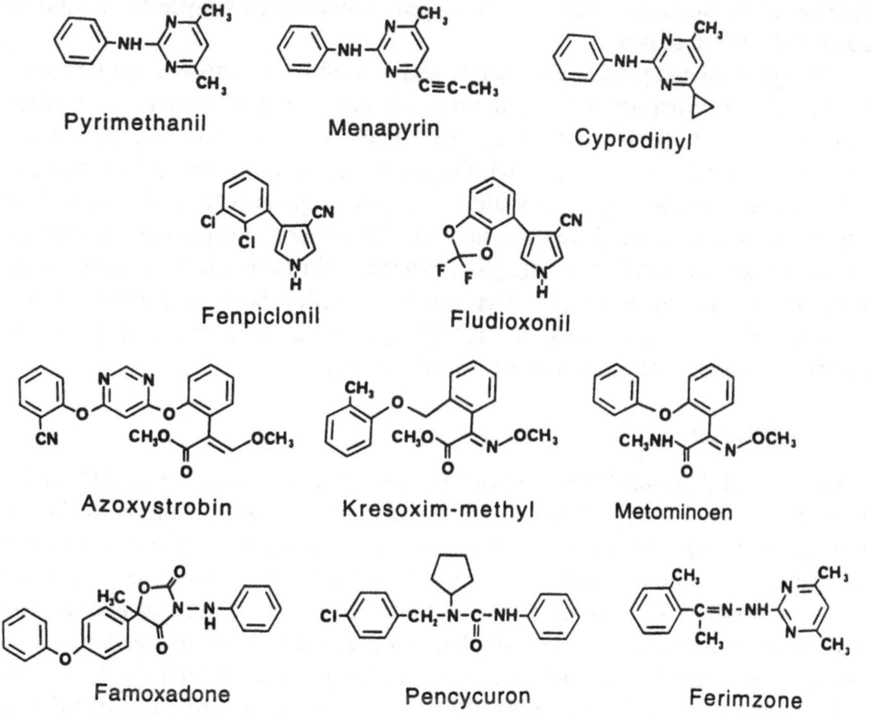

Figure 11 Anilinopyrimidines, phenylpyrroles, strobilurins, and miscellaneous fungicides with specific modes of action.

phasis on *Botrytis* and *Venturia* spp., and on several cereal diseases. The anilinopyrimidines are therapeutic fungicides with effects on disease establishment and sporulation but not on spore germination [38,51]. The mode of action has been related to the inhibition of extracellular enzyme secretion required in pathogenesis [51, 52], but convincing evidence for a mode of action in methionine biosynthesis has also been presented [18].

A second class of recently developed fungicides are the phenylpyrroles, with fenpiclonil and fludioxonil (Fig. 11) developed commercially. The lead structure of synthesis efforts was the natural antifungal product pyrrolnitrin produced by *Pseudomonas pyrocinia*. In principle, phenylpyrolles have a very broad spectrum of activity against all classes of pathogens. At present, however, this broad spectrum is not reflected in an equally broad introduction of commercial products. Fenpiclonil is currently used for dressing seeds and potato tubers, with high activity against *Fusarium* spp. and good activity against many other diseases. Fludioxonil was introduced in mixture with the anilinopyrimidine cyprodinyl as a foliar fungicide in the control of gray mold. The mode of phenylpyrrole action involves the inhibition of sugar and amino acid uptake [33], but details of inhibitor action remain unknown.

Recent attention has focused on a new group of fungicides referred to as strobilurins, or methoxyacrylates (MOAs). The development was based on the natural antifungal inhibitors strobilurin A, oudemansin A, and myxothiazol A, which are all produced by Basidiomycetes [54]. The three inhibitors shared structural features and were shown specifically to inhibit mitochondrial respiration by binding to cytochrome bc_1. The same mode of action has been verified for the three fully synthetic structural analoges azoxystrobin, kresoxim-methyl, and metominoen (SSF-126) (Fig. 11) [4,20,54]. All three fungicides are in early stages of commercial introduction. The structurally different cytochrome bc_1 inhibitor famoxadone has been introduced more recently [35].

The strobilurins exhibit a very broad disease spectrum including Oomycetes, Ascomycetes/Deuteromycetes, and Basidiomycetes [4,20]. The primary target diseases for azoxystrobin and kresoxim-methyl are currently powdery mildews on a variety of crops, several cereal diseases, and several tree fruit diseases, as well as several oomycete diseases. Azoxystrobin and SSF-126 are also under development for the control of rice blast. All strobilurins are very potent inhibitors of spore germination but have limited activity in completely inhibiting the mycelial growth of pathogens. This limitation is probably due to the induction of an alternative oxidase that circumvents part of the inhibitor action [53]. Therapeutic value of the strobilurins is inherent to their antisporulant activity [20].

2. Resistance

For the anilinopyrimidines and *Botrytis cinerea*, it was observed that frequencies of highly resistant isolates at an experimental vineyard with several years of exposure had increased to high levels and that disease control had become insufficient [17,26]. This finding signals a considerable risk of resistance. The restriction of seasonal applications or the use of a cyprodinyl/fludioxonil mixture were recommended as an antiresistance strategy in the control of gray mold [17]. No information is currently available for other pathogens controlled with anilinopyrimidines.

For the class of phenylpyrroles, the situation appears more complex. As described above, laboratory mutants of *Botrytis cinerea* resistant to fludioxonil were also resistant to dicarboxides [27]. A similar situation was reported for laboratory mutants of *Ustilago maydis* [15]. This type of resistance appears irrelevant to field populations, because all dicarboximide-resistant field isolates of *B. cinerea* were fully sensitive to the phenylpyrrole fludioxonil [17,26,27].

The risk of resistance to the strobilurins is similarly undetermined at present. Mutants of *Saccharomyces cerevisiae* resistant to this type of respiration inhibitor have been described [13,19], and resistance was based on a target site mutation. The level of resistance in these yeast mutants was low, and similar studies for mutants of filamentous fungi or plant pathogens are not available at present. A resistance mechanism based on the induction of an alternative oxidase was described for a laboratory mutant of *Septoria tritici*, but this in vitro mechanism had no impact on the in vivo control of the disease [76].

I. Miscellaneous Fungicides with Specific Modes of Action

In the past, numerous other fungus-specific fungicides were introduced into the market for plant disease control. They include edifenfos, iprobenfos, and isoprothiolane, which are used exclusively for the control of rice blast and with modes of action in the synthesis of phospholipids. Pyrazophos and dodine must be mentioned as fungicides previously used to control powdery mildews and apple scab, respectively. Similarly, the fermentatively produced natural fungicides with practical application in plant disease control, blasticidine S, kasugamycin, polyoxins, validamycin A, and mildiomycin, primarily used in Japan, had specific modes of antifungal action. However, the value of these fungicides in current disease control programs has declined either because of resistance or because alternatives became available.

Two more recently introduced fungicides are interesting in the context of specific fungicidal action. For example, pencycuron (Fig. 11), introduced in the 1980s, is highly specific against *Rhizoctonia solani*, and within this group, sensitivity is even restricted to certain anastomosis groups. Although the mode of action has been investigated [37], the amazing specificity of this fungicide is not understood. Pencycuron is currently in use for dressing of seed potatoes.

A second specific fungicide introduced in 1989 and yet not fully explored is ferimzone (Fig. 11), which was developed for the control of several rice diseases. Inhibition of spore germination is not pronounced, and although an action on membrane functions was suggested [58], mechanistic details remain to be clarified. Interestingly, ferimzone was recently shown to induce salicylic acid synthesis and to protect against viral diseases [55], a property that might link the mode of action to an indirect effect of triggering host defenses described below.

IV. FUNGICIDES WITH INDIRECT MODES OF ACTION

A. Introduction

All fungicides described above inhibit the growth of fungal pathogens directly, be it by the inhibition of numerous metabolic steps or by the inhibition of a single and often highly fungus-specific target site. For almost two decades, it has been suggested that interference with the steps crucial in fungal infections rather than the direct inhibition of metabolic steps should yield plant protection agents with an elevated level of specificity. Thus far, only a few commercial products with such indirect modes of action have been introduced.

B. Fungicides Preventing Host Penetration

Penetration into host plants subsequent to attachment of pathogenic spores to host surfaces is one of the most crucial events in plant infection. Commercial penetration inhibitors have only been developed for the control of rice blast caused by *Pyricularia oryzae*. The three rice blast fungicides fthalide, tricyclazole, and pyriquilon (Fig. 12) have no effects on spore germination or mycelial growth of the pathogen. They specifically inhibit the synthesis of the pigment melanin in walls of appressoria and are known as melanin biosynthesis inhibitors (MBIs). Without the strengthening of appressorial cell walls by melanin, a sufficiently high internal turgor pressure necessary for the penetration of hard surfaces will not be reached. Consequently, attempts of host penetration become abortive in the presence of MBIs. A fourth rice blast fungicide acting as an MBI, carpropamid (KTU 3616) (Fig. 12), was introduced recently [25].

The commercial acceptance of MBIs in rice blast control was triggered by their lasting residual activities. All MBIs are highly systemic and impregnate host cuticles with inhibitor concentrations sufficient for durable inhibition of penetration. Although the inhibitors act protectively and would not provide any control when applied after infections, they lack target sites in plants. They can, thus, move systemically, impregnate plant cuticles from the inside and provide protection over a long period of host development. In laboratory studies, MBIs

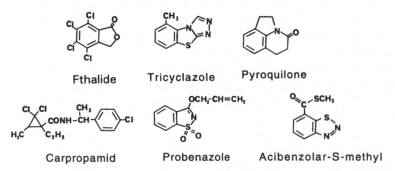

Fthalide   Tricyclazole   Pyroquilone

Carpropamid   Probenazole   Acibenzolar-S-methyl

Figure 12 Fungicides with indirect modes of action.

were also active against *Colletotrichum* spp.; commercial fungicides are not available for the control of *Colletotrichum* diseases. The MBIs also inhibit melanin biosynthesis in numerous other pathogens, but penetration from appressoria is not dependent on the presence of melanin in the majority of pathogens.

The first MBIs, fthalide and tricyclazole, were discovered in the early 1970s and have been used on a broad scale for 20 years against rice blast. Resistance has not become a problem thus far, and although the reasons for this lack of resistance are not understood, the MBIs remain effective fungicides despite their specific mode of action.

C. Fungicides Inducing Host Defense Reactions

A second strategy to chemically controlling plant diseases without interference in fungal metabolism is the triggering of plant defense reactions. A dogma in plant pathology implies that the establishment of disease after attachment of a potentially pathogenic spore is the exception rather than the rule. Plants either are nonhosts and ward off attempted infections by a nonpathogen, or specific resistance genes present in certain host cultivars prevent infections by avirulent pathogen races. In both cases, the plant responds actively, and susceptibility implies that this active response of the host is overcome by the pathogen. Therefore, the induction of a successful resistance response triggered by chemicals appeared feasible. Such fungicides would lack, by definition, any direct target site in fungal organisms.

Until recently, probenazole (Fig. 12), a broadly used rice blast fungicide introduced in 1979, was the only known fungicide triggering defense reactions in rice against infection by *Pyricularia oryzae*. Although the induction of a pathogenesis-related protein in rice by probenazole was reported recently [50], a detailed mechanistic analysis has not been provided thus far.

A novel approach to triggering plant defense reactions for disease control is related to the long known phenomenon of systemic acquired resistance (SAR). The concept implies that the challenge of a lower leaf by a pathogen or an avirulent race of a pathogen triggers resistance responses in upper leaves. The induced synthesis of salicylic acid was identified as a key step in this SAR reaction [11]. The new benzothiadiazole fungicide, acibenzolar-S-methyl (BTH, CGA- 245704) (Fig. 12), acts as a structural analogue of salicylic acid and triggers defense reactions effective against a variety of diseases [21]. The fungicide was commercially introduced for the control of wheat diseases, in particular powdery mildew, and is currently being tested for commercial control of several other diseases such as downy mildews and black sigatoka [66].

REFERENCES

1. Albert, G., & H. Heinen. 1996. How does dimethomorph kill fungal cells ? - A time lapse video study with *Phytophthora infestans*. In *Modern Fungicides and Antifungal Agents*. H. Lyr, P.E. Russel, & H.D. Sisler, eds. Andover, Massachusetts: Intercept, pp 141–146.

2. Akagi, T., S. Mitani, T. Komyoji, & K. Nagatani. 1995. Quantitative structure-activity relationships of fluazinam and related fungicidal N-phenylpyridiamines: Preventive activity against *Botrytis cinerea*. *J. Pestic. Sci.* 20:279–290.

3. Akagi, T., S. Mitani, T. Komyoji, & K. Nagatani. 1996. Quantitative structure-activity relationships of fluazinam and related fungicidal N-phenylpyridiamines: Preventive activity against *Sphaerotheca fuliginea*, *Pyricularia oryzae* and *Rhizoctonia solani*. *J. Pestic. Sci.* 21:23–29.

4. Baldwin, B.C., J.M. Clough, C.R.A. Godfrey, J.R. Godwin, & T.E. Wiggins. 1996. The discovery and mode of action of ICIA5504. In *Modern Fungicides and Antifungal Agents*. H. Lyr, P.E. Russel, and H.D. Sisler, eds. Andover, Massachusetts: Intercept, pp 69–78.

5. Brandt, U., J. Schubert, P. Ceck, & G. von Jagow. 1992. Uncoupling activity and physicochemical properties of derivatives of fluazinam. *Biochim. Biophys. Acta* 1101:41–47.

6. Broomfield, P.L.E., & J. Hargreaves. 1992. A single amino-acid change in the iron-sulphur protein subunit of succinate dehydrogenase confers resistance to carboxin in *Ustilago maydis*. *Curr. Genet.* 22:117–121.

7. Brown, J.K.M., S. Le Boulaire, & N. Evans. 1996. Genetics of responses to morpholine-type fungicides and of avirulences in *Erysiphe graminis* f. sp. *hordei*. *Eur. J. Plant Pathol.* 102:479–490.

8. Buchenauer, H. 1990. Physiological reactions in the inhibition of plant pathogenic fungi. In *Chemistry of Plant Protection*. Vol. 6. G. Haugh & H. Hoffmann, eds. Berlin: Springer-Verlag, pp 217–392.

9. Chabane, K., P. Leroux, N. Maia, & G. Bompeix. 1996. Resistance to dimethomorph in laboratory mutants of *Phytophthora parasitica*. In *Modern Fungicides and Antifungal Agents*. H. Lyr, P.E. Russel, & H.D. Sissler, eds. Andover, Massachusetts: Intercept, pp 387–392.

10. Cohen, Y., A. Baider, & B.H. Cohen. 1995. Dimethomorph activity against fungal plant pathogens. *Phytopathology* 85:1500–1506.

11. Conrath, U., Z. Chen, J. Malamy, J. Durner, J. Hennig, P. Sanchez-Casas, H. Silva, J. Ricigliano, & D.F. Klessig. 1996. The salicyclic acid signal for the activation of plant disease resistance: Induction, modification, perception and transduction. In *Modern Fungicides and Antifungal Agents*. H. Lyr, P.E. Russel, & H.D. Sisler, eds. Andover, Massachusetts: Intercept, pp 467–474.

12. DeWaard, M.A., J.G.M. van Nistelrooy, C.R. Langveld, J.A.L. van Kan, & G. del Sorbo. 1996. Multidrug resistance in filamentous fungi. In *Modern Fungicides and Antifungal Agents*. H. Lyr, P.E. Russel, eds. H.D. Sisler, eds. Andover, Massachusetts: Intercept, pp 293–300.

13. Di Rago, J.-P., J.-Y. Coppée, & A.-M. Colson (1989). Molecular basis for re-
     sistance to myxothiazol, mucidin (strobilurin A), and Stigmatellin. *J. Biol. Chem.*
     264:14543–14548.
14. Dutzmann, S., D. Berg, N.E. Clausen, W. Krämer, K.H. Kuck, R. Pontzen, R.
     Tiemann, & J. Weissmüller. 1996. KWG 4168: A novel foliar fungicide with a
     particular activity against powdery mildew. *Proc. Brighton Crop Prot. Conf.*
     1996:47–52.
15. Eberle, A., & K. Schauz. 1996. Effects of the phenylpyrrole fungicide fludioxonil
     in sensitive and resistant *Ustilago maydis* strains. In *Modern Fungicides and Anti-
     fungal Agents.* H. Lyr, P.E. Russel, and H.D. Sisler, eds. Andover, Massachu-
     setts: Intercept, pp 393–402.
16. Engels, A.J.G., B.C. Mantel, & M.A. de Waard. 1996. Effects of split applica-
     tions of fenpropimorph-containing fungicides on sensitivity of *Erysiphe graminis*
     f. sp. *tritici. Plant Pathol.* 45:636–643.
17. Forster, B., & T. Staub. 1996. Basis for use strategies of anilinopyrimidine and
     phenylpyrrole fungicides against *Botrytis cinerea. Crop Prot.* 15:529–537.
18. Fritz, R., C. Lanen, V. Colas, & P. Leroux. 1997. Inhibition of methionine bio-
     synthesis in *Botrytis cinerea* by the anilinopyrimidine fungicide pyrimethanil. *Pestic.
     Sci.* 49:40–46.
19. Gaier, M., H. Schägger, U. Brandt, A.-M. Colson, & G. von Jagow. 1992. Point
     mutation in cytochrome b of yeast ubihydroquinone: cytochrome-c oxidoreductase
     causing myxothiazol resistance and facilitated dissociation of the iron-sulfur sub-
     unit. *Eur. J. Biochem.* 208:375–380.
20. Gold, R.E., E. Ammermann, H. Köhle, G.M.E. Leinhos, G. Lorenz, J.B.
     Speakman, M. Stark-Urnau, & H. Sauter. 1996. The synthetic strobilurin BAS
     490F: Profile of a modern fungicide. In *Modern Fungicides and Antifungal Agents.*
     H. Lyr, P.E. Russel, and H.D. Sisler, eds. Andover, Massachusetts: Intercept, pp
     79–92.
21. Görlach, J., S. Voirath, G. Knauf-Beiter, G. Hengy, U. Bechhove, K.-H. Kogel,
     M. Oostendorp, T. Staub, E. Ward, H. Kessmann, & J. Ryals. 1996. Benzo-
     thiadiazole, a novel class of inducers of systemic acquired resistance, activates gene
     expression and disease resistance in wheat. *Plant Cell* 8:629–643.
22. Grayson, B.T., D.M. Batten, & D. Walter. 1996. Adjuvant effects on the thera-
     peutic control of potato late blight by dimethomorph wettable powder formulations.
     Pestic. Sci. 46:355–359.
23. Griffith, J.M., A.J. Davis, & B.R. Grant. 1992. Target sites of fungicides to control
     Oomycetes. In *Target Sites of Fungicide Action.* W. Koeller, Boca Raton, Florida:
     CRC Press, pp 60–100.
24. Gubler, W.D., H.L. Ypema, D.G. Ouimette, & L.J. Bettiga. 1996. Occurrence
     of resistance in *Uncinula necator* to triadimefon, myclobutanil, and fenarimol in
     California grapevines. *Plant Dis.* 80:902–909.
25. Hattori, T., K. Kurahashi, S. Kagabu, J. Konze, & U. Kraatz. 1994. KTU 3616:
     A novel fungicide for rice blast control. *Proc. Brighton Crop Prot. Conf.* 1994:
     517–525.

26. Hilber, U.W., & H. Schuepp. 1996. A reliable method for testing the sensitivity of *Botryotinia fuckeliana*. *Pestic. Sci.* 47:241–247.

27. Hilber, U.W., F.J. Schwinn, & H. Schuepp. 1995. Comparative resistance patterns of fludioxinil and vinclozolin in *Botrytinia fuckeliana*. *J. Phytopathology* 143:423–428.

28. Hollomon, D.W. 1992. Resistance to azole fungicides in the field. *Biochem. Soc. Trans.* 21:104-7-1051.

29. Holloway, P.J. 1993. Structure and chemistry of plant cuticles. *Pestic. Sci.* 37:203–206.

30. Huang, R., & Y. Levy. 1995. Characterization of iprodione-resistant isolates of *Alternaria brassicicola*. *Plant Dis.* 79:828–833.

31. Ioriatti, C., E. Pasqualini, & A. Toniolli. 1992. Effects of the fungicides mancozeb and dithianon on mortality and reproduction of the predatory mite *Amblyseius andersoni*. *Exp. Appl. Acarol.* 15:109–116.

32. James, D.G., & M. Rayner. 1995. Toxicity of viticultural pesticides to the predatory mites *Amblyseius victoriensis* and *Typhlodromus doreenae*. *Plant Protect. Q.* 10:99–102.

33. Jespers, A.B.K., L.C. Davidse, & M.A. De Waard. 1993. Biochemical effects of the phenylpyrrole fungicide fenpiclonil in *Fusarium sulphreum* Schlecht. *Pestic. Biochem. Physiol.* 45:116–129.

34. Joseph-Horne, T., D. Hollomon, N. Manning, & S.L. Kelly. 1996. Investigation of the sterol composition and azole resistance in field isolates of *Septoria nodorum*. *Appl. Environ. Microbiol.* 62:184–190.

35. Joshi, M.M., & J.A. Sternberg. 1996. DPX-JE874: A broad-spectrum fungicide with a new mode of action. *Proc. Brighton Crop Prot. Conf.* 1996:21–26.

36. Keon, J.P.R., & J.A. Hargreaves. 1996. An *Ustilago maydis* mutant partially blocked in $P450_{14DM}$ activity is hypersensitive to azole fungicides. *Fung. Genet. Biol.* 20:84–88.

37. Kim, H.T., T. Kamakura, & T. Yamaguchi. 1996. Effect of pencycuron on the osmotic stability of protoplasts of *Rhizoctonia solani*. *J. Pestic. Sci.* 21:159–163.

38. Knauf-Beiter, G., H. Dahmen, U. Heye, & T. Staub. 1995. Activity of cyprodinil: Optimal treatment timing and site of action. *Plant Dis.* 79:1098–1103.

39. Koenraadt, H., & A.L. Jones. 1992. Resistance to benomyl conferred by mutations in codon 198 or 200 of the beta-tubulin gene of *Neurospora crassa* and sensitivity to diethofencarb conferred by codon 198. *Phytopathology* 83:850–854.

40. Koenraadt, H., S.C. Sommerville, & A.L. Jones. 1992. Characterization of mutations in the beta-tubulin gene of benomyl-resistant field strains of *Venturia inaequalis*. *Phytopathology* 82:1348–1354.

41. Köller, W. 1991. Fungicide resistance in crop protection. In *CRC Handbook of Pest Management in Agriculture*. 2nd ed. Vol. 2. D. Pimentel, ed. Boca Raton, Florida: CRC Press, pp 679–720.

42. Köller, W., ed. 1992. *Target Sites of Fungicide Action*. Boca Raton, Florida: CRC Press.

43. Köller, W., W.F. Wilcox, J. Barnard, A.L. Jones, & P.G. Braun. 1997. Detec-

tion and quantification of resistance of *Venturia inaqeualis* populations to sterol demethylation inhibitors. *Phytopathology* 87:184–190.

44. Komyoji, T., K. Sugimoto, & K. Suzuki. 1995. Effect of fluazinam, a new fungicide, on infection processes of several plant pathogenic fungi. *Ann. Phytopathol. Soc. Jpn.* 61:145–149.

45. Komyoji, T., K. Sugimoto, S. Mitani, N. Matsuo, & K. Suzuki. 1995. Biological properties of a new fungicide, fluazinam. *J. Pestic. Sci.* 20:129–135.

46. Kovach, J., C. Petzold, J. Degni, & J. Tette. 1992. A method to measure the environmental impact of pesticides. *NY Food Life Sci. Bull.* 139.

47. Latoree, B.A., V. Flores, A.M. Sara, & A. Roco. 1995. Dicarboximide-resistant isolates of *Botrytis cinerea* from table grape in Chile: Survey and characterization. *Plant Dis.* 7:990–994.

48. Lyr, H., ed. 1995. *Modern Selective Fungicides.* Jena: Gustav Fischer Verlag.

49. Lyr, H. 1995. Selectivity in modern fungicides and its basis. In *Modern Selective Fungicides.* H. Lyr, ed. Jena, Germany: Gustav Fischer Verlag, pp 13–22.

50. Midoh, N., & M. Iwata. 1994. Cloning and characterization of a probenazole-inducible gene for an intracellular pathogenesis-related protein in rice. *Plant Cell Physiol.* 35:9–18.

51. Milling, R.J., & A. Daniels. 1996. Effects of pyrimethanil on the infection process and secretion of fungal cell wall degrading enzymes. In *Modern Fungicides and Antifungal Agents.* H. Lyr, P.E. Russel, & H.D. Sisler, eds. Andover, Massachusetts: Intercept, pp 53–60.

52. Miura, I., T. Kamakura, S. Maeno, S. Hayashi, & I. Yamaguchi. 1995. Inhibition of enzyme secretion in plant pathogens by mepanipyrim, a novel fungicide. *Pest, Biochem. Physiol.* 48:222–228.

53. Mizutani, A., N. Miki, H. Yukioka, H. Tamura, & M. Masuko, 1996. A possible mechanism of control of rice blast disease by a novel alkoxyiminoacetamide fungicide. *Phytopathology* 86:295–300.

54. Mizutani, A., H. Yukioka, H. Tamura, N. Miki, M. Masuko, & R. Takeda. 1995. Respiratory characteristics in *Pyricularia oryzae* exposed to a novel alkoxyiminoacetamide fungicide. *Phytopathology* 85:306–311.

55. Nakayama, M., K. Matsuura, & T. Okuno. 1996. Production of salicylic acid in tobacco and cowpea plants by a sytemic fungicide ferimzone and induction of resistance to virus infection. *J. Pestic. Sci.* 21:69–72.

56. Nuninger, C., G. Watson, N. Leadbitter, & H. Ellgehausen. 1996. CGA329351: Introduction of the enantiomeric form of the fungicide metalaxyl. *Proc. Brighton Crop Prot. Conf.* 1996:41–46.

57. Oerke, E.-C., H.W. Dehne, F. Schönbeck, & A. Weber. 1994. *Crop Production and Crop Protection.* Amsterdam: Elsevier.

58. Okuno, T., I. Furusawa, K. Matsuura, & J. Shishiyama. 1989. Mode of action of ferimzone, a novel systemic fungicide for rice diseases. Biological properties against *Pyricularia oryzae in vitro. Phytopathology* 79:827–832.

59. O'Riley, P., S. Kobayashi, S., Yamane, W.G. Phillips, P. Raymond, & B.

Castanho, 1992. Mon 24000, a novel fungicide with broad-spectrum disease control. *Proc. Brighton Crop Prot. Conf.* 1992:427–434.

60. Orth, A.B., M. Rzhetskaya, E.J. Pell, & M. Tien. 1995. A serine (threonine) protein kinase confers fungicide resistance in the phytopathogenic fungus *Ustilago maydis. Appl. Environ. Microbiol.* 61:2341–2345.

61. Orth, A.B., A. Sfarra, E.J. Pell, & M. Tien. 1992. An investigation into the role of lipid peroxidation in the mode of action of aromatic hydrocarbon and dicarboximide fungicides. *Pestic. Biochem. Physiol.* 44:91–100.

62. Ragsdale, N.N., & H.D. Sisler. 1972. Inhibition of ergosterol synthesis in *Ustilago maydis* by the fungicide triarimol. *Biochem. Biophys. Res. Commun.* 46:81–96.

63. Raposo, R., J. Delcan, V. Gomez, & P. Melgarejo. 1996. Distribution and fitness of isolates of *Botrytis cinerea* with multiple fungicide resistance in Spanish greenhouses. *Plant Pathol.* 45:497–505.

64. Reiter, B., M. Wenz, H. Buschhaus, & H. Buchenauer. 1996. Action of proparmocarb against *Phytophthora infestans.* In *Modern Fungicides and Antifungal Agents.* H. Lyr, P.E. Russel, & H.D. Sisler, eds. Andover, Massachusetts: Intercept, pp 147–156.

65. Romero, R.A., & T.B. Sutton. 1997. Sensitivity of *Mycosphaerella fijiensis,* causal agent of black sigatoka of banana, to propiconazole. *Phytopathology* 87:96–100.

66. Ruess, W., K. Mueller, G. Knauf-Beiter, W. Kunz, & T. Staub. !996. Plant activator CGA 245704: An innovative approach for disease control in cereals and tobacco. *Proc. Brighton Crop Prot. Conf.* 1996:53–60.

67. Schwinn, F., & T. Staub. 1995. Phenylamide and other fungicides against Oomycetes. In *Modern Selective Fungicides.* H. Lyr, ed. Jena, Germany: Gustav Fischer Verlag, pp 323–346.

68. Sijpestein, A.K. 1985. Mode of action of some traditional fungicides. In *Mode of Action of Antifungal Agents.* A.P.J. Trinci & J.F. Ryley, eds. Cambridge, UK: Cambridge University Press, pp 135–153.

69. Siegel, M.R., & H.D. Sisler, eds. 1977. *Antifungal Compounds.* New York: Marcel Dekker.

70. Steel, C.C. 1996. Catalase activity and sensitivity to the fungicides, iprodione and fludioxinil in *Botrytis cinerea. Lett. Appl. Microbiol.* 22:335–338.

71. Steel, C.C., & N.G. Nair. 1995. Oxidative protective mechanisms and resistance to the dicarboximide fungicide, iprodione, in *Alternaria alternata. J. Phytopathol.* 143:531–535.

72. Sundin, G.W., D.H. Demezas, & C.L. Bender. 1994. Genetic and plasmid diversity within natural populations of *Pseudomonas syringae* with various exposure to copper and streptomycin. *Appl. Environ. Microbiol.* 60:4421–4431.

73. Tokutake, N., H. Miyoshi, & T. Fujita. 1991. Electron transport inhibition of the cytochrome bc1 complex of rat liver mitochondria by phenolic uncouplers. *Biochim. Biophys. Acta* 105-7:377–383.

74. Torgeson, D.C., ed. 1967. *Fungicides—An Advanced Treatise.* New York: Academic Press.

75. Vali, R.J., & G.W. Moorman. 1992. Influence of selected fungicide regimes on frequency of dicarboximide-resistant and dicarboximide-sensitive strains of *Botrytis cinerea*. *Plant Dis.* 76:919–924.
76. Ziogas, B.N., B.C. Baldwin, & J.E. Young. 1997. Alternative respiration: a biochemical mechanism of resistance to azoxystrobin (ICIA 5504) in *Septoria tritici*. *Pestic. Sci.* 50:28–34.

14

Arthropods as Pests of Plants: An Overview

Robert L. Metcalf†
University of Illinois, Urbana-Champaign, Illinois

I. INTRODUCTION

The earliest known fossil records of land plants date from the Silurian period (about 420 million years before the present [MY BP]), and flowering plants, the angiosperms, were well established by the Triassic period (about 225 MY BP). The oldest known insect fossil is that of a bristle-tail (Thysanura), *Gaspeya paloventognathae*, from Devonian rock dated about 390 MY BP. By the Carboniferous period, about 300 MY BP, the Insecta had begun to dominate the terrestrial world and at least 10 orders were present. Unmistakable evidence of insect-damaged plant leaves is found in the fossil record of the Permian period (about 270 MY BP) [29]. The Insecta, therefore, have been engaged in herbivory for more than 100-fold years longer than the ascendancy of the hominoid ancestors of humans which appeared in Africa about 3 million years ago. Thus, the direct competition between insect species and humans for food and fiber is readily explainable from this extensive period of plant/insect interactions, from the vast diversity of insect species, and from the enormous number of individuals involved. World populations of individual insects are estimated at about 1×10^{18} by C.B. Williams compared to about 6×10^9 humans. The present-day diversity of the Insecta is summarized by their classification into 26 orders and as many as 937 families [3]. Insect herbivory is concentrated in species of 10 orders and these, together with important families, are shown in Table 1.

---

†Deceased.

Table 1 Orders and Families of Insects of Importance as Pests of Plants

Order/Family	Number of described species	
	North America	World
Orthoptera	1100	28,000
Locustidae (grasshoppers, locusts)		
Gryllidae (crickets, mole crickets)		
Tettigoniidae (katydids)		
Dermaptera (earwigs)	18	1200
Isoptera (termites)	41	2100
Thysanoptera (thrips)	606	4700
Homoptera	6700	33,000
Aleyrodidae (whiteflies)		
Aphididae (aphids)		
Cicadellidae (leafhoppers)		
Membracidae (treehoppers)		
Fulgoridae (planthoppers)		
Coccidae (scales, mealybugs)		
Hemiptera	4600	23,500
Pentatomidae (stink bugs)		
Miridae (leaf bugs)		
Tingidae (lace bugs)		
Lygaeidae (seed bugs)		
Coreidae (squash bugs)		
Coleoptera	30,000	300,000
Cucujidae (flat bark beetles)		
Buprestidae (metallic woodborers)		
Elateridae (wireworms)		
Meloidae (blister beetles)		
Scarabaeidae (white grubs)		
Cerambycidae (long-horned beetles)		
Chrysomelidae (leaf beetles)		
Bruchidae (weevils)		
Curculionidae (snout weevils)		
Lepidoptera	11,000	113,000
Cossidae (carpenter moths)		
Psychidae (bagworms)		
Lyonetiidae (ribbed cocoon makers)		
Gracillariidae (leaf blotch miners)		
Coleophoridae (casebearers)		
Oecophoridae (webworms)		
Gelechiidae		
Tortricidae (leafrollers)		
Olethreutidae		
Pyralidae		

Table 1 Continued

Order/Family	Number of described species	
	North America	World
Sphingidae (hornworms)		
Geometridae (measuring worms)		
Notodontidae		
Lymantriidae (tussock moths)		
Noctuidae		
Arctiidae (wooly bears)		
Lasiocampidae (tent caterpillars)		
Papilionidae (swallowtails)		
Pieridae		
Nymphalidae		
Diptera	18,000	120,000
Cecidomyiidae (gall gnats)		
Tephritidae (fruit flies)		
Chloropidae (fruit flies)		
Anthomyiidae (root maggots)		
Hymenoptera	17,400	120,000
Cephidae (stem sawflies)		
Tenthredinidae (sawflies)		

Source: Data from refs. 2 and 20.

II. IMPORTANCE OF THE PHYLUM ARTHROPODA

There are an estimated 885,000 described species of arthropods, of which about 762,000 species are insects [2,20,25]. Insects comprise about 72% of all described species of animals, and a single order, the Coleoptera, with more than 300,000 species, outnumbers all described species of plants. The fauna of North America contains approximately 94,000 described species of insects [2,20,25]. Of these insects, about two-thirds of the species are direct feeders on plants. Weiss [32], for example, in a catalogue of the 10,000 species of insects identified in New Jersey, found that 62% were phytophagous, 28% entomophagous, and about 10% were phlebotomous, feeding on higher animals.

A. Insects as Pests

Webster's Unabridged New International Dictionary defines *pest* as "any particularly injurious or destructive insect," an unrealistically narrow definition that nevertheless is pleasing to entomologists. It must be emphasized that the term *pest* is an arbitrary label and has no ecological validity. Some insect species are pests under certain conditions and beneficial at other times. An insect is usually considered a pest when it is in competition with humans for a resource, and when significant numbers are present [12]. If we accept as a working definition that an insect pest is one that occasionally requires an applied control measure, we estimate that about 1% of the total insect fauna can be defined as pests. Therefore, in the United States, there are approximately 10,000 species of insects, mites, and ticks that cause losses to agriculture. However, only about 600 species of arthropods require annual applied control measures, and about 400 of these feed directly on plants [20].

Insect pests can be further categorized as *occasional pests*, *minor pests*, and *major pests*. The numbers of insect species thus involved in attacking major crops in North America are shown in Table 2, which provides estimates of major and minor pests of agricultural crops. These numbers aggregate to about 260 major pests and as many as 750 minor pests [20].

A substantial proportion of North American pests are exotic species that were introduced during the past 300 years of formal agriculture. Plant cultivars grown in the United States are largely those introduced from other regions of the world

Table 2 Numbers of Major, Minor, and Exotic Insect Pest Species Attacking Important U.S. Crops

Crop (origin)	Major pest species[a]	Minor pest species[a,b]	Exotic pest species[a]
Corn (Mexico)	8	30	3
Wheat (Iraq)	15	45	5
Rice (Southeast Asia)	4		2
Sorghum (Africa)	3		1
Sugarcane (India)	4		2
Soybean (China)	11	40	3
Alfalfa (Transcaucasus)	24	75	7
Cotton (South America)	15	85	6
Tobacco (South America)	11		1
Apple (Eurasia)	10	43	7
Pear (Southeastern Europe)	10		6
Peach (China)	17		6
Citrus (China)	25	30	17
Potato (Americas)	5		1
Tomato (Tropical America)	6		1
Turf (U.S.)	6		4

[a]Data from ref. 20.
[b]Data from ref. 18.

where agriculture developed from 8000 to 5000 years before colonists discovered North America. Corn was introduced from Mexico, wheat from Iraq, soybean from China, and cotton from South America. Nearly all of our common deciduous fruits and most of our common vegetables were introduced by colonists from Europe. Therefore, about 99% of the agricultural land in the United States is planted with exotic crops, and this type of disturbed ecosystem is highly vulnerable to attack by insect pests [23].

A large number of exotic insect species have been brought to North America since 1700. Sailer [26] catalogued 1683 insect species introductions, of which 630 were considered minor pests and 235 were important pests. The number of exotic pest species attacking the principal crops grown in the United States is estimated in Table 2 and totals about 72 of the 174 major pest species enumerated, or about 1% of the total insect fauna. Most of these exotic pests were inadvertently introduced together with cultivars or other foodstuffs and arrived in the New World without the associated fauna of natural enemies that regulated their populations in their native habitats. As a consequence, exotic insect pest species have caused far more extensive crop damage than their relative numbers would suggest. This is demonstrated in Table 3, which summarizes the most important exotic agricultural pest introductions into North America over the past three centuries. The species included are all capable of producing annual damages ranging from $10 to $100 million or more [8,9,20]. The introduction of such exotic pests continues at an accelerated rate because of burgeoning air transportation, tourism, immigration from semitropical and tropical countries, and expanded shipments of food commodities under free trade agreements. It has been estimated that 11 new exotic insects are introduced into the United States annually, 7 are likely to be of some importance as pests, and that about every third year, a major pest is introduced [26].

III. CROP LOSSES FROM INSECT PEST ATTACKS

Crop losses from insect pests in the United States are generally considered to amount to 10–15% of annual crop production [8,27]. The U.S. Department of Agriculture has repeatedly surveyed annual losses for major crops heavily treated with insecticides [9,13,30,31], as summarized in Table 4. The average losses over the first 60 years of the twentieth century ranged from 9.0 to 15.5% for the individual crops surveyed and the overall mean loss was 12.8%. This information is in agreement with Pimentel's [22] estimate of 13% annual loss from insect pest attacks and aggregated to about $6.6 billion in 1990 dollars. A more detailed estimate for 1988 based on specific loss estimates for 63 important agricultural crops showed an aggregate loss of approximately $8 billion. The

Table 3 Major Exotic Insect Pests of Agriculture Introduced into North America

Discovery date and location	Pest	Geographical origin
1779 Long Island	Hessian fly, *Mayetiola destructor*	Europe
1800?	Codling moth, *Cydia pomonella*	Europe
1832 Connecticut	Pear psylla, *Cacopsylla pyricola*	Europe
1860 Quebec	Imported cabbageworm, *Pieris rapae*	Europe
1870 California	San Jose scale, *Quadraspidiotus perniciosus*	China
1872 California	California red scale, *Aonidiella aurantii*	China
1872?	Peach twigborer, *Anarsia lineatella*	Europe
1879?	Pea aphid, *Acrythosiphon pisum*	Europe
1882 Virginia	Greenbug, *Schizaphis graminum*	Europe
1885 Ottawa	Carrot rust fly, *Psila rosae*	Europe
1892 Texas	Boll weevil, *Anthonomus grandis*	Mexico
1904 Utah	Alfalfa weevil, *Hypera postica*	Europe
1916 New Jersey	Japanese beetle, *Popillia japonica*	Japan
1917 Massachusetts	European cornborer, *Ostrinia nubilalis*	Italy
1917 Texas	Pink bollworm, *Pectinophora gossypiella*	Mexico
1922 Mississippi	Vegetable weevil, *Listroderes difficilis*	Brazil
1924 Canada	Sweet clover weevil, *Sitona cylindricollis*	Europe
1952 Nova Scotia	Face fly, *Musca autumnalis*	Europe
1954 New Mexico	Spotted alfalfa aphid, *Therioaphis maculata*	North Africa
1962 Michigan	Cereal leaf beetle, *Oulema melanopus*	Europe
1975 California	Blue alfalfa aphid, *Acyrthosiphon kondoi*	Japan
1975 California	Mediterranean fruit fly, *Ceratitis capitata*	Central America
1986 Texas	Russian wheat aphid, *Diuraphis noxia*	Europe

Source: Data from refs. 4 and 20.

component losses were: staple crops $5.55 billion (9.1%), fruit crops $0.81 billion (8.4%), truck crops $0.51 billion (12.3%), and greenhouse and nursery crops $1.11 billion (15%) [20].

The combined loss information suggests that there has been no overall decrease in U.S. crop losses from insect pest attacks over the present century despite the massive use of insecticides [14,15]. Increases in crop protection have been offset by the planting of more susceptible varieties, reduced crop sanitation, reduced tillage, reduced crop rotations, pest resistance to insecticides, and increased cosmetic standards for produce quality [14,15,22].

A variety of direct damage assessments from individual insect pest attacks are presented in Table 5. These values do not include indirect effects (e.g.,

Table 4 Estimated Annual Losses from Insect Pests Attacking Major U.S. Crops

Crop	% of production lost[a]				
	1900–1904	1910–1933	1942–1951	1951–1960	Average
Cotton	10	15	15	19	14.7
Corn	8	12	4	12	9.0
Apple	20	10	14	13	14.2
Potato	10	22	16	14	15.5
Cabbage	10	20	8	17	13.7
Alfalfa	10	5	9	15	9.7
Average	11.3	14.0	11.0	15.0	

[a]Data from refs. 9, 13, 30, and 31.

resurgences, development of secondary pests, environmental disruptions) associated with the management of these pests.

IV. CHARACTERISTIC BIOLOGY OF ARTHROPOD PESTS

A. Dispersal of Insect Pests

In general, most insect pests disperse from oviposition sites or larval habitats only as far as necessary to acquire satisfactory food supplies. Nearly all Lepidoptera (moths and butterflies) lay their eggs on suitable host plants after stimulation by host-produced kairomones. Examples include the cabbage looper, *Trichoplusia ni*, and the diamondback moth, *Plutella xylostella*, on crucifers; the codling moth, *Cydia pomonella*, on apples; the European corn borer, *Ostrinia nubilalis*, and the corn earworm, *Helicoverpa zea*, on corn; and the alfalfa caterpillar, *Colias eurytheme*, on alfalfa [20].

For the Coleoptera, the boll weevil, *Anthonomus grandis*; the plum curculio, *Conotrachelus nenuphar*; the alfalfa weevil, *Hypera postica*; and the Mexican bean beetle, *Epilachna varivestis*, are monophagous or oligophagous feeders that both oviposit and feed on plants of a narrow taxonomical range, returning to the same general locality each year. The same is true of the northern corn rootworm, *Diabrotica barberi*, and the western corn rootworm, *D. virgifera virgifera*; which, however, oviposit directly into soil adjacent to corn plants [20].

Migratory species represent the other extreme and their journeys in search of niches for feeding and reproduction may cover hundreds to thousands of kilometers (see Chap. 3). The most notorious North American migratory species is the Rocky Mountain grasshopper, *Melanoplus spretus*, which invaded

Table 5 Estimated Annual U.S. Crop Losses from Selected Major Insect Pests

Pest	Crop	Year(s)	Loss $Million
Boll weevil, *Anthonomus grandis*	Cotton	1907–1949	203
Bollworm and Budworm, Heliothine spp.	Cotton	1951–1960	100
California red scale, *Aonidiella aurantii*	Citrus	1943–1944	10
Chinch bug, *Blissus leucopterus*	Corn, grain	1934	55
Codling moth, *Cydia pomonella*	Apple	1940–1944	25
Corn earworm, *Helicoverpa zea*	Corn	1945	140
European cornborer, *Ostrinia nubilalis*	Corn	1949	350
Grasshoppers, *Melanoplus* spp.	Many	1936	102
Greenbug, *Schizaphis graminum*	Small grain	1950	25
Hessian fly, *Mayetiola destructor*	Wheat	1944	47
Hornworms, *Manduca* spp.	Tobacco	1944	84
Mediterranean fruit fly, *Ceratitis capitata*	Fruits	1981–1987	50
Onion thrips, *Thrips tabaci*	Onion	1944	14
Pea aphid, *Acyrthosiphon pisum*	Pea, alfalfa	1944	35
Potato aphid, *Macrosiphum euphorbiae*	Potato	1944	66
Russian wheat aphid, *Diuraphis noxia*	Wheat	1988–1989	100
Spotted alfalfa aphid, *Therioaphis maculata*	Alfalfa	1956	42
Whiteflies, *Bemisia* spp.	Winter vegs.	1991	137

Source: Data from refs. 8, 18, 20–22, and 27.

the middle-western croplands in huge swarms during the period of 1850–1900, migrating from oviposition sites on the eastern slopes of the Rocky Mountains. Encroaching cultivation of these arid breeding grounds destroyed the eggs and disrupted the migration ecology, and such extreme migrations no longer occur [20]. However, the closely related migratory grasshopper, *Melanoplus sanguinipes*, a general feeder on cultivated crops, often flies many miles to new breeding sites.

The armyworm, *Pseudaletia unipuncta*, often migrates in huge numbers during destructive peak population cycles, destroying vegetation over hundreds of square kilometers. The beet leafhopper, *Circulifer tenellus*, is a notorious migratory insect that overwinters on, for example, Russian thistle, saltbush, and wild mustard, in southwestern deserts. As these plants senesce in spring, the leafhoppers migrate hundreds of kilometers to feed on cultivated tomatoes and sugar beets in northern California, Idaho, and Montana, inoculating these crops with the destructive curly top virus [20].

Migrations from subtropical and tropical overwintering areas to summer croplands in the Mississippi Valley are encouraged by prevailing winds. The black cutworm, *Agrotis ipsilon*, overwinters as a pupa in the Gulf states. Newly emerged female moths mate during migratory flights of 1300–1600 km to lay their eggs on weeds adjacent to newly planted corn fields in Iowa, Missouri, and Illinois [7]. The spotted cucumber beetle, or southern corn rootworm, *Diabrotica undecimpunctata howardi*, is a nondiapausing species that hibernates in the southern states as an adult and migrates northward up the Mississippi Valley in the Spring. The distances and directions of these migrations are controlled by prevailing winds, and this species often reaches Canada.

Two tropical insects that migrate into the United States are the soybean looper, *Pseudoplusia includens*, and the velvetbean caterpillar, *Anticarsia gemmatalis*. The soybean looper overwinters in Central America and migrates northward to cause major damage to soybeans in the southern states in August and September. The velvetbean caterpillar also regularly migrates from Central America into the gulf states and has been found feeding on soybean as far north as Ontario, Canada [20].

B. Invasion Biology

Once established as a pest through exotic invasion, by a change in host preference, or by a mutation resulting in a new biotype, the rates and patterns of spread of an insect pest determine its ecological and economic impact. Increased understanding of the complexities of the invasion process is fundamental to efforts in containment, eradication, and control of introduced pests. The following brief sketches of the invasions of several notorious pest species into North America illustrate some of the problems involved. Establishing the date and place of invasion and the route by which it occurred are common historical uncertainties. The basic pattern of invasion biology common to these examples consists of three phases: (a) a relatively static period of consolidation marked by mutual accommodation between the invading species and the agroecosystem, (b) radial and relatively symmetrical spread as permitted by local geographical features, and (c) rapid colonization as influenced by the physiology and ecology of the insect pest, agricultural crop patterns, climatic extremes, geographical barriers, and patterns of transportation [1].

Phase 1 is illustrated by the gypsy moth, *Lymantria dispar*, which remained confined to eastern Massachusetts for more than 20 years following its inadvertent release. During the next 100 years, it spread throughout the states east of the Mississippi River. Another example, the cotton boll weevil, remained confined to southeastern Texas for approximately 10 years after its introduction near Brownsville before spreading eastward to the Atlantic Coast over the next 14 years. In Phase 2, the weevil spread over a series of relatively concentric circles covering the Cotton Belt [20]. This invasion response is also clearly demonstrated by the spread of the cereal leaf beetle, *Oulema melanopus*, from southwestern Michigan [1], and by the spread of the mutant cyclodiene-resistant western corn rootworm from its origin in southeastern Nebraska [18].

Understanding the invasion biology of any species is complicated by (a) historical uncertainties as to the date and area of introduction and of the area of origin of the invading species (e.g., the codling moth and the cereal leaf beetle), (b) the results of repetitive multiple introductions versus endemicity (e.g., the frequent invasions of the Mediterranean fruit fly, *Ceratitis capitata*, in California) [21]; (c) the genetic constitution of multiple biotypes (e.g., the European cornborer with its univoltine, bivoltine, and multivoltine biotypes) [28]; (d) host preferences of monophagous or polyphagous species (e.g., cotton boll weevil, *Anthonomus grandis* [monophagy], or the Japanese beetle, *Popillia japonica* [polyphagy]); (e) changes in host preference (e.g., the Colorado potato beetle, *Leptinotarsa decemlineata*, and the apple maggot, *Rhagoletis pomonella*); (f) changes in and proliferation of agriculture such as irrigation, conservation tillage, new plant varieties or genotypes; and (g) capricious human behavior relating to the introduction and dispersal of invading species.

The gradual and regular geographical progressions of infestations by the boll weevil which required 92 years, and the pink bollworm, *Pectinophora gossypiella*, which required 48 years to reach southern California from southern Texas [20,21], are examples of manmade invasions controlled by the westward progress of irrigated cotton culture. The almost uncontrollable spread of infestations of tephritid fruit flies, such as the apple maggot and the Mediterranean fruit fly [21], are primarily the consequence of the undetectable maggots tunneling in edible fruits either imported to distant markets or discarded by thoughtless travelers.

A tabulation of exotic insect species invading California over the period of 1769–1994 [4] indicated that 62% of the exotic species belong to the order Homoptera and demonstrated that shipments of nursery stock harboring these small and often sessile scales, whiteflies, mealybugs, and aphids are the major way in which these insects have been spread about the world. The spread of severe outbreaks in the United States began with the accidental importations of the cottony cushion scale, *Icerya purchasi*, from Australia in 1868; the San Jose scale, *Quadraspidiotus perniciosus*, from China in 1870; the red and yellow scales, *Aonidiella* spp., from China in 1872; and the black scale, *Saisettia oleae*, from California to Florida in 1874.

V. PRINCIPLES OF HOST PLANT SELECTION

Insect herbivores have been classified as monophagous (single host), stenophagous (restricted hosts), oligophagous (few hosts), and polyphagous (many hosts). However, with our present knowledge of the chemical factors (allelochemics) present in plant tissues that determine host plant selection, these distinctions should be viewed as portions of a continuum of insect responses to allomones and kairomones that provide neurological stimuli resulting in decision making during the phases of the host selection process [10]. These phases are delineated as (a) host habitat finding, (b) host finding, (c) host recognition, (d) host acceptance, and (e) host suitability.

A. Host Finding

Host finding involves phototaxis, anemotaxis, geotaxis, and temperature and humidity preferences (10). It results from long-range sensory inputs, both visual and olfactory. The characteristic reflectance spectrum of green plants is in the green-yellow-orange region and this has its counterpart in the maximal sensitivity of the insect visual pigments or rhodopsins, which is in the region of 500–580 nm. Highly polyphagous herbivores, such as migratory locusts, are attracted to patches of vegetation largely by visual stimuli. However, chemosensory inputs from the characteristic emanations of the "green volatiles" of growing plants are important in specific host finding [19]. Mobile insects characteristically fly upwind (anemotaxis) toward plant sources of volatile attractants (kairomones).

B. Host Recognition

Host recognition is facilitated by specific allelochemics, and oviposition is characteristically triggered by a combination of olfactory and gustatory inputs. Thus, the female apple maggot fly is attracted to maturing apple fruits through a combination of visual imaging of orange-red, fruit-sized spheres together with the emanation of butyl hexanoate and related volatiles from apple fruits [24]. The onion maggot fly, *Delia antiqua*, is attracted to its host plant by the characteristic odor of propyl disulfide; and the cabbage maggot fly, *Delia brassicae*, is drawn to allyl isothiocyanate liberated from the glucosinolates characteristic of the Cruciferae [19].

C. Host Acceptance

Host acceptance is usually determined by "test biting" or by ovipositional probing. The presence of phagostimulant kairomones is of definitive importance in promoting continuous feeding by larvae of the silkworm, *Bombyx mori*, on *Morus* spp. in response to the kairomone morin; by larvae of the catalpa sphinx, *Ceratomia catalpae*, in response to catalposide in the leaves of *Catalpa* spp.; and by *Chrysolina* spp. beetles that feed on *Hypericum* spp. in response to the kairomone hypericin. Many other examples of such specific monophagy or stenophagy are known [5,19].

D. Host Suitability

Host suitability is determined by the nutritional value of the plant through its content of sugars, amino acids, lipids, proteins, and vitamins and by the absence of deleterious factors such as toxic allomones [10].

Consideration of these factors suggests that true polyphagy is found only occasionally, as with migratory locusts and armyworms that devour almost any green vegetation, although there are marked preferences. Other insects with very wide host ranges are recognized; the gypsy moth larva is known to feed on more than 500 species of plants, the Japanese beetle adult, *Popillia japonica,* has been recorded as feeding on more than 350 plant species, and the Mediterranean fruit fly has a host range of at least 250 species. At the opposite extreme of host acceptance, true monophagy is rare and perhaps nonexistent, with species such as the silkworm and catalpa sphinx perhaps qualifying. The boll weevil feeding largely on *Gossypium* spp. and the asparagus beetles, *Crioceris* spp., feeding on *Asparagus* are monophagous from the viewpoint of agriculture.

The great majority of insect herbivores have a restricted stenophagous or oligophagous host range despite the presumed advantages of utilizing many different food sources. Appropriate examples of stenophagy are found in the aphids (Aphididae), where the majority of the species have a very restricted host range as indicated by their common names: bean aphid, chrysanthemum aphid, columbine aphid, melon aphid, rose aphid, violet aphid, walnut aphid, for example. [20].

VI. BIOLOGICAL FACTORS THAT CHARACTERIZE INSECT PEST STATUS

A major goal in the study of applied entomology is to understand the physiological and ecological factors that define successful pest status. Although much has been learned over the past half century, as outlined above, the complexities of insect/plant interrelations are so convoluted as presently to defy precise analysis. Consideration of several groups of host-related, sympatric, and often sibling species, where one or more species are major pests and the others are of little significance, offers grist for future research and contemplation.

A. *Diabrotica* Rootworms

The Nearctic *virgifera* group of diapausing *Diabrotica* spp. rootworms (Coleoptera: Chrysomelidae) includes five species widely distributed across North

America. The larvae of the northern corn rootworm, *D. barberi*, and of the western corn rootworm, *D. virgifera virgifera*, feed almost exclusively upon the roots of corn, where these two species have become major pests throughout the Corn Belt [20]. *D. longicornis* is a sibling species of *D. barberi* that appears to be primarily associated with buffalo gourd, *Cucurbita foetidissima*, a plant of western deserts. The more distantly related *D. lemniscata* is also routinely collected from buffalo gourd. *D. cristata* is also a sibling species of *D. barberi* and *D. longicornis*, but its larvae feed almost exclusively on the roots of big-bluestem, *Andropogon gerardi*. This species is now confined to relict prairies where it is of no economic importance [11]. The males of each of these five species respond to female pheromone esters of 8-methyl-2-decanol [11], and both males and females of all five species are compulsive feeders on the cucurbitacin kairomones that are characteristic phagostimulants of the Cucurbitaceae [17]. The physiology of diapause, oviposition, fecundity, and reproduction seem almost identical for all five species; the ecology and behavior of the adults are remarkably similar, and it seems highly unlikely that natural enemies play major roles in regulating the populations of any of the species. Our present knowledge is clearly not adequate to express the factors responsible for major pest status.

B. *Bactrocera (=Dacus)* Fruit Flies

The Oriental fruit fly, *Bactrocera dorsalis*, is known to feed on at least 173 species of fruits and is one of the world's major pests [19,20]. Sibling species include the banana fruit fly, *B. musae*, which breeds almost exclusively in banana, *Musa* spp.; and the solanum fly, *B. cacuminatus*, which breeds only in *Solanum* spp. The biology of all three species is remarkably similar, and males of all three species characteristically respond to the kairomonal attractant and phagostimulant methyl eugenol [16,19].

C. *Pieris (=Ascia)* spp. as Stenophagous Feeders on Cruciferae

The cabbage butterfly, *Pieris rapae*, is a major pest that is distributed worldwide. Related species such as the southern cabbage butterfly, *Pieris protodice*, the western white, *P. occidentalis*, the California white, *P. sisymbri*, the potherb butterfly, *P. oleracae*, and the Gulf white butterfly, *P. monusta*, are much more restricted in distribution and appear to breed largely on wild Cruciferae [20]. The several species are very similar in appearance and biology, and their stenophagy on the Cruciferae indicates that host selection is controlled by the kairomonal glucosinolates that characterize this plant family.

Examples of these biological conundrums that determine major, minor, or nonpest status for very closely related insect species can be amplified many-fold. Prokopy's [21] conclusions about the current state of understanding of the Mediterranean fruit fly are generally applicable:

> It seems unlikely that progress can be made toward developing safer or more effective approaches to controlling Medfly without developing a) a much firmer understanding of how . . . behavior is organized in space and time in natural habitats, and b) much more complete appreciation of how variations in environmental factors and fly physiological information, or genetic state factors affect patterns of behavioral organization.

VII. INSECT PEST PROBLEMS OF DISTURBED AGROECOSYSTEMS

Agroecosystems where crops are produced are highly disturbed and of little plant diversity as compared with natural ecosystems. Contrast, for example, the original tall grass prairie characteristic of the Midwestern United States and containing as many as 100 associated plant species, with its present-day successor the cultivated field of wheat, corn, soybean, or alfalfa, where the optimum is to obtain a single and totally dominant plant species. There is every reason to believe that human intervention in terms of modern agriculture is the single most important factor in determining whether or not an insect species becomes an important pest.

No better argument could be presented than that of the Colorado potato beetle, *Leptinotarsa decemlineata*, which Thomas Say described in 1824 as a rare denizen of the eastern slopes of the Rocky Mountains where it fed on the buffalo burr, *Solanum rostratum*, and other wild *Solanum* spp. For 30 years longer, this insect continued to live in obscurity until pioneer settlers brought the cultivated potato, *Solanum tuberosum*, into the region. The insect rapidly deserted its weed hosts for the cultivated potato and began its notorious march eastward spreading from potato patch to potato patch, reaching Nebraska in 1859, Illinois in 1864, Ohio in 1869, and the Atlantic Coast in 1874. Its annual rate of spread was about 137 km per year [20].

The western corn rootworm, *Diabrotica virgifera*, was first described by Joseph LeConte in 1868 from collections on *Cucurbita foetidissima* in Southwestern Kansas and was first recorded as a pest of corn near Fort Collins, Colorado in 1909. This insect is an example of a manmade pest, as its larvae develop almost exclusively on the roots of corn, *Zea mays*, and there is but a single generation annually. The western corn rootworm cannot survive on any other crop plant and is effectively controlled by crop rotation with soybean or other legume. It is largely because of economic pressure to grow continuous corn throughout the Corn Belt that the western corn rootworm has become one of the most damaging insects in North America [7].

The pink bollworm, *Pectinophora gossypiella*, is an exotic pest of irrigated desert-grown cotton. The larva attacks developing bolls, feeding in the seed and damaging the lint. Damage from this pest can be largely minimized by planting early-maturing varieties of cotton and by area-wide stalk destruction to destroy the food resources of the pest and largely to eliminate overwintering populations. Producer demands for two sets of bolls each year in an extended cotton growing season in the desert area of southern California have resulted in heavy infestations of the pink bollworm and insecticide costs that have sometimes exceeded $400 per acre. As a result, cotton acreage in California's Imperial Valley has declined from 125,000 acres to less than 20,000 acres over a recent 5 year period [6].

VIII. DEVELOPMENT OF PESTICIDE RESISTANCE, RESURGENCES, AND SECONDARY PESTS

The wholesale application of broad-spectrum pesticides to agroecosystems has resulted in the development of many new arthropod pests through the combined phenomena of selection for insecticide-resistant biotypes, destruction of key natural enemies, resurgences in target pest populations, and development of secondary pests [14,15]. Insecticide use in agriculture in the United States increased about 10-fold over the 40-year period from 1940 to 1980, yet as shown in Table 4, there has been no appreciable reduction in crop losses caused by insect pests. There has, however, been a dramatic increase in the selection of insecticide-resistant races of arthropods from 11 known examples in 1946 to more than 500 in 1995 [14,15]. High levels of pesticide resistance have been demonstrated to all known classes of insecticides: fumigants, arsenicals, organochlorines, organophosphates, carbamates, pyrethroids, insect growth regulators, and microbial insecticides. Many pest species now exhibit multiple resistance to most or all of these types of insecticides and have become almost uncontrollable by chemical applications [14,15,18]. The development of the cyclodiene-resistant corn rootworm, *D. virgifera virgifera*, at a single locus in southeastern Nebraska in 1959, resulted in a new and very destructive biotype which then invaded the entire U.S. Corn Belt between 1959 and 1980.

The development of pesticide-resistant races of arthropods, together with destruction of their natural enemies, has resulted in the appearance of a new class of secondary pests. The classic example is the emergence of resurgent epidemics of red spider mites (Tetranychidae). These have been particularly damaging on citrus and deciduous fruit crops, and their impact on pest control programs is demonstrated by the sudden appearance in the entomological literature of the terms *acaricide* and *miticide* in 1947, following the proliferation of the new organochlorine insecticides [14,15]. Before that time, red spider mites were rarely abundant enough to require special applied control measures. As an example of the drastic changes brought about in insect control, Circular 212 of the Illinois Agricultural Experiment Station (1918) "Directions for Spraying Fruit in Illinois," recommended eight spray treatments for apples using lime sulfur, Bordeaux mixture, oil, and nicotine. The revised Circular 1151 (1968) recommended 43 pesticides, including the selective acaricides chloropropylate, dicofol, ovex, propargite, tetradifon, oxythioquinos, and organotins [15,18].

Today, the use of broad-spectrum insecticides has almost inevitably been followed by pest resistance, pest resurgence, and by the development of secondary pests. A 1977 inventory of the 25 most important insect pests of agriculture in California disclosed that 17 were resistant to one or more insecticides, and that 24 were either pest resurgences aggravated by the use of insecticides or secondary pest outbreaks [14,15]. The disastrous results of the development of secondary pests in cotton agroecosystems following massive insecticide applications for the control of the boll weevil and pink bollworm were demonstrated by the emergence of the bollworm and budworm, *Helicoverpa zea* and *Heliothis virescens*, respectively, as voracious secondary pests. Their pest status has been exacerbated by the high levels of resistance in the budworms to many insecticides [14,15]. By 1992, these two secondary pests were reported to cause a yield reduction of 2.21% in U.S. cotton production, exceeding that of the 2.12% reduction credited to the major key pest, the boll weevil [6]. In some cotton-growing areas, these secondary cotton pests have required from 10 to 20 applications of insecticides, with control costs of $300–$400 per acre. As a result, there have been catastrophic declines in cotton production in many areas [6,18]. In contrast, the successful eradication of the boll weevil in the southeastern United States has resulted in reduced applications of insecticides and expansion of cotton acreage.

Further discussion of the insect pest problems created by the injudicious use of insecticides is beyond the scope of this chapter. It suffices to state that the combined effects of insecticide resistance, pest resurgences, and secondary pest development are major factors in the development of the present-day emphasis on Integrated Pest Management (IPM) [14,15,18]. Nevertheless, the development of effective IPM programs for insect pest management must consider the ecology, behavior, genetics, and host plant relationships of the target pests. Continued strict reliance on chemical insect controls while disregarding the life history of the target pest is doomed to failure over the long run.

REFERENCES

1. Andow, D.A., P.M. Kareiva, S.A. Levin, & A. Okubo. 1993. Spread of invading organisms: patterns of spread. In *Evolution of Insect Pests*. K.C. Kim & B.A. McPheron, eds. New York: Wiley, pp 219–242.

2. Borror, D.J., D.M. DeLong, & C.A. Triplehorn. 1981. *An Introduction to the Study of Insects*. 6th ed. Philadelphia: Saunders.

3. Brues, C.T., A.L. Melander, & F.M. Carpenter. 1954. *Classification of Insects*. Bull. Harvard Mus. Comp. Zool. 108. Cambridge, Massachusetts: Harvard University.

4. Dowell, R.V. 1995. List of exotic invertebrates that became established in California between 1600 and 1994. In *The Mediterreanean Fruit Fly in California: Defining Critical Research*. J.G. Morse, R.L. Metcalf, & J.R. Carey. Riverside, California: University of California, College of Agricultural Natural Sciences, pp. 22–39.

5. Fraenkel, G. 1969. Evaluation of our thoughts on secondary plant compounds. *Entomol. Exp. Appl.* 12:473–86.

6. Frisbie, R.E., H.T. Reynolds, P.L. Adkisson, & R.F. Smith. 1994 Cotton insect pest management. In *Introduction to Insect Pest Management*. 3rd ed. R.L. Metcalf & W.H. Luckmann, eds. New York: Wiley, pp 421–468.

7. Gray, M.E., & W.H. Luckmann. 1994. Integrating the cropping system for corn insect pest management. In *Introduction to Insect Pest Management*. 3rd ed., R.L. Metcalf & W.H. Luckmann, eds. New York: Wiley, pp 507–541.

8. Haeussler, G.J. 1952. Losses caused by insects. In *Insects, the Yearbook of Agriculture*, Washington, DC: U.S. Department of Agriculture, pp 141–146.

9. Hyslop, J.A. 1938. Losses occasioned by insects, mites, and ticks in the U.S. In *USDA Bureau of Entomology*, E-444, Washington, DC: U.S. Department of Agriculture.

10. Kogan, M. 1994. Plant resistance in pest management. In *Introduction to Insect Pest Management*. 3rd ed. R.L. Metcalf & W.H. Luckmann, eds. New York: Wiley, pp 73–128.

11. Krysan, J.L., I.C. McDonald, & J.H. Tumlinson. 1989. Phenogram based on allozymes and its relationship to classical bio-systematic structure among eleven Diabroticites (Coleoptera: Chrysomelidae). *Ann. Entomol. Soc. Am.* 82:574–581.

12. Luckmann, W.H., & R.L. Metcalf. 1994. The pest management concept. In *Introduction to Insect Pest Management*. 3rd ed., R.L. Metcalf & W.H. Luckmann, eds. New York: Wiley, pp 1–34.

13. Marlatt, C.L. 1904. Annual losses caused by destructive insects in the U.S. In *USDA Yearbook of Agriculture*. Washington, DC: U.S. Department of Agriculture, pp 461–474.

14. Metcalf, R.L. 1980. Changing role of insecticides in crop protection. *Annu. Rev. Entomol.* 25:219–256.

15. Metcalf, R.L. 1986. The ecology of insecticides and the chemical control of insects. In *Ecological Theory and Integrated Pest Management Practice*. M. Kogan, ed. New York: Wiley, pp 251–297.

16. Metcalf, R.L. 1990. Chemical ecology of Dacinae fruit flies (Diptera: Tephritidae). *Ann. Entomol. Soc. Am.* 83:1017–1030.

17. Metcalf, R.L. 1994. Chemical ecology of Diabroticites. In *Novel Aspects of the Biology of Chrysomelidae*. P.H. Jolivet, M.L. Cox, & E. Petitpierre, eds. The Netherlands: Kluwer, pp 153–169.

18. Metcalf, R.L., & W.H. Luckmann, eds. 1994. *Introduction to Insect Pest Management*. 3rd ed. New York: Wiley.

19. Metcalf, R.L., & E.R. Metcalf. 1992. *Plant Kairomones and Insect Ecology and Control*. New York: Chapman & Hall.
20. Metcalf, R.L., & R.A. Metcalf. 1993. *Destructive and Useful Insects: Their Habits and Control*, 5th ed. New York: McGraw-Hill.
21. Morse, J.G., R.L. Metcalf, J.R. Carey, & R.V. Dowell, eds. 1995. *The Mediterranean Fruit Fly in California: Defining Critical Research*. Riverside, California: University of California College of Agricultural Natural Sciences.
22. Pimental, D. 1986. Agroecology and economics. In *Ecological Theory and Integrated Pest Managment Practice*. M. Kogan, ed. New York: Wiley, pp 299–319.
23. Pimentel, D. 1993. Habitat factors in new pest invasions. In *Evolution of Insect Pests*. K.C. Kim & B.A. McPheron, eds. New York: Wiley, pp 165–181.
24. Prokopy, R.A., & B.A. Croft. 1994. Apple insect pest management. In *Introduction to Insect Pest Management*. 3rd ed. R.L. Metcalf & W.H. Luckmann, eds. New York: Wiley, pp 543–585.
25. Sabrosky, C.W. 1952. How many insects are there? In *Insects, the Yearbook of Agriculture*. Washington, DC: U.S. Department of Agriculture, pp 1–7.
26. Sailer, R.I. 1983. History of insect introductions. In *Exotic Plant Pests and North American Agriculture*. C.T. Wilson & C.L. Graham, eds. New York: Academic Press, pp 15–38.
27. Schwartz, P.H., & W. Klassen. 1981. Estimates of losses caused by insects and mites to agricultural crops. In *Handbook of Pest Management in Agriculture*. D. Pimentel, eds. Boca Raton, Florida: CRC Press, pp 15–17.
28. Showers, W.V. 1993. Diversity and variation of European corn borer populations. In *Evolution of Insect Pests*. K.C. Kim & B.A. McPheron, eds. New York: Wiley, pp 287–310.
29. Smart, J., & N.F. Hughes. 1973. The insect and the plant: progressive paleontological introduction. In *Insect/Plant Interrelationships*. H.F. van Emden, ed. Entomological Society of Lond., Symposium No. 6. London: Blackwell, pp. 148–156.
30. U.S. Department of Agriculture. 1954. Losses in agriculture. *Agricultural Research Service Report* 20-1, Washington, DC.
31. U.S. Department of Agriculture. 1965. Losses in agriculture. *Agricultural Handbook* 291, Agr. Res. Ser., Washington, DC.
32. Weiss, H.B. 1925. Notes on ratios of insect food habits. *Proc. Biol. Soc. Wash.* 38:1–14.

15
Cultural Approaches to Managing Arthropod Pests

John N. All
University of Georgia, Athens, Georgia

I. INTRODUCTION

Cultural control is any farm practice that is used with some consideration for insect management. In this context, a *farm practice* is a multipurpose agricultural operation used for profitable crop production as a fundamental priority, such as choosing seed, planting, cultivating, or harvesting. Farming operations can have positive, negative, or neutral influences on insect populations, and knowledge of all three impacts is useful in making cultural control decisions. Traditionally, cultural control has referred to the use of farming practices to manage insects [14]. This simple idea is too restrictive and does not include concepts associated with risk management. In the current context, cultural control also includes methods that avoid creating hazardous crop environments for insect infestations (i.e., preventative pest management).

Terms such as *environmental control* [52], *ecological management* [50,55], and *habitat management* [4,24] have been used in connection with an "ecosystem management" philosophy for integrated pest management (IPM) that includes traditional cultural control methods, but may also incorporate modern technologies such as relative insect sampling methods, IPM decision models requiring regular monitoring of environmental parameters, or the use of menu-driven interactive computer systems. These systems often require entomological capability at a professional level and emphasize insect impact and biological interaction rather than agronomic efficiency of the cultural control practices. On the other hand, successful farmers and pest management specialists know that simplicity is often essential for business success in real-world agriculture, and they tend to actually use uncomplicated cultural control practices with a view that any pest suppression technology they adopt has to be profitable.

In Sun Tzu's *Art of War*, the author presents an admonition highly applicable to cultural control of arthropod pests: "Know the enemy and know yourself: in a hundred battles, you will never be defeated." The effective use of cultural controls likewise requires detailed knowledge of the pest's ecology and how the manmade cropping system affects its ecology. The principal environmental factors affecting insect/crop interactions and cultural control are the suitability of plants as insect food and the physical habitat. Proliferation of insect populations is fundamentally determined by crop and cultivar selection, and this important farm decision is the foundation of cultural control [4,61,68]. Many environmental conditions influence the quality of insect habitats, but the principal parameters are temperature, moisture (rainfall and humidity), solar characteristics (day length, sun exposure), air movement, and soil characteristics. Ambient and accumulated temperature and moisture levels of air and soil are important physical parameters used for cultural control decisions, because they can be measured relatively easily and accurately on site. Soil texture, organic matter content, and pH, for example, can be analyzed at moderate cost in public and private laboratories, and can thus be readily used for cultural control decisions.

II. PREVENTATIVE AND RESPONSIVE CONTROL PHILOSOPHY

Cultural control is preventative pest management and emphasizes avoidance of damaging insect populations rather than suppression of pest outbreaks. Preventative control usually requires patience and relies on detailed knowledge of the interaction between the crop, the crop's pests, and the pests' natural enemies. The goal of preventative control is to inhibit or avoid a build-up of pests or damage to economic injury levels using procedures of cultivar selection, biological control, habitat management, or the prophylactic use of insecticides [4]. The deployment of preventative control strategies is usually incorporated with farming operations conducted more for crop production than pest control, and the nonchemical procedures usually involve little or no extra cost [4].

Responsive, or therapeutic, control is concerned with suppressing insect outbreaks that threaten economic damage to the commodity [4]. The components of responsive control are insect monitoring, the use of action thresholds, and deployment of insecticides or releases of natural enemies [5]. These are costly operations that are required for rapid elimination of a high proportion of an insect population that threatens economic damage. Cultural control operations typically are not rapid or suppressive enough of insect populations for use in responsive control programs. Insect suppression with cultural operations often takes weeks, months, or even years for fruition. On the other hand, cultural control can accumulate over time to produce sustained repression of pest populations below economic injury levels at little or no extra cost for the grower [14].

III. HAZARD ASSESSMENT OF AGRICULTURAL ENVIRONMENTS

Hazard assessment for pest problems associated with a proposed farming practice is an important aspect of cultural control, but it is often neglected in the planning process of agricultural programs [30]. Farmers face an array of decisions for crop production on each section of land, including crop and cultivar options. Additionally, decisions on purchase, calibration, deployment, and maintenance of a variety of implements for tillage, planting, pesticide application, fertilization, irrigation, and harvest are routinely made by farmers, often with an accompanying debt repaid from, or levied or borrowed against, the current year's crop. Each cultural practice is chosen for a variety of agronomic and economic reasons relevant to crop production, often with little consideration of the hazard the technology has for promoting insect problems. In retrospect, growers often realize that a high-yielding cultivar or tillage practice used for soil conservation is also vulnerable to insect infestations.

Hazard assessment is a consideration of risk probabilities [77] for increased pest problems under a given scenario of cultural practices for crop production. Presently, hazard assessment is mostly intuitive and, as such, is quantitative only in the sense of a high, intermediate, or low assessment of hazard for pest problems associated with adoption of a farming practice. In the future, if sufficient quantitative information becomes available on the positive or negative influences of cropping technologies on pest populations, more sophisticated mathematical procedures (or improved sampling methodology) may be used to model and predict the pest hazard of specified cropping programs. For example, Mack et al. [45] used simple weather parameters of temperature and lack of rainfall in the development of "borer days," which are essentially hazard levels to indicate when to initiate scouting and insecticide applications for the lesser cornstalk borer, *Elasmopalpus lignosellus*, infestations in conventionally tilled peanut fields.

A crop life table can provide useful information for hazard assessment of cropping practices for cultural control. Crop life tables are tabulations of the major pest problems during the life of the crop, and several have been published for different commodities [29]. To be highly useful for cultural control, a crop life table needs to be tailored to local pest conditions and field environments.

Unfortunately, the biological details of pest/crop interactions needed for localized crop life tables usually are lacking. Nevertheless, the intuitive process of characterizing crop development in a series of discrete cohorts with assessment of the general hazard for infestations of major resident or transient pests provides a valuable conceptual framework for planning cultural programs to manage pests. Life tables for insects are the counterpart of crop life tables, and they can be an invaluable source of clues in developing cultural control strategies for pests [17,32,60].

A cultural control method never operates independently of other cropping practices that themselves can influence the population potential of a targeted pest [3,55]. This is important to consider in making cost/benefit projections for using specified farming practices and evaluating their hazard or control potential as cultural control practices. Often a profitable agronomic method with undesirable pest hazard attributes can be offset by other cultural practices that have insect control characteristics. For example, southern corn billbug, *Sphenophorus callosus*, infestations are increased in no-tillage corn when an alternate weed host, nut sedge, is present in or around fields. However, the cultural practices of in-row subsoiling, irrigation, and fertilization can be used in no-tillage systems to enhance crop tolerance of billbug injury [34].

IV. DISRUPTION OF PLANT HOST SUITABILITY

Insect populations thrive in crop monocultures that are easily accessible, have low natural defenses, and will nutritionally support one or more generations of a pest. Avoiding highly susceptible plant hosts for pest infestations is a foundation operation of cultural control [6]. Farmers can avoid problems beginning with the process of developing a field plan for a farm by considering the field history for insect problems with a particular crop and with an appreciation of potential pest build-up and movement between adjoining fields. For example, the hazard of problems arising from chinch bug, *Blissus leucopterus*, and cereal leaf beetle, *Oulema melanopus*, infestation in seedling corn is increased when winter grain fields are located nearby [22], and cotton growers in the southeastern United States are aware that increased bollworm damage is associated with moth populations issuing from tobacco fields and maturing fields of corn [66].

A. Crop Rotation

Crop rotation is seasonally alternating one or more crops within a field to disrupt insects that are biologically dependent on one of the crops. Crop rotation sometimes effectively controls insects referred to as *resident-type pests* [4]. These insects are a recurring problem, and they often have a restricted host range and more than a 1-year life cycle. White grubs of various species are examples of resident pests that build over time in monocropped fields and can be effectively managed by rotating crops [47]. Crop rotation has generally not been effective for *transient-type pests* such as the corn earworm, *Helicoverpa zea*, which seasonally moves into fields from adjoining and distant areas and usually has multiple generations per year. Transient-type pests are often polyphagous, and this type of feeding behavior tends to negate the benefits of crop rotation.

The northern (*Diabrotica barberi*) and western (*D. virgifera virgifera*) corn rootworms are insects that determine their offspring's diet (corn) before it is planted by laying eggs for overwintering diapause in monocropped fields. Thus, rootworm populations have been successfully managed by rotating corn with a crop (e.g., soybeans) unsuitable for larval survival [41].

In 1985, reports of northern corn rootworm infestations in corn rotated with soybean in areas of South Dakota, Iowa, and Minnesota prompted studies which revealed rootworm strains which produce eggs that had prolonged diapause for 2 or more years [38]. The rootworm strains capable of extended diapause were apparently genetically selected by crop rotation and became numerous enough to produce economic damage. Insects like rootworms that preinfest fields endure similar ecological stresses as certain nematode, bacterial, and fungal species that produce encysted or otherwise durable reproductive structures capable of prolonged inactivity in soil [1]. Options such as longer rotational cycles (using each of the crops for at least 2 years before rotating) and insecticide deployment have been suggested as management strategies to counteract the extended diapause characteristic of western corn rootworm populations in these areas [41,49].

A corn/soybean rotation for rootworm control has not been profitable enough for many Midwestern farmers who continue to use a continuous corn program and apply a planting time insecticide for pest management. In order for crop rotation to be a feasible choice preferred over the use of a prophylactic insecticide: (a) the rotation crops should be culturally compatible (e.g., use the same equipment for production, grow well in the field), (b) both crops should be profitable and in harmony with short- and long-term farm objectives, (c) using the crops should not increase hazard for other pests (including nematodes, diseases, and weeds), (d) crop rotation should approach similar suppression of insect populations as a prophylactic insecticide, (e) chemical control should become less effective against the pest owing to development of insecticide resistance or enhanced microbial degradation of the insecticides, (f) the cost/benefit relationship for use of insecticides must become marginal or prohibitive, or (g) the use of insecticides might be more stringently regulated because of concern for adverse environmental effects.

The crop rotation concept also can be applied in multiple cropping systems and for continuous plant-growing operations in greenhouses and plant nurseries. Southern corn billbug infestations of corn were reduced in fields where canola (used as a green manure in fields) and corn were alternated during the season [57], and damage in no-tillage corn by southern corn rootworm larvae was less abundant in fields in which winter wheat had been used as a cover crop as compared with hairy vetch [20]. Rotating plant species within and between greenhouses is recommended to avoid infestations of whiteflies, mites, leafminers, and thrips, and it is common practice periodically to leave greenhouse units vacant for several weeks to purge a variety of pests [71].

B. Cultivar Selection

Choosing a cultivar is one of the most important farming decisions, and many factors are considered with the goal of achieving optimum crop performance. Crop yield, produce quality, determinate or indeterminate growth, earliness, harvest characteristics, for example, as well as pest resistance properties, are considered before purchasing seed. For cultural control purposes, cultivar selection is the use of adapted cultivars that are least conducive to the development of insect outbreaks. The concept includes the use of resistant or tolerant varieties if they are available, but it additionally includes the avoidance of highly susceptible varieties in environments with a high hazard for pest infestations.

Developing insect-resistant crops by traditional breeding or by new genetic technologies is an important part of IPM and a topic of another chapter (see Chap. 8). In reality, a grower's first priorities for a cultivar are usually for optimum yield or produce quality, and often insect-resistant varieties with these characteristics are not available. From the point of view of cultural control strategies, cultivar selection is working with the wide assortment of commercial varieties that are available and choosing one that has the best profit potential with the least hazard for insect problems. Knowledge of the differential susceptibility among highly productive crop cultivars to specified insects provides the IPM specialist with information to make risk assessments for using a particular variety. The information also can be a basis to prepare alternative control strategies when the profitability attributes of the cultivar outweigh its hazard for having insect problems.

Unfortunately, it is often difficult to obtain susceptibility and resistance profiles for commercial cultivars to insects. Many states have annual performance evaluations of crop varieties in different locations, and sometimes insect damage ratings are made along with data on yield and produce quality [10]. Seed companies will usually inform buyers of general vulnerability or resistance of cultivars in their sales line. Local cooperative extension agents, independent consultants, and neighboring farmers can be sources of information on what crop varieties to use or avoid. Some farmers make it a practice to have small plantings of several cultivars and conduct their own evaluations of crop performance.

C. Crop Tolerance

Farming practices that aid crop tolerance of insect injury are useful cultural control strategies. For example, optimum fertilization, irrigation, and subsoiling all aid corn growth and have been shown to reduce injury by the western corn rootworm, southern corn billbug, lesser cornstalk borer, *Elasmopalpus lignosellus*, and to reduce disease severity of maize chlorotic dwarf and maize dwarf mosaic viral diseases which are transmitted by insects [2,7,8,64]. On the other hand, these same practices promote robust, succulent growing plants that are considered more vulnerable to aphids [70]. Also, the manner in which a cultural practice is deployed can influence the potential of pest success. Irrigation, optimum fertilization, and weed control undoubtedly aid grape tolerance of the debilitating injury of the grape root borer, *Vitacea polistiformis*, to grape roots. However, drip irrigation tends to maintain moist surface soil conditions under vines, which promote optimum survival of freshly hatched borer larvae. In contrast, overhead sprinkler irrigation results in faster drying of surface soil and encourages crusting which is inhospitable to young larvae [5].

V. CULTURAL PRACTICES THAT INFLUENCE ENVIRONMENTAL HAZARD FOR PESTS

A. Sanitation of Breeding Reservoirs

Sanitation has generally pertained to the destruction of insect breeding sites or other pest reservoirs using farming equipment or other multipurpose machinery [14,50]. Knipling [36] argues that sanitation practices can decrease the hazard of agricultural environments for developing insect infestations in a manner that insecticides cannot. This is due to a proportional relationship between the number of insects eliminated and decrease in the population's reproductive potential that occurs when breeding or overwintering sites are destroyed. Insect suppression with insecticides does not produce a parallel decline in reproductive potential, and surviving populations can theoretically increase more rapidly, because there are fewer competitors for the undisturbed habitat. Sanitation helps prevent a "snowballing" of pest infestation from occurring in fields by removing breeding material that allows proliferation of insect populations.

Removing fallen or culled fruits and vegetables from fields, or shredding, bush hogging, and plowing of postharvest crops destroys insect life stages that may be harbored within or on the materials. For example, destruction of cotton plants following harvest is mandated in the Boll Weevil Eradication Program conducted across the U.S. Cotton Belt, because unharvested squares and bolls are essential for adult weevils to build the fat reserves they require for winter survival [68]. Pink bollworm, *Pectinophora gossypiella*, life stages also are destroyed by postharvest cotton destruction [28].

In corn, the overwintering stages of the European corn borer, *Ostrinia nubilalis;* southwestern corn borer, *Diatraea grandiosella;* and southern corn-stalk borer, *D. crambidoides*, are killed by crop destruction [47]; whereas, in wheat, the Hessian fly, *Mayetiola destructor*, is one of several pests that are disrupted by postharvest crop destruction [76]. Sanitation of fruits, nuts, and vegetables removes major pests such as the codling moth, *Cydia pomonella;* plum curculio, *Conotrachelus nenuphar;* melon flies, *Dacus cucurbitae;* pecan weevils, *Curculio caryae;* pickleworms, *Diaphania nitidalis;* and potato tuber-worms, *Phthorimaea operculella*, from planted areas [42].

Sanitation is a cultural control practice used in forests managed commercially or for recreation. Many species of scolytid bark beetles, such as the southern pine beetle, *Dendroctonus frontalis*, produce aggregating pheromones when colonizing insects successfully attack vulnerable trees. The pheromones rapidly attract high numbers of beetles to a concentrated cluster of trees (brood trees or group kills), and expedient salvage of infested timber can often remove a substantial portion of a beetle population from a forest stand and prevent pro-liferation of infestations [16]. The challenge to foresters has been the logistical problems of locating and removing infested trees before bark beetle emergence. This is often difficult to justify economically; beetle-infested timber is not as valuable, because blue stain (caused by the fungus *Ceratocystus pini* and asso-ciated with bark beetle attacks) degrades the quality of the wood for lumber or paper products. Aerial surveying may be used to rapidly locate group kills. Foresters often schedule cutting of large areas (up to an hectare) around beetle spots to be certain that all infested lumber is removed. Additionally, tree limbs and other postharvest debris are burned or rendered unsuitable for bark beetle brood survival [25].

Burning of trash in agricultural fields or in forest environments is a ques-tionable cultural procedure, because it produces air pollution, but it is an inex-pensive sanitation practice that removes weeds and debris that compete with or otherwise obstruct the establishment of new crops. Burning eliminates breed-ing material for pests and sources of plant pathogens that may be transmitted by insects to crops [54]. Direct exposure to fire is obviously destructive to in-sects, but the use of burning for insect control may not be as effective as would be expected. For example, incineration of small grain stubble is destructive to the puparia of the Hessian fly located in the duff left after harvest [67]. On the other hand, the relatively low heat generated by burning of field stubble is not lethal to Hessian fly puparia in the soil and has a negligible impact on soil pests [23]. Also, burning can increase the hazard for infestations of the lesser cornstalk borer. Smoke and freshly incinerated detritus are attractive to ovipositing moths of the lesser cornstalk borer, and increased infestations by the pest in various crops have been associated with burning [72].

B. Sanitation of Alternate Hosts

Destruction of volunteer crops or weeds that germinate within fields or when crops are planted is often important, because the plants can be reservoirs for pests that build and spread into germinating crops. This can particularly be a problem when insect-transmissible plant pathogens are present in rogue plants that are extant uniformly throughout a field. The plants serve as sources of inoculum for short-distance movement by vectors and can create a widespread epiphytotic more rapidly than normal [53]. For example, two viruses, maize dwarf mosaic (MDMV) and maize chlorotic dwarf (MCDV), overwinter in Johnson grass rhizomes and in the spring are transmitted by aphids (MDMV) and leafhoppers (MCDV) to corn. The economic injury level for Johnson grass as a reservoir for MDMV and MCDV is probably less than is its weed economic threshold. This is especially apparent in no-tillage fields. The aphid and leafhopper vectors can transmit the viruses into germinating corn seedlings sooner in no-tillage fields owing to the presence of Johnson grass in fields when corn is germinating. This results in higher rate of spread and greater severity of the disease on corn in conservation tillage systems than in conventional tillage systems. Even if Johnson grass is not completely controlled by herbicides and tillage in conventional tillage systems, the rhizomes are disrupted by cultivation operations before planting so that weed development is delayed for a few weeks. The delay is important in epiphytotics, because severity and yield loss to both diseases is negatively correlated with the age of the plants at the time they are inoculated with the viruses [2].

Sanitation also is used for preventative control of beet yellows virus, and transmission of tobacco etch virus by the green peach aphid, *Myzus persicae*, is augmented by the presence of volunteer plants in fields [48,73]. Both pathogens infect several wild and domestic hosts which may germinate in freshly plowed fields or reside as perennials in bordering areas. Burning, cultivation, and the use of herbicides to maintain clean fields and border areas have been used with success for control of both diseases by eliminating alternate hosts from which insects can transfer the pathogens to standing crops [74].

C. Benefit/Risk Relationships

When sanitation is considered as a cultural control practice, benefit/risk relationships need to be carefully examined, such as whether the hazard for increased pest problems by not removing an insect breeding source outweighs its benefit for other purposes. For example, it is clearly established that cultivation reduces populations of several pests by either direct destruction or removal of food sources. But does lack of cultivation or the use of no-tillage practices increase the hazard for pest problems and does the risk outweigh the conservation and agronomic benefits of no-tillage? Infestations of certain pests are increased in no-tillage systems, but the problems can usually be managed by other pest control strategies to allow for the profitable use of reduced cultivation [9].

Another example where the benefits of a cultural practice may outweigh its risks is the practice of composting, which may include the presence of culled fruits, vegetables, and plant debris that are infested with insects. If the materials are processed properly, the temperatures within compost piles reach 160°C, which is lethal to insects. However, temperatures in the outer layers of compost piles are not lethal, and if the material is not turned and processed properly, it can be a source of infestations. Insect pests that develop in compost piles can be prevented from escaping by using screens or other coverings or by periodic application of insecticides to the materials [27].

VI. DISRUPTING CONTINUITY OF PEST BIOLOGICAL RHYTHMS: PHENOLOGICAL ASYNCHRONY

If manipulation of crop phenology is culturally feasible, it can be among the most effective control strategies for many pests. Yield potential of many crops is strongly influenced by the timing of planting, and the possibility of reduced production must be weighed against pest hazard when considering this strategy. Other factors related to cultivar selection, such as, for example, maturity date and determinate or indeterminate flowering, can be used in developing a planting time strategy for pest control. Producing phenological asynchrony in pest/ crop interactions can be accomplished by alternating the time of planting (or harvest) or by selecting a cultivar with desirable development traits that avoid pest peaks. The strategy can be applied either to a pest complex on a crop during the season or to specific insects that produce damage at a distinct stage of crop development.

There are many examples of successful manipulation of crop phenology for preventative control of both resident- and transient-type pests [69]. In general, insects that are vulnerable to this method have one or more of the following attributes:

  a. A brief breeding period or a short reproductive life span (e.g., Hessian fly; sorghum midge, *Contarinia sorghicola*)
  b. Narrow crop phenological requirements for reproduction (e.g., sorghum midge)

c. Not adapted for year-long survival in a region and migrate consider-able distances each growing season to infest crops (e.g., corn ear-worm; fall armyworm, *Spodoptera frugiperda*; velvetbean caterpillar, *Anticarsia gemmatalis*; aster leafhopper, *Macrosteles quadrilieatus*)

d. Two or more generations per season, and high populations occur in "flushes" of rapid increase and decline that are closely associated with certain hosts (e.g., three-cornered alfalfa hopper, *Spissistilus festinus*; chinch bugs; soybean looper, *Pseudoplusia includens*)

e. Multiple overlapping generations per year in a region and a tendency to build populations during the season before harvest (e.g., lesser cornstalk borer; boll weevil, *Anthonomus grandis grandis*)

f. Feeding preference for specific maturity stages of a crop (e.g., velvetbean caterpillar, stink bugs)

Examples of insects having these attributes, and how they can be exploited, are presented below.

Attribute 1: The Hessian fly is an example of an insect having a brief reproductive period. The use of fly-free planting dates for managing Hessian flies is an example of manipulating the time of planting to dis-rupt an insect with a short window of breeding activity. This insect lays eggs on germinating winter wheat in the fall, larvae develop on the plants, then pupae diapause during the winter. Three characteristics of Hessian fly reproductive biology lend themselves to manipulation by altering planting dates of winter wheat. First, the fly's requirement for diapause to overwinter in northern states ensures the synchronization of populations in the fall. Second, adults are short-lived (3 or 4 days). Third, vegetative plant stages are required for the fly's oviposition and larval survival. Control is achieved by understanding the adult emer-gence period and delaying planting until most have died [75]. This sce-nario breaks down in southern states where mild winters do not limit adult activity and fail to induce diapause. For example, in the Georgia coastal plain, Hessian flies have up to six broods a year and are present throughout much of the planting period of wheat [21]. In general, de-laying planting dates reduces the incidence of most pests of autumn-sown winter crops, such as aphid vectors of the barley yellow dwarf virus.

Attribute 2: An example of an insect with a narrow host phenological requirement which makes it vulnerable to manipulation of crop phenol-ogy is the sorghum midge. It can successfully oviposit in a floral spikelet of sorghum for only the 1–2 days when it is actively producing pol-len. When sorghum is planted early, it pollinates before high midge populations develop in wild hosts (mainly Johnson grass) and, if the plants flower uniformly in a field as a result of good agronomic practice, the period for oviposition is minimized. Thus, a widespread program of early planting is an effective preventative control of midges in sorghum [69].

Attribute 3: Several insects, including the corn earworm and fall armyworm, exhibit this attribute. These insects have annual migratory flights from southern overwintering areas into northern states where they are unable to survive the winters. Sweetcorn growers from mid Georgia north into eastern Canada know that early planting or the use of an early maturing hybrid avoids high earworm infestations and may completely escape fall armyworm flights.

Attributes 4 and 5: Insects having these attributes are usually indigenous in a region, but differ in population dynamics, with the former having distinct "flushes" in numbers two or more times in a season, whereas the other type typically has a steady build-up in individuals during the year. The sunflower moth, *Homoeosoma electellum*, has two or three generations per year which are synchronized so that "flushes" of ovipositing moths occur at distinct intervals and plantings of sunflower can be adjusted so that the susceptible flowering period escapes peak moth populations [15]. The lesser cornstalk borer has three or four overlapping generations, and populations increase during the season on the many crop and weed hosts that the insect infests. Most row crops are only susceptible in seedling stages, so early planting is recommended [7]. Peanuts are infested throughout development, but damage to pegs is of most concern to growers, and early planting usually favors avoidance of severe infestations by lesser cornstalk borer [44].

Attribute 6: The velvetbean caterpillar (VBC), is an example of insects with this attribute, and whose population dynamics are associated with crop growth patterns. The insect is a serious soybean defoliator in the southeastern United States and has migratory flights that usually reach Georgia, Alabama, and Mississippi in late September and October. Resistant soybean cultivars, such as "Crocket," inhibit several lepidopteran defoliators prior to flowering, but inhibition declines during pod-filling stages of plant development. Since Crocket is a later maturing cultivar (maturity group VIII), the resistance sometimes is not present when high populations of VBC move into soybean fields in Florida and Georgia in late fall. The use of earlier maturing cultivars (maturity group VI) such as "Lamar" and "Lyon" is more effective, because the plants have matured when the VBC populations occur [59].

VII. DIVERSION TACTICS FOR CULTURAL CONTROL

A. Trap Cropping

The use of cultural practices to create environments that divert insects from a crop has been considered for many years and implemented on a small scale, but they usually have not had widespread implementation, because the practices were often impractical or were not as effective as insecticides. This may be changing, probably more from impending restrictions on insecticides than improved capabilities of the diversion strategies. For example, the potential value of trap cropping as a pest management tactic has been known for many years [14,31,65]. Trap cropping is the establishment of a preferred host of a specified pest within or adjacent to the cash crop. The intention is to attract a high percentage of the pest into the trap crop, where it can be destroyed more efficiently than under usual field infestation conditions. The concept has generally been directed at a single dominant pest and may utilize the same crop species (a more preferred cultivar or the same cultivar in a more desirable phenological stage of development) or a different, more preferred, host of the insect. The trap crop must be highly attractive, and the target insect must be mobile enough to aggregate readily in the trap areas. Finally, a reliable method for controlling the insect in the trap crop must be available for rapid deployment when necessary. The whole sequence can seem complicated to growers, and thus they often are reluctant to adopt trap cropping until they are convinced that it is an effective and profitable method for preventing insect infestations.

A pilot IPM program was conducted by the University of Georgia in 1992–1994 using trap cropping of stinkbugs in soybeans to demonstrate the value of the method for reducing insecticide costs and for conserving parasitoids and predators of soybean insects [33]. The program was based on research that demonstrated the economic efficiency of using a border of soybean one to two maturity groups earlier than the main crop as a trap for stinkbugs [46,62]. Eight rows of an early-maturing soybean cultivar were planted around the periphery of fields planted with a late-maturing, high-yielding variety. The trap variety was sprayed one to three times with a truck-mounted airblast sprayer. Stinkbug populations in the trap crop area were up to 10 times higher than the main fields in the program, and insecticide use was reduced by 80% in the field with trap crops. Damage to soybean pods by stinkbugs was substantially reduced in the trap crop fields compared with unsprayed fields. Grower adoption of stinkbug trap cropping has increased by approximately 25% in Georgia during 1992 through 1994.

B. Diversity Cropping

Polyculture, companion planting, intercropping, diversity planting, and *weedy culture* are terms applied to intermixing one or more crops to disrupt normal insect host selection so that fewer plants are attacked [11,13,26,51,56]. The concept is popular with ecologists and proponents of organic farming systems who are concerned about the negative influence of crop monocultures on insect population dynamics, and who can demonstrate that, under certain circumstances, increasing plant diversity may reduce pest populations [18,40,56]. This practice has influenced certain host-specific pests, but it has been ineffective or variable for polyphagous insects. Unfortunately, most diversity programs have not been proven to be practical in large-scale production agricultural systems, and tangible benefits in increased yield or profitability have not been demonstrated [13]. In order for a crop diversity program to be a viable cultural control option in modern agricultural systems, it should not only reduce the hazard for insect infestations, but it also should be as profitable as a monocrop option in the absence of pests.

The basic requirement for economic feasibility of diversity cropping to stimulate farmer adoption is discussed by Stern [65] using intercropping for management of lygus bugs in cotton. Lygus damage in cotton could be reduced by using a mixed cropping system of alfalfa strips planted alternatively with cotton because of the insect's strong preference for alfalfa over cotton. Normally, using a strip row planting system is not as profitable as a monocropping cotton system where insecticides are used to control lygus. However, in the early 1960s, federal regulations for cotton land allotments were assessed in a manner economically favorable to the establishment of skip row planting systems. Many farmers apparently adopted the alfalfa/cotton mixed row system for the management of lygus despite the fact that irrigation requirements for the two crops were not compatible and insecticides were effective. Regulations were changed in the 1970s so that monocropping of cotton became the most profitable cultural practice regardless of the presence of the pest, and farmers soon abandoned the strip row system in favor of the more economically efficient monocropping practice.

Polyculture farming has been promoted for subsistence agriculture in developing countries on small farms where access to insecticides and other pest control technologies are limited [12,18,19,39,43]. Additionally, these technologies may be effective in gardening and small vegetable farm operations [26,35,58]. However, even in these systems, the cost effectiveness of mixed cropping methods as an alternative to conventional agronomic practice must be carefully weighed against the value of pest suppression.

C. Strip Harvesting

Strip harvesting is a procedure that utilizes different maturity states of the same crop in large strips or multiple rows within a field. These are more practical agronomic methods as compared with other polyculture practices, because conventional farming equipment can be used more efficiently in establishing and harvesting the crops. Strip cutting alfalfa for hay so that a reservoir of plants is maintained in uncut strips helps prevent pest migration to adjacent crops and conserves biological control agents [65].

D. Reflective Mulches

Reflective mulches have been used with success for diversion of aphid vectors of viral diseases from plantings of susceptible vegetables and ornamentals [37,63].

VIII. CONCLUSIONS

Cultural control is an old concept that has been updated, probably not to the extent deserving new nomenclature, and as used here, it is *any farm practice used with some consideration for insect management*. Cultural control is the raw reality of applied ecology and operates under rules of economic efficiency. It has the goal of achieving insect management within a noncontaminated and sustainable agricultural environment, but only by using cultural practices that are cost efficient for farmers. Within academic circles, cultural control may entertain sophisticated ecological theories and conjecture about managing agroecosystems in a perfect world, but adoption by farmers usually requires applying ecological principles that agree with the practical realities required for profitable crop production. Cultural control programs have, for the most part, needed to be user friendly and employ simple, economic procedures with proven value.

Cultural control is preventative insect management that manages insect populations by manipulating food suitability or physical habitat. A variety of options, including crop and cultivar selection, crop rotation, and manipulation of planting and harvest dates, are available for use as preventative control tactics for one or several pests. Few cultural control methods should be considered as a single tactic for a given pest, but they ought to be used in tandem with other cropping methods which may or may not be suppressive to insects but do not pose an increased hazard for infestations.

The future of cultural control methods that encourage diversity in agricultural environments, such as companion planting, strip cropping, and spatial arrangements of crops, as large pest management programs is uncertain. Ecological dogma contends that monocultures foster pest problems, because they present a few adapted insects with a uniform food medium for proliferation. Logic would indicate that utilizing methods that increase diversity would result in reduced pest problems. There is some indication that certain methods that promote environmental diversity do prevent infestations, but in general, their pest control impact is outweighed by economic inefficiency for crop production. In order for future progress to occur, agricultural entomologists need to (a) clearly demonstrate the economic value for pest control of a crop diversity program and (b) show (preferably in cooperation with crop production scientists and agricultural economists) that in the absence of pests the diversity methods are as profitable as conventional monocropping technologies.

REFERENCES

1. Agrios, G.N. 1987. *Plant Pathology*. New York: Academic Press.
2. All, J.N. 1984. Integrating techniques of vector and weed host suppression into control programs for maize virus diseases. In *Proceedings of the International Maize Virus Disease Colloquium and Workshop*. D.T. Gordon, J.R. Knoke, L.R. Nault, & R.H. Ritter, eds. Wooster, Ohio: Ohio State University Press, pp 243–247.
3. All, J.N. 1987. Importance of concomitant cultural practices on the biological potential of insects in conservation tillage systems. In *Arthropods in Conservation Tillage Systems*. G.J. House & B.R. Stinner. Lanham, Maryland: Entomological Society of America Miscellaneous Publications, pp 11–18.
4. All, N.J. 1989. Importance of designating prevention and suppression control strategies for insect pest management programs in conservation tillage. In *Conservation Farming, Integrated Pest Management*. I.O. Teare, E. Brown, & C.A. Trimble, pp 1–5. Proceedings of the Southern Conservation Tillage Conference. University of Florida Press, Special Bulletin.
5. All, J.N. 1992. Preventative and responsive control strategies for management of grape pests. In *Viticultural Science. Proceedings of the 15th Viticultural Scientific Symposium*. S. Leong, ed. Tallahassee, Florida: Florida A&M University, p 5.
6. All, J.N. 1994. Interaction of cultivar, tillage practice, and aldicarb on thrips populations in seedling cotton. In *Annual Plant Resistance to Insects Newsletter*. E.E. Ortman & R.H. Ratcliffe, eds. W. Lafayette, Indiana: Purdue Unviersity Press, pp 15–19.
7. All, J.N., W. Gardner, E.F. Suber, & B. Rogers. 1982. Lesser cornstalk borer as a pest of corn and sorghum. *A Review of Information on the Lesser Cornstalk Borer, Elasmopalpus lignosellus* (Zeller). H.H. Tippins, ed. University of Georgia College of Agriculture Special Publication, pp 33–46.
8. All, J.N., R.S. Hussey, & D.G. Cummins. 1984. Influence of no-tillage, coulter-in-row-chiseling, and insecticides on damage severity on corn by the southern corn billbug and plant parasitic nematodes. *J. Econ. Entomol.* 7:178–182.
9. All, J.N., & G.J. Musick. 1986. Management of vertebrate and invertebrate pests.

In *No-Tillage and Surface-Tillage Agriculture: The Tillage Revolution*. M.A. Sprague & G.B. Triplett. New York: Wiley, pp 347–387.

10. All, J.N., G.B. Rowan, & H.R. Boerma. 1993. Field cage and greenhouse ratings of soybean cultivars for resistance to three insect species. In *1992 Field Crops Performance Tests*. P.L. Raymer, J.L. Day, R.B. Bennett, S.H. Baker, W.D. Branch, & M.G. Stephenson, eds. Georgia Agricultural Experimental Station Special Report, p 53.

11. Allardice, P. 1993. *A-Z of Companion Planting*. New York: Harper Collins.

12. Andow, D.A. 1983. Effect of agricultural diversity on insect populations. In *Environmentally Sound Agriculture*. W. Lockeretz, ed. New York: Praeger, pp 91–118.

13. Andow, D.A. 1991. Yield loss to arthropods in vegetationally diverse agroecosystems. *Environ. Entomol.* 20:1228–1235.

14. Anon. 1969. *Insect-Pest Management and Control*. Washington DC: National Academy of Sciences.

15. Aslam, M., G.E. Wilde, T.L. Harvey, & W.D. Stegmeier. 1991. Effect of sunflower planting date on infestation and damage by the sunflower moth (Lepidoptera: Pyralidae) in Kansas. *J. Agric. Entomol.* 8:101–108.

16. Belanger, R.P., R.L. Hedden, & P.L. Lorio. 1993. Management strategies to reduce losses from the southern pine beetle. *South. J. Appl. Forest.* 17:150–154.

17. Bellows, T.S.J., R.G. Van Driesche, & J.S. Elkington. 1992. Life table construction and analysis in the evaluation of natural enemies. *Annu. Rev. Entomol.* 37:587–614.

18. Bohlen, P.J., & G.W. Barrett. 1990. Disperal of the Japanese beetle (Coleoptera: Scarabaeidae) in strip-cropped soybean agroecosystems. *Environ. Entomol.* 19:955–960.

19. Bottenberg, H., & M.E. Irwin. 1992. Flight and landing activity of *Rhopalosiphum maidis* (Homoptera: Aphididae) in bean monocultures and bean-corn mixtures. *J. Entomol. Sci.* 27:143–153.

20. Buntin, G.D., J.N. All, D.B. McCracken, & W.L. Hargrove. 1994. Cover crop and nitrogen fertilizer effects on southern corn rootworm (Coleoptera: Chrysomelidae) damage in corn. *J. Econ. Entomol.* 87:1683–1688.

21. Buntin, G.D., P.L. Bruckner, & J.W. Johnson. 1990. Management of Hessian fly (Diptera: Cecidomyiidae) in Georgia by delayed planting of winter wheat. *J. Econ. Entomol.* 83:1025–1033.

22. Burkhardt, C.C. 1985. Insect pests of corn. In *Fundamentals of Applied Entomology*. R.E. Pfadt, ed. New York: Macmillan, pp 282–309.

23. Chapin, J.W., J.S. Thomas, & M.J. Sullivan. 1992. Spring- and fall-tillage system effects on Hessian fly (Diptera: Cecidomyiidae) emergence from a coastal plain soil. *J. Entomol. Sci.* 27:293–300.

24. Corbet, P.S. 1973. Habitat manipulation in the control of insects in Canada. In *Proceedings of the Tall Timbers Conference on Ecological Animal Control by Habitat Management*. H.T. Gormeratz, ed. Tallahassee, Florida: Tall Timbers, pp 147–171.

25. Coulson, R.N., & J.A. Witter. 1984. *Forest Entomology, Ecology and Management*. New York: Wiley.

26. Cromartie, J.W., Jr. 1981. The environmental control of insects using crop diversity. In *CRC Handbook of Pest Management in Agriculture*. D. Pimentel, ed. Boca Raton, Florida: CRC Press, p 223–251.

27. Flint, M.L. 1990. *Pests of the Garden and Small Farm: A Growers Guide to Using Less Pesticide*. Davis, California: University of California.

28. Frisbie, R.E., & J.K. Walker. 1981. Pest management systems for cotton insects. In *CRC Handbook of Pest Management in Agriculture*. D. Pimentel, ed. Boca Raton, Florida: CRC Press, pp 187–202.

29. Harcourt, D.G. 1970. Crop life tables as a pest management tool. *Can. Entomol.* 102:950–955.

30. Hedden, R.L. 1981. Hazard-rating system development and validation: an overview. In *Hazard-Rating Systems in Forest Insect Pest Management: Symposium Proceedings*. R.L. Hedden, S.J. Barras, & J.E. Coster, eds. Washington, DC: U.S. Department of Agriculture, Forest Service, pp. 9–12.

31. Hokkanen, H.M.T. 1991. Trap cropping in pest management. *Annu. Rev. Entomol.* 36:119–138.

32. Horn, D.J. 1988. *Ecological Approach to Pest Management*. London: Guilford Press.

33. Hudson, R.D., & R.G. McDaniels. 1994. The use of trap crops as a means of controlling stink bugs in soybeans. In *Proceedings of the 1994 Southern Soybean Conference*. W. Flinchum, ed. St. Louis: American Soybean Association, pp 96–99.

34. Javid, A.M., J.N. All, & J.S. Laisa. 1986. The influence of cultural treatments on feeding damage of southern corn billbug (Coleoptera: Curculionidae) to corn. *J. Entomol. Sci.* 21:276–282.

35. Kloen, H., & M.A. Altieri. 1990. Effect of mustard (*Brassica hirta*) as a non-crop plant on competition and insect pests in broccoli (*Brassica oleracae*). *Crop Prot.* 9:90–96.

36. Knipling, E.F. 1979. *The Basic Principles of Insect Population Suppression and Management*. Washington DC: U.S. Department of Agriculture.

37. Kring, J.B., & T.J. Kring. 1990. Aphid flight behavior. In *Aphid-Plant Interactions: Populations to Molecules*. D.C. Peters, J.A. Webster, & C.S. Chlouber, eds. Oklahoma State University: Oklahoma Agricultural Experimental Station, pp 203–214.

38. Krysan, J.L., D.E. Foster, T.F. Branson, K.R. Ostlie, & W.S. Cranshaw. 1986. Two years before the hatch: rootworms adapt to crop rotation. *Bull. Entomol. Soc. Am.* 32:250–253.

39. Kyamanywa, S., & J.K.O. Ampofo. 1988. Effect of cowpea/maize mixed cropping on the incident light at the cowpea canopy and flower thrips (Thysanoptera: Thripidae) population density. *Crop Prot.* 7:186–189.

40. Lamp, W.O. 1991. Reduced *Empoasca fabae* (Homoptera: Cicadellidae) density on oat-alfalfa intercrop systems. *Environ. Entomol.* 20:118–126.

41. Levin, E., & H. Oloumi-Sadeghi. 1991. Management of Diabroticite rootworms in corn. *Annu. Rev. Entomol.* 36:229–255.
42. Liquido, N.J. 1991. Fruit on the ground as a reservoir of resident melon fly (Diptera: Tephritidae) populations in papaya orchards. *Environ. Entomol.* 20:620–625.
43. Litsinger, J.A., V. Hasse, A.T. Barrion, & H. Schmutterer. 1991. Response of *Ostrinia furnacalis* (Guenee) (Lepidoptera: Pyralidae) to intercropping. *Environ. Entomol.* 20:988–1004.
44. Mack, T.P., & C.B. Backman. 1990. Effects of two planting dates and three tillage systems on the abundance of lesser cornstalk borer (Lepidoptera: Pyralidae), other selected insects, and yield in peanut fields. *J. Econ. Entomol.* 83:1034–1041.
45. Mack, T.P., D.P. Davis, & R.E. Lynch. 1993. Development of a system to time scouting for the lesser cornstalk borer (Lepidoptera: Pyralidae) attacking peanuts in the southeastern United States. *J. Econ. Entomol.* 86:164–173.
46. McPherson, R.M., & L.D. Newsom. 1984. Trap crops for control of stink bugs in soybean. *J. Entomol. Sci.* 19:470–479.
47. Metcalf, R.L., & R.A. Metcalf. 1993. *Destructive and Useful Insects: Their Habits and Control,* 5th ed. New York: McGraw-Hill.
48. Padgett, G.B., F.W. Nutter, Jr., C.W. Kuhn, & J.N. All. 1990. Quantification of disease resistance that reduces the state of tobacco etch virus epidemics in bell pepper. *Phytopathology* 80:451–455.
49. Paul, J. 1987. Rootworm hiberation proves disastrous to corn. *Agrichem. Age* 7:12–13.
50. Pedigo, L.P. 1989. *Entomology and Pest Management.* New York: Macmillan.
51. Pimentel, D. 1960. Species diversity and insect population outbreaks. *Ann. Entomol. Soc. Am.* 54:76–86.
52. Pimentel, D. 1981. Introduction. In *CRC Handbook of Pest Management in Agriculture.* D. Pimentel, ed. Boca Raton, Florida: CRC Press, pp 3–11.
53. Pitre, H.N., & F.J. Boyd. 1970. A study of the role of weeds in corn fields in the epidemiology of corn stunt disease. *J. Econ. Entomol.* 63:195–197.
54. Powell, D.M., & R.L. Wallis. 1974. Spraying peach trees and burning weed hosts in managing green peach aphid to reduce the incidence of virus in potatoes and sugarbeets. In *Proceedings of the Tall Timbers Conference on Ecological Animal Control by Habitat Management.* T.G. Gomeratz, ed. Gainesville, Florida: Tall Timbers, pp 87–98.
55. Rabb, R.L., G.K. Defoliart, & G.G. Kennedy. 1984. An ecological approach to managing insect populations. In *Ecological Entomology* C.B. Huffaker & R.L. Rabb, eds. New York: Wiley, pp 697–728.
56. Risch, S.J., D. Andow, & M.A. Altieri. 1983. Agroecosystem diversity and pest control: data, tentative conclusions and new research directions. *Environ. Entomol.* 12:625–629.
57. Roberts, P.M. 1993. *Risk Assessment of Southern Corn Billbug, Lesser Cornstalk Borer, and Fall Armyworm Infestations in Field Crops Using Sustainable Agricultural Practices.* PhD dissertation, University of Georgia, Athens, Georgia.

58. Roltsch, W.J., & S.H. Gage. 1990. Influence of bean-tomato intercropping on population dynamics of the potato leafhopper (Homoptera: Cicadellidae). *Environ. Entomol.* 19:534–543.

59. Rowan, G.B., H.R. Boerma, J.N. All, & J.W. Todd. 1993. Soybean maturity effect on expression of resistance to lepidopterous insects. *Crop Sci.* 33:433–436.

60. Ruesink, W.G. 1982. Analysis and modeling in integrated pest management. In *Introduction to Insect Pest Management.* R.L. Metcalf & W.F. Luckmann, ed. New York: Wiley, pp 353–373.

61. Sailer, R.I. 1981. Extent of biological and cultural control of insect pests of crops. In *CRC Handbook of Pest Management in Agriculture.* D. Pimentel, ed. Boca Raton, Florida: CRC Press, pp 57–67.

62. Schumann, G.W., & J.W. Todd. 1982. Population dynamics of the southern green stink bug (Heteroptera: Pentatomidae) in relation to soybean phenology. *J. Econ. Entomol.* 75:748–753.

63. Smith, F.F., G.V. Johnson, R.P. Kahn, & A. Bing. 1964. Repellency of reflective aluminum to transient aphid virus-vectors. *Phytopathology* 54:748–752.

64. Spike, B.P., & J.J. Tollefson. 1991. Response of western corn rootworm-infested corn to nitrogen fertilization and plant density. *Crop Sci.* 31:776–785.

65. Stern, V. 1981. Environmental control of insects using trap crops, sanitation, prevention, and harvesting. In *CRC Handbook of Pest Management in Agriculture.* D. Pimentel, ed. Boca Raton, Florida: CRC Press, pp 199–207.

66. Stinner, R.E., R.L. Rabb, & J.R. Bradley. 1974. Population dynamics of *Heliothis zea* (Boddie) and *H. virescens* (F.) in North Carolina: A simulation model. *Environ. Entomol.* 3:163–168.

67. Suber, E.S., P.B. Martin, & W.L. Morrill. 1978. *Control of Insects in Small Grains.* Athens, Georgia: Georgia Extension Service.

68. Summy, K.R., & E.G. King. 1992. Cultural control of cotton insect pests in the United States. *Crop Prot.* 11:307–319.

69. Teetes, G.L. 1981. The environmental control of insects using planting times and plant spacing. In *Handbook of Pest Management in Agriculture.* D. Pimentel, ed. Boca Raton, Florida: CRC Press, pp 209–221.

70. van Emden, H.F. 1973. Aphid host plant relationships. In *Perspectives in Aphid Biology.* A.D. Lowe, ed. Christchurch, New Zealand: Caxton Press, pp 54–64.

71. van Lenteren, J., & J. Woets. 1988. Biological and integrated pest control in greenhouses. *Annu. Rev. Entomol.* 33:233–270.

72. Viana, P.A. 1981. *Effect of Soil Moisture, Substrate Color and Smoke on the Population Dynamics and Behavior of the Lesser Cornstalk Borer, Elasmopalpus lignosellus, Zeller (Lepidoptera: Pyralidae).* PhD dissertation, Purdue University, West Lafayette, Indiana.

73. Wallis, R.L. 1967. Some host plants of the green peach aphid and beet western yellows virus in the Pacific Northwest. *J. Econ. Entomol.* 60:904–907.

74. Wallis, R.L., & J.E. Turner. 1969. Burning weeds in drainage ditches to suppress populations of green peach aphids and incidence of beet western yellows disease in sugar beets. *J. Econ. Entomol.* 62:307–309.

75. Walton, W.R., & C.M. Packard. 1930. The Hessian fly and how losses from it can be avoided. In USDA Farmers' Bulletin, Washington DC: U.S. Department of Agriculture.
76. Wilde, G. 1981. Wheat arthropod-pest management. In *CRC Handbook of Pest Management in Agriculture*. D. Pimentel, ed. Boca Raton, Florida: CRC Press,
77. Wilson, R., & E.A.C. Crouch. 1987. Risk assessment and comparisons, an introduction. *Science* 236:267–270.

16
Biological Control of Arthropod Pests

John R. Ruberson
University of Georgia, Tifton, Georgia

James R. Nechols
Kansas State University, Manhattan, Kansas

Maurice J. Tauber
Cornell University, Ithaca, New York

I. INTRODUCTION: DEFINITION AND SCOPE OF BIOLOGICAL CONTROL

Throughout the history of human reliance on plants for food and fiber, native natural enemies have played a major role in suppressing herbivorous arthropods. The concept of biological control reflects the assumption that naturally occurring, biotic mortality factors limit or restrain most populations of living organisms. In addition to these naturally occurring biological controls, humans have intentionally used natural enemies of arthropods as an integral part of crop production since the early development of agriculture.

No single definition of biological control adequately encompasses arthropod pests, weeds, and plant pathogens, although such definitions have been attempted (e.g., see ref. 6). For purposes of this chapter, which deals exclusively with the biological control of arthropods, we have adopted DeBach's [29] definition of biological control as the ". . . action of parasites, predators, or pathogens in maintaining another organism's population density at a lower average than would occur in their absence." This definition neither explicitly includes nor excludes human involvement, and it is also indifferent to the economic value of the suppression. Indeed, the definition applies equally well to naturally occurring biological control of unmanaged systems as well as agricultural ones. This point is important, because much biological control of pest or potential pest populations occurs outside of the crop system in space and time (See Chap. 4).

Our definition also excludes so-called 'biologically based' tactics of pest management, such as pheromone trapping, mating disruption, host plant resistance, genetic controls, and biological toxins. These valuable tactics, which are discussed elsewhere in this text, have little in common with biological control or with each other (see refs. 9 and 47 for discussion).

This chapter focuses on the biological control of arthropod pests in crop production systems. Numerous works have addressed biological control of arthropods (e.g., see refs. 7, 24, 29, 33, 69, 70, 71, 89, 90, 100, 107, 132, 146, and 148); these texts contain more detailed discussions of the subject. In addition, various texts provide valuable information on the biology of predators [8,22,55,131], parasitoids [4,10,22,59,154], and pathogens [15,18,48,50, 109,126,133].

II. HISTORY OF BIOLOGICAL CONTROL OF ARTHROPODS

The history of biological control of arthropods shows that the role of natural mortality factors, such as natural enemies, in suppressing pest populations has been recognized for many centuries [54]. Predaceous ants of the genus *Oecophylla* were used to manage pests of orchard systems as early as A.D. 900 in China (and their use in this manner persists today).

Manipulation of natural enemies for biological control in Europe and North America appears to have developed more slowly. During the eighteenth and nineteenth centuries, Western societies became aware of biological control—specifically the roles of predators, parasitoids, and pathogens—as efforts were made to manage pest populations by manipulating natural enemy populations. These efforts culminated in the highly successful biological control of the cottony-cushion scale in California (and subsequently other areas of the world) by a predaceous lady beetle, the vedalia beetle (*Rodolia cardinalis*), and a parasitoid (the cryptochetid fly, *Cryptochetum iceryae*) imported from Australia [17,36].

The spectacular success of the vedalia beetle project on citrus stimulated the importation of natural enemies for control of other introduced pests (later termed "classical biological control"). An intense but poorly documented period of importation followed, with most of the efforts involving lady beetles [17,54]. Between 1920 and 1940, stringent regulations were enacted in various countries to reduce the free movement of plant and animal material across political boundaries, and classical biological control was conducted more carefully. Subsequently, the advent of synthetic organic insecticides and acaricides in the 1940s and 1950s relegated biological control to a very minor role in crop protection, particularly in the United States and other developed countries where pesticides were readily available. Simultaneously, research in biological control received relatively few resources and attracted relatively little interest, and as a consequence of heavy pesticide use, natural enemy populations were severely disrupted in many agroecosystems.

Several events renewed interest in biological control during the 1950s and 1960s. First, the development of resistance to pesticides in numerous pest species demonstrated serious limitations to pesticide use and stirred considerable alarm among the public and agriculturalists. Second, the publication of Rachel Carson's landmark book *Silent Spring* in 1962 sparked intense environmental awareness and accelerated the incorporation of natural enemies into what was ultimately termed integrated pest management (IPM) [127].

III. APPROACHES TO BIOLOGICAL CONTROL OF ARTHROPODS

In general, three approaches are employed in the biological control of arthropod pests: (1) importation of natural enemies, (b) augmentative releases of natural enemies, and (c) conservation of natural enemies. Below we consider each of these three approaches.

A. Importation ("Classical" Biological Control)

Importation of biological control agents has been widely practiced since the initial spectacular success of the vedalia beetle in 1889–1890 (see Section II and refs. 23 and 86). This approach has yielded some of the best-known examples of biological control [16]. It focuses primarily on the importation of natural enemies for introduced pests that, after invading a new area, have escaped from the regulating action of their natural enemies in their native environments. Because of the high frequency of immigrants among resident pest faunas (see Chap. 14), classical biological control has widespread application [119]. Importation of exotic natural enemies that have not evolved with exotic or native pest species, but attack closely related species, has also been promulgated (see ref. 19; called "new associations" by Hokkanen and Pimentel [64,65]). This approach has yielded some successes, and therefore should not be ignored. However, because release of these natural enemies could pose increased risks to nontarget organisms, their use likely will be restricted.

The success of a classical biological control program depends greatly on practitioners who carefully plan and execute a sequence of procedures (e.g., see refs. 118 and 147). Each of these is addressed below.

1. Procedures in Classical Biological Control Programs

Taxonomic Determination of Pest and Literature Search.   Prior to initiating extensive searches for natural enemies, it is vital to establish the taxonomic identity of the targeted pest and to examine the literature to determine its geographical origins. Improper identification of the targeted pest can seriously hamper exploration efforts and delay success of the program (for examples, see ref. 29 and cassava mealybug case history in Section III.2). Furthermore, it is important to assess the published information on natural enemies of the pest in order to determine whether biological control agents are available and, if so, where they might be acquired.

Foreign Exploration.   When likely geographical sources of natural enemies are identified, appropriate permits must be obtained, and personal and political contacts should be established to permit entry or access into countries where candidate organisms might be available and for subsequent export of candidate organisms. Exploration may involve transport of scientists to the country in question or collaborative work with scientists working in the country to be explored.

In selecting a location for collecting natural enemies, climate is a crucial consideration. To increase the likelihood that the introduced biological control agent will be capable of establishing in its new home, efforts should be made to match the climate of the collection area with the proposed release area.

Among the less resolved questions in the foreign exploration component of importation are (a) how many different sites within the geographical range of the natural enemy should be sampled, and (b) how many individuals of the target natural enemies should be collected from each site to enhance the probability of successful establishment? These questions are rooted in population genetics and population ecology, and they address the complex issue of genetic variability present in the founder stock. No clear patterns between the numbers collected and the success of establishment of a natural enemy have emerged, and collections of large numbers of individuals at a single site may provide little added benefit—the majority of the local population's allelic variability would probably be present in a relatively small sample [66,88,113]. Rather than collecting large numbers of natural enemies in one or two locales, it has been suggested that collections be made at a variety of sites in the natural enemy's geographical range, to increase genetic variability in the sample [21,88].

Importation.   After appropriate candidate organisms have been found, it is necessary that they be properly (and legally) exported and subsequently (again, legally) imported into the targeted country in preparation for release. This requires shipment of natural enemies in appropriate life stages and under proper environmental conditions to permit survival until they arrive at their destination [14]. This problem has been ameliorated greatly with the development of rapid, efficient transportation and delivery systems. Nevertheless, care must be taken to maximize survival of the natural enemies during shipment. Imported organisms are held in quarantine facilities subsequent to their arrival in the country of intended release.

Quarantine/Testing. Imported organisms are held in quarantine, typically for one or several generations, for the removal of unwanted pathogens and hyperparasites, and to permit evaluation of the vulnerability of nontarget organisms to attack by the imported organism [41]. Quarantine facilities are carefully designed to permit efficient propagation of natural enemies while precluding their escape [82]. The quarantine period also provides an opportunity to assess life-history attributes of natural enemies prior to release.

Release and Evaluation. After an imported natural enemy has been deemed safe for release, its numbers may be increased considerably through rearing programs; subsequently it is distributed for release. Efforts are made to release the natural enemies into a variety of regions and habitats to increase the opportunities for establishment. To date, no clear theory has emerged concerning the optimal size and genetic makeup of a released population to enhance the probability of establishment. Mackauer [88] observed that neither the number of natural enemies released nor the number of sites where releases were made had any substantive influence on the likelihood of establishment. Nevertheless, it is generally thought that every effort should be made to maximize genetic variability within the released cohorts; the greater the prevalent variability, the greater the presumed likelihood that a genotype will occur that is appropriate to the released environment. However, no consensus has emerged regarding the means of achieving this variability. Some workers have suggested that the various strains or races of the collected natural enemy should be maintained and released separately to retain the variation [42,66]. Others have advocated pooling all strains to allow the best genotypes to be sorted out after release [142,143].

Evaluations of released natural enemies could be of great value for assessing what factors determine success or failure to establish and suppress the target pest. Unfortunately, such studies are rare because of the limited resources provided for such work. Although some excellent follow-up studies have been done, these efforts are dwarfed by the numerous projects where no evaluations were undertaken. It is clear that some evaluation should accompany each biological control project, and a few selected ones should be studied intensively. Only by doing so can sufficient data be accumulated to place biological control of importation on a sound scientific footing [40,153].

2. Case Histories of Classical Biological Control

Cassava Mealybug. The highly successful, ongoing biological control of the cassava mealybug in Africa provides a valuable case history of classical biological control [62].

The cassava mealybug, *Phenacoccus manihoti*, a native of South America, first appeared on cassava in Africa in 1973, presumably through introduction of plant material [62]. Although a complex of indigenous natural enemies attacked the mealybug in Africa, pest suppression was poor and the problem rapidly escalated. Initial searches for natural enemies in South America and the Caribbean area were seriously handicapped by misidentification of the mealybug—parasitoids were collected from the wrong host species and the natural enemies were unable to reproduce on the cassava mealybug. Finally, in 1981, specimens of the cassava mealybug were located in Paraguay and their natural enemies were collected; among the parasitoids was the encyrtid wasp *Epidinocarsis lopezi*. Since its introduction into Africa, this parasitoid has provided excellent control of the mealybug in areas where it has established. Conservative estimates indicate that the project yielded a benefit-cost ratio of 149 to 1 [104].

E. lopezi possesses several biological attributes that likely contributed to its success [62]. First, it has a short generation time relative to its host; thus, its intrinsic rate of increase is high in relation to that of its host. Second, the parasitoid not only parasitizes hosts, but also kills mealybugs by feeding on their hemolymph. The mutilation resulting from this behavior accounts for considerable mortality in mealybug populations. Third, the parasitoid is highly host specific and very efficient at locating hosts even at low densities. Thus, the parasitoid is able to persist and simultaneously suppress pest populations at low pest densities. The parasitoid is not effective when mealybugs attain high densities, but by being effective at low densities, high pest populations rarely occur. Fourth, *E. lopezi* exhibits a strong capacity to aggregate where host populations are present (i.e., density-dependent aggregation).

Ash Whitefly. The ash whitefly, *Siphoninus phillyreae*, appeared in southern California in 1988 and spread rapidly throughout California and into adjacent states [124]. Because of its broad host range, this pest can cause severe damage to many economically important trees and shrubs. The source of the pest was determined to be the palearctic region and a search for natural enemies was initiated. Two species of natural enemies were imported and released: the coccinellid beetle *Clitostethus arcuatus*, and the aphelinid parasitoid *Encarsia inaron* (identified as *E. partenopea*). *E. inaron* rapidly established and suppressed the whitefly [12]. Although the life history of *E. inaron* is relatively unknown at present, females appear to be highly efficient at locating incipient populations of the whitefly. Furthermore, the parasitoid seems capable of overwintering in deciduous habitats (because of protracted adult longevity and high survival of immature parasitoids in hosts on abscised leaves), whereas the white-fly must seek the foliage of alternate, evergreen hosts during the winter. Thus, when the whitefly returns to its preferred deciduous hosts in the spring, the parasitoid is present and ready to initiate reproduction simultaneously with that of the whitefly [39].

3.   Obstacles to Classical Biological Control

A number of formidable impediments currently limit the application of classical biological control. We address several of these below.

Lack of Predictive Theory. It is difficult to predict whether an introduced natural enemy will establish and suppress the target population [40,96,153]. Lack of success may result from a variety of factors, although it should be noted that in many cases, postrelease evaluations are insufficient to determine the primary causes of failure. A number of factors can prevent or enhance the establishment of an imported natural enemy: its adaptability to the climate in the release area, synchrony of the natural enemy's phenology with that of the target pest, availability of alternate hosts or food, compatibility of the habitat (e.g., overwintering sites) with the introduced natural enemy, competitive ability of the natural enemy, complex of indigenous natural enemies attacking the imported natural enemy, and the genetic structure of the imported population (e.g., see refs. 30, 128, and 146).

Incompatibility of the imported natural enemy with the climate in the introduction area is believed to be responsible for most important failures (e.g., see refs. 63, 92, 128, 138, and 330). In the absence of an effective theoretical base, importation efforts rely on the application of several empirical "rules of thumb" (e.g., see ref. 49): (a) efforts should be made to match the climate of the collection area and the release areas, (b) it is better to collect numerous individuals of a natural enemy species from a wide range of habitats (localities) to increase the probability of obtaining a genotype appropriate to the area of release, (c) host-specific organisms are more effective natural enemies (particularly at low host/prey densities) than are more polyphagous species, and (d) the most important natural enemy in the targeted pest's native habitats will likely be efficacious in the release area. Although numerous exceptions to these "rules" can be found, they serve as useful guidelines.

In recent years, much progress has been made to understand and predict the interactions of natural enemy populations with those of their prey or hosts (e.g., see ref. 95). The extrapolation of these population models to predict outcomes for natural enemies and their prey or hosts in novel environments is far more complex, and as a result, theory in this area has been much slower to develop.

Regulatory Constraints. Importation of natural enemies has come under increasing scrutiny and criticism as the public, environmentalists, and politicians show heightened sensitivity to the conservation of natural species (e.g., see refs. 68, 85, 94, and 122). Various scientists have expressed concerns about the effects that introduced natural enemies may have on nontarget biota; however, the evidence of natural species suffering irreparable damage from importation of arthropod natural enemies in professionally conducted biological control programs is generally highly circumstantial (see ref. 46 for assessment of the situation in Hawaii). Nevertheless, caution must be exercised in introducing natural enemies because of the potential nontarget effects, and these risks should be assessed relative to alternative pest suppression measures. Currently, herbivores introduced for biological control of weeds are tested rigorously. Selective tests should be conducted for biological control of arthropods, particularly in areas, such as Hawaii, that harbor rare or endangered species [116].

Postcollection studies of candidates for introduction should be conducted rigorously during quarantine to determine the safety of the proposed agent to the environment into which it will be released. The relative safety of an organism is generally assessed by evaluating its host range to determine its potential impact on the target and nontarget populations in regions proposed for release. The more polyphagous the organism, the greater its probable risk to nontarget species [108]. To date, these protocols have yielded a fine safety record [45,46,120]. As Howarth [67] pointed out, there is no panacea to managing pests. Thus we are relegated to assessing risks and benefits (environmental, economic, social) of the various management tactics used individually or used in concert with one another. Based on its historical record of environmental, social, and economic impact, biological control, as practiced by professionals, must be regarded highly as a pest management tool (see ref. 114).

Environmentally related concerns have stimulated the development of new regulations addressing importation programs in Australia; presently, analogous regulations are being developed in the United States. The proposed regulations attempt to define a balance between the safety in importing biological control agents relative to relying on alternatives such as chemical controls and the possible adverse effects of introducing exotic entomophagous organisms into native ecosystems (see refs. 75 and 114 for discussion). Indeed, the balance is difficult to define, and it will be difficult to satisfy all parties. Nevertheless, all possible risks and benefits must be carefully considered.

B. Augmentative Releases (Inoculative/Inundative Releases)

Augmentative releases of indigenous or naturalized (exotic species that have been previously released and have become established) biological control agents have long been used to suppress various pest species [12]. Periodic releases may involve native or naturalized species of natural enemies that have shown efficacy or releases of imported species that have not become established but are effective as biological pesticides in suppressing the target pest. Augmentation falls into two categories: (a) inoculative approaches in which natural enemies are released in relatively low numbers to establish or bolster local populations of resident natural enemies for either short- or long-term suppression of a target pest or pest complex, and (b) inundation, in which large numbers of natural enemies are released to obtain rapid pest suppression. These two approaches impose different demands on production and release technologies, and these differences will be addressed below.

1. Requirements for Augmentation

The first requirement for a release program is to select an appropriate natural enemy species or strain for the pest target. Often the natural enemy must be amenable to mass rearing and storage, and it must be matched to the conditions under which it will be used. This need has precluded certain difficult-to-rear groups of natural enemies (e.g., tachinid flies and obligately parasitic pathogens) from being widely used in inundation programs, although some of these may be fine natural enemies. Ease of rearing is somewhat less of a constraint in inoculative release programs, where small quantities of natural enemies are sufficient. The natural enemy must also be adapted to the conditions under which it will be utilized. For example, release of a natural enemy that is adapted to an arboreal habit would be of little value in a row-crop system.

The second requirement of a release program is the development of cost-effective mass-rearing methodology. Large numbers of the natural enemies are necessary and they must be produced economically. If natural enemies are mass reared for commercial distribution to growers, they must be available at prices that are competitive with other pest management tools (e.g., pesticides; see ref. 134). Efficient storage may be a significant factor in this process [139]. The development of artificial diets and artificial media has facilitated the in vitro production of some natural enemies. However, presently there are numerous efficacious species for which no artificial rearing media or procedures for storage are available. Production of these species requires much labor, and thus accrues greater cost. Research is needed to develop diets and artificial media and more efficient and cost-effective rearing and storage methodology.

Another critical issue is the quality of the mass-produced natural enemies (e.g., see refs. 2, 13, and 155). Natural enemies acquired from producers must perform consistently at a high level, commensurate with their cost and with the relative costs of alternative control technologies (e.g., insecticides). Furthermore, growers must have adequate information to conduct releases properly and sometimes to make evaluations. To enhance efficacy, the proper natural enemy species must be selected, and releases must be timed to coincide with the appropriate host stage of the target pest and the appropriate environmental conditions. Likewise, adequate numbers of natural enemies must be released to achieve the desired level of suppression. More research is needed in all of these areas.

2. Examples of Augmentative Releases

Egg parasitoids of the hymenopteran genus *Trichogramma* have been widely used worldwide for inundative releases against key pests [84]. Various species of this genus are easily reared and released, making them excellent candidates for such an application. There are, however, significant pitfalls associated with the use of *Trichogramma* spp. Accurate species determinations are difficult within this genus because of highly conserved morphological characters and the occurrences of biotypes. In addition, those species that are most frequently used (e.g., *T. minutum* and *T. pretiosum*) are polyphagous and widely distributed. The first characteristic presents risks to nontarget species, whereas the second complicates evaluations (was the outcome due to released or naturally occurring parasitoids?). In addition, timing of releases is often critical because of the rapid embryonic development in many insect pests and the resulting changes in suitability for parasitism as the host eggs age [106].

   Trichogramma spp. were released annually in over 2 million hectares of crop and forest systems in China [83]. Furthermore, Li [83,84] suggested that long-term benefits accrued in years subsequent to the releases. Similarly, over 8 million hectares of crops received *Trichogramma* spp. annually in the former Soviet Union [152]. *Trichogramma* spp. have been used in Western Europe and the United States but on a much more limited scale than that of Eastern Europe and Asia—355 thousand hectares in the United States in 1983 and 10 thousand hectares in western Europe in 1985–1986 [149,150]. The efficacy (and economic returns) of all the above *Trichogramma* releases is poorly documented.

   Large-scale trials with *T. pretiosum* to manage populations of the cotton bollworm, *Helicoverpa zea*, and the budworm, *Heliothis virescens*, on cotton in the southern United States failed to provide adequate suppression [76]. Releases with large numbers of parasitoids (120,000–370,000 parasitoids/ha/release) were made at frequent intervals (18 releases from June 10 to August 21 in one trial). The failure of this project was largely attributed to the widespread use of insecticides detrimental to the parasitoid [76], although other factors (e.g., quality control, release methodology) cannot be ruled out. The use of less noxious amounts or types of insecticides, coupled with improvements in the systematics, rearing, delivery and release technology, and quality control of commercially produced parasitoids, will likely open up considerable opportunities to use *Trichogramma* spp. more effectively in the future.

Predators have also been used widely in release programs. For example, green lacewings of the genus *Chrysoperla* (Neuroptera: Chrysopidae) have been utilized worldwide to suppress various pests, including aphids, lepidopteran eggs and larvae, and other soft-bodied insect pests. Several biological features of green lacewings make them excellent agents for release programs: (a) they are generalists capable of developing and reproducing on a wide variety of prey, (b) they tend to be more tolerant of insecticides than many other entomophagous arthropods [27], and (c) they can be mass reared and stored, [20,103]. Lacewings are typically released as eggs or prefed neonate larvae and have demonstrated high efficacy in some production systems when released in high numbers (e.g., see refs. 81 and 105).

Knipling [77,78] advocated large-scale ("areawide"), early-season inoculative releases of natural enemies to improve biological control; this approach would shift the natural enemy/host ratio in favor of the natural enemy early in the season when pest populations are still relatively small. Theoretically, this methodology permits fewer natural enemies to be released, with a greater per capita effect on the target population than late-season releases of large numbers of natural enemies. This approach depends on the selection of an appropriate natural enemy that can efficiently locate the target species and on proper timing of releases to maximize the natural enemy: target ratio when the target is in appropriate stages for the natural enemy to be successful. However, releases of natural enemies over large areas will require governmental regulation and execution; such procedures intrude on the growers' decision making and can have detrimental political effects. Nevertheless, the concept has merit. In pilot projects concluded in Georgia and North Carolina, early-season release of the tachinid fly *Archytas marmoratus* against the corn earworm, *Helicoverpa zea*, yielded substantially increased levels of parasitism [110]. Unfortunately, the season-long effect of these releases on the pest population in the release areas was not monitored.

3. Considerations for Use

Several issues must be considered before releasing natural enemies for augmentation. First, the cost and anticipated benefits of using biological control relative to alternative tactics, such as pesticides, must be weighed carefully. Second, the overall crop production system should be considered; releases may be wasted if crop production practices are inimical to the use of natural enemies. Finally, information must be available so that users may make decisions about which natural enemy species or biotype would be appropriate, and the most effective means of application (e.g., how to release the natural enemies, how many to release per unit area or per pest, how to evaluate efficacy). Reliable taxonomic and ecological information is critical.

C. Conservation of Natural Enemies (or Habitat Management)

Conservation of biological control agents refers to any environmental modification that either reduces or eliminates conditions that are unfavorable to natural enemies or that provides resources that promote population growth, recruitment, or performance. Conservation can constitute the most important approach to biological control for at least two reasons. First, practices conducive to natural enemy conservation influence the success of native, imported, and periodically released natural enemies. Second, conservation of natural enemies enables growers to utilize beneficial species that exist in the agroecosystem. Thus, it serves to strengthen growers' appreciation of biological control and their understanding of the natural enemies' ecology in relation to crop production systems.

Ironically, conservation is the least emphasized of the three major approaches to biological control. This disparity may stem from the growers' reluctance to modify their production practices to accommodate or encourage natural enemies, possibly because such practices conflict with conventional crop production or protection methods. Also, many growers lack confidence in the reliability of natural enemies for pest suppression. Most importantly, often very little is known concerning the environmental factors that either limit or promote the effectiveness of natural enemies. Conservation of natural enemies requires detailed knowledge of the natural enemy's phenology and resource requirements. In addition, understanding the interactions of various management methods (agronomic as well as pest management techniques) with natural enemy populations is indispensable. Currently such data are available for few species and systems.

Agroecoystems can be extremely hostile environments for natural enemies, because agricultural practices cause frequent and regular disturbance (e.g., chemical pesticides or physical disturbances), and because agricultural systems are highly artificial habitats that often lack essential resources for natural enemies. Despite obstacles, biological control researchers have discovered means for conserving natural enemies that have led to very successful biological control programs. Equally important, many of these environmental modifications have been shown to be practical and can be incorporated into agricultural or silvicultural practices. A few examples of successful conservation programs follow.

Modified pesticide use—including the use of selective pesticides, careful timing of applications, and reduced application rates—is an approach for conserving natural enemies that is available to most growers [27,51,117]. It is particularly applicable to cropping systems that have multiple pests, only some of which are subject to effective biological control. Because a dynamically changing, diverse number of arthropods (e.g., see ref. 91), including a large complement of natural enemies (e.g., see ref. 102), may colonize cropping systems, the use of target-specific pesticides or application rates may protect numerous biological control agents.

The ongoing project to eradicate the cotton boll weevil, *Anthonomus grandis*, from the southeastern United States illustrates the problem and potential associated with modified pesticide use. From 1987–1990 repeated, widespread application of organophosphate insecticides to eradicate the boll weevil throughout most of Georgia severely impaired a large complex of generalist natural enemies that ordinarily suppresses the beet armyworm, *Spodoptera exigua*, on cotton [115]. Insecticidal disruption triggered devastating outbreaks of this lepidopteran pest on cotton. Since that time, cotton growers in Georgia have adopted more selective insecticides and modified the rates of application to conserve natural enemies, and in so doing, they have relegated the beet armyworm to minor pest status.

Timing of insecticide applications to reduce the disruption of natural enemies can also conserve them. Thus, understanding the phenology and colonization potential of important beneficial species is important for timing insecticide applications and thus conserving natural enemies. For example, in cotton fields of the southeastern United States, widely separated applications of pyrethroids during the middle and later portions of the growing season are less disruptive than are early-season treatments. During the later part of the season, when pest pressure is typically most intense, large populations of beneficial species from the diverse habitats surrounding the cotton fields rapidly recolonize the cotton fields.

Beneficial species can also be conserved by restricting the area to which insecticides are applied. Fifty years ago, Isely [72] suggested spot treatment for the cotton aphid, *Aphis gossypii*, noting its highly clumped distribution. van den Bosch and Stern [144] developed an effective system for conserving natural enemies of the purple scale (*Lepidosaphes beckii*) on citrus by applying insecticides to every other pair of tree rows. The combination of insecticide and natural enemies effectively suppressed the scale. The use of spot treatments may become more feasible as site-specific farming technology improves.

The manipulation of crop plantings or other vegetation represents another means of protecting or enhancing natural enemies. A classic example involves the interplanting of blackberries [38] and, more recently, French prune trees [156] in California vineyards. Both plant types harbor leafhopper species that remain active during the winter months, and whose eggs serve as alternate hosts for the egg parasitoid *Anagrus epos*, a key parasitoid of the grape leafhopper, *Erythroneura elegantula*. Plant community structure may affect conservation of natural enemies on an even larger spatial scale (see Chap. 4). Parasitization of the European cornborer, *Ostrinia nubilalis*, by the ichneumonid wasp *Eriborus terebrans* and other larval parasitoids was significantly greater near wooded field borders than in the center of fields or at unwooded borders of corn fields in

Michigan [80]. Adult food, alternate host species, or favorable microclimates associated with wooded borders may have enhanced the effectiveness of the natural enemy.

Ground cover crops can provide a continuous source of prey and, perhaps, a more favorable microclimate for generalist and specialist predators that attack spider mites, aphids, and lepidopteran pests on orchard crops such as almonds [61], pecans [140], and apples [5,26]. A general scheme of how cover crops can contribute to improved biological control by serving as a temporal bridge or relay is shown in Fig. 1.

Habitat diversification within and between growing seasons has been advocated for enhancement of entomophagous organisms. For example, some cover crops can provide an early-season habitat from which beneficial organisms may

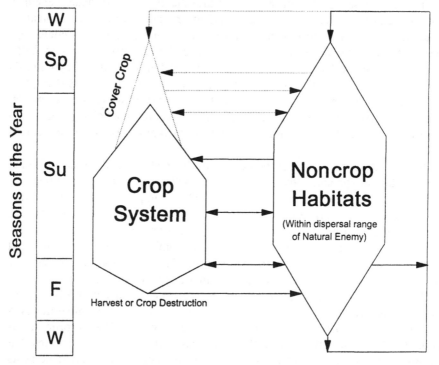

Figure 1 Schematic diagram of the possible role of spring cover crops in relaying natural enemies into target crop systems in the Temperate Zone. Solid lines on arrows denote the movement of natural enemy species between habitats. Dashed lines indicate movement of natural enemies when a cover crop is available in the crop system. The relative widths of the "crop system" and "noncrop system" polygons imply relative availability of resources for natural enemies within the respective system.

subsequently move into the crop (see Fig. 1). This approach seeks to provide alternative hosts, prey, nutrition, and/or favorable habitats for the entomophagous species within or adjacent to the crop system; the primary targets are polyphagous, mobile species that can exploit food and reproductive resources in the alternate plantings and also attack the pests in the cropping area when they become available. However, care must be taken that the alternate plants do not favor pests of the main crop and that they serve as sources for beneficial species and do not compete with the crop for natural enemies [25].

In some cases, reduced tillage appears to increase the diversity and abundance of natural enemy populations by providing food, reducing mechanical injury, eliminating dust, or other deleterious factors (see ref. 145). For example, in low- or no-tillage corn, soil-inhabiting predators, including spiders and ground beetles, had a substantially greater impact on black cutworms than in conventionally tilled corn [129]. An indirect benefit of conservation tillage to natural enemies is to increase the diversity of flowering plants which serve as food for adult parasitoids. The presence of flowering weeds in minimum-till corn increased the longevity, reproduction, and parasitization rates for the cutworm parasitoid *Meteorus rubens* [44]. However, the benefits of reduced tillage for natural enemies are not unequivocal; predation of European cornborer eggs by chrysopids was greater in no-till versus tilled corn, whereas the reverse was found for the coccinellid predator *Coleomegilla maculata* [3].

IV. ATTRIBUTES OF EFFECTIVE NATURAL ENEMIES

In most cases, it is difficult to show a cause-effect relationship between the characteristics of a natural enemy and its effectiveness. However, natural enemies that are considered to be effective usually possess one or more traits that contribute to pest suppression. These may include good searching ability (particularly at low pest population density), high population growth, and the ability to function over a broad spectrum of climatic conditions (see refs. 37 and 93).

Whether one is searching for natural enemies for importation or augmentative releases, it is noteworthy that no one species or biotype will encompass all of the "ideal" attributes, nor is this necessary for a successful biological control program. For example, a natural enemy that lacks vagility may be distributed by biological control practitioners or farmers (see ref. 121).

The intended use of a natural enemy is a key determinant of which biological qualities are most important. For example, amenability to mass rearing generally is required of a natural enemy that will be used in an augmentation program. However, a natural enemy that has been imported for establishment against an exotic pest may be successful without a large-scale rearing effort (although the release of large numbers of genetically diverse natural enemies may increase the likelihood of colonization and establishment). Similarly, the ability to kill pests quickly is desirable in most augmentation programs, whereas qualities that ensure persistence of the enemy and stable suppression of the pest population are often more important in classical biological control programs.

The attributes sought in a natural enemy also depend on the life history of the pest. For example, a pest that occurs sporadically, but with a tendency to undergo explosive population growth, will typically require a natural enemy that is a good searcher and has a high capacity for population growth. These same characteristics may not be as important against a pest that begins its season at low densities, exists in the same area from year to year, and has only moderate population growth.

Natural enemy attributes also should be compatible with the biological and physical makeup of the agroecosystem. In general, the greater the physical extremes (e.g., temperature, moisture), the more important it is to choose natural enemies that possess adaptations to survive and function under rigorous climatic conditions (see refs. 135, 137, and 138). Knowledge of the biological community into which the natural enemies will be released may be important when evaluating candidates for biological control programs. For example, in areas where natural enemies of the proposed biological control agent exist, defense mechanisms, a capability for offsetting population losses (e.g., through reproduction), or caging during establishment may be necessary [101]. If natural enemies of the pest occur, choosing species with high competitive abilities also may be necessary. However, predicting the outcomes of multiple species interactions may be difficult [40,42]. Competition may be reduced, and the overall level of biological control increased, by choosing natural enemies that complement resident enemy species. For example, the spiraling whitefly, an introduced pest in Hawaii, Guam, and other tropical areas, has two key natural enemies: a lady beetle, *Nephaspis oculatus*, and a parasitic wasp, *Encarsia ?haitiensis*. In Guam, lady beetles respond to whiteflies at high densities; whereas the parasitoids function relatively more effectively when whitefly populations are at lower densities [98].

V. EVALUATION OF NATURAL ENEMY EFFICACY

Assessing the role of natural enemies in suppressing pests constitutes a vital aspect of biological control programs. There are a variety of tools and techniques for making these assessments (see refs. 30, 31, 32, 73, 87, and 130). In general, it is desirable to use several different methods when testing the effectiveness of natural enemies. The following is a brief review of the major methods.

A. Direct Observation

The age-old method of direct observation has yielded excellent data on pest natural enemy relationships, but it is highly time consuming and labor intensive; its success also depends on knowing the temporal foraging patterns of natural enemies of the target pest. Direct observation has the advantages of being highly reliable, generally requiring no special equipment, and being compatible with most experimental designs. However, the observer's presence can sometimes affect the arthropods' behavior.

B. Life Tables

Life table studies can be useful for analyzing stage-specific rates of population change within and among generations of the target species [125,151]. Through detailed observations and sampling, the timing and sources of key mortality factors—including natural enemies—can be estimated. Life tables are most effective for evaluating natural enemy efficacy when combined with other techniques (e.g., experimental methods discussed below), because life tables alone generally do not elucidate the role of individual species of natural enemy; this problem is particularly acute when a complex of natural enemies is involved. Nevertheless, when appropriately designed, life tables can help point to important natural enemy species [11].

C. Experimental Methods

In these widely used, diverse approaches using experimental methods, the natural enemy is excluded from or confined to a portion of the pest population; the survival and growth of the enemy-free portion of the pest population is compared with the one in which the natural enemy is present. Such comparisons may involve simple short-term studies of pest population shifts or more detailed life table studies.

Exclusion can be achieved by using selective pesticides (e.g., fungicides, insecticides) or physical barriers (e.g., walls, cages, sticky compounds). Differential levels of exclusion (e.g., using cages with different mesh sizes to shut out natural enemies of various sizes or placing sticky barriers around plants to exclude walking natural enemies) provide evidence for the activity of specific natural enemies or groups of natural enemies. Inclusion is achieved by introducing the natural enemy into a pest-infested cage or other confinement, in an area that is unoccupied by the natural enemy, or where it is present at negligible levels. Depression of the pest population density is considered as evidence of the natural enemy's efficacy. However, confinement may alter enemy behavior and abundance. Whenever possible, the size and design of the exclusion or inclusion should take into account the biology of the target species and its presumed natural enemies. Exclusion and inclusion cage studies have been widely used to demonstrate the impact of predators and parasitoids on a variety of pest species [87,99].

The "addition" method involves the release of a natural enemy into an area where it did not exist (e.g., importation) and comparing the pest population density before and after introduction (see examples in ref. 32). Ideally, studies of an introduced natural enemy's impact are conducted simultaneously in areas with and without the natural enemy; this procedure allows direct comparisons. The method also provides valuable data not only on the role of a particular natural enemy in suppressing a pest but also information pertaining to those life-history attributes of natural enemies that contribute to successful introduction (e.g., see ref. 98).

D. Labeling and Serological Methods

A relatively recent approach for evaluating the activity of natural enemies involves labeling the prey/hosts with radioactive materials, rare elements (such as rubidium), or dyes or using serological [52,53] or electrophoretic techniques to determine which natural enemy species are feeding on a given pest. Handling radioactive materials and rare elements involve hazards and considerable expense, and some of these materials are toxic to both prey and predators. Serological methods, such as the enzyme-linked immunosorbent assay (ELISA) and immunoblot assays, and electrophoresis provide workers with safe, highly sensitive, and relatively inexpensive tools for determining predators of a target pest and estimating predation rates. With a pest-specific monoclonal antibody, predators can be sampled from the field and rapidly assayed to determine if the target pest is present in the predator's gut [57,97,123]. Such a technique can present problems; for example, it may be difficult to develop a monoclonal antibody, the target antigen may decay rapidly in the predator's gut [123]. Moreover, distinguishing whether the presence of antigen in the predator resulting from a single or multiple predation event can be problematic, and secondary predation (preying on a predator that had consumed the target species) cannot always be separated from primary predation.

Another recently developed use of immunotechniques is that of labeling natural enemies with persistent protein markers [56,58]. This inexpensive and easy to use technique should improve our ability to study more closely the movement and prey/host associations of beneficial arthropod species.

E. Economic Evaluations

A very significant, but often neglected, aspect of biological control programs is the economic impact. Accurate appraisals of economic benefits of biological control program require considerable data (e.g., see refs. 60 and 111) and may be challenging to make, because assigning values to some of the relevant parameters (e.g., environmental benefits) is highly complex [141]. Despite these hurdles, biological control has been shown to be highly cost effective [74,146], and demonstration of economic advantages is vital to grower acceptance and adoption of biological control and to continued support for research. On the basis of biological control projects (covering both arthropods and weeds) undertaken by the Australian Commonwealth Science and Industry Research Organization's (CSIRO) Department of Entomology from 1960 to 1975, Tisdell [141] projected that the benefit:cost ratio from the year 1960 to the year 2000 was about 32:1; this compared with a benefit:cost ratio of 2.5:1.0 for work in areas of non-biological control. AliNiazee [1] estimated even higher benefit:cost ratios for four biological control programs in the western United States.

VI. FUTURE NEEDS

A. Biological Control

The number of successful examples of biological control continues to increase worldwide. However, nearly four decades after the concept "integrated control" was first introduced to the literature [127], few programs have been designed actively to use natural enemies rather than emphasizing reductions in pesticide use and resistance management. To increase the role of biological control—especially for high-value food and ornamental crops—additional research is needed. Specific areas for future investigation include the following.

1. Ecological Studies of Natural Enemies

Treatment thresholds that incorporate both pest and natural enemy populations are rare (see ref. 28). This deficiency primarily stems from a poor understanding of natural enemy/pest population dynamics which limits the predictability of biological control. Thus, ecological studies are needed to understand natural enemy (and pest) phenology, factors that influence host/prey searching and reproductive behavior, and conditions that promote or inhibit development, survival, and foraging efficiency of natural enemies.

2. Interactions of Natural Enemies with Other Management Practices

Most production practices in agriculture, forestry, and other human-managed ecosystems where arthropod pests are a concern disregard potentially adverse direct or indirect efforts on naturally occurring or released biological control agents. Consequently, the ability to conserve natural enemies, and to influence and predict their impact on pests, may be hampered severely. Understanding how human activity affects natural enemies is a prerequisite to integrating biological control as a fully operational pest management tactic. Because of the broad range of production activities associated with these systems, research and implementation programs should be interdisciplinary.

3. Systematic Studies

A continuing obstacle for pest management in general and biological control in particular is an inability to identify correctly and classify arthropod pests and their natural enemies (see above). Insufficient resources, coupled with a critical shortage of trained systematists for many important taxa of natural enemies (e.g., the parasitic Hymenoptera), have exacerbated the problem [34,79].

4. Development of Reliable Biological Control Theory

Biological control continues to be an empirical science (e.g., see ref. 40). As such, it is difficult to predict the outcome of biological control programs. However, as data accumulate for different systems, it should be possible to make some general theoretical predictions that will have broad applications for biological control programs. The keystone to reaching this goal is a combination of basic and applied research.

5. Implementation

With increased social and governmental pressure to reduce pesticide use, the benefits of biological control are reaching an ever-growing, receptive audience. Nevertheless, despite a favorable atmosphere, many producers are reluctant to adopt biological control because of their reluctance to take risks, lack of experience, or resistance to make changes from well-established procedures. Ultimately, the failure to adopt biological control is related chiefly either to a lack of economic and other impact data or insufficient, impractical, or inconvenient guidelines for implementation. A significant part of the problem is a shortage of demonstration programs for producers. Also, many extension workers and private pest consultants have limited training and experience in biological control. Thus, educational programs are needed at all levels.

B. Increasing Biological Control's Competitive Edge Within IPM

From an agricultural perspective, the future of biological control within an IPM setting lies in fulfilling growers' needs for environmentally sensitive products and procedures; these products and procedures must be efficacious, readily available, adaptable to production systems, and profitable [35,134,136] (see also ref. 150). Despite biological control's great environmental advantages, and despite its successes, it suffers from some real and perceived deficiencies when it is compared with competing, conventional industrial products or systems in agricultural pest management. Such weaknesses reduce biological control's image and rate of adoption.

What can be done to strengthen the competitiveness and to hasten the increased use of biological control as a component within IPM systems? To answer this question, it is practical to use as a paradigm the development of industrial or agricultural products, because the agricultural industry has overcome some of the hurdles that biological control faces. Thus, following this paradigm, biological control would benefit greatly from increased planning and activity in four crucial areas: research, development, implementation, and evaluation. Each of these must be characterized by well-established short- and long-term goals, realistic timetables, and sustained budgets.

1. Research (development of new products or procedures, improvement of established ones). Industry constantly seeks to develop new products (e.g., seed varieties, pesticides) and new procedures for improving the attractiveness and utility of traditional products. By analogy, to be competitive the biological control community (policy makers and practitioners) must:

   a. Continuously search for effective (native and exotic) natural enemies and biotypes that can be used in classical, augmentative, and inundative biological control (success of these efforts is contingent upon strong systematics; see above)
   b. Characterize the attributes, host ranges, and safety of natural enemies that are currently in agricultural systems and those that are candidates for release
   c. Develop and make available efficient, profitable production systems for mass rearing and marketing natural enemies
   d. Develop procedures for manipulating natural enemies so as to improve their effectiveness

2. Development (fine tuning or adaptation of products to specific crop and growing conditions). Industry expends considerable effort and resources in developing and preparing its products for market; it establishes parameters for the production of uniform products; defines the conditions under which the products can be used, and determines effective marketing strategies and tactics. By analogy, the biological control community should:

   a. Assess the marketing potential and the factors that influence the acceptance and implementation of biological control systems
   b. Ensure that commercial insectaries have incentives for providing standardized, consistently high-quality products (natural enemies) that have a name or number for the particular variety or biotype of natural enemy

c.   Define procedures for use of natural enemies in particular cropping systems

Important considerations include (a) physical factors during crop production, (b) biological factors that influence the natural enemies and their interactions within the tritrophic system, and (c) cultural or management practices associated with crop production.

3.   Implementation. For conventional pest control procedures, industry (in conjunction with federal and state agencies) provides well-developed and packaged systems with follow-up service. Such guidelines are seriously lacking in implementation of biological control, particularly in augmentative programs. The biological control community must develop and implement operational systems that include delivery, marketing, and follow-up services.

a.   As with the seed and pesticide industries, the biological control community must ensure a constant and reliable supply of standardized products (natural enemies) and procedures that are "user friendly."

b.   Industry is very successful in featuring its products to customers. The biological control community also has the ability to promote its products and procedures, but relative to industry, it is far from fully competitive in using modern marketing techniques (advertising, demonstration, education, outreach).

c.   Industry salespersons' livelihoods depend on personally promoting and selling the product and providing advice and follow-up service. In contrast, the biological control community does not yet have a developed infrastructure for training, rewarding, and promoting extension personnel who are involved in implementing biological control.

4.   Evaluation. All too often, the biological control community forgets its greatest assets—the environmental advantages and highly favorable cost:benefit ratio of biological control. (a) Too few resources are devoted to either long- or short-term evaluation and thus the biological effectiveness of biological control is often not assessed quantitatively, and (b) the economic effectiveness of biological control is seldom calculated. By improving these two functions, the biological control community could simultaneously (a) garner favorable publicity and public acceptance as well as (b) develop a strong basis for seeking increased resources.

5.   Recommendations. To move biological control forward in research and development, to increase its acceptance by growers, and to incorporate it into IPM requires:

a.   Careful planning: federal, state, and educational institutions, in cooperation with private sector, must define their goals for biological control and develop comprehensive long- and short-term plans for allocating resources for achieving these goals. When appropriate, the biological control community should use relevant aspects of the industrial paradigm that the chemical, seed, and drug industries have developed; this paradigm should be tailored to the specific needs of biological control.

b. Execution: (a) provide a consistent, reliable, adequate resource base for achieving the defined goals, thereby avoiding annual fluctuations in the investment of resources (b) develop specific full-time state and federal professional positions that work with the private sector to deliver specified biological control systems.

c. Evaluation: all proposed biological control projects should include realistic plans for evaluation, and all funded projects should allocate adequate resources for a predetermined level of evaluation.

REFERENCES

1. AliNiazee, M.T. 1995. The economic, environmental, and sociopolitical impact of biological control. In *Biological Control in the Western United States: Accomplishments and Benefits of Regional Research Project W-84, 1964-1989.* J.R. Nechols, L.A. Andres, J.W. Beardsley, R.D Goeden, & C.G. Jackson, eds. Oakland, California: ANR, Pub. No. 3361, pp 47–56.

2. Anderson, T.E., & N.C. Leppla. 1992. *Advances in Insect Rearing for Research and Pest Management.* Boulder, Colorado: Westview Press.

3. Andow, D.A. 1992. Fate of eggs of first-generation *Ostrinia nubilalis* (Lepidoptera: Pyralidae) in three conservation tillage systems. *Environ. Entomol.* 21:388–393.

4. Askew, R.R. 1971. *Parasitic Insects.* London: Heinemann.

5. Asquith, D. 1971. The Pennsylvania integrated control program for apple pests—1970. *Penn. Fruit News* 50:43–47.

6. Baker, R. 1985. Biological control of plant pathogens: definitions. In *Biological Control in Agricultural IPM Systems.* M.A. Hoy & D.C. Herzog, eds. New York: Academic Press, pp 25–39.

7. Baker, R.R., & P.E. Dunn (eds.) 1990. *New Directions in Biological Control: Alternatives for Suppressing Agricultural Pests and Diseases.* New York: Liss.

8. Balduf, W.V. 1939. *The Bionomics of Entomophagous Insects.* Part II. New York: Swift.

9. Barbosa, P., & S. Braxton. 1993. A proposed definition of biological control and its relationship to related control approaches. In *Pest Management: Biologically Based Technologies.* R.D. Lumsden & J.L.Vaughn, eds. Washington, DC: American Chemical Society, pp 21–27.

10. Beckage, N.E., S.N. Thompson, & B.A. Federici (eds.) 1993. *Parasites and Pathogens of Insects, vol. 2.* San Diego, California: Academic Press.

11. Bellows, T.S., Jr., R.G. Van Driesche, & J.S. Elkinton. 1992. Life table con-

struction and analysis in the evaluation of natural enemies. *Annu. Rev. Entomol.* 37:587–614.

12. Bellows, T.S., Jr., T.D. Paine, J.R. Gould, L.G. Bezark, & J.C. Ball. 1992. Biological control of ash whitefly: a success in progress. *Calif. Agric.* 46:24–28.

13. Bigler, F. (ed.). 1991. *Proceedings of the Fifth Workshop on the IOBC Global Working Group "Quality control of mass reared arthropods"* (25–28 March 1991, Wageningen, The Netherlands). Zurich: Swiss Federal Research Station for Agronomy.

14. Boldt, P.E., & J.J. Drea. 1980. Packaging and shipping beneficial insects for biological control. *FAO Plant Prot. Bull.* 28:64–71.

15. Burges, H.D., & N.W. Hussey.1981. *Microbial Control of Insects and Mites.* New York: Academic Press.

16. Caltagirone, L.E. 1981. Landmark examples in classical biological control. *Annu. Rev. Entomol.* 26:213–232.

17. Caltagirone, L.E., & R.L. Doutt. 1989. The history of the vedalia beetle importation to California and its impact on the development of biological control. *Annu. Rev. Entomol.* 34:1–16.

18. Cantwell, G.E., ed. 1974. *Insect Diseases.* New York: Marcel Dekker.

19. Carl, K.P. 1982. Biological control of native pests by introduced natural enemies. *Biocontrol News and Inform.* 3:191–200.

20. Chang Y.-F., M.J. Tauber, & C.A. Tauber. 1995. Storage of the mass-produced predator *Chrysoperla carnea* (Neuroptera: Chrysopidae): Influence of photoperiod, temperature, and diet. *Environ. Entomol.* 24:1365–1374.

21. Clausen, C.P. 1936. Insect parasitism and biological control. *Ann.Entomol. Soc. Am.* 29:201–223.

22. Clausen, C.P. 1940. *Entomophagous Insects.* New York: McGraw-Hill.

23. Clausen, C.P., ed. 1978. *Introduced Parasites and Predators of Arthropod Pests and Weeds: A World Review.* U.S. Department of Agriculture, Agriculture Handbook No. 480.

24. Coppel, H.C., & J.W. Mertins. 1977. *Biological Insect Pest Suppression.* New York: Springer-Verlag.

25. Corbett, A., & R.E. Plant. 1993. Role of movement in the response of natural enemies to agroecosystem diversification: a theoretical evaluation. *Environ. Entomol.* 22:519–531.

26. Croft, B.A. 1975. *Integrated Control of Apple Mites.* Bull. E-825, Michigan State Cooperative Extension Service.

27. Croft, B.A. 1990. *Arthropod Biological Control Agents and Pesticides.* New York: Wiley.

28. Croft, B.A., & D.L. McGroarty, 1977. The role of *Amblyseius fallacis* in Michigan apple orchards. *Michigan Agricultural Experimental Station Research Report* No. 333.

29. DeBach, P., ed. 1964. *Biological Control of Insect Pests and Weeds.* New York: Chapman & Hall.

30. DeBach, P., & B.R. Bartlett. 1964. Methods of colonization, recovery and evaluation. In *Biological Control of Insect Pests and Weeds.* P. DeBach, ed. New York: Chapman & Hall, pp 402–426.

31. DeBach, P., & C.B. Huffaker. 1971. Experimental techniques for evaluation of the effectiveness of natural enemies. In *Biological Control*. C.B. Huffaker, ed. New York, NY: Plenum Press, pp 113–140.

32. DeBach, P., C.B. Huffaker, & A.W. MacPhae. 1976. Evaluation of the impact of natural enemies. In *Theory and Practice of Biological Control*. C.B. Huffaker & P.S. Messenger, eds. New York: Academic Press, pp 255–285.

33. Debach, P., & D. Rosen. 1991. *Biological Control by Natural Enemies*. 2nd ed. New York: Cambridge University Press.

34. Delucchi, V., D. Rosen, & E.I. Schlinger. 1976. Relationship of systematics to biological control. In *Theory and Practice of Biological Control*. C.B. Huffaker & P.S. Messenger, eds. New York: Academic Press, pp 81–91.

35. Dietrick, E.J. 1989. Commercialization of biological control in the United States. In *Internatl. Symp. on Biological Control Implementation (McAllen TX. 4–6 April 1989). Proceedings and Abstracts*. NAPPO Bull. No. 6, pp 71–87.

36. Doutt, R.L. 1958. Vice, virtue, and the vedalia. *Bull. Entomol. Soc. Am.* 4:119–123.

37. Doutt, R.L. 1964. Biological characteristics of entomophagous adults. In *Biological Control of Insect Pests and Weeds*. P. DeBach, ed. New York: Chapman & Hall, pp 145–176.

38. Doutt, R.L., & J. Nakata. 1973. The *Rubus* leafhopper and its egg parasitoid: an endemic biotic system useful in grape-pest management. *Environ. Entomol.* 2:381–386.

39. Dreistadt, S.H., & M.L. Flint. 1995. Ash whitefly (Homoptera: Aleyrodidae) overwintering and biological control by *Encarsia inaron* (Hymenoptera: Aphelinidae) in Northern California. *Environ. Entomol.* 24:459–464.

40. Ehler, L.E. 1990. Introduction strategies in biological control of insects. In *Critical Issues in Biological Control*. M. Mackauer, L.E. Ehler, & J. Roland, eds. Andover, Hants, UK: Intercept, pp 111–134.

41. Fisher, T.W., & G.L. Finnery. 1964. Insectary facilites and equipment. In *Biological Control of Insect Pests and Weeds*. P. Debach, ed. London: Chapman & Hall, pp 381–401.

42. Force, D.C. 1967. Genetics in the colonization of natural enemies for biological control. *Ann. Entomol. Soc. Am.* 60:722–729.

43. Force, D.C. 1974. Ecology of insect host-parasitoid communities. *Science* 184:624–632.

44. Foster, M.A., & W.G. Ruesink. 1984. Influence of flowering weeds associated with reduced tillage in corn on a black cutworm (Lepidoptera: Noctuidae) parasitoid, *Meteorus rubens* (Nees von Esenbeck). *Environ. Entomol.* 13:664–668.

45. Frank, J.H., & E.D. McCoy, 1995. Introduction to insect behavioral ecology: The good, the bad, and the beautiful: Non-indigenous species in Florida. Invasive adventive insects and other organisms in Florida. *Florida Entomol.* 78:1–15.

46. Funasaki, G.Y., P. Lai, L.M. Nakahara, J.W. Beardsley, & A.K. Ota. 1988. A review of biological control introductions in Hawaii: 1890–1985. *Proc. Hawaiian Entomol. Soc.* 28:105–160.

47. Garcia, R., L.E., Calatagirone, & A.P. Gutierrez. 1988. Comments on a redefinition of biological control. *BioScience* 38:692–694.

48. Gaugler, R. & H.K. Kaya, eds. 1990. *Entomopathogenic Nematodes in Biological Control.* Boca Raton, Florida: CRC Press.

49. Gonzalez, D., & F.E. Gilstrap. 1992. Foreign exploration: assessing and prioritizing natural enemies and consequences of preintroduction studies. In *Selection Criteria and Ecological Consequences of Importing Natural Enemies.* W.C. Kauffman & J.R. Nechols, eds. Proceedings Thomas Say Publications in Entomology. Hyattsville, Maryland: Entomological Society of America, pp 53–70.

50. Granados, R.R., & B.A. Fedrici, eds. 1986. *Biology of Baculoviruses* Boca Raton, Florida: CRC Press.

51. Greathead, D.J. 1995. Natural enemies in combination with pesticides for integrated pest management. In *Novel Approaches to Integrated Pest Management.* R. Reuveni, ed. Boca Raton, Florida: Lewis, pp 183–197.

52. Greenstone, M.H. 1996. Serological analysis of arthropod predation: past, present, and future. In *The Ecology of Agricultural Pests: Biochemical Approaches.* W.O.C. Symondson & J.E. Liddell, eds. London: Chapman & Hall, pp 267–321.

53. Greenstone, M.H., & C.E. Morgan. 1989. Predation on *Heliothis zea* (Lepidoptera: Noctuidae): an instar-specific ELISA assay for stomach analysis. *Ann. Entomol. Soc. Am.* 82:45–49.

54. Hagen, K.S., & J.M. Franz. 1973. A history of biological control. From *The History of Entomology.* Palo Alto, California: Annual Reviews.

55. Hagen, K.S., S. Bomborosch, & J.A. McMurtry. 1976. The biology and impact of predators. In *Theory and Practice of Biological Control.* C.B. Huffaker & P.S. Messenger, eds. New York: Academic Press, pp 93–142.

56. Hagler, J.R. 1997. Field retention of a novel mark-release-capture method. *Environ. Entomol.* 26:1079–1086.

57. Hagler, J.R., & S.E. Naranjo. 1996. Using gut content immunoassays to evaluate predaceous biological control agents: a case study. In *The Ecology of Agricultural Pests.* W.O.C. Symondson & J.E. Liddell, eds. London: Chapman & Hall, pp 383–399.

58. Hagler, J.R., A.C. Cohen, E. Bradley-Dunlop, & F.J. Enriquez. 1992. New approach to mark insects for feeding and dispersal studies. *Environ. Entomol.* 21:20–25.

59. Hawkins, B.A., & W. Sheehan. 1994. *Parasitoid Community Ecology.* New York: Oxford University Press.

60. Headley, J.C. 1985. Cost-benefit analysis: defining research needs. In *Biological Control in Agricultural IPM Systems.* M.A. Hoy & D.C. Herzog, eds. New York: Academic Press, pp 53–62.

61. Hendricks, L.C. 1995. Almond growers reduce pesticide use in Merced County field trials. *Calif. Agric.* 49:5–10.

62. Herren, H.R., & P. Neuenschwander. 1991. Biological control of cassava pests in Africa. *Annu. Rev. Entomol.* 36:257–283.

63. Hokkanen, H.M.T. 1985. Success in classical biological control. *CRC Crit. Rev. Plant Sci.* 3:35–72.

64. Hokkanen, H., & D. Pimentel. 1984. New approaches for selecting biological control agents. *Can. Entomol.* 116:1109–1121.

65. Hokkanen, H., & D. Pimentel. 1989. New associations in biological control: theory and practice. *Can. Entomol.* 121:829–840.
66. Hopper, K.R., & R.T. Roush. 1993. Management of genetics of biological-control introductions. *Annu. Rev. Entomol.* 38:27–51.
67. Howarth, F.G. 1983. Classical biocontrol: panacea or Pandora's box. *Proc. Hawaiian Entomol. Sci.* 24:239–244.
68. Howarth, F.G. 1991. Environmental impacts of classical biological control. *Annu. Rev. Entomol.* 36:485–509.
69. Hoy, M.A., & D.C. Herzog, eds. 1985. *Biological Control in Agricultural IPM Systems.* New York: Academic Press.
70. Huffaker, C.B. ed. 1971. *Biological Control.* New York: Plenum Press.
71. Huffaker, C.B., & P.S. Messenger, eds. 1976. *Theory and Practice of Biological Control.* New York: Academic Press.
72. Isely, D. 1946. *The Cotton Aphid.* University of Arkansas Agricultural Experimental Station Bulletin No. 462. Fayetteville, Arkansas.
73. Jervis, M., & N. Kidd. 1996. *Insect Natural Enemies: Practical Approaches to Their Study and Evaluation.* New York: Chapman & Hall.
74. Jetter, K., K. Klosky, & C. Pickett. 1997. Cost benefit analysis of the ash whitefly biological control program in California. *J. Arboricult.* 23:65–71.
75. Kauffman, W.C., & J.R. Nechols, eds. 1992. *Selection Criteria and Ecological Consequences of Importing Natural Enemies.* Proceedings Thomas Say Publications in Entomology. Lanham, Maryland: Entomological Society of America.
76. King, E.G., L.F. Bouse, D.L. Bull, R.J. Coleman, W.A. Dickerson, W.J. Lewis, J.D. Lopez, R.K. Morrison, & J.R. Phillips. 1986. Management of *Heliothis* spp in cotton by augmentative releases of *Trichogramma pretiosum* Ril. *J. Appl. Ent.* 101:2–10.
77. Knipling, E.F. 1977. The theoretical basis for augmentation of natural enemies. In *Biological Control by Augmentation of Natural Enemies.* R.L. Ridgway & S.B. Vinson, eds. New York: Plenum Press, pp 79–123.
78. Knipling, E.F. 1992. *Principles of Insect Parasitism Analyzed from New Perspectives: Practical Implications for Regulating Insect Populations by Biological Means.* U.S. Department of Agriculture, Agriculture Handbook No. 693.
79. Knutson, L. 1981. Symbiosis of biosystematics and biological control. In *Biological Control in Crop Production.* G.C. Papavizas, ed. London: Allanheld, Osmun, pp 61–78.
80. Landis, D.A., & M.J. Haas. 1992. Influence of landscape structure on abundance and within-field distribution of European corn borer (Lepidoptera: Pyralidae) larval parasitoids in Michigan. *Environ. Entomol.* 21:409–416.
81. LaRock, D.R., & J.J. Ellington. 1996. An integrated pest management approach emphasizing biological control, for pecan aphids. *Southwest. Entomol.* 21:153–166.
82. Leppla, N.C. & T.R. Ashley, eds. 1978. *Facilities for Insect Research and Production.* U.S. Department of Agriculture Technical Bulletin 1576.
83. Li, L. 1984. Research and utilization of *Trichogramma* in China. In *Proceedings of the Chinese Academy Science—United States National Academy of Science Joint*

Symposium on Biological Control of Insects. P.L. Adkisson & M. Shijun, eds. Beijing: Science Press, pp 204–223.

84. Li, L. 1994. Worldwide use of *Trichogramma* for biological control on different crops: a survey. In *Biological Control with Egg Parasitoids.* E. Wajnberg & S.A. Hassan, eds. Wallingford, UK: CAB International, pp 37–53.

85. Lockwood, J.A. 1993. Environmental issues involved in biological control of rangeland grasshoppers (Orthoptera: Acrididae) with exotic agents. *Environ. Entomol.* 22:503–518.

86. Luck, R.F. 1981. Parasitic insects introduced as biological control agents for arthropod pests. In *CRC Handbook of Pest Management in Agriculture.* Vol. 2. D. Pimentel, ed. Boca Raton, Florida: CRC Press, pp 125–284.

87. Luck, R.F., B.M. Shepard, & P.E. Kenmore. 1988. Experimental methods for evaluating arthropod natural enemies. *Annu. Rev. Entomol.* 33:367–391.

88. Mackauer, M. 1976. Genetic problems in the production of biological control agents. *Annu. Rev. Entomol.* 21:369–385.

89. Mackauer, M., L.E. Ehler, & J. Roland, eds. 1990. *Critical Issues in Biological Control.* Andover, Hants, UK: Intercept.

90. Maxwell, F.G., & F.A. Harris (eds.). 1974. *Proceedings of the Summer Institute on Biological Control of Plant Insects and Diseases.* Jackson, Mississippi: University Press of Mississippi.

91. Mayse, M.A., & P. W. Price. 1978. Seasonal development of soybean arthropod communities in east central Illinois. *Agro-Ecosyst.* 4:387–405.

92. Messenger, P.S. 1971. Climatic limitations to biological control. In *Proceedings of the Tall Timbers Conference on Ecological Animal Control by Habitat Management 3.* Tallahassee, Florida: Tall Timbers Research Station, pp 97–114.

93. Messenger, P.S., F. Wilson, & M.J. Whitten. 1976. Variation, fitness, and adaptability of natural enemies. In *Theory and Practice of Biological Control.* C.B. Huffaker & P.S. Messenger, eds. New York: Academic Press, pp 209–231.

94. Miller, M., & G. Aplet. 1993. Biological control: a little knowledge is a dangerous thing. *Rutgers Law Rev.* 45:285–334.

95. Murdoch, W.W. 1990. The relevance of pest-enemy models to biological control. In *Critical Issues in Biological Control.* M. Mackauer, L.E. Ehler, & J. Roland, eds. Andover, Hants, UK: Intercept, pp 1–24.

96. Murdoch, W.W., & C.J. Briggs. 1996. Theory for biological control: recent developments. *Ecology* 77: 2001–2013.

97. Naranjo, S.E., & J.R. Hagler, 1998. Characterizing and estimating the effect of heteropteran predation. In *Predatory Heteroptera in Agroecosystems: Their Ecology and Use in Biological Control.* M. Coll & J.R. Ruberson, eds. Proceedings Thomas Say Publications in Entomology. Lanham, Maryland: Entomological Society of America, pp 171–197.

98. Nechols, J.R., & D.M. Nafus. 1995. Spiraling whitefly. In *Biological Control in the Western United States: Accomplishments and Benefits of Regional Project W-84, 1964–1989.* J.R. Nechols, L.A. Andres, J.W. Beardsley, R.D. Goeden, & C.G. Jackson, eds. Oakland, California: ANR Publication No. 3361, pp 113–114.

99. Nechols, J.R., & T.F. Seibert. 1985. Biological control of the spherical mealybug, *Nipaecoccus vastator* (Homoptera: Pseudococcidae): assessment by ant exclusion. *Environ. Entomol.* 14:45–47.

100. Nechols, J.R., L.A. Andres, J.W. Beardsley, R.D. Goeden, & C.G. Jackson, eds. 1995. *Biological Control in the Western United States: Accomplishments and Benefits of Regional Project W-84, 1964–1989*. Oakland, California: ANR Publication No. 3361.

101. Nechols, J.R., J.J. Obrycki, C.A. Tauber, & M.J. Tauber. 1996. Potential impact of native natural enemies on *Galerucella* spp (Coleoptera: Chrysomelidae) imported for biological control of purple loosestrife: a field evaluation. *Biol. Control* 7:60–66.

102. Neuenschwander, P., R.D. Hennessey, & H.R. Herren. 1987. Food web associated with the cassava mealybug, *Phenacoccus manihoti* Matile-Ferrero (Hemiptera: Pseudococcidae), and its introduced parasitoid, *Epidinocarsis lopezi* (De Santis) (Hymenoptera: Encyrtidae), in Africa. *Bull. Entomol. Res.* 77:177–189.

103. Nordlund, D.D., & R.K. Morrison. 1992. Mass-rearing *Chrysoperla*. In *Advances in Insect Rearing for Research and Pest Management*. T.E. Anderson & N.C. Leppla, eds. Boulder, Colorado: Westview Press, pp 427–439.

104. Norgaard, R.B. 1988. The biological control of cassava mealybug in Africa. *Amer. J. Agr. Econ.* 70:366–371.

105. Olkowski,W.E., D. Dietrick, & H. Olkowski. 1992. The biological control industry in the United States. Part II. *IPM Practitioner* 14:12–14.

106. Pak, G.A. 1986. Behavioural variations among strains of *Trichogramma* spp. A review of the literature on host-age selection. *J. Appl. Entomol.* 101:55–64.

107. Papavizas, G.C., ed. 1981. *Biological Control in Crop Production*. London: Allanheld, Osmun.

108. Pimm, S.L. 1987. Determining the effects of introduced species. *Trends Ecol. Evol.* 2:106–108.

109. Poinar, G.O., Jr., & G.M. Thomas. 1984. *Laboratory Guide to Insect Pathogens and Parasites*. New York: Plenum Press.

110. Proshold, F.I., H.R. Gross, Jr., & J.E. Carpenter. 1998. Inundative release of *Archytas marmoratus* (Diptera: Tachinidae) against the corn earworm and fall armyworm (Lepidoptera: Noctuidae) in whorl-stage corn. *J. Entomol. Sci.* (in press).

111. Reichelderfer, K.H. 1981. Economic feasibility of biological control of crop pests. In *Biological Control in Crop Production*. G.C. Papavizas, eds. London: Allanheld, Osmun, pp 403–417.

112. Ridgway, R.L. & S.B. Vinson, eds. 1977. *Biological Control by Augmentation of Natural Enemies*. New York: Plenum Press.

113. Roush, R.T. 1990. Genetic variation in natural enemies: critical issues for colonization in biological control. In *Critical Issues in Biological Control*. M. Mackauer, L.E. Ehler, & J. Roland, eds. Andover, Hants, UK: Intercept, pp 263–288.

114. Ruberson, J.R., J.J. Obrycki, & J.R. Nechols. 1992. Arthropod panel recommendations and summary of discussions. In Proc. USDA/CSRS Natl. Workshop,

Regulations and Guidelines: Critical Issues in Biological Control. R. Charudattan
& H.W. Browning, eds. Gainesville, Florida: IFAS, University of Florida, pp
159–167.

115. Ruberson, J.R., G.A. Herzog, W.R. Lambert, & W.J. Lewis. 1994. Manage-
      ment of the beet armyworm (Lepidoptera: Noctuidae) in cotton: role of natural
      enemies. *Fla. Entomol.* 77:440–453.

116. Ruberson, JR, CA Tauber, & MJ Tauber. 1995. Developmental effects of host
      and temperature on *Telenomus* spp (Hymenoptera: Scelionidae) parasitizing
      chrysopid eggs. *Biol. Control* 5:207–220.

117. Ruberson, J.R., H. Nemoto, & Y. Hirose. 1998. Pesticides and conservation of
      natural enemies in pest management. In *Conservation Biological Control.* P.
      Barbosa, ed. Academic Press, New York (in press).

118. Sailer, R.I. 1974. Foreign exploration and importation of exotic arthropod para-
      sites and predators. In *Proceedings of the Summer Institute on Biological Con-
      trol of Plant Insects and Diseases.* F.G. Maxwell, & F.A. Harris, eds. Jackson,
      Mississippi: University Press of Mississippi, pp 97–109.

119. Sailer, R.I. 1983. History of insect introductions. In *Exotic Plant Pests and North
      American Agriculture.* L. Wilson & L. Graham, eds. New York: Academic Press,
      pp 15–38.

120. Samways, M.J. 1988. Classical biological control and insect conservation: are
      they compatible? *Environ. Conserv.* 15: 349–354, 348.

121. Schuster, M.F., J.C. Boling, & J.J. Marony, Jr. 1971. Biological control of
      rhodesgrass scale by airplane releases of an introduced parasite of limited dis-
      persing ability. In *Biological Control.* C.B. Huffaker, ed. New York: Plenum
      Press, pp 227–250.

122. Simberloff, D. 1992. Conservation of pristine habitats and unintended effects of
      biological control. In *Selection Criteria and Ecological Consequences of Import-
      ing Natural Enemies.* W.C. Kauffman & J.R. Nechols, eds. Proceedings Thomas
      Say Publications in Entomology. Lanham, Maryland: Entomological Society of
      America. pp 103–117.

123. Sopp, P.I., & K.D. Sunderland. 1989. Some factors affecting the detection of
      aphid remains in predators using ELISA. *Entomol. Exp. Appl.* 51: 11–20.

124. Sorenson, J.T., R.J. Gill, R.V. Dowell, & R.W. Garrison. 1990. The introduc-
      tion of *Siphoninus phillyreae* (Haliday) (Homoptera: Aleyrodidae) into North
      America: niche competition, evolution of host plant acceptance, and a prediction
      of its potential host range in the Nearctic. *Pan-Pac. Entomol.* 66:43–54.

125. Southwood, T.R.E. 1978. *Ecological Methods With Special Reference to the Study
      of Insect Populations.* 2nd ed. New York: Chapman & Hall.

126. Steinhaus, E.A., ed. 1963. *Insect Pathology: An Advanced Treatise.* New York:
      Academic Press.

127. Stern, V.M., R.F. Smith, R. van den Bosch, & K.S. Hagen. 1959. The integra-
      tion of chemical and biological control of the spotted alfalfa aphid: the integrated
      control concept. *Hilgardia* 29:81–101.

128. Stiling, P. 1983. Why do natural enemies fail in classical biological control pro-
      grams? *Am. Entomol.* 39:31–37.

129. Stinner, B.R., D.A. McCartney, & D.M. Van Doren. 1988. Soil and foliage arthropod communities in conventional, reduced and no-tillage corn (Maize, *Zea mays* L.) systems: a comparison after 20 years of continuous cropping. *Soil Tillage Res.* 11:147–158.

130. Sunderland, K.D. 1987. A review of methods of quantifying invertebrate predation occurring in the field. *Acta Phytopathol. Entomol. Hung.* 22:13–34.

131. Swan, L.A. 1964. *Beneficial Insects.* New York: Harper & Row.

132. Sweetman, H.L. 1958. *The Principles of Biological Control.* Dubuque, Iowa: Brown.

133. Tanada, Y., & H. Kaya. 1992. *Insect Pathology.* New York: Academic Press.

134. Tauber, M.J., & R.J. Helgesen. 1978. Implementing biological control systems in commercial greenhouse crops. *Bull. Entomol. Soc. Am.* 24:424–426.

135. Tauber, M.J., & C.A. Tauber. 1983. Life history traits of *Chrysopa carnea* and *Chrysopa rufilabris* (Neuroptera: Chrysopidae): influence of humidity. *Ann. Entomol. Soc. Am.* 76:282–285.

136. Tauber, M.J., M.A. Hoy, & D.C. Herzog. 1985. Biological control in agricultural IPM systems: a brief overview of the current status and future prospects. In *Biological Control in Agricultural IPM Systems.* M.A. Hoy & D.C. Herzog, eds. Orlando, Florida, Academic Press, pp 3–9.

137. Tauber, M.J., C.A. Tauber, & S. Masaki. 1984. Adaptations to hazardous seasonal conditions: dormancy, migration, and polyphenism. In *Ecological Entomology.* C.B. Huffaker & R.L. Rabb, eds. New York: Wiley, pp 149–184.

138. Tauber, M.J., C.A. Tauber, & S. Masaki. 1986. *Seasonal Adaptations of Insects.* New York: Oxford University Press.

139. Tauber, M.J., C.A. Tauber, & S. Gardescu. 1993. Prolonged storage of *Chrysoperla carnea* (Neuroptera: Chrysopidae). *Environ. Entomol.* 22:843–848.

140. Tedders, W.L. 1983. Insect management in deciduous orchard ecosystems: habitat manipulation. *Environ. Manage.* 7:29–34.

141. Tisdell, C.A. 1990. Economic impact of biological control of weeds and insects. In *Critical Issues in Biological Control.* M. Mackauer, L.E. Ehler, & J. Roland, Andover, Hants, UK: Intercept, pp 301–316.

142. Unruh, T.R., W. White, D. Gonzalez, G. Gordh, & R.F. Luck. 1983. Heterozygosity and effective population size in laboratory populations of *Aphidius ervi* (Hymen.: Aphidiidae). *Entomophaga* 28:245–258.

143. Unruh, T.R., W. White, D. Gonzalez, & R.F. Luck. 1986. Electrophoretic studies of parasitic Hymenoptera and implications for biological control. *Misc. Publ. Entomol. Soc. Am.* 61:150–163.

144. van den Bosch, R., & V.M. Stern. 1962. The integration of chemical and biological control of arthropod pests. *Annu. Rev. Entomol.* 7:367–386.

145. van den Bosch, R., & A.D. Telford. 1964. Environmental modification and biological control. In *Biological Control of Insect Pests and Weeds.* P. DeBach, ed. New York: Chapman & Hall, pp 459–488.

146. van den Bosch, R., P.S. Messenger & A.P. Gutierrez. 1992. *An Introduction to Biological Control.* New York: Plenum Press.

147. Van Driesche, R.G., & T.S. Bellows, Jr., eds. 1993. *Steps in Classical Arthropod Biological Control*. Thomas Say Publications in Entomology. Hyattsville, Maryland: Entomology Society of America.

148. Van Driesche, R.G., & T.S. Bellows, Jr. 1996. *Biological Control*. New York: Chapman & Hall.

149. van Lenteren, J.C. 1987. Environmental manipulation advantageous to natural enemies of pests. In *Protection integree: quo vadis?—"Parasitis 86"*. V. Delucchi, ed. pp 123–163.

150. Lenteren, J.C. van. 1989. Implementation and commercialization of biological control in West Europe. In *International Symposium* on *Biological Control Implementation (McAllen, Texas, 4–6 April 1989), Proceedings and Abstracts*. NAPPO Bull. No. 6, pp 50–70.

151. Varley, G.C., G.R. Gradwell, & M.P. Hassell. 1973. *Insect Population Ecology: An Analytical Approach*. Berkeley, California: University of California Press.

152. Voronin, K.E., & A.M. Grinberg. 1981. The current status and prospects of *Trichogramma* utilization in the U.S.S.R. In *Proceedings of a Joint American-Soviet Conference on Use of Beneficial Organisms in the Control of Crop Pests*. J.R. Coulson, ed. College Park, Maryland: Entomological Society of America, pp 49–51.

153. Waage, J. 1990. Ecological theory and the selection of biological control agents. In *Critical Issues in Biological Control*. M. Mackauer, L.E. Ehler, & J. Roland, eds. Andover, Hants, UK: Intercept, pp 135–157.

154. Waage, J.K., & D.J. Greathead, eds. 1986. *Insect Parasitoids*. Orlando, Florida: Academic Press.

155. Williams, D.W., & N.C. Leppla. 1992. The future of augmentation of beneficial arthropods. In *Selection Criteria and Ecological Consequences of Importing Natural Enemies*. W.C. Kauffman & J.R. Nechols, ed. Proceedings Thomas Say Publications in Entomology. Lanham, Maryland: Entomological Society of America, pp 87–102.

156. Wilson, L.T., C.H. Pickett, D.L. Flaherty, & T.A. Bates. 1989. French prune trees: refuge for grape leafhopper parasite. *Calif. Agric.* 43:7–8

17

Chemical Approaches to Managing Arthropod Pests

Jerry B. Graves,* B. Rogers Leonard, and James A. Ottea
Louisiana State University Agricultural Center, Baton Rouge, Louisiana

I. INTRODUCTION

A. Role of Chemicals in Integrated Pest Management

Integrated pest management (IPM) is ecologically, as well as environmentally, the most logical approach to arthropod pest management. Insecticides and acaricides are generally used in IPM systems only when other control measures (e.g., biological, cultural, host-plant resistance, physical, regulatory) fail to keep pest populations below economic thresholds, and it is critical that effective chemicals be available for most IPM systems to succeed. Considerable research efforts and funds have been expended to improve nonchemical strategies for managing arthropod pest populations and numerous examples of successes can be cited (see Chapters 15, 16, 18, and 19). However, chemical control retains a primary and indispensable role in most IPM systems, especially in annual cropping systems with multiple key pests. Furthermore, chemicals are likely to remain an integral and necessary component of IPM systems for the foreseeable future [70,78,133,134].

The availability of chemicals is decreasing due to difficulties in discovering new chemistry, the exorbitant cost of registration and reregistration, cancellation of uses, and the development of pest resistance to pesticides. Insecticide resistance management (IRM), an integral part of IPM, offers a realistic opportunity to forestall resistance development, which is the only one of the aforementioned difficulties directly manageable by the ultimate users of these products. All IPM and commodity production practices that directly or indirectly affect the need for and use of insecticides and acaricides must be focused in a

---

*Retired.

holistic manner toward reduced reliance on chemical control. It is critical that the use of chemicals be viewed as a single component of an ecologically and biologically based pest management system in order to conserve and prolong the utility of the declining pool of pesticides [57,66,70,133,134].

B. Attributes and Disadvantages of Chemical Control

Chemicals continue to be the principal means of managing numerous arthropod pests. Furthermore, the use of effective chemicals has made possible the profitable production of many commodities in the presence of potentially devastating pests. Some rather obvious advantages of chemical control are (a) satisfactory potency, (b) rapid activity, (c) low cost, (d) availability, and (e) the opportunity afforded individual users to take unilateral action. Disadvantages of chemical control are (a) risks and hazards to humans and other nontarget vertebrates from broad-spectrum action of chemicals, (b) decimation of valuable biological pest control agents, (c) transient nature of chemical control, (d) ability of pests to develop resistance to chemicals, (e) pest resurgence, (f) induction of new arthropod pest problems, and (g) the fact that some economic losses still occur when chemicals are used. The merits and limitations of chemical control are more completely discussed by Newsom and Brazzel [123] and Herzog et al. [78]. When several chemicals are available to ameliorate pest problems, selection of the actual chemical to be used should be based on all factors affecting the benefit/risk equation [79]. Although cost and efficacy tend to be overriding factors, emphasis also should be placed on IPM, IRM, and environmental considerations, for it is these factors which will determine long-term efficacy and availability.

C. Insecticide Use Trends

Insecticide use by the agricultural sector has declined from about 29 million kg/year in 1979–1980 to about 14 million kg/year in 1991–1992 (Table 1): a 51% decrease [60]. Insecticide use on field corn, which accounts for over half of all insecticides used, has decreased 52% during that period. Similarly, insecticide use on cotton, which accounts for almost one third of all insecticides used, has dropped by 47%. Of the 17 commodities listed in Table 1, insecticide use has declined on 15 and increased on only 2 of them: lettuce and watermelon.

Development and refinement of IPM programs were only partially responsible for the reductions in insecticide use over this 12-year period. Other important factors were (a) registration of new insecticides which are effective at significantly lower rates than older chemicals, (b) loss of older insecticides either through voluntary cancellations by registrants or bans by the U.S. Environmental Protection Agency (EPA), (c) reduced acreages of some commodities, and (d) reduced rates of application for older chemicals.

Table 1   Insecticide Use Trends in U.S. Agriculture, 1979–1992

Crop	Kilograms per year		% Change
	1979–1980	1991–1992	
Cabbage	140,525	55,203	-61
Cantaloupes	16,012	11,703	-27
Carrots	41,686	3,720	-91
Celery	30,890	14,062	-54
Cotton	8,503,186	4,477,486	-47
Cucumber	11,612	9,843	-15
Field corn	16,032,945	7,713,468	-52
Grapes	60,918	40,370	-34
Green peas	77,112	15,805	-80
Lettuce	92,217	132,179	+43
Onions	113,264	73,937	-35
Potatoes	1,409,743	1,124,928	-20
Snap beans	267,034	73,166	-73
Soybeans	1,569,910	47,628	-97
Sweet corn	825,098	361,383	-56
Tomatoes	229,703	156,311	-32
Watermelon	32,070	45,179	+41
Total	29,453,925	14,356,371	-51

Source:Adapted from ref. 60.

II.   EARLY HISTORY OF CHEMICAL CONTROL

Various types of naturally occurring chemicals have been used to manage arthropod pests for thousands of years. Since the beginning of recorded history, accounts of using materials such as sulfur, ashes, soot, and dust to alleviate pest problems have appeared [80]. These substances are primarily physical poisons that affect the cuticle of arthropods. Present-day counterparts exist in the form of talc, silica gels, and boric acid.

A.   Inorganic Insecticides

Arsenic mixed with honey or other attractants was used as a stomach poison beginning in the mid 1600s for controlling ants. However, it was not until the mid 1800s that the arsenical-based compound Paris Green was effectively used to control Colorado potato beetle and other chiefly foliage feeding pests [80]. With the demonstration that calcium arsenate (Table 2) controlled the boll weevil [25] and that aircraft could be used to apply it in a dust formulation [26,129], extensive use of the arsenicals on cotton occurred during the 1920s to the 1940s. This approach resulted in outbreaks of secondary pests (cotton aphids, boll-

Table 2 Structure, Mode of Action, Mode of Entry, and Spectrum of Activity of Typical Insecticides and Acaricides from Different Classes of Chemicals

Typical Chemical	Structure	Mode of Action	Mode of Entry	Activity
Inorganics				
Calcium arsenate	$Ca_3(AsO_4)_2$	Inhibition of respiratory enzymes	Stomach	Insecticide Herbicide
Botanicals				
Nicotine		Agonist of nicotinic acetylcholine receptors	Contact	Insecticide
Pyrethrum: a mixture of insecticidal esters (e.g.,pyrethrin I)		Disruption of sodium channel function	Contact	Insecticide
Chlorinated Hydrocarbons				
DDT		Disruption of sodium channel function	Contact Stomach	Insecticide
Dicofol		Disruption of sodium channel function	Contact	Acaricide
Endosulfan		Functional disruption of GABA-gated chloride channels	Contact Stomach	Insecticide

Organophosphates

Name	Structure	Mode of action	Type	Use
Acephate	CH_3S, CH_3O—P(=O)—NHCCH$_3$(=O)	Inhibition of acetylcholinesterase at nerve synapse	Contact Systemic	Insecticide
Diazinon	EtO, EtO—P(=S)—O— (isopropyl pyrimidine ring)	Inhibition of acetylcholinesterase at nerve synapse	Contact	Insecticide Nematicide
Malathion	MeO, MeO—P(=S)—S—CHCOOEt—CH$_2$COEt(=O)	Inhibition of acetylcholinesterase at nerve synapse	Contact	Insecticide Acaricide
Methyl parathion	MeO, MeO—P(=S)—O— (C$_6$H$_4$—NO$_2$)	Inhibition of acetylcholinesterase at nerve synapse	Contact	Insecticide
Profenofos	EtO, PrS—P(=O)—O— (aryl, Cl, Br)	Inhibition of acetylcholinesterase at nerve synapse	Contact	Insecticide Acaricide

Carbamates

Name	Structure	Mode of action	Type	Use
Aldicarb	CH$_3$S—C(CH$_3$)(CH$_3$)—CH=NO—C(=O)—NHCH$_3$	Inhibition of acetylcholinesterase at nerve synapse	Contact Stomach Systemic	Insecticide Acaricide Nematicide
Carbaryl	(naphthyl)—O—C(=O)—NHCH$_3$	Inhibition of acetylcholinesterase at nerve synapse	Contact Stomach	Insecticide

(continued)

Table 2 Continued

Typical Chemical	Structure	Mode of Action	Mode of Entry	Activity
Methomyl	$CH_3-C=NO-C-NHCH_3$ (with SCH_3, O)	Inhibition of acetylcholinesterase at nerve synapse	Contact	Insecticide
Thiodicarb	$CH=NO-C-N-S-N-C-ON=C$ (with SCH_3, CH_3, CH_3, CH_3, O, O)	Inhibition of acetylcholinesterase at nerve synapse	Contact Stomach	Insecticide
Pyrethroids				
Bifenthrin	(structure)	Disruption of sodium channel function	Contact Stomach	Insecticide Acaricide
Cypermethrin	(structure)	Disruption of sodium channel function	Contact Stomach	Insecticide
Deltamethrin	(structure)	Disruption of sodium channel function	Contact Stomach	Insecticide
Permethrin	(structure)	Disruption of sodium channel function	Contact Stomach	Insecticide

Tefluthrin — Disruption of sodium channel function — Contact, Stomach — Insecticide

Insect Growth Regulators

Diflubenzuron — Interferes with chitin synthesis — Stomach — Insecticide

Methoprene — Disrupts molting process — Stomach — Insecticide

Others

Amitraz (formamidine) — Octopamine agonist — Contact — Insecticide, Acaricide

Abamectin (avermectin B1) $C_{48}H_{72}O_{14}$ — Functional disruption of GABA-gated chloride channels — Contact, Stomach — Insecticide, Acaricide, Nematicide

Pirate® (pyrrole) — Uncouples oxidative phosphorylation — Contact, Stomach — Insecticide, Acaricide

(continued)

Table 2 Continued

Typical Chemical	Structure	Mode of Action	Mode of Entry	Activity
Fipronil (phenyl pyrazole)		Functional disruption of GABA-gated chloride channels	Contact Stomach Systemic	Insecticide
Imidacloprid (chloronicotinyl)		Acetylcholine receptor agonist	Contact Systemic	Insecticide
Bacillus thuringiensis toxin	crystalline endotoxin	Cytolytic effects on midgut endothelium	Stomach	Insecticide
Spinosad (spinosyn)	mixture of $C_{41}H_{65}NO_{16}$ and $C_{42}H_{67}NO_{16}$	Activation of nicotinic acetylcholine receptors	Stomach Contact	Insecticide

worms, and tobacco budworms) due to disruption of biological control agents [50,89,157], virtual elimination of the cotton leafworm as a pest of cotton [123], and contamination of soils to the extent that soybean and other arsenic-sensitive crops could not produce optimal yields for several years on these soils. This marked the first appearance of problems associated with the widespread use of persistent pesticides; that is, changes in pest status, induction of new pests, destruction of beneficial species, and environmental contamination. Other inorganic pesticides worth mentioning include, for example, such compounds as fluorides, sulfur, selenium, antimony, mercury, lead, and tin [109,155].

B. Botanical Insecticides

Nicotine, in the form of water extracts from tobacco leaves, was first used as an insecticide in 1763, but it was not until 1828 that the pure alkaloid (see Table 2) was isolated [109]. It was used extensively in the 1930s for controlling aphids, especially outbreaks of cotton aphids following applications of the arsenicals [123]. The primary disadvantage of nicotine was its acute mammalian toxicity (50–60 mg kg^{-1}), which resulted in numerous human fatalities [109,155]. During the years 1930–1934, there were 106 accidental deaths and 182 suicides involving nicotine in the United States [76]. Nevertheless, it was the forerunner of modern synthetic insecticides inasmuch as it was a contact poison acting on the nervous system [80].

Pyrethrum (see Table 2) is found in the flowers of a chrysanthemum now commercially grown in Africa and South America [178]. The ground flower heads were used to control human body lice during the Napoleonic Wars in the 1800s, although much earlier use is known. Pyrethrum was found to be particularly effective against flies, mosquitoes, stored-grain pests, and other household and garden pests. Because of its fast knockdown and broad-spectrum activity against insects, as well as its safety to humans and domestic animals, pyrethrum formulated as sprays, aerosols, and dusts is still extensively used in and around homes and gardens. Pyrethrum, which is a contact nerve poison, has served as the model for the synthesis of the synthetic pyrethroids, several of which are of considerable current agricultural importance [80]. Other botanical insecticides include rotenone, sabadilla, ryania, hellebore, strychnine, nornicotine, and anabasine [109,155].

III. MODERN SYNTHETIC ORGANIC CHEMICALS

A. Chlorinated Hydrocarbon Insecticides

The insecticidal properties of DDT were discovered by Paul Müller of Switzerland in 1939, although it was first synthesized by Zeidler in 1874 [109]. The advent of DDT (see Table 2) and other chlorinated hydrocarbon insecticides in the mid to late 1940s revolutionized insect control. Compared with the inorganics and the botanicals, these chemicals were more efficacious against a broader spectrum of arthropod pests, provided longer residual control due to environmental persistence, and were more economical. As a result, DDT became the most widely used insecticide ever introduced. By 1961, there were 1200 formulations of DDT in the United States directed against 240 different arthropod pests. Peak production of DDT occurred in 1963 with 81,646,200 kg of the material being formulated [80]. Other chlorinated hydrocarbons such as dicofol (see Table 2), methoxychlor, BHC, lindane, toxaphene, aldrin, dieldrin, heptachlor, endrin, chlordane, endosulfan (see Table 2), kepone, and mirex were introduced in the 1940s and 1950s.

Widespread adoption of the chlorinated hydrocarbons resulted in induced pest problems (especially spider mites), resurgence of targeted pests following applications, development of resistance in numerous arthropod pests, and residues that caused acute and chronic toxicity problems in nontarget species. Because of environmental and other problems, DDT was banned in the United States in 1972 [80]. Subsequently, nearly all chlorinated hydrocarbons have had their labels banned or severely restricted. Today, DDT is available only as an insecticide for louse control on humans through a physician's prescription [178].

The chlorinated hydrocarbons act primarily as contact poisons, although they are also excellent stomach poisons. The DDT subgroup of chlorinated hydrocarbons (e.g., DDT, methoxychlor, dicofol) consists of nerve poisons that are believed to exert their toxicity by disrupting the functioning of sodium channels on nerve axons [160,162]. The cyclodiene subgroup (e.g., aldrin, dieldrin, endrin, chlordane, endosulfan) are nerve poisons that exert their toxicity by disrupting the action of gamma-aminobutyric acid (GABA), an inhibitory neurotransmitter in the central nervous system of insects [110,162].

B. Organophosphorus Insecticides

A German chemist, G. Schrader, is credited with the discovery of the first successful organophosphorus insecticide, schradan, in 1941 [80]. Parathion followed in 1944 and became one of the most widely used organophosphates despite high mammalian toxicity. Numerous other organophosphates such as malathion (see Table 2), phorate, dimethoate, trichlorfon, naled, diazinon (see Table 2), chlorpyrifos, ethion, disulfoton, demeton, monocrotophos, dichlorvos, dicrotophos, acephate (see Table 2), methyl parathion (see Table 2), and more recently profenofos (see Table 2) and sulprofos, followed. The organophosphates are much less persistent than the chlorinated hydrocarbons, but they are relatively more toxic to mammals [178]. Just as with the chlorinated hydrocarbons, widespread and indiscriminate use of the organophosphates led to changes in pest status, resurgence of target species, destruction of beneficial species, and numerous instances of resistance. They are all nerve poisons that primarily exert their toxicity by inhibition of acetylcholinesterase, the enzyme responsible for inactivation of the excitatory neurotransmitter acetylcholine [128,162]. The organophosphates, which are relatively nonpersistent and do not bioaccumulate, are widely used in all aspects of arthropod pest management as contact, stomach, systemic, and fumigant poisons. For example, some of the more water-soluble organophosphates such as dimethoate, acephate (see Table 2), phorate, disulfoton, and demeton are translocated to other parts of a plant from the points of contact (systemic activity). Similarly, the organophosphate dichlorvos is sufficiently volatile to be used successfully as a fumigant.

C. Carbamate Insecticides

Originally discovered by H. Gysin of the Geigy Company in 1947, the carbamates are essentially synthetic analogues of the plant alkaloid physostigmine [109]. Introduced in 1956, carbaryl (see Table 2) was among the first truly successful carbamates, and it became one of the most widely adopted owing to its low mammalian toxicity and its broad-spectrum activity against insects. Other carbamates such as methomyl (see Table 2), aldicarb (see Table 2), thiodicarb (see Table 2), propoxur, carbofuran, bendiocarb, methiocarb, and bufencarb soon followed. Like the organophosphates, the carbamates are nerve poisons that inhibit the activity of acetylcholinesterase, are relatively nonpersistent, and do not bioaccumulate [162]. With the exception of the oxime carbamates such as aldicarb (see Table 2), this class tends to have lower mammalian toxicity than the organophosphorus insecticides [80]. Carbamates exert their toxicity via contact, stomach, and systemic activity.

D. Pyrethroid Insecticides

Natural pyrethrins and early synthetic analogues such as allethrin were recognized as excellent insecticides with a broad spectrum of activity against insects but with relatively low toxicity to mammals [11,41,43]. However, they were too unstable and expensive to manage pests of agricultural crops economically and efficiently. Beginning with the synthesis of permethrin (see Table 2) by Elliott et al. [42] in 1972, photostable pyrethroids became available for use on agricultural crops. Furthermore, pyrethroids were efficacious at rates 10–100 times less than organophosphates and carbamates [78]. Permethrin and fenvalerate became available for field use in the United States in 1978, and they were extremely effective in controlling pests such as the tobacco budworm (*Heliothis virescens*), which in some areas had developed resistance to all available classes of insecticides [78]. Several other pyrethroids such as bifenthrin (see Table 2), cypermethrin (see Table 2), tefluthrin (see Table 2), deltamethrin (see Table 2), cyfluthrin, esfenvalerate, lambda-cyhalothrin, fenpropathrin, and zetamethrin became commercially available and widely used against a multitude of arthropod pests.

Pyrethroids are axonic nerve poisons with a mode of action similar to DDT; that is, disruption of sodium channel function [160,162]. As a group, they are primarily contact poisons but are also active as stomach toxicants. Because of their broad spectrum of activity against arthropods, pyrethroid use on a widespread basis often induces other pest problems, such as aphid and spider mite outbreaks, although some of the newer pyrethroids such as bifenthrin (see Table 2) are efficacious against these species. Resistance to pyrethroids in some pest species such as the tobacco budworm, horn fly, pink bollworm, Colorado potato beetle, and diamondback moth has developed as a result of widespread use. In addition, preexisting resistance to DDT in some species conferred cross resistance to pyrethroids, because both classes of insecticides share a common site of action.

E. Insect Growth Regulators

Three decades have passed since the chemical structure of the first juvenile hormone was elucidated in 1967 [167]. However, only a few chemicals have been commercialized as insect growth regulators (IGRs). Presently, only the juvenile hormone mimics, such as methoprene (see Table 2) and fenoxycarb, and the benzoylphenylureas, diflubenzuron (see Table 2), and chlorfluazuron, are being used to control arthropod pests of agricultural and medical importance [117]. Methoprene and other juvenile hormone analogues kill insects by inhibiting the molting process [162]. The benzoylphenylureas cause toxicity by interfering with chitin synthesis [162]. Insect mortality associated with the IGRs is typically slow relative to most other insecticides, which has greatly limited their acceptance and use.

F. Formamidine Insecticides

Since the discovery of their pesticidal activity [34], the formamidines have attracted considerable attention because of their novel effects on insects and acarines. Only two of the formamidines, amitraz (see Table 2) and chlordimeform, have been commercially successful. Only amitraz is commercially available, because all uses of chlordimeform have been canceled by the U.S. EPA because of chronic toxicity concerns in humans. Unique properties of the formamidines include ovicidal activity via contact and vapor routes [35], synergism of many classes of insecticides [135], and modifications of insect behavior [99,132,164,166]. The mechanism by which these chemicals protect plants and animals from arthropod pests is complex and results from a combination of lethal and sublethal effects. Sublethal effects include disruption of feeding and reproduction [31]. The formamidines appear to function as octopamine agonists [81,162], causing a variety of detrimental behavioral effects [81,92].

IV. NOVEL CHEMICALS TO MANAGE ARTHROPODS

Several relatively new classes of chemicals such as the avermectins, chloronicotinyls, pyrroles, phenylpyrazoles, spinosyns, and novel forms of IGRs and *Bacillus thuringiensis* offer great promise for the future of arthropod pest management. Generally, these toxicants are more environmentally acceptable because they (a) offer more specificity with regard to arthropod pests, (b) achieve the desired level of arthropod control at low field use rates, and (c) degrade rapidly in the environment.

A. Avermectin Insecticides

The avermectins are among the most potent insecticidal, acaricidal, and antihelminthic agents known [48,93]. Originally isolated in 1976 by scientists at Merck & Co., Inc. (Rahway, NJ), from a culture of *Streptomyces avermitilis* from Japan, the avermectins seem to exert their toxicity by disrupting the action of both ligand-gated (e.g., GABA) and voltage-gated chloride channels [23,109,162]. The end result is functional disruption of GABA-gated chloride channels. Ivermectin and abamectin (avermectin B_1) are the only avermectins to be successfully commercialized thus far. Ivermectin has been successfully used to control animal parasites. Its agricultural analogue, abamectin (see Table 2), has been less successful, because it degrades so rapidly. Its primary use thus far has been for mite control. Synthesis of avermectin derivatives has yielded a number of compounds that possess insecticidal and acaricidal activity against major arthropod pests of agricultural crops with improvements in environmental stability. Promising chemicals are MK-243 [45] and MK-244 (emamectin benzoate) [37,171,180].

B. Pyrrole Insecticides

Pirate® (AC 303,630) is the lead compound of a novel class of insecticides/acaricides, the pyrroles, which were recently discovered by the American Cyanamid Company (Princeton, NJ) [104]. This broad-spectrum insecticide/acaricide (see Table 2) is highly active by ingestion, exhibits contact activity, and provides moderate residual activity [104]. Because of a different mode of action (uncoupler of oxidative phosphorylation), Pirate® is effective against insects such as the tobacco budworm and diamondback moth that have developed resistance to carbamates, cyclodienes, organophosphates, and pyrethroids [162,176]. In field studies, Pirate® has provided acceptable control of 35 agricultural pest species from the orders Acari, Coleoptera, Diptera, Heteroptera, Homoptera, and Lepidoptera without causing phytotoxicity. It is classified as moderately toxic to mammals based on acute toxicity studies and is nonmutagenic in the modified Ames and Chinese hamster ovary tests [104]. Pirate® is presently under full-scale global development for use against pest insects and mites on cotton, vegetables, fruit, and ornamentals.

C. Phenylpyrazole Insecticides

Fipronil (see Table 2) is a new phenylpyrazole insecticide that has excellent activity against many soil and foliar insect pests on a wide variety of commodities. Its insecticidal properties were discovered by Rhone-Poulenc Ag. Co. (Research Triangle Park, NC) in 1987 [27]. This broad-spectrum insecticide is highly active via ingestion, contact, and systemic routes. It shows great promise as both a soil and foliar insecticide against several important pests of cotton, including boll weevil, tarnished plant bugs, and thrips [18]. Its unique action as a potent blocker of the GABA-gated chloride channel makes it effective against insects resistant to carbamate, organophosphate, and pyrethroid insecticides [27,162]. Based on available data on mammalian toxicity and ecotoxicity, fipronil appears to be acceptable for registration by the U.S. EPA. Fipronil is currently in worldwide development by Rhone-Poulenc Ag. Co. against piercing-sucking and chewing insects, including aphids, leafhoppers, planthoppers, chewing Lepidoptera and Coleoptera, flies, and soil-inhabiting Coleoptera. It currently has received registration and use permits in several Asian and South American countries.

D. Chloronicotinyl Insecticides

Imidacloprid (see Table 2), a nitromethylene derivative [161] synthesized in 1985 by Nihon Bayer Agrochem K.K. (Tokyo, Japan), is a systemic and contact insecticide exhibiting low mammalian toxicity [38,39]. With superior activity against sucking insects such as aphids, leafhoppers, planthoppers, thrips, and whiteflies, it also is effective against some Coleoptera, Diptera, and Lepidoptera. Imidacloprid has a novel mode of action that is similar to nicotine: it acts as an agonist of nicotinic acetylcholine receptor [10,154,162]. No cross resistance to imidacloprid has been observed in insects that possess resistance to other classes of insecticides [120]. With excellent systemic and good residual characteristics, imidacloprid is especially appropriate for seed treatment and soil application.

application. Effective early season control with long-lasting protection has been demonstrated in crops such as cereals, corn, cotton, potatoes, rice, sorghum, and vegetables. Foliar applications successfully control numerous insects attacking these and other commodities later in the season [120]. Imidacloprid has been registered recently for use in the United States on apples, cotton, and potatoes. It also has been registered and is now being used in France, Spain, Japan, and South Africa [118].

E. Spinosyn Insecticides

In 1994, DowElanco (Indianapolis, IN) announced a new class of insect control molecules called the spinosyns [36]. Naturally derived from a new species of Actinomycetes, *Saccharopolyspora spinosa*, they are very active against many pests of crops, ornamentals, forestry, greenhouse, garden, and households. Spinosad (see Table 2), a mixture of spinosyn A and spinosyn D, is the lead compound and it has shown contact and stomach activity against the insect orders Coleoptera, Diptera, Hymenoptera, Isoptera, Lepidoptera, Siphonoptera, and Thysanoptera. Current data indicate that spinosad causes persistent activation of nicotinic acetylcholine receptors in the insect nervous system, a unique mode of action with no known cross resistance with other insecticides [149]. Because of its very favorable mammalian toxicity and environmental profiles, spinosad has already been registered by the EPA for use on lepidopteran pests of cotton [121,172].

F. *Bacillus thuringiensis* Insecticides

Bacillus thuringiensis (*Bt*) is an aerobic, spore-forming bacterium which produces toxins (see Table 2) pathogenic to more than 182 species of insects [105]. Numerous subspecies have been isolated, but only a few have been commercialized: *kurstaki* for lepidopteran pests, *israelensis* for dipterans, *tenebrionis* for coleopterans, and *aizawai* as well as several transconjugates for lepidopterans. The mode of action appears to be a disruption of the midgut endothelium [162]. Although commercially available for more than 30 years, *Bt* products have not been used widely, because they were slow to kill and lacked persistence, characteristics that resulted in poor efficacy in comparison with synthetic organic insecticides. However, several recent developments have vaulted *Bt* products into prominence [46]. First, the efficacy and economy of traditional chemicals have been reduced by the development of resistance to these chemicals in numerous arthropod pests. Second, novel changes in the *Bt* toxin and in formulations have resulted in numerous new *Bt* products that are more efficacious. And third, the *Bt* gene has been successfully inserted into several agronomic crops such as cotton [131], tomatoes [47], and corn [46] (see Chapter 10). Initial studies have shown that cotton plants containing the gene for *Bt* toxins provide excellent control of tobacco budworm and pink bollworm and good control of bollworm [13,90,108,115,182].

G. Insect Growth Regulators

Tebufenozide (RH 5992) is a novel insecticide discovered by Rohm & Haas (Philadelphia, PA) that is highly specific against the larval stages of lepidopteran insects [77]. It belongs to a new class of selective and safe insecticides acting as ecdysone agonists (mimics the action of the natural insect hormone 20-hydroxyecdysone, the physiological inducer of the molting and metamorphosis process in insects [162]). Although not yet registered by the EPA, tebufenozide has been successfully used since 1994 against beet armyworm via a Section 18 emergency exemption. Azadirachtin, an extract from the neem tree (*Azadirachta indica*), also disrupts insect molting by antagonizing the action of ecdysone [162]. Already used commercially in India and other countries, it is under development in the United States. Pyriproxyfen [8], a juvenile hormone mimic, and buprofezin [130], an inhibitor of chitin synthesis, were successfully used against silverleaf whiteflies on cotton during 1996 and 1997 via Section 18 registrations. Other promising new active compounds known as the acetogenins and pyridazinones have been isolated from tropical plants and are being considered for development [117].

V. FORMULATION AND APPLICATION OF INSECTICIDES

A. Formulation

Insecticides and acaricides are seldom used in their initial isolated or synthesized form. Formulation is the processing of a chemical by any method that will improve its properties of storage, handling, application, efficacy, or safety [179].

Prior to 1950, most insecticides and acaricides were applied as dusts, primarily because they were easier to formulate and apply as dusts than as sprays. With the advent of the synthetic organic chemicals in the late 1940s, sprays (liquid or wettable powder formulations) came into general use. Most synthetic insecticides and acaricides are readily soluble in organic solvents but most often are insoluble in water. Emulsifiable concentrates, wettable powders, and flowable suspensions can be easily formulated with the addition of effective emulsifying agents. By mixing with water at the site of application, speed of formulation is increased, and storage and transportation costs are reduced relative to bulky dust formulations. Also, liquid applications can be used at times when wind velocities are too high for effective application of dusts [2,54]. However, in terms of insect control, sprays and dusts are equally effective provided they are applied properly. More recently, granular and slow-release formulations have been effectively employed to make safe and lengthen the residual activity of several pesticides. Granular formulations can be applied under windy conditions and are particularly useful as soil applications at planting to introduce systemic chemicals to roots for transport throughout seedling plants. Water dispersible granules represent an improvement in wettable powder formulations by increasing applicator safety while introducing more efficient mixing properties.

Systemic insecticides, which are primarily derived from the carbamate and organophosphorus classes, have been very useful in controlling pests with piercing-sucking mouthparts such as thrips, aphids, spider mites, true bugs, and some beetles. Additionally, soil-applied systemics are less detrimental to beneficial arthropods than are foliar applications.

B. Application

At present, most agricultural applications of insecticides are made with high-clearance ground machines or aircraft using liquid sprays or granules. Developments in spray nozzles and equipment have made sprays practical and effective at volumes of 9.4–94 L ha^{-1}.

Ultra–low–volume (ULV) refers to sprays of 1.89 L/hectare or less. Arthropod pests may be controlled by proper application of undiluted pesticide concentrates at the rate of 0.47–1.89 L ha^{-1} [1,24]. The major problems in applying volumes as low as 0.47 L ha^{-1} are atomization, uniformity of dispersal, and drift to nontarget areas [88]. Most ULV application of insecticides has been done with airplanes fitted with nozzles having orifices of a very small diameter or with rotary atomizing devices [2,139]. ULV application by ground machines has been less successful, primarily because of the relatively slow ground speed.

For ULV application, fine droplets (30–100 μm diameter) are used because of the reduced volume/hectare. This droplet size is required to obtain the necessary coverage for effective insect pest control. Drift is a serious problem with fine droplets (less than 50 μm diameter), but drift may be less with ULV application than with water-diluted formulations [123,183] under some conditions. Under most conditions, droplets of water-based formulations lose water rapidly through evaporation before they reach the target, whereas ULV droplets of the organic compounds (insecticide and solvent) evaporate at a much slower rate.

The efficacy of some insecticides is enhanced when they are formulated and applied in ULV. For example, malathion has been shown to be much more effective when applied as a ULV concentrate rather than as an emulsifiable concentrate in 9.4–18.8 L of water hectare^{-1} [17,24,101]. More recently, Treacy et al. [175] reported longer residual efficacy of azinphosmethyl and cyfluthrin when applied ULV in oil as compared with conventional application in water.

Chemigation (pesticide application through irrigation) provides a novel way of dispensing pesticides, and it has been successfully used for management of several insects, particularly the bollworm/tobacco budworm complex [21,22, 156,173]. Advantages of chemigation are (a) decreased application costs, (b) improved pesticide coverage, (c) decreased human contact with pesticides during application process, (d) decreased energy consumption, and (e) more flexibility in time of day for application. Disadvantages are (a) lack of formulations approved for chemigation by the EPA, (b) need for flow valves and check systems to prevent contamination of water source, and (c) the fact that water-soluble pesticides are ineffective when applied via chemigation [20].

A promising new approach for arthropod control is the development of attract-and-kill devices [111,158]. For example, polyvinyl chloride (PVC) has been used to combine and slowly release a mixture of grandlure (sex and aggregating phermone for the boll weevil), feeding stimulants, and a toxicant for boll weevil control. The PVC device is in the form of a cap that is placed on top of wooden stakes (commonly referred to as boll weevil bait sticks) that are placed in or around cotton fields. Field tests conducted during 1990 during early season indicated a 15–67% reduction in boll weevil trap captures in treated fields versus control fields [112]. Field tests in Tennessee during 1991, where early season boll weevil populations were low, also proved to be successful [159].

VI. INTEGRATION OF CHEMICALS WITH OTHER APPROACHES IN ARTHROPOD PEST MANAGEMENT

A. Compatibility of Chemical Control with Other Approaches

Insecticides and acaricides are generally used in IPM programs when other approaches (e.g., biological, cultural, physical, regulatory) fail to prevent arthropod pest populations from reaching defined economic thresholds [78]. An important question arises: What is the best way to use these chemicals with the least disruption and detriment to the environment, particularly to biological control agents? Newsom's [125,126] discussions of using selective chemicals, ecological selectivity, and genetic manipulation of beneficial species to increase compatibility with other components of IPM programs remain appropriate today.

Although insecticides and miticides are targeted at arthropod pests, many of them have profound deleterious effects on predators and parasites [26,91]. However, individual chemicals vary in their toxicity among species of pests, predators, and parasites. From a physiological standpoint, species specificity is due to differences in penetration of the cuticle, penetration at the site of action, metabolism (both activation and degradation), storage, and excretion. This allows for selection of an insecticide or acaricide (usually several chemicals are registered for a specific use to control specific pests) that is least detrimental to beneficial species, yet still effective against the targeted species. The compatibility of commercially available insecticides and acaricides with natural enemies or arthropods has been reviewed by several investigators [28,29, 113,119,126].

Development of selective pesticides favoring arthropod natural enemies is presently receiving more attention from the chemical industry as environmental and human impacts from pesticide misuse force cancellation or restricted use of pesticides [119]. However, a biorational approach to chemical development suffers from a lack of understanding of the physiological and biochemical bases for responses of pests and natural enemies to toxicants.

Ecological selectivity is based primarily on behavior and habitat differences among pest and beneficial arthropods. Practicing ecological selectivity in IPM programs is accomplished through judicious use of pesticides based on critical selection, timing, dosage, placement, and formulation of pesticides [86,91]. Thus, evaluations of new pesticides for use in IPM programs should include not only effectiveness against target species, minimum effective dosages, and phytotoxicity, but also survival of beneficial arthropods at pest-effective dosages [85,141], as well as the impact on species constituting the food resources for beneficial arthropods [61,141].

Selection for increased resistance of the phytoseiid mite, *Metaseiulus occidentalis*, to insecticides and acaricides serves as one of the best examples of genetic manipulation to increase the compatibility of chemicals in IPM programs [12]. This mite is the most important predator of spider mites in deciduous orchards and vineyards in western North America [49,84]. Carbaryl-organophosphate-permethrin–resistant and carbaryl-organophosphate-sulfur–resistant strains have been developed through laboratory crosses and selections, and these strains have performed well in field trials [82,83]. Techniques of genetic engineering offer much hope for the future [12]. Introducing desirable traits such as, for example, insecticide resistance, increased fecundity, changes in host or habitat selection, and dispersal into beneficial arthropods is becoming a reality.

B. Facilitating IPM Through Crop Production Practices

Crop production practices greatly influence the growth and development of pest and beneficial arthropod populations. By proper manipulation of such practices as planting date, variety selection, fertilization, irrigation, conservation tillage, and crop residue destruction, arthropod pest populations often can be kept below economic thresholds. The need for chemical control is reduced, which in turn forestalls insecticide resistance development [70]. In certain situations where key pests such as the tobacco budworm in cotton are resistant to all commercially available insecticides, manipulation of crop production practices may become the keystone to successful and profitable crop production [6,177]. For example, production of an early maturing cotton crop is critical for managing the tobacco budworm; early maturity reduces the period of time that the harvestable crop is susceptible to this insect pest.

Conservation tillage provides an excellent example of some of the effects of crop production practices on arthropod pests [98] (see Chapters 15 and 25). There are numerous benefits associated with conservation tillage practices that promote sustainable agriculture by minimizing soil erosion, more efficiently utilizing energy resources, reducing water pollution, improving the timeliness of cultural practices, and, in some instances, enhancing crop performance [16,87, 174]. However, conservation tillage systems provide a favorable microenvironment for arthropods by increasing host plant biomass and moderating soil moisture and temperature extremes. Increased problems with some pests (e.g., cutworms, wireworms, corn rootworms, aphids, white grubs, seed pests, armyworms, corn borers) have been reported for several crops in conservation tillage [3,33,55, 72,98].

VII. ARTHROPOD RESISTANCE TO CHEMICALS

A. The Resistance Problem

Insects and other arthropods are amazingly adaptable. Their remarkable adaptability (i.e., genetic plasticity) is the primary reason that insects are the most successful of all animals in terms of both numbers of species and the number of individuals. Thus, it is not surprising that arthropods quickly adapt to toxic environments. Hence, the development of resistance to insecticides and acaricides is simply an accelerated example of the process of evolution [62].

Resistance is defined as the ability of strains of a species to withstand exposure to dosages of chemicals that exceed that toxic to a normal susceptible population—and this ability is inherited by subsequent generations of the strain [4]. The development of resistance is a population phenomenon (individuals do not develop resistance in their own lifetime) that occurs over time through the selection of those individuals best suited to survive exposure to a chemical [62]. Genes for resistance are present (preexisting) in low frequencies in most species and exposure to a chemical simply increases the frequency of genes conferring resistance by eliminating those individuals lacking these genes from the parent population. The problem of cross resistance, the phenomenon where the development of resistance to a given chemical results in the simultaneous development of resistance to other chemicals to which the species has not been exposed, provides further complications.

There were only a few reported instances of insecticide and acaricide resistance prior to the introduction and widespread use of the synthetic organic chemicals in the late 1940s. However, the number of confirmed cases of resistance and cross resistance has been steadily increasing over the past 50 years such that there are now more than 500 insect and mite species that are resistant to one or more chemicals [59]. Resistance to chemicals used to control arthropod pests has become the limiting factor in the successful population management of several species of insects and mites, which have developed resistance or cross resistance to all commercially available chemicals [116]. The current situation with the tobacco budworm typifies this problem [165].

B. Conserving Arthropod Susceptibility to Chemicals

The declining arsenal of chemicals makes it imperative that development of resistance to existing and new chemicals be slowed or prevented [116]. Historically, a class of insecticides has been lost about every 10 years [133]. Insecticides are too important a resource for this trend to continue. The speed of development of resistance in an arthropod population is dependent on the interaction of genetic (frequency, number, and dominance of genes for resistance and their fitness), biological (e.g., generation turnover, number of progeny, refugia, migration), and operational (e.g., chemical nature of toxicant, toxicant persistence, life stages selected) factors [56]. Genetic and biological factors (excluding refugia) represent inherent qualities of a species that are mostly beyond the control of humans. However, operational factors can and must be managed to prevent or delay resistance development. Thus, resistance would be delayed the most if (a) the selecting chemical is not very persistent, (b) the chemical is not related to a previously used chemical, (c) the formulation does not result in prolonged release, (d) selection by the chemical is directed toward only one stage of the life cycle, (e) the area of chemical application is localized rather than area wide, (f) some generations are not exposed, and (g) refugia are present [62].

Insecticide resistance management (IRM) is not a new concept. Early proponents were such eminent scientists as A. W. A. Brown [15], G. P. Georghiou [56,57,58], and R. M. Sawicki [150,151,152]. These scientists made eloquent and logical arguments on the use of insecticides in such a fashion (e.g., sequential use of different classes of insecticides, use based on their potential to cause resistance development and/or cross-resistance) as to slow or prevent resistance development. More recently, Forrester [51,52], Luttrell [106], Phillips [133], Plapp [137,138], Roush [144–147], and Tabashnik [169,170] have not only provided basic research and theoretical models on which to base IRM programs, but they have also been involved in the actual implementation of IRM programs directed specifically at *Heliothis* and *Helicoverpa* spp., key pests of various crops worldwide.

1. IRM Programs for Tobacco Budworm in the United States

Resistance to the pyrethroids in tobacco budworm was confirmed following field control failures in Texas in 1985 and in the mid South states of Arkansas [137],

Louisiana [94,95], and Mississippi [145] in 1986. IRM programs for Texas [137] and mid-South [5] were designed based on the Australian experience [32,116] with a similar species, *Helicoverpa* (=*Heliothis*) *armigera*. Current IRM programs employ a "windows" approach in which the available classes of insecticides are recommended for use during specific time periods based on available information. Basic information, which was available concerning cross-resistance patterns [96], mechanisms of resistance [127,136,163], genetics of the primary resistance mechanisms [145], susceptibility of the various life stages, seasonal history, alternate hosts, and refugia as well as movement of the pest, was incorporated into the IRM strategies.

Additionally, a rapid and inexpensive monitoring system utilizing a discriminating dose was developed by Plapp et al. [137]. Levels of resistance monitored in Arkansas, Louisiana, Mississippi, Oklahoma, and Texas over the past several years suggest that IRM for the tobacco budworm has been relatively successful [63–65,67–69,138,142]. The frequency of pyrethroid-resistant tobacco budworm genotypes appears to have stabilized in Texas and part of the mid-South [71]. However, overall levels of resistance in tobacco budworms in Louisiana have slowly increased over the past decade [9]. Meanwhile, the tobacco budworm has developed resistance to carbamate, cyclodiene, and organophosphorus insecticides [44,97], which has further complicated IRM.

VIII. FUTURE OF CHEMICAL APPROACHES TO MANAGING ARTHROPODS

A. Constraints

The primary constraints affecting the future use of chemicals to manage arthropods are availability and societal acceptance. The availability of chemicals is decreasing owing to development of resistance, environmental regulations, and cost of development. Societal concerns regarding the deleterious effects of chemicals are being expressed in increased regulation of their development, registration, manufacture, formulation, transportation, storage, and application. IPM offers the best opportunity for responsible and judicious use of chemicals, but IPM systems for most commodities have not been fully developed and refined. Furthermore, application technology must be improved to increase the efficiency of the chemical application process.

1. Availability of Chemicals

The continued availability of effective and economical insecticides is in question because of (a) the rapid development of resistance by arthropods to chemicals used for their control, (b) the increasingly stringent and costly federal and state registration and reregistration requirements, (c) the relatively short patent life of new chemicals, and (d) the difficulty in discovering new leads for insecticides with novel modes of action. These developments have increased the cost of registering new insecticides to such an extent (current estimates range up to $180 million [168]) that some companies, which have historically developed and produced insecticides, are no longer active in insecticide research and development. In the future, it appears that only a few very large companies will be able to compete financially in the agricultural chemical arena. This trend is already underway and the expected outcome is fewer but more expensive new insecticides [134].

2. Societal Concerns

The fact that insecticides have been remarkably effective and relatively cheap ensured their misuse [124]. Although Paul Müller, the discoverer of its insecticidal properties, was awarded the Nobel prize, DDT has become one of the most reviled words in the English language. The deleterious effects of DDT and other pesticides were brought to national and international attention with the publication of Rachel Carson's *Silent Spring* in 1962 [19]. Subsequent regulation of insecticides and other pesticides at state, national, and international levels has increased to the extent that very few insecticides are now being registered and the uses of older chemicals have been greatly curtailed [179]. This plethora of laws and regulations and the numerous agencies, departments, and entities involved in their interpretation and enforcement threaten arthropod pest management systems including IPM.

Societal concerns about insecticides are focused primarily on acute poisoning (resulting from handling and application) and chronic exposure (long-term exposure to minute residues). Based on the examination of statistics on fatalities due to insecticides [100] and the frequency of occurrence rates in humans for various types of cancers and tumors, societal fears appear to be ill founded. In ranking substances most frequently involved in human poison exposure for example, pesticides (including rodenticides) were ranked well below cleaning substances, analgesics, cosmetics, and cough and cold preparations and only slightly above vitamins [100]. Similarly U.S. Department of Agriculture surveys of pesticide residues in the United States food supply indicate that dietary intake is well within the limits determined by the EPA as safe and that the general public is not at risk [153]. Nevertheless, public perception of hazards associated with pesticides of all kinds will continue to be a major factor affecting their future availability and use [73,79,103].

B. Partial Solutions

Despite severe constraints that will limit the development, registration, and use of insecticides and acaricides, chemical control will continue to be an indispensable component of arthropod pest management systems for the foreseeable future. Partial solutions to the many problems associated with chemicals can be found in IPM, biorational development of selective insecticides, prescription use, and novel application methodology.

1. IPM

The panacea philosophy associated with the curative action of insecticides and miticides has severely handicapped arthropod pest management [125]. The discovery of the extraordinary biological activity of DDT and other synthetic organic insecticides was hailed as the solution to many pest problems [107]. Similarly, the successful eradication of the screwworm, *Cochliomyia hominivorax*, from the United States using the sterile male technique raised false hopes that insect sterility was the solution to arthropod pest management problems. In reviewing singular approaches to arthropod pest management, Newsom [122] documented that neither the chemical, sterile insect, nor other methods (e.g., regulatory and quarantine, cultural, biological, varietal resistance) were panaceas. It should be obvious by now that most arthropod pest species have the genetic plasticity to overcome unilateral control techniques. Equally obvious is that IPM, which employs all available population management strategies and which is biologically based, offers the best hope for arthropod pest management in the future.

IPM programs have been developed and successfully implemented for a number of arthropod pests over the past 2–3 decades. One of the best examples is that for the sugarcane borer, *Diatraea saccharalis*, in Louisiana. This IPM program, which is practiced on 70% of the sugarcane acreage [14], emphasizes the balanced use of cultural, biological, and chemical pest control tactics [140]. Developers of the program ascribe 25% of within-season suppression to beneficial arthropods, particularly predators; 25% to resistant varieties of sugarcane; and 50% to selective use of insecticides [140].

Despite demonstrated successes against some arthropod pests of several commodities, IPM is severely handicapped by lack of knowledge concerning such fundamental components as, for example, sampling methodology, economic thresholds, biology and ecology of pest and beneficial species, host-plant resistance, and interactions among pest complexes. Lack of resources such as funds for research and development of new and/or refined programs, sufficient trained personnel for implementation, interdisciplinary training, and research programs also has limited IPM.

2. Biorational Chemicals

Biorational pesticides were initially defined as microbials and biochemical substances such as insect pheromones and hormones that are somewhat species specific [45]. A broader definition that includes all chemicals that affect the growth, development, biology and ecology of pest species differently from that of beneficial species is more appropriate [40,53,75,117,119,143,148]. Croft [30] viewed the limited availability of biorational insecticides to be a problem of policy more than of technology. In most cases, improvements have involved the use of nonselective insecticides in ecologically selective ways rather than the development of physiologically selective chemicals. Strong evidence that the development of biorational insecticides is technologically feasible is the fact that selective acaricides and herbicides have been the norm rather than the exception. It is encouraging that chemical companies are giving more priority to the biorational development of selective pesticides. In fact, some of the newer chemical companies are completely focused on the development of biorational pesticides, and this bodes well for the future of arthropod pest management.

3. Novel Application Methodology

One area of technology that has greatly limited IPM programs and has resulted in considerable environmental damage is the process of applying insecticides and acaricides. When one considers that only about 18.5 mg of cypermethrin is actually required to kill 95% of an extremely high population of 100,000 tobacco budworms (about one larva per plant) per hectare if it is applied directly on the larvae, it seems incomprehensible that we must apply over 3500 times this much to the cotton foliage to achieve approximately 95% field control [134]. Drift, or off-target movement, has been and continues to be a major problem with the application of insecticides [74,156]. In fact, it is not unusual for 50% of a pesticide applied with an airplane to fail to reach the intended target [181]. In addition to the actual application process, the effectiveness of insecticides is often negatively affected by the practice of adding, for example, other pesticides, adjuvants, fertilizers, and minor elements [102]. Occasionally, synergism or other interactions among pesticides results in phytotoxicity. Characterizing current foliar application technology as inefficient, imprecise, and wasteful is not an exaggeration. It is imperative that additional research funds and effort be directed toward improving the pesticide application process, which has remained almost unchanged over the past several decades.

Novel pesticide-delivery systems offer great promise in enhancing the efficiency of the application process and dose targeting [74]. Examples are controlled-release technology [74], the use of baits and/or lures to attract and kill pest species [111,114], and the use of genetic engineering technology to incorporate toxicants directly in plants [131].

4. Prescription Use of Chemicals

The idea of permitting the use of restricted pesticides or restricted uses of pesticides via prescription by certified personnel has been the subject of much discussion [7]. A Congressional bill is currently being developed that would authorize the EPA to permit the use of specific dangerous pesticides by prescription only. This approach to managing the use of highly toxic and persistent pesticides, which are also exceedingly valuable for arthropod pest management, holds much promise for retaining the use of pesticides that are vital for the success of IPM but otherwise might be banned.

IX. CONCLUSIONS

Chemicals are likely to remain an integral and necessary component of IPM systems for the foreseeable future, especially in annual cropping systems with multiple pests. Furthermore, the use of effective chemicals has made possible the profitable production of many commodities in the presence of potentially devastating pests. However, it is critical that the use of chemicals be viewed as a single component of an ecologically and biologically based system in order to conserve and prolong the utility of the declining pool of pesticides. Some advantages of chemical control are (a) satisfactory potency, (b) rapid activity, (c) low cost, (d) availability, and (e) the opportunity afforded individual users to take unilateral action. Disadvantages include (a) risks and hazards to humans and other nontarget vertebrates from broad-spectrum action of chemicals, (b) decimation of valuable biological pest control agents, (c) transient nature of chemical control, (d) ability of pests to develop resistance to chemicals, (e) pest resurgence, (f) induction of new arthropod pest problems, and (g) the fact that some economic losses still occur when chemicals are used.

REFERENCES

1. Adair, H.M., R.T. Kincade, M.L. Laster, & J.R. Brazzel. 1967. Low-volume aerial spraying of several insecticides for cotton insect control. *J. Econ. Entomol.* 60:1121–1127.
2. Akesson, N.B., & W.E. Yates. 1964. Problems relating to application of agricultural chemicals and resulting drift residues. *Annu. Rev. Entomol.* 9:285–318.
3. All, J.N., & G.J. Musick. 1986. Management of vertebrate and invertebrate pests. In *No-Tillage and Surface-Tillage Agriculture: The Tillage Revolution.* M.A. Sprague & G.B. Triplett, eds. New York: Wiley, pp 347–387.
4. Anonymous. 1982. Cotton-pest resistance to insecticides and miticides. *Proceedings of the Beltwide Cotton Conferences.* Memphis, Tennessee: National Cotton Council, pp 29–32.
5. Anonymous. 1986. Cotton entomologists seek to delay pyrethroid resistance in insects. *MAFES Res. Highlights* 49:8.
6. Anonymous. 1993. *Producing Early Maturing Cotton for Management of Insecticide Resistance.* Louisiana Cooperative Extension Service Publication 2513.

7. Anonymous. 1994. Association executives discuss future regulations of pesticides in ESA-sponsored roundtable. *ESA Newsletter* 17:1.

8. Ansolabehere, M.J. 1996. KNACK insect growth regulator. *Proceedings of the Beltwide Cotton Conferences.* Memphis, Tennessee: National Cotton Council, p 53.

9. Bagwell, R.D., J.B. Graves, J.W. Holloway, B.R. Leonard, E. Burris, S. Micinski, & V. Mascarenhas. 1997. Status of insecticide resistance in tobacco budworm and bollworm in Louisiana during 1996. *Proceedings of the Beltwide Cotton Conferences.* Memphis, Tennessee: National Cotton Council, pp. 1282–1289.

10. Bai, D., S.C. Lummis, V. Leicht, H. Breer, & D.B. Sattelle. 1991. Actions of imidacloprid and related nitromethylenes on cholinergic receptors of an identified insect motor neurone. *Pestic. Sci.* 33:197–204.

11. Barthel, W.F. 1961. Synthetic pyrethroids. *Adv. Pest Control Res.* 4:33–41.

12. Beckendorf, S.K., & M.A. Hoy. 1985. Genetic improvement of arthropod natural enemies. In *Biological Control in Agricultural IPM Systems.* M.A. Hoy & D.C. Herzog, eds. New York: Academic Press, pp 167–187.

13. Benedict, J.H., E.S. Sachs, D.W. Altman, W.R. Deaton, R.J. Kohel, D.R. Ring, & S.A. Berberich. 1996. Field performance of cotton expressing transgenic CryIA insectidal proteins for resistance to *Heliothis virescens* and *Helicoverpa zea* (Lepidoptera: Noctuidae). *J. Econ. Entomol.* 89:230–238.

14. Bessin, R.T., E.B. Moser, & T.E. Reagan. 1990. Integration of control tactics for management of the sugarcane borer (Lepidoptera: Pyralidae) in Louisiana sugarcane. *J. Econ. Entomol.* 83:1563–1569.

15. Brown, A.W.A. 1977. Epilogue: resistance as a factor in pesticide management. *Proceedings of the XV International Congress on Entomology,* pp 816–824.

16. Brown, S.M., T. Whitwell, J.T. Touchton, & C.H. Burmester. 1985. Conservation tillage systems for cotton production. *Soil Sci. Am. J.* 49:1256–1260.

17. Burgess, E.D. 1965. Control of the boll weevil with technical malathion applied by aircraft. *J. Econ. Entomol.* 58:414–415.

18. Burris, E., B.R. Leonard, S.H. Martin, C.A. White, & J.B. Graves. 1994. Fipronil: evaluation of soil and foliar treatments for control of thrips, aphids, plant bugs and boll weevils. *Proceedings of the Beltwide Cotton Conferences.* Memphis, Tennessee: National Cotton Council, pp 838–844.

19. Carson, R.L. 1962. *Silent Spring.* Boston: Houghton Mifflin.

20. Chalfant, R.B., & J.R. Young. 1982. Chemigation, or application of insecticide through overhead sprinkler irrigation systems, to manage insect pests infesting vegetable and agronomic crops. *J. Econ. Entomol.* 75:237–241.

21. Chandler, L.D., G.A. Herzog, H.R. Sumner, & C.C. Dowler. 1991. Chemigation methodology for management of cotton insect pests. *Proceedings of the Beltwide Cotton Conferences.* Memphis, Tennessee: National Cotton Council, pp 757–759.

22. Chandler, L.D., G.A. Herzog, & H.R. Sumner. 1992. Residual activity of pyrethroid insecticides applied using chemigation methodology for control of bollworm. *Proceedings of the Beltwide Cotton Conferences.* Memphis, Tennessee: National Cotton Council, pp 856–857.

23. Clark, J.M., J.G. Scott, F. Campos, & J.R. Bloomquist. 1994. Resistance to avermectins: extent, mechanisms and management implications. *Annu. Rev. Entomol.* 40:1–30.

24. Cleveland, T.C., W.P. Scott, T.B. Davich, & C.R. Parencia, Jr. 1966. Control of the boll weevil on cotton with ultra-low-volume (undiluted) technical malathion. *J.Econ Entomol.* 59:973–976.

25. Coad, B.R. 1918. Recent experimental work on poisoning cotton boll weevils. *USDA Bull. 731.*

26. Coad, B.R. 1924. Dusting cotton from airplanes. *USDA Bull.1204.*

27. Colliot, F., K A. Kukorowski, D.W. Hawkins, & D.A. Roberts. 1992. Fipronil: a new soil and foliar broad spectrum insecticide.*Proceedings of the Brighton Crop Protection Conference, Pests and Diseases*, pp 29–34.

28. Croft, B.A., & A.W.A. Brown.1975. Responses of arthropod natural enemies to insecticides. *Annu. Rev. Entomol.* 20:285–335.

29. Croft, B.A., & M.E. Whalon.1982. Selective toxicity of pyrethroid insecticides to arthropod natural enemies and pests of agricultural crops. *Entomophaga* 27:3–21.

30. Croft, B.A. 1990. *Arthropod Biological Control Agents and Pesticides*. New York: Wiley.

31. Davenport, A.P., & D.J. Wright.1985. Toxicity of chlordimeform and amitraz to the Egyptian cotton leafworm (*Spodoptera littoralis*) and the tobacco budworm (*Heliothis virescens*). *Pestic.Sci.*16:81–87.

32. Daly, J.C. 1988. Insecticide resistance in *Heliothis armigera* in Australia. *Pestic. Sci.* 23:165–176.

33. DeSpain, R.R., J.H. Benedict, J.A. Landivar, B.R. Eddleman, S.W. Goynes, R.D. Parker, & M.F. Treacy.1990. Cropping systems and insect management. *Proceedings of the Beltwide Cotton Conferences*. Memphis, Tennessee: National Cotton Council, pp 256–262.

34. Dittrich, V. 1966. N-(2-methyl-4-chlorophenyl)-N', N'-dimethylformamidine (C-8514/Schering 36268) evaluated as an acaricide. *J.Econ.Entomol.* 59:889–893.

35. Dittrich, V. 1967. A formamidine acaracide as an ovicide for three insect species. *J.Econ.Entomol.* 60:13–15.

36. DowElanco. 1994. *Spinosad Technical Guide*. Form No.200-03-001 (GS).

37. Dunbar, D.D., R.D. Brown, J.A. Norton, & R.A. Dybas.1996. Proclaim: a new insecticide for use on cotton. *Proceedings of the Beltwide Cotton Conferences*. Memphis, Tennessee: National Cotton Council, pp 756–758.

38. Elbert, A., H. Overbeck, K. Iwaya, & S. Tsuboi. 1990. Imidacloprid, a novel systemic nitromethylene analogue insecticide for crop protection. *Proceedings of the 1990 Brighton Crop Protection Conference, Pest and Diseases*, pp 21–28.

39. Elbert, A., B. Becker, J. Hartwig, & C. Erdelen.1991. Imidacloprid—a new systemic insecticide. *Pflanzenschutz - Nachrichten Bayer* 44:113–136.

40. Eldefrawi, M., & A. Eldefrawi.1991. Philanthotoxin as a model for a new generation of insecticides with a novel mechanism of action. In *Pesticides and the Future: Toxicological Studies of Risks and Benefits*. E. Hodgson, R.M. Roe, & N. Motoyama, eds. Raleigh, North Carolina: North Carolina State University, pp 229–238.

41. Elliott, M. 1971. The relationship between the structure and the activity of pyrethroids. *Bull WHO* 44:315–317.

42. Elliott, M., A.W. Farnham, N.F. James, P.H. Needham, D.A. Pulman, & J.H. Stevenson.1973. A photostable pyrethroid. *Nature* 246:169–170.

43. Elliott, M.1976. Properties and applications of pyrethroids. *Environ. Health Perspect.* 14:3–13.

44. Elzen, G.W., B.R. Leonard, J.B. Graves, E. Burris, & S. Micinski.1992. Resistance to pyrethroid, carbamate and organophosphate insecticides in field populations of tobacco budworm (Lepidoptera: Noctuidae) in 1990. *J. Econ. Entomol.* 85:2064–2072.

45. Falcon, L.A. 1985. Development and use of microbial insecticides. In *Biological Control in Agricultural IPM Systems*. M.A. Hoy & D.C. Herzog, eds. New York: Academic Press, pp 229–242.

46. Ferguson, J. 1989. New Bt options. *Ag Consult.* 45:1–4.

47. Fischhoff, D.A., K.S. Bowdish, F.J. Perlak, P.G. Marrone, S.M. McCormick, J.G. Niedermeyer, D.A. Deon, K. Kusano-Kretzmer, E.J. Meyer, D.E. Rochester, S.G. Rogers, & R.T. Fraley. 1987. Insect tolerant transgenic tomato plants. *Bio/Technology* 5:807–813.

48. Fisher, M.H. 1989. Novel avermectin insecticides and miticides. In *Recent Advances in the Chemistry of Insect Control II*. L.Crombie, ed. Cambridge, UK: Royal Society of Chemistry, pp 52–68.

49. Flaherty, D.L., & C.B. Huffaker. 1970. Biological control of Pacific mites and Willamette mites in San Joaquin Valley vineyards. I. Role of *Metaseiulus occidentalis*. II. Influence of dispersion patterns of *Metaseiulus occidentalis*. *Hilgardia* 40:267–330.

50. Folsum, J.W. 1928. Calcium arsenate as a cause of aphid infestation. *J. Econ. Entomol.* 21:174.

51. Forrester, N.W. 1990. Designing, implementing and servicing an insecticide resistance management strategy. *Pestic. Sci.* 28:167–179.

52. Forrester, N.W., M.Cahill, L.J. Bird, & J.K. Layland.1993. Management of pyrethroid and endosulfan resistance in *Helicoverpa armigera* (Lepidoptera: Noctuidae) in Australia. *Bull. Entomol. Res.: Suppl. Ser. No. 1.*

53. Frazier, J.L., & P.Y-S Lam. 1989. Three-dimensional mapping of insect taste receptor sites as an aid to novel antifeedant development. In *Recent Advances in the Chemistry of Insect Control II*. L. Crombie, ed. Cambridge: Royal Society of Chemistry, pp 247–255.

54. Gaines, J.C. 1957. Cotton insects and their control in the United States. *Annu. Rev. Entomol.* 2:319–338.

55. Gaylor, M.J., S.J. Fleisher, D.P. Muehleisen, & J.V. Edelson. 1984. Insect populations in cotton produced under conservation tillage. *J. Soil Water Conserv.* 39:61–64.

56. Georghiou, G.P., & C.E. Taylor. 1976. Pesticide resistance as an evolutionary phenomenon. *Proceedings of the XV International Congress of Entomology*, p 759.

57. Georghiou, G.P. 1983. Management of resistance in arthropods. In *Pest Resis-*

tance to Pesticides. G.P. Georghiou & T. Saito, eds. New York: Plenum Press, pp 769–792.

58. Georghiou, G.P. 1986. The magnitude of the resistance problem. In *Pesticide Resistance: Strategies and Tactics for Management.* Commission on Strategies for the Management of Pesticide Resistant Pest Populations, ed. Washington: National Academy Press, pp 14–43.

59. Georghiou, G.P. 1990. Overview of insecticide resistance. In *Managing Resistance to Agrochemicals: From Fundamental Research to Practical Strategies.* M.B. Green, H.M. LeBaron, and W.K. Moberg, eds. Washington, DC: American Chemical Society, pp 18–41.

60. Gianessi, L.P., & J. E. Anderson. 1993. Pesticide use trends in U.S. agriculture, 1979–1992. *National Center for Food and Agricultural Policy Discussion Paper PS-93-1.*

61. Gonzales, D., & L.T. Wilson. 1982. A food-web approach to economic thresholds: a sequence of pests/predaceous arthropods on California cotton. *Entomophaga* 27:31–43.

62. Graves, J.B. 1987. An illustrated look at insecticide resistance and how it develops. *Proceedings of the Beltwide Cotton Conferences.* Memphis, Tennessee: National Cotton Council, pp 29–31.

63. Graves, J.B., B.R. Leonard, A.M. Pavloff, G. Burris, K. Ratchford, & S. Micinski. 1988. Monitoring pyrethroid resistance in tobacco budworm in Louisiana during 1987: resistance management implications. *J. Agric. Entomol.* 5:109–115.

64. Graves, J.B., B.R. Leonard, A.M. Pavloff, S. Micinski, G.Burris, & K. Ratchford. 1989. An update on pyrethroid resistance in tobacco budworm and bollworm in Louisiana. *Proceedings of the Beltwide Cotton Conferences.* Memphis, Tennessee: National Cotton Council, pp 343–346.

65. Graves, J.B, B.R. Leonard, S. Micinski, & E. Burris. 1990. Status of pyrethroid resistance in tobacco budworm in Louisiana, *Proceedings of the Beltwide Cotton Conferences.* Memphis, Tennessee: National Cotton Council, pp 216–219.

66. Graves, J.B., B.R. Leonard, G. Burris, & S. Micinski. 1991. Insecticide resistance management: an integral part of IPM. *Proceedings of the Beltwide Cotton Conferences.* Memphis, Tennessee: National Cotton Council, pp 23–24.

67. Graves, J.B., B.R. Leonard, S. Micinski, & E. Burris. 1991. A three year study of pyrethroid resistance in tobacco budworm in Louisiana: resistance management implications. *Southwest. Entomol.* 15(Suppl):33–41.

68. Graves, J.B., B.R. Leonard, S. Micinski, S.H. Martin, D.W. Long, E. Burris, & J.L. Baldwin.1992. Situation on tobacco budworm resistance to pyrethroids in Louisiana during 1991, *Proceedings of the Beltwide Cotton Conferences.* Memphis, Tennessee: National Cotton Council, pp 743–746.

69. Graves, J.B., B.R. Leonard, S. Micinski, E. Burris, S.H. Martin, C.A. White, & J.L. Baldwin.1993. Monitoring insecticide resistance in tobacco budworm and bollworm in Louisiana. *Proceedings of the Beltwide Cotton Conferences.* Memphis, Tennessee: National Cotton Council, pp 788–794.

70. Graves, J.B. 1994. Insecticide resistance management strategies. *Proceedings of*

the Beltwide Cotton Conferences. Memphis, Tennessee: National Cotton Council, pp 43–45.

71. Graves, J.B., G.W. Elzen, M.B. Layton, R.H. Smith, & M.L. Wall. 1995. Budworm/bollworm management: insecticide resistance and population trends in the Mid-south. *Proceedings of the Beltwide Cotton Conferences.* Memphis, Tennessee: National Cotton Council, pp 136–140.

72. Gregory, W.W., & G.J. Musick. 1976. Insect management in reduced tillage systems. *Bull. Entomol. Soc. Am.* 22:302–304.

73. Guillebeau, L.P. 1994. Risk-benefit analysis of pesticides: the U.S. Environmental Protection Agency perspective. *Am. Entomol.* 40:173–179.

74. Hall, F.R. 1991. Pesticide targeting: improving the dose transfer process. In *Pesticides and the Future: Toxicological Studies of Risks and Benefits.* E. Hodgson, R.M. Roe, & N. Motoyama, eds. Raleigh, North Carolina. North Carolina State University, pp 305–315.

75. Hammock, B.D., A. Szekacs, T. Hanzlik, S. Maeda, M. Philpott, B. Bonning, & R. Possee. 1989. Use of transition state theory in the design of chemical and molecular agents for insect control. In *Recent Advances in the Chemistry of Insect Control II.* L. Crombie, ed. Cambridge, UK: Royal Society of Chemistry, pp 256–277.

76. Hayes, W.J., Jr. 1982. *Pesticides Studied in Man.* Baltimore: Williams & Wilkins.

77. Heller, J.J., H. Mattioda, E. Klein, & A. Sagenmuller.1992. Field evaluation of RH 5992 on lepidopterous pests in Europe. *Proceedings of the Brighton Crop Protection Conference, Pests and Diseases*, pp 59–65.

78. Herzog, G.A., J.B. Graves, J.T. Reed, W.P. Scott, & T.F. Watson. 1996. Chemical control. In *Cotton Insects and Mites: Characterization and Management.* E.G. King, J.R. Phillips, & R.J. Coleman, eds. Memphis, Tennessee: Cotton Foundation, pp 447–469.

79. Higley, L.G., M.R. Zeiss, W.K. Wintersteen, & L.P. Pedigo. 1992. National pesticide policy: a call for action. *Am. Entomol.* 38:139–146.

80. Hodgson, E. 1991. Pesticides: past, present and future. In *Pesticides and the Future: Toxicological Studies of Risks and Benefits.* E. Hodgson, R.M. Roe, & N. Motoyama, eds. Raleigh, North Carolina: North Carolina State University, pp 3–12.

81. Hollingworth, R.M., & A.E. Lund. 1982. Biological and neurotoxic effects of amidine pesticides. In *Insecticide Mode of Action.* J.R. Coats, ed. New York: Academic Press, pp 189–227.

82. Hoy, M.A. 1984. Genetic improvement of a biological control agent: Multiple pesticide resistances and nondiapause in *Metaseiulus occidentalis* (Nesbitt) (Phytoseiidae). *Proceedings of the VI International Congress on Acaralogy.* Vol. 2, pp 673–679.

83. Hoy, M.A., W.W. Barnett, L.C. Hendricks, D. Castro, D. Cahn, & W.J. Bentley. 1984. Managing spider mites in almonds with pesticide-resistant predators. *Calif. Agric.* 38:18–20.

84. Huffaker, C.B., M. Van de Vrie, & J.A. McMurty. 1970. Ecology of tetranychid

mites and their natural enemies: A review. II. Tetranychid populations and their possible control by predators: An evaluation. *Hilgardia* 40:391–458.

85. Hull, L.A., & D.M. Baldwin. 1982. Evaluation of insecticides for use in an integrated pest management program for apple. *Down to Earth* 38:9–11.

86. Hull, L.A., & E.H. Beers. 1985. Ecological selectivity: modifying chemical control practices to preserve natural enemies. In *Biological Control In Agricultural IPM Systems*. M.A. Hoy & D.C. Herzog, eds. New York: Academic Press, pp 103–122.

87. Hutchinson, R.L. & W.L. Shelton. 1991. Alternative tillage systems and cover crops for cotton production on the Macon Ridge. *La. Agric.* 33:6–8.

88. Isler, D.A. 1966. Atomization of low-volume malathion aerial spray. *J. Econ. Entomol.* 59:688–690.

89. Isley, D. 1946. The cotton aphid. *Arkansas Agricultural Experiment Station Bull.* 462.

90. Jenkins, J.N., W.L. Parrott, J.C. McCarty, F.E. Callahan, S.A. Berberich, & W.R. Deaton. 1993. Growth and survival of *Heliothis virescens* (Lepidoptera: Noctuidae) on transgenic cotton containing a truncated form of the delta endotoxin gene from *Bacillus thuringiensis*. *J. Econ. Entomol.* 86:181–185.

91. King, E.G., R.J. Coleman, J.A. Morales-Ramos, K.R. Summy, M.R. Bell, & G.L. Snodgrass. 1996. Biological control. In *Cotton Insects and Mites: Characterization and Management*. E.G. King, J.R. Phillips, & R.J. Coleman, eds. Memphis: Cotton Foundation, pp 511–538.

92. Knowles, C.O. 1982. Structure-activity relationships among amidine acaricides and insecticides. In *Insecticide Mode of Action*. J.R. Coats, ed. New York: Academic Press, pp 243–277.

93. Lasota, J.A., & R.A. Dybas. 1991. Avermectins, a novel class of compounds: implications for use in arthropod pest management. *Annu. Rev. Entomol.* 36:91–117.

94. Leonard, B.R., J.B. Graves, T.C. Sparks, & A.M. Pavloff. 1987. Susceptibility of bollworm and tobacco budworm larvae to pyrethroid and organophosphate insecticides. *Proceedings of the Beltwide Cotton Conferences*. Memphis, Tennessee: National Cotton Council, pp 320–324.

95. Leonard, B.R., J.B. Graves, T.C. Sparks, & A.M. Pavloff. 1988. Variation in resistance of field populations of tobacco budworm and bollworm (Lepidoptera: Noctuidae) to selected insecticides. *J.Econ. Entomol.* 81:1521–1528.

96. Leonard, B.R., J.B. Graves, T.C. Sparks, & A.M. Pavloff. 1988. Insecticide cross-resistance in pyrethroid-resistant strains of tobacco budworm (Lepidoptera: Noctuidae*). J. Econ. Entomol.* 81:1529–1535.

97. Leonard, B.R., E. Burris, J.B. Graves, & G.W. Elzen. 1991. Tobacco budworm: insecticide resistance and field control in the Macon Ridge Region of Louisiana, 1990. *Proceedings of the Beltwide Cotton Conferences*. Memphis, Tennessee: National Cotton Council, pp 642–648.

98. Leonard, B.R., R.L. Hutchinson, J.B. Graves & E. Burris. 1993. Conservation-tillage systems and early-season cotton insect pest management. In *Conservation-Tillage Systems for Cotton*. M.R. McClelland, T.D. Valco, & R.E. Frans, eds. Arkansas Agricultural Experiment Station Special Report 160, pp 80–85.

99. Linn, C.E., Jr. & W.L. Roelofs. 1984. Sublethal effects of neuroactive compounds on pheromone response thresholds in male oriental fruit moths. *Arch. Insect Biochem. Physiol.* 1:331–334.

100. Litovitz, T.L., B.F. Schmitz, N. Matgunas, & T.G. Martin. 1988. Annual report of the American Association of Poison Control Centers national data collection system. *Am. J. Emerg. Med.* 6:479–515.

101. Lloyd, E.P., J.R. McCoy, W.P. Scott, E.C. Burt, D.B. Smith, & F.C. Tingle. 1972. In-season control of the boll weevil with ultra-low-volume sprays of azinphosmethyl or malathion. *J. Econ. Entomol.* 65:1153–1156.

102. Long, D.W., J.A. Ottea, J.B. Graves, B.R. Leonard, G.E. Church, E. Burris & L.M. Southwick. 1992. Physical incompatibility of insecticides co-applied with foliar urea fertilizer. *Proceedings of the Beltwide Cotton Conferences*. Memphis, Tennessee: National Cotton Council, pp 850–855.

103. Loughner, G.E. 1996. Environmental issues. In *Cotton Insects and Mites: Characterization and Management*. E.G. King, J.R. Phillips, & R.J. Coleman, eds. Memphis, Tennessee: Cotton Foundation, pp 831–842.

104. Lovell, J.B., D.P. Wright, Jr., I.E. Gard, T.P. Miller, M.F. Treacy, R.W. Addor, & V.M. Kamki. 1991. AC 303, 630 - an insecticide/acaricide from a novel class of chemistry. *Proceedings of the Beltwide Cotton Conferences*. Memphis, Tennessee: National Cotton Council, pp 736–737.

105. Luthy, P., J.L. Cordier, & H.M. Fischer. 1982. *Bacillus thuringiensis* as a bacterial insecticide: basic consideration and application. In *Microbial and Viral Pesticides*. E. Kurstak, ed. New York: Marcel Dekker, pp 35–74.

106. Luttrell, R. & R T. Roush. 1987. Strategic approaches to avoid or delay development of resistance to insecticides. *Proceedings of the Beltwide Cotton Conferences*. Memphis, Tennessee: National Cotton Council, pp 31–33.

107. Lyle, C. 1947. Achievements and possibilities in pest eradication. *J. Econ. Entomol.* 40:1–8.

108. Mahaffey, J.S., J.S. Bacheler, J.R. Bradley, Jr., & J.W. Van Duyn. 1994. Performance of Monsanto's transgenic Bt cotton against high populations of lepidopterous pests in North Carolina. *Proceedings of the Beltwide Cotton Conferences*. Memphis, Tennessee: National Cotton Council, pp 1061–1063.

109. Matsumura, F. 1985. *Toxicology of Insecticides*. New York: Plenum Press.

110. Matsumura, F., K. Tanaka, & Y. Ozoe. 1987. GABA-related systems as targets for insecticides. In *Sites of Action for Neurotoxic Insecticides*. R.M. Hollingworth, & M.B. Green, eds. Washington, DC: American Chemical Society, pp 44–70.

111. McKibben, G.H., J.W. Smith, & J.E. Leggett. 1990. Field tests with an attract-and-kill device for the boll weevil (Coleoptera: Curculionidae). *Proceedings of the Beltwide Cotton Conferences*. Memphis, Tennessee: National Cotton Council, pp 303–304.

112. McKibben, G.H., J.W. Smith, & E.J Villavaso. 1991. Field research results on the boll weevil bait stick. *Proceedings of the Beltwide Cotton Conferences*. Memphis, Tennessee: National Cotton Council, pp 622–623.

113. Metcalf, R.L. 1980. Changing role of insecticides in crop protection. *Annu. Rev. Entomol.* 25:219–256.

114. Metcalf, R.L., J.E. Ferguson, R. Lampman, & J.F. Anderson. 1987. Dry

cucurbitacin-containing baits for controlling diabroticite beetles (Coleoptera: Chrysomelidae). *J. Econ. Entomol.* 80:870–875.

115. Micinski, S., & W.D. Caldwell. 1991. Field performance of cotton genetically modified to express insecticidal protein from *Bacillus thuringiensis. Proceedings of the Beltwide Cotton Conferences.* Memphis, Tennessee: National Cotton Council, p 578.

116. Miller, T.A. 1996. Resistance to pesticides: mechanisms, development and management. In *Cotton Insects and Mites: Characterization and Management.* E.G. King, J.R. Phillips, & R.J. Coleman, eds. Memphis, Tennessee: Cotton Foundation, pp 323–378.

117. Mitsui, T., S. Atsusawa, K. Ohsawa, I. Yamamoto, T. Miyake, & T. Umehara. 1991. Search for insect growth regulators. In *Pesticides and the Future: Toxicological Studies of Risks and Benefits.* E. Hodgson, R.M. Roe, & N. Motoyama, eds. Raleigh, North Carolina: North Carolina State University, pp 239–248.

118. Moffat, A.S. 1993. New chemicals seek to outwit insect pests. *Science* 261:550–551.

119. Mullin, C.A. & B.A. Croft. 1985. An update on development of selective pesticides favoring arthropod natural enemies. In *Biological Control in Agricultural IPM Systems.* M.A. Hoy & D.C. Herzog, eds. New York: Academic Press, pp 123–150.

120. Mullins, J.W. & C.E. Engle. 1993. Imidacloprid (BAY NTN 33893): a novel chemistry for sweetpotato whitefly control in cotton. *Proceedings of the Beltwide Cotton Conferences.* Memphis, Tennessee: National Cotton Council, pp 719–720.

121. Muzzi, D. 1997. New in 1997. *Mid-South Farmer* 4:13.

122. Newsom, L.D. 1966. Essential role of chemicals in crop protection. *Proceedings of the FAO Symposium on Integrated Pest Control.* Vol.2, pp. 95–108.

123. Newsom, L.D., & J.R. Brazzel. 1968. Pests and their control. In *Advances in Production and Utilization of Quality Cotton: Principles and Practices.* F.C. Elliot, M. Hoover, & W.K. Porter, Jr., eds. Ames, Iowa: Iowa State University Press, pp 367–405.

124. Newsom, L.D. 1970. The end of an era and future prospects for insect control. *Proceedings of the Tall Timbers Conference.* pp. 117–136.

125. Newsom, L.D. 1975. Pest management: concept to practice. In *Insects, Science and Society.* D. Pimentel, ed. New York: Academic Press, pp 257–277.

126. Newsom, L.D., R.F. Smith, & W.H. Whitcomb. 1976. Selective pesticides and selective use of pesticides. In *Theory and Practice of Biological Control.* C.B. Huffaker & P.S. Messenger, eds. New York: Academic Press, pp 565–591.

127. Nicholson, R.A. & T.A. Miller. 1985. Multi-factorial resistance to *trans*-permethrin in field collected strains of the tobacco budworm *Heliothis virescens* (F.). *Pestic. Sci.*16:561–562.

128. O'Brien, R.D. 1967. *Insecticides, Action and Metabolism.* New York: Academic Press.

129. Parencia, C.R., Jr. 1978. One hundred twenty years of research on cotton insects in the United States. *USDA-ARS Handbook 515.*

130. Perez, J.J., A. Obando, & N. Darby. 1994. Evaluation of the growth regulator

buprofezin mixed with conventional insecticides for whitefly control. *Proceedings of the Beltwide Cotton Conferences.* Memphis, Tennessee: National Cotton Council, pp 904–905.

131. Perlak, F.J., R.W. Deaton, T.A. Armstrong, R.L. Fuchs, S.R. Sims, J.T. Greenplate, & D.A. Fischhoff. 1990. Insect resistant cotton plants. *Biotechnology* 8:939–943.

132. Phillips, J.R. 1971. Bollworm control with chlorphenamidine. *Ark. Farm Res.* 21:9.

133. Phillips, J.R., J.B. Graves, & R.G. Luttrell. 1989. Insecticide resistance management: relationship to integrated pest management. *Pestic. Sci.* 27:459–464.

134. Phillips, J.R. & J.B. Graves. 1996. Cotton insect management: a look to the future. In *Cotton Insects and Mites: Characterization and Management.* E.G. King, J.R. Phillips, & R.J. Coleman, eds. Memphis, Tennessee: Cotton Foundation, pp 853–860.

135. Plapp, F.W., Jr. 1976. Chlordimeform as a synergist for insecticides against the tobacco budworm. *J. Econ. Entomol.* 69:91–92.

136. Plapp, F.W., Jr., & C. Campanhola. 1986. Synergism of pyrethroids by chlordimeform against susceptible and resistant *Heliothis. Proceedings of the Beltwide Cotton Conferences.* Memphis, Tennessee: National Cotton Council, pp 167–169.

137. Plapp, F.W., G.M. McWhorter, & W.H. Vance. 1987. Monitoring for pyrethroid resistance in the tobacco budworm. *Proceedings of the Beltwide Cotton Conferences.* Memphis, Tennessee: National Cotton Council, pp 324–326.

138. Plapp, F.W., Jr., J.A. Jackson, C. Campanhola, R.E. Frisbie, J.B. Graves, R.G. Luttrell, W.F Kitten, & M. Wall. 1990. Monitoring and management of pyrethroid resistance in the tobacco budworm (Lepidoptera: Noctuidae) in Texas, Mississippi, Louisiana, Arkansas and Oklahoma. *J. Econ. Entomol.* 83:335–341.

139. Polles, S.G., & S.B. Vinson. 1969. Effect of droplet size on persistence of ULV malathion and comparison of toxicity of ULV and EC malathion to tobacco budworm larvae. *J. Econ. Entomol.* 62:89–94.

140. Reagan, T.E. 1980. A pest management system for sugarcane insects. *La. Agric.* 24:12–14.

141. Reagan, T.E. 1981. Sugarcane borer pest management in Louisiana: leading to a more permanent system. *Proceedings of Second Inter-American Sugarcane Seminar: Insect and Rodent Pests,* pp. 100–110.

142. Riley, S.L. 1989. Pyrethroid resistance in *Heliothis virescens*: Current U.S. management programs. *Pestic. Sci.* 26:411–421.

143. Roe, R.M., K. Venkatesh, D.D. Anspaugh, R.J. Linderman, & D.M. Graves. 1991. Chemical and biological approaches to juvenile hormone esterase based insecticides. In *Pesticides and the Future: Toxicological Studies of Risks and Benefits.* E. Hodgson, R.M.Roe, & N. Motoyama, eds. Raleigh, North Carolina: North Carolina State University, pp 249–260.

144. Roush, R.T., & G.L. Miller. 1986. Considerations for design of insecticide resistance monitoring programs. *J. Econ. Entomol.* 79:293–298.

145. Roush, R.T., & R.G. Luttrell. 1987. The phenotypic expression of pyrethroid

resistance in *Heliothis* and implications for resistance management. *Proceedings of the Beltwide Cotton Conferences.* Memphis, Tennessee: National Cotton Council, pp 220–224.

146. Roush, R.T., & J.A. McKenzie. 1987. Ecological genetics of insecticide and acaricide resistance. *Annu. Rev. Entomol.* 32:361–380.

147. Roush, R.T. 1989. Designing resistance management programs: how can you choose? *Pestic. Sci.* 26:423–441.

148. Saito, T., T. Miyata, & P. Kienmeesuke. 1991. Selective toxicities of insecticides between insect pests and natural enemies. In *Pesticides and the Future: Toxicological Studies of Risks and Benefits.* E. Hodgson, R.M. Roe, & N. Motoyama, eds. Raleigh, North Carolina: North Carolina State University, pp 131–136.

149. Salgado, V.L., G.B. Watson, & J.J. Sheets. 1997. Studies on the mode of action of spinosad, the active ingredient in Tracer® insect control. *Proceedings of the Beltwide Cotton Conferences.* Memphis, Tennessee: National Cotton Council, pp 1082–1086.

150. Sawicki, R.M., & I. Denholm. 1986. Evaluation of existing resistance management against arthropod pests on cotton. *British Crop Protection Conference.* pp 933–941.

151. Sawicki, R.M., & I. Denholm. 1987. Management of resistance to pesticides in cotton pests. *Trop. Pest Manag.* 33:262–272.

152. Sawicki, R.M. 1989. Current insecticide management practices in cotton around the world–short term successes as template for the future. *Pestic. Sci.* 26:401–410.

153. Schreiber, A.A. 1994. Pesticide residues on fruits and vegetables: nation and state. *Agrichem. Environ. News* 99:15.

154. Schroeder, M.E., & R.F. Flattum. 1984. The mode of action and neurotoxic properties of the nitromethylene heterocycle insecticides. *Pestic. Biochem. Physiol.* 22:148–160.

155. Shepard, H.H. 1951. *The Chemistry and Action of Insecticides.* New York: McGraw-Hill.

156. Smith, D.B., & R.G. Luttrell. 1996. Application technology. In *Cotton Insects and Mites: Characterization and Management.* E.G. King, J.R. Phillips, & R.J. Coleman, eds. Memphis, Tennessee: Cotton Foundation, pp 379–404.

157. Smith, G.D., & J.A. Fontenot. 1942. Notes on the effect of arsenicals upon the cotton aphid, predators and other insects. *J. Econ. Entomol.* 34:587.

158. Smith, J.W., W.L. McGovern, & G.H. McKibben. 1990. Design of an attract-and-kill device for the boll weevil. *J. Entomol. Sci.* 25:581–586.

159. Smith, J.W., E. Villavaso, G.H. McKibben, & W.L. McGovern. 1992. Boll weevil suppression using bait sticks in Tennessee. *Proceedings of the Beltwide Cotton Conferences.* Memphis, Tennessee: National Cotton Council, pp 716–717.

160. Soderlund, D.M., & J.R. Bloomquist. 1989. Neurotoxic actions of pyrethroid insecticides. *Annu. Rev. Entomol.* 34:77–96.

161. Soloway, S.B., A.C. Henry, W.D. Kollmeyer, W.M. Padget, J.E. Powell, S.A. Roman, C.H. Tieman, R.A.Corey, & C.A. Horne. 1979. Nitromethylene insecticides. In *Advances in Pest Science, Fourth International Congress on Pesticide Chemistry 1978.* Part 2, pp 206–217.

162. Sparks, T.C. 1996. Toxicology of insecticides and acaricides. In *Cotton Insects and Mites: Characterization and Management*. E.G. King, J.R. Phillips, & R.. Coleman, eds. Memphis, Tennessee: Cotton Foundation, pp 283–322.

163. Sparks, T.C., B.R. Leonard, & J.B. Graves. 1988. Pyrethroid resistance and the tobacco budworm: interactions with chlordimeform and mechanisms of resistance *Proceedings of the Beltwide Cotton Conferences*. Memphis, Tennessee: National Cotton Council, pp 366–370.

164. Sparks, T.C., B.R. Leonard, & J.B. Graves, 1991. Pyrethroid-formamidine interactions and behavioral effects in pyrethroid susceptible and resistance tobacco budworms. *Southwest. Entomol.* 15(Suppl):111–119.

165. Sparks, T.C., J.B. Graves, & B.R. Leonard. 1993. Insecticide resistance and the tobacco budworm: past, present, and future. In *Reviews in Pesticides Toxicology*. Vol. 2. R.M. Roe & R.J. Kuhr, eds. Raleigh, North Carolina: Toxicology Communications, pp 149–183.

166. Sparks, T.C., B.R. Leonard, F. Schneider, & J.B. Graves. 1993. Ovicidal activity and alteration of octopamine titers: effects of selected insecticides on eggs of the tobacco budworm (Lepidoptera: Noctuidae). *J. Econ. Entomol.* 86:294–300.

167. Staal, G.B. 1975. Insect growth regulators with juvenile hormone activity. *Annu. Rev. Entomol.* 20:417–460.

168. Szczepanski, C.V. 1990. Today's research for tomorrow's markets or: how to hit a moving target. In *Recent Advances in the Chemistry of Insect Control II*. L. Crombie, ed. Cambridge, UK: Royal Society of Chemistry, pp 1–16.

169. Tabashnik, B.E. 1986. Computer simulation as a tool for pesticide resistance management. In *Pesticide Resistance: Strategies and Tactics for Management*. Commission on Strategies for the Management of Pesticide Resistant Pest Populations, ed. Washington, DC: National Academy Press, pp 194–206.

170. Tabashnik, B.E. 1989. Managing resistance to more than one pesticide: theory, evidence and recommendations. *J. Econ. Entomol.* 82:1263–1269.

171. Thomas, J.D., J.S. Mink, D.J. Boethel, A.T. Weir, & B.R. Leonard. 1994. Activity of two novel insecticides against permethrin-resistant *Pseudoplusia includens*. *Pestic. Sci.* 40:239–243.

172. Thompson, G.D., J.D. Busacca, O.K. Jantz, P.W. Borth, S.P. Nolting, J.R. Winkle, R.L. Gantz, R.M. Huckaba, B.A. Nead, L.G. Peterson, D.J. Porteous, & J.M. Richardson. 1995. Field performance in cotton of spinosad: a naturally derived insect control system. *Proceedings of the Beltwide Cotton Conferences*. Memphis, Tennessee: National Cotton Council, pp 907–910.

173. Threadgill, E.D. 1985. Current status and future of chemigation. *Proceedings of the Third National Symposium on Chemigation*, pp 1–12.

174. Touchton, J.T., & D.W. Reeves. 1988. A beltwide look at conservation tillage for cotton. *Proceedings of the Beltwide Cotton Conferences*. Memphis, Tennessee: National Cotton Council, pp 36–41.

175. Treacy, M.F., J.H. Benedict, & K.M. Schmidt. 1986. Toxicity of insecticide residues to the boll weevil (Coleoptera: Curculionidae): comparison of ultra-low-volume/oil vs. conventional/water and water-oil sprays. *Proceedings of the*

Beltwide Cotton Conferences. Memphis, Tennessee: National Cotton Council, pp 180–181.

176. Treacy, M.F., T.P. Miller, I.E. Gard, J.B. Lovell, & D.P. Wright, Jr. 1991. Characterization of insecticidal properties of AC 303,630 against tobacco budworm, *Heliothis virescens* (Fabricius), larvae. *Proceedings of the Beltwide Cotton Conferences*. Memphis, Tennessee: National Cotton Council, pp 738–740.

177. Walker, J.K., & C.W. Smith. 1996. Cultural control. In *Cotton Insects and Mites: Characterization and Management*. E.G. King, J.R. Phillips, & R.J. Coleman, eds. Memphis, Tennessee: Cotton Foundation, pp 471–510.

178. Ware, G.W. 1980. *Complete Guide to Pest Control With and Without Chemicals*. Fresno, California: Thomson.

179. Ware, G.W. 1989. *The Pesticide Book*. Fresno, California: Thomson.

180. White, S.M., D.M. Dunbar, R. Brown, B. Cartwright, D. Cox, C. Eckel, R.K. Jansson, P.K. Mookerjee, J.A. Norton, R.F. Peterson, & V.R. Starner. 1997. Emamectin benzoate: a novel avermectin derivative for control of lepidopterous pests of cotton. *Proceedings of the Beltwide Cotton Conferences*. Memphis, Tennessee: National Cotton Council, pp. 1078–1082.

181. Willis, G.H., & L.L. McDowell. 1987. Pesticide persistence on foliage. *Rev. Environ. Contam. Toxicol.* 100:24–73.

182. Wilson, F.D., H.M. Flint, W.R. Deaton, D.A. Fischoff, F.J. Perlak, T.A. Armstrong, R.L. Fuchs, S.A. Berberich, N.J. Parks, & B.R. Stapp. 1992. Resistance of cotton lines containing a *Bacillus thuringiensis* toxin to pink bollworm (Lepidoptera: Gelechiidae) and other insects. *J. Econ. Entomol.* 85:1516–1521.

183. Wolfenbarger, D.A., & R.L. McGarr. 1971. Low-volume and ultra-low-volume sprays of malathion and methyl parathion for control of three Lepidopteran cotton pests. *USDA Producers Research Report 126*.

18
Genetic Approaches to Managing Arthropod Pests

James E. Carpenter
United States Department of Agriculture, Tifton, Georgia

Alan C. Bartlett
United States Department of Agriculture, Phoenix, Arizona

I. INTRODUCTION

Genetic approaches for managing arthropod pests have been considered by scientists for more than half a century [59,107]. Yet, because of recent and current emerging technologies, especially in the field of molecular genetic techniques, genetic approaches are often regarded as novel and unproven. For example, in a recent review of integrated pest management of cotton, Lyon [79] did not mention a single genetic approach which could or should be considered for the control of the insect pests of cotton. Certainly, current advances and theory in this field of study remain to be implemented, but some applications of these methods, such as the sterile insect technique, are well known in their effectiveness and have great potential for managing arthropod pests. Emerging technologies also hold great promise as future pest control methods.

Genetic pest control strategies generally involve alterations of a target pest species' ability to reproduce or the insertion of some deleterious character into a pest population. These control strategies are referred to as "autocidal," because they involve the release of genetically modified insects to control intraspecific pests. Other genetic methods for managing arthropod pests involve (a) the genetic improvement of natural enemies [48], and (b) the development of genetic sexing technology to reduce costs and improve the effectiveness of autocidal control programs. Compared with chemical control methods, all genetic approaches are highly specific to the pest species, safe to pest control workers, and nonpolluting to the environment.

Although previous successes using genetic techniques have been encouraging, candidate species for future pest-suppression programs should be selected carefully. Research and development for genetic autocidal programs demand a long-term commitment before potential benefits can be realized. Therefore, genetic approaches are usually developed only for the most economically important pest species. However, future costs of developing, implementing, and operating genetic approaches for managing arthropod pests may be reduced substantially as a result of the technological gains from current programs using genetic techniques, from research in the integration of genetic and nongenetic approaches, and from recent developments in molecular genetic techniques. In the past, arthropod pest managers have had the means to alter the environment to the disadvantage of the pest and the advantage of the pest's natural enemies. These specialists now may have the tools necessary to alter the genetic future of both a pest and its natural enemies [9].

II. THE STERILE INSECT TECHNIQUE

A. History

> An occasional step backward within the road that led forward is needed, however, for consolidation of our gains, for better utilization of our acquisitions, for improved perspective and orientation, for avoiding the repetition of mistakes, and to bring one an awareness of methods of thinking and investigating that may be re-adapted [89].

Many discoveries in radiation biology, entomology, and insect genetics have contributed greatly to our current ideas concerning the use of genetics to control insects. Only a few of these significant advances can be considered here. For example, in 1895, W.K. Roentgen reported his discovery of the emanations (which he called X-rays) from an electrically excited Crookes tube. As early as 1903, the sterilizing effects of radium rays upon reproductive systems and the development of insects were recognized [16]. Bergonié and Tribondeau [15] reported in 1906 that germinal cells are more sensitive to X-ray irradiation than are interstitial cells, and formulated a hypothesis (subsequently called the law of Bergonié and Tribondeau), which states: "the radiosensitivity of cells is directly proportional to their reproductive activity and inversely proportional to their degree of differentiation" [4].

In examining the application of the Bergonié and Tribondeau law, G.A. Runner [103] found that certain doses of X-rays decreased the reproductive capacity of the cigarette beetle, whereas higher doses killed the beetle. By 1927, H.J. Muller [85] had demonstrated that X-ray radiation causes heritable changes (mutations) in the germ line of *Drosophila*. Dominant lethal mutations were found to be one of the most frequent types of induced mutation. The "partial sterility" of X-ray–treated males was extensively discussed by Muller [86], because this effect tended to reduce the number of visible mutations which could be recovered for a given dose of radiation. As early as 1937, perhaps following the law of Bergonié and Tribondeau, E. F. Knipling conceived a new approach for insect control in which the pests' natural reproductive processes would be disrupted by physical or chemical means [59]. This concept, as it was applied to the screwworm fly starting in the early 1950s, eventually came to be known as the sterile insect technique (SIT), or the sterile insect release method (SIRM). Interestingly, about the time that Knipling was formulating his concept of insect control, a Russian geneticist, A. Serebrovsky [107], suggested the use of chromosomal translocations to reduce reproduction in harmful species. Unfortunately, Serebrovsky's suggestion was lost to entomological research in the West until it was discovered by Curtis [36]. We shall discuss the philosophy of these different approaches to insect control more thoroughly under the heading of classic genetic control procedures. An interesting and detailed discussion of the history and development of genetic methods in insect control procedures can be found in Whitten [123].

B. Sterilization Techniques

Studies of insect reproduction, from the early 1900s to the present time, have demonstrated that insects treated with X-ray or gamma radiation (or with certain mutagenic chemicals) are unable to produce a normal number of living progeny [57]. Treated insects (which may be completely or only partially sterile) are released in large numbers into a field environment and are expected to mate with the feral insects, thus interfering with reproduction. If matings between treated insects and normal insects are successful, reproduction of the field populations will be disrupted, and the population will decline.

1. Chemical Sterilization

By 1936, Melvin and Bushland [84] had developed a successful technique for mass rearing the screwworm fly [59]. A dependable way to interrupt the reproductive processes of the insect was needed. Chemosterilants appeared to be promising sterilization tools, but initial attempts using several chemicals were unsuccessful in sterilizing the screwworm [23]. The concept of chemosterilization still intrigues investigators, because it appears to be an efficient method for causing sterility. For example, chemical sterilants could be added to a diet, or they could be applied to eggs or pupae at times when they are being handled in normal rearing procedures [109].

Many chemicals have been screened for their ability to sterilize insects. Some of these chemicals can cause complete sterility and have been considered for use in sterile insect release programs [20,35,61]. In fact, most of the plans for eradicating the boll weevil in the United States recommend the use of hexametopol (HEMPA) plus a low dose of radiation to produce weevils whose sterility will last for their full lifetime.

However, the use of chemical sterilants is presently very limited because of concerns over environmental contamination with carcinogenic materials. Disposal of contaminated diets, or other carriers used to administer chemosterilants, is indeed problematic. Also, there are problems with chemosterilized insects being consumed by other biological entities, such as predators, birds, mammals, and (indirectly) humans in the environment. McDonald [82] concluded that, although chemosterilants such as aphoxide (TEPA), methapoxide (METEPA), HEMPA, and Apholate were very effective, they must be used cautiously. We are not aware of any sterile insect technique (SIT) programs presently using chemicals to sterilize released insects.

2. Radiation Sterilization

The SIT concept remained dormant from 1937 to 1950 except in the minds of E. F. Knipling and R. C. Bushland. However, in about 1950, the work of H. J. Muller [87,88] came to Knipling's attention, and he suggested that Bushland examine the use of X-ray radiation to sterilize the screwworm fly [59]. Once the correct radiation dosage for inducing screwworm sterility was determined [25], and the ability to rear a large number of adult flies was more or less assured [42], other problems, such as, for example, release methods, release ratios, and field evaluation, were worked out after the initial eradication program was in operation [59].

The requirements for the successful use of a sterility principle have not changed since the original formulation [55]. These requirements can be organized into separate program components [9]:

a. Techniques for rearing large numbers of a target insect species (rearing component)
b. Techniques for sterilizing large numbers of a target insect species (treatment component)
c. Biologically competitive insects that can be released after sterilization (release component)
d. Techniques for assessing field populations accurately before and after release of the treated insects (evaluation component)

e. A treatment area large enough (or sufficiently isolated) to exclude or significantly reduce the immigration of inseminated females into a release area (reinfestation component, for eradication programs only)

It is obvious from a study of these components that no detailed knowledge of genetic principles is necessary for the success of a sterile insect release program, and according to Bushland [21], none was employed. However, some new methods of control demand a detailed knowledge of the genetics of the pest insect, its host (plant or animal), and its associated microorganisms.

SIT was extremely successful for controlling the screwworm fly, and that success led to investigations of the effects of radiation on the reproductive performance of many other economically important insect species. Unfortunately, difficulties are often encountered when SIT is applied to Lepidoptera, to certain Coleoptera, or to species whose noxious behavior precludes release of a large number of insects in the vicinity of people. Whereas a dose of 5500 rad (55 Gy) is sufficient to guarantee 100% sterility of the screwworm fly, some lepidopteran pests can still produce F_1 progeny after exposure to doses exceeding 50,000 rad (500 Gy) . These progeny are not necessarily fertile, but they are viable and active and must be considered in the evaluation component of a SIT program. In the boll weevil, doses of radiation higher than 5000 rad (50 Gy) kill certain dividing gut cells and cause early mortality owing to infection and septicemia [98].

C. Inherited Sterility

Radiation can be used to transfer a piece of one chromosome to another in many insects. Some of the resulting chromosomal configurations are known as translocations. When an insect carrying one chromosomal translocation breeds with a normal insect, 50% of their progeny die because of unbalanced chromosome constitutions. The number of translocations carried by an individual is directly related to the percentage of mortality of that individual's progeny. The percentage of F_1 sterility varies with species and with the configuration of the translocations [123].

Radioresistance in lepidopterans, as well as in some hemipterans, allows these species to be exposed to radiation doses high enough to produce translocations but not high enough to produce full sterility. When these species are exposed to doses of radiation which do not produce full sterility, the resulting F_1 progeny exhibit levels of sterility equal to or higher than those of their treated parents [91]. This F_1 (or delayed) sterility may be useful in control programs, because (a) insects treated at lower doses of radiation would be more competitive than fully sterile insects; (b) the number of F_1 progeny produced in nature would exceed the rearing capabilities of most laboratories, thereby increasing the cost effectiveness of a release program; and (c) sterility effects will last more than one generation. Theoretical models comparing F_1 sterility with SIT [29,56] have shown that the release of partially sterilized insects offers far greater suppressive potential than the release of fully sterile insects.

The use of radiation-induced F_1 sterility for areawide control of Lepidoptera was addressed in a special meeting in 1991 [3]. Studies of the production of F_1 sterility were examined for diverse insect species; that is, diamondback moth (*Plutella xylostella*), European corn borer (*Ostrinia nubilalis*), Asian corn borer (*Ostrinia furnacalis*), pink bollworm (*Pectinophora gossypiella*), tropical armyworm (*Spodoptera litura*), wild mulberry silkworm (*Bombyx mandarina*), gypsy moth (*Lymantria dispar*), and corn earworm (*Helicoverpa zea*). It was concluded that doses of radiation in the range of 100–200 Gy could produce a useful level of inherited (F_1) sterility after the release of partially sterile insects. Currently, research on the use of radiation-induced F_1 sterility is being conducted in more than 20 different countries. However, only a few field trials have been conducted to test the ability of low doses of radiation to induce sufficient inherited sterility to control an economically important insect species.

A field study was conducted in small mountain valleys in North Carolina to assess the influence of released, substerilized (100 Gy) males of the corn earworm, *Helicoverpa zea*, on wild populations and to measure the infusion rate of inherited sterility into wild populations [28]. Results from this study revealed that the number of wild males captured per hectare was positively correlated with the distance from the release site of irradiated males. Analyses of seasonal population curves of wild *H. zea* males calculated from mark-recapture data suggested that seasonal increases of wild *H. zea* males were delayed and/or reduced in mountain valleys where irradiated males were released. The incidence of larvae with chromosomal aberrations (progeny of irradiated, released *H. zea* males [26] collected from the test sites indicated that irradiated males were very competitive in mating with wild females, and they were successful in producing F_1 progeny which further reduced wild populations.

The loss of sperm competitiveness of fully sterile lepidopteran males has been reported by many researchers [40,95,110]. Although sperm competition is improved by using a partially sterilizing dose of radiation, the resulting F_1 males often exhibit reduced sperm competitiveness. In the pink bollworm, F_1 males of parents treated with partially sterilizing doses of radiation often fail to transfer sperm to their untreated mates, and females paired with these F_1 males mate more often than females paired with normal males [63]. Carpenter et al. [29] found that *H. zea* females mated to progeny from irradiated (100 Gy) males outcrossed to normal females exhibited the same attractiveness and mating propensity as virgin females. Females mated to these F_1 males were able to detect the quality of a sperm complement and reduce their refractory period (the postcopulatory period during which females are averse to subsequent mating) if the sperm quality was not satisfactory. Although sperm from these F_1 males were less competitive than sperm from normal males, reduced sperm competitiveness in F_1 males had little effect on the expected population suppression following the release of partially sterile moths.

The release of partially sterile females (compared with fully sterile females) would increase the number of destructive larvae in the field. If the presence of these larvae would cause economic loss, a higher dose of radiation may be required to increase female sterility. Another alternative would be to develop a sexing system (mechanical or genetic) that would make possible the release of males only or the treatment of males and females with different radiation doses.

D. Future Possibilities

Although most technical limitations in inducing sterility in insects have been overcome during the 40 years that SIT has been under investigation, some significant problems still exist. For example, mass rearing has been a very expensive component of SIT programs and has only been successfully accomplished with a limited number of insect species (e.g., the screwworm fly, pink bollworm, boll weevil, tobacco budworm, several fruit fly species, and diamondback moth). The achievement of successful mass rearing programs for other economically important species appears to be more dependent upon fiscal limitations than on scientific constraints.

One of the more interesting ways suggested to reduce the expense and time of developing a mass rearing program is to sterilize insects where they live. This technique, sometimes called autosterilization, involves the use of either a pheromone or other type of attractant (e.g., feeding stimulant, attractive color) that will bring the insects to a point where they can be automatically exposed to a sterilant (usually a chemosterilant). This technique has been examined for the tsetse fly [116]. One problem with this technique is holding the insects in contact with a chemosterilant long enough to accumulate a dose sufficient to produce permanent sterility. Another problem is finding a sterilant that will not be toxic or mutagenic to nontarget species. Juvenile hormone analogues or other growth inhibitors might be good candidates for this technique, since they are quite species specific, but they also are often very specific to stage of development. This is an area of insect control that requires further investigation and development.

The incorporation of SIT into areawide management systems (AMSs) must also be considered. In many such programs, practically every aspect of insect management is incorporated (e.g. cultural control, biological control, insecticidal control, pheromone disruption, or trapping techniques) but SIT is ignored. E.F. Knipling [58] considered SIT to be an integral part of any AMS, because SIT is most effective when used on very low populations such as those that could be expected after a successful program of integrated control. Sterile insects could be effective in locating the few individuals that remain in an area after other control programs have been applied. The release of sterile insects could also be more economical than the use of other control procedures, since SIT effectiveness increases with decreasing populations, whereas most other procedures become less efficient as populations decline. The integration of genetic and nongenetic approaches will be discussed further in Section VII below.

III. CLASSIC GENETIC TECHNIQUES POTENTIALLY USEFUL FOR PEST SUPPRESSION

Some insect species are not good candidates for control by SIT either because some of their somatic tissues are too sensitive to the effects of radiation (e.g., the boll weevil) or because the radiation dose rates necessary for their complete sterility are so high that detrimental effects on behavior and reproduction occur (e.g., codling moth). To overcome these deficiencies in the use of SIT for insect control, entomologists and geneticists initiated elegant genetic schemes. These techniques, along with SIT, ultimately came to be known as autocidal control techniques.

We previously pointed out that sterilization by the use of radiation or chemicals primarily involves the induction of dominant lethal mutations as demonstrated by Muller [85]. However, many other genetic phenomena can be used to affect the fitness of an organism. Some classic genetic techniques which have been considered candidates for use in autocidal control schemes are examined here.

A. Chromosome Translocations

Perhaps the earliest proposed use of genetic manipulation for insect control was by Serebrovsky [107] in 1940. In this scheme, chromosomal translocations are induced and recovered in laboratory populations of the pest species. The best translocations are made homozygous by appropriate breeding schemes. The best translocations are defined as those that are essentially fully fertile when they are made homozygous. Fertile translocation homozygotes are released into natural populations where, hopefully, they will interbreed with native insects carrying normal chromosomes. The resulting F_1 translocation heterozygotes will be partially sterile. The incidence of F_1 sterility increases with the number of translocations carried by the released insect populations. This scheme is similar to the concept of F_1, or delayed, sterility considered in an earlier section, but in this case, genetic manipulation of stocks must take place in the laboratory before release [37]. In the case of F_1 sterility, reared insects are treated with a nonsterilizing dose of radiation (or chemical) and then released. Delayed sterility is primarily suggested for insects with holocentric chromosomes (also called polycentric chromosomes, which refers to chromosomes with diffuse centromeres as opposed to chromosomes which have a single centromeric region to which spindle fibers attach during meiosis), such as Lepidoptera and Hemiptera, whereas the use of recovered translocations is applicable to most orders of insects.

Curtis [36] suggested the use of translocations in pest control programs as mechanisms to introduce deleterious genes into populations, thus generating high and persistent genetic loads. Translocations between autosomes and sex chromosomes also have been proposed for the construction of strains which genetically eliminate one sex during the rearing process. The release of genetically sexed insects is important in genetic control programs to avoid the release of fully fertile female insects into areas where additional damage to crops or animals cannot be tolerated. Release of only genetically altered males for population control would be especially important in certain species of mosquitoes or flies where released females would be particularly noxious while feeding on hosts.

The pink bollworm has been exposed to irradiation for the production of chromosomal translocations in a number of experiments (A.C. Bartlett, unpublished data). Visible eye-color genetic markers have been used to recover reciprocal translocations. However, radiation of the pink bollworm induces a number of detrimental mutations (e.g., recessive lethals, deletions, duplications) with induced reciprocal translocations. A single heterozygous translocation produces about 50% sterility in the insect carrying the translocation. Thus, the reduced fertility due to the translocation and fertility problems caused by induced detrimental mutations leads to rapid loss of translocation-bearing strains. Insect control using translocations thus awaits further experimentation with agents that will induce translocations without causing other reproductive problems.

B. Simply Inherited Traits

Simply inherited detrimental traits have been used very little in genetic control programs, although some have been examined experimentally. For example, a simply inherited conditional lethal mutation, *salmon*, has been examined in the tsetse fly, *Glossina morsitans* [41]. Likewise, the dark body color mutation (*D*) in the cabbage looper, *Trichoplusia ni*, is a dominant allele with recessive lethal effects [7,8,10] similar to several other melanic mutants in moths [12]. Theoretically, a 10 dark : 1 feral release program could reduce the feral population by 40% in a single growing season. Unfortunately, unless only males are released, an unacceptable population increase would occur as a result of reproduction of the released mutant females.

A number of simply inherited morphological mutations have been incorporated into mass reared strains of several species to differentiate males from females (genetic sexing) during rearing [99,120]. As pointed out previously, the presence of females in released populations is generally detrimental either because of their production of additional progeny in the field, because they may be pestiferous to humans or animals, or because they may be disease vectors. Automatic elimination of females during rearing has been attempted with a number of different species [101], most successfully with the mosquito, *Anopheles albimanus* [106], medfly, *Ceratitis capitata* [100], and stable fly, *Stomoxys calcitrans* [105].

C. Hybrid Sterility

Hybrid sterility refers to sterility that occurs when certain strains, races, or closely related species are crossed and either one or both sexes of F_1 progeny are viable but cannot produce viable progeny. This phenomenon has been investigated and used in the management of several pest species, (e.g., between species of the tsetse flies, *Glossina morsitans* and *G. swynnertoni*, between races of the gypsy moth, *Lymantria dispar*, and between the moth species *Heliothis virescens* and *H. subflexa*).

The hybridization of *H. virescens* and *H. subflexa* [65] stimulated interest in reducing field populations of *H. virescens* using released backcross insects. In an experiment on St. Croix, U.S. Virgin Islands, using both male and female backcross insects, sterility was infused into a field population with release ratios of <20 sterile backcross : 1 feral insect [94]. The frequency of sterile male progeny increased for one generation after release, and the distribution of backcross frequencies became homogenous throughout the population. During a 6-week period, 94% of trapped males were sterile progeny from released or field-reared backcross females and native males. Isolated populations of *H. virescens* probably can be eradicated using this method given a sufficient number of released hybrid individuals.

Many attempts have been made to produce sterile hybrids from other pest species. For example, *Helicoverpa zea* has been mated with *H. armigera* from Australia, Russia, and China and *H. assulta* from Pakistan and Thailand [66–72]. These hybridization attempts failed to produce a sterile hybrid. Similarly, no measurable hybrid sterility has been found in crosses between the pink bollworm from areas within the United States, Mexico, Puerto Rico, or St. Croix (A.C. Bartlett, unpublished results). Raina et al. [96] reported no incompatibility between a strain of the pink bollworm from southern India and two strains (one was a long-term laboratory strain, the other a newly colonized strain) from Arizona.

LaChance and Ruud [64] crossed strains of the pink bollworm, *Pectinophora gossypiella*, from Australia and Arizona and observed no loss of fertility. They also made reciprocal crosses between both strains of the pink bollworm and a strain of *Pectinophora scutigera* from Australia. These crosses were characterized by reduced interspecific mating, low fecundity, and low fertility. Some fertile F_1 progeny were produced, especially when *P. scutigera* females were crossed with *P. gossypiella* males. The F_1 individuals were fertile in backcrosses to *P. scutigera* but infertile in backcrosses to the pink bollworm. The investigators suggested that interspecific hybrids between these two species will not be obtained easily and that these methods may not be useful in control procedures. It might be possible to improve the rate of interspecific mating, fecundity, and fertility by artificial selection procedures so that an increased number of F_1 progeny could be produced and released against the pink bollworm without the debilitating effects of radiation. However, the development of hybrid sterility in the pink bollworm may require more research effort and more expense than is justified, especially if releases of radiation-sterilized insects continue to be as efficacious as has been the case in the San Joaquin Valley release program (see below).

D. Sex-Ratio Modifiers

It was previously noted that for some programs it may be necessary to release only males. Similarly, at times, it may be beneficial to separate males and females prior to release to avoid assortative mating. Rearing costs can also be reduced if one sex is eliminated in the egg stage (prezygotic sexing)—twice as many release insects could be produced for a given expenditure of diet and labor. Insect sex ratios can be altered by a number of genetic phenomena.

Strunnikov [114] proposed the use of strains with balanced recessive lethal mutations on the sex chromosomes of the male as a method for controlling lepidopteran pests. Lepidopteran males carry two X chromosomes (homogametic), whereas the females have only one (heterogametic). Males carrying balanced recessive lethal mutations have two different recessive lethal genes; one on each X chromosome at different loci. When such males are crossed with any female, no female progeny are produced unless crossing over occurs between the two loci. For example, Marec [81] constructed a balanced lethal strain of the Mediterranean flour moth, *Ephestia kuehniella*. Males of this strain are balanced for two nonallelic sex-linked recessive lethal mutations, *sl-2* and *sl-15*. Females carry either *sl-2* or *sl-15* in their Z chromosome and the T(W;Z)2 translocation on their W chromosome. Because the translocation includes wild-type alleles of both lethal loci, the females are viable. However, matings between the balanced lethal strain males and normal females of the wild-type strain produce >99% male progeny. Exceptional females resulted from recombination between the sl-1 and sl-15 loci. Strunnikov [114] postulated that the use of such genetically altered insects would be 1.3 times as effective in the F_2 as a single release of fully sterile males and that the effect would increase over generations.

In addition to the direct usefulness of balanced sex-linked recessive lethal mutations for control procedures, such genetic stocks would be useful in genetic sexing of strains where only males should be released to drive a detrimental character (e. g., a potential conditional lethal, such as nondiapause) into a field population. In fact, the two systems (conditional lethal and balanced lethal) would act in concert to reduce pest populations during the growing season and the host-free season.

Bartlett [8] demonstrated that sex-linked recessive lethal mutations can be induced readily in the pink bollworm. The recessive sex-linked eye color mutation, *purple*, is used as a marker in these stocks to help maintain the sex-linked lethal mutations. The lethal mutation was induced in males which were homozygous for brown (wild-type) eye color. These males were crossed to purple females and their F_1 heterozygous wild-type sons were again crossed to purple females. If a recessive lethal is present, then no wild-type F_2 females will be produced in the progeny of the cross. The lethal mutation can then be maintained by using only wild-type (brown-eye) males and mating them to their purple-eye sisters. The presence of any brown-eye females in such a cross is an indication of crossing over between the induced sex-linked recessive lethal and the locus for the purple-eye gene. The production of a balanced sex-linked lethal strain has not yet been accomplished in the pink bollworm nor in the codling moth [2]. However, the development of such strains is being investigated at this time. Sex-linked recessive lethal mutations have also been induced in strains of the codling moth [2], but, as in the pink bollworm, balanced sex-linked lethal strains have not yet been produced.

In another type of scheme for modifying the sex ratio, alleles resistant to specific toxic chemicals (called selectable alleles), such as ethyl alcohol, endrin, purine, potassium sorbate, dieldrin, cyromazine, and propoxur, are identified and purified. If it is shown that inheritance of resistance to a chemical is due to a single locus, an investigator tries to induce a translocation between a sex chromosome (usually the Y chromosome) and a chromosome bearing the resistance locus. This scheme has been successful for manipulating the sex ratio in three species of mosquitoes and is being actively investigated for the medfly. Lepidopteran pests would be ideal for this type of approach if any single locus resistance traits could be identified. This scheme is discussed further in Section IV on molecular techniques.

E. Conditional Lethal Mutations

Strains of insects can be manipulated by artificial selection to carry traits that impair field survival but do not affect the insect's ability to exist in a laboratory. For example, where diapause is necessary for survival of the pink bollworm during host-free or environmentally unsuitable periods, its inability to enter diapause would be a conditional lethal trait. A nondiapausing (ND) strain could be reared readily in the laboratory, but progeny produced by this strain in the field would not diapause, and it could not reproduce during the host-free period. Bartlett and Lewis [11] selected nondiapausing strains of the pink bollworm in the laboratory. The nondiapause character of the pink bollworm is controlled by dominant or partially dominant alleles and is polygenic.

The nature of the inheritance of a nondiapause character in pink bollworm suggests that single releases of the nondiapausing strain should be made in large populations near the end of a reproductive season. In common with other genetic control procedures, it would be most beneficial if only males carrying the deleterious trait could be released to prevent increased larval populations. In this way, crop losses due to the addition of fertile females to field populations would be avoided because released males would mate only with feral females.

F. Cytoplasmic Incompatibility and Meiotic Drive

Laven [73] discovered that crosses between individuals from certain populations of the same species of mosquito (*Culex pipiens*) are sterile. Crosses between other populations are sometimes completely fertile or partially sterile. This sterility appears to be due to a cytoplasmic factor and not to chromosomal incompatibilities. This phenomenon also has been discovered in a number of other mosquito species and has been used as a control measure in experimental releases. Cytoplasmic incompatibility also has been observed in the plum curculio, *Conotrachelus nenuphar* [113], and in *Ephestia cautella* [21] leading to the hope that it might occur in many economically important insects. The use of sterility induced by cytoplasmic incompatibility has one serious drawback. Released strains of insects must be separated by sex before release so that they will not interbreed and become established in the field.

Another genetic phenomenon, called meiotic drive, has been demonstrated in several dipteran species [37]. Meiotic drive occurs when individuals heterozygous for particular gene combinations are crossed with an unequal recovery of alleles from the heterozygotes. In *Drosophila melanogaster*, certain chromosomes are passed to more than 95% of the offspring when only 50% of its offspring would be expected to receive a specific chromosome. Unfortunately, few cases of either cytoplasmic incompatibility or meiotic drive have been observed, because very few simply inherited visible mutations are presently available as chromosome markers in most economically important species. Thus, it is difficult to follow cases of segregation distortion. The lack of other genetic tools, such as crossover suppressors, inversion stocks, and other chromosome balancing factors has made the recovery and maintenance of such strains very difficult. However, the occurrence of certain factors, called "transposable elements," do have some interesting possibilities that will be discussed in the next section.

IV. MOLECULAR GENETIC TECHNIQUES

The development of genetic-sexing schemes has been slow because of our inability to find appropriate selectable genes in pest species. Additionally, the induction of translocations between a sex chromosome and a selectable locus is often an obstacle to successful development of genetic sexing. Recent advances in molecular genetics have led to the speculation that genetic engineering techniques could be used in insect control programs. Transformation or transvection of foreign DNA (genes) into pest species is being investigated for control purposes.

It has been known for some time that genomes of both prokaryotes and eukaryotes are not as stable as they were once thought to be. Genomes can be changed by several different types of forces, such as recombination, chromosomal breakage, and mutation. However, we also now find that certain genetic sequences are able to move from one site to another on a chromosome site. These sequences have come to be known as "transposable elements," or "transposons." These sequences carry genes required for their own transposition, but they also may carry other unrelated gene sequences with them as they move. This phenomenon was demonstrated by the genetic transformation of *Drosophila melanogaster* with transposable vectors (P-elements) and has suggested new approaches which may be used with other insect species.

For example, the genetic marker used by Rubin and Spradling [102] to demonstrate transformation in *Drosophila melanogaster* was a gene for wild-type expression at the *rosy* (ry^+) locus. This gene is the structural locus for the enzyme xanthine dehydrogenase (XDH). Flies lacking the active XDH gene have *rosy* (*ry*) eyes and are sensitive to purine in the rearing diet. The XDH locus has been cloned in *D. melanogaster* and is available for hybridization with Southern blotted genomic DNA from other insects. Saul [104] analyzed a similar locus in the medfly and suggested that this locus could be used for genetic sexing if a suitable vector could be found.

Unfortunately, attempts at using *Drosophila* P-elements as transformants have not been successful in nondrosophilids. Investigators are therefore searching in other insect species for transposable elements analogous to the *Drosophila* P-element. For example, a transposable element family, called mariner, has been found in a number of insect species while searching for gene sequence homology to the mariner element. An interesting approach for controlling insects through the use of a mariner vector is described by Warren and Crampton [118]. Another possibility for gene vectors are various insect viruses, such as baculoviruses, which can be engineered to contain selectable genes and used to carry those constructs into an insect genome to form stable transformations.

Although the technology for transvection of foreign genes into insect pest species is still in the developmental stage, its possibilities are fascinating. Only one of several possibilities is considered here. In most insects, considerable yolk protein (YP) is packaged in an oocyte for use by a developing embryo. The precursors to yolk proteins, called vitellogenins, are synthesized by the fat body of adult females. Thus, this synthesis is sex, stage, and tissue specific. YP genes have been cloned for *Drosophila melanogaster* [6]. In addition, the function and control of YP genes is well understood for *Drosophila*. The genes consist of both promoter and regulatory DNA sequences.

One interest in obtaining cloned YP genes is to utilize the promoter sequences that control the specific expression of the genes. If promoter sequences from YP genes which cause sex, stage, and tissue specificity can be linked to structural sequences of other genes, then genes with desirable activities (such as chemical resistance) can be introduced into a target species. The location on a chromosome of the introduced construct would not be important, since appropriate chemicals in a diet could be used to select for a specific sex or stage of development. This technology would be very useful in the development of a genetic-sexing system, because females could be eliminated early in rearing processes, thereby allowing twice as many males to be reared on the same amount of diet.

Specific obstacles to progress in the use of molecular genetics for insect control include the lack of (a) a general vector for gene transfer, (b) suitable selective criteria for useful loci (i.e., uncertainty about what genetic modifications will prove most detrimental to a given species), and (c) knowledge of the basic genetic biology of most insect pests. None of these obstacles is insurmountable. The basic technology is in place but must be explored thoroughly for each species.

V. RESEARCH AND DEVELOPMENT REQUIREMENTS FOR AUTOCIDAL PROGRAMS

Pest control programs using genetic methods are generally very large and comprise many elements. Because of the complexity of these programs, and because the control method is usually applied over a broad area, nearly all autocidal programs are managed by large organizations (i.e., government agency, international organization, grower organization) which are responsible for obtaining pertinent information needed for decision making and for coordinating a vast array of activities. Fortunately, the implementation of many successful autocidal programs has provided a foundation of knowledge upon which future programs can build. The principal components of a genetic control program, as listed by Knipling [55], have changed very little. Components common to most genetic control programs will be discussed briefly.

A. Mass Rearing

An economical method for mass rearing a pest species is a key requirement for autocidal control programs. Most mass-rearing methods rely on a nutritious artificial diet and mechanization. Methods for mass rearing insect pests have been developed for numerous species [54,75,108]. Still, substantial modification of existing methods may be required before a new species can be successfully mass produced.

Once mass-rearing technology has been developed for an insect species, it is critical for a genetic control program that the fitness of released insects approach that of the feral insects [119]. However, producing quality insects in a mass-rearing facility is no easy task. Artificial diets may be nutritionally incomplete, or diseases can infect an insect colony. Laboratory strains that become adapted to insectary conditions may perform well in mass-rearing facilities, but they also may undergo selection when maintained over a long period [62]. Such laboratory selection may result in behavioral or physiological attributes that cause released insects to be reproductively isolated from feral insects. Also, certain characteristics important for survival in the field may lose their selective advantage under laboratory conditions [119]. A selective breeding program within a laboratory colony may be useful in reducing inbreeding [127]. Another method for improving the performance of an insect colony to be used for a genetic control program is to build a laboratory colony from feral insects indigenous to the area in which releases will be made [126]. This technique is not without risk, however, because without proper precaution, feral insects sometimes introduce diseases, as well as new genes, into laboratory colonies [44].

Providing a continuous supply of a predetermined number of high-quality insects for a genetic control program is a difficult requirement. Lindquist et al. [78] observed that, "Mass-rearing of insects on a continuous basis requires ingenuity, a 7-day week work-force, and a good bit of luck." Nevertheless, because a mass-rearing facility is a factory and can be managed as a factory, quality control can be researched and incorporated into production design. A good foundation of knowledge is available for researchers studying quality control of mass-reared insects [17,18,24,33,48,51,54,74,75,83,121].

B. Population Dynamics, Ecology, and Behavior

An understanding of the population dynamics, ecology, and behavior of a candidate species is prerequisite to the application of a genetic control method. Information crucial to the economic assessment and the development of a genetic control method includes (a) an estimate of the number of insects in the native population and how this number changes over time and space; (b) time of the year when the pest population is at its lowest level, and methods of reducing the native pest population to a minimal number; (c) intrinsic rate of increase and long-range movement capabilities of a pest during various seasons and in different cropping systems; (d) all host plants capable of supporting a pest species, and the relative abundance and economic importance of each host plant; and (e) whether secondary pests will benefit or be deterred by the reduction or absence of a target pest [62]. Researching information needed for genetic control programs often requires a shift in the type of ecological and behavioral studies normally conducted. Research on total pest populations, pest movement, and pest reproductive rates is difficult and expensive. Nevertheless, knowledge gained from such research can prevent costly mistakes once a control program has been initiated.

C. Distribution, Dispersal, and Release Technologies

Methods for collecting and releasing mass-reared insects must be designed to minimize loss of competitiveness due to handling and must be congruent with dispersal capabilities of insects. The developmental stage of an insect to be collected will depend upon the species, the type of genetic method in use, and whether collected insects will receive a radiation treatment. Packaging, transporting, and releasing methods must be developed appropriate to the species and developmental stage. Decisions concerning the best distribution system must be determined by field tests on candidate species. Program effectiveness will depend upon a homogeneous distribution of released insects throughout a test location.

D. Program Monitoring

All successful genetic control programs must have adequate methods for making progressive assessments of release methods and effectiveness of released insects. Extensive trapping can provide data on the ratio of released to native insects, and the degree of homogeneity and dispersal of released insects. Capture of native and released females can provide data on their number, and the type of matings and egg hatchability.

The benefit of a good monitoring system cannot be overemphasized; its development must be well planned to avoid potential pitfalls [62]. Although egg hatchability is a reliable monitoring tool for SIT programs, it is less reliable as a monitoring tool for some other genetic control methods. For example, the use of inherited sterility and backcross sterility can complicate monitoring because native females mated with irradiated (substerilizing dose) males, and field-reared backcross females may be fertile. Also, unmarked males that are the sterile progeny of released insects can be trapped one generation after inherited sterility or backcross sterility releases begin. Cytological examination or laboratory tests of fertility can determine whether these trapped males are progeny of native pairs or the progeny of released and native insects [62]. Additionally, dominant morphological genetic markers in the released strain would allow investigators to monitor the progeny of released insects over several generations [7].

E. Economic Analysis

Before any genetic control program is implemented, an in-depth economic analysis should be conducted. Such an analysis should consider damage estimates by the pests, current cost of their control, cost estimate of a genetic control program, and potential benefits to the environment [62]. The advantages of various methods and combinations of methods of control should be analyzed and compared for multiple years. Knipling [57] observed that reducing pest species to low populations may be difficult and costly by the use of insecticides or by a combination of several methods. But, once this is accomplished, continuous pest population management by genetic approaches may be the most economical, effective, and ecologically acceptable method available. The cost/benefit ratio is not a static figure, but it should continue to decrease with each passing year. Dyck [38] described the thorough economic assessment conducted prior to the start of the codling moth eradication program in British Columbia, Canada, which could be used as a model for future programs. Previous control programs have demonstrated that genetic approaches can be very cost effective. Cost/benefit ratios, generated from increased production and the absence of conventional pest control, estimate that the screwworm program saved $4 billion through 1987, and that the melon fly program will save more than $100 million per year [125].

F. Organizational Structure

Large insect control and eradication programs using genetic methods require an effective team of scientists and technicians with a hierarchy of delegated authority. An organized communications network must be established to coordinate field operations with program officials. Often decisions based on laboratory and field data will call for changes in program operations, such as the distribution of insects or the location of ground personnel. With an appropriate organizational structure for making, communicating, and implementing needed changes, program operations can be managed efficiently [62].

The areawide application of a genetic control program usually depends upon strong financial and political support. Growers often provide much of the financial support, while government organizations coordinate and implement the programs. These program partners should, however, also be aware of the general public's interest and concern. Genetic control programs are very visible to the public operating in large areas, involving many people and frequently using airplanes for release of biological material and insecticides. Therefore, an effective program to inform the public about control projects is essential [78].

VI. CURRENT PROGRAMS USING GENETIC TECHNIQUES

A. Screwworm

The first major use of SIT was for the eradication of the primary screwworm from the island of Curaçao in 1954 [13,76]. Following this success, the pest was then eradicated from the southeastern United States in 1959 and from Puerto Rico in 1975 [122]. Eradication of the screwworm was initiated in the southwestern United States in 1962 and completed in 1982. Because of the extensive damage that the screwworm causes in Mexico, and because it was impossible to prevent reinfestation into the United States, a Mexican-American commission was created in 1972 to eradicate the pest from northern and western Mexico and to establish a "sterile fly barrier" at the Isthmus of Tehuantepec [43]. In 1986, the commission extended its eradication activities to the Yucatan Peninsula and countries of Central America. As a result of this program, Mexico, Belize, and most of Guatemala are now free of the screwworm. Operations are in progress to eradicate the screwworm from Honduras and El Salvador [52]. The present goals of the program are to eradicate the pest from Central America and Panama and to establish a sterile fly barrier at the Darien Gap to prevent its reinfestation. Also, efforts will be made to eradicate the screwworm from Caribbean islands which are still infested [78].

The screwworm invaded North Africa in the late 1980s and became established in the Libyan Arab Jamahiriya. After detecting the screwworm in 1988 and confirming its presence in 1989, a joint decision was made by the Food and Agriculture Organization/International Atomic Energy Agency (FAO/IAEA) and the government of the Libyan Arab Jamahiriya to initiate an eradication campaign using the sterile insect technique. The campaign was implemented late in 1990 and completed in October of 1991. North Africa was declared free of the screwworm in June 1992 [77,78].

B. Fruit Flies

Genetic control programs using the SIT have been successful against several species of fruit flies (Diptera: Tephritidae). Limited field application of the SIT has resulted in population suppression or eradication of the Caribbean fruit fly, *Anastrepha suspensa*, in Florida [22]; cherry fruit fly, *Rhagoletis cerasi*, in Switzerland [19]; Oriental fruit fly, *Dacus dorsalis*, in the Mariana Islands [112]; Queensland fruit fly, *Dacus tyroni* [1]; and Chinese citrus fly, *Dacus citri*, in China [117]. However, areawide use of the SIT has been successful in the eradication or control of several species of fruit flies. The most spectacular successes have been with eradication of the Mediterranean fruit fly, *Ceratitis capitata*, from southern Mexico, and of the melon fly, *Dacus cucurbitae* from Japan, and the prevention of Mexican fruit fly from invading California and Texas [39].

After the first detection of the Mediterranean fruit fly in Mexico, the Ministries of Agriculture of Mexico and Guatemala and the United States Department of Agriculture formed the Moscamed program to combat this pest [97]. The objectives of the program were to stop the northern advance of the pest, to eradicate it from southern Mexico and Guatemala, and in the long term, to eradicate it from Central America and Panama. By combining the discrete use of malathion bait spray with the release of sterile flies, the medfly was eradicated from Mexico in 1982 [46]. The program also has been successful in Guatemala [47,92].

The Japanese government initiated a project to eradicate the melon fly in 1972. Using the SIT following one or more treatments of a lure/toxicant, the eradication effort began on Kume Island and was expanded until the melon fly had been eradicated from all the infested islands of the Kagoshima and Okinawa Prefectures. Japan was declared free of the melon fly in 1992 [125]. Although the eradication program required an investment of about $100 million [125], the benefits from eradication of the melon fly should be over $100 million per year [78].

C. Pink Bollworm

The pink bollworm, *Pectinophora gossypiella*, is a serious pest of cotton in many parts of the world. Although the pink bollworm is occasionally recovered from states east of Texas, this pest is most destructive in Arizona, southern California, and the adjacent northwest Mexican desert. After its introduction and establishment in central Arizona in the mid 1950s and in the Colorado River Basin of western Arizona, southern California, and northwestern Mexico in 1965, a program was initiated in 1968 to protect the 500,000 ha of cotton in the San Joaquin Valley. This program has prevented the establishment of the pink bollworm through the use of sterile insect–release technology, minor use of pheromones as a mating disruptant, and adequate cultural control. The California cotton industry considers the 27-year old San Joaquin Valley Exclusion Project to be a major success. The cotton industry and the United States Department of Agriculture are investigating an expanded program in the Imperial Valley [111].

D. Codling Moth

A sterile insect release program to eradicate the codling moth, *Cydia pomonella* (L.), from the Okanagan region of British Columbia, Canada, by the year 2000 was initiated in 1992. The British Columbia Fruit Growers' Association and the British Columbia and Canadian governments jointly developed an implementation plan which divided the program into three distinct phases. A prerelease sanitation phase was designed to reduce wild populations as much as possible using insecticide applications and cultural control methods for 2 years. The second phase was rearing and release of sterile moths for 3 years. After eradication, the third phase will be protection against reinfestation by monitoring for the presence of wild moths, releasing sterile moths at border sites to prevent invasion of wild moths, and controlling the transfer of infested fruit containers. This phase of the program will last indefinitely [38].

VII. INTEGRATION OF GENETIC AND NONGENETIC APPROACHES

The advantage of combining genetic control methods with other pest control methods has been recognized from the inception of genetic control [55]. All the successful genetic control programs mentioned above have nongenetic components that have served vital roles in the control or eradication of pests, as well as preventing reinfestation of pests. Most often genetic control methods have been integrated with insecticide applications, cultural controls, and quarantines. However, population models constructed to predict the potential advantages of combining genetic control methods with other methods, such as inundative releases of parasites, host plant resistance, pheromones for mating disruption, and insect pathogens, have suggested that these combinations would yield synergistic effects [5,27,57,60]. Because the ratio between irradiated and nonirradiated insects is the most critical factor in regulating the efficacy of SIT or F_1 sterility release programs, any mortality agent (i.e., resistant host plants, insecticides) applied during a continuous release of genetically altered insects would benefit a release program. Although the mortality agent would reduce the number of both released and wild insects, it would not change the ratio. Therefore, subsequent to the application of the mortality agents, wild populations would be lower and the ratio of released to wild insects would be increased by continual releases of genetically altered insects [27].

Integration of genetic techniques and inundative releases of parasitoids may be more complementary than most other pest control combinations because their optimal actions are at opposite ends of the host density spectrum and do not interfere with each other [5]. Although the use of parasitoids and sterile insect techniques have different modes of action, the effectiveness of the sterile insect technique increases the ratio of adult parasitoids to adult hosts, and the effectiveness of the parasitoids increases the ratio of sterile to fertile insects [60]. Greater pest suppression could be expected if parasitoid releases were combined with the F_1 sterility technique. Not only is F_1 sterility in lepidopterans more effective than full sterility in reducing population increases, the F_1 sterility technique produces eggs and sterile F_1 larvae that would provide an increased number of hosts for the parasitoids [27].

Several laboratory and field studies have been conducted to determine the compatibility and effectiveness of combining different types of genetic control techniques with other control techniques. Carpenter and Young [32] found that progeny from irradiated parents and progeny from nonirradiated parents demonstrated the same level of insecticide resistance. Carpenter and Wiseman [30,31] investigated the effects of F_1 sterility and host plant resistance on *Helicoverpa zea* and *Spodoptera frugiperda* development, and they found that larvae resulting from irradiated male by nonirradiated female crosses were equally competitive with normal larvae for all measured parameters. Compatibility between F_1 sterility and parasites prompted Mannion et al. [80] to suggest that the use of combinant strategies, including F_1 sterility and the tachinid parasitoid, *Archytas marmoratus*, may be feasible for managing early-season populations of *H. zea*. Partially sterile *H. zea* adults could be released to produce large populations of sterile larvae on early-season annual plants that would then become hosts for *A. marmoratus* and other parasites. The next generation of parasitoids would be increased and any surviving sterile larvae would become sterile adults. Tillman and Laster [115] conducted ovipositional acceptance tests and parasitism studies with *Microplitis croceipes* in the laboratory and field using *Heliothis virescens* (F.) and *H. virescens–H. subflexa* backcross (see Section III.C) larvae as hosts. They concluded that augmentative releases of *Microplitis croceipes* during or following a sterile backcross release should not adversely affect the backcross release ratio, and that the two control techniques possibly could be used effectively together in an areawide management program for controlling *H. virescens*. Wong et al. [124] investigated the effect of concurrent parasite and sterile fly releases on wild *Ceratitis capitata* populations in the Kula area of Maui, Hawaii. They concluded that concurrent SIT and parasite augmentation programs may interact synergistically, producing a greater suppression in target insect population than either method used alone.

VIII. GENETIC IMPROVEMENT OF NATURAL ENEMIES

Biological control through importation and release, population augmentation, and conservation of natural enemies offers one of the best long-term strategies to help reduce or eliminate the use of synthetic chemical pesticides [90]. The use of natural enemies is an environmentally sound and publicly acceptable approach to pest management that can provide large economic savings to agriculture [60]. However, many attempts to control pests with natural enemies have been met with limited or no success. Recent advances in genetic selection and biotechnology may greatly improve the effectiveness of natural enemies through the development of genetic-sexing strains, disease and pesticide-resistant strains, and colonies with greater genetic variability [90].

Genetic selection of desirable traits in natural enemies has been limited largely to laboratory selection for pesticide resistance. Selection methods employed by researchers vary greatly among species of natural enemies [53]. Successes and achievements of selection programs have been noteworthy [14,34,49,50,53]. However, few selection projects have produced strains of natural enemies that were implemented in the field. In the future, advances in molecular biology may provide the technology needed to obtain desirable genes and insert them into natural enemies.

Selectable traits that could be exploited by genetic engineering efforts targeted at improving the effectiveness and economic value of natural enemies are pesticide resistance, disease resistance, environmental hardiness, and sex-ratio manipulation [45]. The genetic engineering approach will require (a) identification of genes controlling desirable traits, (b) cloning and characterizing genes, and (c) modifying and introducing genes into a target organism's DNA so that desired traits will be transmitted in a Mendelian fashion [90]. Development of methods for genetically transforming or inserting modified genes into chromosomes of target species is currently limiting the use of genetic engineering approach [45].

IX. CONCLUSIONS

Controlling insect pests is required for efficient and successful agricultural production of food, energy, and fiber. Insecticides are currently the major defense against most crop pests and will likely continue to be in the near future. But changes in pest management may be on the horizon [78]. Concerns regarding pesticide pollution and insecticide resistance have increased investigations of alternative management strategies. However, no single strategy should be expected to solve all pest problems. It is important to define and develop strategies that can be combined without deleterious effects and while perhaps creating a complementary pest management system [80]. Genetic methods for pest control are quite advantageous for this approach to pest management. Compared with chemical control methods, genetic methods are highly specific to a pest species, safe to pest control workers, nonpolluting to the environment, long lasting and areawide in approach, and compatible with most other pest control strategies.

Some obstacles are present which will slow the rate of progress in the use of genetic methods for controlling pests. Research and development requirements for genetic approaches often demand a long-term commitment before benefits can be realized. Also, technological constraints, such as the difficulty and high cost of insect rearing, may impede progress. But none of these obstacles are insurmountable. Even the major roadblock to further advances in the use of biotechnology for insect control, development of a genetic transformation method [45], appears to be relenting. For example, Presnail and Hoy [93] have reported the insertion of genetic material from *Drosophila* spp. *Escherichia coli* into female *Metaseiulus occidentalis*, a predatory mite, by direct injection of DNA.

Perhaps the most formidable obstacle to the successful use of genetic methods for pest control is our historical concept of pest management. Although evolving for the past 35 years, concepts of pest management have not kept pace with, nor fully used, rapidly developing genetic technologies. Before genetic control methods can be more fully implemented for pest management, the concepts of preventative and areawide pest management must gain wider acceptance [62]. Future pest management strategies for major pests and pest complexes should be developed with definable, long-term goals. These exercises in strategic planning should consider the long-range economic and ecological advantages that genetic approaches may provide as components of integrated pest management.

ACKNOWLEDGMENTS

We thank Drs. Leo E. LaChance, Sudhir K. Narang, and Charlie E. Rogers for their valuable comments on the original manuscript.

REFERENCES

1. Andrewartha, H.G., J. Monro, & N.L. Richardson. 1967. The use of sterile males to control populations of Queensland fruit fly, *Dacus tryoni* (Frogg.) (Diptera: Tephritidae). II. Field Experiments in New South Wales. *Aust. J. Zool.* 15:475–499.
2. Anisimov, A.I. 1988. Investigations on sex linked recessive lethal mutations as

a possible mechanism for the genetic control of lepidopterous pests. In *Modern Insect Control: Nuclear Techniques and Biotechnology* (Proc. Symp. Vienna). IAEA, Vienna, 1987, pp 65–76.

3. Anonymous. 1993. *Radiation Induced F_1 Sterility in Lepidoptera for Area-wide Control*. Panel Proceeding Series. IAEA, Vienna.

4. Arena, V. 1971. *Ionizing Radiation and Life*. St. Louis, Missouri: Mosby.

5. Barclay, H.J. 1987. Models for pest control: complementary effects of periodic releases of sterile pests and parasitoids. *Theoret. Pop. Biol.* 32:76–89.

6. Barnett, T., C. Pachl, J.P. Gergen, & P.C. Wensink. 1980. Transcription and translation of yolk protein mRNA in the fat bodies of *Drosophila. Cell* 21:729.

7. Bartlett, A.C. 1982. Genetic markers. In *Sterile Insect Technique and Radiation in Insect Control*. IAEA-SM-255, IAEA, Vienna, pp 451–465.

8. Bartlett, A.C. 1988. Induction and use of sex linked lethal mutations in the pink bollworm. In *Modern Insect Control: Nuclear Techniques and Biotechnology* (Proc. Symp. Vienna). IAEA, Vienna, 1987, pp 85–96.

9. Bartlett, A.C. 1990. Insect sterility, insect genetics, and insect control. In *Handbook of Pest Management in Agriculture*. Vol. II. D. Pimentel, ed. Boca Raton, Florida: CRC Press, pp 279–287.

10. Bartlett, A.C., & G.D. Butler, Jr. 1975. Genetic control of the cabbage looper by a recessive lethal mutation. *J. Econ. Entomol.* 68:331–335.

11. Bartlett, A.C., & L.J. Lewis. 1987. Response of the pink bollworm (Lepidoptera: Gelechiidae) to long-term selection for the inability to diapause. *Ann. Entomol. Soc. Am.* 80:797–803.

12. Bartlett, A.C., & J.R. Raulston. 1982. The identification and use of genetic markers for population dynamics and control studies in *Heliothis. Proceedings of the International Workshop on Heliothis Management*. Patancheru, AP: India, ICRISAT, pp 75–85.

13. Baumhover, A.H., A.H. Graham, B.A. Bitter, D.E. Hopkins, W.D. New, F.H. Dudle, & R.C. Bushland. 1955. Screwworm control through release of sterilized flies. *J. Econ. Entomol.* 48:462–466.

14. Beckendorf, S.K., & M.A. Hoy. 1985. Genetic improvement of arthropod natural enemies through selection, hybridization, or genetic engineering techniques. In *Biological Control in Agricultural IPM Systems*. M.A. Hoy & D.C. Herzog, eds. New York: Academic Press, pp 167–187.

15. Bergonié, J., & L. Tribondeau. 1906. Interprétation de quelques résultats de radiothérapie et assai de fixation d'une technique rationelle. *Compt. Rend. Acad. Sci. (Paris)* 143:983 (also published as an English translation in *Rad. Res.* 11:587–588, 1959).

16. Bohn, C. 1903. Influence des rayons du radium sur les animaux en voie de croissance; Influence des rayons du radium sur les oeufs vierges et fécondés, et sur les premiers stades du développement. *Compt. Rend. Acad. Sci. (Paris)*. 136:1012–1085.

17. Boller, E.F. 1972. Behavioral aspects of mass-rearing insects. *Entomophaga* 17:9–25.

18. Boller, E.F., & D.L. Chambers. 1977. Quality aspects of mass-reared insects.

In *Biological Control by Augmentation of Natural Enemies*. R.L. Ridgway & S.B. Vinson, eds. New York: Plenum Press, pp 219–235.

19. Boller, E.F., & U. Remund. 1983. Field feasibility study for the application of SIT in *Rhagoletis cerasi* L. in Northwest Switzerland. (1976–1979). In *Fruit Flies of Economic Importance*. Proceedings of an International Symposium organized by Greek Authorities/CEC/IOBC, Athens. Balkema, Rotterdam: pp 366–370.

20. Borkovec, A.B., & C.W. Woods. 1963. Aziridine chemosterilants: sulfur-containing aziridines. In *New Approaches to Pest Control and Eradication*. S.A. Hall, ed. Washington DC: American Chemical Society, pp 47–55.

21. Brower, J.H. 1976. Cytoplasmic incompatibility: Occurrence in a stored-product pest, *Ephestia cautella*. *Ann. Entomol. Soc. Am.* 69:1011–1015.

22. Burditt, A.K., Jr., D.F. Lopez, L.F. Steiner, D.L. von Windeguth, R. Baranowski, & M. Anwar. 1975. Application of sterilization techniques to *Anastrepha suspensa* Loew in Florida, USA. *In Sterility Principle for Insect Control, Proc. International Symposium IAEA/FAO*. Innsbruck. IAEA STI/PUB/337, pp 93–101.

23. Bush, G.L. & R.W. Neck. 1976. Ecological genetics of the screwworm fly, *Cochliomyia hominivorax* (Diptera: Calliphoridae) and its bearing on the quality control of mass-reared insects. *Environ. Entomol.* 5: 821–826.

24. Bushland, R.C. 1971. Historical development and recent innovations. In *Sterility Principle for Insect Control*. IAEA/SM138/47 IAEA, Vienna, pp 3–14.

25. Bushland, R.C., & D.E. Hopkins. 1951. Experiments with screw-worm flies sterilized by X-rays. *J. Econ. Entomol.* 44:725–731.

26. Carpenter, J.E. 1991. Effect of radiation dose on the incidence of visible chromosomal aberrations in *Helicoverpa zea* (Lepidoptera: Noctuidae). *Environ. Entomol.* 20:1457–1459.

27. Carpenter, J.E. 1993. Integration of inherited sterility and other pest management strategies for *Helicoverpa zea*: status and potential. *Proceedings of FAO/IAEA International Symposium on Management of Insect Pests: Nuclear and Related Molecular and Genetic Techniques*. IAEA, Vienna, pp 363–370.

28. Carpenter, J.E. & H.R. Gross. 1993. Suppression of feral *Helicoverpa zea* (Lepidoptera: Noctuidae) populations following the infusion of inherited sterility from released substerile males. *Environ. Entomol.* 22:1084–1091.

29. Carpenter, J.E., A.N. Sparks, & H.L. Cromroy. 1987. Corn earworm (Lepidoptera: Noctuidae): influence of irradiation and mating history on the mating propensity of females. *J. Econ. Entomol.* 80:1233–1237.

30. Carpenter, J.E. & B.R. Wiseman. 1992. Effects of inherited sterility and insect resistant dent-corn silks on *Helicoverpa zea* (Lepidoptera: Noctuidae) development. *J. Entomol. Sci.* 27:413–420.

31. Carpenter, J.E. & B.R. Wiseman. 1992. *Spodoptera frugiperda* (Lepidoptera: Noctuidae) development and damage potential as affected by inherited sterility and host plant resistance. *Environ. Entomol.* 21:57–60.

32. Carpenter, J.E. & J.R. Young. 1991. Interaction of inherited sterility and insecticide resistance in the fall armyworm (Lepidoptera: Noctuidae). *J. Econ. Entomol.* 84:25–27.

33. Chambers, D.L. 1977. Quality control in mass-rearing. *Annu. Rev. Entomol.* 22:289–308.
34. Croft, B.A. 1990. *Arthropod Biological Control Agents and Pesticides.* New York: Wiley.
35. Crystal, M.M. 1963. The induction of sexual sterility in the screw-worm fly by antimetabolites and alkylating agents. *J. Econ. Entomol.* 56:468–473.
36. Curtis, C.F. 1968. Possible use of translocation to fix desirable genes in insect pest populations. *Nature* 218:368–369.
37. Davidson, G. 1974. *Genetic Control of Insect Pests.* New York: Academic Press.
38. Dyck, V.A., S.H. Graham, & K.A. Bloem. 1993. Implementation of the sterile insect release programme to eradicate the codling moth, *Cydia pomonella* (L.) (Lepidoptera: Olethreutidae), in British Columbia, Canada. *Proc. Management of Insect Pests: Nuclear and Related Molecular and Genetic Techniques.* IAEA/FAO, Vienna, pp 285–297.
39. Economopoulos, A.P. 1990. Progress in the sterile insect technique against fruit flies. In *Pesticides and Alternatives: Innovative Chemical and Biological Approaches to Pest Control.* J.E. Casida, ed. New York: Elsevier, pp 89–92.
40. Flint, M., & E.L. Kressin. 1968. Gamma irradiation of the tobacco budworm: sterilization, competitiveness, and observations on reproductive biology. *J. Econ. Entomol.* 61:477–483.
41. Gooding, R.H. 1982. Laboratory evaluation of the lethal allele *salmon* for genetic control of the tsetse fly, *Glossina morsitans.* In *Sterile Insect Technique and Radiation in Insect Control.* IAEA/STI/PUB/595, Vienna, pp 267–278.
42. Graham, A.J. & F.H. Dudley. 1959. Culture methods for mass rearing of screwworm larvae. *J. Econ. Entomol.* 52:1006–1008.
43. Graham, O.H. 1985. Eradication of the screwworm from the United States and Mexico. *Entomology Society of America Miscellaneous Publication* No. 62.
44. Hamm, J.J., R.L. Burton, J.R. Young, & R.T. Daniel. 1971. Elimination of *Nosema heliothidis* from a laboratory colony of the corn earworm. *Ann. Entomol. Soc. Am.* 64:624–627.
45. Heilmann, L.D. DeVault, R.L. Leopold & S.K. Narang. 1994. Improvement of natural enemies for biological control: a genetic engineering approach. In *Applications of Genetics to Arthropods of Biological Control Significance.* S.K. Narang, A.C. Bartlett, & R.M. Faust, eds. Boca Raton, Florida: CRC Press, pp 167–189.
46. Hendrichs, J., G. Ortiz, P. Liedo & A. Schwarz. 1983. Six years of successful medfly program in Mexico and Guatemala. In *Fruit Flies of Economic Importance.* R. Cavalloro, ed. Proceedings of an International Symposium organized by Greek Authorities/CEC/IOBC, Athens. Rotterdam: Balkema, pp 353–365.
47. Hentze F. & R. Mata. 1986. Mediterranean fruit fly eradication programme in Guatemala. In *Fruit Flies.* A.P. Economopoulos, ed. Proceedings of the II International Symposium, Crete. Amsterdam: Elsevier, pp 533–539.
48. Hoy, M.A. 1976. Genetic improvement of insects: fact or fantasy. *Environ. Entomol.* 5:833–839.
49. Hoy, M.A. 1990. Genetic improvement of arthropod natural enemies: becoming a conventional tactic? In *New Directions in Biological Control: Alternatives for*

Suppressing Agricultural Pests and Diseases. R. Baker & P. Dunn, eds. New York: Liss, pp 405–418.

50. Hoy, M.A. 1990. Pesticide resistance in arthropod natural enemies: variability and selection responses. In *Pesticide Resistance in Arthropods*. R.T. Roush & B.E. Tabashnik, eds. New York: Chapman & Hall, pp 203–236.

51. Huettel, M.D. 1976. Monitoring the quality of laboratory-reared insects: a biological and behavioral perspective. *Environ. Entomol.* 5:807–814.

52. Irastorza, J.M., C. Bajatta, J. Ortega, & S.J. Martinez. 1993. Erradicacion del gusano barrenador del nuevo mundo. *Proceedings on Management of Insect Pests: Nuclear and Related Molecular and Genetic Techniques*. IAEA/FAO, Vienna, pp 313–318.

53. Johnson, M.W. & B.E. Tabashnik. 1994. Laboratory selection for pesticide resistance in natural enemies. In *Applications of Genetics to Arthropods of Biological Control Significance*. S.K. Narang, A.C. Bartlett, & R.M. Faust, eds. Boca Raton, Florida: CRC Press, pp 91–105.

54. King, E.G., & N.D. Leppla, eds. 1984. *Advances and Challenges in Insect Rearing*. New Orleans: Agricultural Research Service (Southern Region), U.S. Department of Agriculture.

55. Knipling, E.F. 1955. Possibilities of insect control or eradication through the use of sexually sterile males. *J. Econ. Entomol.* 48:459–462.

56. Knipling, E.F. 1970. Suppression of pest Lepidoptera by releasing partially sterile males: a theoretical appraisal. *BioScience* 20:465–470.

57. Knipling. E.F. 1979. The *Basic Principles of Insect Population Suppression and Management*. Washington, DC: U.S. Department of Agriculture Agriculture Handbook No. 512.

58. Knipling, E.F. 1982. Present status and future trends of the SIT approach to the control of arthropod pests. In *Steile Insect Technique and Radiation in Insect Control*. IAEA/STI/PUB/595, International Atomic Energy Agency, Vienna, pp 3–23.

59. Knipling, E.F. 1985. Sterile insect technique as a screwworm control measure: the concept and its development. In *Symposium on Eradication of the Screwworm from the United States and Mexico*, O.H. Graham, ed., *Misc. Publ. Entomol. Soc. Am.* No. 62, pp 4–7.

60. Knipling, E.F. 1992. *Principles of Insect Parasitism Analyzed from New Perspectives: Practical Implications for Regulating Insect Populations by Biological Means*. U.S. Department of Agriculture, Agriculture Handbook No. 693.

61. LaBrecque, G.C. 1963. Chemosterilants for the control of houseflies, In *New Approaches to Pest Control and Eradication*. S.A. Hall, ed. Washington, DC: American Chemical Society, pp 42–46.

62. LaChance, L.E. 1985. Genetic methods for the control of lepidopteran species. USDA Agriculture Research Service ARS. 28.

63. LaChance, L.E., R.A. Bell, & R.D. Richard. 1973. Effect of low doses of gamma irradiation on reproduction of male pink bollworms and their F_1 progeny. *Environ. Entomol.* 2:653–658.

64. LaChance, L.E., & R.L. Ruud. 1979. Interstrain and interspecific crosses between *Pectinophora gossypiella* and *P. scutigera*. *J. Econ. Entomol.* 72:618–620.

65. Laster, M.L. 1972. Interspecific hybridization of *Heliothis virescens* and *H. subflexa*. *Environ. Entomol.* 16:682–687.
66. Laster, M.L. 1979. Coping with the tobacco budworm/bollworm problem: status of hybrid insect approach. *Proceedings of the Beltwide Cotton Production Conference.* National Cotton Council: Memphis, pp. 41–42.
67. Laster, M.L., C.E. Goodpasture, E.G. King, & P. Twine. 1985. Results from crossing the bollworms, *Helicoverpa armigera x H. zea*, in search of backcross sterility. *Proceedings of the Beltwide Cotton Production Research Conference.* National Cotton Council: Memphis, pp 146–147.
68. Laster, M.L., & D.D. Hardee. 1995. Intermating compatibility between the North American *Helicoverpa zea* and *H. armigera* (Lepidoptera: Noctuidae) from Russia. *J. Econ. Entomol.* 88:77–80.
69. Laster, M.L., D.D. Hardee, & J.C. Schneider. 1996. *Heliothis virescens* (Lepidoptera: Noctuidae): Influence of sterile backcross releases on suppression. *Southwest Entomol.* 21:433–444.
70. Laster, M.L., E.G. King, N.R. Spencer, & R.E. Furr, Jr. 1995. Response to hybridization of the North American *Helicoverpa zea* and *H. assulta* (Lepidoptera: Noctuidae) from Pakistan and Thailand. *Trends Agric. Sci.* (Entomology) 2:135–140.
71. Laster, M.L. & C.F. Sheng. 1995. A search for hybrid sterility for *Helicoverpa zea* in crosses between the North American *H. zea* and *H. armigera* (Lepidoptera: Noctuidae) from China. *J. Econ. Entomol.* 88:1288–1291.
72. Laster, M.L., N.R. Spencer, E.G. King & P. Twine. 1987. Current status of crossing exotic *Heliothis* spp with *H. zea* in search of hybrid sterility. *Proceedings of the Beltwide Cotton Production Research Conference.* National Cotton Council: Memphis, pp 315–316.
73. Laven, H. 1967. Eradication of *Culex pipiens fatigans* through cytoplasmic incompatibility. *Nature* 216:383–384.
74. Leppla, N.C. & T.R. Ashley, eds. 1978. Facilities for insect research and production. USDA *Agric. Tech. Bull.* 1576.
75. Leppla, N.C., W.R. Fisher, J.R. Rye, & C.W. Green. 1982. Lepidopteran mass-rearing. *Proceedings of the Sterile Insect Technique and and Radiation in Insect Control.* IAEA, Vienna, pp 122–133.
76. Lindquist, A.W. 1955. The use of gamma irradiation for control or eradication of the screw-worm. *Econ. Entomol.* 48:467–469.
77. Lindquist, D.A., M. Abusowa, & W. Klassen. 1993. Eradication of the new world screwworm from the Libyan Arab Jamahiriya. *Proceedings of the Management of Insect Pests: Nuclear and Related Molecular and Genetic Techniques.* IAEA/FAO, Vienna, pp 319–330.
78. Lindquist, D.A., B. Butt, U. Feldmann, R.E. Gingrich, & A. Economopoulos. 1990. Current Status and Future Prospects for Genetic Methods of Insect Control or Eradication. In *Pesticides and Alternatives: Innovative Chemical and Biological Approaches to Pest Control.* J.E. Casida, ed. Amsterdam, Elsevier, pp 69–88.
79. Lyon, J.D.J. de B. 1994. Integrated pest management in cotton. In *Challenging*

the Future: Procedings of the World Cotton Research Conference-1. G.A. Constable & N.W. Forrester, eds. CSIRO, Melbourne, pp 456–465.

80. Mannion, C.M., J.E. Carpenter, & H.R. Gross. 1994. Potential of the combined use of inherited sterility and a parasitoid, *Archytas marmoratus* (Diptera: Tachinidae), for managing *Helicoverpa zea* (Lepidoptera: Noctuidae). *Environ. Entomol.* 23:41–46.

81. Marec, F. 1 991. Genetic control of pest Lepidoptera: construction of a balanced lethal strain in *Ephestia kuehniella. Entomol. Exp. Appl.* 61:271–283.

82. McDonald, F. J. 1974. The future of chemical sterilants in entomology, their use against insect populations in the field and their mammalian toxicity and hazards. In *The Use of Genetics in Insect Control.* R. Pal & M.J. Whitten, eds. North Holland: Elsevier, pp 225–238.

83. McDonald, I.E. 1976. Ecological genetics and the sampling of insect populations for laboratory colonization. *Environ. Entomol.* 5:815–820.

84. Melvin, R., & R.C. Bushland. 1936. A method of rearing *Cochliomyia americana* C. and P. on artificial media. *USDA Bur. Entomol. Plant Q. Rep.* ET-88.

85. Muller, H.J. 1927. Artificial transmutation of the gene. *Science.* 66:84–87. (Reprinted in H.J. Muller. 1962. *Studies in Genetics. The Selected Papers of H.J. Muller,* pp 245–251. Indiana University Press: Bloomington.)

86. Muller, H.J. 1927. The problem of genic modification. *Verhandl. V. Internatl. Kongr. Vererbungswiss.,* Berlin, pp. 234–260. (Reprinted in, H.J. Muller. 1962. *Studies in Genetics. The Selected Papers of H.J. Muller,* pp 252–276. Indiana University Press: Bloomington.)

87. Muller, H.J. 1950. Radiation damage to the genetic material (Part I). *Am. Scientist* 38:35–59.

88. Muller, H.J. 1950. Radiation damage to the genetic material (Part II). *Am. Scientist* 38:399–425.

89. Muller, H.J. 1962. Explanatory note no. 1. *Studies in Genetics. The Selected Papers of H.J. Muller,* Indiana University Press: Bloomington, pp 3–6.

90. Narang, S.K., A.C. Bartlett & R.M. Faust, eds. 1994. *Applications of Genetics to Arthropods of Biological Control Significance.* Boca Raton, Florida: CRC Press.

91. North, D.T. 1975. Inherited sterility in Lepidoptera. *Annu. Rev. Entomol.* 22:167–182.

92. Ortiz, G., P. Liedo, A. Schwartz, A. Villasenor, & J. Reyes. 1986. Mediterranean fruit fly *(Ceratitis capitata)*: status of the eradication programme in southern Mexico and Guatemala. *II International Symposium on Fruit Flies.* Crete, pp 523–532.

93. Presnail, J.K. & M.A. Hoy. 1992. Stable genetic transformation of a beneficial arthropod, *Metaseiulus occidentalis* (Acari: Phytoseiidae), by a microinjection technique. *Proc. Natl. Acad. Sci. USA* 89:7732.

94. Proshold, F.I. 1983. Release of backcross insects on St. Croix, U.S. Virgin Islands, to suppress the tobacco budworm (Lepidoptera: Noctuidae): infusion of sterility into a native population. *J. Econ. Entomol.* 76: 1353–1359.

95. Proverbs, M.D., & J.R. Newton. 1962. Some effects of gamma radiation on the reproductive potential of the codling moth, *Carpocapsa pomonella* (L.) (Lepidoptera: Olethreutidae). *Can. Entomol.* 94:1162–1170.

96. Raina, A.K., Bell, R.A., & Klassen, W. 1981. Diapause in the pink bollworm: preliminary genetic analysis. *Insect Sci. Appl.* 1:231–235.

97. Reyes, J., A. Villasenor, A. Schwartz, & J, Hendrichs. 1988. *Proceedings of the International Symposium on Modern Insect Control: Nuclear Techniques and Technology.* IAEA/FAO, Vienna, pp. 107–116.

98. Riemann, J.G. & M.F. Hollis. 1967. Irradiation effects on midguts and testes of the adult boll weevil, *Anthonomus grandis*, determined by histological and shielding studies. *Ann. Entomol. Soc. Am.* 60:298–308.

99. Rössler, Y. 1979. "69-apricot" a synthetic strain of the Mediterranean fruit fly, *Ceratitis capitata* (Diptera: Tephritidae) with sex-limited pupal color and eye color marker. *Entomophaga* 25:275–281.

100. Rössler, Y. 1979. Automatic sexing of the Mediterranean fruit fly: the development of strains with inherited, sex-limited pupal colour dimorphism. *Entomophaga* 24:411 416.

101. Rössler, Y. 1988. Selection for resistance in the Mediterranean fruit fly for genetic sexing in sterile insect technique programmes. *Proceedings of the Modern Insect Control: Nuclear Techniques and Biotechnology.* IAEA/STI/PUB/763, IAEA, Vienna, pp 229–240.

102. Rubin, G.M., & A.C. Spradling. 1982. Genetic transformation of *Drosophila* with transposable elements. *Science* 218:348.

103. Runner, G.A. 1916. Effect of Roentgen rays on the tobacco, or cigarette beetle and the results of experiments with a new form of Roentgen tube. *J. Agric. Res.* 6:383–388.

104. Saul, S.H. 1984. Genetic sexing in the Mediterranean fruit fly, *Ceratitis capitata* (Wied.) (Diptera:Tephritidae): conditional lethal translocations that preferentially eliminate females. *Ann. Entomol. Soc. Am.* 77:280–283.

105. Seawright, J.A., B.K. Birky, & B.J. Smittle. 1986. Use of a genetic technique for separating the sexes of the stable fly (Diptera:Muscidae). *J. Econ. Entomol.* 79:1413.

106. Seawright, J.A., P.E. Kaiser, S.G. Suguna, & D.A. Focks. 1981. Genetic sexing strains of *Anopheles albimanus*. *Mosq. News.* 41:107.

107. Serebrovsky, A.S. 1940. On the possibility of a new method for the control of insect pests. *Zool. Zh.* 19:618–630. (English translation in *Sterile-Males Technique for Eradication of Harmful Insects*. 1969. IAEA, Vienna, pp 123–127.

108. Singh, P. 1977. *Artificial Diets for Insects, Mites and Spiders*. New York: Plenum Press.

109. Smith, C.N. 1963. Chemosterilants as a potential weapon for insect control. In *New Approaches to Pest Control and Eradication*. S.A. Hall, ed. Washington, DC: American Chemical Society, pp 36–41.

110. Snow, J.W., R.L. Jones, D.T. North, & G.G. Holt. 1972. Effects of irradiation on ability of adult male corn earworms to transfer sperm, and field attractiveness of females mated to irradiated males. *J. Econ. Entomol.* 65:906–908.

111. Staten, R.T., R.W. Rosander, & D.F. Keaveny. 1993. Genetic control of cotton insects. *Proceedings of the Management of Insect Pests: Nuclear and Related Molecular and Genetic Techniques*. IAEA/FAO, Vienna, pp 49–60.

112. Steiner, L.F., W.G. Haart, E.J. Harris, R.T. Cunningham, K.Ohinata, & D.C.

Kamakahi. 1970. Eradication of the oriental fruit fly from the Mariana Islands by the method of male annihilation and sterile insect release. *J. Econ. Entomol.* 63:131–135.

113. Stevenson, J.O., & E. H. Smith. 1961. Fecundity and fertility of two strains of the plum curculio, reciprocal crosses and the F$_1$ generations. *J. Econ. Entomol.* 54:283–284.

114. Strunnikov, V.A. 1979. On the prospects of using balanced sex-linked lethals for insect pest control. *Theoret. Appl. Genet.* 55:17–21.

115. Tillman, P.G., & M.L. Laster. 1995. Parasitization of *Heliothis virescens* and *H. virescens–H. subflexa* backcross (Lepidoptera: Noctuidae) by *Microplitis croceipes* (Hymenoptera: Braconidae). *Biol. Control* 24:409–411.

116. Vale, G.A. 1982. Prospects for using stationary baits to control and study populations of tsetse flies (Diptera: Gollinidae) in Zimbabwe. *Proceedings of the Sterile Insect Technique and Radiation in Insect Control*, IAEA/STI/PUB/595. IAEA, Vienna, pp 191–203.

117. Wang, H. & Zhang H. 1993. Control of the Chinese citrus fly *Dacus citri* (Chen), using the sterile insect technique. *Proceedings of the Management of Insect Pests: Nuclear and Related Molecular and Genetic Techniques*. IAEA/FAO. Vienna, pp 505–512.

118. Warren, A.M. & J.M. Crampton. 1994. Mariner—its prospects as a DNA vector for the genetic manipulation of medically important insects. *Parasitol. Today*. 10:59–63.

119. Waterhouse, D.F., L.E. LaChance, & M.J. Whitten. 1976. Use of autocidal methods. In *Theory and Practice of Biological Control*. C.B. Huffaker & P.S. Messenger, eds. New York: Academic Press, pp 637–659.

120. Whitten, C.J. 1969. Automated sexing of pupae and its usefulness in control by sterile insects. *J. Econ. Entomol.* 62:272–273.

121. Whitten, C.J. 1980. Use of the isozyme technique to assess the quality of mass-reared sterile screwworm flies. *Ann. Entomol. Soc. Am.* 73:7–10.

122. Whitten, C.J. 1982. The sterile insect technique in the control of the screwworm. *Proceedings of the Sterile Insect Technique and Radiation in Insect Control*. IAEA, Vienna, pp 79–84.

123. Whitten, M.J. 1985. Conceptual basis for genetic control. In *Comprehensive Insect Physiology, Biochemistry and Pharmacology*. G.A. Kerkut & L.I. Gilbert, eds. Oxford, UK: Pergamon Press, pp 465–528.

124. Wong, T.T.Y., M.M. Ramadan, J.C. Herr, & D.O. McInnis. 1992. Suppression of a mediterranean fruit fly (Diptera: Tephritidae) population with concurrent parasitoid and sterile fly releases in Kula, Maui, Hawaii. *J. Econ. Entomol.* 85:1671–1681.

125. Yamagishi, M., H. Kakinohana, H. Kuba, T. Kohama, Y. Nakamoto, Y. Sokei & K. Kinjo. 1993. Eradication of the melon fly from Okinawa, Japan, by means of the sterile insect technique. *Proceedings of the Management of Insect Pests: Nuclear and Related Molecular and Genetic Techniques*. IAEA/FAO, Vienna, pp 49–60.

126. Young, J.R., J.J. Hamm, R.L. Jones, W.D. Perkins, & R.L. Burton. 1976. *Development and Maintenance of an Improved Laboratory Colony of Corn Earworms.* USDA ARS, ARS-S-110:1–4.
127. Young, J.R., J.W. Snow, J.J. Hamm, W.D. Perkins, & D.G. Haile. 1975. Increasing the competitiveness of laboratory-reared corn earworm by incorporation of indigenous moths from the area of sterile release. *Ann. Entomol. Soc. Am.* 68:40–42.

120.

121.

19

Behavior-Modifying Chemicals in Management of Arthropod Pests

Alan L. Knight and Thomas J. Weissling*
United States Department of Agriculture, Yakima, Washington

I. INTRODUCTION

Arthropods live in a world of complex chemical stimuli affecting every aspect of their lives. Chemical messages, or *semiochemicals* [86], closely regulate interactions of arthropods with plants and other arthropods. Behavioral responses of arthropods to semiochemicals can be broadly classified as attraction, arrestment, repulsion, stimulation, or deterrence [42]. Chemical signals comprise single- or multiple-component compounds produced by one species that modify the behavior of perceiving organisms of the same species (*pheromones*) or different species (*allelochemics*) [109,110]. Pheromones affect behaviors associated with mating, aggregation, and dispersal (Table 1). Allelochemics are categorized according to the benefit of the message to the receiver and sender, respectively, as: *allomones* (-,+), *kairomones* (+,-), and *synomones* (+,+) (Table 1). In addition, the male lures of the tephritid flies are another important category of semiochemicals used in pest management (Table 1).

The foundation of chemical ecology has been built from basic studies of arthropod olfaction, physiology and biochemistry, behavior, and ecology, as is apparent in numerous recent reviews [1,7,9,10,115,120,122,135). Detailed reviews on aspects of the chemical ecology of several major agriculturally important arthropod groups have also appeared: forest Coleoptera [162], tephritid

---

Current affiliation: University of Florida, Ft. Lauderdale, Florida.

Table 1 Categories of Semiochemicals Used in the Management of Arthropod Pests

Category	Used in associated behavior	Manipulation of
Pheromone		
Sex	Mating	Lepidoptera
Aggregation	Mass attack of host	Bark beetles
Alarm	Predator avoidance	Aphids
Epideictic	Regulate host exploitation	Bark beetles
Allelochemic		
Allomone	Oviposition and feeding deterrents	General
Kairomone	Host selection, feeding stimulants	Parasitoids, *Diabrotica*
Synomone	Pollination, attraction of natural enemies	Bees, Parasitoids
Other		
Male lure	Male attraction	Tephritid flies

flies [131], Lepidoptera [97], mites [146], aphids [118], Heteroptera [2], bees [15], and parasitic Hymenoptera [88]. Since the identification of the first insect sex pheromone, bombykol, in 1959 [23] improvement of analytical tools [68] has led to the commercial availability of hundreds of pheromones and allelochemics for pest management [70]. Although this rapid development of chemical ecology has not led to the creation of a "magic, semiochemical bullet" [17] for managing pest and natural enemies, it has produced a steady acquisition of knowledge of arthropod behavior and greater sophistication of tactics that use semiochemicals for pest management. Although the following supposition of Shorey [142] has not been proven, it likely remains true, and it clearly provides a goal for future applied studies in chemical ecology: "Much of the sensory world of the insect involved in stimulation or inhibition of such behaviors as mating, feeding, and egg-laying is chemical. The reactions of the insects to these chemicals are so predictable that if man could learn enough about the attendant behaviors, he could literally make the insects jump through a hoop" [142].

In this chapter, we present a summary that demonstrates the importance of semiochemicals in contemporary pest management programs. We follow this broad review with specific examples of eight case histories where the use of semiochemicals has contributed extensively to the management of pest species. In concluding, we emphasize our belief that semiochemicals will play an increased role in pest management and outline some of the constraints affecting their development.

II. USE OF SEMIOCHEMICALS IN PEST MANAGEMENT

Management of pests in the latter half of the twentieth century has been achieved largely by the widespread application of pesticides [99]. However, recognition of the impacts of pesticides on the stability of the agroecosystem, spillover effects outside of agriculture, and the continuing spin of the pesticide treadmill have been some of the forces driving the development of integrated pest management (IPM) programs [28]. The importance of semiochemicals in pest management programs during the past 35 years has been covered in numerous reviews [12,14,25,60,75,83,101,104a,111,124,130,143,144,165].

Incorporation of semiochemicals into IPM programs has allowed major restructuring of the established pesticide-based control programs in a number of crop systems, such as tomatoes [72] and cotton [90]. Semiochemicals have several advantages compared with pesticides: (a) they are generally species specific, (b) they tend not to affect secondary pests or natural enemy populations, and (c) they have low mammalian toxicity [145]. The use of semiochemicals directly to control pest species or to improve monitoring has resulted in a significant decrease in pesticide use [25,82,133]. In addition, semiochemicals serve as important tools in studies of an arthropod's ecology, population dynamics, and behavior. Such studies provide grist for development of new approaches for pest control.

There are several ways that behavioral responses have been developed to aid pest management (Table 2). These are addressed below.

A. Monitoring

The use of pheromones or kairomones to attract insects to a trap is probably the most exploited behavioral response in pest management [25]. Attractants are

Table 2 Selective Examples of How Arthropod Semiochemicals Have Been Used in Pest Management

1. Sex pheromones and kairomones for monitoring populations
   A. Detection
   B. Estimating density
   C. Determining phenology
   D. Surveying insecticide resistance
2. Sex pheromone–mediated mating disruption
3. Kairomone and pheromone-based mass trapping
4. Allelochemic augmentation of natural enemies
5. Antiaggregation epideictic pheromones
6. Pheromones to improve pollination
7. Alarm pheromones to disrupt populations
8. Plant allomone feeding deterrents

often highly selective and inexpensive, making them useful tools for pest detection, establishing pest thresholds, for determining the appropriate timing for pesticide applications, and monitoring insecticide resistance. The designs of the lure and the trap must both be crafted to meet the objective of monitoring. For example, the attractiveness of a lure can be adjusted by varying the release rate and composition of attractant [85]. The reliability of interpreting catch data is largely dependent on chemical purity and release rate, but trap density, placement, and maintenance are also important considerations [128].

1. Detection of the Pest

Semiochemicals are widely used by government and commodity groups to detect new pest infestations. There are presently three primary uses of survey traps: (a) trap grids are used to detect introductions of chronic pests such as fruit flies and gypsy moth, and they are maintained at high-risk areas such as urban centers, airports, and harbors; (b) a Cooperative Agricultural Pest Survey program has been established in the United States between federal and individual state agencies to detect the introduction of new exotic pests [74]; and (c) monitoring programs have been established to demonstrate pest-free production and storage zones as a prerequisite for exporting various agricultural commodities to specific countries [129].

2. Assessment of Pest Density

Traps baited with sex pheromones are commonly used to monitor the population density of pests. Traps baited with the sex pheromones of lepidopteran species generally catch only males, and establishing high correlations between moth catch and the population density of other life stages or crop injury has proven to be difficult for most species [79]. Yet, traps have been very useful for setting conservative action thresholds for pesticide use [78,91]. Trap grids have also been used in areawide studies of pest distribution and dispersal [77].

3. Assessment of Pest Phenology

Combining data from sex pheromone–baited traps with predictive phenology models has been widely adopted in the management of some pests. First moth catch (referred to as a "biofix") has been used as a predictor of the beginning of adult emergence [126]. Traps have generally been less effective in predicting peak emergence and the emergence of later generations probably because of trap saturation, pesticide use, female moth–trap competition, and the variability associated with the degree of trap and lure maintenance.

4. Assessment of Insecticide Resistance

Pheromone-baited traps have been used to monitor insecticide resistance in several lepidopteran species [20,58,127]. This use of sex pheromone–baited traps allows rapid collection and determination of resistance for large samples of the pest without having to rear the pest.

B. Mating Disruption with Sex Pheromones

The use of sex pheromones to disrupt mating was first attempted for *Trichoplusia ni* [50], and this has now become an important control method for a number of lepidopteran species [27]. Sex pheromones may disrupt mating through one or more of several mechanisms: adaptation and habituation of the nervous system to the pheromone, false trail following, camouflage of an emitting female's chemical signal, and by creating an imbalance in the male's sensory reception [26]. More than a single mechanism may operate in the same system. For example, attempts to the disrupt pink bollworm, *Pectinophora gossypiella*, with a number of dispenser systems and pheromone components have likely involved all of these mechanisms [5].

There are many factors that affect the development of a successful mating disruption program [25]. Of primary importance is the correct identification of the compound and determination of its physiological and behavioral effects on the species [133]. The sex pheromone of most species comprises a blend of components [4]. Each component of the blend can have a specific role in mate location and sexual behavior [89,106]. Also basic to the use of mating disruption is knowledge of the distribution of pheromone in the air, as well as the moth's response to the pheromone's plume structure [106]. Recent progress [139,156] in measuring the concentrations of pheromone in the air will advance our understanding of these factors.

Optimizing the use of mating disruption is dependent on a variety of factors. The supposition that the natural pheromone would be the best disruptant has not been sufficiently tested or substantiated [8,104]. Ecological factors such as host specificity, reproductive characteristics, and dispersal characteristics also affect the success of mating disruption. Pests with a limited host range, low reproductive potential, and limited dispersal capabilities should be the best candidates [25]. In addition, the importance of the pest within the overall management program, the level of disruption needed, and the impact of changing from a primarily insecticide-based to a pheromone-based management program on secondary pests and their natural enemies will all determine how widely this technology will become adopted.

C. Mass Trapping with Attractants

The utility of sex and aggregation pheromones, kairomones, and male lures for mass trapping of pests has been reviewed [25,85]. Traps or targets may employ semiochemicals in combination with toxicants, sterilants, or pathogens to reduce pest densities [143]. Trapping is generally directed at the adult stage, and it may target either or both sexes. Mass trapping is considered to be most effective for pests which are geographically isolated and/or at low densities [85]. Mass trapping is also often appropriate in urban areas where the use of insecticides is unacceptable [84].

Management of pest populations by mass trapping has varied both in approach and success. Many attempts directed at lepidopteran pests have been unsuccessful because of the requirement of an initial low pest population density and the cost of maintaining traps [66,92]. In general, mass trapping tends to be ineffective in reducing pest population densities if only males are removed [80]. In contrast, mass trapping of the Japanese beetle, *Popillia japonica*, has been effective, because the lure is a combination of a kairomone plus the sex pheromone; thus the trap attracts both sexes [84]. Trap cropping has been enhanced with the use of semiochemical baits [64]. This approach was used against the boll weevil, *Anthonomus grandis*, in cotton in the 1970s with some success [56]. However, more recently, bait sticks treated with the male-produced pheromone, grandlure, plus insecticides have been used for mass trapping boll weevils [96]. Trees baited with aggregation pheromones are used in the management of some bark beetle species [17]. Point sources baited with pheromone plus insecticide have been used to combine mating disruption with mass trapping [57]. The effectiveness of this approach is likely achieved by a combination of disruption and both lethal and sublethal effects from pesticide exposure [57].

D. Augmentation of Natural Enemies

Chemical information provided by the arthropod host (kairomones) and the plant (allomones) play important roles in both the long-range and short-range location of hosts by parasitoids [43,157]. Successful parasitism involves a series of interactions often mediated by allelochemics, including host habitat location, host location, and host acceptance [160]. Host kairomones can be found in moth scales, host feces, silk, cuticular components, gland secretions, and host sex pheromones [157]. Volatiles may also be released by the plant in response to herbivore damage or be released by the plant in response to the presence of the herbivore ("herbivore-induced synomone" [44]).

Semiochemicals may be useful in pest management programs for manipulating densities of the natural enemy or by manipulating the source of the chemical [54,161]. The criteria of how a natural enemy responds to a host's kairomones or plant allomones could be used to preselect promising species for introduction in classic biological control [71,108]. Likewise, differences among strains of parasitoids in their use of semiochemicals in specific cropping systems could be considered [114]. The most commonly used manipulation of semiochemicals in augmenting biological control is through exposing the parasitoid to allelochemics prior to release [158]. Semiochemicals which are attractive to or which arrest the parasitoid could be sprayed on the crop to maintain or attract the parasitoid in the crop [32,54]. This must be applied in a manner, however, that does not confuse or disrupt the functioning of the natural enemies. At present, the use of semiochemicals has played a small role in the augmentation of parasitoids, probably due to the chemical complexity and our rudimentary knowledge of these multitrophic interactions [88].

E. Epideictic Pheromones

Semiochemicals that influence the distribution of insects are produced by a broad range of insect taxa [121]. Epideictic pheromones reduce intraspecific competition by disrupting landing, feeding, or oviposition of pests on their host plants [121]. The epideictic pheromones of the bark beetles, *Dendroctonus* spp., such as verbenone for *D. frontalis* and 3,2-methylcyclohexenone (MCH) for *D. pseudotsugae*, are the most well known. These pheromones are released by beetles that have successfully colonized a host; in combination with aggregation pheromones, these compounds serve to keep beetle densities within an optimal range (six to eight per 1000 cm^2 of bark surface [121]).

Oviposition deterrents have been reported for a number of tephritid flies [121]. The identification of the epideictic pheromone for the European cherry fruit fly, *Rhagoletis cerasi*, has led to small, successful field tests of its use in managing this pest [3]. The possibility that the pest might habituate to the deterrent requires that this strategy be combined with an attractive bait [121].

F. Pollination

Pollination by the honeybee, *Apis mellifera*, is of tremendous importance in the production of numerous high-value crops [94]. However, in a number of crops, blossoms may be either unattractive or less attractive than other noneconomic species (e.g., dandelion, *Taraxacum officinalis*), and this problem can lead to poor pollination of the crop. In addition, poor weather during the bloom period may adversely affect pollinators.

Pheromones isolated from honeybees have been used to improve pollination. Honeybee queens produce a five-component pheromone blend from their mandibular gland (queen mandibular pheromone [QMP]) which has a number of roles in the life cycle of the hive [166]. When applied at 1000 queen equivalents per hectare in several fruit crops, bee foraging and fruit set and size increased [38,39]. The largest improvement in yields occurred when conditions were otherwise poor for pollination [39]. QMP is also used by beekeepers in a variety of hive-management chores such as shipping colonies [107] and queen management [117]. Another pheromone released from the bee's Nasanov gland affects orientation of worker bees and is used to improve bee foraging [93].

G. Alarm Pheromone

Many aphid species release an alarm pheromone when attacked by predators or parasitoids which elicits defensive or avoidance behaviors [118]. Early uses of (E)-β-farnesene were unsuccessful in managing aphid populations [24,62]. However, the use of the alarm pheromone has improved pest control in other studies by increasing the aphids' contact with insecticides [53] or fungal pathogens [63]. Increased residual activity for control of an aphid-vectored disease was obtained by directly applying chemical derivatives instead of β-farnesene [40]. Further studies of aphid chemical ecology may contribute to new approaches for using aphid pheromones in pest management [41].

H. Feeding Deterrents

A large variety of plant allomones protect plants from herbivory, and these compounds offer an opportunity for new approaches in pest management [11,73]. In contrast, the potential of defensive allomones of arthropods has received little attention [16]. The success of pest control tactics using these types of chemicals would likely be improved with compounds that produce both behavioral and physiological effects [73]. For example, most recent work in pest management using plant allomones has been with azadirachtin and related analogues isolated from the fruits of the neem tree, *Azadirachta indica* [140]. Azadirachtin has a number of deterrent effects on the behavior of arthropod species, modifying, for instance, oviposition and feeding [140]. In addition, azadirachtin may cause physiological effects, including the disruption of ecdysone biosynthesis, reduced fecundity, and decreases in egg hatch. These physiological effects occur at lower concentrations than those necessary for deterrent effects.

Deterrent compounds generally have short residual activity and require multiple applications to provide adequate control. This factor combined with the expense of producing plant-derived chemicals makes their use expensive. Deterrents that can function systemically, such as azadirachtin, would have improved persistence and coverage [140].

III. SELECTED CASE HISTORIES

A. Fruit Flies

Chronic introductions of tropical and subtropical tephritid fruit flies into quarantine areas threaten agricultural production in the United States in several ways: via infestation of crops, reduced acceptability of crops for exporting, and disruption of IPM programs [37]. Early detection of pest introductions using traps baited with male lures allows quick eradication of incipient populations [37]. New infestations of the oriental fruit fly, *Bactrocera dorsalis*, in California have been eradicated 12 times since 1966 using methyleugenol in an insecticide-laced bait [101]. Trimedlure was developed empirically for monitoring the Mediterranean fruit fly (medfly), *Ceratitis capitata*, and has been used in traps to pinpoint the location and size of infestation sites [36,37]. Medflies can then be eradicated through the use of a sprayable formulation of protein hydrolysate and insecticide [134].

The standard monitoring program for the medfly in California is five traps baited with trimedlure per square mile in urban areas [116]. After flies are detected an additional 1700 traps are baited in a 4.5-mile radius for three generations. In Los Angeles from 1986 to 1988, releases of sterile flies in combination with helicopter spraying of malathion mixed with protein hydrolysate was successful in eradicating infestations of the medfly, but this became very unpopular because of perceived health and environmental risks [116]. Since 1989, the insecticide bait has only been used in California to eradicate infestations of the medfly outside the Los Angeles basin.

B. Gypsy Moth

Introduced from Europe in 1869, the gypsy moth, *Lymantria dispar*, is one of most devastating pests of forest and shade trees in the eastern United States. Gypsy moths have been managed in the United States primarily through aerial application of insecticides to areas that are considered to be of high monetary value [81]. The identification of a female-produced sex pheromone (disparlure [13]) has led to great advances in the management of the gypsy moth. Mating disruption programs were established on about 7000 ha from 1979 to 1987 in the eastern United States [81]. This approach has not been widely adopted, because mating disruption works best for low population densities and most control programs are targeted at outbreaks of high population density. However, in 1993, a federal program, "Slow the Spread," was initiated to determine the feasibility of slowing the rate of gypsy moth expansion by incorporating IPM strategies over a large geographical area. This program encompasses 3,000,000 ha in four states. Each year, about 13,000 ha are treated, and mating disruption is being used on 15% of this area [123].

The greatest use of disparlure has been in surveys to determine the need for management tactics. Trap-based surveys are used to determine whether the gypsy moth is present in previously uninvested areas. Despite the flightlessness of their females, the mobility of gypsy moths on the possessions of migrating humans is legendary. In the western United States, incipient infestations of gypsy moths were detected in the mid 1970s. In Washington state, gypsy moth adults were trapped in only a few sites, but by the early 1990s, 13–35 new outbreak sites were located each year [167]. These infestations coincided with the dramatic increase in human immigration into the Pacific Northwest from gypsy moth–infested states during a major pest outbreak period.

The standard protocol for monitoring gypsy moth in Washington state each year is placing about 10,000 traps at a density of one trap per square mile in all high-risk areas, which are defined as areas with high human population growth and suitable pest habitat [167]. When a moth is captured, a grid of traps is placed within a 3-mile radius around the catch site. If additional moths are caught and other life stages are found, then three sprays of the bacterium *Bacillus thuringiensis* are applied. The delimiting trap density serves to trap any remaining male moths and is maintained for an additional 2 years to ensure eradication. With this protocol, the gypsy moth has failed to become established in the western United States and British Columbia [167]. Since 1991, another race of *L. dispar*, the asian gypsy moth, has been repeatedly introduced into Washington state and British Columbia from Siberia. This race has a potentially larger host range, and the female can fly. Currently, the same monitoring and treatment program, but at a much larger scale, has been successful in preventing establishment of this pest [167].

C. Pink Bollworm

The introduction of the pink bollworm, *P. gossypiella*, to Arizona in the mid 1950s and to southern California in the mid 1960s drastically changed cotton production in these regions [22]. For the next 20 years, growers relied heavily on insecticides to maintain yields, but the development of insecticide resistance and continual outbreaks of secondary pests forced yearly reductions in cotton acreage.

A commercial formulation of the pink bollworm's sex pheromone, gossyplure, was the first product registered in the United States for the use in mating disruption [50]. Over the past 17 years, the use of mating disruption to control the pink bollworm has matured [5]. Currently, several products are available which are either sprayed or hand applied. Mating disruption for control of bollworms has become important in several cotton-growing areas. For example, about 42,000 and 125,000 ha were treated with pheromone in Arizona and Egypt in 1994, respectively [27].

Implementation of mating disruption for the pink bollworm has varied among geographical areas in the western United States. In the San Joaquin Valley, an eradication program for the pink bollworm has been in effect since 1968 using releases of sterile moths coupled with sex pheromone–baited traps for monitoring [61]. Mating disruption is used to supplement this program when larvae are found or if the program fails to achieve a 50:1 ratio of sterile to fertile males caught in pheromone-baited traps [51]. The pink bollworm became a major pest of cotton in the Imperial Valley of California in the mid 1960s despite applications of 10–15 insecticide sprays per season [5]. A grower-mandated mating disruption program in 1982 was successful but was discontinued [5]. In 1985, demonstration tests using a hand-applied dispenser were also successful, but again, this technology has not been widely adopted [147]. Continual immigration of pink bollworms into the Imperial Valley and severe outbreaks of whiteflies, *Bemisia* spp., have reduced cotton production and the adoption of mating disruption in this region [148]. Grower organizations in Arizona have widely adopted mating disruption to treat entire valleys. For example, growers in Parker Valley formed an informal pest control district in 1990 and adopted the use of mating disruption on over 12,000 ha of cotton [149]. The program has been very successful over the past 7 years, but problems controlling whiteflies have increased and a more expensive pheromone-insecticide program is now required.

D. Codling Moth

An important pest in tree fruits, the codling moth, *Cydia pomonella* L., has historically been controlled by several applications of broad-spectrum insecticides [65]. However, insecticide resistance [155] and the lack of effective alternative materials has prompted a search for new management approaches.

The major sex pheromone component of codling moth (codlemone [132]) has been used extensively for monitoring and for the timing of pesticide applications [159]. Codlemone was used in some of the earliest evaluations of mass trapping [92]. Mating disruption with codlemone has been tested in several apple-growing regions, including parts of Europe [30], Australia [137], and the United States [105]. The effectiveness of mating disruption in these trials depended on moth density, moth immigration, the amount of pheromone released, and the number and positioning of dispensers used per area. A three-component blend housed in a polyethylene dispenser was registered in the United States in 1991 [6] and was used to treat about 10,000 ha in 1994. Further adoption of mating disruption for codling moths will require a restructuring of pest management programs for a number of secondary pests in some fruits [21].

E. Tomato Pinworm

Fresh market tomatoes in Mexico, Florida, and California are attacked by a number of lepidopteran pests, including the tomato pinworm, *Keiferia lycopersicella* [154]. A number of characteristics of this pest have made it amenable to control with mating disruption: (a) unlike other lepidopteran pests of tomatoes, the host range of tomato pinworm is restricted to plants in the family

Solanaceae [112]; (b) insecticides have not been highly effective in controlling tomato pinworm, because it feeds inside the calyx of the fruit; and (3) overuse of insecticides has led to high levels of resistance to many compounds [20].

Tomato pinworm sex pheromone was isolated in 1979 [29], and since 1992 pheromone dispensers produced by several companies have been registered for mating disruption. The first large-scale tests (500 ha) were conducted in Culican, Mexico, in 1981 [72]. Results of this trial demonstrated that programs using mating disruption for pinworms used less insecticide, had reduced fruit injury, and were less expensive than conventional insecticide programs. These clear results have led to a broad adoption of a truly integrated pest management program in tomatoes using pheromones, the biopesticide *Bacillus thuringiensis*, biological control, and cultural practices [153].

F. Tsetse Flies

Tsetse flies, *Glossina* spp., are found only in Africa and are vectors of the parasite that causes sleeping sickness in humans and nagana in cattle [35]. The presence of these parasites makes large areas of land uninhabitable. Management of tsetse flies has traditionally been accomplished by clearing suitable habitat and by areawide insecticide applications. Early experimentation with semiochemicals used livestock treated with insecticides to attract and kill flies [55]. Specific attractants, such as 1-octen-3-ol, were isolated from cattle; and when combined with carbon dioxide and acetone, these compounds were similar to ox odor in attractiveness to flies. The addition of cattle urine further increased attraction. Today, a commonly used bait is a combination of acetone, 1-octen-3-ol, 4-methylphenol, and 3-propylphenol [35].

Baited traps are being used to monitor for the presence of the tsetse fly. Baited targets are also used in combination with insecticides to kill responding flies. In addition, odor-baited targets treated with a sterilant have been tested. The tsetse fly has been eradicated with this approach from a number of areas, but continual reintroductions of flies require that a yearly trapping program be maintained [151].

G. Corn Rootworms

The genus *Diabrotica* includes many economically important species. The western corn rootworm, *D. virgifera virgifera*, and the northern corn rootworm, *D. barberi*, are widely distributed pests of field corn in the midwestern United States [31]. The southern corn rootworm, *D. undecimpunctata howardi*, has a broader host range and is an economic pest of peanuts, sweet potatoes, and corn in the southeastern United States [69]. Root feeding by larvae of all three species causes reduced plant growth and yield. The traditional management tactic for *Diabrotica* in field corn has been broadcast insecticide applications in late summer to reduce oviposition and subsequent larval damage or soil insecticides applied at planting or first cultivation to reduce larval numbers in the root zone [87].

Two aspects of diabroticite chemical ecology make manipulation of their behavior feasible. First, the association between diabroticite beetles and a potent arrestant and feeding stimulant, cucurbitacin, found in the Cucurbitaceae [100] can be so strong that other behaviors such as sex attraction are suppressed [67]. Second, host selection by *Diabrotica* beetles appears to be mediated by several species-specific, plant-derived volatile attractants [101]. Powdered squash fruits and gourd roots have been tested in combination with insecticides as baits for improving the capture of beetles in monitoring programs [141]. In addition, palatable baits containing cucurbitacins and an insecticide have been useful in the short-term suppression of rootworm populations [102,164]. Suppression of the southern corn rootworm was enhanced by the addition of a volatile plant-derived attractant to insecticide-treated baits [103].

In 1993, the first insecticide-laced bait was registered for use against corn rootworms, and it was applied to about 12,000 ha in 1994. This approach reduced insecticide use in corn by 90–95% [98]. However, three factors may limit further adoption: (a) adult management is only feasible in corn-corn rotations, (b) adult management requires more intensive scouting and better timing than conventional programs; and (c) this approach is costly because of difficulties in extracting cucurbitacin from the buffalo gourd, *Cucurbita foetidissima*.

H. *Dendroctonus* Bark Beetles

Mass attacks of living trees by bark beetles of the genus *Dendroctonus* have caused catastrophic losses of timber [48]. Control of these pests has generally been by sanitation-salvage clearcuts. But this approach interferes with management of old growth forests, riparian zones, and recreational uses [136].

Pitman et al. [119] identified *trans*-verbenol, an oxidation product of the host tree compound α-pinene, as a female-produced aggregation pheromone for *Dendroctonus* spp. Maximum aggregation usually involves the addition of one or more host plant constituents to the pheromone. Semiochemicals are widely used to monitor and control bark beetles; two examples of their use are discussed here.

The attractiveness of the aggregation pheromone, *trans*-verbenol to male mountain pine beetles, *D. ponderosae,* is increased when it is combined with the host volatile, myrcene [18]. An additional compound, *exo*-brevicomin, is produced by males, and when this compound is combined with additional volatiles, it functions as an aggregation pheromone at low concentrations and as an antiaggregation pheromone at high concentrations [19]. A blend of these compounds has been marketed for use in monitoring and trapping mountain pine beetle [17]. This program has four operational components: (a) population monitoring, (b) containment and concentration of beetle populations, (c) postlogging mop-up, and (d) eradication of isolated infestations. Traps are used to establish restrictions on when logging can occur in specific areas. Baits are widely used in spot and grid management programs in mature lodgepole pine, *Pinus contorta*, forests in British Columbia [18]. The management of this program is unique in that personnel from the manufacturer of the bait and traps are hired [17].

D. *pseudotsugae* produces a multicomponent aggregation pheromone that is important in host colonization [138]. However, as the number of colonizing beetles increases, an antiaggregation pheromone, 3-methyl-2-cyclohexen-1-one (MCH), released by females reducing intraspecific competition [76]. Colonization of trees by D. *pseudotsugae* can be prevented by application of MCH to trees [47]. A plastic bead formulation containing MCH was developed and applied to plots by aircraft, and this approach greatly decreased attacks by D. *pseudotsugae* on susceptible trees [95]. However, concern about the impact of the plastic beads on wildlife has hindered its registration. Point source application of MCH for preventing attack of high-value trees is being evaluated and results have been encouraging [136].

IV. CURRENT CONSTRAINTS AND FUTURE EXPECTATIONS

Development of semiochemical-based techniques for pest management will continue to face a number of constraints. Developing long-lived controlled-release dispensers for chemicals that are often highly volatile and susceptible to degradation is difficult [163]. In addition, dispensers must release blends in proper ratios. Commercialization of semiochemical-based products is strongly affected by the size of the potential market, the cost of registration, and the product's price competitiveness. The lack of commercial interest by the major agrichemical firms has clearly hampered the development of semiochemical-based products. Commercial successes achieved with semiochemicals should energize the level of commitment by industry in developing new products.

In the United States, registration of semiochemical products has been severely constrained by the Environmental Protection Agency's (EPA) refusal to separate behavior-modifying chemicals from insecticides in their tier-testing registration process [152]. Until 1994, only chemicals used as attractants for survey and detection were excluded from this process. In 1994, the EPA ruled that all arthropod pheromones would be exempt from the requirement of establishing a tolerance if applied in a solid dispenser [46]. This ruling also allowed for testing of up 375 g of each active ingredient per hectare on up to 100 ha without destroying the crop [45]. This breakthrough in the unwieldy registration process [113] should greatly accelerate the process of product development and registration.

The wide-scale use of semiochemicals, such as sex pheromones for the pink bollworm and the tomato pinworm, raises the possibility that resistance may develop. Resistance could occur through elevated release of pheromone by females or changes in the ratio of the blend's components [59]. Laboratory studies with pink bollworm have shown the potential to select for both mechanisms [34]. Selection for changes in the behavior of pest species to attractants or deterrents may also lead to resistance [52].

A combination of factors has propelled pest management rapidly toward the incorporation of more sophisticated and more integrated tactics [125]. The pesticide legacy wherein pest managers were accustomed to immediate, inexpensive, and highly effective pest control is disappearing. Resistance of pests, even to new selective chemicals [33,150], the rising costs of registering new pesticides, and an escalating amount of litigation and lobbying on environmental and health issues have seriously constrained the array of tools available to manage agricultural pests. These constraints may promote further adoption of semiochemical-based tactics.

This chapter has attempted to provide an overview of the various ways by which semiochemicals impact arthropod pest management. Their use is clearly significant. Semiochemical use in established approaches such as monitoring and mating disruption will likely grow in importance. Further exploration of all aspects of arthropod chemical ecology should produce new and more elegant tools to manage pest populations in the future.

REFERENCES

1. Admad, S., ed. 1983. *Herbivorous Insects: Host-Seeking Behavior and Mechanisms*. New York: Academic Press.
2. Aldrich, J.R. 1988. Chemical ecology of the Heteroptera. *Annu. Rev. Entomol.* 33:211–238.
3. Aluja, M., & E.F. Boller. 1992. Host marking pheromone of *Rhagoletis cerasi*: field deployment of synthetic pheromone as a novel cherry fruit fly management strategy. *Entomol. Exp. Appl.* 65:141–147.
4. Arn, H., M. Toth, & E. Priesner. 1992. *List of Sex Pheromones of Lepidoptera and Related Attractants*. Paris: OILB-SROP.
5. Baker, T.C., R.T. Staten, & H.M. Flint. 1990. Use of pink bollworm pheromone in the southwestern United States. *In Behavior-Modifying Chemicals for Insect Management*. R.L. Ridgway, R.M. Silverstein, & M.N. Inscoe, eds. New York: Marcel Dekker, pp 417–436.

6. Barnes, M.M., J.G. Millar, P.A. Kirsch, & D.C. Hawks. 1992. Codling moth (Lepidoptera: Tortricidae) control by dissemination of synthetic female sex pheromone. *J. Econ. Entomol.* 85:1274–1277.

7. Barton Browne, L. 1993. Physiologically induced changes in resource-oriented behavior. *Annu. Rev. Entomol.* 38:1–25.

8. Beevor, P.S., D.G. Campion. 1979. The field use of inhibitory components of lepidopterous sex pheromones and pheromone mimics. In *Chemical Ecology: Odour Communication in Animals.* F.J. Ritter, ed. Amsterdam: Elsevier, pp 313–325.

9. Bell, W.J. 1990. Searching behavior patterns in insects. *Annu. Rev. Entomol.* 35:447–467.

10. Bell, W.J., & R.T. Cardé. 1984. *Chemical Ecology of Insects.* London: Chapman & Hall.

11. Bernays, E.A. 1983. Antifeedants in crop protection. In *Natural Products for Innovative Pest Management.* D.L. Whitehead & W.S. Bowers, eds. New York: Pergamon Press, pp 259–271.

12. Beroza, M., ed. 1976. *Pest Management with Insect Sex Attractants and Other Behavior Controlling Chemicals.* Washington, DC: American Chemical Society.

13. Bierl, B.A., M. Beroza, & C.W. Collier. 1970. Potent sex attractant of the gypsy moth: its isolation, identification, and synthesis. *Science* 170: 87–89.

14. Birch, M.C., ed. 1974. *Pheromones.* Amsterdam: North-Holland.

15. Blum, M.S. 1992. Honey bee pheromones. In *The Hive and the Honey Bee.* J. Graham, ed. Chelsea: Book Crafters, pp 269–372.

16. Blum, M.S., D.M. Everett, T.H. Jones, & H.M. Faler. 1991. Arthropod natural products as insect repellents. In *Naturally-Occurring Pest Bioregulators.* P.A. Hedin, ed. Washington, DC: American Chemical Society, pp 14–26.

17. Borden, J.H. 1990. Use of semiochemicals to manage coniferous tree pests in western Canada. In Behavior-Modifying Chemicals for Insect Management. R.L. Ridgway, R.M. Silverstein, & M.N. Inscoe, eds. New York: Marcel Dekker, pp 281–315.

18. Borden, J.H., J.E. Conn, L.M. Friskie, B.E. Scott, L.J. Chong, H.D. Pierce, & A.C. Oehlschlager. 1983. Semiochemicals for the mountain pine beetle, *Dendroctonus ponderosae* (Coleoptera: Scolytidae) in British Columbia: baited trees studies. *Can. J. For. Res.* 13:325–333.

19. Borden, J.H., L.C. Ryker, L.J. Chong, H.D. Pierce, Jr., B.D. Johnston, & A.C. Oehlschlager. 1987. Response of the mountain pine beetle, *Dentroctonus ponderosae* Hopkins (Coleoptera: Scolytidae) to five semiochemicals in British Columbia lodgepole pine forests. *Can. J. For. Res.* 17:118–128.

20. Brewer, M.J., D.J. Schuster, J.T. Trumble, & B. Alvarado-Rodriguez. 1993. Tomato pinworm (Lepidoptera: Gelechiidae) resistance to fenvalerate from localities in Sinaloa, Mexico and California, USA. *Trop. Agric.* 70:179–184.

21. Brunner, J.F., L.J. Gut, & A. Knight. 1992. Transition of apple and pear orchards to a pheromone-based pest management system. *Proc. Wash. Hort. Assoc.* 88:169–175.

22. Burrows, T.M., V. Sevacherian, H. Browning, & J. Baritelle. 1982. History and

cost of the pink bollworm (Lepidoptera: Gelechiidae) in the Imperial Valley. *Bull. Entomol. Soc. Am.* 28:286–290.

23. Butenandt, A., R. Beckman, D. Stamm, & E. Hecker. 1959. Über den Sexuallockstoff des Seidenspinners *Bombyx mori*; Reindarstellung und Konstitution. *Z. Naturforsch.* 14b:283–284.

24. Calabrese, E.J., & A.J. Sorensen. 1978. Dispersal and recolonization by *Myzus persicae* following aphid alarm pheromone exposure. *Ann. Entomol. Soc. Am.* 71:181–182.

25. Campion, D.G. 1984 Survey of pheromone uses in pest control. In *Techniques in Pheromone Research*. H.E. Hummel & T.A. Miller, eds. New York: Springer-Verlag, pp 405–450.

26. Cardé, R.T. 1990. Principles of mating disruption. In *Behavior-Modifying Chemicals for Insect Management*. R.L. Ridgway, R.M. Silverstein, & M.N. Inscoe, eds. New York: Marcel Dekker, pp 47–72.

27. Cardé, R.T. & A.K. Minks. 1995. Control of moth pests by mating disruption: successes and constraints. *Annu. Rev. Entomol.* 40:559–586.

28. Carson, R. 1962. *Silent Spring*. Boston: Houghton Mifflin.

29. Charlton, R.E., J.A. Wayman, J.R. McLaughlin, J.W. Du, & W.L. Roelofs. 1991. Identification of sex pheromone of tomato pinworm, *Keiferia lycopersicella* (Wals.). *J. Chem. Ecol.* 17:175–183.

30. Charmillot, P.J. 1990. Mating disruption technique to control codling moth in western Switzerland. In *Behavior-Modifying Chemicals for Insect Management*. R.L. Ridgway, R.M. Silverstein, & M.N. Iscoe, eds. New York: Marcel Dekker, pp 165–182.

31. Chiang, H.C. 1973. Bionomics of the northern and western corn rootworms. *Annu. Rev. Entomol.* 18:47–72.

32. Chiri, A.A., & E.F. Legner. 1983. Field application of host-searching kairomones to enhance parasitization of the pink bollworm (Lepidoptera: Gelechiidae). *J. Econ. Entomol.* 76:254–255.

33. Clark, J.M., J.G. Scott, F. Campos, & J.R. Bloomquist. 1995. Resistance to avermectins: extent, mechanisms, and management implications. *Annu. Rev. Entomol.* 40:1–30.

34. Collins, R.D., & R.T. Cardé. 1985. Variation in and heritability of aspects of pheromone production in the pink bollworm moth *Pectinophora gossypiella. Ann. Entomol. Soc. Am.* 78:229–234.

35. Colvin, J., & G. Gibson. 1992. Host-seeking behavior and management of tsetse. *Annu. Rev. Entomol.* 37:21–40.

36. Cunningham, R.T. 1989. Parapheromones. In *Fruit Flies. Their Biology, Natural Enemies and Control*. Vol. 3A. A.S. Robinson & G. Hooper, eds. Amsterdam: Elsevier, pp 221–230.

37. Cunningham, R.T., R.M. Kobayashi, & D.H. Miyashita. 1990. The male lures of tephritid fruit flies. In *Behavior-Modifying Chemicals for Insect Management*. R.L. Ridgway, R.M. Silverstein, & M.N. Inscoe, eds. New York: Marcel Dekker, pp 255–268.

38. Curie, R.W., M.L. Winston, K.N. Slessor, & D.F. Mayer. 1992. Effect of syn-

thetic queen mandibular pheromone sprays on pollination of fruit crops by honey bees (Hymenoptera: Apidae). *J. Econ. Entomol.* 85:1293–1299.

39. Curie, R.W., M.L. Winston, & K.N. Slessor. 1992. Effect of synthetic queen mandibular pheromone sprays on honey bee (Hymenoptera: Apidae) pollination of berry crops. *J. Econ. Entomol.* 85:1300–1306.

40. Dawson, G.W., D.C. Griffiths, J.A. Pickett, R.T. Plumb, C.M. Woodcock, & Z.N. Zhamy. 1988. Structure/activity studies on aphid alarm pheromone derivatives and their field use against transmission of barley yellow dwarf virus. *Pesticide Sci.* 22: 17–30.

41. Dawson, G.W., D.C. Griffiths, L.A. Merritt, A. Mudd, J.A. Pickett, L.J. Wadhams, & C.M. Woodcock. 1990. Aphid semiochemicals - a review and recent advances on the sex pheromone. *J. Chem. Ecol.* 16:3019–3030.

42. Dethier, V.G., L.B. Browne, & C.N. Smith. 1960. The designation of chemicals in terms of the responses they elicit from insects. *J. Econ. Entomol.* 53:134–136.

43. Dicke, M., & M.W. Sabelis. 1988. Infochemical terminology: based on cost-benefit analysis rather than origin of compounds. *Funct. Ecol.* 2:131–39.

44. Dicke, M., M.W. Sabelis, J. Takabayashi, J. Bruin, & M.A. Posthumus. 1990. Plant strategies of manipulating predator-prey interactions through allelochemicals: prospects for application in pest control. *J. Chem. Ecol.* 16:3091–3118.

45. *Federal Register.* 1994. Vol. 50, p. 93681, January 26.

46. *Federal Register.* 1994. Vol. 59, p. 14757, March 30.

47. Furniss, M.M., G.E. Daterman, L.N. Kline, M.D. McGregor, G.C. Trostle, L.F. Pettinger, & J.A. Rudinsky. 1974. Effectiveness of the Douglas-fir beetle antiaggregative pheromone methylcyclohexenone at three concentrations and spacings around felled host trees. *Can. Entomol.* 106:381–392.

48. Furniss, R.L., & V.M. Carolin, eds. 1977. *Western Forest Insects.* USDA Misc. Publ. No. 1339. Washington, DC: USDA.

49. Gaston, L.K., H.H. Shorey, & C.A. Saario. 1967. Insect population control by the use of sex pheromones to inhibit orientation between the sexes. *Nature* 213:1155.

50. Gaston, L.K., R.S. Kaae, H.H. Shorey, & D. Sellers. 1977. Controlling the pink bollworm by disrupting sex pheromone communication between adult moths. *Science* 196: 904–905.

51. Goodell, P.B., & W. Bentley. 1990. Pink bollworm in the San Joaquin Valley - 1990 update. *Calif Cotton Rev.* 18:1–4.

52. Gould, F. 1991. Arthropod behavior and the efficacy of plant protectants. *Annu. Rev. Entomol.* 36:305–330.

53. Griffiths, D.C. & J.A. Pickett. 1980. A potential application of aphid alarm pheromones. *Entomol. Exp. Appl.* 27:199–201.

54. Gross, H.R., Jr. 1981. Employment of kairomones in the management of parasites. In *Semiochemicals: Their Role in Pest Control.* D.A. Nordlund, R.L. Jones, & W.J. Lewis, eds. New York: Wiley, pp 137–152.

55. Hall, D.R. 1990. Use of host odor attractants for monitoring and control of tsetse flies. In *Behavior-Modifying Chemicals for Insect Management.* R.L. Ridgway, R.M. Silverstein, & M.N. Inscoe, eds. New York: Marcel Dekker, pp 517–530.

56. Hardee, D.D. 1982. Mass trapping and trap cropping of the boll weevil *Anthonomus grandis* Boheman. In *Insect Suppression with Controlled Release Pheromone Systems*, Vol. II. A.F. Kydonieus & M. Beroza, eds. Boca Raton, Florida: CRC Press, pp 66–71.

57. Haynes, K.F., W.G. Li, & T.C. Baker. 1986. Control of pink bollworm (Lepidoptera: Gelechiidae) with insecticides and pheromones (attracticide) lethal and sublethal effects. *J. Econ. Enotomol.* 79:1466–1471.

58. Haynes, K.F., T.A. Miller, R.T. Staten, W.G. Li, & T.C. Baker. 1987. Pheromone trap for monitoring insecticide resistance in the pink bollworm moth (Lepidoptera: Gelechiidae): new tool for resistance management. *Environ. Entomol.* 16:84–89.

59. Haynes, K.F., & T.C. Baker. 1988. Potential for evolution of resistance to pheromones: worldwide and local variation in chemical communication system of pink bollworm *Pectinophora gossypiella*. *J. Chem. Ecol.* 14:1547–1560.

60. Hedin, P.A., ed. 1991. *Naturally-Occurring Pest Bioregulators*. Washington, DC: American Chemical Society.

61. Henneberry, T.J. 1994. Pink bollworm sterile moth releases: suppression of established infestations and exclusion from noninfested areas. In *Fruit Flies and the Sterile Insect Technique*. C.O. Calkins, W. Klassen, & P. Liedo, eds. Boca Raton, Florida: CRC Press, pp 181–207.

62. Hille Ris Lambers, D., & D. Schepers. 1978. The effect of trans-β-farnesene used as a repellent against landing aphid alatae in seed potato growing. *Potato Res.* 21:23–26.

63. Hockland, S.H., G.W. Dawson, D.C. Griffiths, B. Marples, J.A. Pickett, & C.M. Woodcock. 1986. The use of aphid alarm pheromone (E)-β-farnesene to increase effectiveness of the entomophilic fungus *Verticillium lecanii* in controlling aphids on chrysanthemums under glass. In *Fundamental and Applied Aspects of Invertebrate Pathology*. R.A. Samson, J.M. Vlak, & and R. Peters, eds. Amsterdam: The Netherlands Society of Invertebrate Pathology, p 252.

64. Hokkanen, H.M.T. 1991. Trap cropping in pest management. *Annu. Rev. Entomol.* 36:119–138.

65. Hoyt, S.C., J.R. Leeper, G.C. Brown & B.A. Croft. 1983. Basic biology and management components for insect IPM. In *Integrated Management of Insect Pests of Pome and Stone Fruits*. B.A. Croft & S.C. Hoyt, eds. New York: Wiley, pp 93–152.

66. Huber, R.T., L. Moore, & M.P. Hoffman. 1979. Feasibility study of area-wide pheromone trapping of male pink bollworm in a cotton insect pest management program. *J. Econ. Entomol.* 72:222–227.

67. Hummel, H.E., & J.F. Andersen. 1982. Secondary plant factors of *Cucurbita* suppress sex attraction in the beetle *Diabrotica undecimpunctata howardi*. *Proc. Int. Symp. Insect-Plant Relationships* 5:163–167.

68. Hummel, H.E., & T.A. Miller. 1984. *Techniques in Pheromone Research*. New York:Springer-Verlag.

69. Hunt, T.N., & J.R. Baker. 1982. Insect and related pests of field crops: some important, common and potential pests in North Carolina. *N.C. Agric. Ext. Publ.* No. 271.

70. Inscoe, M.N., B.A. Leonhardt, & R.L. Ridgway. 1990. Commercial availability of insect pheromones and other attractants. *In Behavior-Modifying Chemicals for Insect Management.* R.L. Ridgway, R.M. Silverstein, & M.N. Inscoe, eds. Amsterdam: Elsevier, pp 631–715.

71. Janssen, A., C.D. Hofker, A.R. Braun, N. Mesa, M.W. Sabelis, & A.C. Bellotti. 1990. Preselecting predatory mites for biological control: the use of an olfactometer. *Bull. Entomol. Res.* 80:177–181.

72. Jenkins, J.W., C.C. Doane, D.T. Schuster, J.R. McLAughlin, & M.J. Jimenez. 1990. Development and commercial application of sex pheromones for control of the tomato pinworm. In *Behavior-Modifying Chemcials for Insect Management.* R.L. Ridgway, R.M. Silverstein, & M.N. Inscoe, eds. Amsterdam: Elsevier, pp 269–280.

73. Jermy, T. 1990. Prospects of antifeedant approach to pest control - a critical review. *J. Chem. Ecol.* 16:3151–3166.

74. Johnson, R.L., & R.A. Schall. 1989. Early detection of new pests. In *Plant Protection and Quarantine*, Vol. III. R.P. Kahn, ed. Boca Raton, Florida, CRC Press, pp 105–116.

75. Jutsum, A.R. & R.F.S. Gordon, eds. 1988. *Insect Pheromones in Plant Protection.* New York: Wiley.

76. Kinzer, G.W., A.F. Fentiman, Jr., R.L. Foltz, & J.A. Rudinsky. 1971. Bark beetle attractants: 3-methyl-2-cyclohexen-1-one isolated from *Dendroctonus pseudotsugae. J. Econ. Entomol.* 64:970–971.

77. Knight, A.L., & L.A. Hull. 1988. Areawide population dynamics of *Platynota idaeusalis* (Lepidoptera: Tortricidae) in southcentral Pennsylvania pome and stone fruits. *Environ. Entomol.* 17:1000–1008.

78. Knight, A.L, & L.A. Hull. 1989. Predicting seasonal apple injury by tufted apple bud moth (Lepidoptera: Tortricidae) with early-season sex pheromone trap catches and brood I fruit injury. *Environ. Entomol.* 18:939–944.

79. Knight, A.L., & B.A. Croft 1991. Modelling and prediction technology. In *Tortricid Pests, Their Biology, Natural Enemies, and Control.* L.P.S. van der Geest & H.H. Evenhuis, eds. Amsterdam: Elsevier, pp 301–312.

80. Knipling, E.F., & J.U. McGuire, Jr. 1966. *Population Models to Test Theoretical Effects of Sex Attractants Used for Insect Control.* USDA Agric. Info. Bull. No. 308. Washington, DC: USDA.

81. Kolodny-Hirsch, D.M., & C.P. Schwalbe. 1990. Use of disparlure in the management of gypsy moth. In *Behavior-Modifying Chemicals for Insect Management.* R.L. Ridgway, R.M. Silverstein, & M.N. Inscoe, eds. New York: Marcel Dekker, pp 363–386.

82. Kydonieus, A.F., & M. Beroza. 1982. Pheromones and their use. In *Insect Suppression with Controlled Release Pheromone Systems.* Vol I. A.F. Kydonieus & M. Beroza, eds. Boca Raton, Florida: CRC Press, pp 3–12.

83. Kydonieus, A.F., & M. Beroza, eds. 1982. *Insect Supression with Controlled Release Pheromone Systems.* Vols. I and II. Boca Raton, Florida: CRC Press.

84. Ladd, T.L., & M.G. Klein. 1986. Japanese beetle (Coleoptera: Scarabidae) response to color traps with phenethyl propionate + eugenol + geraniol (3: 7: 3) and japonilure. *J. Econ. Entomol.* 79:84–86.

85. Lanier, G.N. 1990. Principles of attraction-annihilation: mass trapping and other means. In *Behavior-Modifying Chemicals for Insect Management*. R.L. Ridgway, R.M. Silverstein, & M.N. Inscoe, eds. Marcel Dekker, pp 25–46.
86. Law, J.H. & F.E. Regnier. 1971. Pheromones. *Annu. Rev. Biochem.* 40:533–548.
87. Levine, E. & H. Oloumi-Sadeghi. 1991. Management of diabroticite rootworms in corn. *Annu. Rev. Entomol.* 36:229-255.
88. Lewis, W.J, & W.R. Martin. 1990. Semiochemicals for use with parasitoids: status and future. *J. Chem. Ecol.* 16:3067–3090.
89. Linn, C.E., Jr., M.G. Campbell, & W.L. Roelofs. 1986. Male moth sensitivity to multicomponent pheromones: critical role of female-released blend in determining the functional role of components and active space of the pheromone. *J. Chem. Ecol.* 12:659–668.
90. Luttrell, R.G. 1994. Cotton pest management. Part 2, A U.S. perspective. *Annu. Rev. Entomol.* 39:527–542.
91. Madsen, H.F., & J.M. Vakenti. 1973. Codling moth: use of codlemone baited traps and visual detection of entries to determine need of sprays. *Environ. Entomol.* 2:677–679.
92. Madsen, H.F., & B.E. Carty. 1979. Codling moth (Lepidoptera: Olethreutidae) suppression by male removal with sex pheromone traps in three British Columbia orchards. *Can. Entomol.* 111:627.
93. Mayer, D.F., R.L. Britt, & J.D. Lunden. 1989. Evaluation of Beescent® as a honey bee attractant. *Am. Bee J.* 129:41–42.
94. McGregor, S.E. 1976. Insect Pollination of Cultivated Crop Plants. Agriculture Handbook No. 496. Washington, DC: USDA.
95. McGregor, M.D., M.M. Furniss, R.D. Oaks, K.E. Gibson & H.E. Meyer. 1984. MCH pheromone for preventing Douglas-fir beetle infestations in windthrown trees. *J. For.* 82:613–616.
96. McKibben, G.H., D. Rainer & J.W. Smith. 1994. Boll weevil bait stick use in Nicarauga in 1993. *Proceedings of the Beltwide Cotton Production Research Conference.* Vol 2. National Cotton Council: Memphis, pp 984–986.
97. McNeil, J.N. 1991. Behavioral ecology of pheromone-mediated communication in moths and its importance in the use of pheromone traps. *Annu. Rev. Entomol.* 36:407–430.
98. Meinke, L.J. 1994. Evaluation of a semiochemical-based bait for potential use in corn rootworm management programs. *Proceedings of the National Integrated Pest Management Symposium Workshop* 2:212.
99. Metcalf, R.L. 1975. Insecticides in pest management. In *Introduction to Insect Pest Management*. R.L. Metcalf & W. Luckman, ed. New York: Wiley, pp 235–273.
100. Metcalf, R.L., & R. L. Lampman. 1989. The chemical ecology of Diabroticites and Cucurbitaceae. *Experientia* 45:240–247.
101. Metcalf, R.L., & E.R. Metcalf. 1992. *Plant Kairomones in Insect Ecology and Control*. New York: Chapman & Hall.
102. Metcalf, R.L., J.E. Ferguson, D. Fischer, R. Lampman, & J. Andersen. 1983.

Controlling cucumber beetles and corn rootworm beetles with baits of bitter cucurbit fruits and roots. *Cucurbit Genet. Coop.* 6:79–81.

103. Metcalf, R.L., J.E. Ferguson, R. Lampman, & J.E. Andersen. 1987. Dry cucurbitacin-containing baits for controlling Diabroticite beetles (Coleoptera: Chrysomelidae). *J. Econ. Entomol.* 80:870–875.

104. Minks, A.K., & R.T. Cardé. 1988. Disruption of pheromone communication in moths: is the natural blend really most efficacious. *Entomol. Exp. App.* 49:25–36.

104a. Mitchell, E.R. 1981. *Management of Insect Pests with Semiochemicals: Concepts and Practice.* New York: Plenum Press.

105. Moffitt, H.R., & P.H. Westigard. 1984. Suppression of codling moth (Lepidoptera: Tortricidae) population on pears in southern Oregon through mating disruption with sex pheromone. *J. Econ. Entomol.* 77:1513–1519.

106. Murliss, J., J.S. Elkinton, & R.T. Cardé. 1992. Odor plumes and how insects use them. *Annu. Rev. Entomol.* 37:505–532.

107. Naumann, K., M.L. Winston, M.H. Wyborn, & K.N. Slessor. 1990. Effects of synthetic honey bee (Hymenoptera: Apidae) queen mandibular-gland pheromone on workers in packages. *J. Econ. Entomol.* 83:1271–1275.

108. Noldus, L.P.J.J. 1989. Semiochemicals, foraging behaviour, and quality of entomophagous insects for biological control. *J. Appl. Entomol.* 108:425–451.

109. Nordlund, D.A., & W.J. Lewis. 1976. Terminology of chemical releasing stimuli in intraspecific and interspecific interactions. *J. Chem. Ecol.* 2:211–220.

110. Nordlund, D.A. 1981. Semiochemicals: a review of the terminology. In *Semiochemicals: Their Role in Pest Control.* D.A. Nordlund, R.L. Jones, & W.J. Lewis, eds. New York: Wiley, pp 13–28.

111. Nordlund, D.A., R.L. Jones, & W.J. Lewis, eds. 1981. *Semiochemicals: Their Role in Pest Control.* New York: Wiley.

112. Oatman, E.T. 1970. Ecological studies of the tomato pinworm on tomato in southern California. *J. Econ. Entomol.* 63:1531–1534.

113. O'Connor, C.A. 1990. Registration of pheromones in practice. In *Behavior-Modifying Chemicals for Insect Management.* R.L. Ridgway, R.M. Silverstein, & M.N. Inscoe, eds. New York: Marcel Dekker, pp 605–618.

114. Pak, G.A., J.W.M. Kaskers, & E.J. DeJong. 1990. Behavioural variation among strains of *Trichogramma* spp. host-species selection. *Entomol. Exp. Appl.* 56:91–102.

115. Payne, T.L., M.C. Birch, & C.E.J. Kennedy, eds. 1986. *Mechanisms in Insect Olfaction.* Oxford, UK: Clarendon Press.

116. Penrose, R. 1993. The 1989/1990 Mediterranean fruit fly eradication program in California. In *Fruit Flies: Biology and Management.* M. Aluja & P. Liedo, New York: Springer-Verlag, pp 401–406.

117. Pettis, J.S., M.L. Winston, M. Malyon, & K.N. Slessor. 1993. The use of honey bee queen mandibular gland pheromone in mating nuclei management. *Am. Bee J.* 133:725–727.

118. Pickett, J.A., L.J. Wadhams, C.M. Woodcock, & J. Hardie. 1992. The chemical ecology of aphids. *Annu. Rev. Entomol.* 37:67–90.

119. Pitman, G.B., J.P. Vité, G.W. Kinzer, & A.F. Fentiman, Jr. 1968. Bark beetle attractants: *trans*-verbenol isolated from *Dendroctonus*. Nature 218:168–169.

120. Prestwich, G.D., & G.J. Blomquist, eds. 1987. *Pheromone Biochemistry*. New York, Academic Press.

121. Prokopy, R.L. 1981. Epideictic pheromones that influence spacing patterns of phytophagous insects. In *Semiochemicals Their Role in Pest Control*. D.A. Nordlund, R.L. Jones, & W.J. Lewis, eds. New York: Wiley, pp 181–214.

122. Raina, A.K. 1993. Neuroendocrine control of sex pheromone biosynthesis in Lepidoptera. *Annu. Rev. Entomol.* 38:329–349.

123. Ravlin, W., K. Swain, & R. Wolfe. 1992. *A Strategic Plan to Slow the Spread of Gypsy Moth*. USDA Forest Service Internal report. Asheville: Forest Health Unit.

124. Ridgway, R.L., R.M. Silverstein, & M.N. Inscoe, eds. 1990. *Behavior-Modifying Chemicals for Insect Management*. New York: Marcel Dekker.

125. Ridgway, R.L., M.N. Inscoe, & K.W. Thorpe. 1993. Advances and trends in managing insect pests. In *Management of Insect Pests: Nuclear and Related Molecular and Genetic Techniques*. Vienna: International Atomic Energy Agency, pp 3–15.

126. Riedl, H., B.A. Croft, & A.J. Howitt. 1976. Forecasting codling moth phenology based on pheromone trap catches and physiological-time models. *Can. Entomol.* 108:449–460.

127. Riedl, H., A. Seaman, & F. Henrie. 1985. Monitoring susceptibility to azinphosmethyl in field populations of the codling moth (Lepidoptera: Tortricidae) with pheromone traps. *J Econ. Entomol.* 78:692–699.

128. Riedl, H., J.F. Howell, P.S. McNally, & P.H. Westigard. 1986. *Codling Moth Management Use and Standardization of Pheromone Trapping Systems*. Berkeley, California: University of California Experimental Station.

129. Riherd, C., R. Nguyen, & J.R. Brazzel. 1994. Pest free areas. In *Quarantine Treatments for Pests of Food Plants*. J.L. Sharpe & G.J. Hallman, eds. Boulder, Colorado: Westview Press, pp 213–223.

130. Ritter, F.J., ed. 1979. *Chemical Ecology: Odour Communication in Animals*. Amsterdam: Elsevier.

131. Robinson, A.S., & G. Hooper, eds. 1989. *Fruit Flies: Their Biology, Natural Enemies, and Control*, Vols. 3A and 3B. Amsterdam: Elsevier.

132. Roelofs, W., A. Comeau, A. Hill, & G. Milicevic. 1971. Sex attractant of the codling moth: characterization with electroantennogram technique. *Science* 174:297–299.

133. Roelofs, W.L. 1981. Attractive and aggregating pheromones. In *Semiochemicals Their Role in Pest Control*. D.A. Norlund, R.L. Jones, & W.J. Lewis, eds. New York: Wiley, pp 215–235.

134. Roessler, Y. 1989. Insecticidal bait and cover sprays. In *Fruit Flies, Their Biology, Natural Enemies and Control*. Vol. 3B. A.S. Robinson & G. Hooper, ed. Amsterdam: Elsevier, pp 329–336.

135. Roitberg, B.D., & M.B. Isman, eds. 1992. *Insect Chemical Ecology*. New York: Chapman & Hall.

136. Ross, D.W., & G.E. Daterman. 1997. Reduction of Douglas fir beetle infestations of high risk stands by antiaggregation and aggregation pheromones. *Can. J. For. Res.* 24:2184–2190.

137. Rothschild, G.H.L. 1982. Suppression of mating in codling moth with synthetic sex pheromone and other compounds. In *Insect Suppression with Controlled Release Pheromone Systems*. A.F. Kydonieus & M. Beroza, eds. Boca Raton, Florida: CRC Press, pp 117–134.

138. Rudinsky, J.A., M.M. Furniss, L.N. Kline, & R.F. Schmitz. 1972. Attraction and repression of *Dendroctonus pseudotsugae* (Coleoptera: Scolytidae) by three synthetic pheromones in traps in Oregon and Idaho. *Can. Entomol.* 104:815–822.

139. Sauer, A.E., G. Karg, U.T. Koch, & J.J. de Kramer. 1992. A portable EAG system for the measurement of pheromone concentrations in the field. *Chem. Senses* 17:543–553.

140. Schmutterer, H. 1990. Properties and potential of natural pesticides from the neem tree, *Azadirachta indica*. *Annu. Rev. Entomol.* 35:271–298.

141. Shaw, J.T., W.G. Ruesink, S.P. Briggs, & W.H. Luckmann. 1984. Monitoring populations of corn rootworm beetles (Coleoptera: Chrysomelidae) with a trap baited with cucurbitacins. *J. Econ. Entomol.* 77:1495–1499.

142. Shorey, H.H. 1977. Interaction of insects with their chemical environment. In *Chemical Control of Insect Behavior: Theory and Application*. H.H. Shorey & J.J. McKelvey, eds. New York: Wiley, pp 1–5.

143. Shorey, H.H. 1981. The use of chemical attractants in insect control. In *Handbook of Pest Management in Agriculture*. Vol. II. D. Pimentel, ed. Boca Raton, Florida: CRC Press, pp 307–314.

144. Shorey, H.H., J.J. McKelvey Jr., eds. 1977. *Chemical Control of Insect Behavior: Theory and Application*. New York: Wiley.

145. Silverstein, R.M. 1981. Pheromones: background and potential use in insect pest control. *Science* 213:1326–1332.

146. Sonenshine, D.E. 1985. Pheromones and other semiochemicals of the acari. *Annu. Rev. Entomol.* 30:1–28.

147. Staten, R.T., H.M. Flint, R.C. Weddle, E. Quintero, R.E. Zarate, C.M. Finnell, M. Hernandes, & A. Yamamoto. 1987. Pink bollworm (Lepidoptera: Gelechiidae): largescale field trials with a high-rate gossyplure formulation. *J. Econ. Entomol.* 80:1267–1271.

148. Staten, R.T., R.W. Rosander & D.F. Keaveny. 1993. Genetic control of cotton pests: the pink bollworm as a working programme. In *International Symposium on Management of Insect Pests: Nuclear and Related Molecular and Genetic Techniques*, Vienna: International Atomic Energy Agency, pp 269–283.

149. Staten, R.T., L. Antilla & O. El Lissy. 1993. Using sex pheromones to disrupt insect population development through mating disruption. Workshop Great Lakes Fisheries Commission, Oct. 29–31, St. Paul, Minnesota.

150. Tabashnik, B.E. 1994. Evolution of resistance to *Bacillus thuringiensis*. *Annu. Rev. Entomol.* 39:47–80.

151. Takken, W., M.A. Oladunmade, L. Dengwat, H.U. Feldman, J.A. Onah, S. O. Tenabe, & H.J. Hamann. 1986. The eradication of *Glossina palpalis palpalis*

(Diptera: Glossinidae) using traps, insecticide-impregnated targets, and the sterile insect technique in central Nigeria. *Bull. Entomol.* 76:275–286.

152. Tinsworth, E.F. 1990. Regulation of pheromones and other semiochemicals in the United States. In *Behavior-Modifying Chemicals for Insect Management.* R.L. Ridgway, R.M. Silverstein, & M.N. Inscoe, eds. New York: Marcel Dekker, pp 569–604.

153. Trumble, J.T., & B. Alvarado-Rodriguez. 1993. Development and economic evaluation of an IPM program for fresh market tomato production in Mexico. *Agric. Ecosys. Environ.* 43:267–284.

154. Trumble, J. T. 1994. Sampling arthropod pests in vegetables. In *Handbook of Sampling Methods for Arthropods in Agriculture.* L.P. Pedigo & G.D. Buntin, eds. Boca Raton, Florida: CRC Press, pp 603–626.

155. Varela, L.G., S.C. Welter, V.P. Jones, J.F. Brunner, & H. Riedl. 1993. Monitoring and characterization of insecticide resistance in codling moth (Lepidoptera: Tortricidae) in four western states. *J. Econ. Entomol.* 86:1–10.

156. Van der Pers, J.N.C., & A.K. Minks. 1993. Pheromone monitoring in the field using single sensillum recording. *Entomol. Exp. Appl.* 68:237–245.

157. Vet, L.E.M., & M. Dicke. 1992. Ecology of infochemical use by natural enemies in a tritrophic context. *Annu. Rev. Entomol.* 37:141–172.

158. Vet, L.E.M., & A.W. Groenwold. 1990. Semiochemicals and learning in parasitoids. *J. Chem. Ecol.* 16:3119–3135.

159. Vickers, R.A., & G.H.L. Rothschild. 1991. Use of sex pheromones for control of codling moth. In *Tortricid Pests Their Biology, Natural Enemies and Control.* L.P.S. van der Geest & H.H. Evenhuis, eds. Amsterdam: Elsevier, pp 339–370.

160. Vinson, S.B. 1976. Host selection by insect parasitoids. *Annu. Rev. Entomol.* 21:109–133.

161. Vinson, S.B. 1977. Behavioral chemicals in the augmentaion of natural enemies In *Biological Control by Augmentation of Natural Enemies.* R.L. Ridgway & S.B. Vinson, eds. New York: Plenum Press, pp 237–279.

162. Vite, J.P., & E. Baader. 1990. Present and future use of semiochemicals in pest management of bark beetles. *J. Chem Ecol.* 16:3031–3042.

163. Weatherston, I. 1990. Principles of design of controlled-release formulations. In *Behavior-Modifying Chemicals for Insect Management.* R.L. Ridgway, R.M. Silverstein, & M.N. Inscoe, eds. New York: Marcel Dekker, pp 93–112.

164. Weissling, T.J., & L.J. Meinke. 1991. Potential of starch encapsulated semiochemical-insecticide formulations for adult corn rootworm (Coleoptera: Chrysomelidae) control. *J. Econ. Entomol.* 84:601–609.

165. Whitehead, D.L., & W.S. Bowers, eds. 1983. *Natural Products for Innovative Pest Management.* New York: Pergamon Press.

166. Winston, M.L., & K.N. Slessor. 1992. The essence of royalty: honey bee queen pheromone. *Am. Sci.* 80:374–385.

167. Wood, J. 1994. *Gypsy Moth Program Summary Report.* Washington State Department of Agriculture internal report. Olympia: WSDA.

20
General Overview of Weeds in Crop Systems

David C. Bridges
University of Georgia, Griffin, Georgia

I. INTRODUCTION

Weeds, and their management, have been an important part of agriculture since humans progressed from wandering and gathering to sowing and reaping; for it was the concept of relative value that was the basis for agricultural development. Smith and Secoy [17] found that many books from the time of Theophrastus (ca. 300 B.C.) have mentioned weeds. Biblical accounts of the Creator's encounter with Adam and Eve after they had eaten the forbidden fruit mention thistles and thorns (Genesis 3: 17–18). It is in these passages that God sentences woman to the pain of childbirth and man to the labor of feeding himself among the "weeds." Weeds affect humans and agricultural practices in many ways and are among agriculture's most important pests. The general types of impacts by weeds are economic, esthetic, ecological, and/or health related.

A. What Is a Weed?

This is an important question to answer if one is to have a meaningful understanding of weeds, their impacts, and their management. Much debate and academic discussion has failed to yield consensus on this question. Depending on one's background and interests, the definitions may be widely divergent, but typically belong to one of two camps. Definitions in the first camp reflect a social sway. These definitions often use subjective terms like *undesirable, unwanted,* or *objectionable.* Much like the old adage that beauty is in the eye of the beholder, many believe that the sole requirement for a plant to be designated is

547

for it to be in some way undesirable or unwanted. Another interpretation is that one person's weed is another person's crop; as the poet Emerson commented, "A weed is a plant whose virtues have not yet been discovered" [11]. Definitions of this sort have been widely adopted among agriculturalists. Indeed, the sole criterion often used to establish weediness by many practitioners of weed control has been unwantedness. Given that many modern agricultural systems are monocultures, in which maintaining the monoculture is paramount, it is not surprising that a premium is placed on weed control.

In the other camp, definitions of weeds reveal an ecological perspective rather than a social or anthropogenic sway. These definitions often include references to weeds as pioneer plants, colonizers, or otherwise opportunistic plants. It is true that for many agricultural communities, "weeds" are most common among disturbed or highly manipulated habitats. In fact, among tilled fields, the most common weeds are often annual species that have relatively high reproductive rates. It has been suggested that weeds can be separated as a special class of plants based on certain characteristics. The idea is that weeds share a common set of characteristics that distinguish them from other plants. Table 1 includes a list of some the characteristics that are often ascribed as attributes of the "ideal" weed. Although many important weeds do share these characteristics, so do many crops. This is not surprising. Weeds have coevolved with many of our important agricultural crops. They are equally as well adapted to the conditions of modern agriculture as are crops.

Definitions of a weed that recognize the relative importance of unwantedness, impact, and biological characteristics are generally more useful. Navas [15] proposed a definition that includes an appropriate balance of the perspectives presented by both camps. He defined weeds as plants that, "form populations that are able to enter habitats cultivated, markedly disturbed or occupied by man, and potentially depress or displace the resident plant populations which are deliberately cultivated or are of ecological and/or aesthetic interest." This definition acknowledges the fact that weeds often are especially adapted to establishing and thriving in highly disturbed sites, but it also recognizes the poten-

Table 1 Characteristics Often Associated with Weediness

Attribute	Relevance to weediness
High relative growth rates	Typically associated with competitiveness
Aggressive and rank growth	Able to establish and spread
High reproductive rates	Important for annual species
Multiple seed dispersal mechanisms	Important for spread
Seed dormancy and seed longevity	Ensures survivability
Efficient scavengers and users of resources	Growth under suboptimum conditions

tially suppressive effects of weeds. The recognition that weeds can depress, suppress, and sometimes competitively displace the resident, or preferred, population is important in both agricultural and nonagricultural habitats. These effects, along with economic impact, are the basis for weeds being unwanted. Furthermore, the definition allows for the inclusion of feral, or escaped, crop plants as weeds. It also allows for the consideration of exotic species as weeds.

B. Selection Processes for Weeds

Disturbance and stress typically are primary selective processes in most habitats. For example, production systems that include tillage one or more times each year are generally dominated by annual species. This occurs because tillage prohibits the establishment of perennial species and generally favors species that produce large numbers of viable seed. Furthermore, tillage produces many potential germination sites in the upper soil profile where aeration, light stimulation, and good seed-to-soil contact promote seed germination.

The ultimate goal of "cash" agriculture is economic profit, which requires consistent, efficient, and profitable production. Producing record yields 1 year followed by catastrophe the following year is not financially sustainable. Therefore, late 20th century agriculture has focused on both production efficiency and production consistency. Edaphic and year-to-year environmental variation are major contributors to variations in production. One way to minimize variability is to provide water, nutrients, light, and growth conditions at near-optimum levels at all times. Reducing variability in production while trying consistently to push yield limits leads to systems where resources and conditions are highly manipulated and stress is typically low. However, maintaining these conditions requires constant and intensive management. Undesirable plants that can exploit these resources are often very productive and become the "weeds" within that system. Heterogeneity within weed populations ensures evolution of populations that are highly adapted to the conditions of modern agriculture. They are often so fit and competitive that they competitively suppress or replace the crop or more desirable species. Selection ensures that over time weeds that persist in a production system are as well or better adapted to the environment in which they grow as is the crop. This is why maintaining a monoculture of plants is so difficult.

The selection process is constantly at work and often unforgiving. Reducing or eliminating tillage favors perennials and disfavors annuals. Cultivation selects for species that are not easily uprooted or which regenerate from roots, buds, or crowns. Applying herbicides selects for herbicide-tolerant species and against susceptible species. Repeated use of the same herbicide or different herbicides that share a common mode of action can select a resistant subpopulation from an otherwise susceptible population. As a result, crop production and weed control practices dictate which weed species persists over time. Generally, weeds that are not controlled in 1 year become an increasing problem the next year, assuming that practices are not changed.

II. WEEDS AS PESTS

Which plants are weeds? How many weedy species are there in the world? Are they just unwanted plants? To date, approximately 250,000 species of terrestrial plants have been identified and recognized worldwide. How many of these are weeds? This question was asked by weed scientist LeRoy Holm and his colleagues many years ago. In the book entitled *The World's Worst Weeds* [13], Holm et al. report that about 200 species appear to cause 95% of the world's crop losses due to weeds. Holm et al. listed 80 primary weeds and 120 secondary weeds depending on prevalence and importance on a worldwide basis. According to Holm et al., 12 plant families contain 68% of the most important weeds, with three families, Poaceae (Graminae), Cyperaceae, and Asteraceae (Compositae), containing 43% of the species (Table 2). Holm and coworkers considered 18 species to be of the highest importance worldwide, naming them the most serious weeds of the world. These species are universally recognized by agriculturalists worldwide (Table 3).

The taxonomic distribution of these most serious worldwide weeds is not totally surprising. Weeds, by definition, are well adapted to the sites in which

Table 2 Plant Families that Contain the World's Worst Weeds[a] and the Number of Species Defined as Weeds Contained Within Each Family

Family name	Colloquial name	Number of species
Poaceae	Grasses (Graminae)	44
Asteraceae	Asters (Compositae)	32
Cyperaceae	Sedges	12
Polygonaceae	Smartweeds	8
Amaranthaceae	Amaranths (pigweeds)	7
Brassicaceae	Crucifers (Cruciferae)	7
Fabaceae	Legumes (Leguminosae)	6
Convolvulaceae	Morning glories	5
Euphorbiaceae	Spurges	5
Chenopodiaceae	Goosefoots	4
Malvaceae		4
Solanaceae	Nightshades	4

[a]According to ref. 13.

Table 3 The 18 Most Serious Weeds of the World[a]

Scientific name	English common name
Cyperus rotundus	Purple nutsedge
Cynodon dactylon	Common bermudagrass
Echinochloa crusgali	Barnyard grass
Echinochloa colonum	Junglerice
Eleusine indica	Goosegrass
Sorghum halepense	Johnsongrass
Imperata cylindrica	Cogon grass
Chenopodium album	Lambsquarters
Digitaria sanguinalis	Large crabgrass
Convolvulus arvensis	Field bindweed
Portulaca oleracea	Common purslane
Eichornia crassipes	Water hyacinth
Avena fatua	Wild oat
Amaranthus hybridus	Smooth pigweed
Amaranthus spinosus	Spiny pigweed
Cyperus esculentus	Yellow nutsedge
Paspalum conjugatum	Sour paspalum
Rottboellia exaltata	Itchgrass

[a]According to ref. 13.

they prevail. They are often as well adapted as the crops with which they are associated. A careful study of the origin and evolution of crop and weed species suggests some degree of parallelism. Conditions that favor domesticates (crops) often also favor the weeds. So, in many cases, weeds may have co-evolved with the crop. It is not incidental that many weeds come from families that contribute many crop plants. Therefore, it is not surprising to find that many of these most important weeds of the world come from the same five plant families that contain 12 plants species that provide more than 75% of the world's food. These five families are Poaceae (barley, maize, millet, oats, rice, sorghum, sugarcane, wheat); Solanaceae (eggplant, tobacco, tomato, white potato); Convolvulaceae (sweet potato); Euphorbiaceae (cassava); and Fabaceae (beans, peas, soybean). The association of crops and weeds was studied in great detail by Harlan and de Wet [11], revealing some of the associations found in Table 4.

Weeds occur in every major agricultural production system whether it be annual row or drill crops, orchard crops, fruits, vegetables, or aquatic production. They can be herbaceous or woody, annual or perennial, monocots or dicots. Life form and life cycle are often determined by factors of disturbance that are associated with the habitat in which they grow. Obviously, in winter

Table 4 Families Contributing Crops and Weeds

Many major crops, many important weeds:
Poaceae (Graminae), Fabaceae (Leguminosae)

A few crops, many important weeds:
Asteraceae (Compositae), Euphorbiaceae, Labiatae, Convolvulaceae,
  Chenopodiaceae, Cruciferae, Polygonaceae, Umbelliferae,
  Amaranthaceae

Many minor crops, some weeds:
Rosaceae, Cucurbitaceae

A few crops, some weeds:
Solanaceae, Malvaceae, Palmaceae, Dioscoreaceae, Rubiaceae

No important crops, some weeds:
Cyperaceae, Ranunculaceae, Cactaceae, Caryopyllaceae, Portulacaceae,
  Onagraceae, Asclepidaceae, Plantaginaceae, and others

annual crops such as wheat, one is more likely to find winter annual weeds than other forms, summer annuals, woody perennials, etc. Aquatic weeds occur in ponds, lakes, rivers, ditches, canals, and sometimes in rice paddies. A recent analysis of data obtained from a 1992 Weed Science Society of America survey [8] provides some insight regarding the most important weeds in the United States. Cooperative extension weed scientists from each state were surveyed regarding weeds in their state in each of 47 agricultural commodities. They were asked to identify the 10 most common and 10 most troublesome weeds in each commodity within their state. Common weeds are those species which are widely existing, or prevalent. These species can be found abundantly infesting a significant portion of the acreage for a given crop throughout the state, especially where no weed management intervention has occurred. Weeds listed as troublesome are those species which, despite weed control efforts, are inadequately controlled and interfere with crop production and/or yield, crop quality, or harvest efficiency. Dividing weeds into these two categories allows for some inference about the occurrence and relative importance of species. For example, a species that is referred to as *common* may be widely distributed, but it may not be competitive, or it may be readily controlled with usual and customary control practices. Other species may be as prevalent, or less so, but cause significant losses, reduce crop quality, hamper harvest, or require extraordinary efforts to control them. These species are referred to as *troublesome* in this survey. Compositing reports from all states and across 47 commodities produced a list of the 10 most common and 10 most troublesome weeds in the United States (Table 5). The relative frequency reported in the table indicates the total

Table 5 The 10 Most Common and 10 Most Troublesome Weeds in 47 Agricultural Commodities in the United States, Including Relative Reporting Frequencies

Species	Relative frequency
Most common	
Common lambsquarters	240
Redroot pigweed	204
Barnyardgrass	150
Pigweed species	124
Common ragweed	120
Kochia	111
Quackgrass	110
Large crabgrass	108
Johnsongrass	105
Common cocklebur	99
Most troublesome	
Canada thistle	137
Field bindweed	123
Johnsongrass	117
Quackgrass	115
Yellow nutsedge	113
Common lambsquarters	96
Morning glory species	96
Redroot pigweed	77
Kochia	75
Common cocklebur	72

Source: Compiled from ref. 8.

number of states by commodity reports that occurred within the survey. The 10 most common and troublesome weeds in U.S. corn, cotton, soybeans, and wheat are shown in Table 6.

III. THE IMPACT OF WEEDS

Weeds affect humans and their activities in many ways (Table 7). They reduce the value and yield of agricultural crops. They interfere with the use of recreational facilities. They diminish the value and appearance of real estate. They affect environmental quality and the health and productivity of humans, domesticated animals, livestock, and wildlife [6].

Table 6a   Most Frequent Weeds (Common, Troublesome) in Selected Crops in the United States[a]

Crops	Common	States	Troublesome	States
Corn	CHEAL	19	SORHA	21
	AMARE	18	ABUTH	14
	XANST	18	CIRAR	13
	ABUTH	16	AGRRE	12
	AMAZZ	16	IPOZZ	12
	PANDI	16	CHEAL	11
	SORHA	16	SORVU	11
	IPOZZ	14	PANMI	10
	SETFA	13	XANST	10
	AMBEL	11	APCCA	09
Cotton	IPOZZ	10	IPOZZ	11
	AMAZZ	09	CYPES	08
	SORHA	09	SORHA	07
	ELEIN	08	CYNDA	06
	SIDSP	08	CYPRO	06
	XANST	08	ELEIN	06
	CYPES	06	XANST	06
	CASOB	05	ABUTH	05
	CYNDA	05	CASOB	05
	DIGSA	04	SIDSP	05
Soybeans	XANST	21	SORHA	15
	ABUTH	16	ABUTH	14
	AMAZZ	14	XANST	14
	IPOZZ	14	IPOZZ	13
	CHEAL	13	AMAZZ	07
	SETFA	13	AMBTR	07
	SORHA	13	SOLPT	07
	AMBEL	12	CIRAR	06
	AMARE	08	APCCA	05
	CASOB	08	SIDSP	05
Wheat	LAMAM	19	ALLVI	15
	ALLVI	17	BROSE	14
	STEME	13	CONAR	14
	CIRAR	10	CIRAR	12
	CONAR	10	LOLMU	11
	KCHSC	10	AEGCY	10
	SASKR	10	BROTE	10
	BROTE	09	AVEFA	09
	SINAR	09	KCHSC	08
	BROSE	08	LAMAM	08

[a]See also Table 6b.
Source: Compiled from ref. 8.

Table 6b Appendix for Weeds listed in Table 6

Code	Genus	Species	Common name
ABUTH	*Abutilon*	*theophrasti*	Velvetleaf
AEGCY	*Aegilops*	*cylindrica*	Goatgrass, jointed
AGRRE	*Elytrigia*	*repens*	Quackgrass
ALLVI	*Allium*	*vineale*	Garlic, wild
AMAZZ	*Amaranthus*	spp.	Pigweeds
AMARE	*Amaranthus*	*retroflexus*	Pigweed, redroot
AMBEL	*Ambrosia*	*artemisiifolia*	Ragweed, common
AMBTR	*Ambrosia*	*trifida*	Ragweed, giant
APCCA	*Apocynum*	*cannabinum*	Dogbane, hemp
AVEFA	*Avena*	*fatua*	Oat, wild
BROSE	*Bromus*	*secalinus*	Cheat
BROTE	*Bromus*	*tectorum*	Brome, downy
CASOB	*Cassia*	*obtusifolia*	Sicklepod
CHEAL	*Chenopodium*	*album*	Lambsquarter, common
CIRAR	*Cirsium*	*arvense*	Thistle, Canada
CONAR	*Convolvulus*	*arvense*	Bindweed, field
CYNDA	*Cynodon*	*dactylon*	Bermudagrass
CYPES	*Cyperus*	*esculentus*	Nutsedge, yellow
CYPRO	*Cyperus*	*rotundus*	Nutsedge, purple
DIGSA	*Digitaria*	*sanguinalis*	Crabgrass, large
ELEIN	*Eleusine*	*indica*	Goosegrass
IPOZZ	*Ipomoea*	spp.	Morning glories
KCHSC	*Kochia*	*scoparia*	Kochia
LAMAM	*Lamium*	*amplexicaule*	Henbit
LOLMU	*Lolium*	*multiflorum*	Ryegrass, Italian
PANDI	*Panicum*	*dichotomiflorum*	Panicum, fall
PANMI	*Panicum*	*miliaceum*	Millet, wild-proso
SASKR	*Salsola*	*iberica*	Thistle, Russian
SETFA	*Setaria*	*faberi*	Foxtail, giant
SIDSP	*Sida*	*spinosa*	Sida, prickly
SINAR	*Brassica*	*kaber*	Mustard, wild
SOLPT	*Solanum*	*ptycanthum*	Nightshade, eastern black
SORHA	*Sorghum*	*halepense*	Johnsongrass
SORVU	*Sorghum*	*bicolor*	Shattercane
STEME	*Stellaria*	*media*	Chickweed, common
XANST	*Xanthium*	*strumarium*	Cocklebur, common

A. Economic Impact

Among the most important impact of weeds is the economic impact. It was previously reported that the economic impact of weeds in the United States exceeds $25 billion annually [6]. This very conservative estimate shows that almost 80% of the documented impact occurs within the agricultural produc-

Table 7 Potential Impacts of Weeds on Humans and Their Activities

---

Economic impact—e.g., agriculture, forestry
  Yield reductions—competitive and noncompetitive effects
  Quality reductions—e.g., dockage
  Operational efficiency—planting, maintenance, and harvest
  Cost of control—chemical, biological, cultural, and mechanical
Environmental impact
  Ecological impact—suppression or replacement of preferred or native vegetation
  Environmental quality—effects of tillage and/or herbicide use
Aesthetic impact
  Unsightliness
  Perceived devaluation of property
Health/safety
  Worker safety—direct effects from weeds and direct effects from weed control
    technologies
  Worker efficiency—lost and unproductive work time due to weeds (e.g., allergies,
    thorns)
  Toxicities/poisonings—animal and human
  Impacts of weed control technologies on human health—herbicide exposure to nonusers;
    i.e., residues in air, food, and water and nontarget exposures

---

tion sector of the economy. The economic impact of weeds on other sectors of the U.S. economy, such as forestry, hay, pasture, and range enterprises, is very poorly documented because of the difficulties related to assessing impact in these systems. Recently, several groups have attempted to assess the impact of weeds, especially nonindigenous, invasive species such as melaleuca (*Melaleuca* spp.), and purple loosestrife (*Lythrum salicaria*) [2].

Economic impact consists of two primary components: lost revenues and costs. Within agricultural enterprises, lost revenues result from yield and quality reductions caused by weeds. These are often referred to as *losses* due to weeds. Inputs, or costs, accrue owing to the use of herbicides, tillage, mowing, grazing, cultural inputs, and biological inputs for weed management and control. These are usually referred to as *costs* due to weeds.

Crop yield and quality losses attributed to weeds vary tremendously by crop. For some crops, the primary component affected by weeds may be the yield component. For example, losses in corn and soybean are often due to yield reductions. For many crops, yield losses due to uncontrolled weeds can be as high as 50–75%. In fact, a 1992 Weed Science Society of America study, *Crop Losses Due to Weeds in the U.S.–1992* [7], reports that despite the use of current best management practices, including herbicides, yield losses due to weeds range from about 5 to 10%. Annual U.S. losses due to weeds with best management practices averaged from 1989 through 1991 were approximately $1 billion, $805 million, and $396 million for corn (*Zea mays*), soybeans (*Glycine max*), and wheat (*Triticum aestivum*), respectively (Table 8). So despite tillage, herbicides, cultivation, and other weed management practices, yield losses often still occurred. This is not surprising given the potential competitiveness of some weeds. For example, one common cocklebur (*Xanthium strumarium*) per 3 m of peanut row can reduce peanut (*Arachis hypgaea*) yield by 10%.

Weeds often reduce crop quality quite dramatically. For example, grasses are often green when cotton (*Gossypium* spp.) is harvested with a mechanical cotton picker. Grass is drawn through the spindles resulting in stained cotton fibers, thus greatly reducing the value of the cotton. Contamination of grain with weed seed or weed chaff diminishes grain value. Wheat that is contaminated with aerial bulblets of wild garlic (*Allium* spp.) is virtually unmarketable for milling. Where pigweeds (*Amaranthus* spp.) are uncontrolled in mint (*Mentha* spp.), crop quality suffers greatly; only small proportions of pigweed biomass in the harvested mint renders mint oil unmarketable (Table 9).

These examples illustrate some of the yield and quality effects that result in lost crop revenues due to weeds. In other cases, the primary economic impact may occur as a result of the costs associated with controlling weeds. For example, in high-value, minor-acreage crops, like herbs, strawberries, endive, and others, few or no herbicides are registered for use, and as a result weed con-

Table 8 Estimated Yield Losses Due to Weeds in Selected Crops and Production Regions of the United States

Production region	Estimated losses with current best management practices ($ × 1000)		
	Corn	Soybeans	Wheat
Appalachian	41,245.86	86,119.52	6,833.19
Corn Belt	541,715.26	456,928.70	31,540.46
Delta States	1,381.80	74,413.89	2,054.26
Lake States	147,836.53	69,698.28	20,388.50
Mountain	18,835.76		65,654.16
Northeast	62,973.36	19,679.32	6,631.17
Northern Plains	161,804.53	69,073.94	103,249.33
Pacific	6,322.10		78,891.26
Southeast	20,252.42	18,922.67	6,279.48
Southern Plains	38,309.31	9,476.45	74,341.54
U.S. Total	1,040,676.94	804,312.77	395,863.35

Source: Adapted from *Crop Losses Due to Weeds in the United States–1992*, Weed Science Society of America.

Table 9 Effect of Various Weed Control Practices on Mint Biomass, Mint Oil Yield, Mint Oil Quality, and Oil Value

Treatment	Spearmint biomass (ton/acre)	Spearmint oil (lb/acre)	Organoleptic quality (index[a])	Oil value ($/acre)
Hand weeded	11.8	26.7	2.0	388
Weedy	2.8	3.9	4.4	0
Chemical weed control	9.5	26.6	1.7	367

[a]Organoleptic quality is a determination of aromatic quality. It is reported as an index where: 1 = acceptable in commerce, 2 = acceptable to marginally acceptable, 3 = marginally acceptable, 4 = marginally unacceptable to unacceptable, and 5 = unacceptable in commerce.
Source: Personal communication, Stephen Weller, from mint weed control research conducted by Purdue University.

trol must be accomplished via hand weeding or mulching. In these cases, weed control costs can be very high owing to labor costs. Weeds also can impair planting and harvesting operations. Reduced harvest efficiency translates into increased production costs. For example, the mechanical harvest of tomatoes and snap beans can be adversely affected by weeds. Mechanical cotton harvest can be hampered by weeds that interfere with the operation of cotton picker spindles. Weedy vines and green biomass can plug grain combines, reducing harvest efficiency and increasing harvest losses.

B. Environmental Impact

The environmental impact of weeds is substantial. They displace native species, alter habitats, and affect wildlife, fish, and other aquatic life. For example, the uncontrolled encroachment of nonindigenous species has adversely affected many natural and/or public lands. Species like melaleuca (*Melaleuca* spp.), water hyacinth (*Eichornia crassipes*), and purple loosestrife (*Lythrum salicaria*) are now a focus of public lands management.

Another environmental cost of weeds is the impact associated with weed control agents and practices, such as herbicides, tillage, and fire. Although there is much debate regarding the precise environmental impact of herbicides, effects clearly are common. Contemporary regulatory practices are designed to minimize the potential of environmental damage resulting from herbicide use, but there are occurrences of off-site movement to sensitive, nontarget species every year. Herbicides are often detected in ground and surface water.

The principal reason for post–plant tillage is weed control. Cultivation leads to increased surface water runoff, silt and sediment production [4], and possibly increased herbicide deposition in streams and lakes due to herbicides that are adsorbed to mineral and organic soil fractions being transported from fields (see Chap. 26). It also contributes greatly to wind erosion and general loss of top soil.

C. Aesthetic Impact

The aesthetic impact of weeds should not be underestimated. Considerable resources are devoted each year to the elimination of weeds that are unwanted only because of their displeasing appearance. For example, lawns are sprayed, mowed, and hand weeded. More than 12 million hectares of turf are managed in the United States alone [5]. Weeds in turf disrupt the uniformity in appearance. They thin turf by competing for light, water, and nutrients, and they produce seed and other propagules that ensure a persistent problem in future years. Vegetable and flower gardens are sprayed, tilled, and weeded. Homeowners are in a constant battle with weeds. A Kline and Co. report indicates that more than $2.4 billion is spent annually at the retail level to maintain American lawns and gardens, much of which is for weed control (personal communication, Mancer J. Cyr, Senior Associate, Agribusiness Group, Kline & Co., Fairfield, NJ).

D. Health and Safety Impacts

The effect of weeds on humans, domesticated animals, and wildlife is extremely large but poorly documented. Some weed species directly and negatively impact farmers and farm workers. Other workers face exposure while applying herbicides. Allergic reactions to weeds such as common ragweed (*Ambrosia artemisiifolia*), poison ivy (*Rhus* spp.), and many grasses result in lost worker productivity and increased medical costs. Human and animal toxicities to accidental exposure to and consumption of poisonous plants are significant. Plants, not all poisonous, account for approximately 10% of all inquiries to Poison Control Centers [14]. Not all species that cause human poisonings are considered to be weeds, but many are. It has been estimated that no less than 280,000 cattle (about 1% of the total herd) and 264,000 sheep (about 3.5% of the flock) are lost annually in the 17 contiguous western United States where free-ranging livestock are exposed to poisonous plants [21]. Based on 1987 prices, Williams estimated annual cattle and sheep losses in the western United States to exceed $234 million [21]. These estimates did not includes losses of swine, horses, goats, wildlife, or livestock east of the Mississippi River. Nor did they include reproductive losses, birth defects, or veterinary fees.

The impact of weed control technologies on human health has not been well documented. Effects are not limited to those resulting from herbicide use. Many workers are injured or killed mowing, cultivating, burning, and mechanically removing undesirable vegetation.

IV. STRATEGIES FOR MANAGEMENT OF WEEDS

The management, or control, of weeds is often more difficult than that of any other class of pests. Among pests, weeds are more akin to the plants (crops) in which they occur than are arthropods, pathogens, nematodes, or other herbivores. They are anatomically, physiologically, and biochemically almost identical to the plants among which we often want to control them. Weeds have presented a unique challenge from the very inception of agriculture. It must have been obvious early, in fact it was the basis for selection of food, feed, and fiber crops, that some plants were more desirable crops than others. Contemporary agricultural systems are typically artificial, highly disturbed, and manipulated ecosystems. Diversity is typically low. In fact, many are maintained as monocultures, at least with respect to plants. So, by their nature, these systems are not very stable. Inputs are generally high, with nutrients, water, and other resources being maintained at near-optimum levels. Weeds that are common to these agricultural systems are equally as well adapted to the prevailing conditions as are the crops. They respond much the same way to changes in the system as does the crop being grown. In one respect, weed populations have an advantage. The genetic diversity (heterogeneity) of weed populations exceeds that of the crop. Assuming that the weeds are equally as well adapted, increased genetic diversity among the weed population ensures that achieving any degree of permanence in the suppression of weed populations will be difficult.

A. Weed Management Versus Weed Control

Is there really a difference? Weed management suggests a broader approach, one that involves the minimization of the undesirable effects of weeds rather than the elimination of the causal organism. Weed management, in other words, uses a variety of strategies, of which control is one, in an effort to minimize the effect that weeds have on crop yield and quality losses, unsightliness, and interference with humans and their activities, livestock, and wildlife. Fryer [10] indicated that weed management involved the "rational deployment of all available technology to provide systematic management of weed problems in all situations."

Certainly, weed control, the use of a variety of tactics for the suppression or reduction of weed numbers, will be an important component of all weed management systems. However, weed management will hopefully comprise more, including manipulation of the habitat or crop production system to disfavor the weed and result in a minimal yield or other detrimental effect. Zimdahl [22] noted that when the basis for weed management is fully developed, it should include the following components:

   a. Incorporation of ecological principles
   b. Full utilization of plant interference and crop-weed competition
   c. Definition and incorporation of economic and damage thresholds
   d. Full integration of several weed control techniques, including selective use of herbicides
   e. Supervised weed management, probably by a professional pest manager employed to develop a program for each farm

This and other conceptual frameworks for weed management, of which there are many, recognizes the need for understanding the interaction of the weed with the desirable vegetation (crop) and their common environment. It also emphasizes that the real objective is to minimize, or eliminate, the undesirable impact of the weed, which may be more easily and sustainably accomplished than eliminating the weed. Altieri and Liebman [1] listed 10 key steps in achieving weed management. These were recently concisely described by Zimdahl [22]:

   a. *Monitor* seed and vegetative populations.
   b. *Identify* problem weed species and their density.
   c. *Study* the farmer's present methods.
   d. *Assess* the dominant weed species and their interactions.
   e. *Predict* weed populations and population shifts.
   f. *Decide* whether control should be done.
   g. *Choose* the control technology compatible with the system.
   h. *Integrate* weed and other crop protection measures.
   i. *Evaluate* long-term environmental, social, and economic impacts.

B. Tools of Weed Management

Relative to agricultural production, many aspects of crop management can affect how successfully a population of weeds will persist and compete with the crop. For example, adjusting cultural practices such as planting date, row spacing, seeding rate, and variety selection can significantly affect the competitiveness of weeds. Altering planting date may allow a crop to emerge before certain weeds and thus gain a competitive advantage. The same may be true for altering row spacing and seeding rate. Emergence and growth rate, developmental rate, and maturity are all factors that are variety/cultivar dependent. Moderate gains in crop competitiveness may be achieved by carefully selecting the appropriate variety. These management tactics do not necessarily result in lower weed populations, short-term or long-term, but they reduce the impact of the weed and, therefore, they raise the economic or damage threshold.

C. Prevention, Eradication, and Control

Other management tactics involve manipulating the population of weeds. With respect to weed populations, these tactics are directed at either preventing, eradicating, or controlling weed populations. *Prevention* involves tactics that inhibit or delay the introduction and/or establishment of weeds in an area where the particular weed does not already exist. Often preventative tactics are cultural or regulatory in nature. For example, they may include the promotion or requirement for planting agricultural seed that are certified to be free of certain weed seed. Regulatory practices such as quarantine and weed laws are often means of achieving prevention.

Eradication involves the elimination of a weed species from a defined area. Although eradication is theoretically possible, it is achieved very rarely, particularly if the species has become established in the area. Once weeds are established in an area (i.e., they have survived in an area for one or more life cycles), the probability for successful eradication declines dramatically. Eradication is most often attempted where particularly noxious or invasive weeds have been introduced, but have not yet formed stable, reproducing populations. For example, eradication efforts with parasitic weeds such as witchweed (*Striga* spp.) and broomrapes (*Orobanche* spp.) have been moderately successful in some locales, especially when introduction was restricted to a limited and confined area. More recently, quarantine and eradication efforts have been attempted to confine and limit the spread of tropical soda apple (*Solanum virarum*) in the southeastern United States [3]. Eradication usually requires an intensive program sustained over a period of several to many years, because weeds and other propagules must be depleted from the soil seed bank and reintroduction must be prevented for the program to be successful.

Control involves the use of tactics that are directed toward suppressing or limiting the population of an existing weed species within a prescribed area. Control tactics may reduce weed populations to below threshold levels. They may cause weed shifts and they may result in selections for other changes within the weed population, but they rarely have a lasting effect on the weed population unless they are sustained with vigor over a long period of time. Weed control tactics are usually considered to take one of four forms: mechanical, cultural, chemical, or biological.

D. Approaches to Weed Management

1. Mechanical Weed Control

Mechanical, or physical, methods of weed control include pulling, hoeing, mowing, clipping, tilling, cultivating, burning, chaining, flooding, and mulching, among others (see Chap. 21). They share a common mechanism of action; that is, physical disruption of the normal growth and/or development of the weed. These weed control methods are among the most widely used around the world. Along with cultural controls, mechanical weed control is the mainstay of weed control in all developing agricultural civilizations. Likewise, they remain an essential part of weed control in the United States. Despite the common use of herbicides, mechanical weed control remains the most widespread form of weed control in the United States. Wicks et al. [20] provided an excellent review of the use of mechanical weed control methods in the United States, including how these tactics can be integrated with other weed control methods.

2. Cultural Weed Control

Cultural practices often rely on the interplay between prevention, avoidance, and minimizing the impact of weeds (see Chap. 21). Cultural weed control tactics includes sanitation; that is, planting weed seed-free crop seed, using weed seed-free irrigation water, and cleaning farm equipment to prevent movement of weeds from one field to another. Control tactics also may include crop rotation, including fallow periods in which no seed rain is allowed and weed seed numbers are reduced. Adjustment of planting dates, row arrangement and spacing, and seeding rate also may be used to reduce weed numbers. Two excellent reviews on cultural weed control and prevention have been published [18,19]. Most cultural weed control techniques focus on modifying the relationship between the weed, the crop, and the environment in order to reduce the chance for weed emergence, survival, or competition.

3. Biological Weed Control

Biological weed control involves the use of other biological organisms, other plants, insects, herbivores, or microorganisms to lower the population of a weed species (see Chap. 22). Sometimes the objective is to reduce competitiveness or reproductive output rather than to reduce population in the short term. Success with biological weed control has been limited in number but great in impact. Biological control has been most successful where the agent was deployed over a rather large area and where short-term population suppression was not the goal. Establishment of stable, resident populations of a biological control agent takes time. In many cases, sufficient host numbers must remain to support the biological agent, disease-causing organism, arthropod, or other herbivore. In effect, many successful biological control systems simply provide regulation and stability of the weed population over a period of time rather than having episodic cycles of population growth followed by dramatic control with mechanical or chemical methods.

Successful biological control has been achieved with mycoherbicides, insects, birds, fish, and livestock. Cardina [9] recently provided an excellent review of biological weed control, including success stories illustrating the role of wide variety of organisms in suppressing weed growth and/or reproduction.

4. Chemical Weed Control

The use of chemicals, or herbicides, to control weeds dates back at least to the previous century (see Chap. 23). Inorganic salts like sodium nitrate, ammonium sulfate, iron sulfate, sodium chloride, and acids such as sulfuric acid were used for nonselective weed control well before the turn of the century. By 1900, they were being used somewhat selectively for weed control in cereals. The first organic herbicides were apparently petroleum-based oils, solvents, or fuels that were used for nonselective weed control. During the 1930s, phenolic-type herbicides were used in the United States. But, the real birth of chemical weed control technology occurred shortly after World War II, when 2,4-dinitrophenoxyacetic acid (2,4-D) was revealed to the civilian world. Today, herbicides are an important component of the agrichemical business. It has been recently estimated that approximately 200 active ingredients are used worldwide as herbicides [12]. Recent estimates show that approximately 70% of the pesticides used in the United States were herbicides [16].

V. CONCLUSIONS

Weed management in most agricultural production systems requires the integration of several tactics. In most cases, weeds are not eliminated. Rather weeds are managed so that their numbers and spatial and temporal occurrence have a minimal, or acceptable, impact on the crop. This often requires using an integrated approach that includes the use of cultural, mechanical, chemical, and/ or biological controls. For example, in many row crops, herbicides are used along with tillage and cultivation. In certain high-value, minor-acreage crops where herbicide use is limited, weed control is achieved using a combination of mulches and hand weeding.

REFERENCES

1. Altieri, M.A. and M. Liebman. 1988. Weed management: ecological guidelines. In *Weed Management in Agroecosystems: Ecological Approaches*. (M.A. Altieri & M. Liebman, eds. Boca Raton, FL: CRC Press, pp 331–337.
2. Anonymous. 1993. The consequences of harmful non-indigenous species. In *Harmful Non-Indigenous Species in the United States*. Office of Technology Assessment, OTA-F-565. Washington, DC: U.S. Government Printing Office, pp 51–76.
3. Anonymous. 1996. *Tropical Soda Apple Symposium Proceedings*. Gainesville, FL: University of Florida.
4. Beatley, M.T. 1982. Soil erosion: its agricultural, environmental and socioeconomic implications. Report 92, 29. Ames, IA: Council for Agric. Sci. & Tech.
5. Bingham, S.W., W.J. Chism, and P.C. Bowmik. 1995. Weed management systems for turfgrass. In *Handbook of Weed Management*. (A.E. Smith, ed.) New York: Marcel Dekker, pp 603–665.
6. Bridges, D.C. 1994. Impact of weeds on human endeavors. Weed Technol. 8:392–395.
7. Bridges, D.C. and R.L. Anderson. 1992. Crop losses due to weeds in the United States. In *Crop Losses Due to Weeds in the United States-1992*. (D.C. Bridges ed.), Lawrence, KS: Weed Science Society of America, pp 1–74.
8. Bridges, D.C. and P.A. Bauman. 1992. Weeds causing losses in the United States. In *Crop Losses Due to Weeds in the United States-1992*. (D. C. Bridges ed.), Lawrence, KS: Weed Science Society of America, pp 75–147.
9. Cardina, J. 1995. Biological weed control. In *Handbook of Weed Management*. (A.E. Smith, ed.) New York: Marcel Dekker, pp 279–341.
10. Fryer, J.D. 1985. Recent research on weed management: new light on an old practice. In *Recent Advances in Weed Research*. (W.W. Fletcher, ed.) Old Working, Surrency, UK: Gresham Press.
11. Harlan, J.R. and J.M.J. de Wet. 1965. Some thoughts about weeds. *Econ. Bot.* 19:16–24.
12. Harrison, S.K. and M.M. Loux. 1995. Chemical weed management. In *Handbook of Weed Management*. (A.E. Smith, ed.) New York: Marcel Dekker, pp 101–153.
13. Holm, L.G., D.L. Plucknett, J.V. Pancho, and J.P. Herberger. 1992. *The World's Worst Weeds*. Malabar, FL: Krieger.
14. Lampe, K.F. and M.A. McCann. 1985. *AMA Handbook of Poisonous and Injurious Plants*. Chicago: Chicago Review Press.
15. Navas, M.L. (1991) Using plant population biology in weed research: a strategy to improve weed management. *Weed Res.* 31:171–179.
16. Pimentel, D., L. McLaughlin, A. Zepp, B. Lakitan, T. Kraus, P. Kleinman, F. Vancini, W.J. Roach, E. Grapp, W.S. Keeton, and G. Selig. 1991. Environmen-

tal and economic impacts of reducing U.S. agricultural pesticide use. In *Handbook of Pest Management in Agriculture*. Vol. I. (D. Pimentel, ed.) Boca Raton, FL: CRC Press, pp 679–718.

17. Smith, A.E. and D.M. Secoy. 1976. Early chemical control of weeds in Europe. *Weed Sci.* 24:594–597.

18. Walker, R.H. 1995. Preventative weed management. In *Handbook of Weed Management*. (A.E. Smith, ed.) New York: Marcel Dekker, pp 35–50.

19. Walker, R.H. and G.A. Buchanan. 1982. Crop manipulation in integrated weed management systems. *Weed Sci.* 30(suppl):17–24.

20. Wicks, G.A., O.C. Burnside, and W.L. Felton. 1995. Mechanical weed management. In *Handbook of Weed Management*. (A.E. Smith, ed.) New York: Marcel Dekker, pp 51–99.

21. Williams, M.C. 1994. Impact of poisonous weeds on livestock and humans in North America. *Rev. Weed Sci.* 6:1–27.

22. Zimdahl, R.L. 1995. Introduction. In *Handbook of Weed Management*. (A.E. Smith, ed.) New York: Marcel Dekker, pp 1–18.

21
Mechanical and Cultural Approaches to Weed Management

William K. Vencill
University of Georgia, Athens, Georgia

George W. Langdale*
United States Department of Agriculture, Watkinsville, Georgia

I. INTRODUCTION

Cultural weed management refers to systems in which crop competitiveness with coexisting weed species is maximized, thus providing better weed control with fewer chemical inputs. Before the mid-1940s, and to the present day in underdeveloped regions of the world, growers relied primarily on cultural and mechanical weed control, since alternatives were unavailable. Interest in cultural weed control has been renewed recently due to mandatory reductions in pesticide usage in European countries and growing support for similar reductions in North America (63,72,73,78,89).

Cultural weed control can be used to reduce herbicide inputs by delaying, decreasing, or preventing harmful weed emergence or by reducing the vigor of weeds treated with herbicide (64). One of the most common and effective means of cultural weed control is that of optimizing crop competitiveness with weeds (73,89). Enhancement of crop competitiveness depends on several factors, including timing of crop emergence in relation to weeds, availability of moisture and nutrients, crop vigor and health, spatial patterns of crop and weeds, crop morphology in relation to weeds, soil structure, and allelopathic effects (13,33,51,98,99,134,137,145). The competitive ability of crops can be enhanced by exploiting some of these factors, such as timing the sowing of wheat to avoid

---

*Retired.

an autumn weed flush or using close row spacing or cross drilling to improve the smothering effect of the crop on weeds (111). The components of cultural weed management that are discussed in this chapter are tillage systems, crop rotations, seeding rate and crop populations, canopy arrangement, varietal responses, nutrition, and water management. This chapter focuses on how these factors influence weed pressure and competition, as well as their impact on crop growth, particularly crop competitiveness. Discussion of herbicides is limited to how cultural weed management components can be integrated with chemical weed control (see Chapter 23 for discussion of chemical control of weeds).

II. TILLAGE

The primary role of tillage in an agricultural setting is weed control. The type of tillage employed by a grower, whether it is zero tillage, ridge tillage, moldboard plowing, harrowing, or another type, has a profound effect on crop/weed interactions. Tillage affects weed species present and has a direct impact on the crop's competitive ability. However, as a weed control technique, tillage has several major disadvantages. On a practical scale, tillage operations can injure the crop, increase disease incidence, and provide little residual control (142). In addition, selective cultivation is weather dependent, and if it is not done in a timely manner, it is ineffective. Tillage is a critical component of cultural weed management inasmuch as it defines the agroecosystem in which the crop and weed exist (84).

A wide array of research has examined the utility of tillage for weed control (6,23,81,102,103,124). Timeliness is critical to the success of mechanical weed control. For example, Lovely et al. (81) observed that rotary hoeing when performed while weeds were small reduced weed infestations in soybean (*Glycine max*) by 72% and increased soybean yield 671 kg/ha over soybean without weed control. However, rotary hoeing operations conducted on larger weeds reduced weed infestation by 38% and increased soybean yield by 336 kg/ha over soybean not receiving any weed control. Similarly, Buhler et al. (22) found that timeliness of cultivation as well as environmental conditions, weed density, weed species, and weed emergence patterns influenced the effectiveness of mechanical weed control.

Moss (91–93) compared cultivation systems (plowing, tine cultivation, and direct drilling) for control of blackgrass (*Alopecurus myosuroides*) in spring barley (*Hordeum vulgare*). He found that 80–90% of blackgrass seeds in the direct drilled wheat (*Triticum aestivum*) were derived from recently shed seeds. Blackgrass infestations in tine-cultivated wheat were similar to those in direct drilled wheat. In contrast, plowing destroyed blackgrass seeds from the previous crop and reduced blackgrass populations. In a similar experiment, Froud-Williams (57) observed that burning of surface plant residues after harvest destroyed 97% of ungerminated seeds on the soil surface and reduced seedling numbers by 94%. Shallow cultivation reduced weed seed population by 34%. Plowing to a depth of 20 cm eliminated blackgrass infestations.

The stale seedbed technique integrates cultural, mechanical, and chemical weed control methods. It involves disking or harrowing of a field well before seeding, the application of a nonselective herbicide to weeds that germinate, and planting of the crop with minimal soil disturbance, since soil disturbance can promote weed seed germination (45). The effectiveness of stale seedbed techniques include selection of appropriate cultivars (i.e., fast growing), timely planting, irrigation, and effective weed control with preplant applications of herbicides. Standifer and Beste (125) eliminated viable large crabgrass (*Digitaria sanguinalis*) from the upper 2 cm of soil and significantly reduced populations of *Cyperus* and *Poa* spp. using the stale seedbed technique.

Numerous studies have evaluated the effect of conservation tillage systems on weed population dynamics and results have been mixed. In a detailed review, Moyer (94) provided a list of weeds that are reportedly favored by conservation or conventional tillage systems (Tables 1–3). Wicks et al. (140) reported the results of a 40-year study in North Platte, Nebraska, in which tillage shifted from plowing to disking to ridge tillage. In the plow system, kochia (*Kochia scoparia*), redroot pigweed (*Amaranthus retroflexus*), and green foxtail (*Setaria viridis*) were predominant. Annual grasses dominated when disking supplanted plowing as the tillage system, whereas ridge tillage favored the emergence of winter annual broadleaf weeds. In corn (*Zea mays*), 24% more weed seedlings emerged after corn was planted in a ridge tillage system than in conventional tillage systems. Similarly, Donaghy et al. (40) observed in conservation tillage fields an increase in the density of perennial weeds, annual grasses, windblown weeds, and native plant species that were not normally found in cultivated fields. In contrast, in a long-term study of weed populations in a cropping system of spring wheat, winter wheat, flax (*Linum usitatissimum*), field peas (*Pisum sativum*), and summer fallow, annual and biennial weed populations were not affected by tillage (39). Nevertheless, proper crop rotations may alleviate higher weed populations that can occur in conservation tillage systems (117).

Vencill and Banks (135) examined weed population dynamics in minimum tillage and conventional tillage grain sorghum for 4 years at zero, low, medium, and high herbicide inputs and found that weed species composition was affected more by weed management system than by tillage. However, biennial and perennial weeds did increase in the conservation-tillage sorghum. In Brazil and Argentina, annual weed populations declined in zero tillage systems (50), but

Table 1 Influence of Tillage on Annual Weed Species and Density

Scientific name	Common name	Reference
Weeds with greater densities in conservation tillage than conventional-tillage		
Abutilon theophrasti Medicus	Velvetleaf	21,48,110
Aethusa cynapium L.	Foolsparley	59
Anagallis arvensis L.	Scarlet pimpernel	58,59
Atriplex patula L.	Spreading orach	58
Avena fatua L.	Wild oat	40,54
Brassica kaber (DC.) L.C. Wheeler	Wild mustard	59,87,132
Capsella bursa-pastoris (L.) Medicus	Shepherd's purse	58
Chenopodium album L.	Common lambsquarters	130
Fumaria officinalis L.	Fumitory	56,59
Matricaria spp.	NA	132
Papaver rhoeas L.	Corn poppy	59
Polygonum convolvulus L.	Wild buckwheat	40,54,87
Raphanus raphanistrum L.	Wild radish	59,132
Setaria viridis (L.) Beauv.	Green foxtail	40,54,130
Thalaspi arvense L.	Field pennycress	40,130
Stellaria media (L.) Vill.	Common chickweed	56,132
Viola arvensis Murr.	Field violet	59
Xanthum strumarium L.	Common cocklebur	27
Weeds with greater densities in conventional than conservation-tillage		
Alchemilla arvensis (L.) Scop.	Parsley-piert	132
Alopecurus myosuroides Huds.	Blackgrass	59,132
Amaranthus spp.	Pigweeds	110,135
Aristida oligantha Michx.	Prairie threeawn	26
Avena fatua L.	Wild oat	16,38,59
Brachiaria ramosa (L.) Stapf	Browntop millet	44
Bromus spp.	NA	26,49,59,132,146

Capsella bursa-pastoris (L.) Medicus	Shepherd's purse	132
Cenchrus incertus M.A. Curtis	Field sandbur	26
Chenopodium album L.	Common lambsquarters	110,132,135
Conyza canadensis (L.) Cronq.	Horseweed	26,110,135
Croton texensis (Klotzsch) Muell.-Arg. ex DC.	Texas croton	26
Descurainia spp.	NA	26,68
Digitaria sanguinalis (L.) Scop.	Large crabgrass	26,44,83,135
Echinochloa crus-galli (L.) Beauv.	Barnyardgrass	26,68
Eleusine indica (L.) Gaertn.	Goosegrass	26,130
Euphorbia humistrata Engelm. ex Gray	Prostrate spurge	26,68
Helianthus annuus L.	Common sunflower	26,68
Kochia scoparia (L.) Schrad.	Kochia	26,68
Lactuca serriola L.	Prickly lettuce	26,68
Matricaria spp.	N/A	26,147
Mollugo verticillata L.	Carpetweed	130
Panicum dichotomiflorum Michx.	Fall panicum	10,26,48,59,110
Poa annua L.	Annual bluegrass	56,131,146
Polygonum aviculare L.	Prostrate knotweed	68,132
Polygonum erectum L.	Erect knotweed	26
Portulaca oleracea L.	Common purslane	26
Schedonnardus paniculatus (Nutt.) Trel.	Tumblegrass	26
Senecio vulgaris L.	Common groundsel	132
Setaria viridis (L.) Beauv.	Green foxtail	54,87
Sida spinosa L.	Prickly sida	44
Solanum triflorum Nutt.	Cutleaf nightshade	26
Sorghum bicolor (L.) Moench.	Shattercane	26
Stellaria media (L.) Vill.	Common chickweed	147
Tribulus terrestris L.	Puncturevine	26

Source: Adapted from ref. 94.

Table 2 Biennial Weeds that Have Greater Density in Conservation Tillage than Conventional Tillage

Scientific name	Common name	Reference
Artemisia biennis Willd	Biennial wormwood	68
Carduus acanthoides L.	Plumeless thistle	26
Carduus nutans L.	Musk thistle	26,56,68
Cicuta spp.	Waterhemlock	26
Cirsium vulgare (Savi) Tenore	Bull thistle	26
Lactuca serriola L.	Prickly lettuce	59
Melilotus spp.	Sweetclover	26,68
Tragopogon dubius Scop.	Western salsify	26,68

Source: Adapted from ref. 94.

Zentner et al. (148), Donald and Nalewaja (42), and Fay (49) found that annual grass weeds were favored by conservation tillage systems. Peeper and Wiese (100) reported that growers in Oklahoma who have wheat monocultures must use moldboard plowing to control downy brome (*Bromus tectorum*) despite erosion problems.

Froud-Williams (57) hypothesized that wind-blown weed species may be favored by conservation tillage systems, and this hypothesis is supported by several studies. Derksen (38) found that wind-dispersed weeds were more numerous in no-till than in conventionally tilled treatments. Fay (49) reported that weeds of the family Poaceae that have windborne seeds are not well adapted to burial in soil and are favored in conservation tillage systems.

Gill and Arshad (60) examined weed population dynamics among three tillage intensities (zero, minimum, and conventional tillage). Broadleaf weeds declined as a percentage of total weed species present with a reduction in tillage intensity in wheat, barley and canola (*Brassica napus*). Perennial weed intensity was inversely related to tillage intensity. Species diversity of broadleaf and total weed population exhibited a greater percentage of common and rare species with conventional tillage than with zero tillage.

Hartmann and Nezadal (65) observed that weed coverage after 7 years was reduced from 80% coverage when tillage was performed in daylight to 2% coverage when all tillage operations were performed between 1 h after sunset and 1 h before sunrise. The success of this system is based on weed seeds requiring light for germination as per a phytochrome response. However, many crop plants are light independent for germination. The investigators hypothesized that this type of system could reduce herbicide inputs dramatically for appropriate weeds under certain conditions.

Table 3 Perennial Weeds that Have Greater Density in Conservation Tillage than Conventional Tillage

Scientific name	Common name	Reference
Acroptilon repens (L.) DC	Russian knapweed	26,68
Ambrosia gray (A. Nels.) Shinners	Woollyleaf bursage	26
Andropogon virginicus L.	Broomsedge	26
Apocynum cannabinum L.	Hemp dogbane	26,48,132,143
Asclepias syriaca L.	Common milkweed	26,48,132,143
Calystegia sepium (L.) R. Br.	Hedge bindweed	26
Campsis radicans (L.) Seem. ex Bureau	Trumpetcreeper	146
Cardaria draba (L.) Desv.	Hoary cress	26,68
Chloris verticillata Nutt.	Tumble windmillgrass	26
Cirsium arvense (L.) Scop.	Canada thistle	17,26,40,48,54,68,110,132
Convolvulus arvensis L.	Field bindweed	26,68,132,143
Cynodon dactylon (L.) Pers.	Bermudagrass	26,132,143
Cyperus spp.	NA	26,59,143
Agropyron repens L.	Quackgrass	16,38,40,49,68,101,107,128,130
Euphorbia esula L.	Leafy spurge	26,42,68
Helianthus tuberosus L.	Jerusalem artichoke	26
Hordeum jubatum L.	Foxtail barley	16,26,38,42,68,128
Lygodesmia juncea (Pursh) D. Don.	Skeleton weed	26
Muhlenbergia frondosa (Poir.) Fern.	Wirestem muhly	26,48
Paspalum dilatatum Poir.	Dallisgrass	132,143
Physalis spp.	NA	26,132,143
Rosa spp.	NA	26
Rubus spp.	NA	132
Rumex spp.	NA	26,56,68,132,135
Sassafras albidium (Nutt.) Ness	Sassafras	132
Solanum carolinense L.	Horsenettle	26,48,59,143
Sonchus arvensis L.	Perennial sowthistle	15,68
Sorghum halepense (L.) Pers.	Johnsongrass	17,132
Sporobolus cryptandrus (Torr.) Gray	Sand dropseed	26

Source: Adapted from ref. 94.

III. CROP ROTATION

Prior to 1945 and the use of modern herbicides, crop rotation and mechanical methods were the principal methods of weed control. For example, the traditional cure for wild oat (*Avena fatua*) infestations in cereals was to rotate to a grass or sod crop (79). Several factors are critical determinants of weed infestation severity and species composition. Of these, the crop type is the most critical (129). The crop selected determines planting date, harvest date, and other cultural practices that influence associated weeds (27,59,97). Leighty et al. (79) examined 29 crop rotation combinations. In 21 of 29 trials, weed densities were greater in the continuous monoculture system than in the rotational systems. Cereal crops sown in the spring and in the autumn have similar associated weed communities in northern Europe, but the relative proportions of species differ (59). A rotation from rice to a broadleaf crop such as cotton (*Gossypium hirsutum*), tomatoes (*Lycopersicum esculentum*), safflower (*Carthamus tinctorius*), sorghum (*Sorghum bicolor*), corn, or wheat reduced aquatic weed populations that can increase in flooded fields of rice (*Oryza sativa*) (67). Stewart and Pittman (127) reported that rotations of wheat with sugarbeet (*Beta vulgaris*) increased wheat yields 79% over continuous monoculture wheat. Heard et al. (66) evaluated a four-course rotation of barley, fodder beet, barley intersown with grass clover, and a grass-clover ley and observed reduced populations of broadleaf and most grass weeds in wheat that followed this rotation. Liebman and Dyck (77) compared weed populations in continuous monoculture wheat with wheat grown in a rotational scheme of summer fallow followed by winter wheat. Green foxtail populations were 960 plants m^{-2} in the monoculture wheat versus 3.5 plants m^{-2} in the rotation. Monoculture intensifies the simplification of weed-crop communities, usually resulting in a weed flora dominated by one to several difficult to control weed species (77). In a corn monoculture, a single weed species comprised 94% of the weeds present; whereas in a corn-wheat rotation no single species contributed over 43% of the total weeds present. Crop rotations improve weed control by periodically changing the weed community—various crops differ in planting and harvest dates, growth habit, competitive ability, fertility requirements, and associated production practices, thereby favoring different weed complexes (52,53,77).

IV. COVER CROPS, INTERCROPPING, AND MULCHES

Cover crops are another way of minimizing weed populations while maintaining an aseasonal cover to prevent soil erosion (90). In one study, rye (*Secale cereale*) or hairy-vetch (*Vicia villosa*) residue reduced total weed density by 78% when the cover crop biomass exceeded 300 g m^{-2} compared with treatments without a cover crop (130). Various cover crop species and varieties differ in their relative biomass production. Weston (139) compared nine cover crops in Kentucky and found rye, barley, and wheat produced more biomass than tall fescue (*Festuca arundinacea*), red fescue (*Festuca rubra*), or white clover (*Trifolium repens*). In addition to more cover for weed control, additional biomass can provide several other agronomic benefits. In the same study, cereal grain covers provided the most compatible planting situations for seedling establishment of soybean compared with legume covers. Rye and wheat also provided greater suppression of weed growth than did the other cover crops.

Intercropping (growing two crops simultaneously) has been used to reduce weed growth (77). Intercropping combinations allow the exploitation of more available resources than one crop and probably suppress weed growth by preemptive use of these resources by the crops (78). Wilson and Phipps (144) observed that wild oat seed populations greater than 2000 seed m^{-2} were reduced to 2 seed m^{-2} in wheat following a 3-year intercropping with ryegrass and white clover (36), redtop (*Agrostis gigantea*) (85), and wild oat (36).

Unamma et al. (133) examined the effect of intercropping cowpea (*Vigna unguiculata*) and egusi melon (*Citrullus lanatus*) in corn/cassava (*Manihot esculenta*) in Nigeria. Using either two-hand weedings or herbicides as a comparison with the intercrop, they found that weed control with intercrops was not equivalent to herbicides or hand weeding, but that yields were equivalent. Enache and Ilnicki (46) found that corn yields in a system where corn was planted into subterranean clover (*Trifolium subterraneum*) without tillage or herbicides were equivalent to conventional systems. However, intersown low-growing smother crops can reduce yields if competition for water or nutrients is strong. Competition of smother crops with the main crop may be reduced by applications of growth retardants and low herbicide doses to the smother intercrop.

Nonliving mulches are another means of reducing weed infestations via dense cover. These mulches can consist of organic materials, such as fresh tree bark, straw, or litter composts. In intensive cropping systems such as vegetables and ornamentals, plastic mulches have been widely utilized. Plastic mulches prevent light from reaching the soil surface, thereby inhibiting weed growth. Niggli et al. (96) compared many types of mulches and found that mulches of fresh bark from conifer or oak or rape straw were superior to other mulches in suppressing weeds. Black polyethylene mulch can also provide effective weed control for basil (*Ocimum basilicum*) and rosemary (*Rosmarius officinalis*) but not for parsley (*Petroselinum crispum*) (114).

Stale seedbeds (see above) are another means of providing a weed-suppressing mulch by leaving vegetative residue on the surface that reduces light penetration to soil surface.

V. PLANTING DATE

The crop planting date can have a major impact on the severity of weed infestation encountered in any given crop. Indeed, Brenchley and Warington (18) stated that the planting date of a crop is the main factor determining the composition of the weed flora. Rapid and even emergence of the crop is critical to a crop's success and its competitive advantage over associated weeds. A review of weed control is linseed by Lutman (82) noted that early-sown crops tend to emerge slowly and have an uneven establishment making them more susceptible to weed competition. Similar examples can be found in most major row crops such as corn, soybean, cotton, and peanut.

VI. SPATIAL ARRANGEMENT, SEEDING RATE, AND ROW SPACING

Spatial arrangement, seeding rate, and row spacing utilize the architecture of the crop and canopy to inhibit or reduce weed growth physically. Several reports show that cotton (8), corn, and wheat (41) display differential competitive ability based on geographical orientation. Soybeans planted in north-south rows intercept light better than those planted in east-west rows, and subsequently yield better and are more competitive with associated weeds (104). However, in another study Duncan and Schapaugh (43) did not observe an effect of row orientation on yield of soybean.

Population density of a given crop can dramatically improve its competitive ability with associated weed species. Certain cereal crops can suppress weed growth when sown at high densities and in narrow rows between other crops. Dense planting of barley seed has been shown to suppress quackgrass (*Elytrigia repens*) infestations. Increased seeding rates of winter wheat can reduce the number of flowering heads of blackgrass and the total biomass of sterile oat (*Avena sterilus*) (5,59). Increased crop density and narrow row widths may be used to reduce competitive ability of wild oat in spring barley (5). Higher seeding rate of safflower can hasten the formation of dense canopies and improve crop competitiveness with associated weed species (14). Safflower yields increased with increased crop density until an upper limit was reached. Likewise, high safflower densities (>70 plants m^{-2}) improved crop yields fourfold over safflower planted at lower densities (14).

Many studies have examined the influence of row spacing on the competitive ability of a given crop with associated weed species. Overall, these studies have shown that in most crops, narrower row spacing increases the competitiveness of the crop by allowing the crop to form a canopy more quickly and allowing it to intercept more light than associated weeds.

Reduced row crop width has been shown to favor crop development at the expense of weeds in soybean, peanut (*Arachis hypogaea*), transplanted rice, and corn (59). However, seeding rates of dry-seeded rice did not affect weed competition in this crop. Gunsolus (62) reported that soybean was better suited than corn to take advantage of the competitive advantages offered by late planting and narrow row spacing, because soybean normally closes its canopy slower than corn.

In Illinois, soybean yields increased and weed yields decreased as soybean row spacing was decreased (4,138). With 76-, 51-, and 25-cm row spacing soybean yield increased 10, 18, and 20%, respectively, over the yield with standard 102-cm rows. In comparing soybean's competitiveness with pitted morningglory (*Ipomoea lacunosa*) in 1- and 20-cm row spacing, Howe and Oliver (70) found that pitted morning-glory interfered with soybean growth earlier in the wider row spacing. In conventional and conservation tillage systems, reduced row width improved soybean competitiveness with associated weeds (70). Freed et al. (55) found that soybean yield from 45-cm row planting was greater than from those planted in 60- and 90-cm rows.

Koscelny et al. (76) examined wheat planted in rows spaced from 8 to 23 cm apart. Wheat yield was greater in the narrow-row than in wide-row spacing, but when cheat (*Bromus secalinus*) was present, the narrower row spacing did not improve wheat yield. Kirkland et al. (74) examined row spacing in spring barley and found that barley yields were inversely related to row spacing in a system of 11-, 22-, 33-, and 46-cm row spacing. In the narrower row spacing, they found that wild oat, wild mustard (*Brassica kaber*), and volunteer canola populations were reduced.

Narrowing the row spacing in peanut improved its competitive ability with Florida beggarweed (*Desmodium tortuosum*) and sicklepod (*Senna obtusifolia*) (19). Weed growth was less in 20.3- and 40.6-cm rows when compared with standard 81.2-cm rows and peanut yield was 10–30% higher in the 40.6-cm row spacing than in the 81.2-cm row. Buchanan and Hauser (19) estimated that peanut yield was increased 50% owing to reduced weed competition for peanut planted in 40.6-cm rows compared with peanut in 100-cm rows. Similarly, peanut planted in a twin 18-cm row pattern yielded greater than peanut planted in the standard 91-cm row pattern (31).

Cotton also responds well to narrower row spacing. Rogers et al. (116) found that cotton planted in narrower 53-cm rows produced higher yields than cotton in rows of 79 and 106 cm. Cotton planted in the 53-cm row required 6 weeks of weed free maintenance for a maximum yield, whereas cotton planted in 79- and 106-cm rows required 10 and 14 weeks, respectively, of weed-free conditions to obtain optimum yield.

Environmental conditions can influence the efficacy of modified row spacing for weed suppression. For example, in Texas, grain sorghum was more competitive in wide rows when dry conditions prevailed and in narrow rows when moist conditions were present (141).

VII. CULTIVAR SELECTION

Crop cultivars with different rates of maturity and competitive ability may be used to suppress weed populations and weed growth (59,112). Grundy et al. (61) reported that short-straw cereal cultivars might be at a competitive disadvantage with tall vigorous weeds and that long-straw cultivars tend to suppress weed growth.

Crop varieties vary in their relative competitiveness for nitrogen. In competition with yellow foxtail (*Setaria glauca*), early-maturing corn hybrids as a group were more competitive for nitrogen than the late-maturing ones (126). Below 157 kg N/ha, weeds reduced yield of late-maturing varieties 20%, but yield of early-maturing varieties was only reduced 6% relative to late-maturing ones.

Richards and Whytock (113) evaluated the competitive abilities of several wheat varieties with wild oat infestations and found that those varieties that achieved good early ground cover experienced lower subsequent wheat infestations. In spring barley, the erectoid types were less competitive than intermediate types and early ground cover had a greater effect on weed populations than did crop density. Normal-height wheat varieties were more competitive with green foxtail than semidwarf varieties (12). However, others evaluating wheat have not found a varietal response to weed competition (76). Richards (111) observed that wheat varieties that achieved the most rapid ground cover competed better with weeds than did varieties that covered the ground more rapidly.

Several groups have examined differential responses among crop varieties to weed competition (25,80,93,109). McWhorter and Hartwig (86) examined soybean varietal responses to johnsongrass (*Sorghum halepense*) and common cocklebur (*Xanthium strumarium*) competition. Johnsongrass reduced soybean yield 23–42% and common cocklebur reduced yield 53–75% depending on variety. Factors enhancing differences in cultivar competitiveness include early seedling vigor, high relative growth rate, and short time to canopy closure.

Similar responses have been noted for grain sorghum (24). Differences of 26% in yields between the most competitive varieties and the least competitive varieties have been observed. These differences were due mostly to the rapid emergence of sorghum after planting and relative growth rate.

Many studies on barley/wild oat competition for soil nutrients show that increased nutrient availability benefits both species and can increase the competitive ability of wild oats or barley depending upon the study. For example, Siddiqi et al. (121) found potassium to be a nutrient affecting barley/oat competition and studied the variability among barley cultivars in their ability to compete for available potassium. They hypothesized that cultivar-specific differences were due to differential uptake and utilization of potassium.

VIII. FERTILITY

Competition in the rhizosphere for nutrients and moisture is particularly important for crop vigor and for competitiveness with associated weed species. The relative efficiencies of nutrient acquisition by crops and weeds may be responsible for differences in aboveground competition for light (1,121). Soil fertility can affect weed management in one of two ways. First, soil fertility can affect crop vigor that, in turn, can improve the crop's competitiveness with associated weeds. Second, soil fertility can affect species composition and competitive ability of weeds as well as crops. Weeds usually take up fertilizer more rapidly than do crop plants (2,136). Corn plants growing with pigweed contained only 58% as much nitrogen as weed-free corn plants. Competition for immobile nutrients such as phosphorus and potassium is less acute than is competition for nitrogen (1,9).

Much of the research on soil fertility and weed competition shows that recommendations need to be made on a case-by-case basis rather than by sweeping generalities. For example, wheat yields in wild oat–infested plots declined with increased fertilization while wild oat density increased (28), suggesting that wild oat may be able to utilize the additional nitrogen better than wheat. At wild oat densities lower than 1.6% of total plant density, additional fertilizer did improve wheat yields (48). Schipstra (119) reported that oxeye daisy (*Chrysanthemum leucanthemum*) and creeping buttercup (*Ranunculus repens*) infestations in rye were associated with a potassium deficiency in soils, and greater common lambsquarters infestations were observed where phosphorus was deficient. Applications of nitrogen at 70 and 140 kg ha^{-1} as liquid urea and sulphur nitrate affected weed composition in barley in Czechoslovakia (108). Erect weeds similar in height to barley took better advantage of the enhanced nitrogen supply than basal-growing species. The lower nitrogen application also increased species diversity among weeds. In another study, the application of 112–336 kg ha^{-1} N failed to affect populations of common lambsquarters, giant foxtail, velvetleaf, jimsonweed, and redroot pigweed (48), although germination of common lambsquarters (*Chenopodium album*) increased at higher nitrogen rates.

One means of improving cultural weed management is to adjust fertility levels to enhance crop productivity over that of associated weeds. Several studies have examined the response of specific weed species to fertility levels. This information may improve the producers' ability to predict certain weed infestations or to shift growing conditions to favor weeds that are less competitive or easier to control by various methods. In Australia, capewood (*Cryptostemna calendula*) has been observed to be the dominant weed species where no fertilizers were applied, and this dominance is enhanced by the application of nitrogen fertilizer alone (95). Nitrogen and phosphorus amendments can create conditions that lead to a domination by grass species (69,95). The additions of phosphorus alone led to burr medic dominance, whereas additional potassium alone had little effect. Hoveland et al. (69) examined the response in dry weight accumulation of several weed species to phosphorus and potassium supplements. They found that dry weight accumulation of warm-season weeds such as redroot pigweed, jimsonweed (*Datura stramonium*), and Florida beggarweed was responsive to phosphorus, whereas common chickweed (*Stellaria media*) was the most responsive cool-season weed. Dry weight accumulation of showy crotolaria (*Crotolaria spectabilis*), tall morning-glory (*Ipomoea purpurea*), sicklepod, Carolina geranium (*Geranium carolinianum*), and coffee senna (*Cassia occidentalis*) was not affected by low levels of phosphorus. In terms of dry matter accumulation in response to potassium supplements, redroot pigweed, jimsonweed, and Florida beggarweed were the most affected warm-season weeds, whereas wild mustard and annual bluegrass (*Poa annua*) were the most response cool-season weeds. Buckhorn plantain (*Plantago lanceolata*), Carolina geranium, and curly dock (*Rumex crispus*) were unaffected by low levels of potassium. Generally, the weeds examined were more sensitive to low levels of phosphorus than they were to potassium, and weeds were more sensitive than crop plants to low levels of both phosphorus and potassium. Singh and Singh (122) found that weeds belonging to the genera *Amaranthus, Cleome, Chenopodium*, and *Portulaca* are very efficient in phosphorus uptake, whereas most other weed species evaluated did not differ in phosphorus uptake. Foliar tissues of coffee senna and common lambsquarters were high in nitrogen and calcium compared with other weed species and their maximum absorption of nutrients was in the preflowering stage. Yellow foxtail is very efficient at zinc uptake, whereas *Crotolaria* species are not; indeed, poorjoe (*Diodia terres*) and yellow foxtail decomposition yielded enough zinc to alleviate deficiency in corn (115). Banks et al. (9) examined weed populations in winter wheat plots that had received various conventional fertility treatments continuously for 47 years. Six species of weeds dominated these plots. The fewest weeds were found in the low-fertility plots and the highest number of weeds were found in plots receiving nitrogen, phosphorus, potassium, and lime. Evening primrose (*Oenothora speciosa*) predominated in plots without fertility treatments, and evening primrose populations decreased with increasing fertility. Henbit (*Lamium amplexicaule*) and carpetweed (*Mollugo verticillata*) numbers were greatest in plots receiving only phosphorus or nitrogen plus phosphorus treatments.

The nitrogen supply can also affect the competitive ability of ryegrass (7,47). Under regular applications of nitrogen, the ryegrass root/shoot ratio remained constant in the vegetative phase. However, an intermittent nitrogen supply caused a shift in the root/shoot ratio of ryegrass in monoculture from high to low.

Qualitative aspects of fertilization can also influence weed/crop competition. For example, the response of lettuce (*Lactuca sativa*) to phosphorus can vary by placement of nutrient; lettuce growth was optimized when phosphorus was applied in a band at one-third of the rate required for broadcast treatment (118). Soil analysis indicated that banded phosphorus increased available phosphorus in the lettuce root zone compared with broadcast phosphorus applications. It was hypothesized that this phenomenon could be used to provide lettuce with a competitive advantage over associated weeds.

Some research has shown that nutrient management can be used to manipulate weed populations directly. Broadcast or disked applications of nitrogen fertilizer increased gemination of wild oat seed present in the soil (120). Fertilizer treatments also may be used to deplete seed reserves in fallow years by promoting weed seed gemination from the weed seedbank and reduce the amount of seed returned to the soil in crop years by combining fertilizer application with delayed crop seeding.

IX. WATER MANAGEMENT

Water is used most often as a weed management tool to promote healthy crop growth and improve the crop's ability co compete with associated weeds. Water also is used as a weed management tool to control terrestrial weeds in flooded rice fields (71). Water-saturated soil limits oxygen availability, and rice can grow under flooded conditions, whereas many weeds cannot.

X. ALLELOPATHY

Allelopathy is one mechanism of cultural weed management that does not involve increasing the competitiveness of the crop, relative to weeds, for aboveground or belowground resources. The term *allelopathy* has been used since 1937 when Molisch (88) first described the phenomenon of one plant species eliminating competing plant species through the release of toxic chemical agents. The best-known case of allelopathy is the production of juglone by black walnut (*Juglans nigra*) to eliminate surrounding vegetation. Allelopathy has much untapped potential as a weed management tool (3,7,30,35,37,75,105.106,123).

Various crop plants express allelopathy. For example, rye residues left in a conservation tillage system can reduce weeds germinating from the soil surface. In a study by Barnes and Putnam (11), cover crops of lupine (*Lupinus albus*), wheat, rape (*Brassica napus*), radish (*Raphanus sativus*), and oats contained a total weed biomass of 300–1350 g m^{-2} with 83–100% of the weeds being grasses, whereas rye and triticale (*Triticale triticosecale*) contained weed biomasses of 500 and 1100 g m^{-2} with 64–77% being broadleaf species (11). Christian et al. (29) found that wheat, barley, and oat cover crop residue incorporated to a depth of 20 cm reduced weed seedling establishment by 29%. Putnam and DeFrank (107) observed that total weed biomass and the weight of several indicator species were reduced by barley, oat, wheat, and rye residues. Larger seeded vegetables were unaffected by the small grain residues, whereas small seeded vegetables were stunted by the residues (107). These same residues reduced common purslane (*Portulca oleracea*) and smooth crabgrass (*Digitaria ischaemum*) by 70–98%.

XI. FUTURE TRENDS

The use of cultural control methods to improve weed management systems has increased in popularity in the last few years. In the next few years, precision farming techniques will allow growers to manage many of their cultural practices on a much smaller scale than has been possible in the past. Similarly, the advent of transgenic cultivars will provide growers with greater opportunities to select crop genotypes with features that fit into management systems utilizing cultural methods for weed control. These technological innovations will allow cultural weed control methods to be more effective and will improve weed management systems. The growing popularity of conservation tillage will simultaneously present challenges for weed control, and it will also provide opportunities to devise novel methods of managing weeds (e.g., cover crops, allelopathy). With increasing herbicide resistance and growing costs for registration of new herbicides, cultural weed control practices will become increasingly important in the coming years.

REFERENCES

1. Aldrich, R.J. 1984. *Weed-Crop Ecology: Principles in Weed Management*. North Scituate, MA: Breton Publishers.

2. Alkämper, N. 1976. Influences of weed infestation on effect of fertilizer dressings, *Pflanzenschutz-Nachrichten*, 29:191–235.

3. Altieri, M.A., and J.D. Doll. 1978. The potential of allelopathy as a tool for weed management in crop fields. *PANS* 24:495–502.

4. Anaele, A.O., and U.R. Bishnoi. 1992. Effects of tillage, weed control method and row spacing on soybean yield and certain soil properties. *Soil Till. Res.* 23:333–340.

5. Anderson, B. Influence of crop density and spacing on weed competition and grain yield in wheat and barley. *Proceedings of the European Weed Research Symposium*, 1986, pp 121–128.

6. Armstrong, D.L., J.L. Leasure, and M.R. Corbin. 1968. Economic comparison of mechanical and chemical weed control. *Weed Sci.* 16:369–371.

7. Baan Hofman, T., and G.C. Ennik. 1982. The effect of root mass of perennial ryegrass (*Lolium perenne* L.) on the competitive ability with respect to couchgrass (*Elytrigia repens* (L.) Desv.) *Neth. J. Agric. Sci.* 30:275–283.

8. Baker, D.N., and R.E. Meyer. 1966. Influence of stand geometry on light interception and net photosynthesis in cotton. *Crop Sci.* 6:15–19.

9. Banks, P.A., P.W. Santelmann, and B.B. Tucker. 1976. Influence of long term soil fertility treatments on weed species in winter wheat. *Agron. J.* 68:825–827.

10. Barker, M.R., and W.A. Wünsche. 1977. Plantio directo in Rio Grande do Sul, Brazil. *Outl. Agric.* 9:114–120.

11. Barnes, J.P., and A.R. Putnam. 1986. Evidence for allelopathy by residues and aqueous extracts of rye (*Secale cereale*). *Weed Sci.* 34:384–390.

12. Blackshaw, R.E., E.H. Stobbe, and A.R.W. Sturko. 1981. Effect of seeding dates and densities of green foxtail (*Setaria viridis*) on the growth and productivity of spring wheat (*Triticum aestivum*). *Weed Sci.* 29:212–217.

13. Blackshaw, R.E. 1991. Soil temperature and moisture effects on downy brome versus winter canola, wheat, and rye emergence. *Crop Sci.* 31:1034–1040.

14. Blackshaw, R.E., and J.T. O'Donovan. Higher crop seed rates can aid weed management. *Proc. Brit. Crop Prot. Conf. - Weeds -* 1993, pp 1003–1008.

15. Borgo, A. 1979. O controle Das invasoras. *Rev. A Granja* 376:51–58.

16. Brandt, S.A. Zero tillage vs conventional tillage with two rotations: crop production over the last 10 years. *Proceedings of the Soils and Crops Workshop*, University of Saskatchewan, Saskatoon, 1989, pp 330–338.

17. Brandt, S.A., and K.J. Kirkland. Herbicide programs for reduced tillage in Saskatchewan. *Proceedings of the Tillage Symposium*, North Dakota State University, Bismarck, 1980, pp 153–158.

18. Brenchley, W.E., and K. Warington. 1933. The weed seed population of arable soil II. Influence of crop, soil, and methods of cultivation upon the relative abundance of viable seeds. *J. Ecol.* 21:103–127.

19. Buchanan, G.A., and E.W. Hauser. 1980. Influence of row spacing on competitiveness and yield of peanuts. *Weed Sci.* 28:401–409.

20. Buckeridge, D.J., and J. Norrington-Davies. 1986. Competition for phosphate between establishing plants of *Lolium perenne* and *Trifolium repens* under differing cultivation treatments in an upland pasture. *Agric. Sci., Camb.* 106:449–453.

21. Buhler, D.D., and T.C. Daniels. 1988. Influence of tillage systems on giant foxtail *Setaria faberi* and velvetleaf *Abutilon theophrasti*, density and control in corn, *Zea mays*. *Weed Sci*. 36:642–647.

22. Buhler, D.D., J.L. Gunsolus, and D.F. Ralston. 1992. Integrated weed management techniques to reduce herbicide inputs in soybean. *Agron. J.* 84:973–978.

23. Burnside, O.C., and W.L. Colville. 1964. Soybean and weed yields as affected by irrigation, row spacing, tillage, and amiben. *Weeds* 12:109–112.

24. Burnside, O.C., and G.A. Wicks. 1972. Competitiveness and herbicide tolerance of sorghum hybrids. *Weed Sci*. 20:314–316.

25. Burnside, O.C. 1972. Tolerance of soybean cultivars to weed competition and herbicides. *Weed Sci*. 20:294–297.

26. Burnside, O.C. 1984. Requirements for agricultural chemicals in changing agricultural production systems—current state of the art. In *Changing Agricultural Production Systems and the Fate of Agricultural Chemicals*. G.W. Irving, ed. Chevy Chase, MD: Agricultural Research Institute, pp 49–71.

27. Burnside, O.C., G.A.. Wicks, and D. R. Carlson. 1980. Control of weeds in an oat (*Avena sativa*)-soybean (*Glycine max*) ecofarming rotation. *Weed Sci*. 28:46–50.

28. Carlson, H.L., and J.E. Hill. 1985. Wild oat competition in spring wheat: Plant density effects. *Weed Sci*. 33:176–181.

29. Christian, D.G., B.M. Smallfield, and M.J. Gass. Straw disposal on heavy clay soils. *Proc. Brit. Crop Prot. Conf. - Weeds - 1985*, Vol. 2, 1985, p 621.

30. Collison, R.C., and H.J. Conn. 1925. The effect of straw on plant growth. *NY Agric. Exp. Sta. Geneva Tech. Bull.* 114:1–10.

31. Colvin, D.L., G.R. Wehtje, M. Patterson, and R.H. Walker. 1985. Weed management in minimum-tillage peanuts (*Arachis hypogea*) as influenced by cultivar, row spacing, and herbicides. *Weed Sci*. 33:233–237.

33. Cousens, R. 1985. An empirical model relating crop yield to weed and crop density and a statistical comparison with other models. *J. Agric. Sci*. 105:513–521.

34. Covarelli, G., and F. Tei. Effet de la rotation culturale sur la flore adventice du mais. In *VIIIeme Colloque International Sur la Biologie, L'ecologie et la Systematique des Mauvaises Herbes*. Vol. 2. Paris: Comite Francais de Lutte Contre les Mauvaises Herbes, 1988, pp 477–484.

35. Crutchfield, D.A., G.A. Wicks, and O.C. Burnside. 1985. Effect of winter wheat (*Triticum aestivum*) straw mulch level on weed control. *Weed Sci*. 34:110–114.

36. Cussans, G.W. The influence of changing husbandry on weeds and weed control in arable crops. *Proc. Brit. Crop Prot. Conf. - Weeds - 1976*, pp 1001–1009.

37. de Almeida, F.S. Effect of some winter crop mulches on the soil weed infestation. *Proc. Brit. Crop Prot. Conf. - Weeds - 1985*, Vol. 2, p 621.

38. Derksen, D.A., G.P. Lafond, C.J. Swanton, A.G. Thomas, and H.A. Loeppky. 1993. The impact of agronomic practices on weed communities: tillage systems. *Weed Sci*. 41:409–417.

39. Derksen, D.A., A.G. Thomas, G.P. Lafond., C.J. Swanton, and H.A. Loeppky. 1994. The influence of agronomic practices on weed communities: fallow within tillage systems. *Weed Sci*. 42:184–194.

40. Donaghy, D.I. Effects of tillage systems on weed species. *Proceedings of the Tillage Symposium*. Bismarck, ND: Cooperative Extension Service, North Dakota State University, 1980, pp 153–158.

41. Donald, C.M. 1963. Competition among crop and pasture plants. *Adv. Agron.* 15:1–118.

42. Donald, W.W., and J.D. Nalewaja. Northern Great Plains. In *Systems of Weed Control in Wheat in North America*. W.W. Donald, ed. Champaign, IL: Weed Science Society of America, 1990, pp 90–126.

43. Duncan, S.R., and W.T. Schapaugh, Jr. 1993. Row orientation and plating pattern of relay intercropped soybean and wheat. *J. Prod. Agric.* 6:360–364.

44. Elmore, C.D., and T. Moorman. 1988. Tillage related changes in weed species and other soil properties. *Proc. South. Weed Sci. Soc.* 41:240–291.

45. Elmore, C.D., R.A. Wesley, and L.G. Heatherly. 1992. Stale seedbed production of soybeans with a wheat cover crop. *J. Soil Water Conserv.* 47:187–190.

46. Enache, A.J., and R.D. Illnicki. 1990. Weed control by subterranean clover (*Trifolium subterraneum*) weed as a living mulch. *Weed Technol.* 4:534–538.

47. Ennik, G.C., and T. Baan Hofman. 1983. Variation in the root mass of ryegrass types and its ecological consequences. *Neth. J. Agric. Sci.* 31:325–334.

48. Fawcett, R.S. 1978. Overview of pest management for conservation tillage systems. In *Effects of Conservation Tillage on Groundwater Quality: Nitrates and Pesticides*. Logan, T.J., J. M. Davidson, J.L. Baker, and M.R. Overcash, eds. Chelsea, MA: Lewis Publishers, pp 19–37.

49. Fay, P.K. 1990. A brief overview of the biology and distribution of weeds in wheat. In *Systems of Weed Control in Wheat in North America*. W.W. Donald, ed. Champaign, IL: Weed Science Society of America, pp 33–50.

50. Ferrando, J.C., J.E. Smith, L.B. Donato de Cobo, and A. Benefico. Investigacion en labranza reducida en el area de Castelar. In *Seminario labranza reducida en el cono sur*. H. Caballero R. Diaz, ed. Colonia, Uruguay: Instituto Interamericano de Cooperation par la Agricultera/Centro de Investigaciones Agricolas Alberto Boerger, 1982, pp 78–93.

51. Fischer, R.A., and R.E. Miles. 1973. The role of spatial pattern in the competition between crop plants and weeds: a theoretical analysis. *Mathem. Biosci.* 18:335–350.

52. Forcella, F., K. Eradat-Oskoui, and S.W. Wagner. 1993. Application of weed seedbank ecology to low-input crop management. *Ecol. Appl.* 3:74–83.

53. Foster, R.K. The effect of various conservation cropping practices on wheat. *Annual Report of Crop Development Centre/Department of Crop Science*, University of Saskatchewan, Saskatoon, 1984, pp 68–77.

54. Foster, R.K., and C.W. Lindwall. Minimum tillage and wheat production in western Canada. In *Wheat Production in Canada—A Review*. A.E. Slinkard and D.B. Fowler, eds. University of Saskatchewan, Saskatoon, 1986, pp 254–366.

55. Freed, B.E., E.S. Oplinger, and D.D. Buhler. 1987. Velvetleaf control for solid-seeded soybean in three corn residue management systems. *Agron. J.* 79:119–123.

56. Froud-Williams, R.J. 1981. Potential changes in weed floras associated with re-

duced-cultivation systems for cereal production in temperate regions. *Weed Res.* 21:99-109.

57. Froud-Williams, R.J. 1983. The influence of straw disposal and cultivation regimen on the population dynamics of *Bromus sterilis. Ann. Appl. Biol.* 103:139-148.

58. Froud-Williams, R.J., D.S.H. Drennan, and R.J. Chancellor. 1983. Influence of cultivation regime on weed floras of arable cropping systems. *J. Appl. Ecol.* 20:187-197.

59. Froud-Williams, R.J. 1988. Changes in weed flora with different tillage and agronomic management systems. In *Weed Management in Agroecosystems: Ecological Approaches.* M.A. Altieri and M. Liebman, eds. Boca Raton, FL: CRC Press, pp 213-236.

60. Gill, K.S., and M.A. Arshad. 1995. Weed flora in the early growth period of spring crops under conventional, reduced, and zero tillage systems on a clay soil in northern Alberta, Canada. *Soil Till. Res.* 33:65-79.

61. Grundy, A.C., R.J. Froud-Williams, and N.D. Boatman. 1992. The effects of nitrogen rate on weed occurrence in a spring barley crop. Nitrates and farming systems. *Aspects Appl. Biol.* 30:377-380.

62. Gunsolus, J.L. 1990. Mechanical and cultural weed control in corn and soybeans. *Am. J. Alt. Agric.* 5:114-119.

63. Hansen, O.M., and V.J. Zeljkovich. Investigacion en Labranza Reducida en Argentina. In *Seminario Labranza Reducida en el Cono Sur.* H. Caballero and R. Diaz, eds. Colonia, Uruguay: Instituto Interamericano de Cooperacion para la Agricultera/Centro de Investigaciones Agricolas Alberto Boerger, 1982, pp 1-5.

64. Harper, J.L. 1957. Ecological aspects of weed control. *Outlook Agric.* 1:197-205.

65. Hartmann, K.M., and W. Nezadal. 1990. Photocontrol of weeds without herbicides. *Naturwissenschaften* 77:153-158.

66. Heard, A.J. 1963. Weed populations on arable land after four-course rotations and after short leys. *Ann. Appl. Biol.* 52:177-184.

67. Hill, J.E., and D.E. Bayer. Integrated systems for rice weed control. *Proceedings of the California Weed Control Conference,* Vol. 42, 1990, pp 85-89.

68. Holm, F.A. Herbicides for conservation tillage. *Proceedings of the 35th Annual Meeting of the Canadian Pest Management Society:* Calgary, Alberta, 1988, pp 76-98.

69. Hoveland, C.S., G.A. Buchanan, and M.C. Harris. 1975. Response of weeds to soil phosphorus and potassium. *Weed Sci.* 24:194-201.

70. Howe, O.W., and L.R. Oliver. 1987. Influence of soybean (*Glycine max*) row spacing on pitted morningglory (*Ipomoea lacunosa*) interference. *Weed Sci.* 35:185-193.

71. Jones, J.W. 1993. Effect of reduced oxygen pressure on rice germination. *Agron. J.* 25:69-81.

72. Jordan, N. 1993. Prospects for weed control through crop interference. *Ecol. Appl.* 3:84-91.

73. King, L.D., and M. Buchanan. 1993. Reduced chemical input cropping systems in the Southeastern United States. I. Effect of rotations, green manure crops and nitrogen fertilizer on crop yields. *Am. J. Alt. Agric.* 8:58–77.

74. Kirkland, K.J. 1993. Weed management in spring barley (*Hordeum vulgare*) in the absence of herbicides. *J. Sustainable Agric.* 3:95–104.

75. Klein, R.R., and D.A. Miller. 1980. Allelopathy and its role in agriculture. *Commun Soil Sci. Plant Anal.* 11:43–56.

76. Koscelny, J.A., T.F. Peeper, J.B. Solie, and S.G. Solomon, Jr. 1990. Effect of wheat (*Triticum aestivum*) row spacing, seeding rate, and cultivar on yield loss from cheat (*Bromus secalinus*). *Weed Technol.* 4:487–492.

77. Liebman, M., and E. Dyck. 1993. Crop rotation and intercropping strategies for weed management. *Ecol. Appl.* 3:92–122.

78. Liebman, M., and E. Dyck. 1993. Weed management: A need to develop ecological approaches. *Ecol. Appl.* 3:39–41.

79. Leighty, C.E. 1938. Crop rotation. In *Soils and Men: Yearbook of Agriculture 1938.* Washington, DC: U.S. Dept. of Agric., GPO, pp 406–430.

80. Lotz, L.A.P., R.M.W. Groeneveld, B., Habekottee, B and H Van Oene. 1991. Reduction of growth and reproduction of *Cyperus esculentus* by specific crops. *Weed Res.* 31:153–160.

81. Lovely, W.G., C.R. Weber, and D.W. Stantiforth. 1958. Effectiveness of the rotary hoe for weed control in soybean. *Agron. J.* 50:621–625.

82. Lutman, P.J.W. 1991. Weed control in linseed: a review. *Aspects Appl. Biol.* 28:137–144.

83. Magrini, A.D., C. Anchier, and R.M. Diaz. 1983. II. Effecto residual de rastrojos de invierno sobre cultivos de verano sembrados con minimo ye cero laboreo. In *Miscelanea.* Colonia, Uraguay: Ministerio de Agricultura y Pesca Republica Oriental del Uraguay, pp 1–15.

84. Malhi, S.S., G. Mumey, P.A. O'Sullivan, and K.N. Harker. 1988. An economic comparison of barley production under zero and conventional tillage. *Soil Till. Res.* 11:159–166.

85. Mann, H.H., and T.W. Barnes. 1949. The competition between barley and certain weeds under controlled conditions. III. Competition with *Agrostis gigantea*. *Ann. Appl. Biol.* 36:273–281.

86. McWhorter, C.G., and E.E. Hartwig. 1972. Competition of johnsongrass and cocklebur with six soybean varieties. *Weed Sci.* 20:56–59.

87. Miller, S.D., and J.D. Nalewaja. 1985. Weed spectrum change and control in reduced-till wheat. *North Dakota Farm Res.* 43:11–14.

88. Molisch, H. 1937. *Der Einfluss einer Pflanze auf die Andere—Allelopathie.* Jena: Fischer Verlag.

89. Moomaw, R., and A.R. Martin. 1984. Cultural practices affecting season-long weed control in irrigated corn (*Zea mays*). *Weed Sci.* 32:460–467.

90. Moore, M.J., T.J. Gillespie, and C. J. Stanton. 1994. Effect of cover crop mulches on weed emergence, weed biomass, and soybean (*Glycine max*) development. *Weed Technol.* 8:512–518.

91. Moss, S.R. 1979. The influence of tillage and method of straw disposal on the

survival and growth and blackgrass (*Alopecurus myosuroides*) and its control by chlortoluron and isoproturon. *Ann. Appl. Biol.* 91:91-100.

92. Moss, S.R. 1980. A study of populations of blackgrass (*Alopecurus myosuroides*) in winter wheat, as influenced by seed shed in the previous crop, cultivation system and straw disposal method. *Ann. Appl. Biol.* 94:121-126.

93. Moss, S.R. 1985. The effect of drilling date, pre-drilling cultivations and herbicides on *Alopecurus myosuroides* (black-grass) populations in winter cereals. *Aspects Appl. Biol.* 9:31-40.

94. Moyer, J.R. 1994. Weed management in conservation tillage systems for wheat production in North and South America. *Crop Prot.* 13:243-259.

95. Myers, L.F., and R.M. Moore. 1952. The effect of fertilizers on a winter weed population. *J. Aust. Inst. Agr. Sci.* 18:152-155.

96. Niggli, V., F.P. Weibe, and W. Gut. 1990. Weed control from organic mulch materials in orchards. *Acta Hort.* 285:97-102.

97. Oliver, L.R. 1979. Influence of soybean (*Glycine max*) planting date on velvetleaf (*Abutilon theophrasti*) competition. *Weed Sci.* 27:183-188.

98. Orson, J.H. Growing practices: An aid or hindrance to weed control in cereals. *Proc. Brit. Crop Prot. Conf. - Weeds -* 1987, pp 87-96.

99. Orson, J.H. Integrating cultural and chemical weed control in cereals. *Proc. Brit. Crop Prot. Conf. - Weeds-* 1993, pp 977-984.

100. Peeper, T.F. & A.F. Wiese. 1990. Southern Great Plains. In *Systems of Weed Control in Wheat in North America*. W.W. Donald, ed. Champaign, IL: Weed Science Society of America.

101. Perkins, B. 1987. Problems and losses caused by quackgrass in reduced tillage systems. In *Technical Proceedings Supplement* . H. Glick, ed. Winnipeg: Quackgrass Action Committee, pp 1-3.

102. Peters, E.J., D.L. Klingman, and R.E. Larson. 1959. Rotary hoeing in combination with herbicides and other cultivations for weed control in soybeans. *Weeds* 7:449-458.

103. Peters, E.J., F.S. Davis, D.L. Klingman, & R.E. Larson. 1961. Interrelations of cultivations, herbicides, and methods of application for weed control in soybeans. *Weeds* 9:639-645.

104. Philbrook, B.D., & E.S. Oplinger. 1989. Tramlines, row orientation and individual row effects on solid-seeded soybean plot comparisons. *Agron. J.* 81:498-500.

105. Purvis, C.E., R.S. Jessop, & J. V. Lovett. 1985. Selective regulation of germination and growth of annual weeds by crop residues. *Weed Res.* 25:415-421.

106. Putnam, A.R. 1994. Phytotoxicity of plant residues. In *Managing Agricultural Residues*. P.W. Unger, ed. Boca Raton, FL: CRC Press, pp 286-314.

107. Putnam, A.R., and J. DeFrank. 1985. Use of phytotoxic plant residues for selective weed control. *Crop Prot.* 2:173-181.

108. Pyšek, P., and J. Lepš. 1991. Response of a weed community to nitrogen fertilizer: A multivariate analysis. *J. Veg. Sci.* 2:237-244.

109. Ramsel, R.E., and G.E. Wicks. 1988. Use of winter wheat cultivars and herbicides in aiding weed control in an ecofallow corn rotation. *Weed Sci.* 36:394-398.

110. Regnier, E.E., and R.R. Janke. 1990. Evolving strategies for managing weeds. In *Sustainable Agriculture Systems*. C.A. Edwards, R. Lal, P. Madden, R.H. Miller, and G. House, eds. Ankeny, IA: Soil Water Conservation Society, pp 174–202.

111. Richards, M.C. Crop competitiveness as an aid to weed control. *Proc. Brit. Crop Prot. Conf. - Weeds* - 1989, Vol. 2, pp 755–762.

112. Richards, M.C., and D.H.K. Davies. Potential for reducing herbicide inputs/rates with more competitive cereal cultivars. *Proc. Brit. Crop Prot. Conf. - Weeds* - 1991, pp 1233–1240.

113. Richards, M.C., and G. Whytock. 1993. Varietal competitiveness with weeds. *Aspects Appl. Biol.* 34:345–354.

114. Ricotta, J.A., and J.B. Masiunas. 1991. The effects of black plastic mulch and weed control strategies on herb yield. *HortScience* 26:539–541.

115. Rogers, L.H., O.E. Gail, and R.M. Barnette. 1939. The zinc content of weeds of volunteer grasses and planted land covers. *Soil Sci.* 47:237–242.

116. Rogers, N.K., G.A. Buchanan, and W.C. Johnson. 1976. Influence of row spacing on weed competition with cotton. *Weed Sci.* 234:410–413.

117. Roman, E.S. & A.D. Dinonet. 1990. *Controle de Plantas Daninhas no Sistema de Plantio Direto de Trigo e Soja*. Circular Tecnica 2, Empresa Brasileira de Pesquisa Agropecuaria–Centro Nacional de Pesquisa Agropecuaria, Passo Fundo, Rio Grande do Sul.

118. Sanchez, C.A., S. Swanson, and P.S. Porter. 1990. Banding P to improve use efficiency of lettuce. *J. Am. Soc. Hort. Sci.* 115:581–584.

119. Schipstra, K. 1957. Weeds as indicators of nutritional diseases. *Tijdschr. Plziekt.* 63:15–18.

120. Sexsmith, J.J., and U.J. Pittman. 1962. Effect of nitrogen fertilizers on germination and stand of wild oats. *Weeds* 10:99–101.

121. Siddiqi, M.Y., A.D.M. Glass, A.I. Hsiao, and A.N. Minjas. 1985. Wild oat/barley interactions: Varietal differences in competitiveness in relation to K^+ supply. *Ann. Bot.* 56:2–7.

122. Singh, B.N., and L.B. Singh. 1939. Relative absorption of nutrients by weeds of arable land. *Soil Sci.* 47:227–235.

123. Smeda, R.J., and A.R. Putnam. 1988. Cover crop suppression of weeds and influence on strawberry yields. *HortScience* 23:132–134.

124. Snipes, C.E., & T.C. Mueller. 1992. Cotton (*Gossypium hirsutum*) yield response to mechanical and chemical weed control systems. *Weed Sci.* 40:249–254.

125. Standifer, L.C., and C.E. Beste. 1985. Weed control methods for vegetable production with limited tillage. In *Weed Control in Limited Tillage Systems*. A.F. Wiese, ed. Champaign, IL: Weed Science Society of America, pp 93–99.

126. Stantiforth, D.W. 1961. Response of corn hybrids to yellow foxtail competition. *Weeds* 9: 132–136.

127. Stewart, G., and D.W. Pittman. 1931. Twenty years of rotation and manuring experiments at Logan, Utah. *Utah Agric. Exp. Stn. Bull.* Logan, UT: Utah State University.

128. Stobbe, E.H. 1979. Tillage practices on the Canadian prairies. *Outl. Agric.* 10:21–26.

129. Streibig, J.C., C. Andreasen, and W.M. Blacklaw. 1993. Crop management affects the community dynamics of weeds. *Proc. Brit. Crop Prot. Conf. - Weeds -* 1993, pp 487–494.

130. Teasdale, J.R., C.E. Beste, and W.E. Potts. 1991. Response of weeds to tillage and cover crop residue. *Weed Sci.* 39:195–199.

131. Todd, B.G., and D. A. Derksen. 1986. Perennial weed control in wheat in western Canada. In *Wheat Production in Canada–A Review.* A.E. Slinkard and D.B. Fowler, eds. Saskatoon: University of Saskatchewan, pp 391–404.

132. Triplett, G.B., Jr. 1985. Principles of weed control for reduced-tillage corn production. In *Weed Control in Limited Tillage Systems.* ed. A.F. Wiese, eds. Champaign, IL: Weed Science Society of America, pp 26–40.

133. Unamma, R.P.A., L.S.O. Ene, S.O. Odurukwe, and T. Enyinnia. 1986. Integrated weed management for cassava intercropped with maize. *Weed Res.* 26:9–17.

134. Vaughn, D., and B.G. Ord. 1989. Extraction of potential allelochemicals and their effects on root morphology and nutrient contents. *Special Publ. Brit. Ecol. Soc.,* pp 399–422.

135. Vencill, W.K., and P.A. Banks. 1994. Effect of tillage systems and weed management on weed populations in grain sorghum (*Sorghum bicolor*). *Weed Sci.* 42:541–547.

136. Vengris, J., W.G. Colby, and M. Drake. 1955. Plant nutrient competition between weeds and corn. *Agron. J.* 47:213–216.

137. Walker, R.H., and G.A. Buchanan. 1982. Crop manipulation in integrated weed management systems. *Weed Sci.* 30(1):17–23.

138. Wax, L.M., and J.W. Pendleton. 1968. Effect of row spacing on weed control in soybeans. *Weeds* 16:462–464.

139. Weston, L.A. 1990. Cover crop and herbicide influence on row crop seedling establishment in no-tillage culture. *Weed Sci.* 38:166–171.

140. Wicks, G.A., O.C. Burnside, and W.L. Felton. Weed control in conservation tillage systems. In *Managing Agricultural Residues.* P.W. Unger, ed. Boca Raton, FL: CRC Press, 1994, pp 211–244

141. Wiese, A.F., J.F. Collier, L.E. Clark, and U.D. Havelka. 1964. Effects of weeds and cultural practices on sorghum yields. *Weeds* 12:209–211.

142. Wiese, A.F., and J.M. Chandler. 1988. Weed control in conservation tillage systems. In *Conservation Tillage in Texas, Research Monograph 15.* College Station, TX: The Texas A & M University System, pp 41–52.

143. Williams, J.L., Jr., and G.A. Wicks. 1978. Weed control problems associated with crop residue systems. In *Crop Residue Management Systems.* Oschwald, ed. Madison, WI: ASA/CSSA/SSSA, pp 165–172.

144. Wilson, B.J., and P.A. Phipps. 1985. A long term experiment on tillage, rotation and herbicide use for the control of *A. fatua* in cereals. *Proc. Brit. Crop Prot. Conf. - Weeds -* 1985, pp 32–37.

145. Wilson, B.J. Yield responses of winter cereals to the control of broadleaved weeds. *Proc. EWRS Symp.,* 1986, pp 75–82.

146. Wilson, H.P., M.P. Mascianica, T.E. Hines, and R.F. Walden. 1986. Influence

of tillage and herbicides on weed control in a wheat (*Triticum aestivum*) soybean (*Glycine max*) rotation. *Weed Sci.* 34:590–594.

1473 Witt, W.W. 1984. Response of weeds and herbicides under no-tillage conditions. In *No-tillage Agriculture: Principles and Practices*. R.E. Phillips and S.H. Philips, eds. New York: Van Nostrand Reinhold, pp 152–170.

148. Zentner, R.P., C.W. Lindwall, and J.M. Carefoot. 1988. Economics of rotations and tillage systems for winter wheat production in southern Alberta. *Can. Farm Econ.* 22:3–13.

22
Biological Approaches to Weed Management

Dennis L. Isaacson
Oregon Department of Agriculture, Salem, Oregon

Raghavan Charudattan
University of Florida, Gainesville, Florida

I. DEFINITIONS AND CONCEPTS

A. Definition and Perspectives

The intentional use of organisms to control weeds has involved a variety of organisms: insects, plant pathogens, mites, nematodes, vertebrates, and, arguably, even other plant species. In this chapter, we discuss various approaches to biological control of weeds and consider issues related to environmental safety and utilization of biological control agents for weed control. Because of the breadth of this area, this chapter is necessarily restricted in scope. Additional references of value to the reader are: Charudattan and Walker [14] and TeBeest [70] discuss the use of pathogens for biological weed control; Harley and Forno [29], DeBach and Rosen [22], and Cardina [12] provide comprehensive overviews of the subject; Goeden [26] and Rees et al. [61] give examples of projects and their status in biological weed control. Earlier reviews by Huffaker [38] and Goeden and Louda [28] are also good introductions to the theory and practice of biological weed control.

1. Classic Biological Control in Weed Management

There are several approaches to the use of biological control agents for weed suppression. Classic biological control involves introduction of host-specific pathogens or herbivorous organisms to control exotic, invasive weeds. Also called the inoculative method, classic biological control is suitable for introduced weeds that are distributed over vast areas and lack a complement of natural enemies in their new ranges. For example, a plant population that is removed from the selection pressures of certain pathogens tends to lose its resistance to those pathogens over time, and when the latter are reintroduced, the plant population can be vulnerable. Therefore, introduced weeds, especially those that are clonally reproducing, may become easy targets for control by introduced pathogens from which they have been physically separated for a period. It is, however, impossible to predict the success of an agent or to increase or decrease its efficacy after release, and it is also impossible to recall a bioagent once released. Classic biological control is distinguished in this respect from inundative or augmentative biological control (see below) in that, in these latter efforts, control is expected to be temporary, whereas control resulting from introduced agents is expected to be permanent. Therefore, a particularly careful evaluation of efficacy and safety must precede a classical introduction.

Classic biological control of weeds fills a gap among weed management options. Other management options, including chemicals, manual and physical control, and cultural control, typically require larger inputs of resources per unit area than does classic biological control. In crop production systems, projected returns using intensive weed management techniques must be greater than the costs of weed control, but in forest, range, and near-natural systems, returns are so limited that intensive methods are not economical. This is not to say that biological control of weeds is cheap in every sense. Research into target weeds and their potential biological antagonists may cost several hundred thousand dollars, and thus very large areas are usually infested before biological control is considered as a control approach. But in systems where production returns are very limited, or where there may even be no direct marketable returns from the land, classic biological weed control is most appropriate. Nevertheless, importation efforts also may offer promise in ephemeral cropping systems.

The great majority of weed species which are potential targets for classic biological control have been spread between continents, and many countries have undertaken weed control with classic biological agents. The attempt to control *Lantana camara* L. in Hawaii in 1902 is generally regarded as the first classic biological control effort for weeds. Eight insect species found in Lantana's native range in Mexico were intentionally established in Hawaii [26], providing moderate to excellent control. Activity in biological weed control has since increased. In the most recent global accounting of biological weed control projects, Julien [43] lists over 100 weed species as targets of classic biological weed control, for which nearly 400 different agent species have been released worldwide. He lists a world historical total of 729 releases of exotic species in different countries, 130 of which occurred in the 5 years prior to 1992.

Even though biological control of weeds is practiced in many countries, there are relatively few professional workers in the field, and there is not a well-developed body of literature. Much of the work published on biological weed control is found in journals from related fields, such as plant ecology, weed science, range management, entomology, or plant pathology. Although these sources can be located easily and searched, there are several literature sources that may be difficult to find and are not included in some abstracts of biology and agriculture. Some of the most important of these sources are listed in Table 1.

2. Inundative Biological Control of Weeds

Pathogens may be applied in an inundative or augmentative strategy as bioherbicides. The technique of controlling weeds with planned applications of large doses of a pathogen is regarded as the microbial herbicide strategy (also called *inundative control strategy* or the *bioherbicide strategy*). A significant level of research activity is currently underway in several countries to discover, develop, and use plant pathogens as biological control agents of weeds in both classic and inundative weed control programs (Table 2). Researchers in about 20 countries are attempting to develop pathogens to control weeds in major crops such as rice, wheat, sorghum, soybean, sugar beet, citrus, vegetables, oil crops, and plantation crops, as well as in agroforestry, sod and turf, and ranch lands. Weeds of high economic importance such as *Amaranthus* spp., *Chenopodium album, Convolvulus* spp., *Cyperus* spp., *Euphorbia heterophylla, Senna obtusifolia, Xanthium* spp., various grasses and vines, parasitic weeds (*Cuscuta* spp. and *Striga hermonthica*), and several invasive annuals, aquatic weeds, and tree species have been targeted for inundative control by pathogens. In situations where the economics of weed control do not justify the use of costly control methods, such as in rangelands, natural areas, native forests, and certain natural waters, the option has been to use pathogens as classic biocontrol agents.

B. Ecological Concepts of Biological Control of Weeds

If we search among the work of biogeographers, plant geographers, and plant ecologists to find explanations for limits to the distribution and abundance of plants, we are likely to find discussions of light, temperature, soils, moisture, or the correlation of climate to vegetation, but we are not likely to find discussions of the role of natural enemies in limiting their distribution and abundance (e.g., see ref. 25). Yet there are striking examples of how these organisms can limit plant populations. In North America, the gypsy moth, *Lymantria dispar*, through its ability to defoliate its preferred host species, is known markedly to influence the composition of eastern forests [36]. Chestnut blight, *Endothia parasitica*, was so effective in killing its American chestnut host that this tree

Table 1   Annotated List of Some Key Reference Materials in Biological Control of Weeds

1.   Proceedings of the International Symposia on Biological Control of Weeds, I through IX:

Because there is no international journal dealing with biological weed control, much of the literature in this field is widely dispersed in journals dedicated to related topics. The proceedings of these symposia contain information not readily available in other sources and represent a comprehensive account of developments and activities in the field. Attendance at the symposia includes most of the active professionals working in biological weed control. Because they have been published by several different organizations in several different countries, the individual proceedings can be difficult to find, so details on their publication is listed below. The next symposium (X) is scheduled to be held in Bozeman, Montana, in 1999.

No., Location, and Year of Symposium	*Year of Publication, Editor(s), and Publisher*
I. 1969 Delemont, Switzerland	1970; F. J. Simmonds; Commonwealth Agricultural Bureaux, Farnham Royal, Slough, UK. Misc. Publication No. 1.
II. 1971 Rome, Italy	1973; Paul H. Dunn; Commonwealth Agricultural Bureaux, Farnham Royal, Slough, UK. Misc. Publication No. 6.
III. 1973 Montpellier, France	1974; A. J. Wapshere; Commonwealth Agricultural Bureaux, Farnham Royal, Slough, UK. Misc. Publication No. 8.
IV. 1976 Gainesville, Florida	1976; T. E. Freeman; University of Florida, Gainesville, Florida.
V. 1980 Brisbane, Australia	1981; E. S. Delfosse; CSIRO, Melbourne, Australia.
VI. 1984 Vancouver, British Columbia	1985; E. S. Delfosse; Agriculture Canada, Ottawa, Canada.
VII. 1988 Rome, Italy	1990; E. S. Delfosse; Instituto Sperimentale per la Patalogia Vegetale, Rome, Italy.
VIII. 1992 Canterbury, New Zealand	1992; E. S. Delfosse and R. R. Scott; CSIRO, Melbourne, Australia.
IX. 1996 Capetown, South Africa	1992; J. R. Hoffman; University of Capetown, Capetown, South Africa.

2.   Regional reviews:

Nechols, J. R., L. A. Andres, J. W. Beardsley, R. D. Goeden and C. G. Jackson, eds. 1995. *Biological Control in the Western United States.* University of California Press, Oakland, CA. Part 3 of this work gives case histories of biological control projects focusing on 22 different weeds. Weed history and status, and the results of the projects are given for each weed, along with recommendations for further work.

Clausen, C. P. 1978. *Introduced Parasites and Predators of Arthropod Pests and Weeds: A World Review.* Agriculture Handbook No. 480, U.S. Dept. of Agriculture, Wash., D.C.

(continued)

Table 1 Continued

Part II of this book, by Richard D. Goeden, is notable for the wealth of details given about some of the earliest projects in biological control of weeds, with many references to reports published in sources that are difficult to find. Coverage is for projects through 1968.

Rees, N. E., P. C. Quimby, G. L. Piper, E. M. Coombs, C. E. Turner, N. R. Spencer and L. V. Knutson, eds. 1996. *Biological Control of Weeds in the West*. Western Society of Weed Science, Helena, MT.

Distributed in a looseleaf format, this publication includes color photographs of both biological control agents and target weeds. The history, biology and status of weeds and agents are covered in an outline format.

3.   National reviews (listed without annotation):

    a. Australia

        Wilson, F. 1960. *A Review of the Biological Control of Insects and Weeds in Australia and Australian New Guinea*. Commonwealth Institute of Biological Control Technical Communication No. 1.

    b. Canada

        Kelleher, J. S. and M. A. Hulme. 1984. *Biological Control Programmes against Insects and Weeds in Canada 1969–1980*. Commonwealth Agricultural Bureaux, Farnham Royal, Slough, UK.

        McLeod, J. H. 1962. *A review of the biological control attempts against insects and weeds in Canada. Part I. Biological control of pests of crops, fruit trees, ornamentals, and weeds in Canada up to 1959*. Commonwealth Institute of Biological Control Technical Communication No. 2:1–33.

    a. New Zealand

        Cameron, P. J., R. L. Hill, J. Bain and W. P. Thomas. 1989. *A Review of Biological Control of Invertebrate Pests and Weeds in New Zealand 1874 to 1987*. CAB International, Wallingford, Oxon, UK.

    b. South Africa

        Hoffman, J. R. (sic). 1991. Biological Control of Weeds in South Africa. *Agriculture, Ecosystems and Environment* 37(1–3). Elsevier, Amsterdam.

is now hardly mentioned in guidebooks of North American trees. These are clear demonstrations of how important insects and diseases can be influencing species composition or in limiting the distribution and abundance of a plant species.

The gypsy moth and chestnut blight are introduced organisms which attacked native plants, but many of the world's weeds are species which are themselves introduced into new areas. A working assumption of classic biological weed control is that the success of some weeds in new areas is due in part to their freedom from attack by natural enemies, and an extension of this assumption is that if we can find natural enemies to attack them, the weeds will be less successful. By now we know that this is a practical approach to weed control

Table 2 A List of Current Projects on Biological Control of Weeds of Agricultural Importance Using Plant Pathogens

Weed	Pathogen (Registered name)	Target crop(s)	Country[a]	Status[b]
Abutilon theophrasti	*Colletotrichum coccodes*	Soybean	Canada	4*
Acacia mearnsii	*Ceratocystis* sp.	Tree plantations	South Africa	4
	Cylindrobasidium laeve (Stump Out)	Tree plantations	South Africa	4
A. saligna	*Uromycladium tepperianum*	Tree plantations	South Africa	3
Aeschynomene virginica	*Colletotrichum gloeosporioides* f.sp. *aeschynomene* (Collego)	Rice and soybean	USA, Arkansas	5*
Amaranthus spp.	*Phomopsis amaranthicola*	Vegetables	USA, Florida	3
Avena fatua	*Dreschlera avenaceae* (Teleomorph = *Pyrenophora chaetomioides*)	Wheat	P.R. China	2
Broad-leaved trees	*Chondrostereum purpureum*	Tree plantations	Holland; Canada	4
Chenopodium album	*Ascochyta caulina*	Various	Holland	4
	Cercospora spp.	Various	Holland	2
Cirsium arvense	An unidentified fungus	Various	Canada	2
	Fusarium sp.	Pastures	Germany	2
	Puccinia punctiformis	Pastures	Holland	3
	Botrytis sp.	Pastures	Germany	2
	Phoma sp.	Pastures	Germany	2
Cuscuta spp.	*Alternaria* sp.	Cranberries	USA, Florida	4
Cyperus spp.	*Dactylaria higginsii*	Various	USA, Florida	3
C. esculentus	*Puccinia canaliculata* (Dr. BioSedge)	Various	USA, Georgia	5*
C. rotundus	*Cercospora caricis*	Various	Brazil; Israel	3
	Puccinia romagnoliana	Various	India; Israel	2
Cytisus scoparius	*Fusarium tumidum*	Tree plantations	New Zealand	2
Erigeron annuus	*Phoma putaminum*	Pastures; various	Italy	2
Euphorbia heterophylla	*Helminthosporium* sp.	Various	Brazil	3
Grass weeds	Not identified	Cereals	Vietnam-Australia	1
	Drechslera spp. and *Exserohilum* spp.	Citrus	USA, Florida	3
	Pyrenophora semeniperda	Various	Australia	2

Hakea sericea	Colletotrichum gloeosporioides	Tree plantations	South Africa	4
Heliotropium europaeum	Uromyces heliotropii	Pastures	Australia	3
	Cercospora spp.	Pastures	Australia	2
Imperata cylindrica	Colletotrichum caudatum	Various	Malaysia	1
	Ascochyta sp.	Various	Malaysia	1
	Puccinia rufipes	Various	Malaysia	1
	Colletotrichum graminicola	Various	Malaysia	1
	Didymaria sp.	Various	Malaysia	1
	Dinemasporium sp.	Various	Malaysia	1
Malva pusilla	Colletotrichum gloeosporioides f.sp. malvae (BioMal)	Various	Canada	5*
Mikania micrantha	Cercospora mikaniicola	Plantation crops	Malaysia	2
Morrenia odorata	Phytophthora palmivora (DeVine)	Citrus	USA, Florida	5
Poa annua	Xanthomonas campestris pv. poa	Turf grass	USA-Japan	4
Pteridium aquilinum	Ascochyta pteridis	Pastures	UK	3
Rottboellia	Sporisorium ophiuri chochinchinensis	Cereals	UK	3
	Colletotrichum spp. nov. near graminicola	Cereals	Thailand-UK	3
	Puccinia rottboelliae	Cereals	Thailand-UK	2
Rumex spp.	Uromyces rumicis	Pastures	Azores, Portugal	2
	Ramularia rubella	Pastures	Azores, Portugal	2
Sagittaria spp.	Rhynchosporium alismatis	Rice	Australia	2
Senecio vulgaris	Puccinia lagenophorae	Various	UK-Switzerland	4
Senna obtusifolia	Alternaria cassiae	Soybean	Brazil	3
Sesbania exaltata	Colletotrichum truncatum	Soybean and rice	USA, Mississippi	4
Solanum viarum	Pseudomonas solanacearum	Citrus and sod	USA, Florida	3
	Colletotrichum spp.		Brazil-USA, Florida	3
Sphenoclea zeylanica	Alternaria sp.	Rice	Philippines	2
	Colletotrichum gloeosporioides	Rice	Malaysia	2

(continued)

Table 2 Continued

Weed	Pathogen (Registered name)	Target crop(s)	Country[a]	Status[b]
Striga hermonthica	*Fusarium nygamai*	Various	Sudan-Germany	3
	Fusarium oxysporum	Cereals	West Africa-Canada	3
	Fusarium semitectum var. *majus*	Sorghum	Sudan-Germany	3
Ulex europaeus	*Fusarium tumidum*	Plantation crops	New Zealand	2
Various annual weeds	*Myrothecium verrucaria*	Various	USA, Maryland	3
Various broad-leaf trees	*Chondrostereum purpureum*	Agroforestry	Canada, B.C.; Holland	4
			USA, Florida	3
Various composite weeds	*Psuedomonas syringae* pv. *tagetis*	Various	USA; Canada	4
Viola arvensis	*Mycocentrospora acerina*	Not identified	UK-Australia	2
Weed seeds	*Pyrenophora semeniperda*	Various	Australia	2
	Chaetomium globosum	Seed bank	UK	1
	C. spirale	Seed bank	UK	1
Xanthium spp.	*Alternaria zinniae*	Various	Australia	2
	Colletotrichum orbiculare	Various	Australia	4*
	Puccinia xanthii	Various	Australia	3

[a] "-", cooperative work is in progress between the said countries; ";", the said countries are engaged in independent work.

[b] Status: 1, in exploratory phase; 2, laboratory and/or greenhouse testing underway; 3, field trials in progress; 4, under early commercial or practical development; 4*, commercial development tried but registration uncertain; 5, available for commercial or practical use; and 5*, registered as a bioherbicide but currently unavailable for use due to economic reasons.

Source: Compiled from refs. 1-3.

for some species, because we have good examples of success. Some are spectacular, as in the case of insect control of several species of cacti in the genus *Opuntia* in several countries. In Australia alone, hundreds of thousands of hectares are now relatively free from problems associated with introduced *Opuntia* species [82]. Citing the case of biological control of Klamath weed (*Hypericum perforatum*) by the leaf beetle *Chrysolina quadrigemina*, Huffaker [39] suggested that at the current reduced densities of Klamath weed, the beetle is not necessarily responsible for maintaining control, but the beetles are still critical in the event of weed resurgence and would respond to prevent increases in weed density to former levels. This is a classic case illustrating the mechanism of density dependence in limiting a pest, and it is the ideal for successful classic biological control of weeds [40].

Host specialization is another key concept in classic biological control of weeds. Many potential natural enemies specialize in using a narrow range of closely related plants. Because of specialization, we can select and use organisms that present very limited risks to crops or other desirable domestic species, as well as native plants. The likelihood of an approved biological control agent's switching from one host to another outside its predicted range of use is restricted by a combination of factors [32]. Because host selection by phytophagous insects is a sequence of steps, those that specialize have been described as being ". . . in a deepening rut in evolution" [39]. We can devise experiments to test and define the degree of dependence of biological control agents on the targeted weed group.

In contrast to biological control of arthropod pests, or to biological control with nonclassic agents such as large herbivores, safety is of paramount importance in classic biological control of weeds. The intentional importation and release of exotic organisms for weed control involves consideration of risks to desirable and nontarget plant species, and this is undertaken only after precautions have been exercised to ensure that unintended effects are avoided (see below). In most countries, decisions to import and release biological control agents for weeds have been supported by requirements for scientific observation, experimentation, and submission and reviews of proposals.

II. SAFETY IN CLASSIC BIOLOGICAL CONTROL OF WEEDS

In the United States, as in most other countries with classic biological weed control programs, there is an overriding concern with safety. The probability that an agent introduced to control a target weed may attack a crop or other desirable plants must be minimized. The Plant Quarantine Act of 1912 and the Plant Pest Act of 1957 have been interpreted such that biological control agents used in weed control are classified as plant pests, and guidelines for reviewing proposals for their introduction and release into the United States were adopted in 1957 [45]. With the passage of the National Environmental Protection Act of 1970 and the Endangered Species Act (ESA) of 1973, concern was broadened to include species which are listed as threatened, endangered, or those species considered as candidates to be so listed.

The U.S. Department of Agriculture's Animal and Plant Health Inspection Service (USDA-APHIS) oversees a permitting process for exotic biological control agents proposed for control of weeds. Authors of proposals prepare a petition to import and/or release potential agents, and APHIS circulates petitions to a Technical Advisory Group (TAG), comprising representatives of various federal agencies [17,45], the National Plant Board (representatives of State Departments of Agriculture), the Weed Science Society of America, and representatives from Mexico and Canada. The TAG Chairman collects reviews and comments from TAG representatives and forwards a recommendation to APHIS, which is responsible for making the decision. There are no formal rules for reaching consensus on a recommendation, but in practice individual TAG members have near-veto power. In addition to reviewing petitions for agent proposals, TAG reviews some proposals for test plant lists and proposals to target individual weed species for biological control. From 1987 through 1996, TAG reviewed approximately 100 petitions for approval to release or to receive biological control agents in quarantine.

Pathogens used as classic biocontrol agents are subjected to the same safety and host-specificity considerations as are arthropod biocontrol agents. TAG helps to coordinate research in this area by reviewing proposals to seek biocontrol agents and making recommendations with respect to the need, scope, and adequacy of the proposed research. As in the case of exotic arthropods used for biological control of weeds, the approval of USDA-APHIS is required to introduce foreign pathogens into quarantine facilities, as well as for their eventual release and field establishment, usually after additional safety evaluations have been completed and reviewed by TAG.

A. Safety of Arthropod Biological Control Agents

Guidelines for petitions require information on the taxonomy of the target weed, the design and results of host-range testing experiments, the species of plants included in the tests, and the taxonomy of the proposed agent. For arthropod biological control agents, host range experiments focus on two main indicators of host specificity—feeding and oviposition—and usually include both "choice" and "no-choice" tests. Choice tests are those in which the agents under study are provided with opportunity to feed or oviposit upon a specimen of the target plant and also on one or more other specimens of different plants (related and unrelated to the target weed). No-choice tests are those in which only a single test plant species is presented to agents. Specificity information from field experiments, or from field collections in the native areas of proposed agents, often supplement information derived from choice/no-choice testing. Development of lists of test plant species usually takes into account the taxonomic relationship of the target weed species to other plants [79,80], with more intense testing of species that are more closely related to the target weed. The aim of testing is as much to define the range of hosts acceptable to the proposed agent as it is to determine whether individual plant species might be used. With passage of the Endangered Species Act, however, endangered and threatened plants, or closely related sibling species, are now also considered for testing, and including these species in host range experiments improves assessment of risk to endangered and threatened plants.

Most petitions to APHIS and TAG for approval to import and release arthropod biological control agents include evaluation of the proposed agents according to two scoring systems developed to predict the effectiveness of agents. Harris [30] first recommended this approach, and Goeden [27] proposed a different system of scoring. These systems assign scores to a number of different attributes of potential agents such as reproductive capacity, amount and type of damage to the target plants, and specificity of candidate agents, but there has been no evaluation of the predictive power of these approaches.

There have been troublesome unintended effects of some biological control agents [37,65]. The worst cases have not involved weeds, but other organisms, but there are two instances of classic biological control of weeds from which we should learn. Both involve the attack of native plant species closely related to target weeds by biological control agents, one of which was intentionally introduced into the United States, and one which was not.

A Eurasian weevil (*Rhinocyllus conicus*), which attacks a wide range of thistles, was introduced by Canadian workers into North America in 1968 for control of musk thistle, *Carduus nutans*. The weevil was known, by collection records and by host range experiments, to attack other species of *Carduus*, and also other genera, including *Cirsium* and *Silybum* [84,85]. *R. conicus* was distributed throughout Canada and the United States on *C. nutans* and other weedy thistles, and became widely established and exerted effective control of *C. nutans* in some locations. In 1983–1985, Turner et al. [76] collected flowerheads of several species of native *Cirsium* and found that they were attacked by and likely were suitable hosts for *R. conicus*. Three of the native *Cirsium* species were included on the 1985 U.S. Fish and Wildlife Category 2 list, although they are no longer listed under ESA provisions.

First employed as a biological control agent for cactus in the genus *Opuntia* in Australia in the 1920s, the moth *Cactoblastis cactorum* also has been released in other parts of the world, including several Caribbean islands. Never intentionally introduced into the United States, *C. cactorum* was detected within a Nature Conservancy preserve in Florida in 1989 on *Opuntia spinosissimi*, a rare species with only a few individual specimens known in the United States [65]. It is not certain how *C. cactorum* arrived in Florida; it may have flown there from Cuba, but it is known to have been intercepted in shipments of ornamental cacti from the Dominican Republic to the United States from 1981 through 1986 [57]. After the initial detection, *C. cactorum* was found in other Florida locations, and this insect represents a threat to several species of *Opuntia* native to the United States.

Both of these cases demonstrate the need for caution and deliberation in introducing biological control agents for control of weeds. Nevertheless, neither example is a failure in current procedures for reaching decisions to import biological control agents for weeds into North America. In the case of *R. conicus*, the weevil was widely tested, its host range was well known, and it even attacked North American native *Cirsium* spp. in testing. The decision to introduce the weevil was made, however, before passage of either the National Environmental Protection Act of 1970 or the Endangered Species Act of 1973, and concern for native plants was discounted because of the severity of the thistle problem. It is clear that *R. conicus* would not be intentionally released under current rules and procedures.

The case of *C. cactorum* illustrates two points. One is the need for international cooperation in biological control. The decisions of some countries to introduce this moth affected other countries not included in the decision-making process. APHIS maintains two *ex officio* seats on TAG for Canada and Mexico, and Canada actively participates in review of all petitions, but there is no international convention to consider movements of biological control agents. The other point is that regulatory actions to limit introductions of plant pests are insufficient. Entry of plant pests besides *C. cactorum* are common; indeed, it could be argued that conservative regulation of entry of biological control agents for weeds is of little value give the inflow of other potential plant pests. For example, approval was granted for importation and release of the moth *Leucoptera spartifoliella* for control of an introduced leguminous shrub, Scotch broom, in the United States only to find that the moth was already present in California and Washington [24]. Another exotic moth, *Agonopterix nervosa*, apparently specific to Scotch broom and another weedy leguminous shrub, gorse, is common on these plants in the western United States. An exotic seed weevil, *Bruchidius villosus*, is widely established in eastern North America on Scotch broom, and it has been approved for release into New Zealand as a biological control agent [69]. In addition, two exotic insect species of the family Psyllidae, also quite specific for Scotch broom, are present on Scotch broom in the western United States. It could be argued that efforts to enhance the protection of native plants from introduced pests might best focus on understanding and limiting accidental introduction of herbivores rather than increasing restrictions on intentional introductions.

B. Safety of Pathogenic Biological Control Agents

Typically, surveys are made in the native range of the weed to discover and rank pathogens with biocontrol potential. Prospective pathogens, usually rust fungi because of their host specificity, are studied in their homelands to determine their host specificity and virulence toward the target weeds. Extensive biosystematic and pathological studies may be done to confirm the identity and suitability of individual pathogens. Host-range studies are conducted to establish the safety of the agents. Any gaps in our understanding of the pathogens' life cycles must be filled at this stage. Only those considered safe to nontarget plants, including major and minor crops and ecologically important plants, are approved for introduction into quarantine in the United States. Once in quarantine, additional testing is done on plants of economic and ecological importance in North America. Release of the agents and assessment of their effectiveness are typically coordinated with appropriate federal, state, and regional agencies.

When used as bioherbicides, according to the current regulations in the United States, plant pathogens are classified as "pesticides" under U.S. Federal Insecticide, Fungicide, and Rodenticide Act (FIFRA), and the U.S. Environmental Protection Agency (EPA) has set regulations and guidelines for their registration and safe use (see ref. 13). Under these guidelines, detailed testing of toxicology, environmental impact, product performance, and other aspects are undertaken by the potential registrant, generally a private industry. This process assures that bioherbicides will follow an established and scientifically valid testing scheme, and only those that are considered safe to the public and the environment will be allowed to be used.

Under this regulatory scheme, bioherbicidal agents are tested against nontarget plants and animals and are subjected to environmental risk analysis, toxicology, and other tests as deemed appropriate. It is possible to negotiate exemptions from certain types of tests if scientific evidence warrants such exemptions. The testing is typically done in a tiered format, with Tier I tests being the least stringent. Pathogens that pass Tier I tests with no adverse indications are likely to proceed toward further industrial development as bioherbicides; those that fail Tier I tests (i.e. they are considered risky in one or more aspects) still may be subjected to Tier II and Tier III tests. However, since these tests are very expensive, it may be economically prohibitive to sponsor and test a bioherbicidal agent beyond Tier I tests.

Plant pathogens intended to be used as biological weed control agents are tested against a variety of plants selected to represent ecological, economic, and pathological relevance in an attempt to determine their host range. The centrifugal-phylogenetic testing scheme proposed by Wapshere [79,80] is followed to select plants that are taxonomically and phylogenetically related to the target weed. In addition, due to the EPA regulatory requirements mentioned above, it is essential to screen many economically important crop plants and plants that occupy the same ecological habitats as the target weed and to consider potential impacts on endangered and threatened plants. Generally, 100–150 different plant species must be tested in order to obtain a fairly comprehensive assessment of the host range of a given pathogen species. Since pathogens intended for use as classic biocontrol agents are usually nonnatives, they are subjected to more stringent host-range testing than native pathogens developed as bioherbicides. The principles and protocols involved in the search, selection, evaluation, regulation, and use of microbial weed control agents have been discussed in detail in two treatises edited by Charudattan and Walker [14] and TeBeest [70] and in several excellent reviews.

C. Major Differences Between Arthropod and Plant Pathogens as Weed Control Agents

The similarities and differences between arthropods and plant pathogens used as biological weed control agents are summarized in Table 3. In general, there are few procedural differences between arthropods and pathogens imported and utilized as classic biological control agents for weeds. Both types of agents increase in a density-dependent manner, and, if conditions are conducive to their survival, reproduction, and rapid population build-up, they may reach population densities at which they impose sufficient stress on the target weed to control it. On the other hand, pathogens used as bioherbicides must undergo a technological development and regulatory testing phase that is highly contingent on several economic and commercial realities. Since the process of industrial development, safety testing under the EPA protocols, registration, and marketing is costly and capital intensive, only those agents that can provide economically viable returns on investment will be developed [7,8,44].

III. THE PROCESS OF BIOLOGICAL CONTROL OF WEEDS

A methodical approach to classic biological control of weeds requires inquiry into and management of several different issues. It would be wrong to suggest that resolving these issues are "steps," if that implies the process is sequential and that the issues are discrete, but there is a logical problem-solving structure

Table 3 Similarities and Differences Between Requirements for Insects and Plant Pathogens Used as Biological Control Agents for Weeds

---

Both insects and pathogens
  As classic biological control agents:
    surveys and biological studies
    efficacy and host-range determinations
    determination of effectiveness, host specificity, and safety
    release
    establishment
    density-dependent multiplication
    further spread
    biocontrol success
Pathogens only
  As bioherbicides:
    discover and determine efficacy in small-scale trials
    determine safety on the basis of host-range
    confirm efficacy under field conditions
Regulatory Requirements to Register Bioherbicides as Pesticides and/or Commercial and Practical Necessities
  patent and establish a system for technology transfer
  develop methods for large-scale production of the agent
  develop a suitable formulation of the agent
  test under a variety of field conditions
  conduct toxicology tests
  develop a commercial product and a product label
  obtain EPA registration (United States)
  establish a network for distribution and product promotion

---

to the organization of biological control projects aimed at weeds: defining the problem, developing and evaluating different solutions, selecting the best solution(s), and implementing and evaluating them.

A. Defining the Problem

The investment in research and oversight in obtaining approval for release of a single agent may be several hundred thousand U.S. dollars [31]. Besides the direct costs of obtaining and utilizing one or more biological control agents, there are intangible costs of biological control, such as those associated with the environmental risks involved in the use of an exotic organisms. Overall costs for even a modest project easily could exceed US$1 million. Biological control projects should be directed at weeds where costs can, at a minimum, be reasonably expected to be recovered, and, given the expectation for investment, where a net profit can be returned. It follows that there may be weeds unsuitable for biological control because their current or projected costs do not meet a minimum threshold. For these species, the resources, or costs, to initiate biological control might be greater than any projected benefits. Peschken and McClay [58] proposed a scoring system based on the attributes of weedy plants to predict their suitability as targets for biological control. In many instances, biological control projects have been launched because the costs of the targeted weed clearly exceeded any such threshold even in the absence of quantitative data on costs. *Optunia* spp. in Australia and leafy spurge in North America are examples, although there are now studies of leafy spurge in northern Midwest states of the United States which quantify spurge impacts [5,47].

B. Understanding the Target Weed

If a weed is selected as a target for biological control, the weed's distribution and ecology in its native range and its taxonomic or phylogenetic relationship to other plants must be clearly defined. The richest source of potential biological control organisms for a weed is predicted to occur at the evolutionary center of its distribution [80]. For regions where there are reliable catalogs and descriptions of the flora, searches for potential agents can be planned and conducted by determining centers of diversification of target weed genera.

C. Searching for Candidate Agents

Only organisms specifically adapted to the target weed or its close relatives are likely to meet standards of specificity. These potential agents are identified by field searches, inspection of museum collections, maintenance of "trap" plants, and surveys of existing literature. Often extensive lists of candidate species of potential biological control agents are developed, then subsequently reduced and prioritized by excluding species thought or shown to be too broad in their range of acceptable hosts and by observing the nature and extent of damage caused to target weeds.

D. Selection for Testing

The end objective of this sorting and processing of lists is to identify candidate agents which merit more intensive study and to attempt to expend resources only on species likely to be effective and to pass specificity requirements. Selection of agents with respect to predicted effectiveness may be based on detailed observations in the field or upon scoring schemes like those proposed by Harris [30] and Goeden [27].

E. Testing for Host Suitability and Specificity

Some control agents are very specific, to the point that, even though they are collected from the host within its native range, they may not attack the biotypes or strains present in the area where control is needed. In the case of the rust fungus *Puccinia chondrillina*, used against rush skeletonweed (*Chondrilla juncea*), there are different strains of both the rust and the weed, and virulence and even suitability vary greatly depending on the match of biotypes. Failing to include a range of plant material in prerelease testing from the area where the weed is to be controlled could lead to false conclusions about the suitability of a candidate agent. Tests may be conducted either in laboratory situations or in the field and in or near the country of origin or in the country where agents are to be used. Testing is sometimes complicated by the pest and plant protection regulations of different countries; many do not allow free movement of weedy plant material and impose systems of permitting and quarantine, and a few have outright prohibitions of movements of some live materials, whether plants or potential biological control agents.

F. Review and Conflict of Interest Issues

Safety of an introduction is a primary concern, but there are other issues considered during the review of proposals to import and release an exotic organism for control of a weed. Although researchers may identify potential conflicts of interest, decisions to use agents whose use will predictably have both beneficial and adverse impacts must involve public input (e.g., ref. 23). The possible targeting of wild blackberries in North America provides an example. Blackberries have been successfully controlled in Chile with a rust fungus in the genus *Phragmidium* [56]. This agent also has been introduced into Australia for blackberry control, and prospects for satisfactory control are good [10]. It is likely that a strain or strains of this fungus could be found and tested which would attack the exotic and problematic species of blackberries without harming native or related domestic blackberries. Wild blackberries dominate extensive areas in pastures and in riparian areas in the western United States, where they displace native and desirable vegetation. Herbicide use by homeowners and small landowners is common, and it is not often efficient and is sometimes improper. Industrial and road maintenance organizations expend considerable resources for control of wild blackberries. However, the fruit of wild blackberries is picked and used by the public, and if there were there a proposal to control wild blackberries, all concerned public interests should be addressed and evaluated.

G. Shipment and Handling

One of the primary concerns with shipments of biological control agents from the country of origin to the areas where they are to be introduced is the exclusion of any parasites and diseases associated with them. In Canada, the first introductions of the cinnabar moth (*Tyria jacobaeae*) for control of tansy ragwort failed to establish, and this failure was attributed to infections of the introduced herbivore by both a virus and a microsporidian [11]. Subsequent introductions of disease-free stock resulted in successful establishment. Elimination of parasites and diseases is often accomplished within quarantine facilities, sometimes by rearing a complete generation of the agents to ensure exclusion, sometimes by testing or examining a sample of individual agents. The inability to exclude parasites can be the basis of a decision not to import and release, as in the case of *Nanophyes marmoratus*, a seed-feeding weevil proposed for control of purple loosestrife in North America, where attempts to exclude a parasitic nematode have been unsuccessful.

The physical transport of biological control agents over long distances has been improved with the proliferation of express parcel delivery services and the development of specialized containers and shipping cartons which provide for cooling. But it is still important that care be taken to ship organisms in the appropriate containers and to handle them properly to avoid losses.

The need to document introductions and to maintain records of introductions has been discussed by Coulson [16], who points out that reliable records can be valuable to practitioners of biological control, researchers in related fields, and to regulators in identifying patterns related to successes and failures in biological weed control. Voucher specimens are an essential part of record keeping, and there is a constant need to upgrade, standardize, and integrate systems for tracking releases and their outcomes.

H. Site Selection and Release

Field establishment is the first goal of postintroduction management for biological control of weeds, and some factors such as predation, parasitism, or adverse weather may not be controllable. Competition with other natural enemies and poor climatic matching between native and introduced areas also may be factors in failure, and may be somewhat more under the control of practitioners. Other controllable factors are clearly related to successful establishment. The number of individuals of the agent released is directly related to the probability of successful establishment [51]. Caging of initial releases can improve chances of success. Ensuring the stability and continuity of management at a site can be a critical factor if the sites are not under control of regulatory agencies or biological control practitioners. Locating sites for convenience of monitoring and research is important, but some limitations to easy public access may be advised in some settings to avoid pilferage or vandalism.

In classic biological control programs, the introduced natural enemy is simply related or inoculated into small weed infestations relative to the total infestation. If conditions are favorable, the natural enemy multiplies on the weed host and spreads, causing a high level of damage that may kill the weed, severely limit its growth, or reduce its competitive ability. The weed population may then begin to decline. Since this process depends on a gradual increase in disease or herbivore populations, it may take several months or years to obtain significant levels of weed control. Detailed evaluation is critical to determine success, as well as to document possible nontarget impacts.

For inunduative programs, massive doses of inoculum are applied to the weed population to encourage rapid rates of disease development and disease spread. A high level of epidemic disease occurs if conditions conducive to disease development and weed susceptibility to pathogens are present. Inundative microbiological agents are manufactured, formulated, standardized, packaged, and registered as herbicides according to FIFRA. The pathogens are applied to weeds by methods and tools similar to those used for chemical herbicides, and they can be integrated with other pesticide and crop management schedules [66]. Because an inundative agent can produce rapid and high levels of weed control, it can be used in intensively managed agroecosystems. There is usually a need for annual applications of the inundative agent, since the pathogen typically does not survive in sufficient numbers, nor does it multiply between cropping sequences to initiate a fresh epidemic on new weed infestations. Moreover, agricultural soils hold considerable seed reserves to initiate new weed infestations each growing season, making it necessary to reapply the inundative agent.

I. Monitoring

Establishment of introduced biological control agents, population increase, and damage to the target weed are three measures of success which are routinely monitored. Monitoring usually is an informal process, and observations are often qualitative and subjective. More detailed studies typically are planned and conducted (Section III.K below).

J. Collection and Redistribution

Some organisms used as biological control agents disperse on their own so well that minimal effort is required on the part of practitioners to further distribute agents. The rust fungus, *P. chondrillina,* used against rush skeletonweed in Australia, is one such example [19]. Similarly, the tephritid fly, *Urophora quadrifasciata*, which attacks spotted knapweed, dispersed substantial distances on its own [67]. Others organisms disperse slowly of their own accord; for example, starvation was the major mortality factor at the site of the release of the cinnabar moth, *Tyria jacobaeae*, 13 years after its release, but natural dispersal was only about 10 km from the release point [42]. With most species of agents, planned collection and redistribution will substantially reduce the time taken until dispersal of the agent over the entire range of the target host is realized.

K. Evaluation

If biological weed control is to be truly scientific in it use, there is a need for scientific evaluation of projects to establish clearly the impacts of the biological control agents and to document that changes in populations of the target weeds are in fact attributable to biological control and not to other factors. There are several factors which limit interest in evaluation. First, field evaluations are difficult and expensive, and establishing and maintaining control over treatment plots present challenges that may require extraordinary measures. Also, there may be a bias against evaluating projects in which results have not met expectations either because the project may be considered incomplete or because the further investment of resources is considered a waste. There are also cases where the impacts of biological control agents are apparently evident beyond argument. McClay [49], in a review of 57 cases of reported evaluations, found that a minority incorporated experimental methods and only six measured weed populations. He recommended inclusion of four guiding elements in evaluation projects: (a) to assess the weed, not just the agent; (b) to assess weed populations rather than individual plants; (c) to conduct the work in the field; and (d) to prove the responsibility of the agent.

Few biological control projects have been evaluated for their economic impact. Several good studies exist on the economic impacts of weeds, and the results of these studies are impressive in terms of the magnitude of the documented losses due to weeds (e.g., refs. 5 and 47). Benefits of weed control, though, are difficult to quantify, and the persons responsible for the conduct of biological control programs are usually biologists, pathologists, or entomologists, with a prejudice for documenting results in biological rather than economic terms. Even where benefits have been quantified, it is difficult to capture them all. In the case of the poisonous weed, tansy ragwort, it was relatively easy to document that fewer cattle and horses were being poisoned [15], but the environmental benefits of reduced herbicide use, particularly by small landowners with limited experience and knowledge of proper uses of herbicides, are intangible.

IV. EXAMPLES

A. Classic Biocontrol Agents

1. *Puccinia chondrillina* and Rush Skeletonweed

One of the most remarkable successes in classic biological control of weeds resulted from the introduction of the rust fungus, *Puccinia chondrillina*, from the Mediterranean region into Australia to control rush skeletonweed (*Chondrilla juncea*; Compositae). It has been estimated that this successful biocontrol project has resulted in a cost:benefit ratio of 1:100 in Australia [18,64]. Skeletonweed is of Mediterranean origin and was introduced into Australia, where it became a serious weed in cereal crops and rangelands. After considerable research by the Australian Commonwealth Scientific and Industrial Research Organization, *P. chondrillina* was introduced, along with several insects, in a classic biocontrol attempt. The rust was very effective; following inoculative releases, it rapidly spread, created high levels of disease epidemics, and in the process infected, stressed, and killed the most common and susceptible biotype of the weed. As a result, after the successful establishment of the pathogen, the weed density in cereal crops decreased in less than 10 years to be less than 10 plants m^{-2} from the level of about 200 plants m^{-2} that existed before the rust was introduced. Good control also was obtained in pastures, suggesting an overall projected savings of $25.96 million per annum when the biocontrol system reaches an equilibrium [18].

The rust attacks one of three forms of the weed. Initially, following the successful reduction in the density of the susceptible form due to biocontrol, two other forms that are more resistant to the rust became more widespread [18]. Therefore, subsequent attempts were made to find and introduce additional rust strains virulent on these relatively resistant forms [33-35]. These strains have since been introduced into Australia to control the resistant forms of the weed [35]. This case illustrates one of the potential difficulties in using biological control, particularly with highly specific natural pathogens; namely, the emergence of resistant weed biotypes as problems. However, this example also illustrates the possibility to counter the emergence of resistant weeds by introducing pathogenic strains that are effective against the resistant weed biotypes.

Puccinia chondrillina was also introduced into the United States to control a skeletonweed biotype in the western United States. However, unlike in Australia, the rust was only partially successful, probably because of a poor match between the introduced pathogen and the U.S. skeletonweed forms. Hence, under these conditions of less than expected efficacy, the rust has been utilized along with chemical herbicides and the insect biocontrol agents, *Cystiphora schmidti* (a gall-forming midge) and *Aceria chondrillinae* (a gall-forming mite), in an integrated weed management program to maximize its benefits [46]. As in Australia, the rust has been the most successful of the three introduced biocontrol agents in California and other western states in the United States [68].

2. *Entyloma compositarum* and *Hamakua pamakani*

Another successful example of a classic biocontrol program involved the use of a smut fungus, *Entyloma compositarum*, from Jamaica to control pamakani, *Hamakua pamakani* (*Argeratina riparia;* Compositae), in Hawaiian forests and rangelands [75]. The fungus, originally named as *Cercosporella* sp., was introduced into Hawaii in 1974. Extensive host range testing established this fungus to be highly host specific and safe as a biocontrol agent for pamakani, and it was introduced into Hawaii from Mexico in 1925 as an ornamental plant. By the 1970s, it had spread to an estimated 62,500 ha on the island of Hawaii and 10,00 ha on Oahu. The weed also was present on the island of Maui. About 2–3 months after the pathogen was released in the field, devastating epidemics occurred in dense stands of *A. riparia* in cool, high-rainfall sites on Oahu, Hawaii, and Maui. The weed populations were reduced 80% in a 9-month period. Similar reductions in weed populations were recorded 3–4 years after the pathogen was released at sites with adequate moisture. At sites with low temperatures and low rainfall, there was greater than 50% reduction in the weed population in 8 years after the pathogen's release. It is estimated that more than 50,000 ha of pasture land have been rehabilitated to their full potential by this pathogen. No evidence of host resistance or the presence of mutant strains of the pathogen has been encountered since the release of this pathogen [74].

3. *Puccinia carduorum* and Musk Thistle

Another rust fungus, *Pucccinia carduorum*, has been imported from Turkey and released into the northeastern United States to control musk thistle, *Carduus thoermeri* (Compositae). The fungus was tested for pathogenicity to 16 accessions of *C. thoermeri,* 10 accessions of related weedy *Carduus* spp., 22 native and 2 weedy *Cirsium* species, and *Cynara scolymus* (artichoke). It was determined that species outside the genus *Carduus* are unsuitable hosts for this rust fungus. This group of unsuitable hosts includes 25% of the *Cirsium* species from North America tested, artichoke, and representatives from two other genera, *Saussurea* and *Silybum*. Considering the host-range data and the very susceptible nature of *C. thoermeri*, a judgment was made that species other than *C. thoermeri* would not be threatened by the introduction of *P. carduorum* into North America to control musk thistle. Accordingly, permission was granted in 1987 for a limited field release of the rust in Virginia [9].

Musk thistle plots were inoculated successfully in the fall and spring of each year in 1987–1989. Artichoke and selected *Cirsium* spp. (nontarget plants) were planted between plots of inoculated musk thistle in 2 of 3 years. Only one rust pustule was found on an artichoke plant in 1989; all other nontarget plants remained rust free despite severe disease on surrounding musk thistle plants. This indicated that *P. carduorum* poses no threat to these nontarget plants. The rust overwintered on musk thistle. Its spread was limited during the rosette stage, and the disease became severe only during bolting stage (i.e., stem elongation). Senescence of rust-infected musk thistles was accelerated and their seed production was reduced by 20–57%. Thus, the results confirmed that *P. carduorum* can contribute significantly to the control of musk thistle [6].

4. *Uromycladium tepperianum* and *Acacia saligna*

Morris [53] has described another highly successful example of a classic biological control program from South Africa, where an introduced gall-forming rust fungus, *Uromycladium tepperianum,* has been used against an invasive tree species, *Acacia saligna.* The fungus was introduced from Australia into the Western Cape Province. About 8 years after introduction and establishment at specific sites, the number of infected trees, rust galls per tree, and severity of rust infections increased at all sites. Tree density decreased by at least 80% in rust-established sites. The seed number in the soil seed bank stabilized in most sites. Thus, *U. tepperianum* is proving to be a very effective biocontrol agent, as shown by the greatly reduced population densities of *A. saligna* in South Africa. The rust forms extensive galls on branches and twigs, costing a significant energy loss to the tree, disrupting photosynthesis, causing heavily infected branches to droop, and eventually killing the tree.

B. Bioherbicides

1. DeVine (*Phytophthora palmivora*)

Between 1980 and 1994, four bioherbicides were registered in the United States and Canada. More recently, three other pathogens have been either registered as a bioherbicide or as cut-stump treatments. (Details of these agents are given in Table 4.) The first bioherbicide registered in the United States was DeVine, based on *Phytophthora palmivora* to control stranglervine (= milkweed vine; *Morrenia odorata*) in citrus in Florida. Stranglervine, imported into Florida as a potential ornamental plant, became a serious weed in about 120,000 ha of Florida's citrus groves. A native isolate of *Phytophthora palmivora* was found to attack stranglervine in a citrus grove in Florida, and it was determined to be a suitable bioherbicide agent by Ridings and others [62,63]. It was developed and registered in 1980 by Abbott Laboratories (Chicago, IL) as the bioherbicide DeVine. DeVine yields 90–100% control of the vine with just one application, and the control lasts for at least 2 years following treatment [44].

Table 4 Registered or Approved Bioherbicides—1997

Commercial name	Pathogen name	Weed		Made/sold by/for use in
		Common name	Botanical name	
DeVine	*Phytophthora palmivora*	Milkweed vine	*Morrenia odorata*	Abbott Laboratories, Chicago, IL, for use in Florida in citrus groves.
Collego	*Colletotrichum gloeosporioides* f.sp. *aeschynomene*	Northern jointvetch	*Aeschynomene virginica*	Encore Technologies, Minnetonka, MN, and University Arkansas, Fayetteville, for use in Arkansas, Mississippi, and Louisiana in rice.
BioMal	*Colletotrichum gloeosporioides* f.sp. *malvae*	Round-leaved mallow	*Malva pusilla*	PhilomBios and Agriculture and Agri-Food, Saskatoon, Canada, for use in Canada in various row crops.
Dr. BioSedge	*Puccinia canaliculata*	Yellow nutsedge	*Cyperus esculentus*	No commercial product or seller; this registered agent has never been sold commercially.
Camperico	*Xanthomonas campestris* pv. *poae*	Annual bluegrass	*Poa annua*	Japan Tobacco, Yokohama, Japan, for use in Japan in golf course turfs (Kentucky bluegrass, *Poa pratensis*, and zoysia grass, *Zoysia tenuifolia*).
BioChon	*Chondrostereum purpureum*	Black cherry	*Prunus serotina*	This product is not registered but approved for use as a wood decay promotor for control of *P. Serotina* in the Netherlands in forests.
Stump Out	*Cylindrobasidium laeve*	Weedy trees	Various	This product is approved for use as a stump treatment to prevent resprouting of felled trees in South Africa.

2. Collego (*Colletotrichum gloeosporioides* f.sp. *aeschynomene*)

Collego, based on the fungus *Colletotrichum gloeosporioides* f.sp. *aeschynomene*, was registered in 1981 by The Upjohn Company (Kalamazoo, MI). This product was subsequently sold to other companies and is now marketed by Encore Technologies (Minnetonka, MN) under an agreement with the University of Arkansas. This native fungus was discovered and developed in Arkansas by Templeton and colleagues [72,73] to control northern jointvetch (*Aeschynomene virginica*), a leguminous weed in rice and soybean crops of Arkansas, Mississippi, and Louisiana. In addition to its competition with the crops, the weed produces black, hard seeds that are difficult to separate from harvested rice and soybean. The presence of the contaminant seeds lowers the quality and price of the harvest.

C. *gloeosporioides* f.sp. *aeschynomene* is a highly virulent pathogen that causes an anthracnose disease that is lethal to northern jointvetch. The bioherbicide preparation, consisting of dried spores of the fungus, is hydrated before application and is applied by fixed-wing aircraft [71]. In nearly two decades of combined experimental and commercial use, Collego has consistently yielded 90–100% weed control with no apparent adverse ecological impacts. It has been effectively integrated into rice and soybean pest management programs involving fungicides, herbicides, and insecticides [66].

3. BioMal (*Colletotrichum gloeosporoides* f.sp. *malvae*)

The third bioherbicide that was registered in North America was BioMal, based on the fungus *Colletotrichum gloeosporioides* f.sp. *malvae*, a pathogen of round-leaved mallow (*Malva pusilla*). It is registered for use in several crops in Canada, but owing to technical and economic considerations, it has not been commercially sold since registration. However, attempts are underway to further develop and market this bioherbicide [50,54,55].

4. Dr. BioSedge (*Puccinia canaliculata*)

Dr. BioSedge, a bioherbicide based on the rust fungus *Puccinia canaliculata*, was registered for use against yellow nutsedge (*Cyperus esculentus*) in the United States. Several years of research by Phatak and coworkers [59,60] established the efficacy and the potential to use this native rust fungus as an augmentative agent to control the weed. However, since the fungus is an obligate parasite, it could not be mass produced, and therefore presently there is no mechanism in place for its commercial sales and use.

5. BioChon (*Chondrostereum pupureum*)

A wound-invading, wood-rot–causing basidiomycete, *Chondrostereum purpureum,* has been approved for use as a wood-decay promoter to prevent resprouting of black cherry (*Prunus serotina*) and to control this weedy tree in conifer forests in the Netherlands. The fungus is widespread in the temperate regions of the world and is normally a weak parasite of wounded or weakened trees. It has a fairly wide host range, being capable of infecting several broad-leaved trees. It causes a disease, known as the silverleaf disease, on several rosaceous plants. However, in spite of the lack of a high degree of host specificity and the potential for damage to some fruit crops, it was determined that it is possible to use this pathogen without any extra harm to cultivated plants [20].

In a recent assessment of the potential for risk from this fungus, de Jong et al. [21] estimated the natural occurrence of *C. purpureum* in the forests on southern Vancouver Island (British Columbia, Canada) in relation to the proposed bioherbicidal use of this fungus to control unwanted hardwood trees. Surveys were done during two winters in randomly located, 1000-m^{-2} plots of fructifications (basidiocarps) and compared with potential added fructifications that might occur as a result of using the fungus to control hardwood weeds. In addition, surveys were made for fructifications in forests as well as in urban or agricultural areas by estimating the surface areas of woody substrates covered with basidiocarps. Estimates also were made in locations where the fungus would be expected to occur, such as woodpiles, silvicultural thinnings, and killed trees. The amount of added fructification through the use of this fungus as a biological control agent was determined from inoculated plots as well as from calculated stump-surface areas. From the various calculations, it was determined that the added fructification of *C. purpureum* is of the same order of magnitude as naturally occurring levels or even lower. In addition, it was determined that there is a distinct geographical separation between predominantly forested areas, where the fungus will be used, and predominantly settled areas where fruit and ornamental trees are cultivated. Accordingly, it was calculated that using this fungus as a biological control agent in forestry is not likely to pose a significant threat to fruit growing and commercial forests.

C. purpureum is being developed as BioChon, a stump-treatment product by Koppert B. V., The Netherlands. A similar product may be developed and registered in Canada to control unwanted hardwood trees in conifer forests.

6. Camperico (*Xanthomonas campestris* pv. *poae*)

An isolate of *Xanthomonas campestris* pv. *poae*, a wilt-causing bacterium, isolated in Japan from annual bluegrass (*Poa annua*) has been developed and registered in Japan as the bioherbicide Camperico to control annual bluegrass in golf courses. Fourteen isolates of this bacterial pathogen were isolated from annual bluegrass from throughout Japan and compared for pathogenicity and safety toward desirable turf grasses such as creeping bentgrass (*Agrostis palustris*) and Kentucky bluegrass (*Poa pratensis*). One isolate, JT-P482, was the most effective pathogen of annual bluegrass, and it did not cause any symptoms on desirable turf grasses. By applying a suspension of the bacterium to precut annual bluegrass, greater than 70% control was obtained, determined as weight loss due to plant wilting and death, compared with uninoculated controls. The optimum inoculum concentration was more than 108 colony-forming units (CFU) in an application volume of 100–400 mL m^{-2}. The infection rates were accelerated by applying larger volumes of inoculum. At 25°C/20°C (day/night) annual bluegrass wilted severely in 7–10 days, but lower temperatures caused a loss of efficacy, showing that temperature is an important factor for effective control. Field tests in zoysia (*Zoysia tenuifolia*) greens in the fall with an inoculum rate of 109 CFU mL^{-1} at 400 mL m^{-2} resulted in over 90% disease severity the following spring. Thus, this bacterium was determined to have great potential as a bioherbicide for controlling annual bluegrass without harming the desirable grasses, including the closely related species Kentucky bluegrass. Accordingly, the bacterium has been registered for use as a bioherbicide [41].

7. Stump Out (*Cylindrobasium laeve*)

Morris [53] has developed a wood-infecting basidiomycetous fungus, *Cylindrobasidium laeve*, as a stump-treatment product to control resprouting of cut trees, including weedy trees. A laboratory-based inoculum production system and an oil-based paste formulation of the fungus have been developed, and the fungal product has been approved for use in South Africa under the name Stump Out. Several other pathogens are undergoing evaluation for possible commercial development and registration in different countries. These are listed in Table 4.

V. CONCLUSIONS

There is an upsurge of interest in biological control of weeds using both arthropods and plant pathogens as evidenced by the number of currently active research projects worldwide. It is expected that research into and development of biological weed control agents will continue with support from the public. It is anticipated that three or four additional herbicides and one or two classic biological control projects using pathogens and 20–30 arthropod agents will be developed and implemented in the coming decade.

The U.S. Congressional Office of Technology Assessment (OTA) conducted two studies in which biological control of weeds was identified as an important concern. The first of these dealt with harmful nonindigenous species [77], and the importance of biological control was recognized as a key in managing invading noxious plant species. The second was a review of biologically based technologies for pest control [78] and presented several options to Congress for increasing the benefits of biological weed control. Both of these studies, and the growing concerns expressed by many landowners and managers about invading weeds, have led to a recognition of the seriousness of noxious weeds. As one of the solutions to this developing problem, biological control is receiving more support and attention. At a recent national weed symposium, U.S. Interior Secretary Babbitt reported that he had been directed to prepare an action plan that establishes goals and steps that can be taken to reduce problems associated with invasive aliens, and at the symposium support for biological control was one of 12 priority action items identified by participants [4]. As a result, considerably more effort in biological control of weeds is anticipated.

One important issue relevant to the future of biological control of weeds is the safety of agents. Concern over attacks of native plant species by biological control agents is increasing, and this will predictably slow the rate of importation of new agents and limit the importation of potentially valuable agents. The work by Turner et al. [76] on the attack of a biological control agent on rare native thistle species, published in an entomological journal, spurred increased concern about safety within the biological control community, but not with the general public. Louda et al. [48] documented much the same phenomenon as did Turner and published the result in the journal *Science*, attracting much more public interest. A predictable public policy response to this developing concern will be the issuance of stricter rules for approval of agents. More conservative approaches to selection of targets and potential agents by biological control workers is also very likely. Some very troublesome weeds, such as the introduced pepperweed, *Lepidium latifolium*, have sibling species that are uncommon, and even rare, North American natives, and the concern for native plants will predictably limit interest and progress in biological control of such targets.

Continued and enhanced use of pathogens as bioherbicides will depend on the ability of industrial concerns to develop efficient and economical means for producing pathogens, and being able to deliver consistent results with pathogenic products. In addition, the high degree of specificity of pathogens used as bioherbicides mitigates against their use by growers, unless product costs are sufficiently low that growers can afford to use multiple bioherbicides against the array of weed species typically encountered in crop production systems. More detailed studies of pathogen and herbivore ecology and biology are critically needed to advance further the biological control of weeds using these agents.

REFERENCES

1. Anonymous (ed.). 1995. *Abstracts and Posters of Papers. European Weed Research Society Workshop on Biological Control of Weeds.* February 8–10, 1995, Montpellier, France.
2. Anonymous (ed.). 1996. *Programme and Abstracts. III International Bioherbicide Workshop. International Bioherbicide Group.* January 19–21, 1996, Stellenbosch, South Africa.
3. Anonymous (ed.). 1996. *Programme and Abstracts. 9th International Symposium on Biological Control of Weeds.* January 21–26, 1996, Stellenbosch, South Africa. Agricultural Research Council of South Africa.
4. Anonymous. 1998. *Proceedings: Science in Wildland Weed Management.* April 10–18, 1998; Denver (Babbitt's remarks are on: www.blm.gov/weeds/symp98/addrbabb.html).
5. Bangsund, D. A., and F. L. Leistritz. 1991. *Economic Impacts of Leafy Spurge on Grazing Lands in the Northern Great Plains.* Economic Report No. 275-S, Dept. of Agricultural Economics, North Dakota State University, Fargo.
6. Baudoin, A. B. A. M., R. G. Abad, L. T. Kok, and W. L. Bruckart. 1993. Field evaluation of *Puccinia carduorum* for biological control of musk thistle. *Biol. Control* 3:53–60.
7. Bowers, R. C. 1982. Commercialization of microbial biological control agents. In *Biological Control of Weeds with Plant Pathogens.* R. Charudattan & H. L. Walker, eds. New York: Wiley, pp 157–173.
8. Bowers, R. C. 1986. Commercialization of Collego--an industrialist's view. *Weed Sci.* 34 (suppl 1):24–25.
9. Bruckart, W. L., D. J. Politis, G. Defago, S. S. Rosenthal, and D. M. Supkoff. 1996. Susceptibility of *Carduus, Cirsium,* and *Cynara* species artificially inoculated with *Puccinia carduorum* from musk thistle. *Biol. Control* 6:215–221.
10. Bruzzese. E. 1992. Present status of biological control of European blackberry (*Rubus fruticosus* Aggregate) in Australia. *Proceedings of 8th International Symposium on Biological Control of Weeds.* E. S. Delfosse, & R. R. Schott, eds.. February 2–7, Lincoln University, Canterbury, New Zealand. DSIR/CSIRO, Melbourne, Australia, pp 297–299.
11. Bucher, G. E., and P. Harris. 1961. Food-plant spectrum and elimination of disease of Cinnabar moth, *Callimorpha jacobaeae* (L.) (Arctiidae: Lepidoptera). *Can. Entomol.* 93:931–936.
12. Cardina, J. 1995. Biological weed control. In *Handbook of Weed Management.* A. E. Smith, ed. New York: Marcel Dekker, pp 279–341.
13. Charudattan, R., and H. W. Browning (eds.). 1992. *Regulations and Guidelines: Critical Issues in Biological Control.* Proceedings of a USDA/CSRS National Workshop, June 10–12, 1991, Vienna, Virginia. Institute of Food and Agricultural Sciences, University of Florida, Gainesville.
14. Charudattan, R., and Walker, H. L. 1982. *Biological Control of Weeds with Plant Pathogens.* New York: Wiley.

15. Coombs, E. M., H. Radtke, D. L. Isaacson, and S. P. Snyder. 1996. Economic and regional benefits from the biological control of tansy ragwort, *Senecio jacobaea*, in Oregon. *Proceedings of 9th International Symposium on Biological Control of Weeds*. V. C. Moran and J. H. Hofman, eds. January 19–26, 1996, Stellenbosch, South Africa, University of Capetown, pp 489–494.

16. Coulson, J. R. 1992. Documentation of classical biological control introductions. *Crop Protect.* 11:195–205.

17. Coulson, J. R. 1992. The TAG: Development, functions, procedures, and problems. *Regulations and Guidelines: Critical Issues in Biological Control.* R. Charudattan & H. W. Browning, ed. Proceedings of a USDA/CSRS National Workshop, June 10–12, 1991, Vienna, VA. Institute of Food and Agricultural Sciences, University of Florida, Gainesville, pp 53–60.

18. Cullen, J. M. 1985. Bringing the cost benefit analysis of biological control of *Chondrilla juncea* up to date. *Proceedings of 6th International Symposium on Biological Control of Weeds*. E. S. Delfosse, ed. August 19–25, 1984, Vancouver, Canada, Agriculture Canada, Ottawa, pp 145–152.

19. Cullen, J. M., P. F. Kable, and M. Catt. 1973. Epidemic spread of a rust imported for biological control. *Nature* 244:462–464.

20. de Jong, M. D., P. C. Scheepens, and J. C. Zadoks. 1990. Risk analysis for biological control: A Dutch case study in biocontrol of *Prunus serotina* by the fungus *Chondrostereum purpureum*. *Plant Dis.* 74:189–194.

21. de Jong, M. D., E. Sela, S. F. Shamoun, and R. E. Wall. 1996. Natural occurrence of *Chondrostereum purpureum* in relation to its use as a biological control agent in Canadian forests. *Biol. Control* 6:347–352.

22. DeBach, P., and D. Rosen. (eds.). 1991. *Biological Control by Natural Enemies*, 2nd ed. New York: Cambridge University Press.

23. Delfosse, E. S., and J. M. Cullen. 1985. CSIRO Division of Entomology submission to the inquiries into biological control of *Echium plantaggineum* L., Paterson's curse/salvation Jane. *Plant Prot. Q.* 1:24–40.

24. Frick, K. E. 1964. *Leucoptera spartifoliella*, an introduced enemy of Scotch broom in the Western United States. *J. Econ. Entomol.* 57:589–591.

25. Gleason, H. A., and A. Cronquist. 1964. *The Natural Geography of Plants*. New York: Columbia University Press.

26. Goeden, R. D. 1978. Part II: Biological control of weeds. In *Introduced Parasites and Predators of Arthropod Pests and Weeds: A World Review*. C. P. Clausen, ed. U. S. Dept. Agric. Handbook No. 480, pp 357–414.

27. Goeden, R. D. 1984. Critique and revision of Harris' scoring system for selection of insect agents in biological control of weeds. *Prot. Ecol.* 5:287–301.

28. Goeden, R. D., and S. M. Louda. 1976. Biotic interference with insects imported for weed control. *Annu. Rev. Entomol.* 21:325–342.

29. Harley, K. L. S., and I. W. Forno (eds.). 1992. *Biological Control of Weeds: A Handbook for Practitioners and Students*. Melbourne and Sidney: Inkata Press.

30. Harris, P. 1973. The selection of effective agents for the biological control of weeds. *Can. Entomol.* 105:1495–1503.

31. Harris, P. 1991. Classical biocontrol of weeds: its definition, selection of effective agents, and administrative-political problems. *Can. Entomol.* 123:827–849.

32. Harris, P. & P. McEvoy. 1992. The predictability of insect host plant utilization from feeding tests and suggested improvements for screening weed biological control agents. *Proceedings of 8ᵗʰ International Symposium on Biological Control of Weeds*. Delfosse, E. S. & R. R. Scott, eds. February 2-7, Lincoln University, Canterbury, New Zealand, DSIR/CSIRO, Melbourne, Australia, pp 213-219.

33. Hasan, S. 1981. A new strain of the rust fungus *Puccinia chondrillina* for biological control of skeleton weed. *Ann. Appl. Biol.* 99:119-124.

34. Hasan, S. 1985. Search in Greece and Turkey for *Puccinia chondrillina* strains suitable to Australian forms of skeleton weed. *Proceedings of 6ᵗʰ International Symposium on Biological Control of Weeds*, E. S. Delfosse, ed. August 19-25, 1984, Vancouver, Canada, Agriculture Canada, Ottawa, pp 625-632.

35. Hasan, S., P. Chaboudez, and C. Espiau. 1992. Isozyme patterns and susceptibility of North American forms of *Chondrilla juncea* to European strains of the rust fungus *Puccunia chondrillina*. In *Proceedings of the 8th Symposium on Biological Control of Weeds*. E. S. Delfosse and R. R. Scott, eds. February 2-7, Lincoln University, Canterbury, New Zealand, DSIR/CSIR, Melbourne, Australia, pp. 367-373.

36. Houston, D. R., J. Parker, and P. M. Wargo. 1981. Effects of defoliation on trees and stands. In *The Gypsy Moth: Research Toward Integrated Pest Management*. C. C. Doane & M. L. McManus, eds. U.S. Dept. Agric. Tech. Bull. 1584, pp 217-297.

37. Howarth, F. G. 1991. Environmental impacts of classical biological control. *Annu. Rev. Entomol.* 36:485-509.

38. Huffaker, C. B. 1959. Biological control of weeds with insects. *Annu. Rev. Entomol.* 4:251-276.

39. Huffaker, C. B. 1962. Some concepts on the ecological basis of biological control of weeds. *Can. Entomol.* 94:507-514.

40. Huffaker, C. B., and C E. Kennett. 1959. A ten-year study of vegetational changes associated with biological control of Klamath weed. *J. Range Manag.* 12(2):69-82.

41. Imaizumi, S., T. Nishino, K. Miyabe, T. Fujimori, and M. Yamada. 1997. Biological control of annual bluegrass (*Poa annua* L.) with a Japanese isolate of *Xanthomonas campestris* pv. *poae* (JT-P482). *Biol. Control* 8:7-14.

42. Isaacson, D. L. 1973. A life table for the cinnabar moth in Oregon. *Entomophaga* 18:291-303.

43. Julien, M. H. (ed.). 1992. *Biological Control of Weeds: A World Catalogue of Agents and Their Target Weeds*. Wallingford, Oxon, U.K: CAB International.

44. Kenney, D. S. 1986. DeVine--the way it was developed--an industrialist's view. *Weed Sci.* 34 (suppl 1):15-16.

45. Klingman, D. L., and J. R. Coulson. 1982. Guidelines for introducing foreign organisms into the United States for biological control of weeds. *Weed Sci.* 30:661-667.

46. Lee, G. A. 1986. Progress on classical biological and integrated control of rush skeletonweed (*Chondrilla juncea*) in the western U.S. *Weed Sci.* 34 (suppl 1):2-6.

47. Leistritz, F. L., D. A. Bangsund, N. M. Wallace, and J. A. Leitch. 1992. *Eco-*

nomic Impact of Leafy Spurge on Grazingland and Wildland in North Dakota. Report AE92005 presented at the 12th Annual Meeting of International Association for Impact Assessment, August 19–22, Washington, D.C.

48. Louda, S. M., D. Kendall, J. Connor and D. Simberloff. 1997. Ecological effects of an insect introduced for the biological control of weeds. *Science* 277:1088–1090.

49. McClay, A. S. 1992. Beyond "before-and-after:" experimental design and evaluation in classical weed biological control. *Proceedings of 8ᵗʰ International Symposium on Biological Control of Weeds.* E. S. Delfosse & R. R. Scott, eds. February 2–7, Lincoln University, Canterbury, New Zealand. DSIR/CSIRO, Melbourne, Australia, pp 213–219.

50. Makowski, R. M. D. 1992. Regulating microbial pest control agents in Canada: the first mycoherbicide. *Proceedings of 8ᵗʰ International Symposium on Biological Control of Weeds.* E. S. Delfosse, & R. R. Scott, eds. February 2–7, Lincoln University, Canterbury, New Zealand, DSIR/CSIRO, Melbourne, Australia, pp 641–648.

51. Memmot, J., S. V. Fowler, H. M. Harmon & L. M. Hayes. 1996. How best to release a biological control agent. *Proceedings of 9ᵗʰ International Symposium on Biological Control of Weeds.* V. C. Moran & J. H. Hoffman, eds. January 19–26, 1996, Stellenbosch, South Africa, University of Capetown, pp 291–296.

52. Morris, M. J. 1996. The development of mycoherbicides for an invasive shrub, *Hakea sericea,* and a tree, *Acacia mearnsii,* in South Africa. *Proceedings of 9ᵗʰ International Symposium on Biological Control of Weeds.* V. C. Moran & J. H. Hoffman, eds. January 19–26, 1996, Stellenbosch, South Africa, Univerisity of Capetown, p 547.

53. Morris, M. J. 1997. Impact of the gall-forming rust fungus *Uromycladium tepperianum* on the invasive tree *Acacia saligna* in South Africa. *Biol. Control* 10:75–82.

54. Mortensen, K. 1988. The potential of an endemic fungus, *Colletotrichum gloeosporioides,* for biological control of round-leaved mallow (*Malva pusilla*) and velvetleaf (*Abutilon theophrasti*). *Weed Sci.* 36:473–478.

55. Mortensen, K. 1988. Biological control of weeds using microorganisms. In *Plant-Microbe Interactions and Biological Control.* G. J. Boland & L. D. Kuykendall, eds. New York: Marcel Dekker, pp 223–247.

56. Oehrens, E. B. 1977. Biological control of the blackberry through the introduction of rust, *Phragmidium violaceum,* in Chile. *FAO Plant Prot. Bull.* 25:26–28.

57. Pemberton, R. W. 1992. *Cactoblastis cactorum* in the United States of America: an immigrant biological control agent or an introduction of the nursery industry? *Am. Entomol.* 41:30–32.

58. Peschken, D. P., and A. S. McClay. 1992. Picking the target: a revision of McClay's scoring system to determine the suitability of a weed for classical biological control. *Proceedings of 8ᵗʰ International Symposium on Biological Control of Weeds.* E. S. Delfosse, & R. R. Scott, eds. February 2–7, Lincoln University, Canterbury, New Zealand. DSIR/CSIRO, Melbourne, Australia, pp 137–143.

59. Phatak, S. C., M. B. Callaway, and C. S. Vavrina. 1987. Biological control and its integration in weed management systems for purple and yellow nutsedge (*Cyperus rotundus* and *C. esculentus*). *Weed Technol.* 1:84–91.

60. Phatak, S. C., D. R. Sumner, H. D. Wells, D. K. Bell & N. C. Glaze. 1983. Biological control of yellow nutsedge with the indigenous rust fungus *Puccinia canaliculata*. *Science* 219:1446–1447.

61. Rees, N. E., P. C. Quimby, G. L. Piper, E. M. Coombs, C. E. Turner, N. E. Spencer, and L. V. Knutson (eds.). 1996. *Biological Control of Weeds in the West*. Bozeman, Montana: Western Society of Weed Science.

62. Ridings, W. H. 1996. Biological control of stranglervine in citrus—a researcher's view. *Weed Sci.* 34 (suppl 1):31–32.

63. Ridings, W. H., D. J. Mitchell, C. L. Schoulties, and N. E. El-Gholl. 1976. Biological control of milkweed vine in Florida citrus groves with a pathotype of *Phytophthora citrophthora*. *Proceedings of 4th International Symposium on Biological Control of Weeds*. E. Freeman, ed. August 30–September, 1976, Gainesville, Florida: Center for Environmental Programs, IFAS, University of Florida, Gainesville, pp 224–240.

64. Room, P. M. 1980. Biological control of weeds—modest investments can give large returns. In *Proceedings, Australian Agronomy Conference, Pathways to Productivity*. I. M. Wood, ed. Australian Society of Agronomy, Australia.

65. Simberloff, D. 1992. Conservation of pristine habitats and unintended effects of biological control. In *Selection Criterion and Ecological Consequences of Importing Natural Enemies*. W. C. Kauffman & J. R. Nechols, eds. Proceedings, Thomas Say Publications in Entomology, Entomological Society of America, Lanham, Maryland: pp 103–117.

66. Smith, R. J., Jr. 1992. Integration of biological control agents with chemical pesticides. In *Microbial Control of Weeds*. D. O. TeBeest, ed. New York: Chapman & Hall, New York, pp 189–208.

67. Story, J. 1985. First report of the dispersal into Montana of *Urophora quadrifasciata* (Diptera: Tephritidae), a fly released in Canada for biological control of spotted and diffuse knapweed. *Can. Entomol.* 117:1061–1062.

68. Supkoff, D. M., D. B. Joley, and J. J. Marois. 1988. Effect of introduced biological control organisms on the density of *Chondrilla juncea* in California. *J. Appl. Ecol.* 25:1089–1095.

69. Syrett, P., and D. J. O'Donnell. 1987. A seed-feeding weevil for biological control of broom. *Proceedings of 40th New Zealand Weed and Pest Control Conference*. A. J. Popay, ed. August 11–13, 1987, Nelson, New Zealand, The New Zealand Pest Control Society, Palmerston North, pp 19–22.

70. TeBeest, D. O. 1991. *Microbial Control of Weeds*. New York: Chapman and Hall.

71. TeBeest, D. O., and G. E. Templeton. 1985. Mycoherbicides: progress in the biological control of weeds. *Plant Dis.* 69:6–10.

72. Templeton, G. E. 1986. Mycoherbicide research at the University of Arkansas--Past, present, and future. *Weed Sci.* 34(suppl 1):35–37.

73. Templeton, G. E., D. O. TeBeest, and R. J. Smith, Jr. 1984. Biological weed control in rice with a strain of *Colletotrichum gloeosporioides* (Penz.) Sacc. used as a mycoherbicide. *Crop Prot.* 3:409–422.

74. Trujillo, E. E. 1995. Biological control of *Hamakua pamakani* with *Cercosporella* sp. in Hawaii. *Proceedings of 6th International Symposium on Biological Control*

   of Weeds. E. S. Delfosse, ed. August 19–25, 1984, Vancouver, Canada, Agriculture Canada, Ottawa, pp 66–671.
75. Trujillo, E. E., M. Aragaki, and R. A. Shoemaker. 1988. Infection, disease development, and axenic culture of *Entyloma compositarum*, the cause of *Hamakua pamakani* blight in Hawaii. *Plant Dis.* 72:355–357.
76. Turner, C. E., R. W. Pemberton, and S. S. Rosenthal. 1987. Host utilization of native *Cirsium* thistles (Asteraceae) by the introduced weevil *Rhinocyllus conicus* (Coleoptera: Curculionidae) in California. *Environ. Entomol.* 16:111–115.
77. U.S. Congress, Office of Technology Assessment. 1993. *Harmful Non-indigenous Species in the United States*. OTA-F-565. Washington, DC: U.S. Government Printing Office.
78. U.S. Congress, Office of Technology Assessment. 1995. *Biologically Based Technologies for Pest Control*, OTA-ENV-636. Washington, DC: U.S. Government Printing Office.
79. Wapshere, A. J. 1974. A strategy for evaluating the safety of organisms for biological weed control. *Ann. Appl. Biol.* 77:201–211.
80. Wapshere, A. J. 1974. Host specificity of phytophagous organisms and the evolutionary centres of plant genera or sub-genera. *Enotomophaga* 19:301–309.
81. Wapshere, A. J. 1975. A protocol for programmes for biological control of weeds. *PANS* 21:295–303.
82. White, G. G. 1981. Current status of prickly pear control by *Cactoblastis cactorum* in Queensland. *Proceedings of the 5th Symposium on Biological Control of Weeds*. E. S. Delfosse, ed. Brisbane Australia, CSIRO; Melbourne Australia, pp 609–616.
83. Zwölfer, H. 1965. Preliminary checklist of phytophagous insects attacking wild Cynareae (Compositae) species in Europe. *Commonwealth Institute of Biological Control Technical Bulletin* No. 6:81–154.
84. Zwölfer, H. 1973. Competitive co-existence of phytophagous insects in the flowerheads of *Carduus nutans*. *Proceedings of 2nd International Symposium on Biological Control of Weeds*. Paul H. Dunn, ed. October 4–7, 1971, Rome. Miscellaneous Publication No. 6, Commonwealth Institute of Biological Control, Commonwealth Agricultural Bureaux, Farnham Royal, Slough, U.K., pp 74–77.
85. Zwölfer, H., and P. Harris. 1984. Biology and host specificity of *Rhinocyllus conicus* (Froel.) (Col., Curculionidae), a successful agent for biological control of the thistle, *Carduus nutans*. L. *Z. Angew. Entomol.* 97:36–62.

23

Chemical Approaches to Weed Management

John W. Wilcut and Shawn D. Askew
North Carolina State University, Raleigh, North Carolina

I. INTRODUCTION

A. History of Chemical Weed Control

Since at least the mid 1800s, humans have used a variety of chemicals, including copper salts, iron sulfate, and sulfuric acid, as herbicides to control unwanted vegetation. However, modern chemical weed management began with the discovery of the phenoxy herbicides. This discovery ushered in the modern era of chemical weed management [47,72]. Modern selective chemical management of weeds increased rapidly during the 1950s and 1960s, because it was extremely successful and cost effective. This success and cost effectiveness is best illustrated by the fact that selective chemical management of unwanted vegetation is now used on over 95% of the agronomic hectarage in the United States.

The term *selective* is used to describe a herbicide that is toxic to some plants (e.g., example, broadleaf weeds) and much less toxic to others (e.g., monocot crop plants) at a given dosage. Some plants may be more tolerant to a herbicide than others, because less herbicide enters these plants or mobility of the herbicide is reduced within these plants [11]. Other plants may be more tolerant, because they possess the ability to metabolize the herbicide rapidly to an inactive form after its absorption. Yet other plants may withstand exposure to some herbicides by having biochemical differences or alterations in the site normally attacked by the herbicide in susceptible plants [11].

Since the initial discovery of the phenoxy herbicides several hundred herbicidal products have been commercialized successfully in the United States by foreign and domestic agrochemical companies. The herbicide industry must synthesize and screen thousands of compounds in order to find a commercially viable new herbicide [48]. It takes this many experimental compounds to develop one commercial product for several reasons. First, current commercially available herbicides provide very good weed control and will only be displaced from the market by an exceptional product. Second, the cost of registration has become quite substantial owing to the number of studies that must be conducted to ensure consumer, food, environmental, and applicator safety according to guidelines established by the U.S. Environmental Protection Agency and other regulatory authorities. Third, these increased safety requirements make it even more difficult to find a new herbicide that is simultaneously more effective for weed management and also safer environmentally and to the applicator.

It has been estimated that in the near future 70,000–80,000 compounds must be screened to find one successful herbicide [5]. Between 1984 and 1988, 170 novel compounds were synthesized and evaluated for phytotoxic properties. Only 21% eventually reached the market as commercial products [28]. Once a compound has been discovered to have herbicidal or pesticidal properties, a search is initiated to find other active compounds that are similar in structure. Thus, several herbicides often are developed within a chemical family that are structurally related and have essentially the same mode and/or site of action in plants [2,72]. However, it cannot be stated that all members of a chemical family have the same site and mode of action, translocation patterns, and crop tolerances. Also, herbicides within a chemical family often vary in selectivity as a result of physicochemical differences that cause them to behave differently in the soil and plant system.

B. Purpose of the Chapter

The purpose of this chapter is to discuss the thought processes necessary systematically to develop an integrated and comprehensive weed management program and to present a systematic overview of the herbicides currently available. The overview of herbicides is organized according to herbicide modes and sites of action. Several comprehensive books and numerous scientific articles are available on herbicide modes and sites of action and weed management in various crops (see, e.g., Refs. 2, 5, 11, 16, 24, 42, 47, 48, 62, 63, 66, 72, 73, and 74).

II. DEVELOPING A CHEMICAL WEED MANAGEMENT SYSTEM

Today there are over 130 compounds worldwide that are used as selective herbicides, and there are approximately 30 others that are used as nonselective herbicides [72]. In contrast, there are several hundred trademarked herbicide products commercially available, since there may be more than one manufacturer of a particular herbicide and/or multiple formulations that all contain the same active ingredient. When one considers the fact that most herbicides are applied in combinations or "tank mixtures" containing two or more herbicides, the variety of application methods available, and the fact that generally one application method will not provide adequate full-season control, the development of a weed management system that relies on herbicides becomes almost endless (considering, e.g., the list herbicide options) and frequently is considered a daunting task by many producers and agribusiness professionals [66]. In determining the most appropriate and cost-effective herbicide program to use, one must consider the: spectrum of weeds present, tillage system, method of application, crop tolerance relative to edaphic conditions, crop varieties, possible rotation restrictions, potential for development of herbicide-resistant weeds, and cost of the herbicides. Selecting a herbicide program is not a simple matter. However, following logical steps may eliminate confusion in this process.

Step 1

To plan a weed management program wisely, one must accurately identify and/ or predict what weeds will be present in a given field. Herein lies the value of scouting during the production season, as well as between seasons, and weed mapping. Weeds present in the fall (or at the end of the growing season) have set seed and most likely will be the predominant problems for at least the following season. Knowledge of the weeds present allows producers and consultants to choose from preplant-incorporated (PPI), preemergence (PRE), postemergence (POST), and POST-directed herbicide options. *Preplant incorporated* is defined as when herbicide application is made prior to planting and the herbicide is physically mixed in the soil (incorporated) with some type of tillage implement. *Preemergence* is defined as when the herbicide is applied after planting but prior to crop and, generally, weed emergence. The term *postemergence* (POST) usually implies herbicide application after crop emergence, but it is just as appropriate to use the term to denote after weed emergence. *POST-directed* application is defined as when the herbicide is sprayed postemergence but the herbicide spray is directed toward the base of the crop plant and does not come into contact with the foliage of the crop plant, especially the apical meristematic region. Depending on the weed complex and crop selectivity, herbicide options may be limited to one method of application. Timeliness of application and management practices are extremely important for postemergence management.

Most fields, regardless of crop, contain a complex of annual broadleaf and grass weeds. For example, in corn (*Zea mays*), peanuts (*Arachis hypogaea*), and soybeans (*Glycine max*), numerous herbicides are available for POST control of most species including annual broadleaf weeds. In cotton (*Gossypium hirsutum*), however, POST herbicide options for annual broadleaf weed control are limited to pyrithiobac for nontransgenic and transgenic cotton cultivars, whereas glyphosate and bromoxynil can only be used on transgenic glyphosate-tolerant cotton and bromoxynil-resistant cotton varieties, respectively [66,67]. These herbicides have only been registered in the last few years, and before these registrations, cotton producers had to rely on soil-applied PPI and PRE herbicides for early-season weed control [66].

Although numerous POST herbicides are available for use in many agronomic crops, certain weeds are controlled better with soil-applied herbicides or a combination of soil and POST applications. For example, with few exceptions, POST herbicides are generally inadequate for sicklepod (*Senna obtusifolia*) control in soybeans. A program utilizing soil-applied PPI and/or PRE herbicides plus a POST herbicide is more effective. Without knowledge of sicklepod as a problem, producers may not use soil-applied herbicides; consequently adequate weed control may be more difficult to obtain. Similarly, Florida beggarweed (*Desmodium tortuosum*) is difficult to control in peanuts, and a complete herbicide program employing PPI, PRE, and POST herbicides often is necessary for season-long control [65]. However, Florida beggarweed can be easily controlled in corn, cotton, and soybeans with soil-applied or postemergence-applied herbicides [66].

Some weeds can be effectively controlled with POST herbicides, and if a given field contains patches of weeds [68,69], producers may forego soil application of herbicide to the complete field in favor of treating only infested areas. Control of annual grasses POST is an alternative to soil-applied PPI and/ or PRE herbicide programs in corn, cotton, peanuts, and soybeans [8,21,33,36,64]. Total POST programs also are preferred in high organic soils where soil-applied herbicides are either not effective or are cost prohibitive because of the rate required.

After determining which weeds will most likely cause problems the following season, herbicide options can be narrowed significantly. Consultants, extension personnel, and dealers usually have weed-response charts available with recommendations of herbicides that are most effective for a specific weed problem. It is valuable to compile a list of herbicides and tank mixtures that will control the various weeds anticipated in a given field. Ideally, this compilation should be developed on a field-by-field basis. The ultimate goal is to determine the best herbicide program for each individual field. If the use of specific herbicide programs for individual fields is not practical, fields with similar weed problems should be grouped. Herbicides are too expensive to try and develop a single blanket treatment to handle all the different weeds likely to be found in different fields on large farms. Additionally, no one management system will control all weed species.

In developing a list of possible herbicide options, one should initially include all registered products and tank mixtures for PPI, PRE, POST, and POST-directed application regardless of cost, carryover potential, or soil-type limitations. The presence of certain weeds may limit herbicide options to only one application method or only one or two products or tank mixtures. For most situations, however, several herbicide options are available.

Step 2

Modify the list of herbicide options depending on whether the planted varieties have enhanced herbicide tolerance or are transgenic varieties that are resistant to previously nonselective herbicides.

Step 3.

Reduce the list of herbicide options by considering the rotational restrictions on certain products. Eliminate those herbicides that have the potential to carry over and damage the crop to be planted the following season. This information can be found on the herbicide labels and/or in extension and educational publications.

Step 4

Consider your soil texture and organic matter content. Use of certain soil-applied herbicides can be risky on coarse-textured soils or low–organic matter soils or fields with variable soils (see Chap. 26). For example, linuron and metribuzin can cause severe injury on coarse-textured soils. In contrast, trifluralin and vernolate are not cost effective on soils with high organic matter (> 5%). Eliminate those choices that do not fit your soil type. On high–organic matter soils, graminicides (POST herbicides that control annual and perennial grasses), POST herbicides for broadleaf weeds such as acifluorfen, bentazon, chlorimuron, imazethapyr, or glyphosphate for grass and broadleaf weed control may be more effective and economical. This information can be obtained from herbicide labels or from extension publications.

Step 5

If you are using insecticides or other pesticides in your production system, you will need to make sure that no adverse interactions between the pesticides will occur. For example, the use of disulfoton as a soil-applied insecticide will increase the likelihood of injury from the use of bentazon in peanuts. Additionally, in a few isolated instances, soil-applied insecticides may safen a crop (improve crop tolerance) to a previously nonselective herbicide (e.g., use of the insecticide disulfoton safens cotton to clomazone applied PPI or PRE).

Step 6.

The method of herbicide application is important, as herbicides can be applied PPI, PRE, POST, or POST-directed. Some herbicides have more than one application method and the weeds may be susceptible to more than one method of application. If the list of herbicide options includes more than one application method, decide which method is best for a particular situation. The method of application choice will vary with the producer depending on equipment and management input limitations. With numerous broad-spectrum soil-applied broadleaf herbicides and prepackaged mixtures now available for use in many agronomic crops, application of herbicides for grass and broadleaf weed control at planting may be most appropriate. By correctly matching herbicides to specific weed complexes, the need for POST or POST-directed herbicides may be eliminated in some cases. Depending upon the weed spectrum, herbicides applied to the soil at planting as well as POST or POST-directed herbicides may be necessary. Additionally, environmental conditions affecting herbicide activation may limit the effectiveness of PRE soil-applied herbicides. For example, in some cases where soil-applied herbicides are not activated by rainfall or irrigation, POST herbicides may be necessary. Activation of PRE soil-applied herbicides is necessary for good weed control. Precipitation, whether from irrigation or rainfall, places the herbicide into the soil solution so that it is available for plant uptake. If the herbicide is not in soil solution, it is not available for plant uptake, and consequently weed control will be reduced.

Consider the benefits and limitations of PPI versus PRE application. PRE herbicides require rainfall or irrigation for activation, whereas PPI herbicides are frequently activated by the physical mixing of the herbicide into the soil (the incorporation method activates the herbicide unless the soil is extremely dry). PPI application provides insurance against a performance failure due to lack of rainfall or irrigation. On the other hand, incorporating herbicides requires additional equipment and more time [39]. Also, proper incorporation is critical. This would include incorporating the herbicide to the correct depth (refer to each respective herbicide label), proper soil conditions for incorporation, specialized equipment operated at the current speed, uniform distribution of the herbicide throughout the soil profile, and so forth. Poor incorporation (such as improper depth of incorporation, incorporating on wet soil, incorrect speed for tillage equipment to incorporate the herbicide uniformly in the soil profile, making only one pass versus needing to make two passes through the field to distribute the herbicide, and so forth may result in poor weed control and/or crop injury. Some PPI herbicides cost less than PRE herbicides; however, the extra expenses of incorporation should be considered. Also, consider the savings possible with banded PRE herbicides (spraying the herbicide only on the drilled seed row and not on the row middles) used in conjunction with cultivation to control emerged weeds in the row middles [41].

For some soils and some weeds, POST herbicides may be more economical than soil-applied herbicides. This is especially true for the good manager who can get acceptable control with reduced application rates [10,64]. However, unless weeds are small, weather conditions are ideal, and application variables such as pressure, volume, nozzle selection, and adjuvant usage are optimal, reduced rates may lead to disappointing results.

When applying POST herbicides, one should consult the label for adjuvant selection. Most POST and POST-directed herbicide labels recommend the use of a spray adjuvant. Spray adjuvants are nonpesticidal materials added to an herbicide to improve its effectiveness. This improvement is the result of one or more of the following: increased absorption and penetration into the plant, better and more uniform coverage, prevention of photodegradation, or other mechanisms [40]. The most commonly used adjuvants are nonionic surfactants and crop oil concentrates.

The major advantage of using POST herbicides is the ability to wait and see what weed problems are present before determining whether the herbicide treatment is justified. Proper timing of application is one of the keys to a successful POST herbicide program. About 5–7 days after planting, start checking the fields and continue on a 3- to 5-day interval. If ensuring timely scouting and application will be difficult, inclusion of a soil-applied herbicide program at planting may be the best option.

Annual weeds can be controlled with a total POST program. If the herbicides are properly applied (at the correct rates and times), POST management can be successful. The major disadvantage of a total POST program is the need for a higher level of management ability. Identification of seedling weeds so that the proper herbicide or tank mixture can be selected is critical. Timing of application is also critical, and adverse weather may delay application beyond the proper time.

In many cases, the optimal times to apply both the graminicide and broadleaf herbicides will coincide. When this occurs, a tank mixture application can save a trip across the field. With most tank mixture applications of graminicide and broadleaf herbicides, reduced grass control (antagonism) is possible [26,30,61]. To overcome this problem, the herbicide labels for some POST graminicides suggest the use of increased rates, which can reduce antagonism. The increased rate may not be necessary under ideal conditions (small grass that is actively growing with good soil moisture) but is suggested for consistent control. Sequential applications can eliminate or reduce antagonism. If sequential applications are made, the graminicide rate generally need not be increased.

If sequential applications are made, the recommended waiting interval between applications of the graminicide and broadleaf herbicides varies depending upon which herbicides are used and which herbicide is applied first (see herbicide label for details). Always refer to the respective herbicide labels for specific guidelines.

Occasionally, the use of spray additives can reduce antagonism. For example, addition of ammonium sulfate and/or replacement of crop oil concentrate with BCH 815085 (DASH) reduced or eliminated the antagonism observed between sethoxydim and bentazon [30]. However, there are few other herbicide combinations in which adjuvants overcome antagonism.

Step 7

An additional process that is now prudent is carefully to assess the potential for herbicide resistant weeds to occur or, if they are present, how to control them. Awareness of herbicide-resistant weeds prompted university and agrochemical company personnel to develop guidelines to educate growers on minimizing the risks of selecting for herbicide-resistant weeds [45]. These guidelines include similar recommendations: (a) Scout fields prior to application of any herbicide to determine the species and whether economical population levels are sufficient to justify an herbicide application. (b) Use alternative weed management practices, such as mechanical cultivation, delayed planting, and weed-free crop seeds [62,63]. (c) Rotate crops with an accompanying rotation of herbicides varying in site of action to avoid using herbicides with the same site of action on the same field. (d) Limit the number of applications of a single herbicide or herbicides with the same site of action in a single growing season. (e) Use mixtures or sequential treatments of herbicides that each control the weeds in question, but have a different site of action. (f) Scout fields after application to detect weed escapes or shifts. If a potentially resistant weed or weed population has been detected, use available control methods to minimize seed deposition in the field. (g) Clean equipment before leaving fields infested with or suspected to have resistant weeds. An additional suggestion in our viewpoint is to (h) restrict the number of herbicide treatments with the same site of action in a cropping system. One needs carefully to analyze and keep a written record of herbicide use patterns in a cropping system over several years in addition to monitoring use patterns in a single growing season.

Step 8

The final consideration is the price of a particular herbicide or mixture of herbicides that effectively controls specific weed complexes. Choosing the most cost-effective option will potentially provide the greatest profit. Additionally, when considering the price of the herbicide treatment, one needs accurately to assess the value of the crop and the benefit to cost ratio associated with the herbicide treatment or lack of treatment [3].

III. HERBICIDE CLASSIFICATION

Herbicides are often grouped or classified according to their basic chemical structure and biochemical mode of action in plants. The term *mode of action* may be defined as the complete sequence of events leading to plant injury, and therefore includes all areas of interaction between a herbicide and a crop or weed species [72]. The imidazolinones, diphenyl ethers, dinitroanilines, and chloroacetamides are examples of chemical families that each contain several commercial herbicides having the same mode and/or site of action in susceptible plants. The term *site of action* is the location of the molecular interaction which triggers the events that hopefully lead to the death of the unwanted weeds. Until recently knowing the mode and/or site of action was important only from an academic viewpoint to university research and extension personnel and agrochemical research company personnel. But the recent increase in the occurrence of herbicide-resistant weeds has meant that producers and pest management personnel also must be familiar with the different types of herbicidal modes and sites of action to prevent or preclude the development and/or establishment of herbicide-resistant weed populations [27,42,45].

There are numerous ways to classify herbicides, but classifying based on site of action is a critical component for successful management of herbicide-resistant plant populations. The Weed Science Society of America in conjunction with the Herbicide Resistance Action Committee has recently developed a listing of herbicides based on the site of action [45]. A number of the herbicides that are available for commercial use in the United States, grouped by site of action, are presented in Table 1 in order of likelihood of resistance development.

The seven major modes of action categories along with the sites of action are shown in Table 1 and are described in the following sections. A description of each general mode of action category is followed by a brief discussion of the site of action, if known, for each chemical family.

A. Amino Acid Synthesis Inhibitors

Since the early 1970s, a new generation of herbicides of major agronomic importance has been developed that inhibits the biosynthesis of amino acids. Proteins are made up of long chains of interconnected amino acids [44]. Therefore, amino acids are the building blocks of proteins, and the failure to produce certain amino acids not only has an effect on the biosynthesis of enzymes but also leads to the failure of plant metabolism in general. There are 20 amino acids

Table 1 Risk of Resistance Evolution and Sites of Action for Commonly Used Herbicides in Agronomic Crops

Risk category	Herbicide family	Site of action	WSSA[a] group number	Common names	Trade names
High	Imidazolinone	Inhibitor of acetolactate synthase (ALS) and also called acetohydroxyacid synthase (AHAS)	2	Imazamox	Raptor
			Imazapic	Cadre	
			Imazapyr[b]	Prepackage only	
			Imazaquin[c]	Scepter	
				Imazethapyr[c]	Pursuit
	Sulfonylurea	See Imidazolinone	2	Chlorimuron[c]	Classic, Skirmish
				Halosulfuron	Permit
				Nicosulfuron[c]	Accent
				Primisulfuron[c]	Beacon
				Prosulfuron[c]	Peak
				Rimsulfuron[b]	Prepackage only
				Thifensulfuron[c]	Pinnacle
				Tribenuron[c]	Express
	Triazolopyrimidine	See Imidazolinone	2	Cloransulam	FirstRate
				Flumetsulam[3]	Python
High	Pyrimidinylthiobenzoate	See Imidazolinone	2	Pyrithiobac	Staple
Moderate to high	Aryloxyphenoxy propionate	Inhibitor of acetyl-CoA carboxylase (ACCase)	1	Diclofop	Holeon
				Fenoxaprop[b]	Prepackage only
				Fluazifop[c]	Fusilade DX
				Quizalofop	Assure II

Persistence	Chemical family	Mechanism of action	No.	Common name	Trade name
	Cyclohexanedione	see Aryloxyphenoxy propionate	1	Clethodim	Select
				Sethoxydim	Poast
Moderate	Dinitroaniline	Microtubule assembly inhibitor	3	Ethalfluralin	Sonalan
				Pendimethalin[c]	Prowl
				Trifluralin[c]	Treflan, Trilin[d]
	Organoarsenical	Unknown	17	DSMA	Many brands
				MSMA	Many brands
	Triazine	Inhibitor of photosynthesis at photosystem II	5	Ametryn	Evik[d]
				Atrazine[c]	AAtrex, Atrazine[d]
				Cyanazine[c]	Bladex, Cy-pro
Moderate	Triazine	Inhibitor of photosynthesis at photosystem II	5	Prometryn	Caparol[d]
				Simazine	Princep[d]
Low	Triazinone	See Triazine	5	Metribuzin[c]	Lexone, Sencor
	Acetamide	Unknown	15	Napropamide	Devrinol
	Amino acid derivative	EPSP synthase inhibition	9	Glyphosate[c]	Roundup Ultra[d]
				Sulfosate	Touchdown
		Glutamine synthase (GS) inhibition	10	Glufosinate	Liberty
	Benzothiadiazole	(see Triazine, differs in binding behavior)	6	Bentazon[c]	Basagran, Pledge
	Benzoic acid	Synthetic auxin	22	Dicamba[c]	Banvel, Clarity
	Bipyridilium	Electron diversion in photosystem I		Paraquat	Gramoxone Extra, Starfire
	Chloroacetamide	Unknown	15	Acetachlor[c]	Harness, Surpass
				Alachlor[c]	Lasso, Micro-Tech

(*continued*)

Table 1 Continued

Risk category	Herbicide family	Site of action	WSSA[a] group number	Common names	Trade names
Low	chloroacetamide	Unknown	15	Dimethenamid[c]	Frontier
				Metolachlor[c]	Dual, Dual II
	Diphenyl ether	Inhibitor of protoporphyrinogen oxidase (PPO)	14	Acifluorfen[c]	Blazer, Status
				Fomesafen[c]	Reflex, Flexstar
				Lactofen	Cobra
				Oxyfluorfen	Goal
	Isoxazolidinone	Inhibits all diterpenes	13	Clomazone	Command
	N-phenylphthalimide	See Diphenylether	14	Flumiclorac[c]	Resource
	Nitrile	See Triazine, differs in binding behavior)	6	Bromoxynil	Buctril
	Phenoxy	See Benzoic acid	4	2,4-D[c]	Many brands
				2,4-DB	Many brands

Low	Phenylpyridazine	6	Pyridate	Tough	See Triazine, differs in binding behavior)
	Pyridazinone	12	Norflurazon	Zorial	Inhibitor of carotenoid biosynthesis at the phytoene desaturase step (PDS)
	Carboxylic acid	4	Clopyralid[c]	Stinger	See Benzoic acid
	Thiocarbamate	8	Butylate EPTC Pebulate Vernolate	Sutan +[d] Eradicane[d] Tillam Vernam	Inhibitor of lipid synthesis—not ACCase inhibition
	Triazolinone	14	Sulfentrazone[c]	Spartan	See Diphenylether
	Urea	7	Diuron Fluometuron Linuron	Karmex, Direx[d] Cotoran, Meturon[d] Lorox, Linex[d]	See Triazine, differs in binding behavior

[a] Herbicide classification system developed by the Weed Science Society of America.
[b] A formulation containing only this active ingredient is not registered for use on agronomic crops. However, this active ingredient is present in prepackaged herbicides mixtures for agronomic crops.
[c] This active ingredient is also present in prepackaged herbicide mixtures.
[d] Other brands available.

necessary for normal plant growth, and plants must synthesize all of their amino acids. Consequently, the risks of mammalian toxicity from these herbicides are slight, because animals cannot synthesize histidine, isoleucine, leucine, methionine, threonine, tryptophan, phenylalanine, and valine, which are termed essential amino acids, and which have considerable dietary significance [44]. Visible injury symptoms on susceptible plants are rather slow to develop for these compounds in comparison with the action of some other classes of herbicides and usually do not appear for a period of several days, although growth may be stopped almost immediately. There are two reviews available on herbicides that inhibit amino acid synthesis [32,52]. Only three molecular sites of action in amino acid synthesis have been shown to be the primary sites of action. The herbicides with this mode of action are extensively used on virtually every crop in the world and are also used extensively in a number of noncropland, turf, and other specialty areas. The array of target weeds is quite extensive and crop selectivity and weeds controlled vary greatly even within the same herbicide family.

1. Glyphosate and Glyphosate Analogues

Herbicides in this class are nonselective and are extremely effective in controlling a broad spectrum of annual and perennial grass, broadleaf, and sedge weeds [18,31,51]. Glyphosate and its analogues (Fig. 1) are rapidly inactivated in soil by biological and chemical reactions and thus are effective only when applied to plant foliage. Glyphosate, the primary member of this class of herbicides, specifically inhibits the enzyme 3-phospho-5-enolpyruvate synthase (EPSP synthase), which is involved in the synthesis of the three essential aromatic amino acids (phenylalanine, tryptophan, and tyrosine) [13]. These three amino acids are products of the shikimate pathway where many aromatic secondary plant products (e.g., lignins, alkaloids, flavonoids, benzoic acids) that are important in plant growth and development are produced. No other significant primary sites of action of glyphosate other than those in the shikimate pathway have been

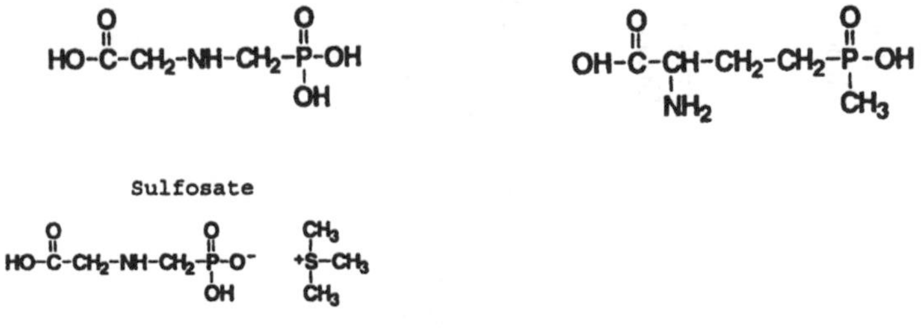

Glyphosate   Glufosinate

$$HO-\overset{\overset{O}{\|}}{C}-CH_2-NH-CH_2-\overset{\overset{O}{\|}}{\underset{\underset{OH}{|}}{P}}-OH \qquad OH-\overset{\overset{O}{\|}}{C}-\underset{\underset{NH_2}{|}}{CH}-CH_2-CH_2-\overset{\overset{O}{\|}}{\underset{\underset{CH_3}{|}}{P}}-OH$$

Sulfosate

$$HO-\overset{\overset{O}{\|}}{C}-CH_2-NH-CH_2-\overset{\overset{O}{\|}}{\underset{\underset{OH}{|}}{P}}-O^- \qquad \overset{\overset{CH_3}{|}}{\underset{\underset{CH_3}{|}}{+S}}-CH_3$$

Figure 1 Glyphosate and glyphosate analogues.

identified. Some weeds or weed biotypes are more tolerant of glyphosate than others. Differences in spray retention [19], absorption [37], and vegetative reproductive potential [35] have been reported as tolerance mechanisms.

Glyphosate and related compounds are readily translocated in plants through the symplast and apoplast to actively growing shoot tissue, as well as to actively growing underground vegetative propagules of perennial species. As a result, these compounds are effective in preventing vegetative reproduction and overwintering of perennial weeds if applied at the correct stage of plant growth.

Toxicity symptoms of treated plants normally begin to appear 3–10 days after application and include stunting and chlorosis of newly developing leaf tissue in the shoot meristematic areas. Purple or reddish discoloration of older shoot tissue also may appear on perennial grasses in the early stages. Within 10–14 days treated plants turn totally necrotic and vegetative propagules of perennial species shrivel, turn brown, and become desiccated [24].

The recent advances in biotechnological research have allowed for the development of glyphosate-tolerant crops, including cotton and soybeans, and current development efforts also include corn and other grassy crops [18,51,67]. The registration of glyphosate-tolerant cotton and soybeans have radically changed herbicide use patterns for these crops in the United States and worldwide. It is estimated that glyphosate-tolerant soybeans will soon be planted on 60% or greater of the U.S. hectarage. To illustrate how quickly this technology is being adopted by U.S. soybean producers, it is estimated that 80% of the U.S. soybean hectarage was treated with an imidazolinone and/or sulfonylurea herbicide in the mid 1990s [71]. As seed availability increases, it would not be surprising to have glyphosate-tolerant soybean varieties being planted on 80% or more of the U.S. soybean hectarage by the year 2000. If this management practice comes even close to 80% of the hectarage, it will represent the most massive change in herbicide use ever in U.S. agriculture and in a shorter period of time. A similar trend toward planting more glyphosate-tolerant cotton varieties will continue as seed becomes more widely available. For example, within two years of registration, glyphosate-tolerant cotton was planted on more than 50% of the North Carolina hectarage in 1998 and is expected to increase further.

2. Glufosinate

Glufosinate (phosphinothriein [PPT]) is nonselective on most plants (crops and weeds) and only used postemergence [55] (see Fig. 1). Glufosinate does not readily translocate in the symplast or apoplast; thus thorough spray coverage is required for good weed control. Injury symptoms usually develop in 2–3 days. Injury is somewhat similar to that caused by membrane-disrupting herbicides (diphenyl ethers, bipyridiliums; see below). However, the speed of membrane disruption (necrosis) is slower. Phytotoxicity is rapid; leaf chlorosis, desiccation, and necrosis may be observed within 2 days after treatment, and plant death results 3 days later. Glufosinate disrupts important nitrogen metabolism in plants and indirectly inhibits electron flow in photosynthesis through effects on amino donors caused by the enzyme glutamine synthetase [34]. Glutamine synthetase is the initial enzyme in the pathway that assimilates inorganic nitrogen into organic compounds. It is a pivotal enzyme in nitrogen metabolism in that, in addition to assimilating ammonia produced by nitrile reductase, it recycles ammonia produced by other plant processes, including photorespiration and deamination reactions [34]. Although in the long run this would lead to a general failure to produce the amino acid glutamate, death is apparently more rapid owing to the accumulation of the toxic ammonia. In the presence of light, inhibition of electron flow in photosynthesis causes induction of lipid peroxidation (membrane damage). As with glyphosate, some crop plants, including corn, cotton, rice, and soybeans, have been developed with resistance to glufosinate by classic plant breeding techniques or with genetic engineering [1,50].

3. Imidazolinones, Sulfonylureas, Triazolopyrimidines, and Pyrimidinylthiobenzoate

During the 1980s, three new herbicide families (Figs. 2–4) were developed that were potent, selective, broad-spectrum inhibitors of plant growth at field rates measured in grams rather than kilograms per hectare. These herbicides are some of the most selective herbicides available and provide broad-spectrum control of annual and perennial grasses, broadleaf plants, and sedges [29,33,38,46,60,65–67]. Although these herbicide families are chemically different from one another, they inhibit the same enzyme and generally produce similar phytotoxic symptoms in susceptible plants. The effectiveness of these herbicides resulted in large-scale usages which rapidly challenged and replaced older herbicides in most commodities and markets.

All of these herbicides have activity as foliar- and/or soil-applied treatments. They are translocated in the apoplast and symplast to all actively growing plant parts, especially the meristematic areas, and are most effective on small rapidly growing annual weeds [56,57]. These herbicides are particularly noteworthy in that they often are used at extremely low rates, sometimes at rates below 5 g/ha. Because of their high degree of specificity for certain weeds and crops, these herbicides must be matched carefully with the crop and its target weed spectrum. Additionally, a number of these herbicides have significant rotational restrictions, and rotational requirements for the producers' cropping systems need to be carefully assessed if one uses these chemicals.

These herbicides inhibit acetohydroxyacid synthase (AHAS or ALS) [57], which is necessary for the production of the three branched-chain amino acids:

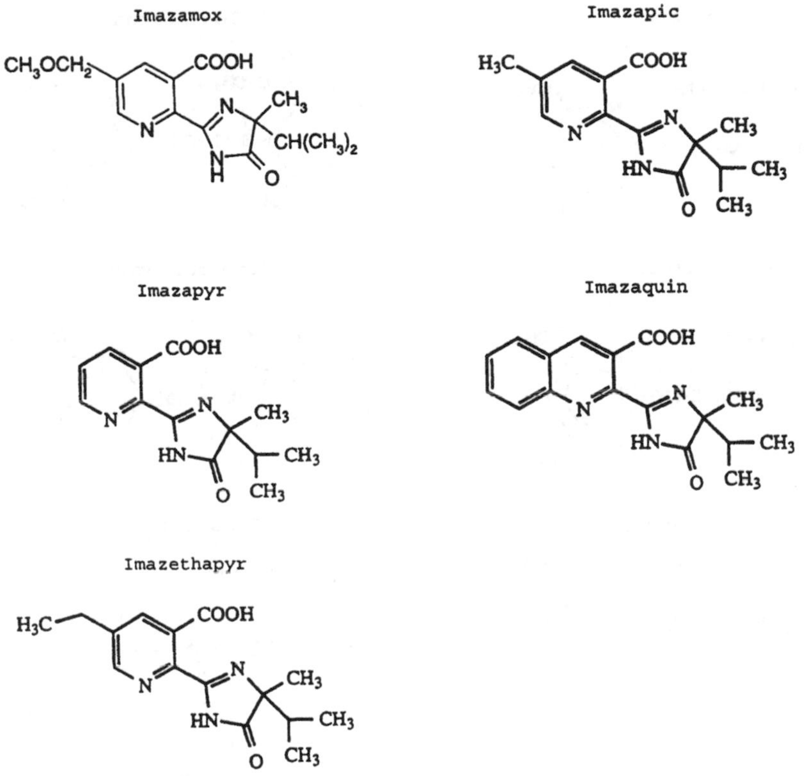

Figure 2 Imidazolinone herbicides.

isoleucine, leucine, and valine. The three classes of ALS inhibitors are structurally diverse and each herbicide class has a unique set of interactions with the enzyme [56]. Susceptible plants die very slowly from these herbicides with meristematic tissue showing the first symptoms. All of these herbicides can be considered growth inhibitors and cause plant death within a period ranging from several days to more than a week.

Susceptible grass plants are stunted and develop intervenous chlorosis and/or purpling of the foliar tissue within 7–10 days after herbicide application [24]. Newly emerging grass leaves may appear to be bleached and malformed with additional time. Lateral roots of susceptible plants may be pruned or poorly developed if herbicide is present in the soil, thus creating the so-called "bottle-brushed" appearance of primary roots. Shoot apical meristems turn necrotic and die back before older plant parts completely die [24].

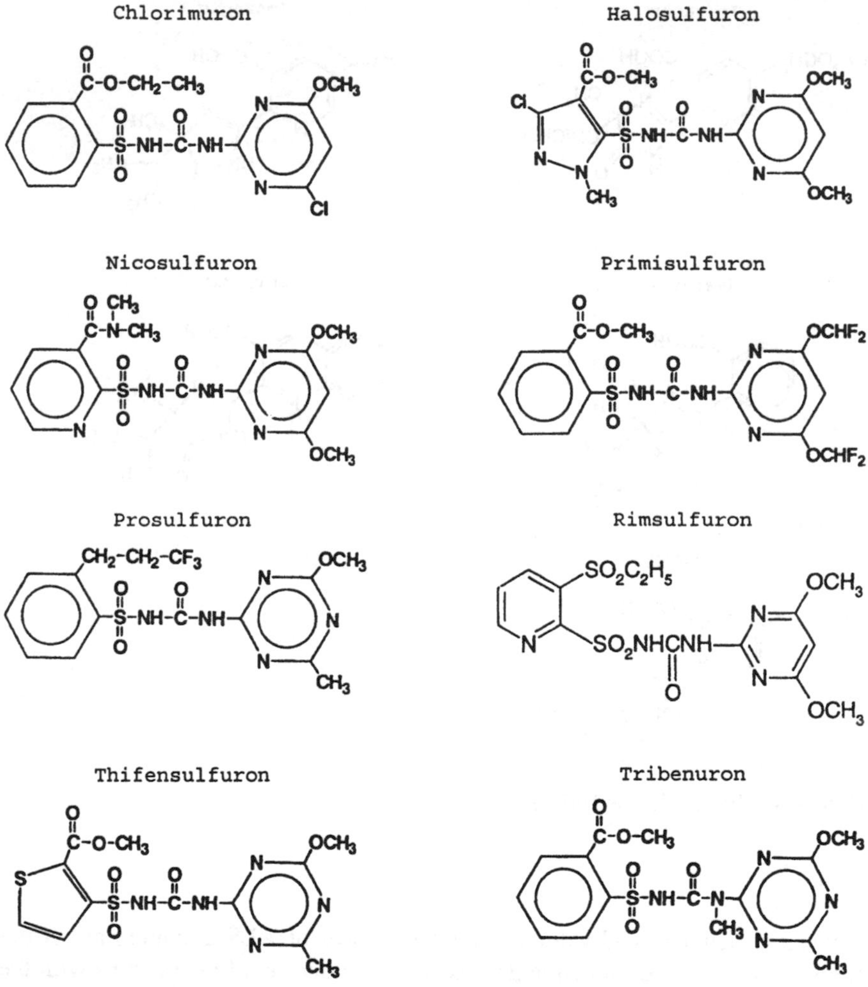

Figure 3 Sulfonylurea herbicides.

B. Contact Disrupters of Membranes

Contact herbicides have only limited translocation, exhibit little or no herbicidal activity in soil, and are generally applied postemergence. After foliar tissue of susceptible plants absorbs the herbicides, secondary reactions generate toxic free radicals that disrupt cell membranes [14]. Membrane disruption causes intracellular contents (water and cytoplasm) to leak out into the intercellular spaces and surrounding tissue. This phenomenon is initially manifested at the whole plant level by a "water-soaked" appearance and wilting of the foliage. As wa-

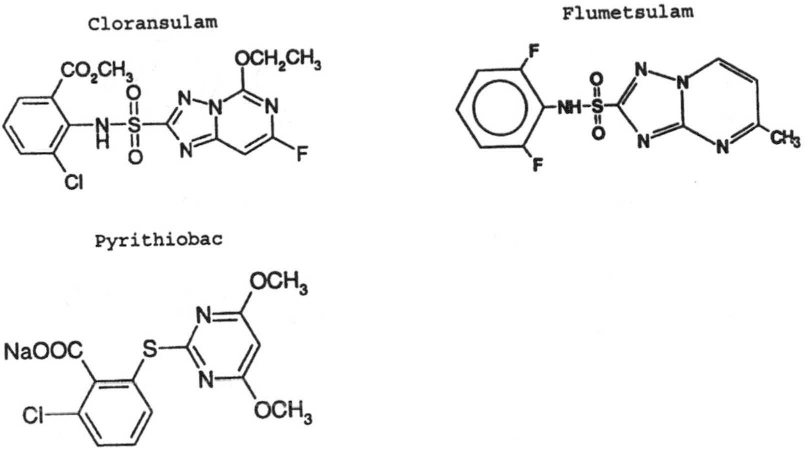

Figure 4 Triazolopyrimidine and pyrimidinylthiobenzoate herbicides.

ter from the spilled cellular contents evaporates, the foliage desiccates and turns necrotic, leading to an appearance similar to that observed several days after a killing frost. Initial symptoms of contact herbicides develop rapidly, usually within hours after application. The rapid action and lack of soil activity make some contact herbicides ideal for "burndown" of vegetation prior to no-till crop planting [24]. Other selective contact herbicides are used to control annual weeds in tolerant annual and perennial crops.

Because contact herbicides have limited translocation in plants, thorough spray coverage and a proper adjuvant are required for maximum activity. Also because of limited translocation, contact herbicides generally have little direct effects on roots of weedy plants or the underground vegetative propagules of perennial weeds, and they are less effective on large established annual weeds. The activity of contact herbicides increases with sunlight, temperature, and humidity and is maximized on young plants that are unstressed and growing actively [14,24].

1. Bipyridiliums

Bipyridiliums (diquat and paraquat) (Fig. 5) are the fastest acting herbicides (even faster than the diphenyl ethers), and under strong light, symptoms of wilting may be observed 20–30 min after treatment. Bipyridilium herbicides are nonselective herbicides that are used as burndown herbicides or harvest-aid desiccants. Bipyridilium molecules are positively charged (cations) that bind strongly to negatively charged soil clay and organic matter particles; hence they are generally not available for root uptake and exhibit no residual activity at

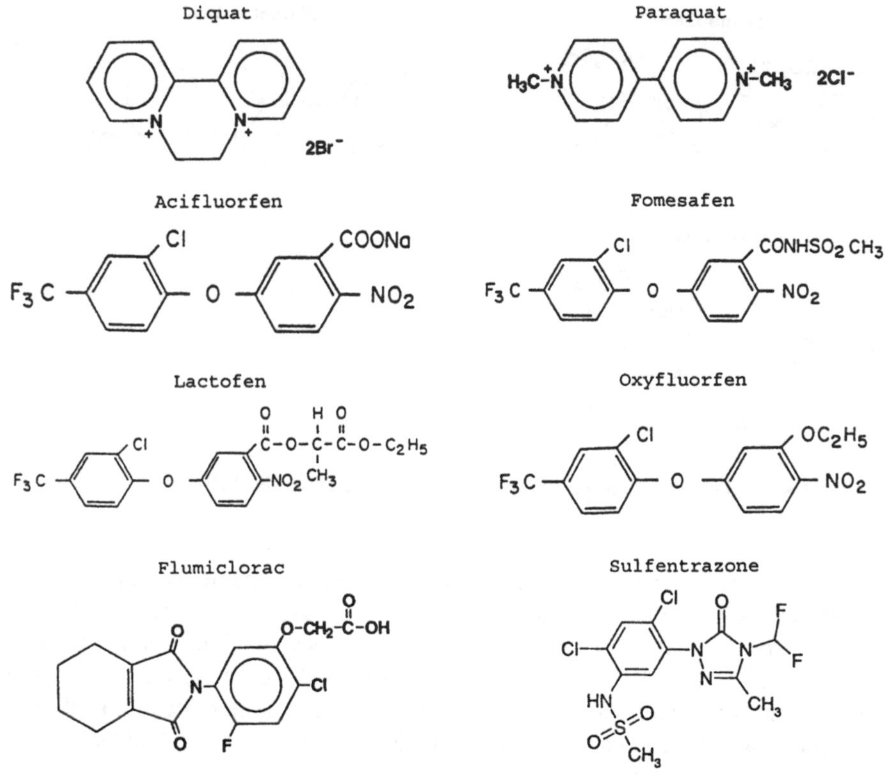

Figure 5 Contact disrupters of membranes (bipyridiliums, diphenyl ethers, and others).

normal use rates in most agricultural soils. These herbicides penetrate plant cells and enter chloroplasts, the intracellular cellular organelles that contain chlorophyll and carry on photosynthesis [17]. The bipyridilium molecules intercept electrons as they flow through their normal pathway in the chlorophyll-mediated light reactions of photosynthesis [9]. The extra electrons that reside in the newly activated paraquat radicals are transferred to molecular oxygen, creating superoxide radicals and subsequently forming other toxic free radicals which react spontaneously with plant cell membranes and other oxidizable cellular components [9,17].

Maximum herbicidal activity is observed under high sunlight and temperatures conducive to maximum rates of photosynthesis. Extreme care should be used when handling paraquat or diquat, because if ingested, they can also act as electron scavengers and free radical generators in mammalian systems, resulting in high acute mammalian toxicity. Intelligent and careful use of these herbicides is prudent.

2. Diphenyl Ethers

The diphenyl ethers (see Fig. 5) are an important group of broad-spectrum herbicides that have been used widely for selective weed control in major world crops, including cotton, peanuts, rice, and soybeans [12,66]. There are several herbicides belonging to this herbicide family, including acifluorfen, bifenox, fomesafen, lactofen, and oxyfluorfen (see Table 1) [65,66].

Diphenyl ether herbicides are used to control annual grass and broadleaf weeds following preemergence or early postemergence applications. A majority of the time in agronomic crops, they are used for selective postemergence control of broadleaf weeds [12,65,66]. Diphenyl ether herbicides inhibit the enzyme protoporphyrinogen oxidase (PROTOX) in susceptible plants [14]. This enzyme plays a key role in the synthesis of chlorophyll in green plants. When a diphenyl ether inhibits PROTOX, there is an abnormal build-up of chlorophyll precursors that are extremely light sensitive. Upon exposure to light, the unstable precursors undergo photoreactions that result in formation of destructive free radicals that oxidize lipids in cell membranes and other oxidizable cell components [14]. Thus, as in the bipyridilium herbicides, herbicidal activity is greatest under high light intensity and warm temperatures.

The initial injury symptoms in susceptible plants usually take several hours to develop. Injury is first expressed as foliage necrosis after 4–6 h of sunlight following postemergence application. The first symptoms are a water-soaked appearance (dark green spots on the foliage) followed by necrosis of the water-soaked area. Preemergence treatments damage the tissue as the herbicide comes in contact with the emerging seedling. As with postemergence applications, the damage is tissue necrosis. Tolerant crops are able to metabolize diphenyl ethers to a nontoxic form, although there is usually some desiccation of the mature foliage before the crop plants resume growth and completely recover.

C. Hormonelike Herbicides

Hormonelike herbicides are used on more land area worldwide than any other group. Some of these herbicides are used extensively on the three leading world crops (corn, rice, and wheat) [12], and there is substantial use on brush, rangeland, turf, and other grass crops [4,6,54]. Historically, the discovery and development of 2,4-D and MCPA in the mid 1940s immediately after World War II revolutionized weed control. These herbicides signaled a new era of selective chemical control of weeds. These herbicides demonstrated that synthetic organic compounds could be developed and used to control weeds in crops selectively and economically. Following introduction of the hormonelike herbicides, the chemical industry began major synthesis and evaluation programs which led to the development of the wide array of herbicides that are available today. The discovery of the phenoxy herbicides came directly from basic research on plant growth regulators.

Hormonelike herbicides are synthetic growth regulators that mimic the activity of natural plant hormones, especially auxins. They are phytotoxic to many plants, because they are much more potent than natural auxins and cause secondary effects that inhibit plant growth. As a group, the growth regulator–type herbicides are usually applied postemergence, but they also may be effective through the soil. They move systematically throughout the plant following absorption by roots and shoots. They induce similar symptoms in plants, but their precise modes and sites of action remain elusive.

Hormonelike herbicides can exhibit activity after absorption by either roots or shoots, but some have much longer residual activity in soils than others. Hormonelike herbicides are highly systemic in susceptible plants and are translocated via the apoplast and symplast to meristematic regions and vegetative propagules of perennial weeds if applied at the proper stage of growth. They are primarily effective for controlling broadleaf plants and many are used for selective control of broadleaf weeds in grass crops; however, they can injure grasses if applied during sensitive stages of growth, including early seedling development and reproduction.

The chemical families that comprise the hormonelike herbicides include the phenoxy acids, benzoic acids, and 2-pyrinecarboxylic acids (see Table 1; Fig. 6). All of these herbicides are believed to cause phytotoxicity by interfering with nucleic acid metabolism [16]. Some hormonelike herbicides are known to cause a hormonal imbalance in plant tissues that induces a rapid increase in the biosynthesis of ribonucleic acid (RNA). Overproduction of RNA leads initially to a temporary stimulation of protein synthesis, photosynthesis, ion uptake, and cell division. However, subsequent tissue swelling prevents translocation of photosynthate to vital areas leading to aberrant growth of developing plant tissues and organs. The earliest injury symptom of hormonelike herbicides in broadleaf plants usually occurs within a few hours and is characterized by epinasty, a downward twisting of the stem and leaf petioles. These herbicides produce profound effects upon the growth and structure of plants; symptoms include malformed leaves, epinastic bending and swelling of stems, deformed roots, and tissue decay. Hormonelike herbicides cause parenchymal cells to divide resulting in callus tissue formation, excessive vascular tissue in young leaves, plugging of the phloem, and root growth inhibition. Young meristematic tissues are more affected than mature tissue with cambium, endodermis, pericycle, and phloem parenchyma tissues being particularly sensitive [11].

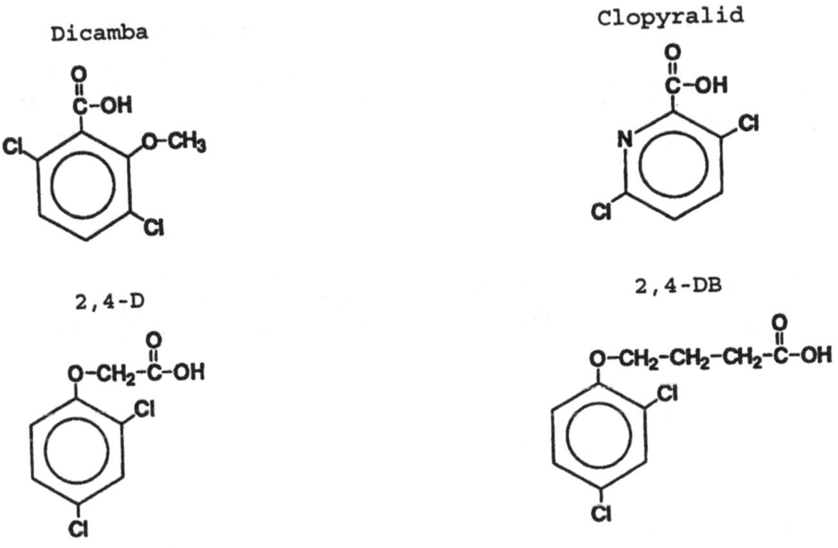

Figure 6 Hormonelike herbicides (benzoic acids, phenoxys, and carboxylic acids).

Hormonelike herbicides generally do not translocate well in most grasses and are metabolized over time. Grasses that are affected may develop tightly rolled leaves, brittle stems, and/or malformations of vegetative or reproductive structures [24].

D. Lipid Synthesis Inhibitors

There are several herbicides from different chemical families that inhibit lipid synthesis in plants. The herbicide families included as lipid synthesis inhibitors in the classification scheme presented here are the cyclohexanediones, aryloxyphenoxy propionates, and thiocarbamates (Fig. 7). These herbicides inhibit fatty acid biosynthesis as their primary mode of action [22]. Lipids (waxes, oils) are composed of fatty acids. Lipids are essential plant components, not only as major seed storage compounds but also as constituents of membranes and cuticular waxes [58]. Changes in membrane fatty acid content result in changes in the functionality of chloroplasts and mitochondria.

The cyclohexanediones and aryloxyphenoxy propionates are highly selective herbicides that control many annual and perennial grasses, whereas broadleaf weeds and crops are very tolerant to these herbicides [24,36,61,65,66]. Some turfgrass species exhibit varying degrees of tolerance [4]. In general, broadleaf crops and weeds are highly tolerant of or resistant to these herbicides, and some cereal and turfgrass species also exhibit varying degrees of tolerance.

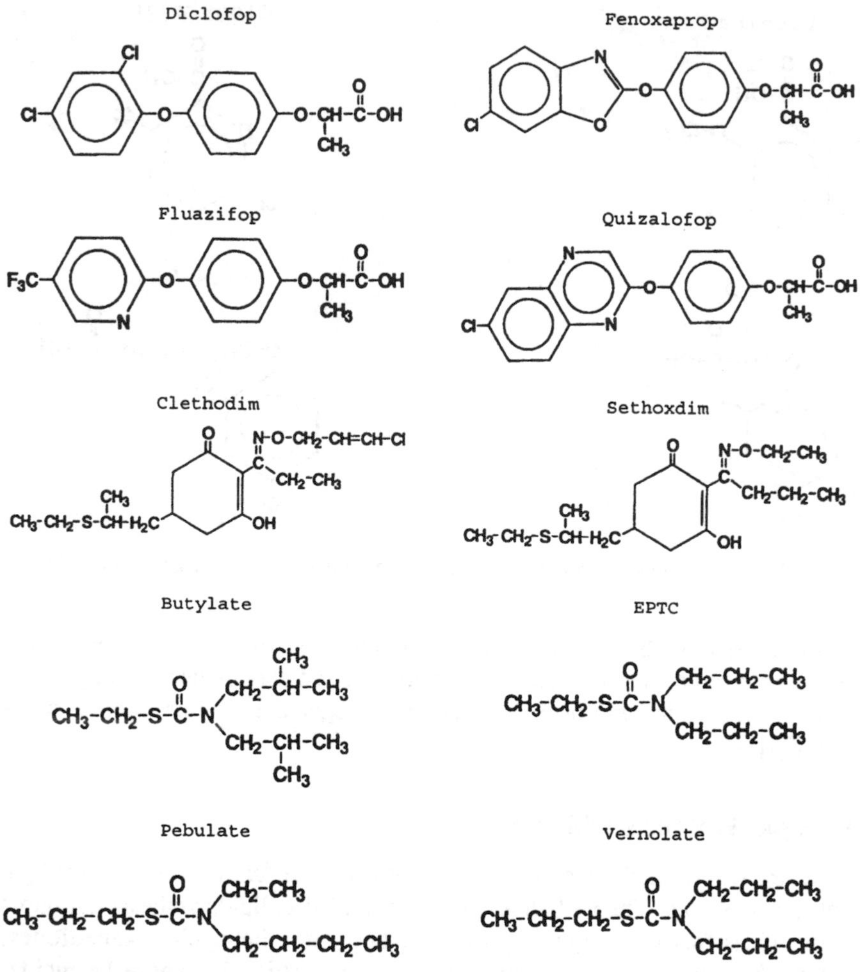

Figure 7 Aryloxyphenoxy propionate, cyclohexanedione, and thiocarbamate herbicides.

Aryloxyphenoxy propionates and cyclohexanediones are foliar-applied, although some do have limited residual activity in soils for a short period. Translocation occurs in the apoplast and symplast with the herbicide tending to concentrate in the meristematic tissue and other areas of high metabolic activity [11]. The amount of herbicide translocated appears to be very low in relation to the amount applied.

Fatty acid synthesis in susceptible plants is blocked when aryloxyphenoxy propionates and cyclohexanediones inhibit acetyl-coenzyme A (acetyl-CoA) carboxylase (ACCase) [22]. This enzyme is responsible for the first committed step in fatty acid biosynthesis [11,22]. Injury symptoms observed on grasses begin to appear several days after application. Root and shoot growth ceases rapidly (within hours), long before visual signs of injury appear. Plants then develop red coloration in the leaves (anthocyanin formation) and necrotic areas in the nongreen meristematic leaf tissue just above the growing point (intercalary meristem). These emerging new leaves can be pulled easily from the rest of the plant after 5–7 days and will be necrotic at the base. Total plant necrosis usually occurs within 10–14 days. Sublethal dosages in grasses can cause stunting along with chlorosis and translucent "banding" of new leaves as they emerge from the whorl. Differential metabolism and/or insensitivity of ACCase (i.e., genetic resistance) appears to be the major mechanism of selectivity for the aryloxyphenoxy propionates and cyclohexanediones [22].

As previously mentioned, antagonism has been reported when the postemergence grass herbicides are applied in a mixture with postemergence broadleaf-controlling herbicides. Applicators should consult the respective herbicide labels when considering joint application of postemergence grass and broadleaf herbicides.

The thiocarbamate herbicides (see Fig. 7) are applied preplant incorporated and they provide selected control of annual grass and some annual small-seeded broadleaf weeds during germination and early seedling growth. The thiocarbamate herbicides control susceptible weeds through their inhibition of the biosynthesis of very long chain fatty acids which are important components of leaf surface lipids (waxes, cutin, suberin) [70]. The site of action of the thiocarbamate herbicides is generally assumed to be in the developing shoot, but their exact site of action has not been identified.

E. Photosynthesis Inhibitors

Photosynthesis is the process in plants by which light energy is utilized to shift energy in the form of energized electrons into chemical forms required for plant metabolism [58]. Photosynthesis inhibitors block the light reactions of photosynthesis [17]. Most photosynthesis inhibitors are more toxic to dicots (broadleaf plants) than grasses at selective dosages, but some herbicides in this class are also quite toxic to grasses (Fig. 8). Some photosynthesis inhibitors are used only as foliar-applied treatments (e.g., bentazon and pyridate; Fig. 8), whereas others are effective as both soil- and foliar-applied treatments (e.g., atrazine).

Photosynthesis inhibitors are translocated mainly in the apoplast, so internal movement within the plant is upward via the transpirational stream from the site of absorption. The exceptions are hydroxybenzonitriles and bentazon, which have limited symplastic translocation. More photosynthesis inhibitors are soil applied than are postemergence applied, so that root uptake will ensure their translocation to all foliar plant parts along with water and minerals in the tran-

Ametryn

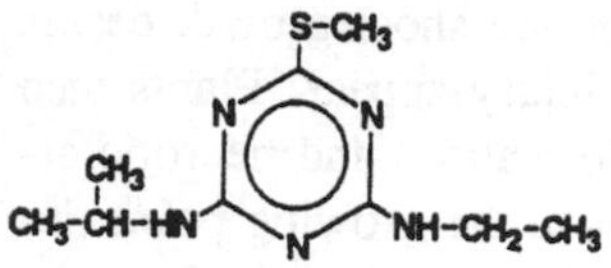

Atrazine

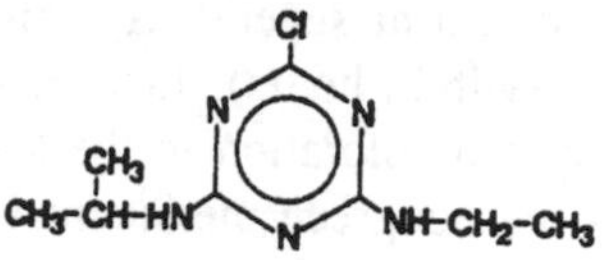

Cyanazine

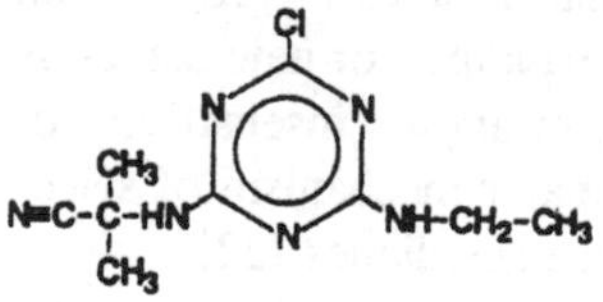

Prometryn

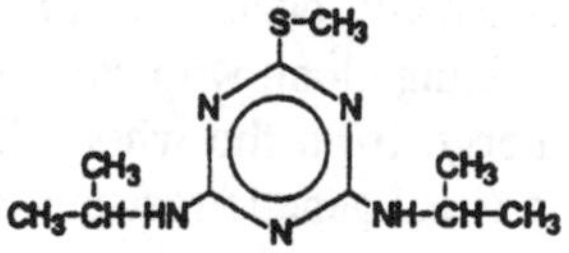

Simazine

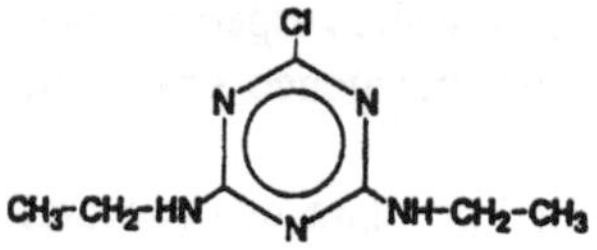

Metribuzin

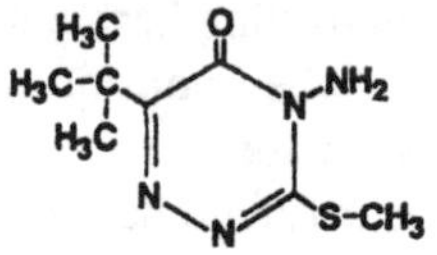

Bentazon

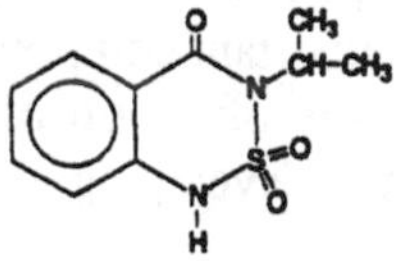

Pyridate

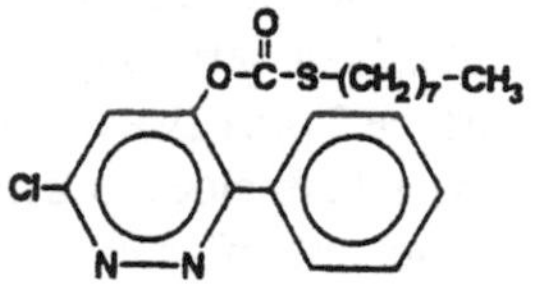

Diuron

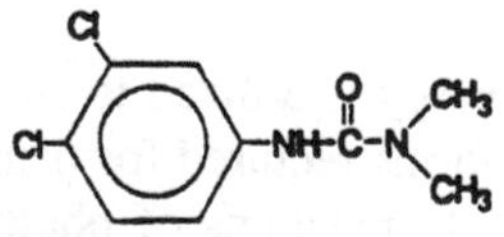

Fluometuron

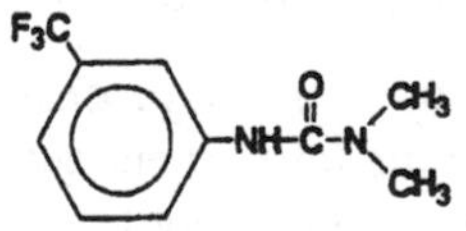

Linuron

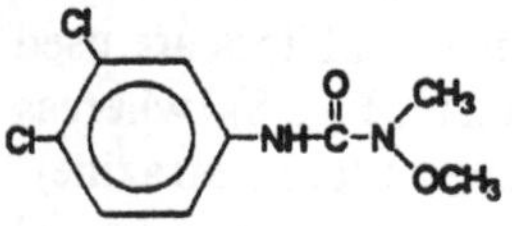

Figure 8 Photosynthesis inhibitors (triazine types, benzothiadiazole, phenylpyridazine, and urea).

spiration stream. Foliar applications of photosynthesis inhibitors can be effective but require uniform spray coverage of the foliage and the addition of an adjuvant in the spray mixture for acceptable weed control [24].

Inside the plant cell, photosynthesis inhibitors enter the chloroplast and bind to a specific protein in the photosynthetic apparatus [11,17]. Herbicide binding of the protein site blocks the flow of electrons in photosynthesis. Consequently, light energy cannot be converted and "trapped" into chemical forms required for normal metabolism and growth. Additionally, sunlight-induced secondary reactions can cause photo-oxidation of chlorophyll resulting in contact-like herbicide symptoms and more rapid destruction of plant tissue than is caused by simple starvation [17]. The most common symptoms produced by photosynthesis inhibitors in susceptible plants is chlorosis (yellowing) of the leaf tissue.

F. Carotenoid Synthesis Inhibitors

The orange-colored carotenoid pigments are not generally a conspicuous feature of leaves, for their presence is masked by an approximately eight times greater concentration of chlorophylls. Carotenoids are auxiliary pigments in green plants that have two main functions: (a) augmentation of the light reactions of photosynthesis by "harvesting" and transferring of sunlight energy to chlorophyll and (b) protection of chlorophyll and the photosynthetic apparatus from photooxidation [49,58]. Unlike the photosynthesis inhibitors, carotenoid synthesis inhibitors (Fig. 9) do not directly block the light reactions of photosynthesis but rather inhibit the production of carotenoid pigments. The most striking symptom resulting from treating plants with herbicides that inhibit carotenoid biosynthesis is the almost totally white foliage (albino growth) following treatment. Plants will continue to grow for a time but without production of green photosynthetic tissue; growth of affected plants cannot be maintained. Susceptible plants develop bleached white and/or chlorotic foliage after absorption of the herbicide. Otherwise susceptible plants appear normal for a short period with no apparent growth malformations or other morphological symptoms. Starvation and overall disruption of plant processes ensue, eventually

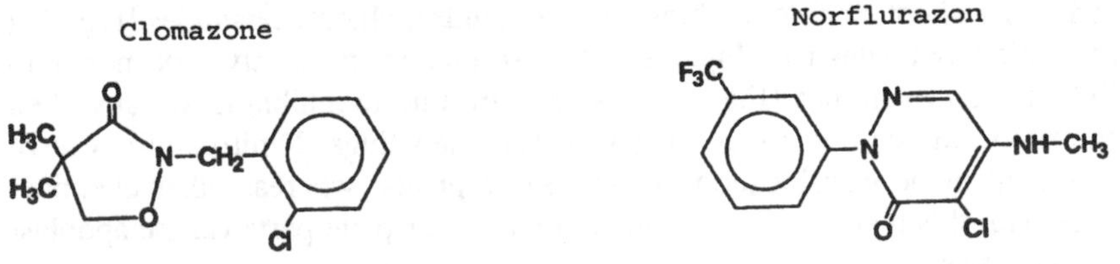

Figure 9 Carotenoid-inhibiting herbicides.

resulting in tissue necrosis and plant death [49]. Herbicides that inhibit caro-tenoid biosynthesis do not affect preexisting carotenoids. Thus, plant tissues formed before treatment do not show albino growth. There is a turnover of carotenoid pigments; thus tissue formed prior to treatment will eventually show chlorosis and die.

Even though new growth in treated plants is white, these herbicides do not directly inhibit green pigment (chlorophyll) biosynthesis. The loss of chloro-phyll is the result of destruction of chlorophyll by light (photooxidation), or it is perhaps due to the absence of carotenoids disrupting normal chlorophyll bio-synthesis and chloroplast development [7].

Herbicides that are known to inhibit biosynthesis of carotenoids are translo-cated in the apoplast and with a few exceptions are used primarily as soil-ap-plied treatments. As with other apoplastically translocated herbicides, activity from foliar applications is maximized with thorough spray coverage and the use of a spray adjuvant. The row-crop herbicides in this class vary widely in se-lectivity and are active on various annual grass and broadleaf weeds. Aminotriazole and isoxazolidinones are believed to interfere with other related enzymes in the carotenoid biosynthesis pathway [11,49].

G.   Meristematic Mitotic Inhibitors

Active cell division in plants occurs in regions called meristems, and these re-gions are where new primary growth occurs [15]. Meristematic tissue occurs in plant shoot and root tips [15]. Meristem mitotic inhibitors affect fundamen-tal process(es) involved in mitosis, the stepwise process of cell division. As a result, these inhibitors are soil-applied herbicides that are most effective when placed in direct contact with germinating seedlings. These herbicides do not inhibit the onset of mitosis but rather disrupt the mitotic sequence (prophase-metaphase-anaphase-telophase) once initiated [59]. All of these compounds in-terfere with the normal movement of chromosomes during the mitotic sequence. The spindle apparatus, which is composed of protein structures called micro-tubules, is the framework responsible for moving chromosomes to the various stages of mitosis [59].

Meristem mitotic inhibitors are often further subdivided into herbicides that are primarily shoot meristem inhibitors (acetamides, chloroacetamides [Fig. 10] and the thiocarbamates [see Fig. 7] and those that are primarily root meristem inhibitors (dinitroanilines [Fig. 10]). Meristem mitotic inhibitors are absorbed by the shoot and/or root tissue of germinating seedlings. Dinitroanilines have very limited or no translocation in susceptible plants, whereas other chemical families with this mode of action translocate to other plant parts via the apoplast and/or symplast.

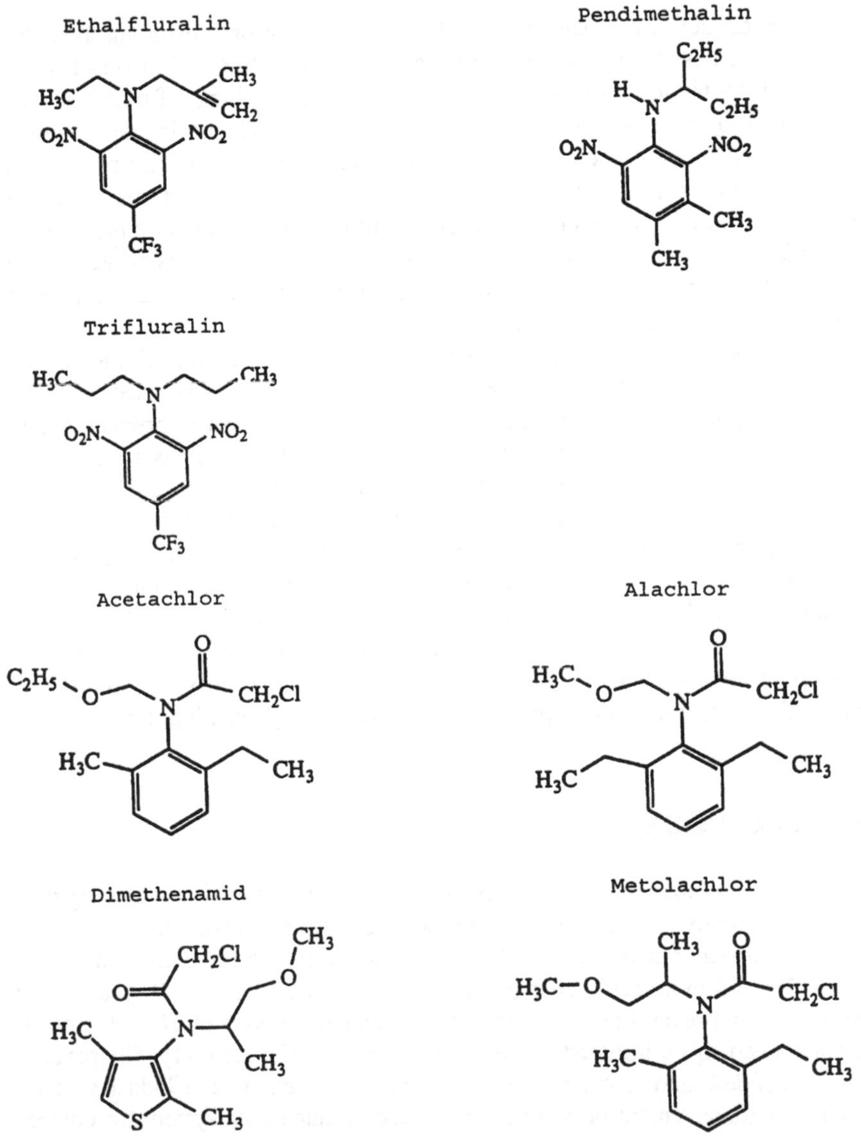

Ethalfluralin

Pendimethalin

Trifluralin

Acetachlor

Alachlor

Dimethenamid

Metolachlor

Figure 10 Mitotic inhibitors (dinitroanilines chloroacetamides, and acetamides).

The principal use of the dinitroaniline herbicides is for preemergence control of seedling grasses and certain small-seeded annual broadleaf weeds in selected crops [65,66]. In general, seedling grasses are more susceptible to these herbicides than are seedling broadleafs. Dinitroaniline herbicides inhibit root meristematic activity in susceptible plants by slowing or preventing assembly of microtubules during mitosis [59]. A secondary effect often observed in dinitroaniline-treated roots is abnormal enlargement (swelling) of root tips. This so-called club root symptom on root tips is often used in diagnosing dinitroaniline injury. This injury is the result of microtubules being additionally required in cell wall development and elongation [58]. Other general symptoms of susceptible plants include nonemergence of germinating seedlings or stunting of plants with a lack of lateral roots or severely pruned roots in the herbicide-treated zone Broadleaf plants may develop a swollen and cracked hypocotyl and callus tissue at the base of the stem [24].

Herbicide families that are primary shoot meristem inhibitors include the acetamides, thiocarbamates (see Fig. 7), and chloroacetamides (see Fig. 10). The specific site of action is unknown, but they are known to prevent mitotic entry of meristem cells and cause various cellular abnormalities in meristematic tissue [24,25,59]. General symptoms include nonemergence of germinating seedlings and malformation and stunting of shoots. Affected grasses may leaf out underground or not unfurl properly, and injured broadleaf plants develop puckered leaves with leaf buds that do not open (thiocarbamates) and/or a shortened mid vein with a notched distal leaf margin (acetamides). Small-seeded grass and broadleaf weeds are most susceptible to meristem mitotic inhibitors in general, and selectivity in crops is at least partially due to differential seed size and herbicidal placement relative to crop seedling depth.

V. CONCLUSIONS

There are a number of very effective herbicides available for controlling most weeds selectively and economically in most agronomic crops. There have been a number of major changes in the last decade concerning weed management with herbicides and more rapid changes will likely occur in the next 10 years. New strategies for preventing or delaying the further development of herbicide resistance need to be initiated in the near future [20,27,42,67,71]. Cooperation will be needed among research and extension, academia, and industry to develop new management programs that include an integrated system for control. To minimize the impact of herbicides on the environment while maintaining crop productivity, new systems are being developed to integrate computers into the management decision-making process [43,53,68,69]. The use of comprehensive computer programs that integrate herbicide selection based on environmental impact, weed control efficacy, economic thresholds, weed population dynamics, weed biology, and herbicide resistance potential could help improve crop quality and profitability while reducing the impact to the environment [65]. New technological developments will integrate the aforementioned factors with global positioning systems, variable rate sprayers, and weed "sensing" sprayers [23] to further increase the efficiency of chemical weed management.

REFERENCES

1. Agracetus, Inc. 1991. Institutional Biosafety Reports. Construction and Use of Dominant Selectable Markers for Use in Transformation of Plant Cells. Updated Addendum. Middleton, Wisconsin: Agracetus, pp 1–9.
2. Ahrens, W. H., ed. 1994. *Herbicide Handbook*. 7th ed. Champaign, Illinois: Weed Science Society of America.
3. Auld, B. A., K. M. Menz, and C. A. Tisdell. 1987. *Weed Control Economics*. New York: Academic Press.
4. Bingham, S. W., W. J. Chism, and P. C. Bhowmik. 1995. Weed management systems for turfgrass. In *Handbook of Weed Management Systems*. A. E. Smith, ed. New York: Marcel Dekker, pp 603–665.
5. Boger, P., and G. Sandman. 1989. *Target Sites of Herbicide Action*. Boca Raton, Florida: CRC Press, p 295.
6. Bovey, R. W. 1995. Weed management systems for rangeland. In *Handbook of Weed Management Systems*. A. E. Smith, ed. New York: Marcel Dekker, pp 519–552.
7. Bramley, P. M., and K. E. Pallet. 1993. Phytoene desaturase: a bichemical target of many bleaching herbicides. *Proceedings of the Brighton Crop Protection Conference. Weeds* 713–722.
8. Byrd, J. D., Jr., and A. C. York. 1987. Annual grass control in cotton (*Gossyipium hirsutum*) with fluazifop, sethoxydim, and selected herbicides. *Weed Sci.* 35:388–394.
9. Chia, L. S., D. G. McRae, and J. E. Thompson. 1982. Light-dependence of paraquat-induced membrane deterioration in bean plants. *Physiol. Plant* 56:492–499.
10. DeFelice, M. S., W. B. Brown, R. J. Aldrich, B. D. Sims, D. T. Judy, and D. R. Guethle. 1989. Weed control in soybeans (*Glycine max*) with reduced rates of postemergence herbicides. *Weed Sci.* 37:365–374.
11. Devine, M. D., S. O. Duke, and C. Fedtke. 1993. *Physiology of Herbicide Action*. Englewood Cliffs, New Jersey: Prentice Hall.
12. Donald, W. W., E. F. Eastin. 1995. Weed management systems for grain crops. In *Handbook of Weed Management Systems*. A. E. Smith, ed. New York: Marcel Dekker, pp 401–476.
13. Duke, S. O. 1988. Glyphosate. In *Herbicides: Chemistry, Degradation, and Mode of Action*. P. C. Kearney and D. D. Kaufman, eds. New York: Marcel Dekker, pp 1–70.
14. Duke, S. O., J. Lydon, J. M. Becerril, T. D. Sherman, L. P. Lehnen, Jr., and M. Matsumato. 1991. Protoporphyrinogen oxidase–inhibiting herbicides. *Weed Sci.* 39:465–473.

15. Esau, K. 1977. *Anatomy of Seed Plants*. 2nd ed. New York: Wiley.
16. Fedtke, C. 1982. *Biochemistry and Physiology of Herbicide Action*. Berlin: Springer-Verlag.
17. Fuerst, E. P., and M. A. Norman. 1991. Interaction of herbicides with photosynthetic electron transport. *Weed Sci.* 39:458–464.
18. Gimenez, A. E., A. C. York, J. W. Wilcut, and R. B. Batts. 1998. Annual grass control by glyphosate plus bentazon, chlorimuron, fomesafen, or imazethapyr mixtures. *Weed Technol.* 12:134–136.
19. Gototrup, O., P. A. O'Sullivan, R. J. Schraa, and W. H. VandenBorn. 1976. Uptake, translocation, metabolism, and selectivity of glyphosate in Canada thistle and leafy spurge. *Weed Res.* 16:197–201.
20. Gressel, J., and L. A. Segel. 1990. Modeling the effectiveness of herbicide rotations and mixtures as strategies to delay or preclude resistance. *Weed Technol.* 4:186–198.
21. Grichar, W. J., and T. E. Boswell. 1986. Postemergence grass control in peanut (*Arachis hypogaea*). *Weed Sci.* 43:587–590.
22. Gronwald, J. W. 1991. Lipid biosynthesis inhibitors. *Weed Sci.* 39:435–449.
23. Hanks, J. E., and J. L. Beck. 1998. Sensor-controlled hooded sprayer for row crops. *Weed Technol.* 12:308–314.
24. Harrison, S. K., and M. M. Loux. 1995. Chemical weed management. In *Handbook of Weed Management Systems*. A. E. Smith, ed. New York: Marcel Dekker, pp 101–153.
25. Hess, F. D. 1987. Herbicide effects on the cell cycle of meristematic plant cells. *Rev. Weed Sci.* 3:183–203.
26. Holhouser, D. L., and H. D. Coble. 1990. Compatibility of sethoxydim with five postemergence broadleaf herbicides. *Weed Technol.* 4:128–133.
27. Holt, J. S., and H. M. LeBaron. 1990. Significance and distribution of herbicide resistance. *Weed Technol.* 4:141–149.
28. Hopkins, W. L. 1989. A global evaluation of "new" herbicide activity, 1984–1988: its changing dynamics and a look at its future direction. *Proceedings of the Brighton Crop Protection Conference. Weeds* 1:213–236.
29. Jennings, K. M., A. C. York, R. B. Batts, and A. S. Culpepper. 1997. Sicklepod (*Senna obtusifolia*) and entireleaf morningglory (*Ipomoea hederacea* var. *intergriuscula*) management in soybean (*Glycine max*) with flumetsulam. *Weed Technol.* 11:227–240.
30. Jordan, D. L., and A. C. York. 1989. Effects of ammonium fertilizers and BCH 81508 S on antagonism with sethoxydim and bentazon mixtures. *Weed Technol.* 3:450–454.
31. Jordan, D. L., A. C. York, J. L. Griffin, P. A. Clay, P. R. Vidrine, and D. B. Reynolds. 1997. Influence of application variables on efficacy of glyphosate. *Weed Technol.* 11:354–362.
32. Kishore, K. M., and D. M. Shah. 1988. Amino acid biosynthesis inhibitors as herbicides. *Annu. Rev. Biochem.* 57:627–663.
33. Krausz, R. F., and G. Kapusta. 1998. Total postemergence weed control in imidazolinone-resistant corn (*Zea mays*). *Weed Technol.* 12:151–156.

34. Lea, P. J., and S. M. Ridley. 1989. Glutamine synthetase and its inhibition. In *Herbicides and Herbicide Metabolism*. A. D. Dodge, ed. Cambridge, UK: Cambridge University Press, pp 137–170.

35. Marquis, L. Y., R. D. Comes, and C. P. Yang. 1979. Selectivity of glyphosate in creeping red fescue and red canarygrass. *Weed Res.* 17:335–342.

36. Nastasi, P., R. E. Frans, and M. R. McClelland. 1986. Economics and new alternatives in cotton (*Gossypium hirsutum*) weed management programs. *Weed Sci.* 34:634–638.

37. Neal, J. C., W. A. Skroch, and T. J. Manaco. 1985. Effects of plant growth stage on glyphosate absorption and transport in ligustrum (*Ligustrum japonicum*) and blue Pacific juniper (*Juniperus conferta*). *Weed Sci.* 34:115–121.

38. Nelson, K. A., and K. A. Renner. 1998. Postemergence weed control with CGA-277476 and cloransulam-methyl in soybeans (*Glycine max*). *Weed Technol.* 12:293–299.

39. Ozkan, H. E. 1995. Herbicide application equipment. In *Handbook of Weed Management Systems*. A. E. Smith, ed. New York: Marcel Dekker, pp 155–216.

40. Ozkan, H. E. 1995. Herbicide formulations, adjuvants, and spray drift management. In *Handbook of Weed Management Systems*. A. E. Smith, ed. New York: Marcel Dekker, pp 217–244.

41. Poston, D. H., E. C. Murdock, and J. E. Toler. 1992. Cost-efficient weed control in soybean (*Glycine max*) with cultivations and banded herbicide applications. *Weed Technol.* 6:990–995.

42. Powles, S. B., and J. A. M. Holtum, eds. 1994. *Herbicide Resistance in Plants, Biology, and Biochemistry*. Boca Raton, Florida: CRC Press.

43. Rankins, A., Jr., D. R. Shaw, and J. D. Boyd, Jr. 1998. HERB and MSU-HERB field validation for soybean (*Glycine max*) weed control in Mississippi. *Weed Technol.* 12:88–96.

44. Rawn, J. D. 1983. *Biochemistry*. New York: Harper & Row.

45. Retzinger, E. J., Jr., and C. Mallory-Smith. 1997. Classification of herbicides by site of action for weed resistance management strategies. *Weed Technol.* 11:384–393.

46. Richburg, J. S., III, J. W. Wilcut, and G. L. Wiley. 1995. AC 263,222 and imazethapyr rates and mixtures for weed management in peanut (*Arachis hypogaea*). *Weed Technol.* 9:801–806.

47. Robbins, W. W., A. S. Crafts, and R. N. Raynor. 1952. *Weed Control*. New York: McGraw-Hill.

48. Saggers, D. T. 1976. The search for new herbicides. In *Herbicides*. L. J. Audus, ed. Vol. 2. London: Academic Press, pp 447–473.

49. Sandmann, G., A. Schmidt, H. Linden, and P. Boger. 1991. Phytoene desaturase, the essential target for bleaching herbicides. *Weed Sci.* 39:474–480.

50. Sankula, S. M., P. Braverman, and S. D. Linscombe. 1997. Glufosinate-resistant, BAR-transformed rice (Oryza *sativa*), and redrice (*Oryza sativa*) response to glufosinate alone and in mixtures. *Weed Technol.* 11:662–666.

51. Scott, R., D. R. Shaw, and W. L. Barrentine. 1998. Glyphosate tank mixtures with SAN 582 for burndown or postemergence applications in glyphosate-tolerant soybeans (*Glycine max*). *Weed Technol.* 12:23–26.

52. Shaner, D. L. 1989. Sites of action of herbicides in amino acid metabolism: primary and secondary physiological effects. *Rec. Adv. Phytochem.* 23:227–261.

53. Shaw, D. R., A. Rankins, Jr., J. T. Ruscoe, and J. D. Byrd, Jr. 1998. Field validation of weed control recommendations for HERB and SWC herbicide recommendation models. *Weed Technol.* 12:78–87.

54. Smith, A. E., and L. D. Martin. 1995. Weed management systems for pastures and hay crops. In *Handbook of Weed Management Systems.* A. E. Smith, ed. New York: Marcel Dekker, pp 477–517.

55. Steckel, G. J., L. M. Wax, F. W. Simmons, and W. H. Phillips. 1997. Glufosinate efficacy on annual weeds is influenced by rate and growth stage. *Weed Technol.* 11:484–488.

56. Stidham, M. A. 1991. Herbicides that inhibit acetohydroxyacid synthase. *Weed Sci.* 39:428–434.

57. Stidham, M. A., and B. K. Singh. 1991. Imidazolinone–acetohydroxy acid synthase interactions. In *The Imidazolinone Herbicides.* D. L. Shaner and S. L. O'Connor, eds. Boca Raton, Florida: CRC Press, pp 71–90.

58. Taiz, T., and E. Zeigler. 1991. *Plant Physiology.* Redwood City, California: Benjamin/Cummings, pp 179–218.

59. Vaughn, K. C., and L. P. Lehnen, Jr. 1991. Mitotic disrupter herbicides. *Weed Sci.* 39:450–457.

60. Vencill, W. K., J. S. Richburg, III, J. W. Wilcut, L. R. Hawf. 1995. Effect of MON-12037 on purple (*Cyperus rotundus*) and yellow (*Cyperus esculentus*) nutsage. *Weed Technol.* 9:148–152.

61. Vidrine, P. R. Johnsongrass (*Sorghum halepense*) control in soybeans (*Glycine max*) with postemergence herbicides. *Weed Technol.* 3:455–458.

62. Walker, R. H. 1995. Preventive weed management. In *Handbook of Weed Management Systems.* A. E. Smith, ed. New York: Marcel Dekker, pp 35–50.

63. Wicks, G. A., O. C. Burnside, and W. L. Felton. Mechanical weed management. In *Handbook of Weed Management Systems.* A. E. Smith, ed. New York: Marcel Dekker, pp 51–99.

64. Wilcut, J. W., G. R. Wehtje, T. V. Hicks, and T. A. Cole. 1990. Postemergence weed management systems for peanuts. *Weed Technol.* 4:76–80.

65. Wilcut, J. W., A. C. York, and G. R. Wehtje. 1994. Interactions and control of weeds in peanut (*Arachis hypogaea*). *Rev. Weed Sci.* 6:177–205.

66. Wilcut, J. W., A. C. York, and D. L. Jordan. 1995. Weed management systems for oil seed crops. In *Handbook of Weed Management Systems.* A. E. Smith, ed. New York: Marcel Dekker, pp 343–400.

67. Wilcut, J. W., H. D. Coble, A. C. York, and D. W. Monks. 1996. The niche for herbicide-resistant crops in U.S. agriculture. In *Herbicide-Resistant Crops: Agricultural, Environmental, Economic, Regulatory, and Technical Aspects.* S. O. Duke, ed. Boca Raton, Florida: CRC Press, pp 213–230.

68. Wiles, L. J., G. G. Wilkerson, H. J. Gold, and H. D. Coble. 1992. Modeling weed distribution for improved postemergence control systems. *Weed Sci.* 40:546–553.

69. Wiles, L. J., G. W. Oliver, A. C. York, H. J. Gold, and G. G. Wilkerson. 1992.

Spatial distribution of broadleaf weeds in North Carolina soybean (*Glycine max*). *Weed Sci.* 40:554–557.

70. Wilkinson, R. E. 1988. Carbamothioates. In *Herbicides: Chemistry, Degradation, and Mode of Action*. Vol. 3. P. C. Kearney and D. D. Kaufman, eds. New York: Marcel Dekker, pp 245–300.
71. Wrubel, R. P., and J. Gressel. 1994. Are herbicide mixtures useful for delaying the rapid evolution of resistance? A case study. *Weed Technol.* 8:635–648.
72. Zimdahl, R. L. 1993. *Fundamentals of Weed Science*. San Diego: Academic Press.
73. Dodge, A. D., ed. 1989. *Herbicides and Herbicide Metabolism*. Cambridge, UK, Cambridge University Press.
74. Smith, A. E. 1995. *Handbook of Weed Management Systems*. New York: Marcel Dekker.

24

Overview and Management of Vertebrate Pests

Richard A. Dolbeer
United States Department of Agriculture, Sandusky, Ohio

I. INTRODUCTION

Agricultural pest management has traditionally focused on arthropods, weeds, and pathogenic organisms. However, many vertebrate species (primarily wild birds and mammals) at one time or another require management actions to reduce conflicts with agricultural production. In fact, vertebrate pest problems appear to be increasing in the United States [25]; in large part because successful wildlife management and conservation programs, as well as land use changes, have allowed populations of many vertebrate species such as deer, geese, and blackbirds to expand. In a 1989 survey of American farmers, about half of all field crop producers reported wildlife-related losses that totaled $237 million annually [117].

Vertebrate pest management must be approached differently than traditional management of plant and invertebrate pests for four interrelated reasons. First, vertebrate species are sentient organisms with complex, adaptable, and often secretive behaviors. Second, the sociological aspects of vertebrate pest management are multifarious and emotional, particularly the polarized views of society regarding the killing and managing of wildlife species. Third, vertebrate pest species such as deer and Canada geese also are important economic resources because of their status as prominent game species. Finally, the regulatory aspects of vertebrate pest management are intricate, especially regarding the legal status of wildlife species (to be discussed in more detail below).

Vertebrate pest control programs can be thought of as having four parts: (a) problem definition, (b) ecology of the problem species, (c) control methods application, and (d) evaluation of control. Problem definition refers to determining the species and numbers of animals causing the damage, the amount of loss, and other biological and social factors related to the problem. Ecology of the problem species refers to understanding the life history of the species, especially in relation to the crop damage. Control methods application refers to taking the information gained from parts 1 and 2 to develop an appropriate management program to reduce the damage. Evaluation of control permits an assessment of the reduction in damage relative to costs and of the impact of the control on target and nontarget populations. Increasingly, emphasis is being placed on integrated pest management for vertebrates whereby populations are monitored and control methods are used in combination and coordinated with other management practices being used at that time (Fig. 1).

This chapter first reviews the legal status of vertebrate pests as related to crop protection. Second, methods of assessment and characteristics and amounts of crop damage caused by various avian and mammalian species are summarized. Examples of control techniques are then presented. Finally, managing blackbird damage to corn is discussed as an example of a plan that integrates the four components of vertebrate pest management outlined above. Scientific names of vertebrate species mentioned in the text are presented in Table 1.

II. LEGAL REQUIREMENTS FOR CONTROL

Before action is taken to manage vertebrate pests, it is important to understand the laws covering the target species. The management of most wild mammals, reptiles, and amphibians in the United States is the responsibility of the individual states. The capture, possession, or killing of these vertebrates to achieve control of damage is regulated by state laws. Migratory birds, in contrast to these other vertebrates, move freely across political boundaries and are thus managed at the federal level under the Migratory Bird Treaty Act of 1918, which is a treaty that has been amended several times and includes formal agreements with Canada, Mexico, Japan, and Russia. Federal regulations in the United States require that a depredation permit be obtained from the U.S. Fish and Wildlife Service, Department of the Interior, before any person may capture, kill, possess, or transport most migratory birds to control depredations. No federal permit is required merely to scare or herd depredating birds other than endangered species or eagles, but some states require permits for species such as waterfowl.

Introduced avian species in the United States such as house sparrows, rock doves (pigeons), and European starlings have no federal protection. Furthermore, a federal permit is not required to control blackbirds (the term *blackbird*

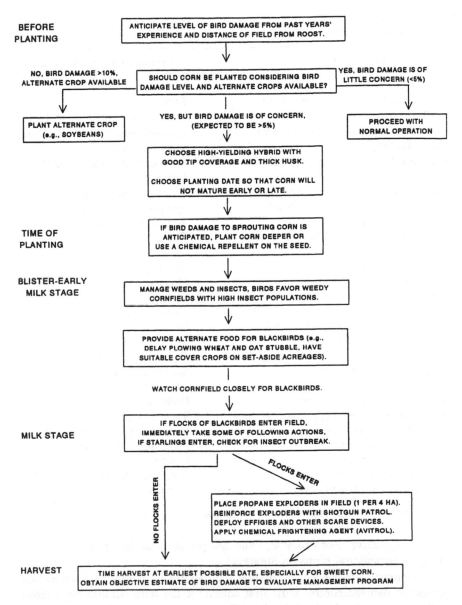

Figure 1 Schematic chart of integrated on-farm management program to reduce black-bird damage to corn. (Adapted from ref. 38.)

loosely refers to a group of about 10 species of North American birds, the most common of which are the red-winged blackbird and common grackle) when they are damaging or about to damage agricultural crops. However, federal provisions do not circumvent any state laws or regulations which may be more, but

Table 1 Common and Scientific Names of the Major Vertebrate Pest Species for Agricultural Crops in North America[a]

Common (scientific) name	Typical crop damage situations
Birds	
American robin (*Turdus migratorius*)	Fruit
Canada goose (*Branta canadensis*)	Winter wheat/sprouting soybeans
Common grackle (*Quiscalus quiscula*)	Sprouting and ripening grain/fruit
Duck (*Anas* spp.)	Ripening grain
European starling (*Sturnus vulgaris*)	Fruit/sprouting wheat/feedlots
House sparrow (*Passer domesticus*)	Ripening grain
Red-winged blackbird (*Agelaius phoeniceus*)	Sprouting and ripening grain
Rock dove (Pigeon) (*Columba livia*)	Sprouting grain/feedlots
Sandhill crane (*Grus canadensis*)	Sprouting and ripening grain
Snow goose (*Chen caerulescens*)	Winter wheat
Mammals	
Bear (*Ursus americanus*)	Ripening corn/apiaries
Beaver (*Castor candensis*)	Flooding cropland
Coyote (*Canis latrans*)	Melons
Deer (*Odocoileus* spp.)	Various crops/orchard trees
Ground squirrel (*Spermophilus* spp.)	Sprouting crops/range vegetation
Kangaroo rat (*Dipodomys* spp.)	Sprouting crops/alfalfa
Meadow vole (*Microtus pennsylvanicus*)	Orchard tree trunks
Mouse (*Peromyscus* spp.)	Sprouting no-till crops
Nutria (*Myocaster coypu*)	Sugarcane/rice
Pine vole (*Microtus pinetorum*)	Orchard tree roots/trunks
Pocket gopher (*Geomys* spp., other genera)	Various crops and trees
Prairie dog (*Cynomys* spp.)	Various crops and range vegetation
Rabbit (*Sylvilagus* spp.)	Sprouting crops/orchard trees
Raccoon (*Procyon lotor*)	Ripening corn
Roof rat (*Rattus rattus*)	Sugar cane/melons
Woodchuck (*Marmota monax*)	Sprouting and ripening crops

[a]See ref. 45 for an account of other vertebrate species that cause damage.

not less, restrictive. For example, Ohio law prohibits the killing of blackbirds damaging agricultural crops on Sundays.

In summary, anyone contemplating the capture or killing of a vertebrate species for damage control must first determine the state wildlife regulations for that species. For birds, federal wildlife regulations must also be followed. In addition, as with any pest control activity, federal and state Environmental Protection Agency (EPA) regulations must be followed whenever chemicals are used. Jacobs [63] provided a comprehensive list of EPA-registered chemicals for wildlife damage control in the United States.

III. METHODS OF ASSESSMENT, CHARACTERISTICS, AND AMOUNTS OF CROP DAMAGE

A. Methods of Assessment

Some pests, such as large flocks of blackbirds, are highly visible, and the damage they inflict in ripening grain fields is usually conspicuous. For these reasons, subjective estimates of blackbird damage sometimes overestimate losses as much as 10-fold [111]. However, rodent and ungulate (deer) damage to orchard trees or sprouting crops often occurs underground or at night by secretive animals, allowing losses to go unnoticed or to be underestimated. Thus, objective estimates of bird and mammal damage to agricultural crops are critical to define accurately the magnitude of problems and to plan appropriate, cost-effective control actions [39].

For many species of vertebrate pests, real crop losses are difficult to assess, because damage is usually highly variable among and within fields. Thus, to estimate losses, an unbiased sampling scheme is needed to select fields that are to be examined and then determine plants or areas to be measured in the selected fields [95]. For example, to estimate objectively the amount of bird and mammal damage in a ripening corn or sunflower field, at least 10 locations widely spaced in the field should be examined. At each location, the estimator should randomly select ears or heads from 10 plants to measure or visually estimate the amount of damage. This damage can then be converted to yield loss per hectare [35,37,116].

Losses of agricultural crops to birds also can be estimated indirectly through avian bioenergetics. By estimating the number of birds of the target species feeding in an area, the percentage of the agricultural crop in the birds' diet (obtained by examination of stomach contents from a sample of collected birds), the caloric value of the crop, and the daily caloric requirements of the birds, one can project the total biomass of crop removed by birds on a daily or seasonal basis [111,112].

Fruit loss to birds in orchards and vineyards can be estimated by counting the numbers of undamaged, pecked, and removed fruits per sampled branch or fruit bunch [34,103,105]. Pearson and Forshey [85] compared yield of apple trees visibly damaged by voles to those not showing damage to determine the dollar losses in gross return per tree. Richmond et al. [88] determined reductions in growth, yield, and fruit size of apple trees damaged by pine vole populations of known size maintained in enclosures around the trees. Forage and macadamia nut losses have been estimated by comparing production on areas with and without rodents [71,106].

An index of rodent damage to sugar cane was developed through sampling at harvest to determine the percentage of stalks damaged [68]. Clark and Young

[19] established transects in cornfields and noted rodent damage to individual seedlings over a 10-day period. Sauer [90] used exclusion cylinders to determine losses of forage to ground squirrels. Likewise, sprouting rice or other small grains removed by birds can be estimated by comparing plant density in exposed plots with that in adjacent plots with wire bird exclosures [82]. These loss estimates must be converted to accurate assessments of final yield reduction to enable cost/benefit evaluation of control programs. This conversion is often difficult given the complexity of factors that affect final yield [106,116].

These examples illustrate the complexity of damage situations and the need for better damage-assessment methods. Lack of methods for determining damage levels, particularly for rodents, has been a serious impediment to the development of cost-effective control strategies. Damage assessment methodology is an area of high priority for future research.

B. Characteristics and Amounts of Damage

1. Birds

Most bird damage occurs during daylight hours, and the best way to identify the species causing damage is by careful observation. However, the presence of a bird species in a crop receiving damage does not automatically prove the species guilty. For example, after careful observation and examination of the stomach contents of large, conspicuous flocks of common grackles in sprouting winter wheat fields, they were found to be eating corn residue from the previous crop. Smaller numbers of starlings were removing the germinating wheat seeds [49]. Conversely, flocks of starlings in ripening corn fields fed on insects, whereas flocks of blackbirds fed on the milk-stage grain [38]. Blackbirds characteristically remove the seed contents of the grain leaving the pericarp on the cob in contrast to deer and raccoons which remove the entire kernel (Fig. 2).

Birds annually destroy many millions of dollars worth of agricultural crops in North America. The greatest loss appears to be from blackbirds feeding on ripening corn; a survey in 1981 indicated a loss of 300,000 tons worth $31 million in the United States [5]. A follow-up survey in 1993 of all wildlife (bird and mammal) damage to ripening field corn indicated a loss of 860,000 tons, valued at $92 million, in the top 10 corn-producing states [118]. Blackbird damage to ripening sunflowers in the upper Great Plains states was estimated at $8 million in 1980 [61]. Damage by various bird species to fruit crops, peanuts, truck crops, and small grains also can be severe in localized areas [4]. For example, bird (primarily starling and robin) damage to blueberries in the United States and Canada was estimated at $8.5 million in 1989 [2].

Damage by ducks and sandhill cranes to swathed or maturing small-grain crops during autumn is a serious, localized problem in the northern Great Plains region [66]. Damage occurs from direct consumption of grain and from trampling, which dislodges kernels from heads [101]. Canada and snow geese grazing and compacting wheat and rye crops in winter can reduce subsequent grain and vegetative yields [24,65]. Canada geese also can be a serious problem to sprouting soybeans in spring and for fields of standing corn in autumn.

2. Mammals

Unlike birds, most wild mammals are secretive and not easily observed; many are nocturnal. Often the investigator must rely on various signs, such as tracks, trails, tooth marks, missing plant parts, scats (feces), or burrows to determine the species doing the damage. Traps may be necessary to identify positively small rodent species.

Characteristics of the damage may also provide clues to the species involved. In orchards, for example, major stripping of roots is usually caused by pine voles, whereas damage at the root collar or on the trunk up to the extent of snow depth is most often caused by meadow voles (Fig. 3). In sugar cane, various species of rats gnaw stalks so that they are hollowed out between the internodes but usually not completely severed. Rabbits, in contrast, usually gnaw through the stalks leaving only the ring-shaped internodes [68,69]. Unlike rodents and lagomorphs (i.e., rabbits), deer and other ungulates do not have upper incisors. Thus, twigs or plants nipped by these species may not show the neat, sharp-cut edge left by most rodents and lagomorphs but instead show a rough, shredded edge and usually a square or ragged break. Dolbeer et al. [45] presented a comprehensive description of damage characteristics by various mammal species.

Ungulate damage to various agricultural, forestry, and ornamental crops caused by feeding, trampling, and antler rubbing is an increasing problem [117], but objective estimates of economic loss are difficult to obtain. For example, losses in yield or tree value may accumulate for many years after damage occurs and vary with other stresses, including rodent damage, inflicted on the plants. In Ohio, growers reported average losses to deer in 1983 of $204/ha for orchards, $219/ha for Christmas tree plantings, and $268/ha in nursery plantings [92]. Fruit and tree losses apparently are in the millions of dollars annually in some U.S. states [6,20,27]. With regard to row crops, Hygnstrom and Craven [62] estimated a mean loss to deer of 2680 kg of corn per hectare for 51 unprotected cornfields in Wisconsin. Yield reductions in soybean fields are most severe when deer feeding occurs during the first week of sprouting [33].

(a)

Figure 2 Damage to corn by (a) blackbirds and (b) raccoons can sometimes be confused. Blackbirds usually slit the husk and peck out the soft contents of kernels leaving
the pericarp. Raccoons and squirrels chew through the husk and bite off the kernels.
(Photographs courtesy of R. A. Dolbeer.)

Assessments of damage caused by rodents and lagomorphs, although limited,
indicate that these mammals also cause tremendous annual losses of food and
fiber in the United States. Forest animal damage in Washington and Oregon,
primarily caused by rodents, was estimated to total $60 million annually to
Douglas fir and ponderosa pine plantings, and the potential reduction in the total
value of forest resources was estimated to be $1.83 billion [6,16]. Miller [79]

(b)

Figure 2 Continued

surveyed forest managers and natural resource agencies in 16 southeastern states and estimated annual wildlife-caused losses to timber and croplands, primarily caused by beavers, to be $11.2 million on 28.4 million hectares. An additional $1.6 million was spent to control wildlife damage on this land. Arner and Dubose [1] estimated that economic loss to beavers exceeded $4 billion over a 40-year period on 400,000 ha in the southeastern United States.

Rats cause substantial losses in sugar cane. Lefebvre et al. [68] estimated annual losses to be about $6 million ($235/ha) in one third of the area produc-

Figure 3 Meadow voles cause reduced apple production, and sometimes loss of trees, in orchards where they tunnel through snow and girdle trees by gnawing bark near the root collar and up the trunk as far as snow cover extends. (Photograph courtesy of M. E. Tobin.)

ing sugar cane in Florida. Hawaiian losses were reported to be in excess of $20 million per year [93]. Ferguson [55] estimated that in 1978, voles caused losses that approached $50 million to apple growers in the eastern United States. Losses of forage on range lands to rodents, rabbits, and hares are also extensive; however, accurate estimates of the monetary losses are difficult to obtain because of the nature of the damage and the wide area over which it occurs [72].

Mice can cause significant losses to corn seedlings in conservation tillage systems, but this damage may be offset by their consumption of harmful insects and weed seeds [19,64]. Ground squirrels, kangaroo rats, prairie dogs, nutrias, pocket gophers, and other rodents can also inflict serious damage to pastures, range lands, grain and bean fields, vegetable gardens, and fruit and nut crops. The burrows of some of these species can cause collapse of irrigation levees, increase erosion, and result in damage to farm machinery [74]. Carnivores such as raccoons and bears feeding on ripening corn [23,98] and coyotes on melons [54] can also cause substantial losses. Annual losses from these miscellaneous rodent and carnivore species are largely unquantified on regional scales but are likely to be in the $100s of millions nationally.

IV. DAMAGE CONTROL TECHNIQUES

A. Modifications of Habitat and Cultural Practices

Habitat and cultural modifications can be implemented in many situations to make roosting, loafing, or feeding sites less attractive to birds and provide long-lasting relief. For example, thinning marsh vegetation (cattails [*Typha* spp.]) can cause roosting blackbirds to move from agricultural areas where they are causing damage to sunflowers [70].

The use of lure crops, where waterfowl or blackbirds are encouraged to feed, is sometimes cost effective in reducing damage to nearby commercial fields of grain and sunflowers where bird-frightening programs are in place [29,100]. Provision of alternative foods (e.g., grain scattered on ground) may reduce loss of corn seedlings to rodents in no-tillage fields [64] and damage to apple trees in winter [102]. Davis [32] reported that pine vole damage in an apple orchard was reduced by mowing three times a year, clearing vegetation from under the trees, removing pruned branches, restricting the distribution of fertilizer, and inspecting and cleaning vulnerable parts of the orchard after harvest.

Bird-resistant cultivars of corn, sunflower, and sorghum have been effective in reducing damage. For example, cultivars of corn with ears having long, thick husks difficult for blackbirds to penetrate incur less damage than do cultivars with ears having short, thin husks [51,52]. Early-maturing cultivars of cherries are generally more susceptible to bird damage than late-maturing cultivars [104]. Planting grain crops so that they do not mature unusually early or late can also reduce damage by blackbirds [15]. Control of insects in corn fields can make those fields less attractive to blackbirds and reduce subsequent damage to the corn crop [42,115].

B. Proofing and Screening

Plastic netting is cost effective in excluding birds from individual fruit trees or high-value crops such as blueberries or grapes [56] (Fig. 4). Many different fence designs have been tested for excluding ungulates, including the 1.5-m Penn State Vertical Electric Deer Fence consisting of five strands of high-tensile steel wire (Fig. 5). Single-strand electric wire fences, 0.6–1.0 m high and baited with peanut butter to entice deer to contact the wire with their muzzles, have been cost effective in reducing damage in orchards and corn fields [62,87]. In England, wire netting and electrified netting fences have excluded rabbits from crop fields [75].

Figure 4 Nylon netting can be a cost-effective means of eliminating bird damage from high-value crops, such as in this vineyard on Long Island, New York. (Photograph courtesy of M. E. Tobin.)

Individual seedling protectors made of photodegradable plastics (e.g., VEXAR tubes; International Reforestation Supplers, Eugene, OR) are effective in reducing ungulate and rodent damage to young conifer trees [17,36]. In orchards, rabbit and aboveground rodent damage can be eliminated by wrapping trees with hardware cloth or burlap that is buried about 10 cm deep around the tree base [107].

C. Frightening Devices

Many devices are marketed, or homemade, to frighten birds and certain mammals such as deer. Target species usually habituate to such devices no matter how effective they may be initially. For example, deer habituated in less than 3 days to propane cannons used to protect piles of corn when the cannons were programmed to detonate systematically every 8–10 min. However, motion-activated cannons were effective for up to 6 weeks [3]. Two important rules are

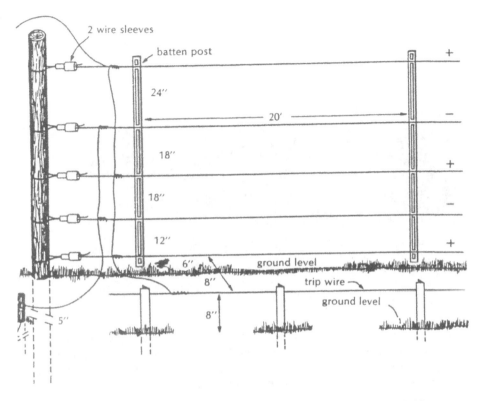

Figure 5 A 2-m high Penn State Vertical Electric Deer Fence. (From ref. 83.)

never to rely solely on one type of device for frightening and to vary the use of devices by altering the timing and location.

Probably the most widely used frightening device for birds and deer is the propane cannon (Fig. 6), which produces a loud explosion at timed intervals. Several models are marketed, including ones with automatic timers and rotating barrels. To be effective in frightening birds from crops, at least one cannon should be used for each 4 ha [30] and cannons should be moved every few days. An occasional shotgun patrol to reinforce the exploders is important for frightening birds [38], using either live ammunition or shell crackers. Shell crackers, fired from a 12-gauge shotgun, shoot a projectile that explodes 50–75 m away. Other pyrotechnic devices for frightening birds include rockets and whistle bombs [13].

Recorded alarm and distress calls of birds broadcast over a speaker system sometimes work well to frighten birds [12]. These calls are commercially avail-

Figure 6 Propane exploders are often used to frighten birds, especially blackbirds, from corn and other crops. For best results, exploders should be elevated above the vegetation, moved around periodically, and occasionally supplemented with a shotgun patrol or other frightening device. (Photograph courtesy of R. A. Dolbeer.)

able for many birds species [91]. Shooting at birds with a shotgun is often used to reinforce the distress calls.

Ultrasonic devices emitting sounds with frequencies above the level of human hearing (20,000 hz) sometimes are marketed for bird control. However, objective field tests have not demonstrated effectiveness of ultrasonic devices in repelling birds [114]. Most birds detect sounds in about the same range of frequencies as do humans.

Flags, helium-filled balloons with and without eyespots, and hawk-kites suspended from balloons or bamboo poles have been used with some success to repel birds from small agricultural fields (e.g., see ref. 21). Mylar flags, 15 cm × 1.5 m in size, are used to keep geese from winter wheat, corn, and alfalfa. Ten flags per 4 ha are recommended [59]. Reflecting tape made of Mylar, strung in parallel lines at 3- to 7-m intervals, has reduced blackbird numbers in small agricultural fields [50] (Fig. 7).

D. Repellents

Numerous odor and taste repellents have been developed to reduce deer and rodent browsing on ornamental plants, fruit trees, and crops. High cost and

Figure 7 Mylar reflecting tape strung above the vegetation can reduce blackbird feeding activity in agricultural fields. (Photograph courtesty of R. A. Dolbeer.)

variable effectiveness during the growing season generally make mammal repellents impractical for use on low-value row crops such as corn [62]. Repellents are most effective on trees and shrubs during the dormant season, but results are inconsistent. Even under optimal conditions, some damage occurs (e.g., see 22,26, and 84).

Birds generally have a poorer sense of smell and taste than do mammals, and repellents based on these senses are usually not effective. One exception may be methyl anthranilate, a taste repellent that has shown effectiveness with birds [31]. In contrast to taste repellents, chemicals that produce illness or adverse physiological response upon ingestion (i.e., conditioned aversion) appear to work well as bird repellents [89]. Methiocarb, a carbamate insecticide, is a condition-aversive repellent that has been effective as a powdered seed treatment for corn [60] and as a spray treatment for ripening cherries, blueberries, and grapes [43]. However, methiocarb is not presently registered for bird control in the United States. There is a need for registered bird repellents, especially to protect sprouting grain and ripening fruits.

E. Shooting and Trapping

Shooting can be effective in reducing local populations of depredating birds if only a few birds are involved. Shooting has little effect on large numbers of birds other than the repelling value [80]. This concept has been promoted in Wisconsin through a hunter referral program in which farmers allow goose hunters to shoot in agricultural fields experiencing chronic damage [58].

The effective use of the hunting season to reduce populations of deer in areas of high damage is one of the best ways to control damage [27]. Some states also have special depredation permits that can be issued to a landowner to remove a specific number of deer at a problem site outside the normal hunting season if sufficient control cannot be achieved during the hunting season.

Various live traps have been used in attempts to reduce populations of starlings near cherry orchards [7] and blackbirds in areas of ripening rice [78] and corn [109]. As with shooting, these methods have generally been successful only when local, small populations of birds are involved [86]. Various live and kill traps are used to capture rodents and crop-damaging predators such as raccoons [45]. Again, these methods can be effective in reducing localized damage by small populations of pests. One important point with live trapping is that euthanasia, not relocation, is the preferred method of disposing of captured animals. Relocated animals, besides having poor chances for long-term survival, may spread diseases or parasites and create additional problems at the relocation site or in attempts to return home. Many states have developed restrictive policies for relocation of vertebrate pests.

F. Toxicants

The use of toxic baits and chemicals to kill pest birds and mammals in agricultural situations presently is greatly restricted compared to historic practices. As one recent example, the wetting agent PA-14 was registered by the U.S. Department of Agriculture for killing blackbirds and starlings in upland roosts from 1974 to 1992. During this time, over 38 million birds were killed to reduce agricultural damage and public health concerns [41,47], but PA-14 is no longer registered or being used. When mixed with water and applied by aircraft or ground spray systems to roosting birds, PA-14 allowed water to penetrate the birds' feathers, cooling the birds so that they died of hypothermia [96].

Two toxic baits still being used on birds in limited agricultural situations are DRC-1339 (USDA, Pocatello, ID) and Avitrol (Avitrol Corp., Tulsa, OK). DRC-1339 is incorporated into poultry pellets and marketed as Starlicide Complete for killing starlings at feedlots and poultry yards. DRC-1339 is also used in limited situations to kill blackbirds damaging sprouting rice [57]. The active ingredient of Avitrol, 4-aminopyridine, when ingested in small doses, causes affected birds to emit distress calls while flying in erratic circles. Affected birds usually die within 0.5 h, but their initial behavior can act to frighten other birds away [38]. In agricultural situations, Avitrol is registered for use on starlings in feedlots and for blackbirds in corn and sunflower fields.

Numerous rodenticide formulations are registered for use in commensal rodent control, around farm buildings, and in noncrop areas, but few rodenticides are registered for in-crop use [67]. Zinc phosphide, one rodenticide with limited in-crop uses, is relatively safe to humans, and its use usually does not result in secondary poisoning of nontarget species. The efficacy of zinc phosphide is poor or inconsistent on some field rodents but often can be improved by prebaiting and proper bait placement [73,107]. Development of registrations for in-crop use of rodenticides, particularly anticoagulants, is a high priority area for research.

Fumigants, another class of toxicants used in rodent control in agriculture, produce gases that are lethal when inhaled. They are placed into individual burrow holes to kill various fossorial mammals, such as pocket gophers, commensal rodents, prairie dogs, ground squirrels, and woodchucks [44].

V. INTEGRATED PEST MANAGEMENT: BLACKBIRD DAMAGE TO CORN AS AN EXAMPLE

As noted earlier, blackbirds feeding on ripening corn cause the greatest agricultural losses from birds in North America. This widespread problem, involving highly mobile, migratory birds with esthetic and other beneficial attributes, is generally not resolvable by direct reduction of populations or exclusion of birds from crops. Thus, damage-reduction programs based on the ecological relationship of blackbirds and the corn crop provide an excellent example of the types of integrated management efforts needed to deal with complex vertebrate pest problems (see Fig. 1). For an example of an integrated management scheme for mammalian pests (rodents in orchards), see Tobin and Richmond [107].

A. Population Ecology of Red-Winged Blackbirds

Red-winged blackbirds (hereafter referred to as redwings), being well adapted to agricultural land uses, are the most abundant bird in North America [47] and the principal avian pest of corn and other grain crops. The continental population, perhaps 165 million at the start of the nesting season in April, doubles to over 350 million birds by July when most young have fledged [46,77]. In July, redwings concentrate in nocturnal roosts containing up to several million birds, usually in marshes within 200 km of their nesting localities [40]. During the day, these birds forage, generally within 10 km of the roosts. The additional energy demands from feather molting [113] coincide with peak numbers of birds at the time the corn crop is ripening, setting the stage for economically significant damage to some corn fields (i.e., >5% yield loss) within 10 km of these late-summer roosts [10,38]. During autumn, after the corn is harvested, redwings and other blackbird species migrate to the southern United States [40]. Residue corn in harvested fields is an important food for redwings from winter [112] to early summer [18].

Environmentally safe reduction of redwing populations causing late-summer damage to corn and other crops has generally proven not to be feasible, both from a population dynamics perspective [41,46] and in practice [108]. Given the localized nature of the damage, the adaptations of redwings to the agricultural environment, and the beneficial attributes and legal status of these birds, population reduction as a routine management action, even if feasible, might not be desirable anyway. For these reasons, most research has focused on nonlethal approaches to reducing damage.

B. Managing Crops and Wetland Vegetation at Roost Sites

Because redwing damage to corn is generally a problem only within a few kilometers of late-summer roost sites, the first strategy to reduce damage is to plant alternate crops not fed upon by blackbirds, such as soybeans, in these high-damage areas. A second strategy is to disperse the large concentrations of birds by thinning the dense, wetland vegetation where the birds roost at night. Dispersing the large concentrations of birds from a single roost in an area of intensive agriculture into smaller concentrations in widely dispersed roosts may reduce overall damage [70].

C. Blackbird Feeding in Relation to Corn Maturation

If a decision is made to grow corn within a few kilometers of a late-summer blackbird roost, an understanding of redwing feeding habits is critical to effective damage management. Corn ear development begins about 60 days after planting when the pistillate flowers, with elongated styles commonly referred to as silks, form in leaf axils. About 15 days after silking (DAS), ears have developed to the point where kernels are in the milk stage and are first vulnerable to damage by birds. Sweet corn ears are usually harvested from 16–20 DAS, when the immature kernels still contain 70–80% water, so this crop is vulnerable to bird damage for only a few days. Field corn is vulnerable to loss for several more weeks as the kernels, accumulating biomass, go through the dough and dent stages of development before harvest (Fig. 8).

Redwings are commonly observed in cornfields near roosting sites during the 2 weeks between silking and initial kernel development. No ripening corn is available in the fields for consumption at this time but insects, such as corn rootworm beetles (*Diabrotica* spp.) and the European corn borer (*Ostrinia nubilalis*), are often abundant. Several studies confirmed that redwings are initially attracted to corn fields in the silking period to feed on insects, and that

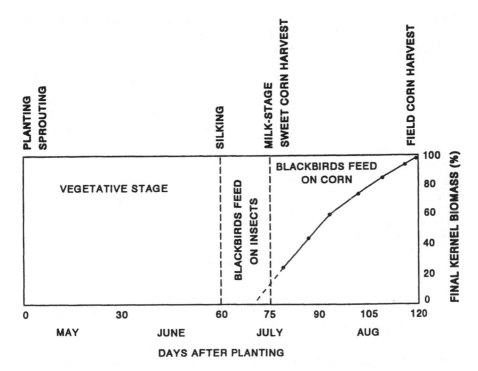

Figure 8 Chronology of typical sweet or field corn planting in midwestern United States showing feeding patterns of blackbirds and rate of biomass accumulation in corn kernels. (From ref. 116.)

control of insects with insecticides can subsequently reduce bird damage to ripening kernels [9,81,94,99,115]. Redwing feeding on beetles during silking in corn fields near roosts may be sufficient to offset some of the subsequent damage caused by feeding on kernels [11].

At the end of the silking period when kernels enter the milk stage, feeding by redwings on insects declines sharply as the birds switch to corn (Table 2). Most damage (kernels eaten) occurs during the milk and early dough stage of kernel development (16–30 DAS) [8,14,53] when kernels have little biomass. For example, kernels contain only about 25% of their final biomass at 20 DAS compared with 70% at 40 DAS (see Fig. 8). Thus, a flock of birds in a field must damage about three times as many kernels at 20 DAS as at 40 DAS to obtain the same corn biomass.

One obvious management implication for sweet corn is that the timing of harvest can have a dramatic influence on the level of bird damage. Damage can increase rapidly once kernels enter the vulnerable milk stage. Advancement of harvest by only 1 day can result in substantially less damage [38], especially if

Table 2 Food (aggregate % volume of gut contents) of Red-Winged Blackbirds Collected in Corn Fields During the Predamage Period and in Corn Fields (New York) or at a Nighttime Roost (Ohio) During the Damage Period

	Predamage (silking) period		Damage (kernel maturation) period	
Food	New York[a] (n = 21)	Ohio[b] (n = 17)	New York[a] (n = 84)	Ohio[b] (n = 66)
Rootworm beetles	59	33	4	<1
Other arthropods	8	35	6	6
Corn	19[c]	23[c]	89	88
Weed seeds	0	6	0	5
Other	14	3	1	1

[a]Data from Bollinger and Caslick [9].
[b]Data from Okurut-Akol [81].
[c]Mature kernels gleaned off of ground from previous year's harvest.

combined with insect control in the silking period to reduce the initial attractiveness of sweet corn fields to birds.

For field corn where damage occurs over a longer period of kernel development, the adjustment of harvest date is less effective in reducing damage. Nonetheless, protective measures to reduce damage (to be discussed below) are more critical in the early stages of kernel development (i.e., 15–30 DAS) than later, because the damage potential is much higher at this time. In addition, bird damage during the early maturational stages is more likely to lead to secondary damage (fungal, insect, sprouting) of kernels than is damage at later stages [116]. Thus, bird numbers being equal, a farmer will obtain a much greater return from his or her efforts by protecting field corn from damaging flocks during the milk stage of maturity than during later stages, Cummings et al. [28] noted a similar pattern of blackbird damage in ripening sunflowers.

D. Cultivar Resistance to Reduce Damage

An ideal resolution of the conflict between corn farmers and redwings would be to couple the beneficial feeding habits of the birds in silking-stage corn, as outlined above, with management techniques to reduce damage once corn enters the milk stage. One means of reducing damage is through the use of cultivars resistant to attack by birds. Aviary and field testing has established that sweet and field corn cultivars with long, heavy husks receive less bird damage than short, thin-husked cultivars [51,52]. In addition, cultivars with an extended period of vulnerability due to variation in maturity among plants are likely to have greater losses to birds than cultivars with synchronized, rapid ear development among plants [110]. No cultivars tested have been completely resistant to damage. However, certain cultivars do have consistently less damage than others, even in no-choice regimens in aviary tests that are more severe than field situations, where birds would always have alternative food sources such as weed seeds and insects. As a corollary, management of the habitat within the birds' foraging range surrounding roosts to provide alternative feeding areas should enhance the effectiveness of resistant cultivars. This will be discussed more fully below.

E. Bird-Frightening Techniques to Reduce Damage

Conventional bird-frightening techniques, such as propane exploiters, hawk-kites and other effigies, shooting, reflective ribbons, and Avitrol can all be of assistance in reducing redwing damage to corn [21,30,38]. The deployment of these devices when corn is most vulnerable to damage, as discussed above, is critical to their success. Their effectiveness should also be enhanced when used in fields of resistant cultivars. One caveat in the use of frightening techniques is that labor and material costs for their proper deployment are often relatively high, and careful assessment of expected benefits in relation to costs is needed before deployment [39].

F. Alternative Foods

The final component of an integrated management program involves the provision of alternative foods outside the corn fields but within the birds' daily foraging range from the roost. The basic tenet is that if the techniques discussed above are to be successful in discouraging birds from feeding in cornfields once kernels enter the milk stage, alternative feeding sites must be available. This hypothesis is supported by a study where blackbird damage to corn was highest in areas of least agricultural diversity [97].

There are numerous opportunities for providing alternative feeding areas in late summer. For example, oats and wheat are typically harvested at the time corn enters the silking stage. Delayed plowing of the stubble until after the corn harvest permits redwings to feed on the waste grains [76], reducing feeding pressure on corn. Delayed plowing of early-harvested sweet corn fields can also serve this purpose. The provision of natural and planted plots of wildlife food crops on wildlife refuges and on areas of farms set aside in price-support programs should also be beneficial. Such "lure crops" have been used for years to reduce waterfowl damage to maturing small-grain crops [66] and have shown promise for reducing redwing damage to nearby commercial sunflower fields [29].

VI. CONCLUSIONS

Vertebrates will be an increasingly important component of pest management in agricultural crops as long as successful wildlife conservation programs are continued among agricultural lands. Much work needs to be done, especially in refining our abilities to estimate costs and benefits associated with vertebrate species that damage agricultural crops and in further refinements of control strategies. However, important steps have been taken in achieving this goal for many species and the future for resolving conflicts appears bright.

Progress must continue in finding ways to accommodate wildlife as a part of agricultural ecosystems so that people can enjoy the positive attributes while simultaneously managing the pest species and crops to reduce damage to tolerable levels. Management techniques include improved abilities to predict when, where, and how much loss will occur so that control is undertaken only when necessary; selective population reductions through hunting and other means; cultural practices such as damage-resistant crops, timing of harvest, and manipulation of attractants; the use of physical barriers and repellents; proper timing in the deployment of frightening devices; and provision of alternate feeding areas. None of these techniques used alone will be sufficient; success will be achieved when they are used together in coordinated programs built upon a foundation of knowledge about the relationship of the vertebrate pest and agricultural crop.

ACKNOWLEDGMENTS

I thank J.L. Cummings, E.P. Hill, and M.E. Tobin for reviewing early drafts of the manuscript.

REFERENCES

1. Arner, D.H., & J.S. Dubose. 1982. The impact of the beaver on the environment and economics in the southeastern United States. *Trans. Int. Congr. Game Biol.* 14:241–247.
2. Avery, M.L., J.W. Nelson, M.A. Cone. 1992. Survey of bird damage to blueberries in North America. *Proc. East. Wildl. Damage Control Conf.* 5:105–110.
3. Belant, J.L., T.W. Seamans, & C.P. Dwyer. 1996. Evaluation of propane exploders as white-tailed deer deterrents. *Crop Prot.* 15:575–578.
4. Besser, J.F. 1986. A guide to aid growers in reducing bird damage to U.S. agricultural crops. *USDA APHIS, Denver Wildl. Res. Cent. Bird Damage Res. Rep.* 377.
5. Besser, J.F., & D.J. Brady. 1986. Bird damage to ripening field corn increases

in the United States from 1971 to 1981. *U.S. Fish Wildl. Serv., Fish Wild. Leafl.* 7.

6. Black, H.C., E.J. Dimock, II, J. Evans, & J.A. Rochelle. 1979. Animal damage to coniferous plantations in Oregon and Washington. Part I. A survey, 1963–75. *Oregon State Univ. For. Res. Lab. Res. Bull.* 25.

7. Bogatich, V. 1967. The use of live traps to remove starlings and protect agricultural products in the state of Washington. *Proc. Vertebr. Pest Conf.* 3:98–99.

8. Bollinger, E.K. 1983. Phenology of red-winged blackbird use of field corn in central New York. *Proc. East. Wildl. Damage Control Conf.* 1:265–270.

9. Bollinger, E.K., & J.W. Caslick. 1985. Red-winged blackbird predation on northern corn rootworm beetles in field corn. *J. Appl. Ecol.* 22:39–48.

10. Bollinger, E.K., & J.W. Caslick. 1985. Factors influencing blackbird damage to field corn. *J. Wildl. Manage.* 49:1109–1115.

11. Bollinger, E.K., & J.W. Caslick. 1985. Northern corn rootworm beetle densities near a red-winged blackbird roost. *Can. J. Zool.* 63:502–505.

12. Bomford, M., & P.H. O'Brien. 1990. Sonic deterrents in animal damage control: a review of device tests and effectiveness. *Wildl. Soc. Bull.* 18:411–422.

13. Booth, T.W. 1994. Bird dispersal techniques. In *Prevention and Control of Wildlife Damage.* S. E. Hygnstrom, R. M. Timm & G. E. Larson, eds. pp E19-E24. Lincoln, Nebraska: University of Nebraska Cooperative Extension Service.

14. Bridgeland, W.T. 1979. Timing bird control applications in ripening corn. *Proc. Bird Control Semin.* 8:222–228.

15. Bridgeland, W.T., & J.W. Caslick. 1983. Relationships between cornfield characteristics and blackbird damage. *J. Wildl. Manage.* 47:824–829.

16. Brodie, J.D., H.C. Black, E.J. Dimock, II, J. Evans, C. Kao, & J.A. Rochelle. 1979. Animal damage to coniferous plantations in Oregon and Washington - Part II. An economic evaluation. *Oregon State Univ. For. Res. Lab. Res. Bull.* 26.

17. Campbell, D.L., & J. Evans. 1975. "Vexar" seedling protectors to reduce wildlife damage to Douglas fir. *U.S. Fish Wildl. Serv. Leafl.* 508.

18. Clark, R.G., P.J. Weatherhead, H. Greenwood, & R.D. Titman. 1986. Numerical responses of red-winged blackbird populations to changes in regional land-use patterns. Can. J. Zool. 64:1944–1950.

19. Clark, W.R., & R.E. Young. 1986. Crop damage by small mammals in no-till cornfields. *J. Soil Water Conserv.* 41:338–341.

20. Connelly, N.A., D.J. Decker, & S. Wear. 1987. Public tolerance of deer in a suburban environment:implications for management and control. *Proc. East. Wildl. Damage Control Conf.* 3:207–218.

21. Conover, M.R. 1984. Comparative effectiveness of Avitrol, exploders, and hawk-kites in reducing blackbird damage to corn. *J. Wildl. Manage.* 48:109–116.

22. Conover, M.R. 1984. Effectiveness of repellents in reducing deer damage in nurseries. *Wildl. Soc. Bull.* 12:399–404.

23. Conover, M.R. 1987. Reducing raccoon and bird damage to small corn plots. *Wildl. Soc. Bull.* 15:268–272.

24. Conover, M.R. 1988. Effect of grazing by Canada geese on the winter growth of rye. *J. Wildl. Manage.* 52:76–80.

25. Conover, M.R. & D.J. Decker. 1991. Wildlife damage to crops: perceptions of

agricultural and wildlife professionals in 1957 and 1987. *Wildl. Soc. Bull.* 19:46–52.

26. Conover, M.R., & G.S. Kania. 1987. Effectiveness of human hair, BGR, and a mixture of blood meal and peppercorns in reducing deer damage to young apple trees. *Proc. East. Wildl. Damage Control Conf.* 3:97–101.

27. Craven, S.R. 1983. New directions in deer damage management in Wisconsin. *Proc. East. Wildl. Damage Control Conf.* 1:65–67.

28. Cummings, J.L., J.L. Guarino, & C.E. Knittle. 1989. Chronology of blackbird damage to sunflowers. *Wildl. Soc. Bull.* 17:50–52.

29. Cummings, J.L., J.L. Guarino, C.E. Knittle, & W.C. Royall, Jr. 1987. Decoy plantings for reducing blackbird damage to nearby commercial sunflower fields. *Crop Prot.* 6:56–60.

30. Cummings, J.L., C.E. Knittle, & J.L. Guarino. 1986. Evaluating a pop-up scarecrow coupled with a propane exploder for reducing blackbird damage to ripening sunflower. *Proc. Vertebr. Pest Conf.* 12:286–291.

31. Cummings, J.L., P.A. Pochop, J.E. Davis, Jr., & H.W. Krupa. 1995. Evaluation of Rejex-it AG-36 as a Canada goose grazing repellent. *J. Wildl. Manage.* 59:47–50.

32. Davis, D.E. 1976. Management of pine voles. *Proc. Vertebr. Pest Conf.* 7:270–275.

33. DeCalesta, D.S., & D.B. Schwendeman. 1978. Characterization of deer damage to soybean plants. *Wildl. Soc. Bull.* 6:250–253.

34. Dehaven, R.W., & R.L. Hothem. 1979. Procedure for visually estimating bird damage to grapes. In *American Society for Testing and Materials, Special Technical Publication* 680. J.R. Beck, ed. Philadelphia: Vertebrate Pest Control and Management Materials, pp 169–177.

35. DeGrazio, J.W., J.F. Besser, J.L. Guarino, C.M. Loveless, & J.L. Oldemeyer. 1969. A method for appraising blackbird damage to corn. *J. Wildl. Manage.* 33:988–994.

36. DeYoe, D.R. & W. Schaap. 1983. Comparison of 8 physical barriers used for protecting Douglas-fir seedlings from deer browse. *Proc. East. Wildl. Damage Control Conf.* 1:77–93.

37. Dolbeer, R.A. 1975. A comparison of two methods for estimating bird damage to sunflowers. *J. Wildl. Manage.* 39:802–806.

38. Dolbeer, R.A. 1980. Blackbirds and corn in Ohio. *U.S. Fish Wildl. Serv. Resour. Publ.* 136.

39. Dolbeer, R.A. 1981. Cost-benefit determination of blackbird damage control for cornfields. *Wild. Soc. Bull.* 9:44–51.

40. Dolbeer, R.A. 1982. Migration patterns for age and sex classes of blackbirds and starlings. *J. Field Ornithol.* 53:28–46.

41. Dolbeer, R.A. 1986. Current status and potential of lethal means of reducing bird damage in agriculture. *Proc. Int. Ornithol. Congr.* 19:474–483.

42. Dolbeer, R.A. 1990. Ornithology and integrated pest management: red-winged blackbirds *Agelaius phoeniceus* and corn. *Ibis* 132:309–322.

43. Dolbeer, R.A., M.L. Avery, & M.E. Tobin. 1994. Assessment of field hazards to birds from methiocarb applications to fruit crops. *Pest. Sci.* 40:147–161.

44. Dolbeer, R.A., G.E. Bernhardt, T.W. Seamans, & P.P. Woronecki. 1991. Efficacy of two gas cartridge formulations in killing woodchucks in burrows. *Wildl. Soc. Bull.* 19:200–204.

45. Dolbeer, R.A., N.R. Holler, & D.W. Hawthorne. 1994. Identification and control of wildlife damage. In *Research and Management Techniques for Wildlife Habitats.* T.A. Bookhout, ed. Bethesda, Maryland: The Wildlife Society, pp 474–506.

46. Dolbeer, R.A., C.R. Ingram, & J.L. Seubert. 1976. Modeling as a management tool for assessing the impact of blackbird control. *Proc. Vertebr. Pest Conf.* 7:35–45.

47. Dolbeer, R.A., D.F. Mott, & J.L. Belant. 1997. Blackbirds and starlings killed at winter roosts from PA-14 applications: Implications for regional population management. *Proc. East. Wildl. Damage Manage. Conf.* 7:77–86.

48. Dolbeer, R.A., & R.A. Stehn. 1983. Population status of blackbirds and starlings in North America, 1966-81. *Proc. East. Wildl. Damage Control Conf.* 1:51–61.

49. Dolbeer, R.A., A.R. Stickley, Jr., & P.P. Woronecki. 1979. Starling (*Sturnus vulgaris*) damage to sprouting wheat in Tennessee and Kentucky, U.S.A. *Prot. Ecol.* 1:159–169.

50. Dolbeer, R.A., P.P. Woronecki, & R.L. Bruggers. 1986. Reflecting tapes repel blackbirds from millet, sunflowers, and sweet corn. *Wildl. Soc. Bull.* 14:418–425.

51. Dolbeer, R.A., P.P. Woronecki, & J.R. Mason. 1988. Aviary and field evaluations of sweet corn resistance to damage by blackbirds. *J. Am. Soc. Hort. Sci.* 113:460–464.

52. Dolbeer, R.A., P.P. Woronecki, & T.W. Seamans. 1995. Ranking and evaluation of field corn hybrids for resistance to blackbird damage. *Crop Prot.* 14:399–403.

53. Dolbeer, R.A., P.P. Woronecki, & R.A. Stehn. 1984. Blackbird (*Agelaius phoeniceus*) damage to maize: crop phenology and hybrid resistance. *Prot. Ecol.* 7:43–63.

54. Emerson, S. 1994. Wildlife vs. growers: who's winning the battle? *Citrus & Vegetable Mag.* Oct:13–19.

55. Ferguson, W.L. 1980. Rodenticide use in apple orchards. *Proc. East. Pine and Meadow Vole Symp.* 4:2–8.

56. Fuller-Perrine, L.D., & M.E. Tobin. 1993. A method for applying and removing bird-exclusion netting in commercial vineyards. *Wildl. Soc. Bull.* 21:47–51.

57. Glahn, J.F., & E.A. Wilson. 1992. Effectiveness of DRC-1339 baiting for reducing blackbird damage to sprouting rice. *Proc. East. Wildl. Damage Control Conf.* 5:117–123.

58. Heinrich, J., & S. Craven. 1987. Distribution and impact of Canada goose crop damage in east-central Wisconsin. *Proc. East. Wildl. Damage Control Conf.* 3:18–19.

59. Heinrich, J. & S. Craven. 1990. Evaluation of three damage abatement techniques for Canada geese. *Wildl. Soc. Bull.* 18:405–410.

60. Heisterberg, J.F. 1983. Bird repellent seed corn treatment: efficacy evaluations

and current registration status. *Proc. East. Wildl. Damage Control Conf.* 1:255–258.

61. Hothem, R.L., R.W. DeHaven, & S.D. Fairaizl. 1988. Bird damage to sunflower in North Dakota, South Dakota, and Minnesota, 1979–1981. *U.S. Fish Wildl. Tech. Rep.* 15.

62. Hygnstrom, S.E., & S.R. Craven. 1988. Electric fences and commercial repellents for reducing deer damage in cornfields. *Wildl. Soc. Bull.* 16:291–296.

63. Jacobs, W.W. 1994. Registered vertebrate pesticides. In *Prevention and Control of Wildlife Damage.* S.E. Hygnstrom, R.M. Timm, & G. E. Larson, eds. Lincoln, Nebraska: University of Nebraska Cooperative Extension Service, pp G1–G22.

64. Johnson, R.J. 1986. Wildlife damage in conservation tillage agriculture: a new challenge. *Proc. Vertebr. Pest Conf.* 12:127–132.

65. Kahl, R.B., & F.B. Samson. 1984. Factors affecting yield of winter wheat grazed by geese. *Wildl. Soc. Bull.* 12:256–262.

66. Knittle, C.E., & R.D. Porter. 1988. Waterfowl damage and control methods in ripening grain: an overview. *U.S. Fish Wildl. Serv. Tech. Rep.* 14.

67. Lefebvre, L.W., N.R. Holler, & D.G. Decker. 1985. Comparative effectiveness of full-field and field-edge bait applications in delivering bait to roof rats in Florida sugarcane fields. *J. Am. Soc. Sugar Cane Tech.* 5:64–68.

68. Lefebvre, L.W., C.R. Ingram, & M.C. Yang. 1978. Assessment of rat damage to Florida sugarcane in 1975. *Proc. Am. Soc. Sugar Cane Tech.* 7:75–80.

69. Lindsey, G.D. 1969. Characteristics of sugarcane damage caused by rodents in Hawaii. *Sugar J.* 39:22–24.

70. Linz, G.M., R.A. Dolbeer, J.J. Hanzel, & L.E. Huffman. 1993. Controlling blackbird damage to sunflower and grain crops in the northern Great Plains. *USDA, APHIS, Ag. Inform. Bull.* 679.

71. Luce, D.G., R.M. Case, & J.L. Stubbendieck. 1981. Damage to alfalfa fields by Plains pocket gophers. *J. Wildl. Manage.* 45:258–260.

72. Marsh, R.E. 1985. Competition of rodents and other small mammals with livestock in the United States. In *Parasites, Pests and Predators.* S.M. Gaafar, W.E. Howard, & R.E. Marsh, eds. Amsterdam, Elsevier, pp 485–508.

73. Marsh, R.E. 1988. Relevant characteristics of zinc phosphide as a rodenticide. *Proc. Great Plains Wildl. Damage Control Conf.* 8:70–74.

74. Marsh, R.E., & W.E. Howard. 1982. Vertebrate pests. In *Handbook of Pest Control.* 6th ed. A. Mallis, ed. Cleveland: Franzak and Foster, pp 791–861.

75. McKillop, I.G., & C.J. Wilson. 1987. Effectiveness of fences to exclude European rabbits from crops. *Wildl. Soc. Bull.* 15:394–401.

76. McNicol, D.K., R.J. Robertson, & P.J. Weatherhead. 1982. Seasonal, habitat, and sex-specific food habits of red-winged blackbirds: implications for agriculture. *Can. J. Zool.* 60:3282–3289.

77. Meanley, B., & W.C. Royall, Jr. 1976. Nationwide estimates of blackbirds and starlings. *Proc. Bird Control Semin.* 7:39–40.

78. Meanley, B. 1971. Blackbirds and the southern rice crop. *U.S. Fish and Wildl. Serv. Resource Publ.* 100.

79. Miller, J.E. 1987. Assessment of wildlife damage on southern forests. In *Proceedings of the Management of Southern Forests for Wildlife and Fish*. J.G. Dickinson & D.E. Maughan, eds. U.S. For. Serv. Gen. Tech. Rep. SO-65, pp 48–52.

80. Murton, R.K., N.J. Westwood, & A.J. Isaacson. 1974. A study of wood-pigeon shooting: the exploitation of a natural animal population. *J. Appl. Ecol.* 11:61–81.

81. Okurut-Akol, F.H. 1989. Relations among blackbird corn damage, pest insects and insecticide applications. MS thesis, Colorado State University, Fort Collins.

82. Otis, D.L., N.R. Holler, P.W. Lefebvre, & D.F. Mott. 1983. Estimating bird damage to sprouting rice. In *American Society for Testing and Materials, Special Technical Report* 817. D.E. Kaukeinen, ed. Philadelphia: Vertebrate Pest Control and Management Materials, pp 76–89.

83. Palmer, W.L., J.M. Payne, R.G. Wingard, & J.L. George. 1985. A practical fence to reduce deer damage. *Wildl. Soc. Bull.* 13:240–245.

84. Palmer, W.L., R.G. Wingard, & J.L. George. 1983. Evaluation of white-tailed deer repellents. *Wildl. Soc. Bull.* 11:164–166.

85. Pearson, K., & C.G. Forshey. 1978. Effects of pine vole damage on tree vigor and fruit yield in New York orchards. *Hort. Sci.* 13:56–57.

86. Plesser, H., S. Omasi, & Y. Yom-tov. 1983. Mist nets as a means of eliminating bird damage to vineyards. *Crop Prot.* 2:503–506.

87. Porter, W.F. 1983. A baited electric fence for controlling deer damage to orchard seedlings. *Wildl. Soc. Bull.* 11:325–327.

88. Richmond, M.E., C.G. Forshey, L.A. Mahaffy & P.N. Miller. 1987. Effects of differential pine vole populations on growth and yield of McIntosh apple trees. *Proc. East. Wildl. Damage Control Conf.* 3:296–304.

89. Rogers, J.G., Jr. 1974. Responses of caged red-winged blackbirds to two types of repellents. *J. Wildl. Manage.* 38:418–423.

90. Sauer, W.C. 1977. Exclusion cylinders as a means of assessing losses of vegetation due to ground squirrel feeding. In *American Society for Testing and Materials, Special Technical Report* 625. W.B. Jackson & R.E. Marsh, eds. Philadelphia: Vertebrate Pest Control and Management Materials, pp 14–21.

91. Schmidt, R.H., & R.J. Johnson. 1983. Bird dispersal recordings: an overview. In *American Society for Testing and Materials, Special Technical Report* 817. D.E. Kaukeinen, ed. Philadelphia: Vertebrate Pest Control and Management Materials, pp 43–65.

92. Scott, J.D., & T.W. Townsend. 1985. Characteristics of deer damage to commercial tree industries of Ohio. *Wildl. Soc. Bull.* 13:135–143.

93. Seubert, J.L. 1984. Research on nonpredatory mammal damage control by the U.S. Fish and Wildlife Service. In *Organization and Practice of Vertebrate Pest Control*. A.C. Dubbock, ed. Surrey, UK: Imperial Chemical Industries PLC, pp 553–571.

94. Stickley, A.R., Jr., & C.R. Ingram. 1976. Methiocarb as a bird repellent for mature sweet corn. *Proc. Bird Control Semin.* 7:228–238.

95. Stickley, A.R., Jr., D.L. Otis & D.T. Palmer. 1979. Evaluation and results of a

survey of blackbird and mammal damage to mature field corn over a large (three-state) area. In *American Society for Testing and Materials, Special Technical Publication* 680. J.R. Beck, ed. Philadelphia: Vertebrate Pest Control and Management Materials, pp 169–177.

96. Stickley, A.R., Jr., D.J. Twedt, J.F. Heisterberg, D.F. Mott, & J.F. Glahn. 1986. Surfactant spray system for controlling blackbirds and starlings in urban roosts. *Wildl. Soc. Bull.* 14:412–418.

97. Stone, C.P. & C.R. Danner. 1980. Autumn flocking of red-winged blackbirds in relation to agricultural variables. *Am. Midl. Nat.* 103:196–199.

98. Stowell, L.R., & R.C. Willging. 1992. Bear damage to agriculture in Wisconsin. *Proc. East. Wildl. Damage Control Conf.* 5:96–104.

99. Straub, R.W. 1989. Red-winged blackbird damage to sweet corn in relation to infestations of European corn borer (Lepidoptera: Pyralidae). *J. Econ. Entomol.* 82:1406–1410.

100. Sugden, L.G. 1976. Waterfowl damage to Canadian grain. *Can. Wildl. Serv. Occas. Pap.* 24.

101. Sugden, L.G., & D.W. Goerzen. 1979. Preliminary measurements of grain wasted by field-feeding mallards. *Can. Wildl. Serv. Prog. Notes* 104.

102. Sullivan, T.P., & D.S. Sullivan. 1988. Influence of alternative foods on vole populations and damage in apple orchards. *Wildl. Soc. Bull.* 16:170–175.

103. Tobin, M.E., & R.A. Dolbeer. 1987. Status of Mesurol as a bird repellent for cherries and other fruit crops. *Proc. East. Wildl. Damage Control Conf.* 3:149–158.

104. Tobin, M.E., R.A. Dolbeer, C.M. Webster, & T.W. Seamans. 1991. Cultivar differences in bird damage to cherries. *Wildl. Soc. Bull.* 19:190–194.

105. Tobin, M.E., R.A. Dolbeer, & P.P. Woronecki. 1989. Bird damage to apples in the Mid-Hudson Valley of New York. *Hort. Sci.* 24:859.

106. Tobin, M.E., A.E. Koehler, R.T. Sugihara, G.R. Ueunten, & A.M. Yamaguchi. 1993. Effects of trapping on rat populations and subsequent damage and yields of macadamia nuts. *Crop Prot.* 12:243–248.

107. Tobin, M.E., & M.E. Richmond. 1993. Vole management in fruit orchards. *U.S. Fish Wildl. Serv., Biol. Rep.* 5.

108. Weatherhead, P.J. 1982. Assessment, understanding and management of blackbird-agriculture interactions in eastern Canada. *Proc. Vertebr. Pest Conf.* 10:193–196.

109. Weatherhead, P.J., H. Greenwood, S.H. Tinker, & J.R. Bider. 1980. Decoy traps and the control of blackbird populations. *Phytoprotection* 61:65–71.

110. Weatherhead, P.J. & S. Tinker. 1983. Maize ear characteristics affecting vulnerability to damage by red-winged blackbirds. *Prot. Ecol.* 5:167–175.

111. Weatherhead, P.J., S. Tinker, & H. Greenwood. 1982. Indirect assessment of avian damage to agriculture. *J. Appl. Ecol.* 19:773–782.

112. White, S.B., R.A. Dolbeer, & T.A. Bookhout. 1985. Ecology, bioenergetics, and agricultural impacts of a winter-roosting population of blackbirds and starlings. *Wildl. Monogr.* 93.

113. Wiens, J.A, & M.I. Dyer. 1975. Simulation modelling of red-winged blackbird impact on grain crops. *J. Appl. Ecol.* 12:63–82.

114. Woronecki, P.P. 1988. Effect of ultrasonic, visual, and sonic devices on pigeon numbers in a vacant building. *Proc. Vertebr. Pest Conf.* 13:266–272.
115. Woronecki, P.P., R.A. Dolbeer, & R.A. Stehn. 1981. Response of blackbirds to Mesurol and Sevin applications on sweet corn. *J. Wildl. Manage.* 45:693–701.
116. Woronecki, P.P., R.A. Stehn, & R.A. Dolbeer. 1980. Compensatory response of maturing corn kernels following simulated damage by birds. *J. Appl. Ecol.* 17:737–746.
117. Wywialowski, A.P. 1994. Agricultural producers' perceptions of wildlife-caused losses. *Wildl. Soc. Bull.* 22:370–382.
118. Wywialowski, A.P. 1996. Wildlife damage to field corn in 1993. *Wildl. Soc. Bull.* 24:264–271.

25
Impact of Tillage Systems on Pest Management

Ronald B. Hammond and Benjamin R. Stinner
The Ohio State University, Wooster, Ohio

I. TILLAGE SYSTEMS

Historically, some form of soil disturbance prior to planting crops has been considered a standard procedure. Conventional tillage, typically performed with a moldboard plow, inverts and mixes soil profiles down to 30 cm. This practice loosens the soil, incorporates crop residues, animal manures and fertilizers, destroys weeds, and creates a nearly bare soil surface for planting a crop. Although conventional or deep-tillage agriculture does create an excellent seedbed, this practice has been criticized for at least the past 50 years as a major causative factor in soil erosion [4,23,26].

With the introduction of herbicides for weed suppression in the 1950s, and further motivated by concerns over soil erosion, researchers began to develop planting systems that relied less and less on major soil disturbance and left crop residues remaining on the soil surface [82]. Approximately 25 years ago, farmers began using reduced or conservation tillage, initially mostly for corn (*Zea mays*) and soybean (*Glycine max*) row crop production. Although the adoption of reduced tillage was slow at first, there were dramatic increases during the late 1970s and early 1980s because of increased concerns over soil erosion and savings in time and energy accrued by not having to plow [26]. More recently, the use of conservation tillage has increased dramatically in some areas to the extent that surveys are reporting up to 50% of all row crops are in some form of reduced tillage in the United States [14]. Adoption of conservation tillage is increasing on a worldwide basis as well. Therefore, much of our discussion in this chapter is centered on the impacts and implications of conservation tillage for pest management.

During the past 20 years, a great deal of effort has focused on developing a diversity of reduced-tillage systems. For the intent of this chapter, we use the following terminology: *conventional tillage*, as indicated above, generally describes moldboard or a type of tillage which inverts the upper soil profile and leaves virtually no crop residues on the soil surface. *Reduced*, or *minimum*, *tillage* encompasses a range of mechanical means using chisels, sweeps, and disks for the purpose of providing adequate seedbed and controlling early season weeds. There are many variations on the minimum tillage theme almost too numerous to mention; however, ridge tillage is one of the more interesting concepts, where crops are planted in permanently maintained ridges and crop residues are concentrated in the valleys between ridges. *No-tillage* is conservation tillage in the extreme where absolutely no physical soil disturbance is done prior to planting, and specially designed equipment is used to obtain soil-seed contact. With no-tillage farming, all plant residues remain on the soil surface. The term *conservation tillage* spans both minimum and no-tillage systems and is officially defined by the Soil Conservation Service as resulting in at least 30% of previous crop residues remaining on the soil surface to ameliorate soil erosion [14].

II. CHANGES IN ENVIRONMENTAL FACTORS

Lack of tillage brings about various changes in environmental factors that can modify soil properties [74]. Changes in soil properties, in turn, often lead to shifts in the status of a pest species either through a pest's potential to occur within a field or in its ability to increase to damaging levels. Two factors associated with conservation tillage that have a great impact on pest ecology and pest status are the mechanical disturbance of the soil and the subsequent placement of residue. Tillage can also alter the ecology of beneficial species as well as pests, as is discussed later.

Conventional tillage, normally associated with a moldboard plow, inverts the top 30 cm of soil; in contrast, there is a minimum of soil disturbance with conservation tillage. This lack of tillage preserves soil structure, including aggregate size, and pore density and distribution, thereby contributing to increased water infiltration and retention. Conservation tillage fields have higher soil moisture levels compared with conventional tillage fields, which can often affect the severity of crop diseases. Soil aeration can also be greater in conservation tillage soils. Numerous indirect effects are possible because of changes in soil properties, and these effects will vary with the pest type and species. Soil biota will be affected by the reduction of soil disturbance, because soil organic matter and nutrients tend to concentrate in the upper soil levels where lower pH values are often recorded. The lack of mechanical disturbance affects the horizontal and vertical placement of that soil biota; microbial populations tend to be more concentrated near the soil surface. Reduced tillage may favor the development of fungus-based soil food webs compared to bacteria-based webs seen in conventional tillage fields. Fungus-based food webs increase decomposition of organic matter and favor greater nutrient mobility in conservation tillage soils [33]. Lack of tillage also serves to increase the density of weed seed near the soil surface.

Residue cover can range from none with conventional tillage to nearly 100% coverage in no-tillage. Residue remaining on the soil surface also influences those factors associated with the lack of soil disturbance. Litter and organic matter tend to concentrate near the soil surface, with the debris being incorporated into the soil slowly by invertebrate action. By providing a surface cover, the residue ameliorates the soil temperature and moisture extremes, providing for a more stable environment for soil- and litter-dwelling biota, including certain soil predators and decomposers. Residue cover also affects the soil moisture levels by reducing the amount of evapotranspiration.

Mechanical disturbance directly affects some arthropod species by exposing certain pest life stages to unfavorable weather conditions; thus, the lack of such soil disturbances might serve to encourage a pest's presence. The presence of a residue cover directly affects some pests by providing a more favorable habitat. Changes in soil parameters and increased residue cover can cause a shift in pest species, as has been observed with weed populations. A shift toward grassy, perennial, wind-disseminated weeds and volunteer crop plants has been observed under conservation tillage [78]. The ecology of many insects and other invertebrates is linked to weeds [61], and thus conservation tillage often results in a shift in certain arthropod pest species because of this interaction with weeds. Lack of adequate weed control, or a change of weed species, often can lead to either increased or decreased populations of certain insect pests.

III. TILLAGE AND PESTS

Over the years, tillage was considered an essential part of pest management for insects, weeds, diseases, and nematodes, because it served to destroy food sources, pest habitats, and the pests themselves. Insects that overwintered within or just below crop residue, pathogens that overwintered in infected crop residue, and weed seeds were buried deep below the soil surface.

Tillage has been considered a major component of mortality for numerous arthropod pests. An excellent example is the pink bollworm, *Pectinophora gossypiella*, in the southwestern United States [21]. Along with defoliation or desiccation of the mature crop, early harvesting, and stalk shredding, plowing was part of the series of tactics targeting this pest. These practices were legislated and have contributed to the pink bollworm's current status as a minor pest in Texas [65]. Management of residue borne diseases in most crops, including those caused by many bacterial and fungal pathogens, typically begins with plowing down the infected residue. Tillage, both plowing and later field cultivation, has always been considered a primary means of weed control.

Growers are now adopting conservation tillage practices throughout much of the United States because of concerns about soil erosion and environmental quality. Because growers and researchers have historically regarded tillage as a way to reduce pest infestations, many thought that conservation tillage would increase pest problems. However, although the potential of certain pest problems is indeed increased under conservation tillage, the widespread problems that were expected have not materialized [74].

IV. PEST GROUPS

A. Insects and Mollusks

The earliest work on the impact of conservation tillage on pests was with insects and other related invertebrates. Indeed, based on a literature search for this chapter, the majority of publications examining pests in conservation tillage systems over the past two decades have addressed arthropods and related invertebrates, followed by weeds, then diseases, and finally nematodes. It was thought, a priori, that there would be an increase in the severity of invertebrate pest problems as growers adopted conservation tillage practices [28]. As a rule, this has not been the case. Depending upon the individual species, the potential for problems can increase, decrease, or remain the same [1]. Generally, conservation tillage will have the greatest influence on soil and litter insects, with the impact on foliar insects often occurring through their ecological relationships with weeds.

The earliest work on the impact of conservation tillage on invertebrates was from corn in Ohio. Studies had been conducted on the northern corn rootworm, *Diabrotica longicornis* [58], true armyworm, *Pseudaletia unipuncta* [55,60], black cutworm, *Agrotis ipsilon* [59], and slugs [59]. The continued widespread adoption of conservation tillage methods during the 1980s was accompanied by numerous studies on the impacts of reduced tillage on numerous arthropod species from many field crops. Some of the better-known examples of tillage modifying the pest-crop damage relationship follow.

Early workers suggested that the European corn borer, *Ostrinia nubilalis*, would become a greater concern for growers with the widespread adoption of conservation tillage in corn production. European corn borer pupae survive the winter within corn stalks, and plowing is considered a management tactic because of the mortality resulting from the burial of the pupae along with the stalks. It was thought that allowing residues to remain on the soil surface would increase the survival rate of the European corn borer and lead to greater problems [59]. Andow and Ostlie [2] pointed out that few data existed suggesting that conservation tillage would increase the probability of European corn borer problems. Indeed, Stinner et al. [76] examined the incidence of borer problems in three tillage systems and found no relationship between tillage practice and European corn borer damage to corn. Andow and Ostlie [2] actually found greater plant injury to corn in chisel-plow systems compared with no-tillage ones. These results, contrary to those expected, were believed to be due to higher European corn borer oviposition in the chisel-plowed corn because of the higher soil temperatures during early evening when oviposition occurred. It should be noted that the use of resistant corn hybrids has reduced the severity of European corn borer infestations, especially of the first generation, during the time period when conservation tillage practices were gaining acceptance. Increased predator activity associated with greater residues could also be a factor in the observed reductions [7].

Slugs have become a serious problem in conservation tillage crops (including corn, soybeans, and alfalfa) in the midwestern and eastern portions of United States where temperatures are cooler and moisture plentiful [10,29]. Slugs are one of the few invertebrate pests that have become more serious pests in conservation tillage systems (especially no-tillage) than in conventional tillage systems (compared to those arthropods with ecological relationships with weeds [e.g., cutworms, stalk borers]). The incidence of slugs is correlated with the amount of residue cover, with few problems occurring when residue is removed through soil incorporation. Four slug species have been associated with conservation tillage crops: the gray garden slug, *Deroceras reticulatum*; the marsh slug, *Deroceras laeve*; the banded slug, *Arion fasciatus*; and the dusky slug, *Arion subfuscus*. Control can be best achieved by the removal of surface residue through sufficient tillage either with a plow or multiple disking/chiseling. When this is not desirable, early planting to allow time for the crop to reach a size sufficient to withstand slug injury or the application of a molluscicide, if available, are possible means of control.

The seedcorn maggot, *Delia platura*, is an insect whose overall pest status was expected to increase with the widespread adoption of conservation tillage. Maggots feed on germinating seeds and can cause large stand reductions in field crops. It was suggested that surface residues and decaying vegetation would provide sites for adult fly oviposition [28]. It was further suggested that lower temperatures provided by the residue cover would slow crop germination, adding to the problem [28]. Studies have shown, however, that seedcorn maggot populations are not increased under most conservation tillage practices where little tillage occurs [30]. Populations are increased because of tillage; specifically, when green, living organic matter such as a rye/wheat cover crop or an old alfalfa stand is incorporated into the soil in early spring. This organic matter begins to decompose immediately, attracting adult flies; oviposition occurs within a short time span following tillage. Normally, chemical control is best achieved with an insecticidal seed treatment. However, Hammond and Cooper [31] suggested that in those fields where living, green organic matter is incorporated in the spring, delaying planting approximately 2–3 weeks after tillage allows the seedcorn maggot larvae to reach the pupal stage. At this time, the population would be in a nonfeeding stage and the crop could be safely planted and permitted to germinate.

There has not been as much attention given to wheat as there has been to some other field crops. Tillage of wheat stubble infested by the Hessian fly, *Mayetiola destructor*, has been considered an important component of pest management, because it removes a significant proportion of the existing pupae. Recent studies from the southeastern United States have compared the effect of spring and/or fall tillage of wheat stubble with conservation tillage practices on Hessian fly emergence in the fall. Chapin et al. [13] found that, although plowing was highly effective at reducing Hessian fly emergence, spring or fall disking only reduced fly emergence about 50%. However, they felt that both these practices have environmental costs that make them increasingly undesirable compared with conservation tillage practices. Zeiss et al. [88] suggested that, although Hessian fly emergence was reduced with some tillage methods, the overall benefits are not sufficient to justify the delays in subsequent soybean planting and the increases in erosion that may accompany tillage.

Shifts in populations of some arthropods under conservation tillage have been linked to the presence of weeds or other crops, with the additional weed/crop serving to increase or decrease the suitability of the insect's habitat. Some of the most severe insect problems under conservation tillage occur because of the insects' relationships with other plant species. For example, the black cutworm has long been associated with weeds in conservation tillage corn. Black cutworm adults migrate into corn fields in early spring where moths oviposit on weeds [9] followed then by the subsequent movement of the larvae to corn seedlings. Showers et al. [72] showed that weed control practices before planting can reduce black cutworm damage to corn. Indeed, reduction of early-season weeds is considered the starting point in the management of black cutworms.

Economic damage to corn from armyworms is greatest in conservation tillage fields, especially those under no-tillage practices where a grassy cover crop is used [55]. Female moths are attracted to and oviposit in the grass cover, such as rye, *Secale cereale*. Armyworm larvae then move to the corn where devastating losses can occur. Preventive tactics against armyworms would include the destruction and/or the removal of the grass cover.

The common stalk borer, *Papaipema nebris*, another insect species attracted to grassy plants (weeds, cover crops, volunteer crops), can cause serious damage to corn grown under conservation tillage [75]. However, stalk borer moths oviposit on these grasses in the fall with the eggs hatch occurring the following spring. When the weeds are killed with a herbicide in the spring, the larvae migrate and tunnel into the corn plant making insecticidal control very difficult. Indeed, as a curative tactic, insecticides must be applied before the larvae enter the plant. Levine [45] suggests that better management could be achieved by controlling grassy weeds in late summer and early fall to prevent oviposition rather than relying on spring control practices.

The presence of weeds can also provide alternate insect feeding sites, often contributing to an increase in an insect's population size. Hammond and Stinner [32] and Hammond (unpublished data) found higher numbers of Japanese beetles (*Popillia japonica*) in weedy conservation tillage soybean fields. This beetle is known to have over 100 species of host plants, and it is thought that the presence of various broadleaf weeds that serve as alternate food sites increases the overall population size of Japanese beetles. Although not directly related to conservation tillage per se, populations of other insects have been shown to increase under weedy situations (often common to conservation tillage fields). For example, populations of flower thrips, *Megalurothrips sjostedti*, and some heteropteran pests of cowpea can increase significantly in the presence of weeds [22].

In a few instances, the presence of a weed or another crop provides an unfavorable environment for the crop pest. A well-known example is the potato leafhopper, *Empoasca fabae*. Potato leafhopper numbers were reduced in alfalfa [3] and soybeans [32] that had a higher grassy weed density compared to conventional tillage situations with fewer weeds. The detrimental effect of the grass crop's presence within a legume cropping system on the leafhopper is being examined as an alternative management tactic (oats/alfalfa [43], wheat/soybeans [73]).

B. Weeds

As indicated previously, weed management has always been a central issue with tillage systems research. With the trend toward reduced tillage, considerable effort has been directed toward controlling weeds with means other than tillage. Although the majority of this effort has been placed on herbicides [15], more recently there has been significant progress in managing weeds with cover crops [34], living mulches [16], and crop rotations [68]. In general, as primary tillage intensity decreases, there is an associated increase in weed problems [53]. Most frequently, shifts in weed community composition toward more perennial species have been observed as production systems move from more to less tillage. Kapusta and Krausz [40], comparing 11 years of continuous conventional, reduced (disk), and no-tillage soybeans, found that the number of broadleaf weed species decreased in all treatments; this decrease was proportionately greater under no-tillage systems. Concomitantly, in the same study, there was an increase in the density of perennial weed species in the no-tillage treatment. Other studies have observed shifts toward more grass-dominated weed communities under reduced and no-tillage management [81,87]. Swanton et al. [78] recently reviewed studies of tillage impacts on weed density and community composition. They pointed out that despite often conflicting results among reported findings, long-term studies did show trends toward higher overall weed densities, increases in grass-dominated communities, and increased perennial species as the degree of tillage is decreased. The investigators also emphasized that both short-term fluctuations in weeds and long-term secondary ecological succession occur as tillage systems are varied, and they presented an hierarchical scheme for succession management in conservation tillage systems.

Crop rotation interacts with tillage systems to play an important role in determining weed densities and species composition. Weed populations have been reported to be lower in reduced and no-tillage wheat management when grown with a fallow period in the rotation compared to continuous wheat growing [37]. Forcella and Lindstrom [24] showed that there were fewer weeds under conservation tillage management when corn was rotated with soybeans as opposed to continuous corn cropping. Similarly, Schreiber [68] found redroot pigweed, *Amaranthus retroflexus*, density to be reduced significantly with crop rotation. There are indications that weed diversity may be higher as crop rotations are increased owing to the association of specific weeds with specific crops [71].

There has been considerable effort placed on using cover crops in conjunction with conservation tillage to help manage weeds with less herbicide, as well as to increase soil organic matter, reduce soil erosion, and increase symbiotic nitrogen fixation with legumes [54]. Although most studies have found considerably less effective weed control with cover crops than with herbicides or tillage, Putnam et al. [64] reported up to 90% suppression of weeds for row crops during a 60-day postplanting period using cover crops. Teasdale et al. [79] showed that the residues from rye and hairy vetch, *Vicia villosa*, planted as cover crops, suppressed total weed density by 78% under no-tillage conditions. Johnson et al. [39], in a Missouri study, found that mowing cover crops of hairy vetch and rye increased their effectiveness in controlling weeds, especially giant foxtail, *Setaria faberii*. Allelopathic chemicals from small grains, particularly rye, have been isolated [70], and legume residues also have been shown to suppress weed seed germination [86].

A number of studies have documented the effects of tillage practices on weed seed bank ecology. Cardina et al. [11] found that after 25 years of continuous no-tillage, reduced tillage (disk), and no-tillage practices across three soil types in Ohio, weed seed numbers to a 15-cm soil depth were highest in the no-tillage treatment with two of the soil types, but otherwise the effects of tillage on seed depth were not consistent across soil types. Species diversity indices indicated that increasing soil disturbance resulted in less weed species diversity. Lueschen et al. [49] found that the number of velvetleaf weed seeds declined more rapidly with tillage compared to no-tillage conditions, and that after 17 years, up to 25% of the original seed population could be recovered in the no-tillage treatment.

C. Diseases

Tillage has been considered an effective means of managing certain crop diseases through the burial of disease propagules that overwinter in crop residue. Plowing and other deep tillage serves to change the vertical distribution of crop residues and disease propagules, placing them in a zone where colonization and degradation is facilitated by saprophytic organisms. The potential exists that conservation tillage, because it allows crop residue to remain on the soil surface, can lead to increases in disease incidence and severity of residue-borne pathogens. There are numerous examples which illustrate this pattern.

Residue removal has been recommended for the reduction of certain wheat diseases. Tan spot, *Pyrenophora tritici-repentis*, is a residueborne pathogen that grows saprophytically on host debris [69,77]. It is considered a major pathogen in continuous, conservation tillage wheat in the central plains of the United States. The number of ascocarps per square meter of field area was reduced by 91% in disked soils compared with no-tillage conditions [89]. Of particular interest, Zhang and Pfender [89] found a negative correlation between ascocarp production and long-period wetness events (leading to wetness duration of wheat straw) in the no-tillage treatments. They concluded that management practices that increase straw-wetness duration could reduce the severity of tan spot by favoring indigenous or applied biocontrol agents. Biocontrol agents applied to wheat straw have been found to reduce the residueborne primary inoculum of tan spot [63]. Although conservation tillage practices increase the inoculum of tan spot, they may provide for greater moisture retention and thereby assist these biocontrol agents.

Other examples of wheat diseases affected by tillage are septoria nodorum blotch, *Phaeosphaeria nodorum*, which was increased, and septoria tritici blotch,

Mycosphaerella graminicola, which was suppressed, in no-tillage systems [77]. Both of these are also residueborne pathogens. The inverse relationship between septoria tritici blotch with the density of wheat residue on the soil was unexpected. It was hypothesized that the causative agent for tan spot antagonized the septoria tritici blotch agent, or perhaps it induced host resistance to the pathogen [77].

Numerous corn diseases caused by residueborne pathogens can also increase under conservation tillage. Griffith et al. [27] earlier reported that northern leaf blight, southern leaf blight, and yellow leaf blight (*Helminthosporium* spp.) were increased on corn under conservation tillage. More recent studies have confirmed these observations for these and other diseases (e.g., corn anthracnose [47] and gray leaf spot [62]). In soybeans, management of numerous diseases can also be achieved through the use of tillage and the burial of crop debris, including many foliage- and stem-infecting bacteria and fungi [38]. Verticillium wilt (*Verticillium dahliae*) and phymatotrichum root rot (*Phymatotrichum omnivorum*) are additional examples from cotton, *Gossypium hirsutum* [21].

This relationship between tillage and disease management is common to many crops where the pathogens are residueborne. Growers must weigh the benefits of conservation tillage against the potential for increased diseases. Consideration should be given to disease severity in previous years and the need for conservation tillage (slope of land and potential soil erosion). Alternative control measures, including crop rotations, resistant varieties or hybrids if available, and seed treatments (or other pesticide applications) can be used to assist in management of such diseases under conservation tillage.

With pathogens that are able to survive in the soil, such as those causing seed decay and root rot (e.g., *Fusarium* and *Pythium*), tillage will usually not be the best direct method of management. These pathogens are always present and turning the soil will not serve to reduce the overall inoculum. However, tillage reduction will often have an indirect effect on these diseases. Soils will generally remain cooler and more moist under conservation tillage in some geographical areas providing more favorable conditions for many of these plant pathogens. Additionally, these same conditions slow crop germination causing the germinating seed and seedlings to remain in the soil for a longer length of time increasing their window of susceptibility to the pathogens.

However, in some drier regions, conservation tillage can reduce the incidence and severity of certain diseases because of greater soil moisture (and thus less stress on the plants). In Texas, for example, common root rot of wheat is a disease that is most severe when plants are experiencing moisture stress [50]. Soil moisture retention from the use of conservation tillage practices might help to alleviate problems from this disease. Therefore, awareness of soil-inhabiting crop pathogens should be considered extremely important when growers use conservation tillage practices. Similar control measures should be employed against soilborne pathogens as those used against residueborne pathogens if available.

D. Nematodes

Nematodes are less affected by tillage than are the other pest types. Literature on nematode management seldom mentions tillage, or the lack thereof, as having any great importance for nematode control. Most recommendations for nematode control suggest the use of resistant varieties, crop rotations, or nematicides where available. Where tillage is discussed, it is usually suggested to alleviate the impact of nematode feeding [46].

However, studies have indicated that tillage can increase, decrease, or have no effect on the population size of nematodes depending upon the species. There is evidence that reduced tillage may help to suppress the build-up of some nematode species, including the soybean cyst nematode, *Heterodera glycines* [51,19,66]. Researchers suggest that conservation tillage reduced the continued movement of the nematodes throughout the field and possibly encouraged other organisms which helped to blunt the build-up of the nematodes. Bergeson and Ferris [5] found that lesion nematode, *Pratylenchus* sp., numbers in corn roots were higher in plowed fields compared with no-tillage fields. They observed that the soil under no-tillage was more compact, and suggested that plowing fields would favor root development and facilitate nematode movement. However, Caveness [12] reported that other species, *Meloidogyne incognita* and *Helicotylenchus pseudorobustus*, had higher populations in no-tillage plots compared with conventional tillage plots.

Although tillage appears to have some impact on nematodes, it does not appear to play a large role in nematode management at the present time. Because researchers have indicated that the impact of tillage on nematodes might take years to have a measurable affect [67], perhaps we might see tillage play more of a role in the future as conservation tillage practices become more widely accepted.

V. BENEFICIAL ORGANISMS—NATURAL ENEMIES

A. Soil-Dwelling Predators

In parallel with the concerns about pest incidence, early on in the development of conservation tillage, it was observed that the environment created by reduced tillage conditions was conducive to soil surface–dwelling predacious arthropods, especially carabid beetles and spiders [17,59]. Subsequent studies have supported the view that reduced soil disturbance and increased surface-maintained crop residues provide habitat for a more numerous and diverse soil predatory fauna in conservation compared to conventional tillage [7,36,74]. Additional studies supporting this concept come from varying geographical regions in the United States and Europe.

Many, if not most, species of ground beetles (Coleoptera: Carabidae) are predators as adults, larvae, or both [80]. These beetles tend to be very generalized in feeding habits and will prey on a wide variety of pest and nonpest invertebrates throughout much of the year. Along with ground beetles, spiders, rove beetles (Coleoptera: Staphylinidae), and centipedes (Chilopoda) are often cited as important soil-dwelling macroarthropod predators. As opposed to more host-specific predators or parasitoids that may exhibit positive numerical responses to increasing pest densities, generalist predators function on a more continuous basis to help maintain pest numbers below an economic threshold level.

It is our contention then that this continuous and sustained predation pressure is increasingly important as tillage is decreased. In the Northern Great Plains of the United States, Weiss et al. [85] found that lower numbers of individual carabid beetle species were associated with conventional tillage compared to either reduced or no-tillage systems. This study also found that tillage type significantly affected species composition of the carabid community. Similarly, House and Rosario-Alzugaray [35] reported that the abundance of ground beetles and spiders was higher in no-tillage compared to conventional tillage corn agroecosystems located in the southeastern United States. Also, in Ohio, Brust et al. [7] showed that macroarthropod predator populations were higher in no-tillage than in conventional tillage corn agroecosystems. Within the same series of studies, these investigators reported that the quantity of black cutworm larvae consumed per predator also was greater in no-tillage compared with conventionally plowed treatments.

Although most of the information on soil predators in relation to tillage concerns macroarthopods, there has been some research on soil and litter microarthropods as predators. Mites (Acarina, Mesostigmata, and Prostigmata) and certain collembolan groups are key predators of other invertebrate pests. Stinner et al. [76] found significantly more predatory mesostigmatid mites in no-tillage and reduced (disk) tillage compared to conventionally plowed soil. In North Carolina, Brust and House [8] reported that the soil-inhabiting mite, *Tyrophagous putrescentiae*, was more abundant in no-tillage versus conventional tillage–grown peanuts, and that this soil microarthropod was an important egg predator of the southern corn rootworm, *Diabrotica undecimpunctata*.

B. Foliage-Inhabiting Natural Enemies

Tillage impacts foliage-dwelling predators and parasitoids either directly through soil disturbance or indirectly through effects on weed communities. On the

Georgia Piedmont, House and Stinner [76] reported higher densities of predatory insects on plant foliage in no-tillage compared with conventional tillage corn. Similarly, in Louisiana, Troxclair and Boethel [83] found more heteropteran predators in no-tillage than conventionally grown soybeans, although this pattern depended upon specific locations. Funderburk et al. [25] concluded that tillage practices can significantly affect common foliage-inhabiting predators of soybean pests, especially predatory bigeyed bugs and damsel bugs.

VI. INFLUENCE OF TILLAGE ON DECOMPOSER FAUNA

There are numerous studies addressing the influences of tillage on soil-inhabiting arthropods and other invertebrates involved in organic matter decomposition and nutrient cycling processes. Although there is much variability in results depending upon geographical conditions, the overall conclusion to be drawn is that invertebrate populations decline as tillage intensity and frequency are increased [6,76]. Earthworm populations can be dramatically affected by tillage and cultivation practices [44]. In England, Low [48] found that earthworm numbers were reduced by 84% in arable tilled versus untilled grassland after 25 years of continuous treatment. Lee [44] argued that the loss of surface litter and decreases in soil organic matter associated with tillage is equally or more important than actual mechanical disturbance in reducing earthworm numbers. In the early 1950s, McCalla [52] reported that earthworm abundance was greater in stubble-mulched row crop systems compared to plowed soils. Subsequent studies from England [18], Nigeria [41], and Germany [20] have demonstrated the beneficial impact of reduced tillage on earthworm populations.

Although, compared to earthworms, there has been less attention given to tillage effects on other groups of decomposer invertebrates, there is sufficient information to conclude that, similar to earthworms, tillage decreases invertebrate abundance either through direct mechanical disturbance or indirectly via the impact on placement of organic material. Stinner et al. [76] observed a greater abundance of oribatid mites and collembola in no-tillage and reduced (disk) tillage soils compared to conventionally plowed soil. They also noted shifts in community composition of microarthopods among tillage treatments. House and Rosario-Alzugaray [35] found that macroarthropod decomposers (primarily Coleoptera larvae) were particularly abundant in no-tillage row crop systems that included leguminous cover crops of hairy vetch, Vicia villosa, and crimson clover, Trifolium incarnatum. Moreover, the investigators indicated that the decomposer fauna was an important prey base for predators in the systems. Appropriate to this latter point, we want to emphasize that there can be important linkages between detrital and predatory food chains in terms of pest management. As indicated previously, many of the generalist predators (ground beetles, spiders, and so forth) do not sustain themselves solely and season long on economically important crop pests. Rather these predators feed on a wide diversity of prey species that typically include organisms involved primarily in decomposition of organic material [84]. Therefore, it is important to consider the maintenance of detrital food webs in designing pest management systems.

VII. CROPPING SYSTEM INTERACTIONS WITH TILLAGE

There has been a number of studies addressing the relationships between tillage systems and cropping systems, and subsequent effects on pest and beneficial invertebrates. As early as the 1960s, Musick [56] observed that whether or not conservation and no-tillage systems led to increases or decreases in pest populations depended upon the presence of cover crops, rotation type, and crop species. As discussed earlier (see Section IV.A), armyworm larvae are usually a serious pest of corn only when following a small grain cover crop that provides oviposition sites for the adult moths [57]. Similarly, Stinner et al. [75] reported that the common stalk borer can become a severely damaging pest of no-tillage corn when grasses have been present the previous year as oviposition sites.

Beneficial arthropods also can be affected significantly by tillage and cropping pattern interactions. In Virginia, Laub and Luna [42] found that compared with herbicide treatments, mowed cover crop treatments in no-tillage corn supported higher populations of predatory ground beetles and wolf spiders (Araneae: Lycosidae). Their studies indicated that these predators played a significant role in reducing damage by armyworm in no-tillage corn.

VIII. TILLAGE AS A CONTROL MEASURE

Management tactics against pests usually fall into two categories depending upon how they relate to the pest in question. Tactics against pests that attempt to reduce their populations once they reach high levels are known as curative, or therapeutic, measures that are initiated a posteriori. These tactics are applied only after the pest is present and economic damage is imminent. This approach constitutes the basis of many integrated pest management programs: monitoring pest populations, using economic thresholds, and using a curative tactic, usually the application of a pesticide. Measures which attempt to reduce the capacity of a pest to exist or to inhibit an increase in their populations are known as preventive tactics, which are done a priori. Examples of preventive tactics include host plant resistance, biological control, and sanitation. Because of the costs and environmental concerns associated with the use of pesticides as curative tactics, there is much interest in alternative tactics that are considered preventive with little or no cost to the grower and environment.

Tillage practices are considered preventive tactics against many insects, weeds, and diseases. Turning the soil to bury insect adults, larvae, and pupae that are on the soil surface just beneath the soil or in crop residue will aid in reducing the insect's initial population. Deep tillage will bury weed seeds deeper in the soil and aid in weed control and also inhibit weeds that allow for the presence of certain insect species (e.g., cutworms and stalk borers). For diseases, the burial of residueborne pathogens will remove the disease propagules preventing them from infecting the new crop. In contrast, the use of conservation tillage as a preventive tactic against pests is limited. The lack of soil disturbance and the presence of crop residue does not usually have the detrimental effects on pests that would be a requisite of a preventive tactic.

An area where conservation tillage might take on a preventive management role is with arthropods, with the increase-in the diversity and population size of many beneficial organisms, especially the soil-inhabiting predators. As discussed, numerous studies suggest that an increase in the diversity and population size of soil predators can assist in reducing the population size of pests. Whether these practices can indeed serve to reduce the severity of pest problems is unclear, but the potential is there.

Growers must weigh the benefits of conservation tillage (e.g., reduced soil erosion, conservation of soil moisture, reduced labor costs) against the potential of a pest becoming more of a problem when determining whether to use tillage as a preventive management tactic. Consideration should be given to pest severity in the previous years and surrounding areas and to the presence of weeds, and these weighed against the need for preventing erosion. When used, conservation tillage cropping practices should include other pest management tactics, including crop rotations, resistant varieties or hybrids, seed treatments, and, when appropriate, therapeutic pesticide applications based on integrated pest management principles of scouting, evaluating, and justifiable use.

REFERENCES

1. All, J. 1987. Importance of concomitant cultural practices on the biological potential of insects in conservation tillage systems. In *Arthropods in Conservation Tillage Systems*. G.J. House & B.R. Stinner, eds. Lanham, MD: Entomological Society of America Miscellaneous Publication 65, pp 11–18.
2. Andow, D.A, & K.R. Ostlie. 1990. First-generation European corn borer (Lepidoptera:Pyralidae) response to three conservation tillage systems in Minnesota. *J. Econ. Entomol.* 83:2455–2461.

3. Barney, R.J, & B.C. Pass. 1987. Influence of no-tillage planting on foliage-inhabiting arthropods of alfalfa in Kentucky. *J. Econ. Entomol.* 80:1288–1290.
4. Bennett, H.H. 1935. Facing the erosion problem. *Science* 81:321–326.
5. Bergeson, G.B, & J.M. Ferris. 1986. Influence of tillage methods on *Pratylenchus* spp. in two soil types. *Plant Dis.* 70:326–328.
6. Blumberg, A.Y, & D.A. Crossley, Jr. 1983. Comparison of soil surface arthropod populations in conventional tillage, no-tillage and old field systems. *Agro-Ecosystems* 8:247–253.
7. Brust, G.E., B.R. Stinner, & D.A. McCartney. 1985. Tillage and soil insecticide effects on predator-black cutworm (Lepidoptera: Noctuidae) interactions in corn agroecosystems. *J. Econ. Entomol.* 78:1389–1392.
8. Brust, G.E., & G.J. House. 1988. A study of *Tyrophagus putrescentiae* (Acari: Acaridae) as a facultative predator of southern corn rootworm eggs. *Exp. Appl. Acarol.* 4:335–344.
9. Bushing, M.K., & F.T. Turpin. 1977. Survival and development of black cutworm (*Agrotis ipsilon*) larvae on various species of crop plants and weeds. *Environ. Entomol.* 6:63–65.
10. Byers, R.A., R.L. Mangan, & W.C. Templeton. 1983. Insect and slug pests in forage legume seedlings. *J. Soil Water Conserv.* 38:224–226.
11. Cardina, J., E. Regnier, & K. Harrison. 1991. Long-term tillage effects on seed banks in three Ohio soils. *Weed Sci.* 39:186–194.
12. Caveness, F.E. 1975. Plant-parasitic nematode population differences under no-tillage and tillage soil regimes in western Nigeria (abstr). *J. Nematol.* 6:138.
13. Chapin, J.W., J.S. Thomas & M.J. Sullivan. 1992. Spring- and fall-tillage system effects on Hessian fly (Diptera: Cecidomyiidae) emergence from a coastal plain soil. *J. Entomol. Sci.* 27:292–300.
14. CTIC. 1994. *Conservation Impact*. Conservation Technology Information Center. Vol. 12, November.
15. Curran, W.S., R.A. Liebl, & F.W. Simmons. 1992. Effects of tillage and application method on clomazone, imazaquin, and imazethapyr persistence. *Weed Sci.* 40:482–489.
16. Echtenkamp, G.W., & R.D. Ilnicki. 1990. Weed control by subterranean clover (*Trifolium subterraneum*) used as a living mulch. *Weed Technol.* 4:534–538.
17. Edwards, C.A., & J.R. Lofty. 1977. The influence of cultivations on soil animal populations. In *Progress in Soil Zoology*. J. Vaneck, ed. Prague: Academia, pp 349–407.
18. Edwards, C.A., & J.R. Lofty. 1977. *Biology Of Earthworms*. London: Chapman & Hall.
19. Edwards, J.H., D.L. Thurlow, & J.T. Eason. 1988. Influence of tillage and crop rotation on yields of corn, soybean, and wheat. *Agron. J.* 80:76–80.
20. Ehlers, W. 1975. Observations on earthworm channels and infiltration on tilled and untilled less soil. *Soil Sci.* 119:242–249.
21. El-Zik, K.M., & R.E. Frisbie. 1991. Integrated crop management systems for pest control. In *Handbook of Pest Management in Agriculture*. 2nd ed. D. Pimentel, ed. Boca Raton, Florida: CRC Press, pp 3–104.

22. Ezueh, M.I., & L.O. Amusan. 1988. Cowpea insect damage as influenced by the presence of weeds. *Agric. Ecosyst. Environ.* 21:255-263.

23. Faulkner, E. 1943. *Plowman's Folly*. Tulsa, Oklahoma: University of Oklahoma Press.

24. Forcella, F., & M.J. Lindstrom. 1988. Weed seed populations in ridge and conventional tillage. *Weed Sci.* 36:500-503.

25. Funderburk, J.E., D.L. Wright,, & I.D. Teare. 1988. Preplant tillage effects on population dynamics of soybean insect predators. *Crop Sci.* 28:973-977.

26. Gebhardt, M.R., T.C. Daniel, E.E. Schweizer, & R.R. Allmaras. 1985. Conservation tillage. *Science* 230:625-630.

27. Griffith, D.R., J.V. Mannering, & W.C. Moldenhauer. 1977. Conservation tillage in the eastern corn belt. *J. Soil Water Conserv.* 32:20-28.

28. Gregory, W.W., & G.J. Musick. 1976. Insect management in reduced tillage systems. *Bull. Entomol. Soc. Am.* 22:302-304.

29. Hammond, R.B. 1985. Slugs as a new pest of soybeans. *J. Kansas Entomol. Soc.* 58:364-366.

30. Hammond, R.B. 1990. Influence of cover crops and tillage on seedcorn maggot (Diptera: Anthomyiidae) populations in soybeans. *Environ. Entomol.* 19:510-514.

31. Hammond, R.B., & R.L. Cooper. 1993. Interaction of planting times following the incorporation of a living, green cover crop and control measures on seedcorn maggot populations in soybean. *Crop Prot.* 12:539-543.

32. Hammond, R.B., & B.R. Stinner. 1987. Soybean foliage insects in conservation tillage systems: effects of tillage, previous cropping history, and soil insecticide application. *Environ. Entomol.* 16:524-531.

33. Hendrix, P.F., R.W. Parmalee, D.A. Crossley, Jr., D.C. Coleman, E.P. Odum & P.M. Groffman. 1986. Detritus food webs in conventional and no-tillage agroecosystems. *BioScience* 36:374-380.

34. Hoffman, M., E. Regnier, & J. Cardina. 1993. Weed and corn (*Zea mays*) responses to a hairy vetch (*Vicia villosa*) cover crop. *Weed Tech.* 7:594-599.

35. House, G.J., & M. Rosario-Alzugaray. 1989. Influence of cover cropping and no-tillage practices on community composition of soil arthropods in a North Carolina agroecosystem. *Environ. Entomol.* 18:302-307.

36. House, G.J., & B.R. Stinner. 1983. Arthropods in no-tillage agroecosystems: community composition and ecosystem interactions. *Environ. Manage.* 7:23-28.

37. Hume, L., S. Tessier, & F.B. Dyck. 1991. Tillage and rotation influences on weed community composition in wheat (*Triticum aestivum* L.) in southwestern Saskatchewan. *Can. J. Plant Sci.* 71:783-789.

38. Jacobsen, B.J., & P.A. Backman. 1989. Soybean Disease Management Strategies. In *Compendium of Soybean Diseases*. 3rd ed. J.B. Sinclair & P.A. Backman. eds. Part IV, Minneapolis, Minnesota: APS Press, pp 94-100.

39. Johnson, G.A., M.S. Defelice, & Z.R. Hensel. 1993. Cover crop management and weed control in corn (*Zea mays*). *Weed Tech.* 7:425-430.

40. Kapusta, G., & R.F. Krausz. 1993. Weed control and yield are equal in conventional, reduced-, and no-tillage soybean (*Glycine max*) after 11 years. *Weed Technol.* 7:443-451.

41. Lal, R. 1974. No-tillage effects on soil properties and maize (*Zea mays* L.) production in western Nigeria. *Plant Soil* 40:321–331.

42. Laub, C.A., & J.M. Luna. 1992. Winter cover crop suppression practices and natural enemies of armyworm (Lepidotera: Noctuidae) in no-till corn. *Environ. Entomol.* 21:41–49.

43. Lamp, W.O., R.J. Barney, E.J. Armbrust, & G. Kapusta. 1984. Selective weed control in spring-planted alfalfa: effect on leafhoppers and planthoppers (Homoptera: Auchenorrhyncha), with emphasis on potato leafhopper, *Empoasca fabae. Entomol. Exp. Appl.* 36:125–131.

44. Lee, K.E. 1985. *Earthworms: Their Ecology and Relationship with Soils and Land Use.* Sidney: Academic.

45. Levine, E. 1993 . Effect of tillage practices and weed management on survival of stalk borer (Lepidoptera: Noctuidae) eggs and larvae. *J. Econ. Entomol.* 86:924–928.

46. Lewis, S.A. 1989. Lance nematodes. In *Compendium of Soybean Diseases.* 3rd ed. J.B. Sinclair & P.A. Backman, eds. Minneapolis, Minnesota: APS Press, pp 67–68.

47. Lipps, P.E. 1985. Influence of inoculum from buried and surface residues on the incidence of corn anthracnose. *Phytopathology* 75:1212–1216.

48. Low, A.J. 1972. The effect of cultivation on the structure and other physical characteristics of grassland and arable soils (1945-1970). *J. Soil Sci.* 23:363–380.

49. Lueschen, W.E., R.N. Anderson, T.R. Hoverstad, & B. K. Kanne. 1993. Seventeen years of cropping systems and tillage affect velvetleaf (*Abutilon theophrasti*) seed longevity. *Weed Sci.* 41:82–86.

50. Mathieson, J.T., D. Bordovsky, L.E. Clark, & O.R. Jones. 1990. Effects of tillage on common root rot of wheat in Texas. *Plant Dis.* 74:1006–1008.

51. Marking, S. 1991. Conservation tillage suppresses SCN. *Soybean Dig.*, Nov.

52. McCalla, T.M. 1953. *Microbiology Sudies of Stubble Mulching.* Nebraska Agricultural Experimental Station Bull. 417.

53. McWhorter, C.R. 1984. Future needs in weed science. *Weed Sci.* 32:850–855.

54. McVay, K.A., D.E. Radcliffe, & W.L. Hargrove. 1989. Winter legume effects on soil properties and nitrogen fertilizer requirements. *Soil Sci. Soc. Am. J.* 53:1856–1862.

55. Musick, G.J. 1987. History, perspective, and overview of entomological research in conservation tillage systems. In *Arthropods in Conservation Tillage Systems.* G.J. House & B.R. Stinner, eds. Lanham, MD: Entomological Society of America Miscellaneous Publication 65, pp 1–10.

56. Musick, G.J. 1970. Insect problems associated with no-tillage. In *Proc. N.E. No-Tillage Conf.* 44–59.

57. Musick, G.J. 1973. Control of armyworm in no-tillage corn. *Ohio Rep.* 58:42–45.

58. Musick, G.J., & D.L. Collins. 1971. Northern corn rootworm affected by tillage. *Ohio Rep.* 56:88–91.

59. Musick, G.J., & H.B. Petty. 1973. Insect control in conservation tillage systems. In *Conservation Tillage: The Proceedings of National Conference.* Ankeny, Iowa: Soil Conservation Society of America, pp 120–125.

60. Musick, G.J., & P.J. Suttle. 1973. Suppression of armyworm damage to no-tillage corn with granular carbofuran. *J. Econ. Entomol.* 66:735–737.
61. Pavuk, D.M., & B.R. Stinner. 1991. Relationship between weed communities in corn and infestation and damage by the stalk borer (Lepidoptera: Noctuidae). *J. Entomol. Sci.* 26:253–260.
62. Payne, G.A., & H.E. Duncan & C.R. Adkins. 1987. Influence of tillage on development of gray leaf spot and number of airborne conidia of *Cercospora zea-maydis*. *Plant Dis.* 71:329–332.
63. Pfender, W.F., W. Zhang, & A. Nus. 1993. Biological control to reduce inoculum of the tan spot pathogen *Pyrenophora tritici-repentis* in surface-borne residues in wheat fields. *Phytopathology* 83:371–375.
64. Putnam, A.R., J. Defrank, & J.P. Barnes. 1983. Exploitation of allelopathy for weed control in annual and perennial cropping systems. *J. Chem. Ecol.* 9:1001–1010.
65. Reynolds, H.T., P.L. Adkisson, R.F. Smith, & R.E. Frisbie. 1982. Cotton insect pest management. In *Introduction to Pest Management*. 2nd ed. R.L. Metcalf & W.H. Luckman, eds. New York: Wiley, pp 379–443.
66. Schmitt, D.P., & L.A. Nelson. 1987. Chemical control of selected plant-parasitic nematodes in soybeans double-cropped with wheat in no-till and conventional tillage systems. *Plant Dis.* 71:323–326.
67. Schmitt, D.P., & R.D. Riggs. 1989. Population dynamics and management of *Heterodera glycines*. *Agric. Zool. Rev.* 3:253–269.
68. Schreiber, M.M. 1992. Influence of tillage, crop rotation and weed management on giant foxtail (*Setaria faberi*) population dynamics and corn yield. *Weed Sci.* 40:645–653.
69. Schuh, W. 1990. The influence of tillage systems on incidence and spatial pattern of tan spot of wheat. *Phytopathology* 80:804–807.
70. Shilling, D.G., R.A. Leibl, & A.D. Worsham. 1985. Rye and wheat mulch: the suppression of certain broadleaved weeds and the isolation and identification of phytotoxins. In *The Chemistry of Allelopathy*. A.C. Thomson, ed. ACS Symp. Ser. 268. Washington DC: American Chemical Society, pp 17–21.
71. Slife, F.W. 1981. Environmental control of weeds. In *Handbook of Pest Management in Agriculture*. Vol. 1. D. Pimentel, ed. Boca Raton, Florida: CRC Press, pp 485–491.
72. Showers, W.B., L.V. Kaster, T.W. Sappington, P.G. Mulder, & F. Whiford. 1985. Development and behavior of black cutworm (Lepidoptera: Noctuidae) populations before and after corn emergence. *J. Econ. Entomol.* 78:588–594.
73. Smith, A.K., R.B. Hammond, & B.R. Stinner. 1988. Rye cover crop management influence on soybean foliage arthropods. *Environ. Entomol.* 17:109–114.
74. Stinner, B.R., & G.J. House. 1990. Arthropods and other invertebrates in conservation-tillage agriculture. *Annu. Rev. Entomol.* 35:299–318.
75. Stinner, B.R., D.A. McCartney, & W.A. Rubink. 1984. Some observations on ecology of the stalk borer (*Papaipema nebris* (GN): Noctuidae) in no-tillage corn agroecosystems. *J. Ga. Entomol. Soc.* 19:229–234.
76. Stinner, B.R., D.A. McCartney, & D.M. Van Doren, Jr. 1988. Soil and foliage arthropod communities in conventional, reduced and no-tillage corn (*Maize, Zea*

mays L.) systems: a comparison after 20 years of continuous cropping. *Soil Tillage Res.* 11:147–158.

77. Sutton, J.C., & T.J. Vyn. 1990. Crop sequences and tillage practices in relation to diseases of winter wheat in Ontario. *Can. J. Plant Pathol.* 12:358–368.

78. Swanton, C.J., D.R. Clements, & D.A. Derksen. 1993. Weed succession under conservation tillage: a hierarchical framework for research and management. *Weed Technol.* 7:286–297.

79. Teasdale, J.R., C.E. Beste, & W.E. Potts. 1991. Response of weeds to tillage and cover crop residue. *Weed Sci.* 39:195–199.

80. Thiele, H.U. 1977. *Carabid Beetles in their Environments.* Berlin: Springer-Verlag.

81. Triplett, G.B., & G.D. Lytle. 1972. Control and ecology of weeds in continuous corn grown without tillage. *Weed Sci.* 20:453–457.

82. Triplett, G.B., & D.M. Van Doren. 1977. Agriculture without tillage. *Sci. Am.* 236:28–33.

83. Troxclair, N.N., & D.J. Boethel. 1984. The influence of tillage practices and row spacing on soybean insect populations in Louisiana. *J. Econ. Entomol.* 77:1571–1579.

84. Wallwork, J.A. 1970. *Ecology of Soil Animals.* London: McGraw-Hill.

85. Weiss, M.J., E.U. Balsbaugh, E.W. French, & B.K. Hoag. 1990. Influence of tillage management and cropping system on ground beetle (Coleoptera:Carabidae) fauna in the Northern Great Plains. *Environ. Entomol.* 19:1388–1391.

86. White, R.H., A.D. Worsham, & U. Blum. 1989. Allelopathic potential of legume debris and aqueous extracts. *Weed Sci.* 37:674–679.

87. Wrucke, M.A., & W.E. Arnold. 1985. Weed species distribution as influenced by tillage and herbicides. *Weed Sci.* 33:853–856.

88. Zeiss, M.R., R.L. Brandenburg, & J.W. Van Duyn. 1993. Effect of disk harrowing on subsequent emergence of Hessian fly (Diptera: Cecidomyiidae) adults from wheat stubble. *J. Entomol. Sci.* 28:8–15.

89. Zhang, W., & W.F. Pfender. 1992. Effect of residue management on wetness duration and ascocarp production by *Pyrenophora tritici-repentis* in wheat residues. *Phytopathology* 82:1434–1439.

26
Reducing Agricultural Pesticide Losses to Surface and Groundwater Resources

James L. Baker
Iowa State University, Ames, Iowa

I. INTRODUCTION

The use of new and improved technology, including pesticides, has allowed U.S. agriculture to provide more than adequate supplies (some for export) of food, feed, and fiber at reasonable costs. However, off-site movement of pesticides to surface and groundwater resources causes concerns for uses of that water. This is particularly true with respect to drinking water and human health, and the viability of the aquatic ecosystem that often receives most, if not all, of its water in the form of agricultural drainage. Water quality problems associated with pesticide transport in surface and subsurface agricultural drainage are part of what is often termed nonpoint source pollution.

The options available for mitigation (or best management) practices to address potential aquatic ecosystem and human health water quality concerns from pesticide use, and the efficiencies of pesticide loss reduction of these practices, is the main topic of this chapter. However, preceding that discussion will be sections on examples of pesticide concerns and exposure found in U.S. surface and groundwater resources presently or within the recent past, and on the current understanding of pesticide fate and transport that is necessary to develop and evaluate potential practices to reduce agricultural pesticide losses to surface and groundwater resources.

With respect to pesticide concerns in the United States, the U.S. Environmental Protection Agency (EPA) is responsible for setting standards for pesticides in drinking water to protect human health against acute and chronic toxicity, as described by Benson [14]. For a few pesticides, standards have been set in terms of a maximum contaminant level (MCL; the maximum permissible level of a contaminant in water which is delivered to any user of a public water system). The MCL for a pesticide is an enforceable concentration, and the water supplier must analyze for the pesticide in "finished" water at least quarterly. If the running annual average of the four quarters exceeds the MCL, the supplier must take action to bring the concentration level back into compliance. For most older herbicides, EPA has at least established what is termed a health advisory (HA; for a lifetime HA, it is the concentration of a chemical in drinking water that is not expected to cause any adverse noncarcinogenic effects over a lifetime of exposure, with a margin of safety).

The setting of MCLs and HAs is based on the toxicity of the pesticides as assessed by various means such as epidemiology and animal tests. The classification scheme used by EPA for categorizing chemicals according to their carcinogenic potential is:

> Group A: Human carcinogen—sufficient evidence in epidemiological studies to support causal association between exposure and cancer,
>
> Group B: Probable human carcinogen—limited evidence in epidemiological studies (Group B1) and/or sufficient evidence from animal studies (Group B2),
>
> Group C: Possible human carcinogen—limited evidence from animal studies and inadequate or no data in humans
>
> Group D: Not classifiable—inadequate or no human or animal evidence of carcinogenicity
>
> Group E: No evidence of carcinogenicity for humans—no evidence of carcinogenicity in at least two adequate animal tests in different species, or inadequate epidemiological animal studies.

Currently, the EPA is in the process of revising cancer guidelines.

From toxicity testing, a reference dose (RFD) is determined, which is an estimate of a daily exposure to the human population that is likely to be without appreciable risk of deleterious effects over a lifetime. From the RFD, a drinking water equivalent level (DWEL) is established, which is a lifetime exposure concentration protective of adverse, noncancer health effects, that assumes all of the exposure to a contaminant is from a drinking water source. Safety factors ranging from 5 to 5000 are then used in establishing HAs (or MCLs), in part because of less-than-lifetime exposures in chronic toxicity testing, because sources other than water may contribute to pesticide exposures, and because of carcinogenicity considerations. Table 1 provides toxicity and HA and MCL data, where they exist, for example pesticides of significant current use.

Recently, concerns for the impact of pesticides in surface water resources relative to the viability of the aquatic ecosystem have received additional attention. In March 1992, a task force within the U.S. EPA Office of Pollution Prevention and Toxic Substances was charged with reviewing and assessing ecological and environmental fate data requirements for registration and reregistration of pesticides. The task force reached policy decisions that dealt with need for improved risk assessment and the use of mitigation practices in a timely manner when a level of concern was expected to be or was being exceeded. To provide input into implementation of this "new paradigm," a group of people representing the EPA (regulators and researchers), academia, agrichemical companies, and environmental and agricultural interest groups produced a document [46] considering methodology for establishing levels of concern for exposure and for implementing mitigation practices (e.g., to reduce pesticide losses in surface runoff when needed). Recently, the EPA has established a similar group (ECOFRRM) to implement changes in the methodology of determining pesticide exposure and effects.

II. PESTICIDES IN SURFACE AND GROUNDWATER RESOURCES

In an extensive 5-year study (1976–1980) of the water quality of drainage in an agricultural watershed, pesticide concentrations and losses were measured from individual corn and soybean fields, as well as from the watershed as a whole [29,30]. The crops in adjacent corn and soybean fields (about 6 ha each) were rotated each year, with the field to be planted to corn receiving a preemergent treatment of propachlor and cyanazine (2.2 kg/ha each) and the field to be planted to soybean receiving a preemergent treatment of alachlor and metribuzin (2.2 and 0.6 kg/ha, respectively). The intensively farmed watershed as a whole (50.5 km^2) was 56% corn, with 99% of that area receiving herbicides in 1980; corresponding values for soybeans were 24 and 100%. Table 2 shows pesticide data for surface runoff from the fields and stream flow from Four-Mile Creek draining the whole watershed; data for atrazine are shown only for the whole watershed, as it was not used on the individual study fields. Average stream flow for the 5-year period was 198 mm, which is slightly above the long-term average.

As shown in Table 2, there was a wide variation in pesticide losses as a result of widely varying flow volumes; during the dry year of 1977, essentially no pesticide loss occurred, whereas during the wet year of 1979, surface runoff losses from the field were as high as 7% and losses with stream flow were as high as 2%. Pesticide concentrations and generally losses were highest for the first runoff event after application, which has been observed in other studies [53]. For the four herbicides, which would be classified as moderately adsorbed with distribution coefficients or ratios of concentrations in soil or sedi-

Table 1 Examples of Pesticide Toxicity and HA and MCL Data

Pesticide name		Cancer group[a]	RfD (mg/kg/day)	DWEL (µg/L)	HA (µg/L)	MCL (µg/L)
Common	Trade					
Herbicides						
2,4-D	-	D	-	400	70	70
acetochlor	Harness/Surpass	-	-	-	-	-
acifluoren	Blazer	B2	0.013	400	-	-
alachlor	Lasso	B2	0.01	400	-	2
atrazine	AAtrex	C	0.035	200	3	3
bentazon	Basagran	D	0.0025	90	20	-
bromoxynil	Buctril	-	-	-	-	-
chlormuron-ethyl	Classic	-	-	-	-	-
cyanazine	Bladex	C	0.002	70	1	-
dicamba	Banvel	D	0.03	1000	200	-
fluazifop-P-butyl	Fusilade	-	-	-	-	-
glyphosate	Roundup	E	0.1	4000	700	700
imazaquin	Scepter	-	-	-	-	-

			RfD	DWEL	HA	MCL
imazethapyr	Pursuit	-	-	-	-	-
metolachlor	Dual	C	0.1	3500	70	-
metribuzin	Sencor/Lexone	D	0.013	500	100	-
nicosulfuron	Accent	-	-	-	-	-
pendimethalin	Prowl	-	-	-	-	-
thifensulfuron-methyl	Harmony	-	-	-	-	-
trifluralin	Treflan	C	0.0075	300	5	-
Insecticides						
chlorpyrifos	Lorsban	D	0.003	100	20	-
fonofos	Dyfonate	D	0.002	70	10	-
methyl parathion	-	D	0.00025	9	2	-
permethrin	Ambush/Pounce	-	-	-	-	-
tefluthrin	Force	-	-	-	-	-
terbufos	Counter	D	0.00013	5	0.9	-

HA, health advisory; MCL, maximum contaminant level; RfD, reference dose; DWEL, drinking water equivalent level.

[a]See text.

Source: EPA Office of Drinking Water, May 1995.

Table 2 Pesticide Concentrations and Losses in Field Surface Runoff and Stream Flow (Four-Mile Creek)

Year/site	Flow volume (mm)	Alachlor (µg/L)[a]	(%)[b]	Metribuzin (µg/L)	(%)	Propachlor (µg/L)	(%)	Cyanazine (µg/L)	(%)	Atrazine (µg/L)	(%)
1976											
fields	51/59[c]	19.5	0.48	7.3	0.72	9.7	0.22	38.0	0.96	-	-
stream	123	0.7	0.10	0.1	0.06	0.1	0.13	0.1	0.07	0.9	0.17
1977											
fields	12/1	<1.0	<0.01	<1.0	<0.0	ND	0.0	1.5	<0.01	-	-
stream	44	ND	0.0	ND	0.0	ND	0.0	ND	0.0	0.1	<0.01
1978											
fields	47/46	11.1	0.27	5.4	0.45	12.6	0.25	46.8	0.82	-	-
stream[d]	197	0.2	0.05	0.1	0.04	0.1	0.11	0.1	0.04	0.5	0.26
1979											
fields	252/199	39.1	2.74	18.8	7.20	5.3	0.56	60.1	5.57	-	-
stream	445	2.7	1.08	0.4	2.01	<0.1	0.65	1.3	1.72	-	-
1980											
fields	120/88	17.0	0.58	6.0	1.15	12.6	0.75	50.8	2.85	-	-
stream	182	10.6	1.89	0.3	0.82	0.8	0.45	3.7	1.63	-	-

ND, not detected; -, no measurement made.
[a]Concentrations are flow-weighted for total flow.
[b]Losses are % of that applied to the field or to the whole watershed (based on annual inventories).
[c]Flows for corn field, then soybean field are both given.
[d]In 1978, extra samples of stream flow during two storms were not taken, possibly resulting in low calculated losses.
Source: From refs. 29 and 30.

ment to those in water (discussed further later) ranging from 2 to 15, the data showed that 90% or more of the transport in stream or in surface runoff was generally in solution with water. The exception was during the wet year of 1979 when erosion from the cropped fields averaged 63.4 t/ha and percentages of measured losses of alachlor, metribuzin, propachlor, and cyanazine that took place in solution were 84, 83, 83, and 58%, respectively. Corresponding values for stream flow from the watershed were 7.6 t/ha and 96, 96, 92, and 85%, respectively. Sediment deposition between field borders and Four-Mile Creek can account for the lower percentages being transported with sediment for stream flow and can decrease pesticide transport and cause some attenuation of losses. However, again considering the 1979 data, the attenuation between the field and stream of 61, 72, and 69% for alachlor, metribuzin, and cyanazine, respectively, implies that other attenuation processes are active beyond just sediment deposition, such as runoff infiltration and pesticide adsorption to in-place soil and living and dead plant tissue in the transport path to the stream. The data for propachlor may indicate an exception; however, it is less reliable as only 1.7% of the watershed was treated with propachlor in 1979 (as opposed to 41.5, 20.4, and 19.7% for alachlor, metolachlor, and cyanazine, respectively).

On a larger regional scale, the U.S. Geological Survey reported in 1994 [45] on the intensive monitoring of selected herbicides and two atrazine metabolites in storm runoff from nine stream basins in five Midwestern states from April through July 1990. Two of these stream basins were selected for further study from April 1991 through March 1992. The four major-use herbicides atrazine, alachlor, metolachlor, and cyanazine were among those studied, and their use in the 12-state Corn Belt region accounted for about 73% of the 65 million kg/ yr of pesticides used in 1987–1989 [24]. Prior to April 1990, concentrations of the triazine herbicides, including atrazine and cyanazine, were less than 1 μg/L; however, in the early May and June 1990 period, concentrations increased sharply in this postplant period to peaks in the 10–75 μg/L range. Alachlor and metolachlor were the major chloroacetanilide herbicides found, with concentration spikes of 20 μg/L. Herbicide concentrations correlated with stream discharge immediately following herbicide application, and thus herbicide transport to streams was seasonal, generally occurring during runoff following chemical application associated with the planting of crops. The repeated sampling at two of the sites in 1991–1992 confirmed the seasonal transport of herbicides in midwestern streams.

On an even larger scale, the U.S. Geological Survey recently released a report [20] on the monitoring of the Mississippi River at three points above Baton Rouge, Louisiana, from April 1991 to September 1992. The herbicide atrazine was detected in almost all the samples taken, with highest concentrations at all locations occurring in the months of May, June, and July, which was likely related to the times of or shortly after most applications, as noted earlier. The maximum atrazine concentration for any individual sample from the Mississippi River was 4.4 µg/L; the overall average concentration was less than 1.0 µg/L. The next three most commonly detected herbicides were alachlor, metolachlor, and cyanazine, which also had their highest concentrations in May, June, and July. However, maximum concentrations for individual samples of alachlor, metolachlor, and cyanazine were lower than for atrazine at 0.9, 2.3, and 3.2 µg/L, respectively. Also, unlike the more persistent atrazine, there were extended periods when concentrations of these three herbicides were below their limits of detection. Because of dilution and attenuation, maximum herbicide concentrations were much lower in the Mississippi River than in flow from the smaller stream systems in the intensively cropped Midwest.

Pesticide concentrations in surface runoff from treated fields can exceed 1 mg/L (1000 µg/L), as discussed earlier, particularly right after application; however, water draining from the bottom of the root zone generally has much lower concentrations, often ≤1 µg/L. This is shown in Table 3 for a study [28] in north central Iowa of pesticides leaching through the root zone to tile lines spaced 7.6 m apart. Detections, concentrations, and losses were greater for more persistent (e.g., atrazine) and less strongly adsorbed (e.g., bentazon) pesticides. Similar low concentrations for atrazine and cyanazine have been measured in other studies [5,31].

In a study of pesticide movement into tile drains on a low organic matter (1.3%) soil in Indiana [33], almost all the samples that had pesticide concentrations ≥1 µg/L were taken in the spring after pesticide application. At that time, typical concentration ranges were: carbofuran 5-150 µg/L, atrazine 1-10 µg/L, cyanazine 1-10 µg/L, and alachlor 1-2 µg/L; no chlorpyrifos was ever detected. These concentrations were inversely related to the degree of soil adsorption, with carbofuran being adsorbed the least.

With water draining from the root zone possibly recharging an aquifer, groundwater contamination with low levels of pesticides is possible. In the recent past, several monitoring studies have been performed at the state, regional, and national levels to assess this contamination. As examples at the state level, surveys have been done in Iowa [32,34] and Ohio [3]. In Iowa, groundwater from 686 rural domestic wells across the state was sampled for 16 pesticides and metabolites in 1988-1989. About 14% of the well water samples contained detectable pesticides, with about 1% exceeding a MCL or HA level for any pesticide. Wells <15 m deep were determined to be more susceptible to contamination. In Ohio, groundwater samples from over 16,000 wells have been tested since 1987. Of those samples tested for alachlor, atrazine, and cyanazine (610 wells), about 1% exceeded a MCL or HA level.

At the regional level, Monsanto Agricultural Company [40] surveyed groundwater from 1430 wells in a 26-state area of extensive herbicide use for alachlor as well as atrazine, simazine, and metolachlor. They found that about 13% of

Table 3 Pesticide Concentrations in Surface Drainage

Pesticide name		Application rate (kg/ha)	No. of samples	Detection (%)	Concentration (μg/L)	
Common	Trade				Maximum	Average[a]
Herbicides						
2,4-D	–	0.14	57	0.0	0.00	0.00
alachlor	Lasso	2.80	66	0.0	0.00	0.00
atrazine	AAtrex	1.68	63	86.4	1.20	0.30
bentazon	Basagram	0.84	133	29.2	6.40	0.63
cyanazine	Bladex	2.24	417	15.1	5.00	0.16
dicamba	Banvel	0.28	54	1.7	0.56	0.01
fluazifop-P-butyl	Fulsilade	0.14	68	0.0	0.00	0.00
metolachlor	Dual	2.80	423	7.2	1.90	0.05
metribuzin	Sencor	0.56	118	11.5	2.20	0.11
pendimethalin	Prowl	1.68	60	0.0	0.00	0.00
trifluralin	Treflan	1.12	133	0.0	0.00	0.00
acifluoren	Blazer	0.41	66	2.3	0.33	0.01
Insecticide						
terbufos	Counter	1.12	4	0.0	0.00	0.00

[a]Flow-weighted average.
Source: From ref. 28.

the well water samples contained detectable levels of one or more herbicides, but less than 0.1% exceeded MCL or HA levels for any pesticide. At the national level, the EPA [51] monitored groundwater from 564 community and 783 rural domestic wells across the United States for 126 pesticides and pesticide metabolites. They found that about 7% of the well water samples contained one or more pesticides, but less than 0.8% exceeded health concerns for any pesticide.

The data for the studies just cited all came from water-supply well samplings, which may not be truly indicative of groundwater quality. In addition to contamination from nonpoint sources (e.g., field leaching), pesticides detected in well water may come from point sources and/or may be a result of poor well construction.

III. CURRENT UNDERSTANDING OF PESTICIDE FATE AND TRANSPORT

As shown in Figure 1, there are three modes of transport or carriers for off-site losses of field-applied pesticides. These are (a) surface runoff water and (b) eroded soil that transport pesticides overland (in solution or adsorbed to sediment) to surface water resources such as streams, rivers, lakes, and reservoirs; and (c) percolating or leaching water that carries pesticide out the bottom of the root zone. This leaching water either may move vertically and recharge an aquifer or move laterally and return to a surface water resource if the hydraulic gradient is in that direction. This lateral flow is known as base flow and can include tile drainage if an artificial drainage system is present.

If the hydraulic gradient is away from a surface water resource, water can move from that source to groundwater. When this happens for a stream, it is called a losing stream. This occurrence, plus special situations involving sinkholes and agricultural drainage wells illustrate why protection of surface and groundwater resources cannot be considered independently. Those resources must be considered as a system involving all aspects of water flow and chemical movement within a watershed.

On a smaller scale, one concept that is important for understanding the fate and transport of soil-applied pesticides is that of the thin "mixing zone" at the soil surface, illustrated in Figure 2. Rainfall and runoff water can mix with soil in this zone and can dissolve/extract pesticide present there and transport it. The amount of applied pesticide remaining in or near the soil surface generally decreases with time; therefore, runoff losses of pesticides are usually greatest for the first runoff event following application when more pesticide is "available" to be lost [4]. There are three sets of factors that interact and determine how

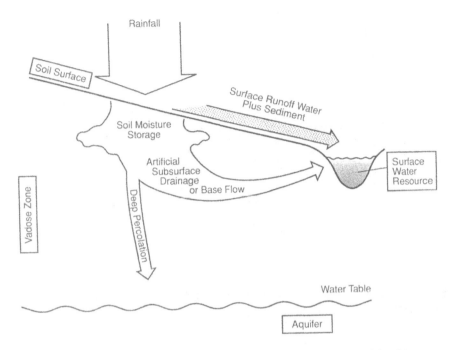

Figure 1 Schematic diagram of surface runoff water, sediment, and leaching water carriers of pesticides to surface and groundwater resources.

and to what degree pesticide losses to water resources will occur: chemical, hydrological, and management.

A. Chemical Factors

Properties of an individual pesticide determine to a large degree its fate, including its potential for off-site loss to water resources, although environmental factors also play a role. The four most important properties are persistence (or resistance to transformation or degradation), soil adsorption, water solubility, and vapor pressure.

1. Pesticide Persistence

Pesticide persistence, especially in the thin "mixing zone" at the soil surface, determines how much of the applied pesticide is present to be lost with surface runoff water and sediment or with leaching water. Persistence is dependent on how resistant the pesticide is to microbiological, chemical, and/or photochemical

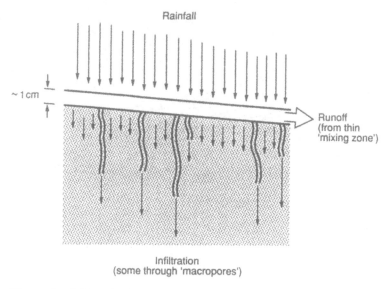

Figure 2 Schematic diagram of the concepts of the thin surface soil "mixing zone" and "macropores" or preferential flow paths.

breakdown. As a first approximation, pesticide degradation is assumed to be a first-order reaction, which means the rate of degradation is proportional to the amount present. The proportionality constant, or rate of reaction, is inversely related to the half-life of the pesticide, or the time it takes for half of the pesticide to degrade. The Natural Resources Conservation Service (NRCS) has maintained a database where half-life data from the literature are listed for most pesticides used in agriculture today [54]. Soil sampling in field studies is one method used to determine half-lives. For example, the half-life of atrazine applied to corn on a silt-loam soil in Iowa was 1–2 months, whereas half-lives determined for other herbicides used were only about 2 weeks [6]. The longer half-life for atrazine, along with its extensive use, are two reasons that it is the pesticide most often detected in surface and groundwater monitoring.

Persistence of a pesticide throughout the soil profile and in a saturated zone (such as a shallow water table) will affect the potential for contamination of leaching water or subsurface drainage and subsequent contamination of a surface or groundwater resource. If the travel time of the lateral movement of shallow groundwater back to a surface water resource, or travel time of vertical movement to an aquifer is long compared to a pesticide half-life, pesticide concentrations can be significantly reduced.

2. Soil Adsorption and Water Solubility

Soil adsorption, in conjunction with water solubility, determines to a large degree how a pesticide will be transported from a treated field. A pesticide added to the soil-water system in the field comes into an equilibrium of being adsorbed to soil or dissolved in the soil water. In general, the degree of soil adsorption and water solubility are inversely related; however, there are exceptions of highly soluble pesticides being strongly adsorbed. These pesticides are positively charged and are adsorbed by the negatively charged clay surfaces. One example is paraquat, a divalent cation, which although soluble enough to be applied in a true solution, is adsorbed so strongly that, in one study [29], it was not found in runoff water at detectable concentrations, although it was measured in sediment. For nonionic pesticides, the organic matter content of the soil is an extrinsic factor important in determining the degree of adsorption.

To quantify adsorption, a partition or adsorption coefficient, often designated K, is used. Batch experiments of pesticide, soil, and water mixtures are run to determine pesticide concentrations in soil (C_s) and in water (C_w) at equilibrium at constant temperature. The resulting curves of C_s plotted against C_w are called "isotherms," and they are often fitted with the Freundlich equation with empirical coefficients K and n:

$$C_s = K(C_w)^{1/n}$$

Generally, the relationship is nearly linear, at least over a limited concentration range, and $1/n$ is assumed to be unity and therefore $K = C_s/C_w$. The higher the K value, the stronger the adsorption.

For strongly adsorbed pesticides (e.g., with $K > 200$), the main mode of transport is with sediment, because pesticide is not readily released to water flowing over or through the soil surface [19]. Thus, although concentrations in water may not decrease during a runoff event because of lack of movement down out of the mixing zone, they are low in surface runoff water. For moderately adsorbed pesticides ($1 < K < 20$), pesticide is more readily released to water flowing over or through the soil surface, and runoff losses with water dominate over losses with sediment [6,17,18]. Even though concentrations may be several times higher in sediment than water, usually the mass of water lost during a surface runoff event is at least 100 times the mass of sediment lost. Concentrations of moderately adsorbed pesticides in runoff water usually decrease with time during a surface runoff event as the amount of pesticide at or near the soil surface decreases with pesticide movement from the mixing zone with runoff and infiltrating water [8,10]. However, because of the moderate adsorption, some of the pesticide will remain at or near the soil surface at the end of the first runoff event to be available for loss with later events.

Although leaching losses of moderately adsorbed pesticides below the root zone will be much less than surface runoff losses (on a percent-of-applied basis) because of adsorption to surface and subsoils, pesticide concentrations in the micrograms per liter range are often found in shallow drainage or water wells [7,28,31,41]. It is believed that rapid flow of water from the soil surface through preferential flow paths or "macropores," bypassing adsorptive soils (see Figure 2), can explain the presence of moderately adsorbed pesticides at depths ≥ 1 m. For nonadsorbed to weakly adsorbed pesticides ($0 < K < 0.1$), pesticide is easily released to water at or near the soil surface and concentrations in and losses with sediment are small. Pesticide concentrations in runoff and leaching water can be initially high but decrease quickly with time as water moving over or through the surface soil removes pesticide present there. As will be discussed below, the amount of infiltration that occurs before surface runoff begins will significantly affect surface runoff losses. For weakly adsorbed pesticides, the major mode of transport will be highly influenced by what percentage of rainfall infiltrates and what percentage runs off. If most of the water infiltrates, the major carrier will be leaching water. The major mode of transport for pesticides with values between 0.1 and 1.0 also depends on relative amounts of runoff and leaching waters; and for K values between 20 and 200, whether the major carrier is sediment or water depends on sediment loss versus runoff water loss.

3. Vapor Pressure

Pesticide vapor pressures are dependent on the pesticide, and can range over several orders of magnitude [56]. Vapor pressure is one measure of the potential for volatilization loss from pesticide-treated fields, and one fate for pesticides lost to the atmosphere is redeposition with rainfall [42]. Because of soil adsorption and low pesticide concentrations (2 kg/ha applied to 10 cm of soil would give a concentration of less than 2 ppb on a dry-soil-weight basis), the pesticide vapor pressure over a treated field is usually much less than the saturated vapor pressure of the pure compound. Thus a pesticide that might volatilize readily from a nonreactive surface such as glass may not volatilize much from soil. However, some pesticides, such as butylate and trifluralin, have such a high vapor pressure that, if applied to soil without incorporation, they could have significant volatilization losses. If the surfaces to which pesticides are applied are living and/or dead plant tissue, or soil with little capacity for adsorption (e.g., soils with low organic matter and/or high sand content), or if there is competition for adsorption sites with something such as water molecules, the decreased adsorption could result in unusually high volatilization losses to the atmosphere.

In one study of washoff of herbicides applied directly to corn residue [12], a mass balance indicated that more than 60% of the applied propachlor was lost, presumably by volatilization, in the 1 day between application and rainfall simulation. In another study [50] where pesticide volatilization from soil and corn residue surfaces was measured in a closed system, about 2% of the propachlor applied to soil escaped into the air in the first 5 days after application, whereas 12% was lost when propachlor was applied to corn residue. Measured losses of cyanazine were much less, because its vapor pressure is over 100,000 times less than that of propachlor.

B. Hydrological Factors

The two primary hydrological factors affecting the off-site transport of pesticides from treated fields are the rate of infiltration (relative to rainfall or irrigation intensity) and the route of water infiltrating into and through the soil. The infiltration rate, relative to the water input rate, determines the time of initiation and the quantity of runoff during a runoff event. The quantity of leaching water can be calculated from input minus runoff and soil water storage. The amount of pesticide in the thin mixing zone at the soil surface (discussed above) combined with infiltration is very important in determining pesticide concentrations and losses. It is assumed that it is from this layer that pesticide present is dissolved in rainfall or irrigation water that in turn becomes runoff or leaching water. For moderately, and especially for weakly to nonadsorbed pesticides, the amount of pesticide remaining in the mixing zone decreases with time and with the amount of water moving through this zone during a rain or irrigation event. Thus the time interval between water input and the initiation of runoff is important, as the pesticide concentration in initial runoff decreases with a delay in runoff. Generally the decrease in concentration during runoff, once it begins, is mostly the result of leaching of the pesticide below the mixing zone. Reduction of pesticide amount in the mixing zone with time following application, due to degradation, possible volatilization loss, diffusion, and movement with non–run–off-producing events, explains why pesticide concentrations and losses with surface runoff (depending on relative runoff volumes) are generally the greatest for the first runoff event after application and decrease with time thereafter [4].

An example of the influence of infiltration rate on pesticide concentrations and losses is provided by a rainfall-simulation study where the effects of the amount of corn residue on a recently disked soil on hydrology and pesticide losses were measured [9]. Residue at the rate of 1500 kg/ha protected the soil against surface sealing and delayed runoff by 18 minutes, reducing surface runoff volume by a factor of 3.6 compared to disked plots without corn residue. As a result of delayed and reduced runoff, measured atrazine concentrations were reduced about 50% and atrazine losses with surface runoff were reduced by a factor of 5.9 by the presence of residue. Another example demonstrates the effects of compaction on infiltration rates and pesticide losses [8]. Under rainfall simulation, runoff from recently disked plots with tractor tracks occurred sooner and was 8% greater than from plots without compaction from tractor traffic. Furthermore, average concentrations of the three surface-applied herbicides (atrazine, propachlor, and alachlor) were more than tripled in runoff from the compacted plots, resulting in losses about four times those from the uncompacted plots.

The potential importance of the routes of infiltrating water on chemical leaching is also shown in Figure 2. When the soil surface becomes saturated, not only can runoff begin, but also some water may move into and through the soil through preferential flow paths or macropores, leaching pesticides deeper and more quickly through the soil than would be expected if the water had to flow through the whole soil profile displacing water below. However, flow through macropores could reduce pesticide leaching if the pesticide was not on the surface but within aggregates, such that water moving through macropores would bypass the pesticides. These macropores resulting from root channels, insect burrows, and/or soil cracks are believed to be responsible for the occurrence of soil-adsorbed pesticides at depths and at times are not generally expected. In a tile-drainage study [41], four triazine herbicides were detected in drainage water at a depth of 1.2 m shortly after application, although at low concentrations. In another tile-drainage study [28], the measured loss of atrazine, one of several herbicides studied, was only 0.14% of that applied and much less than the 1–3% or more that could be expected to be lost with surface runoff [9,53]. The maximum atrazine concentration in any individual sample was less than 2 μg/L, although detectable concentrations (≥0.1 μg/L) were measured in some samples 2 years after the last atrazine application.

C. Management Factors

Total pesticide loss from cropland is equal to the summation of the products of the masses of the carriers times the pesticide concentrations in the respective carriers; that is

$$\text{Total loss} = \sum_{n=1}^{3} \text{Mass}_n \times \text{Concentration}_n$$

where n=1 through 3 represent surface runoff water, sediment, and leaching water. Thus, a management practice that reduces carrier mass and/or pesticide concentration reduces loss. When a single management practice is not sufficient to provide the desired level of control, a system of practices is needed. To devise a single management practice or a system of management practices that are efficient in reducing pesticide transport to water resources, knowledge of the major transport mechanism(s) is needed. This requires information on the pesticide's properties, the source(s) of the pesticide, and the soil and climatic conditions. A system may include a combination of in-field and off-site practices. Although it may be possible to devise several different management practices to use in a system to reduce pesticide losses, practical and economic realities considerably limit the number of practices currently feasible. Major in-field management practices involve cropping, tillage, and the rate, method, and timing of chemical application.

Crops planted on agricultural lands affect surface and subsurface hydrology in addition to influencing pesticide inputs, and thus they affect the volumes of carriers and pesticide concentrations in the carriers from those lands. Although economic considerations play a large role in crop selection, often dictating the use of row crops, close-grown or solid-seeded crops such as perennial grasses and legumes generally result in both lower runoff volumes and sediment concentrations, as well as lower pesticide concentrations (in part from less treated area) in water and sediment compared to row crops. Strip intercropping, where narrow crop strips are grown contiguously in the same field, has the potential to maintain or even increase crop yields while providing surface and/or subsurface water quality benefits. In a rainfall-simulation study of a three-crop system of corn, soybeans, and oats/clover, the presence of the oats/clover strip below corn and soybean strips reduced runoff volume about 40%. This, combined with the reduction in imazethapyr application by one third (applied postemergence to only corn and soybeans) because of the oats/clover strip, resulted in a fourfold reduction in herbicide runoff (J.L. Baker, personal observation). In another study [25], atrazine runoff loss from treated corn plots was reduced from 64 to 91% if the lower 6 m of the 22-m plot was oats instead of corn.

Conservation tillage has played a major role in meeting soil loss limits on highly erodable lands. By definition, conservation tillage systems leave at least 30% of the soil surface covered with crop residue after planting. It has been observed that it is residue cover, and not the tillage tool used, that determines the degree of erosion control for a conservation tillage system relative to the conventional moldboard plow system. Conservation tillage, by its effects on erosion, hydrology, and pesticide applications, affects pesticide concentrations and losses. By reducing erosion and soil loss, conservation tillage can control loss of strongly adsorbed, sediment-transported pesticides (see Chapter 25). However, one drawback with conservation tillage is that it is difficult to incorporate pesticides and still maintain 30% crop residue on the soil surface. The other concern is that increased infiltration often associated with conservation tillage will increase pesticide leaching, particularly for no-till, with more macropores coming to the soil surface. As discussed earlier, water flow through macropores may increase or decrease pesticide leaching.

When pesticides are broadcast sprayed in conservation tillage systems, a significant portion of the soil surface is covered by the previous year's crop residue, which will intercept the spray. There are several possible effects of pesticide being intercepted by crop residue. One involves washoff with precipitation or irrigation (and the fate of that washoff water) and another involves potential volatilization of pesticide from the residue. In one study [35], the washoff of atrazine, alachlor, cyanazine, and propachlor applied to corn residue was measured. It showed that there was little interaction between the four herbicides and the residue, with about one third of the applied herbicides washing off with the first 5 mm of water and about three quarters washing off with 35 mm. In a rainfall-simulation study on the effects of conservation tillage on pesticide runoff losses [10], it was determined that despite reduced runoff and erosion, pesticide losses were not reduced relative to the moldboard plow tillage system. In fact, concentrations in runoff from rain applied shortly after pesticide application increased as the percentage of the soil surface covered with corn residue increased. It was felt that water washoff coming directly from the residue and containing high concentrations of pesticide was becoming part of runoff and resulted in the increased concentrations. Further studies [12,35] have confirmed that the pesticides studied wash off easily from crop residue. Also, it was found that neither formulation nor application carrier affected the amount of washoff with 68 mm of simulated rain, and that by far the largest amount of washoff came with the first increment of rain.

With respect to volatilization loss from crop residue, Burt [16] was one of the first to note that volatility was a major factor for dissipation when herbicides were applied to dried plant material as compared to soil. Taylor et al. [49] found that significant amounts (12 and 45%, respectively) of the insecticides dieldrin and heptachlor volatilized from grass surfaces in the first 12 h after application. Martin et al. [35] also noted that there was an apparent significant loss, in a short time, of herbicide applied to corn residue. Baker and Shiers [12] and Baker and Mastbergen [11] noted the same thing. Tremwel [50] followed the dissipation of five herbicides applied to bare soil or soil covered by corn residue by sampling the surface layers down to about 10 cm periodically after herbicide application. For the least volatile herbicide, cyanazine, the amount recovered in residue plus the soil beneath averaged more than was recovered in the bare soil. After 19 days (and two rains), 26% of the cyanazine originally applied was still on the residue and 16% was in the soil. For the most volatile herbicide, propachlor, after 19 days, 0.6% of the propachlor originally applied was still on the residue and 1.0% was in the soil beneath; 3.8% was recovered from the bare soil. Corresponding numbers after 7 days (no rain) for propachlor were 11, 1.5, and 62%. Tremwel also recirculated air over the herbicide-treated residue and bare soil to measure the relative magnitudes of herbicide trapped from the air streams. Although the absolute calculated percentage loss for propachlor after 7 days was only 8.9% (about 50% extra was missing as calculated from the numbers above after 7 days), for propachlor and each of the other four herbicides studied, volatilization was at least four times greater for herbicide applied to residue compared to bare soil. Because of the potential volatilization of herbicides from crop residue, care must be taken in the choice of herbicides used when application to crop residue occurs. Choice of formulation has some potential to reduce this problem. Although the ease of herbicide washoff with rainfall (or irrigation) is good to move the herbicide to the soil, if washoff becomes part of runoff, increased losses could become a problem. A machine that would allow application beneath crop residue with conservation tillage would reduce potential volatilization problems; in addition, if the herbicide also could be soil incorporated, it would have an additional advantage in reducing runoff losses.

It should be pointed out that in studies of tillage effects on pesticide runoff losses under natural rainfall, conservation tillage generally reduced losses. One reason for this is that small rainfalls could have occurred between herbicide application and the first runoff-producing rainfall event, with the small rainfalls washing most of the herbicide off the residue to the soil below. A second reason could be that herbicide dissipation through volatilization losses from the residue could have reduced the amount left to be lost. However, for the studies of Baker and Johnson [6] and Hall et al. [26], nonvolatile cyanazine was involved and losses were still reduced with conservation tillage. In terms of hydrology, on an annual basis, conservation tillage often results in less surface runoff than for a moldboard plow system; however, as noted in a review by Baker [4], for the first runoff event after tillage, areas with the most recent or most severe tillage usually have the least runoff. Therefore, if tillage and pesticide application are done in conjunction, reduced or conservation tillage may be at a disadvantage in terms of surface runoff pesticide loss, because, as noted previously, it is usually the first runoff event that results in the largest concentrations and losses. The other aspect of the change in field hydrology with conservation tillage is the concern for potentially more leaching because of both increased infiltration and the greater probability of the existence of macropores or preferential flow paths at and near the soil surface, particularly for no-till [31].

The rate of pesticide application will obviously affect the amount of pesticide in the soil, and in particular the mixing zone, and thus the concentrations and losses with runoff water, sediment, and leaching water. With respect to runoff water and sediment, in a natural rainfall study, Hall et al. [27] found that for six atrazine application rates ranging from 0.6 to 9.0 kg/ha, the amount lost with surface runoff was roughly proportional to the amount applied, averaging about 3%. In other studies of herbicide runoff losses, Barnett et al. [13] found that doubling the 2,4-D application rate about doubled losses; and Bovey et al. [15] found that increasing the picloram application rate from 0.56 to 2.24 kg/ha doubled losses.

With respect to concentrations and losses with leaching water, in a study of herbicide banding [28], where one third the area was treated resulting in a three-fold decrease in application rate compared to a broadcast application, the amount of cyanazine in tile drainage was decreased by a factor of three by banding. The same was true for atrazine losses with tile drainage water in another study area [5]. Variable-rate application equipment, utilizing geographical information system/global positioning system (GIS/GPS) technology and spatially varied information such as soil organic matter contents, may also contribute to reductions in overall field application rates.

Timing of pesticide applications can significantly affect off-site pesticide losses. As already discussed, pesticide runoff losses are usually the greatest for the first runoff event after application [4], and increased time intervals between pesticide application and runoff events can significantly decrease pesticide losses if rapid dissipation occurs. In one study of application timing using rainfall simulation, it was shown that atrazine surface runoff losses were reduced 56% when rainfall runoff occurred 96 h after application compared to 1 h. The weather forecasting of intense storms in the short term might be used to avoid application just prior to an expected major runoff event. In the longer term, historical weather records may be used on a statistical basis to adjust application timing to avoid larger losses. One of the reasons for the labeling of metolachlor for fall application in the Corn Belt was that by earlier application, less metolachlor would be present, particularly in the thin surface "mixing zone," when the intense thunderstorms commonly occur in the May, June, and early July period.

In one study with fall application of metolachlor [47] the maximum concentration in runoff water was for snow melt (<120 µg/L), whereas for spring application, the maximum concentration was for the first rainfall runoff event that occurred 22 days after application (>2000 µg/L). At that time, the concentration for fall-applied metolachlor was 20 µg/L. Overall, because of a greater volume of snow-melt runoff than rainfall runoff, losses on the percent-of-applied basis were slightly higher for fall application, being about 1%. In a study of the water quality impact of pre- and postemergence application of atrazine to corn [43], Pantone et al. found that reduced runoff, because of different antecedent conditions due to crop and weed vegetation (planting dates were staggered so that herbicide applications occurred on the same date), was the critical factor in the postemergence application having one third the surface runoff loss of the preemergence application.

Incorporation of pesticides into the soil, by mechanical tillage or some other means, such as sprinkler irrigation, to reduce the amount at or near the surface (i.e., in the mixing zone) is a good way to reduce losses with surface runoff. In a rainfall-simulation study of atrazine, propachlor, and alachlor [8], incorporation of herbicides with a tandem disk reduced surface runoff losses with water and sediment by three times over surface application to the plot surfaces after disking. Furthermore, if the surface application without incorporation was made to a disked surface that was compacted by tractor wheel traffic, the losses were reduced by another factor of three. In another study of preplant incorporation [25], atrazine runoff losses were reduced about a factor of three when the herbicide was applied at 2.2 kg/ha and soil incorporated versus a surface preemergence application, although the reduction was not as great for 4.5 kg/ha applications. In a study of the new John Deere Mulch Master (Deere and Co., Moline, IL) secondary tillage tool [38], which includes incorporation wheels, it was found that although after tillage the Mulch Master left more (nearly all) of the corn residue that had been on the surface, the effect on reduction of herbicide losses was about half that from disk incorporation. There have been other noncommercial attempts to develop pesticide incorporation tools that would protect soil conserving surface crop residue [22,36,48], but more work is needed in this area.

Off-site landscape modifications, such as vegetated buffer strips and wetlands, have the potential to reduce pesticide transport between the field boundary and the water resource of concern. Vegetated buffer or filter strips include bands of vegetation located downslope of cropland with the intent to control erosion in the immediate area and possibly to reduce pesticide transport. Vegetated buffer strips can be situated on field borders or as riparian zones (or within fields themselves on the contour or as grassed waterways). The aboveground vegetation and rooting system of the close-grown plants slow overland flow, and the generally better soil structure and lower moisture content compared to the field draining to the buffer strip increases infiltration and removal of pesticides dissolved in that flow. Surface roughness and reduced flow velocity within the buffer strip reduce the carrying capacity for sediment, and sediment and pesticides adsorbed to it can be deposited. In addition, depending on their chemical and physical properties, some pesticides can be removed from overland flow through adsorption to in-place soil and/or living and dead vegetation.

The differences in reduction of pesticide transport between riparian zones, contour buffer strips, and grassed waterways will result from differences in relative areas of drainage to the vegetated buffer strips, the length of travel through the vegetated buffer strip, and the degree of concentration (or depth) of flow. Thus, the topography and relative geometry of the source area and the vegetated area will be important. In a study of the effects of vegetated buffer strips on runoff, sediment transport, and herbicide removal, Wauchope et al.

[55] reported that one third less rain was needed for bare plots to produce the same runoff as for grassed plots in a rainfall simulation study. Asmussen et al. [2], using simulated rainfall, studied the reduction of 2,4-D load in surface runoff flowing through a grassed waterway. On 24.2-m long waterways, incoming suspended sediment was reduced 98 and 94% for dry and wet antecedent conditions, respectively. The total losses of 2,4-D from the plots were 2.5 and 10.3% for the dry and wet plots, respectively. Only about 30% of the 2,4-D that entered the top of the waterway reached the bottom. In a similar study, Rhode et al. [44] determined percentage of surface runoff losses of trifluralin broadcast applied to fields and how vegetated buffer strips receiving that runoff reduced transport. Annual losses as percentages of that applied were low (0.17 and 0.03%) for the 2 years of measurement under natural rainfall. When runoff caused by rainfall simulation was directed onto vegetated buffer strips, trifluralin losses were reduced 96% if the buffer strip was initially dry and 86% if it was prewetted. Over one half of this reduction was attributed to adsorption on vegetation, organic matter, and soil. Hall et al. [25] determined atrazine surface runoff losses from treated corn fields with and without an oat strip at the slope base and also as affected by preplant herbicide incorporation. Atrazine loss without the oat strip, or incorporation, was 3.5% of that applied. With the oat strip, loss was reduced by more than a factor of ten. Preplant incorporation reduced losses even further over a preemergence application.

In a rainfall-simulation study on the effect of vegetative buffer strip length on atrazine transport, reductions in herbicide transport of 31.7 and 55.4% were observed for 4.6 and 9.1-m lengths, respectively, with no significant difference in reduction whether the runoff contained sediment or not [37]. Arora et al. [1] reported that for the first field runoff event after herbicide application, an average of 13% atrazine, 22% metolachlor, and 15% cyanazine was retained for 15:1 and 30:1 area ratios (drainage area to vegetated buffer strip area); the reductions were less for the higher area ratio, but the differences between 15:1 and 30:1 were not statistically significant. In a rainfall-simulation study where bromide was used to trace the infiltration of runoff entering a grassed buffer strip [39], it was determined that the major mode of reduction of transport through the buffer strip for moderately adsorbed herbicides was infiltration. Dillaha et al. [23] conducted a study to investigate the performance of vegetative buffer strips as an agricultural nonpoint source pollution-control measure. They concluded that vegetative buffer strips are effective on steeper hill slopes; that they become ineffective under concentrated flow conditions with time; and that when sediment was deposited higher than the adjacent field, flow parallel to the strips took place that reduced their effectiveness.

One of the strategies for reducing agricultural chemical contamination of surface and groundwater receiving renewed attention is the construction or restoration of wetlands in agricultural watersheds specifically as sinks for agricultural chemical contaminants [52]. Wetlands are areas of intense biological activity, and there is considerable opportunity for chemical transformation and loss as pollutants dissolved in water move through these systems [21]. Although more research is needed, there is preliminary information that pesticide adsorption to wetland biomass can quickly reduce concentrations (W.G. Crumpton, personal communication) and that wetland plants can enhance pesticide degradation (J.R. Coats, personal communication).

IV. CONCLUSIONS

In reducing or controlling nonpoint source pollution from the use of pesticides, an understanding of the importance of and the relationships between chemical, hydrological, and management factors is necessary. Although water quality concerns relate to pesticide concentrations, pesticide loss reduction is usually the goal as the amount of loss will determine the impact relative to concentrations in a water resource "downstream." Loss reduction is achieved by reducing the concentrations in and/or the volumes of runoff water, sediment, and leaching water carriers. Generally pesticide losses with surface runoff dominate over leaching losses. In-field and/or off-site practices can be used to reduce losses; in the instances when a single management practice is not sufficient to achieve the necessary loss reduction, a system of practices will be needed.

REFERENCES

1. Arora, K., J.L. Baker, S.K. Mickelson, & D.P. Tierney. 1996. Herbicide retention by vegetative buffer strips from runoff under natural rainfall. *Trans. ASAE* 39:2155–2162.
2. Asmussen, L.E., A.W. White, E.W. Hauser, & J.M. Sheridan. 1977. Reduction of 2,4-D load in surface runoff down a grassed waterway. *J. Environ. Qual.* 6:159–162.
3. Baker, D.B., L.K. Wallrabenstein, & R.P. Richards. 1994. Well vulnerability and agricultural contamination: Assessments from a voluntary well testing program. In *New Directions in Pesticide Research, Development, Management, and Policy: Proceedings of the 4th National Conference on Pesticides*. Blacksburg, Virginia: Virginia Polytechnic Institute, pp 470–494.
4. Baker, J.L. 1980. Agricultural areas as nonpoint sources of pollution. *In Environmental Impact of Nonpoint Source Pollution*. M.R. Overcash and J.M. Davidson, eds. Ann Arbor, Michigan: Ann Arbor Science Publishers, pp 275–310.
5. Baker, J.L., T.S. Colvin, D.C. Erbach, & R.S. Kanwar. 1995. Potential water quality and production efficiency benefits from reduced herbicide inputs through

banding. In *Comprehensive Report, Integrated Farm Management Demonstration Program, Rep. No. IFM 16*. Ames, Iowa: Iowa State Extension, pp 5.6–5.9.

6. Baker, J.L., & H.P. Johnson. 1979. The effect of tillage systems on pesticides in runoff from small watersheds. *Trans. ASAE* 22:554–559.

7. Baker, J.L., R.S. Kanwar, & T.A. Austin. 1985. Impact of agricultural drainage wells on groundwater quality. *J. Soil Water Conserv.* 40:516–520.

8. Baker, J.L., & J.M. Laflen. 1979. Runoff losses of surface-applied herbicides as affected by wheel tracks and incorporation. *J. Environ. Qual.* 8:602–607.

9. Baker, J.L., & J.M. Laflen. 1982. Effect of corn residue and herbicide placement on herbicide runoff losses. *Trans. ASAE* 25:340–343.

10. Baker, J.L., J.M. Laflen, & H.P. Johnson. 1978. Effect of tillage systems on runoff losses of pesticides, a rainfall simulation study. *Trans. ASAE* 21:886–892.

11. Baker, J.L., & B. Mastbergen. 1986. *Fate of Broadcast Herbicides Used with Conservation Tillage Systems—Part II*. Completion Report No. NCRPIAP 218, North Central Region Pesticide Impact Assessment Program. Columbus, Ohio: Ohio State University.

12. Baker, J.L., & L.E. Shiers 1989. Effects of herbicide formulation and application method on washoff from corn residue. *Trans. ASAE* 32:830–833.

13. Barnett, A.P., E.W. Hauser, A.W. White, & J.H. Holladay. 1967. Loss of 2,4-D in washoff from cultivated fallow land. *Weeds* 15:133–137.

14. Benson, R.W. 1989. EPA drinking water standards. In *Proceedings of the 5th National Domestic Water Quality Symposium*. ASAE, Dec. 11–12, New Orleans, Louisiana. St. Joseph, Michigan: ASAE, pp 67–76.

15. Bovey, R.W., C. Richardson, B. Burnett, M.G. Merkle, & R.E. Meyer. 1978. Loss of spray and pelleted picloram in surface runoff water. *J. Environ. Qual.* 7:178–180.

16. Burt, G.W. 1974. Volatility of atrazine from plant, soil, and glass surfaces. *J. Environ. Qual.* 3:114–117.

17. Caro, J.H., H.P. Freeman, D.E. Glotfelty, B.C. Turner, & W.M. Edwards. 1973. Dissipation of soil-incorporated carbofuran in the field. *J. Agric. Food Chem.* 21:1010–1015.

18. Caro, J.H., H.P. Freeman, & B.C. Turner. 1974. Persistence in soil and losses in runoff of soil-incorporated carbaryl in a small watershed. *J. Agric. Food Chem.* 22:860–863.

19. Caro, J.H., & A.W. Taylor. 1971. Pathways of loss of dieldrin for soils under field conditions. *J. Agric. Food Chem.* 19:379–384.

20. Coupe, R.H., D.A. Goolsby, J.L. Iverson, D.J. Markovchick, & S.D. Zaugg. 1995. *Pesticide, Nutrient, Water Discharge and Physical-Property Data for the Mississippi River and Some Of Its Tributaries, April 1991-September 1992*. U.S. Geol. Surv. Open file rep. 93-657, Denver, Colorado.

21. Crumpton, W.G., T.M. Isenhart, & S. Fisher. 1993. Transformation and fate of nitrate in wetlands receiving nonpoint source agricultural inputs. In *Constructed Wetlands for Water Quality Improvement*. G.A. Moshiri, ed. Chelsea, Michigan: Lewis Publishers, pp 283–291.

22. Dawelbeit, M.I. 1983. *Design and evaluation of a corn residue managing machine*

for conservation tillage systems. Unpublished PhD dissertation. Iowa State University, Ames, Iowa.

23. Dillaha, T.A., J.H. Renueau, S. Mostaghimi, & D. Lee. 1989. Vegetative filter strips for agricultural nonpoint source pollution control. *Trans. ASAE* 32:513–519.

24. Gianessi, L.P., & C.M. Puffer. 1990. *Herbicide Use in the United States—National Summary Report*, revised April 1991. Washington, DC: Resources for the Future.

25. Hall, J.K., N.L. Hartwig, & L.D. Hoffman. 1983. Application mode and alternate cropping effects on atrazine losses from a hillside. *J. Environ. Qual.* 12:336–340.

26. Hall, J.K., N.L. Hartwig, & L.D. Hoffman. 1984. Cyanazine losses in runoff from no-tillage corn in 'living' and dead mulches vs. unmulched, conventional tillage. *J. Environ. Qual.* 13:105–110.

27. Hall, J.K., M. Pawlus, & E.R. Higgins. 1972. Losses of atrazine in runoff water and soil sediment. *J. Environ. Qual.* 1:172–176.

28. IDALS. 1994. *Agricultural Drainage Well Research and Demonstration Project,* annual report. Ames, Iowa: Iowa Department of Agriculture and Land Stewardship and Iowa State University.

29. Johnson, H.P., & J.L. Baker. 1982. *Field-to-Stream Transport of Agricultural Chemicals and Sediment in an Iowa Watershed: Part I. Data Base for Model Testing (1976–1978).* Report No. EPA-600/53-82-632. Washington, DC: EPA.

30. Johnson, H.P., & J.L. Baker. 1984. Field-to-Stream Transport of Agricultural Chemicals and Sediment in an Iowa Watershed: Part II. Data Base for Model Testing (1979-1980). Report No. EPA-600/53-84-055. Washington, DC: EPA.

31. Kanwar, R.S., T.S. Colvin, & D.L. Karlen. 1995. Tillage and crop rotation effects on drainage water quality. In *Proceedings, Clean Water–Clean Environment–21st Century.* Vol III. St. Joseph, Michigan: ASAE, pp 163–166.

32. Kelley, R.D., G.R. Hallberg, L.G. Johnson, R.D. Libra, C.A. Thompson, R.C. Splinter, & M.G. Detroy. 1986. Pesticides in ground water in Iowa. In *Agricultural Impacts on Ground Water.* Westerville, Ohio: National Water Well Association.

33. Kladivko, E.J., G.E. Van Scoyoc, E.J. Monke, K.M. Oates, & S.W. Pask. 1991. Pesticide and nutrient movement into subsurface tile drains on a silt loam soil in Indiana. *J. Environ. Qual.* 20:264–270.

34. Kross, B.C., G.R. Hallberg, R.D. Libra, L.F. Burmeister, & K.L. Cherryholmes. 1990. *The Iowa State-wide Rural Well-water Survey; Water Quality Data: Initial analysis.* Technical information series 19. Des Moines, Iowa: Department of Natural Resources.

35. Martin, C.D., J.L. Baker, D.C. Erbach, & H.P. Johnson. 1978. Washoff of herbicides applied to corn residue. *Trans. ASAE* 21:1164–1168.

36. Mickelson, S.K., & J.L. Baker. 1991. Band injection of herbicide. Paper no. 91-1542 presented at *ASAE International Winter Meeting*, ASAE, Chicago. St. Joseph, Michigan: ASAE.

37. Mickelson, S.K. & J.L. Baker. 1993. Buffer strips for controlling herbicide runoff losses. Paper No. 93-2084 presented at *ASAE International Summer Meeting*, Spokane, Washington. St. Joseph, Michigan: ASAE.

38. Mickelson, S.K., J.L. Baker, J.A. Baldauf, & P.M. Boyd. 1995. Tillage and herbicide incorporation effects on runoff, erosion and herbicide loss. Paper No. 95-2695 presented at *ASAE International Annual Meeting*. Chicago. St. Joseph, Michigan: ASAE.

39. Misra, A.K., J.L. Baker, S.K. Mickelson, & H. Shang. 1996. Contributing area and concentration effects on herbicide removal by vegetative buffer strips. *Trans. ASAE* 39:2105-2111.

40. Monsanto Agricultural Company. 1990. *The National Alachlor Well Water Survey (NAWWS): Data Summary*. Monsanto Technical Bulletin. St. Louis, Missouri: Monsanto Agricultural Company.

41. Muir, D.C., & B.E. Baker. 1976. Detection of triazine herbicides and their degradation products in tile-drain water from fields under intensive corn (maize) production. *J. Agric. Food Chem.* 24:122-125.

42. Nations, B.K., & G.R. Hallberg. 1992. Pesticides in Iowa precipitation. *J. Environ. Qual.* 21:486-492.

43. Pantone, D.J. R.A. Young, D.D. Buhler, C.V. Eberlein, W.C. Koskinen, & F. Forcella. 1992. Water quality aspects associated with pre-and postemergence applications of atrazine in maize. *J. Environ. Qual.* 21:567-573.

44. Rhode, W.A., L.E. Asmussen, E.W. Hauser, R.D. Wauchope & H.D. Allison. 1980. Trifluralin movement in runoff from a small agricultural watershed. *J. Environ. Qual.* 9:37-42.

45. Scribner, E.A., D.A. Goolsby, E.M. Thurman, M.T. Meyer, & M.L. Pomes. 1994. *Concentrations of Selected Herbicides, Two Triazine Metabolites and Nutrients in Storm Runoff from Nine Stream Basins in the Midwestern United States, 1990-92*. U.S. Geol. Surv. open-file report 94-396, Denver, Colorado.

46. SETAC. 1994. *Aquatic Risk Assessment and Mitigation Dialogue Group-Final Report*. Pensacola, Florida: Society of Environmental Toxicology and Chemistry.

47. Shang, H., J.L. Baker, S.K. Mickelson, & D.P. Tierney. 1996. Water quality aspects of fall metolachlor application. *WSSA Abstracts* 36:62.

48. Solie, J.B., H.D. Wittmuss, & O.C. Burnside. 1983. Improving weed control with a subsurface jet injector system for herbicides. *Trans. ASAE* 26:1022-1029.

49. Taylor, A.W., D.E. Glotfelty, B.C. Turner, R.E. Silver, H.P. Freeman, & A. Wiess. 1977. Volatilization of dieldrin and heptachlor residues from field vegetation. *J. Agric. Food Chem.* 25:542-548.

50. Tremwell, T.K. 1985. Fate of broadcast herbicides used with conservation tillage systems. MS Thesis, Iowa State University, Ames, Iowa.

51. USEPA. 1990. *National Pesticide Survey Phase 1 Report*. EPA 570/9-90-003, Sept. Washington, DC.

52. van der Valk, A.G., & R.W. Jolly. 1992. Recommendations for research to develop guidelines for the use of wetlands to control rural NPS pollution. *Ecol. Eng.* 1:115-134.

53. Wauchope, R.D. 1978. The pesticide content of surface water draining from agricultural fields-a review. *J. Environ. Qual.* 7:459-472.

54. Wauchope, R.D., T.M. Buttler, A.G. Hornsby, P.W.M. Augustijn-Beckers, & J.P. Burt. 1992. The SCS/ARS/CES pesticide properties database for environmental decision-making. *Reviews of Environ. Contam. and Toxicol.* 123:1-164.

55. Wauchope, R.D., G.W. Randall, & L.R. Marti. 1990. Runoff of sulfometuron-methyl and cyanazine from small plots: effects of formulation and grass cover. *J. Environ. Qual.* 19:119–125.
56. Weed Science Society of America. 1994. *Herbicide Handbook.* 7th ed. Champaign, Illinois.

27
Decision Thresholds in Pest Management

Leon G. Higley
University of Nebraska–Lincoln, Lincoln, Nebraska

Larry P. Pedigo
Iowa State University, Ames, Iowa

I. INTRODUCTION

The key issues in pest management are largely matters of sustainability. These include the notion of economic sustainability (maintaining producer profits), environmental sustainability (choosing and using tactics in such a way as to minimize detrimental impacts on the environment), and sustainability of tactics (using tactics in such a way that their effectiveness is maintained through time). At the heart of these issues of sustainability is the recognition that not all pests require management. Defining when management is necessary is the raison d'être of pest management decision making.

Identifying a species as a pest is very much an anthropomorphic designation. Typically, we regard species as pests when their activities impair human health or economic well-being or when they are esthetically displeasing (which in some instances may also have an economic dimension). This is easily seen with agricultural pests, for example, when the injury to a crop is proportional to the number of pests. There are pests, however, for which the relationship between their importance as a pest and their number is less clear. For example, many homeowners would regard one or two cockroaches in a house just as unacceptable as four or five. The importance of pests is a function both of their activities and of their numbers. Consequently, when pest numbers are very low or when the pests' impact is minor, then little or no management action is needed and the pests can be tolerated. Decision making in pest management is important for identifying when pests are tolerable and when they are not.

Fundamentally, the concept of tolerance is one of the defining principles of pest management. By this we mean that not every occurrence of a pest requires management action. A pest *control* philosophy typically focuses on pest presence or absence; if pests are present then action is taken. A *management* philosophy, in contrast, argues that it is only necessary to take action against pests when their activities are sufficient to justify economically intervention. In its simplest statement, this is a break-even philosophy in which the costs of taking actions against pests are weighed against the costs of not taking action.

This issue of identifying which levels of pests can be tolerated depends on the economics of crop production and the economics of managing pests. The challenge to pest management decision making is identifying how much yield loss can be tolerated. The simplest instance, of course, is when pest activity does not result in diminished yield. This represents an intrinsic tolerance of plants to injury and is a function of a specific plant and pest species combination. The more challenging question is to determine how much yield loss is necessary to justify management action. The economic injury level (EIL) is the criterion for identifying this point. In its simplest statement, the EIL represents that level of injury (typically expressed not as injury but as a number of pests, such as numbers of insects or numbers of weeds) producing an economic value equal to the economic cost of management action.

Because the EIL defines what is tolerable, EILs are a cornerstone for pest management. This is true in two ways. First, as an operational criterion, the EIL is extremely valuable in determining when we need to take action against pests and when we need not. To the extent that we are limited in choice of tactics for taking action against existing pest problems, so is the operational use of the EIL limited. In particular, most use of EILs has been associated with pesticides, because pesticides typically are the only tactic available for curing pest problems. Second, EILs can define the nature of a pest-crop relationship. It is easy to recognize that those pest species routinely producing injury in excess of the EIL are more severe pests than those that only occasionally exceed the EIL. Thus, the EIL defines a pest species and provides a ruler for measuring the severity of pests.

In the remainder of the chapter, we briefly discuss the conceptual background underlying EILs and economic thresholds. We also will consider some new adaptations and uses of thresholds, as well as identifying some limitations that continue to exist both in the development and application of thresholds. There are a number of reviews relating to thresholds and their use, including papers by Stern [58], Mumford and Norton [34], and Pedigo et al. [48], and a new volume exclusively on thresholds in pest management [18]. Rather than repeat information available from these other sources, we will try to highlight key issues in threshold development and use and will refer the reader to additional resources as appropriate.

II. HISTORY OF EILs AND THRESHOLDS

A. Stern et al.'s Contributions

The original definition of EILs and thresholds was presented by Stern et al. [59] in their seminal paper entitled "The Integrated Control Concept." In this paper, Stern and coworkers set out a theoretical framework for integrating different types of management tactics and for identifying when the use of a given management tactic is not necessary. Other workers had previously raised questions about whether all pest situations merited control action, but it was not until Stern et al. defined the concepts of the EIL and economic threshold that formal rules were developed for addressing this important question.

Stern et al. [59] defined the EIL as "the lowest population density that will cause economic damage." They defined an economic threshold (ET) as "the population density at which control measures should be determined to prevent an increasing pest population from reaching the EIL." The EIL defines where the cost of management is equal to the benefits of management, and it defines a time when action should be taken to prevent increasing injury (an increasing pest population) from reaching the EIL. The EIL represents the calculated, fundamental value for making management decisions; it represents the operational criterion for taking action against pests based upon an EIL.

Without question, the contributions of Stern et al. in defining the EIL and ET were crucial in providing objective criteria for making management decisions about pests and in contributing to the development of pest management. Nevertheless, there were gaps in the definitions they provided [59]. Pedigo et al. [48] discussed some of these problems in original definitions of the EIL and ET. In brief, although Stern et al. laid the framework for calculating EILs, their original statement did not detail how EILs could be calculated. In particular, Stern et al. defined EILs with respect to economic damage but did not explain how to calculate economic damage. Also, in defining the EIL as a "population density" rather than as a level of injury, they opened the door for some subsequent confusion in exactly what was represented by an EIL. Fortunately, these problems were rectified by subsequent workers; unfortunately, the same cannot be said of confusion arising out of definitions of ET.

Stern et al. defined the ET as a population density, but the population density actually is used as an index of injury or, more specifically, as a predictor of future injury. Although the EIL can be calculated based upon field experimentation, calculating ETs (given their predictive nature) has proven to be an almost intractable problem. This difficulty has been complicated by extraordinary confusion over what constitutes an EIL and what constitutes an ET. The literature has many examples of values called ETs that are clearly EILs. Additionally, many investigators have invented new terms, such as action threshold, action levels, or action threshold levels, to refer to things that are identical to ETs or only negligibly different from them. Most of these new terms do little to advance a real understanding of decision making and do much to confuse that understanding. We have argued [48] that the most logical way to proceed is to employ the original terminology of Stern et al., modifying it for clarity as necessary, and to avoid the construction of new terms unless they represent genuinely novel ideas.

B. Calculating EILs

Although Stern et al. presented their concepts of the EIL and ET in late 1959, it took over a decade for the first published application of these ideas. In 1972, Stone and Pedigo [60] presented calculated EILs for the green cloverworm, *Plathypena scabra* (F.), in which they applied the concepts from Stern et al. to calculate formally an EIL based upon research data. Between 1959 and 1972, there are occasional mentions in the literature of other EILs with the implication that such EILs were being implemented in some situations; however, we are unable to find any published description of how such EILs were calculated. We suspect these reports of EILs actually were for informal management guidelines (nominal thresholds sensu Poston et al. [53]) rather than formally calculated EILs. The key contribution in Stone and Pedigo [60] was their formal definition of economic damage. In principle, economic damage is just that level of damage (expressed in dollars) equal to the cost of managing the pests. Although Stone and Pedigo calculated economic damage as a monetary value, they proposed expressing economic damage in terms of marketable produce, which they called the gain threshold. Following Stone and Pedigo, there was more rapid development of ETs and more sophisticated expression of EILs. Over the next two decades, most of the effort on EILs was focused upon defining EILs for different pests or pest complexes [49]. Most of these efforts were associated with insects, which is not surprising given the origin of the concept in entomology and the common use of therapeutic tactics (insecticides) for insect management. More recently, expansions of the EIL concept for new types of pests and different situations has occurred. Also, greater attention has been given to the determination of ETs.

C. Other Approaches

Although the EIL has been the most influential approach to pest management decision making, other approaches have been proposed. The EIL/ET model is a form of cost-benefit analysis, and other forms of cost-benefit models have been developed based on marginal analysis and optimization approaches. Dynamic programming (e.g., see refs. 39 and 56) and linear programming (e.g., see ref. 29) allow for consideration of multiple factors in determining optimal management decisions. Related approaches include various decision theory and behavioral decision models. These alternative procedures offer more than simple cost-benefit consideration, reflecting factors such as multiple stressors and tactics, as well as individual preferences and noneconomic criteria. Unfortunately, the price of these broader considerations is the addition of considerable complexity. Mumford and Norton [34] and Norton and Mumford [36] provide detailed reviews of these procedures.

III. ECONOMIC INJURY LEVEL

A. Definitions

Although Stern et al. [59] provided a conceptual definition of the EIL, it remained for subsequent workers to provide a detailed mathematical description. Stone and Pedigo [60] provided the first approach to this question, and Norton [35] proposed a somewhat broader model. Later, Pedigo et al. [48] drew on the Norton definition to propose a general model of the EIL, which seems to be widely accepted (e.g., see ref. 31). This definition is

$$\text{EIL} = \frac{C}{\text{CDIK}} \tag{1}$$

where EIL = the EIL in pests per production unit (e.g., insects/ha), C = management costs per production unit ($/ha), V = market value per unit of production ($/kg), D = damage per unit injury (kg reduction/ha/injury), I = injury per pest (injury/insect), and K = proportional reduction in injury with management (under certain circumstances K may be interpreted as the proportional reduction in damage rather than injury). Often K is assumed to be equal to 1 (indicating all damage is prevented with management) and is omitted from the EIL equation. When injury and damage cannot be differentiated, D and I may be replaced by a single variable, D' (= yield loss per pest) and in these instances:

$$\text{EIL} = \frac{C}{\text{VD'K}} \tag{2}$$

In this instance, K represents the proportional reduction in damage (i.e., yield loss).

Other expressions of the EIL are possible. For example, an alternative definition is commonly used for weeds; however, it has the same terms (with slightly different names) as in equation 1 [33]. With weeds, the EIL and ET are equivalent and the weed literature refers to ETs rather than EILS. Also, modifications to the C/VDIK model are possible; for example, a term may be included to reflect the impact of the weed bank [9]. Mortensen and Coble [33] provide a thorough review of thresholds in weed management. Various other modifications of the EIL are possible to reflect additional criteria, such as the inclusion of environmental costs [22], interseasonal effects [67], esthetic value [54], or crop quality [26].

A few points regarding the C/VDIK model (see equation 1) bear mentioning. First, it has been argued that it is unnecessary to differentiate between injury and damage in determining EILs [37,38]. At least three considerations argue against such a view: (a) by ignoring injury, understandings of injury/yield-loss relationships are limited to simple regressions and may be confounded with other factors, (b) pest injury rates may not be constant in some situations (which requires that I be considered separately from D), and (c) explicit considerations of injury can provide a basis for developing multiple-species EILs. Another incorrect notion is that the C/VDIK model implies a linear relationship between injury and damage. Actually, D and I are independent variables with D having a functional relationship with total injury but not injury per individual (I). Finally, the EIL model (and related ETs) are based on preventable injury. If significant injury has occurred before the use of an EIL or ET, the C/VDIK calculation may not be valid. It is possible to reflect past injury in calculating an EIL, but at the cost of considerable additional complexity (Higley and Pedigo [19] present a modified EIL formula that includes past injury). As a practical issue, EILs and ETs have the unstated premise that management decisions will be made before substantial injury has occurred. Taking timely samples and making early decisions seem to be more reasonable approaches than correcting for past injury through involved calculations in a modified EIL.

As calculated, EILs and ETs typically are expressed in numbers of pests per unit area—a pest density; however, more fundamentally, thresholds are based on injury, and pest densities serve as an index to injury. After all, pest numbers per se do not affect yields, it is the injury from pests (such as leaf feeding, light competition from weeds, or disease) that actually impacts plants. It is possible to express EILs in terms of injury, and, in some instances, it is even possible to express ETs through injury. For example, ETs for replanting soybeans after stand loss from insect injury are expressed as a percentage of stand loss [21].

Recognizing the distinction between injury and pests is particularly important in determining ETs. Usually, the ET is thought of as some pest density less than the EIL, with the premise that the pest population will increase to reach the EIL. But the more precise view of an ET is as a level of injury that will increase to reach the EIL. In some instances, pest numbers may not change, but their injury will increase. For example, weed seedlings produce increasing injury as they grow, as do many immature insects. It is even possible to have ETs expressed in pest numbers that are higher than the EIL (also expressed in pest numbers); for instance, an ET for immature insects with high larval mortality (and low injury rates) may be larger than the EIL for mature larvae.

B. Components of the EIL

The EIL embodies considerable economic and biological information. Because economic and biological relationships are variable, so too may EILs vary. Although EILs often are expressed as single values, they actually vary with their components. Some investigators speak of dynamic and static EILs, but these terms typically have little meaning. All EILs are dynamic, because they depend on potentially changing components. It may be possible to include more explicit considerations of this variation in the EIL calculation (e.g., by representing D as a function over different plant stages), but such modifications do not make the EIL more dynamic; they merely represent underlying variability explicitly rather than implicitly.

Significant modifications of the EIL are possible through considerations of factors such as esthetics, multiple stressors, quality effects, environmental risk, or perennial effects [18]. These modifications have greatly expanded the scope and utility of the EIL, and the development of additional modifications seems likely. Because the EIL includes the key economic and biological variables associated with pest problems, focusing on modifying these variables may offer potential for new directions in pest management. For example, we [17,45] have suggested using the elements of the EIL as a framework for improving the environmental sustainability of pest management.

1. Economic Components

The economic variables of the EIL include the management costs, C, and value of the commodity, V. Additionally, the efficacy of management, K, although a biological variable, is strongly related to C, because costs may vary among tactics with different efficacies or even within a given tactic (such as with different rates of a pesticide). Despite the considerable work done on the efficacy of management tactics, K typically receives little attention in the determination of EILs unless the efficacy is poor relative to other tactics. Were sufficient information available, it would be possible to compare different pesticides and rates with corresponding variation in costs and efficacies. However, management costs and efficacies are typically taken as set values in EIL calculations, and EILs are calculated for a standard range of costs. Another possible modification of the management costs is the inclusion of additional factors. For example, adding environmental costs has been proposed as a mechanism for developing more environmentally responsive EILs [22].

Variability in the commodity value, V, typically is accommodated in EIL calculations by determining EILs for a range of prices. In forecasting a final market value of a commodity, some uncertainty is necessarily incorporated into the EIL, but this variability is no different than with other aspects of agricultural production economics. Market value can be a crucial determinant of pest status; the tolerability of a pest. High-value crops are less able to tolerate pests (EILs are lower) than are lower value crops (EILs are higher). This influence of market value on EILs also extends to other aspects of pest management. For instance, management tactics requiring some degree of pest tolerance, such as biological control, may be less applicable on higher value commodities.

2. Biological Variables

The biological variables of the EIL include injury per pest, I, the damage or yield loss per unit injury, D, and the proportion of injury prevented by management, K. These components of the EIL can be difficult to estimate, typically are variable, and may be less well known than the other elements of the EIL. The biological variables of the EIL embody the relationship between pest and host, and understanding those relationships is the most difficult challenge in developing EILs.

Management efficacy is dependent on the relationship between pest biology and the impact of a specific tactic, but, as we indicated previously, K also has an economic aspect. Certainly, K is the most straightforward of the biological variables, but in the context of management failures, particularly through the development of pest resistance, K can be a significant factor. Nevertheless, relatively little attention has been given to modifications of K, and it often is ignored or treated as a constant in EIL calculations.

Most attention in the biological aspects of EIL development involves the D and I components. Characterizing D and I represents a practical application of our knowledge of pest-host relationships. Unfortunately, constraints in our understandings of these relationships also represent constraints in our ability to develop better EILs. For EILs, the relationship between pests and plant stress, expressed as yield loss, is most important. In describing biotic stress, we commonly distinguish between injury, damage, and stress [1,48,61]. Specifically, injury is a stimulus producing an abnormal change in a physiological process; damage is a measurable reduction in plant growth, development, or reproduction resulting from injury; and stress is a departure from optimal physiological conditions [16]. For EILs, injury is represented in I and damage in D. Thus, the relationship described by the combination of D and I represents a practical description of plant stress; specifically, stress as measured by yield reduction.

Injury. Of the two components, injury (I) is the most easily understood. Key factors influencing injury include the type of injury, injury rate, and spatial and temporal patterns of injury. Although quantifying injury is difficult or impossible for some pests (including many pathogens, weeds, and sucking insects), the factors influencing injury are still important considerations in developing EILs for these species.

The type of injury is a fundamental determinant of plant response. Historically, insects have been differentiated based on which plant part they injure. For example, insects have been grouped as leaf feeders, root feeders, fruit feeders, and so forth. This categorization has some usefulness in describing injury, but various workers have recognized that a more meaningful differentiation is provided by recognizing different physiological impacts of injury on plants. Boote [5] proposed a list of categories of physiological response of plants to injury, and his categories have been expanded by subsequent workers [16,48]. For insects, categories of physiological impact include population or stand reduction, leaf-mass reduction, leaf photosynthetic rate reduction, leaf senescence alteration, light reduction, assimilate removal, water balance disruption, seed or fruit destruction, architecture modification, and phenological disruption, with additional categories likely with further research [16].

Characterizing physiological responses to injury is important as a mechanism for developing injury guilds. Injury guilds are groups of species whose physiological impact are equivalent. Injury guilds provide a basis for developing multiple species EILS. Various workers have proposed requirements for establishing an injury guild [16,27,28], and these requirements include such features as similar injury types, common injury phenologies, similar intensities of injury, and injury to the same plant parts. Similarities in physiological response have been identified for different pest species [16,50,51,52,64,65,66]. Injury guilds have been established including those for weeds [69], defoliating caterpillars in soybean [28], and stubble defoliators in alfalfa [51]. For weeds, injury guilds are based on competitive indices and are essential, because weeds invariably occur as a complex of species [33].

Beyond the injury type, rates of injury are another important consideration. Ostlie's [40] research on soybean defoliation demonstrated that plant response to injury is a function of the temporal pattern of injury, as well as the type and amount of injury. This finding has important practical implications, because much initial work in developing EILs for insects was based on hail-simulation studies that impose injury over a single day, unlike actual insect injury. Unfortunately, the legacy of hail studies persists, and many studies relating insect injury to yield loss impose injury on a single day or over a greatly abbreviated interval.

The plant part injured has an obvious influence on plant response. One important distinction is between direct injury (injury to yield-producing structures) and indirect injury (injury to non–yield-producing structures). Generally, plants are less able to tolerate direct injury, and EILs will be lower for pests producing direct injury.

A final issue regarding injury is the amount of injury. For a pest population, the amount of injury is directly related to stress severity, and we will consider the quantitative relationship between injury and yield in our discussion of damage. For individual pests, the amount of injury can be important and variable. For pathogens and many types of sucking insects, the amount of injury per pest cannot be determined, and injury is considered with damage. Similarly, injury per weed is included as a part of weed-competitive indices that can be related to yield loss. For many insects, injury rates are equivalent to consumption rates, so factors influencing consumption will influence injury. Typically, consumption rates of insects have logarithmic growth by immature stage [4,14,20,29,51], so most injury is associated with the older instars (a similar pattern of dramatic increase in injury occurs with weed seedlings once they establish a height differential and can shade the crop). Other factors influencing injury rates may include temperature [14], parasitism of the pest [10], and crowding [47].

The host tissue can also influence injury rates, in that injury to expanding leaf tissue increases as the leaf expands. Hunt et al. [25] documented an eightfold increase in final tissue loss with the bean leaf beetle, *Ceratoma trifurcata* (Forster), injury to seedling soybean and developed a procedure for reflecting the influence of leaf expansion on injury rates. Such a modification is only necessary when expanding tissue represents a significant proportion of total plant tissue, as with seedling plants.

Damage.   Because damage represents the interaction of pest attack with plant physiology, it is the most complex component of the EIL. Damage is a reflection of plant stress as measured through yield reductions, and all the variables influencing yield can influence damage.

Although comprehensive models of abiotic plant stress are being developed [6,32], these do not include a consideration of biotic stress. Nevertheless, the development of these models holds the promise that integrated understandings of many types of stress, including biotic and abiotic plant stress, may some day be possible. Many workers have argued for improved understandings of biotic stress (e.g., see refs. 1,5,12,10, and 66), and Higley et al. [16] have proposed a conceptual framework for building more integrated understandings of biotic stress. Detailed understandings have been developed in specific areas, such as disease physiology, but much work on biotic stress of plants is still in its infancy. Unfortunately, because our understanding of biotic stress is limited, so our characterizations of damage are also limited. We believe a poor understanding of plant stress is the greatest impediment to EIL development.

Considerable differences exist in the understanding of biotic stress for different types of pests. Work on disease physiology is perhaps the most mature of that for any pest, although the emphasis has been on details of the infection process and molecular and cellular responses rather than yield-loss relationships. Research on weed-crop competition and mathematical models of these relationships have received growing attention over the past 10 years. Mortensen and Coble [33] provide a comprehensive review of this literature. Although the literature on the impact of plants on insects is large, the literature on the impact of insects on plants is not extensive. A number of important references address aspects of plant response to insect injury [1,16,20,48,50,61,62,64,66], and in the ecological literature, some important reviews [3,7,30,68] summarize research on the impact of herbivores on plants, which has some practical implications.

Despite limitations in our knowledge of biotic stress, the theoretical relationship between increasing injury and crop yield is well known (Fig. 1). This relationship is called the damage curve, and it represents the array of possible yield responses to increasing levels of injury. Not all pests and crops will display all portions of the damage curve, but it encompasses all possible responses. Increasing levels of injury are associated with increasing numbers of pests; therefore, another way to view the damage curve is as a reflection of how crop yields may be affected by increasing numbers of pests in a generic sense. The damage curve was originally described by Tammes [61] and is an experimentally derived expression of all potential of responses of crops to pest attack. In the context of tolerance, it is noteworthy that at some low levels of injury, there

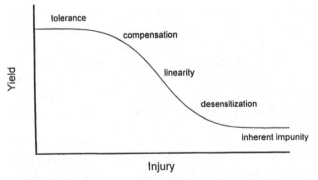

Figure 1 The damage curve representing the theoretical relationship between injury and yield loss.

may be no impact of the pest upon yield. At higher levels of injury, yield decreases but not at a maximal rate; the plant compensates for injury. Next, is a region where the maximum impact of injury occurs followed by portions of the curve where the impact diminishes. In a detailed sense, the damage curve illustrates exactly how injury alters yield.

Pedigo et al. [48] developed a terminology to describe specific portions of the damage curve, and they present examples of different types of damage curves. Higley and Peterson [20] further review examples of the damage curve for different insect-crop combinations, and they critique some inappropriate approaches used in describing these relationships.

To the extent that the damage curve simply relates observed yield to observed levels of injury, it is descriptive and offers little direct insight into plant stress. But, indirectly, the shape of the damage curve does indicate the existence of tolerance, compensation, and even overcompensation to injury. One logical outgrowth of research defining the damage curve is the development of physiologically based explanations for yield loss. Higley [15] and Hunt et al. [24] present examples of ongoing work of this type in describing the importance of altered canopy light interception in mediating yield loss in soybean from insect defoliation.

The quantitative relationship between injury and yield is not a constant, and various factors may influence the damage curve. Pedigo et al. [48] discuss these factors, so we will only briefly review them. As previously discussed, the injury-yield loss relationship is affected by the type of injury, plant part injured, and the duration of injury. Although the damage curve does not explicitly reflect the duration of injury, duration is typically reflected in how the curve is experimentally derived. Consequently, it may be more appropriate to think of the damage curve as reflecting yield loss in response to different intensities of injury, where intensity of injury refers to both the quantity and duration of injury.

The damage curve also is influenced by plant phenology. Responses to injury at one stage of plant development often differ from that at another developmental stage. In determining EILs, this variability commonly is accommodated by calculating EILs specific to different plant developmental stages. As general principles (admittedly with exceptions), younger plants are more susceptible to injury, but if they survive, they have greater time to compensate; and injury during reproductive stages (including early reproductive stages) tends to cause greater yield reductions than injury at other stages.

Among the most important factors influencing the damage curve are the biotic and abiotic environment. Most commonly, this variation is associated with water stress, and dramatic differences in damage functions have been documented between plants in water-stressed and nonstressed conditions [13,42]. Much of the variability in the literature on yield loss relationships undoubtedly stems from environmental interactions. One impetus to developing more physiologically based understandings of biotic stress is to use such understandings to develop considerations of stress interactions, particularly water stress. Ultimately, such an approach may help us improve the accuracy of EILs and better reflect the influence of the environment in our thresholds.

IV. ECONOMIC THRESHOLDS

The economic threshold (ET) is probably the best-known index for making pest management decisions. It differs from the EIL in that it is a practical or operational rule, instead of a theoretical one. Stern et al. [59] defined the ET as "the population density at which control action should be determined (= initiated) to prevent an increasing pest population from reaching the economic injury level." Although expressed in insect numbers, the ET is actually a time parameter, with pest numbers being used to indicate when management tactics should be applied [48]. Therefore, it is also referred to as the *action threshold*.

The ET and the EIL are intimately related. In most instances, the ET is visualized as occurring below the EIL. The basis of this conceptual relationship comes from the original work of Stern et al. [59], who dealt with populations of the spotted alfalfa aphid, *Therioaphis maculata* (Buckton), a pest with overlapping generations and increasing density as the growing season progresses. In this instance and others like it, setting the ET at a lower value than the EIL assumes that the population will increase to exceed the EIL. Hence, taking action before the EIL is reached allows unnecessary losses to accrue. In instances of pests having discrete generations, the ET may be more appropriately placed above the EIL to account for natural mortality occurring within a generation [8,42]. But, if the pest population is expressed in injury equivalents (potential injury of a single pest individual) [57], the ET will always be situated below the EIL [48].

The ET also differs from the EIL in being more complex and, therefore, more difficult to determine objectively. The difficulty arises when attempting to estimate and predict the following: (a) variables of the EIL (this is because the ET is based on the EIL), (b) pest and host phenology, (c) population growth and injury rates, and (d) time delays associated with potential integrated pest management (IPM) tactics. Because of a lack of understanding and the inability to predict most of these variables, ETs usually have been set arbitrarily [17]; therefore, most ETs are crude compared with the more theoretical EIL values.

Several approaches, representing different levels of sophistication, have been used to establish ET values. These approaches can be grouped as subjective determinations or objective determinations.

A. Subjective ETs

The crudest type of determination of the ET is the subjective type. Thresholds in this category, also called *nominal thresholds* [53], are not based on objective criteria, such as an EIL. They are usually based on a practitioner's experience or expert opinion. Unfortunately, these types of thresholds are still the most common type of ET found in IPM recommendations. Most often, these thresholds are necessitated because of the lack of crop loss functions. Although they tend to be static and lack precision, they are still considered to be more progressive than using no threshold at all, and they have resulted in reduced pesticide use [17].

B. Objective ETs

Objective ET values, on the other hand, are established from calculated EILs and change with changes in the primary variables of an EIL (e.g., market values and management costs). For objective ET values, a current EIL is calculated, and projections are made on the potential of pest injury to exceed the EIL. Subsequently, decisions on action to be taken are made based on expected injury rates, delays in applying tactics, and activity rates of the tactics. At least three major types of ETs have been developed for use in IPM, including (a) fixed ETs, (b) descriptive ETs, and (c) dichotomous ETs [46].

1. Fixed ETs

The fixed ET is the most common type of objective ET. This type is used when the rate of pest population growth and injury is unknown. Here, the ET is usually set at a fixed percentage of the EIL, often 25–50% below the EIL. Use of the term *fixed* does not mean an unchanging ET, but rather it refers to the constancy of the percentage. Examples of ETs in this category can be found in many crops [48].

2. Descriptive ETs

Less common than fixed ETs are descriptive ETs. Descriptive ETs are based on predicting pest population growth and/or injury rates to determine the need for management tactics. Both stochastic and deterministic procedures have been used to establish this type of ET.

With stochastic procedures, action decisions and timing are derived by understanding previous pest population growth rates and basing future rates on these. A stochastic ET can be seen in the work of DuToit [11] with the aphid, *Diuraphis noxia* (Mordvilko), on wheat. Here, an EIL was calculated at 14% infested plants for the plant growth stage at maximum aphid infestation (growth stage 59). An earlier plant growth stage (growth stage 31) was found to be the optimum time for insecticide application. Aphid growth rate then was determined for an economic infestation between growth stages 31 and 59. Subsequently, this pest growth rate was used to calculate the ET for growth stage 31 with the formula: ET = (EIL) (C^x), where EIL = economic injury level, C = factor of population increase per week, and x = time in weeks between plant growth stages 31 and 59. This type of ET has the advantage of using current sampling data to assess the injuriousness of a pest population. Its weakness is that, in certain instances, the actual growth rate of the pest may not approximate the derived growth rate. But then this is a criticism of most descriptive ETs.

Deterministic procedures to establish descriptive ETs depend on estimating future pest-population growth and injury rates from age-specific parameters or processes. As opposed to the stochastic approach, this procedure is usually applied to pests with discrete generations, and decisions are made within and for a specific pest generation. Here growth estimates may be based on the probability of age-specific survival from life tables or on mechanistic models of a population process; for example, a predator-prey model. An example of the deterministic type is shown in work with the green cloverworm, *Plathypena scabra* (F.), in soybean [42]. With this pest, stage-specific survival probabilities from life tables were used to compute mortality-adjusted injury equivalents (IEQs). An IEQ is the proportion of potential injury obtained by a damaging stage when premature mortality occurs. IEQs are established for each larval stage or group of stages, and they express the amount of food consumed to date plus expected consumption based on the probability of survival. The calculated IEQs can be used by multiplying them by numbers in each stage obtained in field samples. By tallying the results, realized and future injuriousness of a pest population can be estimated and decisions based on this estimate. ETs of this type perhaps have more precision; however, they tend to be more complicated to use than stochastic types.

3. Dichotomous ETs

Dichotomous ETs can be developed by using a version of the sequential probability ratio test of Wald [63] to determine pest status from field samples taken over a period of time. This procedure has been called time-sequential sampling [44] and can be used with the damaging stage of a pest to determine objectively its ET. With this approach, an EIL is calculated and class limits (m_0, m_1) are established for typical subeconomic and economic populations by using population growth data from field observations. Employing this information, along with dispersion characteristics of the data, a time-sequential sampling plan is developed and displayed in tabular form [43]. Such a plan, developed and tested for green cloverworm larvae in soybean, was both accurate and efficient. In making management decisions, the plan gave an approximate 25% time savings over a plan with a fixed number of sampling periods and provided an objective basis for determining the ET [46].

V. THRESHOLD DEVELOPMENT AND IMPLEMENTATION

A. Current Threshold Development

In a survey of the scientific literature on EILs and ETs from 1959 to 1993, Peterson [49] identified fewer than 250 published papers on thresholds. Of these, over 90% addressed arthropods, with the remainder focused on weeds and pathogens (including nematodes). This disparity in thresholds among pests is not surprising. The practical use of thresholds is for therapeutic tactics and single-event decision making. With pathogens, few therapeutic tactics are available, and management efforts largely focus on preventive approaches. Unless new tactics are developed for therapeutic disease management, it seems likely the applicability of EILs and ETs for plant pathogens will remain limited.

In weeds, research interest in thresholds was coincident with the development of effective postemergence herbicides. Before effective postemergence herbicides were available, a lack of therapeutic tactics may have limited threshold development. Additionally, new concerns about herbicide safety, particularly ground water contamination, also have been an impetus for new approaches in weed science, including thresholds.

That so many calculated EILs were developed for insects is not surprising given the origin of the concept in entomology and the easy applicability of thresholds for many insect pests. What is surprising is how few thresholds have been calculated for insect pests. Peterson found published thresholds for fewer than 100 insect pest species. Undoubtedly, many additional nominal thresholds exist in the extension literature, but given the potential variation even in calculated thresholds, the reliability of nominal thresholds seems suspect. Even in soybean, which has the largest number of calculated EILs (including some highly sophisticated models), calculated EILs are available only for about a third of the potential arthropod pest species. Peterson notes, "If economic thresholds form the keystone of modem pest management, then the structure is still rather unsteady." We have to agree.

Undoubtedly, experimental constraints and incomplete understandings of plant stress have impeded threshold development, but such explanations do not account for the slow progress in threshold development. Possibly, poor threshold development is a reflection of a conflict in research priorities in pest management. Elsewhere [17,18], we have argued that much activity in the guise of pest management actually is directed at control, the reduction of pest populations to the limits of our technology. Thus, we perceive competing philosophies of management, with its emphasis on maintaining pests at tolerable levels, and control, with its emphasis on tactics to reduce (and often eradicate) pest populations, in pest science. From our perspective, the relatively limited development of thresholds is one consequence of this conflict.

Beyond philosophical debates in pest management, a number of practical impediments have retarded threshold development. Foremost of these constraints is the substantial research required to produce an EIL. Particularly in describing the damage curve, substantial experimental challenges exist in working with biotic stressors, and these experimental difficulties undoubtedly impede progress. As more comprehensive understandings of biotic stress are developed, we hope to transcend the current practice of considering every pest on each host as a unique situation requiring individual experimentation. The development of injury guilds and thresholds for these guilds shows that more general approaches are possible, at least for some types of pests.

Other constraints include the limited applicability of thresholds for certain types of pests and management practices. Certainly, thresholds are most applicable for single-event decisions and for therapeutic management tactics, which will limit threshold applications for pests like pathogens. However, many modifications and advances in threshold development are broadening the applications of EILs and ETs. This is especially noteworthy with respect to the development of esthetic injury levels and applications of EILs to consider multiple-year effects of injury.

Undoubtedly, many thresholds in use are nominal thresholds (sensu Poston et al. [53]), and, at least in principle, these should be replaced by calculated thresholds through time. To date, we are unaware of any comprehensive survey of thresholds in the extension literature or of broad comparisons of nominal to calculated thresholds. Consequently, we cannot argue that nominal thresholds are wrong, and we recognize that nominal thresholds are better than no thresholds. Nevertheless, we suspect many of these nominal thresholds are too conservative (indicating unnecessary management action). For example, calculated EILs for bean leaf beetle injury to seedling soybean [25] were not only more than double previous nominal thresholds, but the calculated EILs were so high as to indicate bean leaf beetles would rarely be economic pests on seedling soybean.

B. Implementation

Intrinsic impediments to the implementation of thresholds include the availability and applicability of thresholds for specific pest/host combinations. Beyond these factors, additional biological, economic, and social issues all weigh in the adoption of thresholds and other IPM practices. Bechinski [2] discusses problems in implementation relative to insect sampling, but his observations are equally applicable to thresholds. In particular, Bechinski points out the dilemma in which detailed biological information is needed for making a decision, but it cannot be too complicated to use. Additionally, the labor and costs associated with sampling and other activities surrounding the use of thresholds undoubtedly limit implementation.

A further problem is that the EIL and related ETs are cost/benefit criteria. When a producer's goal is to maximize short-term profits, thresholds can be important management tools. In contrast, if a producer is more interested in risk aversion than in short-term profitability, thresholds are less likely to be used. Although the importance of risk aversion has been recognized in pest management decision making [36], additional emphasis in this area seems essential if we are to improve the adoption of IPM, including the use of thresholds.

VI. FUTURE DEVELOPMENT OF THRESHOLDS

Concerns over threshold development notwithstanding, a number of new developments have broadened the usefulness of the EIL and ET. These approaches include application of EILs to veterinary pests, adaptations of EILs to consider multiple-year effects, and considerations of plant quality in EIL calculations. Higley and Pedigo [18] provide detailed treatment of these and other new approaches. We will briefly highlight two new directions in EIL development here.

Making management decisions for pests on horticultural plantings has been a long-standing problem. Because horticultural plants often do not have readily identifiable, marketable yield, establishing conventional yield-loss relationships necessary to EIL development was not possible. In response to this problem, Raupp et al. [54] developed a procedure for estimating the economic loss associated with esthetic injury in horticultural plantings. Subsequently, various workers have developed additional aesthetic thresholds, and Sadof and Raupp [55] present a comprehensive review of research in this area. The development of esthetic thresholds has opened the door for formal pest management decision making in a number of novel areas.

Higley and Wintersteen [22] proposed including environmental costs associated with management tactics (pesticides) with management costs to produce an environmental EIL. In their procedure, relative risks to different aspects of the environment were determined for different pesticides and a contingent valuation survey of pesticide users was used to set monetary costs associated with different levels of risk. By incorporating environmental risk information into EILs, environmental EILs may be useful in choosing among pesticides based on environmental criteria, as well as cost and efficacy. Higley and Wintersteen [23] review approaches to environmental sustainability through EILs and ETs, including a review and responses to critiques of their approach.

These new developments illustrate that EILs and ETs still have great potential for addressing various novel situations. Although other approaches to pest management decision making promise various advantages to the simple cost/benefit approach of the EIL, none has had the primacy or influence as the EIL. Renewed emphasis on threshold development and novel applications suggest that thresholds will grow increasingly important for pest management systems. Serious challenges remain in research and implementation of thresholds, and in meeting these challenges we see much of the excitement and promise of pest management.

ACKNOWLEDGMENTS

We thank Robert Peterson, DowElanco, and Wendy Wintersteen, Iowa State University, for their reviews of this chapter. This is Journal Paper No. J-16424 of the Iowa Agriculture and Home Economics Experiment Station, Ames, Projects 3207 and 3183 and was supported, in part, by the University of Nebraska Agricultural Experiment Station Projects NEB-17-055 and NEB-17-059.

REFERENCES

1. Bardner, R., & K.E. Fletcher. 1974. Insect infestations and their effects on the growth and yield of field crops: a review. *Bull. Entomol. Res.* 64:141–160.
2. Bechinsky, E.J. 1994. Designing and delivering in-the-field scouting programs. In *Handbook of Sampling Methods for Arthropods in Agriculture*. L.P. Pedigo and G.D. Buntin, eds. Boca Raton, Florida: CRC Press, pp 683–706.
3. Belsky, A.J. 1986. Does herbivory benefit plants? A review of the evidence. *Am. Nat.* 127:870–892.
4. Boldt, P.E., K.D. Biever, & C.M. Ignoffo. 1975. Lepidopterous pests of soybeans: consumption of soybean foliage and pods and development time. *J. Econ. Entomol.* 68:480–482.
5. Boote, K.J. 1981. Concepts for modeling crop response to pest damage. *ASAE Pap. 81-4007*. St. Joseph, Michigan: American Society of Agricultural Engineers.
6. Chapin, F.S. 1991. Integrated responses of plant to stress. *BioScience* 41:29–36.
7. Crawley, M.J. 1989. Insect herbivores and plant population dynamics. *Annu. Rev. Entomol.* 34:531–564.
8. Chiang, H.C. 1979. A general model of the economic threshold level of pest populations. *FAO Plant Prot. Bull.* 27:71–73.
9. Cousens, R., C.J. Doyle, B.J. Wilson, & G.W. Cussans. 1986 Modelling the economics of controlling *Avena fatua* in winter wheat. *Pest. Sci.* 17:1–12.
10. Duodu, Y.A., & D.W. Davis. 1974. A comparison of growth, food consumption, and food utilization between unparasitized alfalfa weevil larvae and those parasitized by *Bathyplectes curculionis* (Thomson). *Environ. Entomol.* 3:705–710.

11. DuToit, F. 1986. Economic thresholds for *Diuraphis noxia* (Hemiptera: Aphididae) on winter wheat in the eastern orange free state. *Phytophylactica* 18:107–109.

12. Fenemore, P.G. 1982. *Plant Pests and Their Control*. Wellington, New Zealand: Butterworths.

13. Hammond, R.B., & L.P. Pedigo. 1982. Determination of yield-loss relationships for two soybean defoliators by using simulated insect-defoliation techniques. *J. Econ. Entomol.* 75:102–107.

14. Hammond, R.B., L.P. Pedigo, & F.L. Poston. 1979. Green cloverworm leaf consumption on greenhouse and field soybean leaves and development of a leaf consumption model. *J. Econ. Entomol.* 72:714–717.

15. Higley, L.G. 1992. New understandings of soybean defoliation and their implications for pest management. In *Pest Management of Soybean*. L.G. Copping, M.B. Green, & R.T. Rees, eds. Amsterdam: Elsevier, pp 56–65.

16. Higley, L.G., J.A. Browde, & P.M. Higley. 1993. Moving towards new understandings of biotic stress and stress interactions. In *International Crop Science*. D.R. Buxton, R. Shibles, R.A. Forsberg, B.L. Blad, K.H. Asay, G.M. Paulson, & R.F. Wilson, eds. Madison, Wisconsin: Crop Science Society of America, pp 749–754.

17. Higley, L.G., & L.P. Pedigo. 1993. Economic injury level concepts and their use in sustaining environmental quality. *Agric., Ecosystems, Environ.* 46:233–243.

18. Higley, L.G., & L.P. Pedigo, eds. 1997. *Economic Thresholds for Integrated Pest Management*. Lincoln, Nebraska: University of Nebraska Press.

19. Higley, L.G., & L.P. Pedigo, 1997. The economic injury level concept. In *Economic Thresholds for Integrated Pest Management*. L.G. Higley & L.P. Pedigo, eds. Lincoln, Nebraska: University of Nebraska Press, pp 9–21.

20. Higley, L.G., & R.K.D. Peterson. 1997. The biological basis of the EIL. In *Economic Thresholds for Integrated Pest Management*. L.G. Higley, & L.P. Pedigo, eds. Lincoln, Nebraska: University of Nebraska Press, pp 22–40.

21. Higley, L.G., & T.E. Hunt. 1994. Early-season soybean insects: past problems and future risks, *Proc. 1994 Integrated Crop Management Conference*. Ames, Iowa: Iowa State University, pp 91–99.

22. Higley, L.G., & W.K. Wintersteen. 1992. A novel approach to environmental risk assessment of pesticides as a basis for incorporating environmental costs into economic injury levels. *Am. Entomol.* 38:34–39.

23. Higley, L.G., & W.K. Wintersteen. 1997. Thresholds and environmental quality. In *Economic Thresholds for Integrated Pest Management*. L.G. Higley & L.P. Pedigo, eds. Lincoln, Nebraska: University of Nebraska Press, pp 249–274.

24. Hunt, T.E., L.G. Higley, & J.F. Witkowski. 1994. Soybean growth and yield after simulated bean leaf beetle injury to seedlings. *Agron. J.* 86:140–146.

25. Hunt, T.E., L.G. Higley, & J.F. Witkowski. 1995. Bean leaf beetle injury to seedling soybean: consumption, effects on leaf expansion, and economic injury levels. *Agron. J.* 87:183–188.

26. Hutchins, S.H. 1997. Thresholds involving plant quality and phenological disruption. In *Economic Thresholds for Integrated Pest Management*. L.G. Higley and L.P. Pedigo, eds. Lincoln, Nebraska: University of Nebraska Press, pp 275–296.

27. Hutchins, S.H., & J.E. Funderburk. 1991. Injury guilds: a practical approach for managing pest losses to soybean. *Agric. Zool. Rev.* 4:1–21.

28. Hutchins, S.H., L.G. Higley, & L.P. Pedigo. 1988. Injury equivalency as a basis for developing multiple-species economic injury levels. *J. Econ. Entomol.* 81:1–8.

29. Hutchins, S.H., L.G. Higley, L.P. Pedigo, & P.H. Calkins. 1986. A linear programming model to optimize management decisions with multiple pests: an integrated soybean pest management example. *Bull. Entomol. Soc. Am.* 32:96–102.

30. McNaughton, S.J. 1983. Physiological and ecological implications of herbivory. *Encycl. Plant Physiol.* New Ser. 12C:657–677.

31. Metcalf, R.L., & W.H. Luckmann, eds. 1994. *Introduction to Insect Pest Management.* New York: Wiley.

32. Mooney, H.A., W.E. Winner, & E.J. Pell. 1991. *Responses of Plants to Multiple Stresses.* San Diego, California: Academic Press.

33. Mortensen, D.A., & H.D. Coble. 1997. Developing economic thresholds for weed management. In *Economic Thresholds for Integrated Pest Management.* L.G. Higley & L.P. Pedigo, eds. Lincoln, Nebraska: University of Nebraska Press, pp 89–113.

34. Mumford, J.D., & G.A. Norton. 1984. Economics of decision making in pest management. *Annu. Rev. Entomol.* 29:157–174.

35. Norton, G.A. 1976. Analysis of decision making in crop protection. *Agro- Ecosystems* 3:27–44.

36. Norton, G.A., & J.D. Mumford, eds. 1993. *Decision Tools for Pest Management.* Oxford, UK: CAB International.

37. Onstad, D.W. 1987. Calculation of economic-injury levels and economic thresholds for pest management. *J. Econ. Entomol.* 80:297–303.

38. Onstad, D.W. 1988. Letter to the editors. *J. Econ. Entomol.* 82:1–2.

39. Onstad, D.W., & R. Rabbinge. 1985. Dynamic programming and the computation of economic injury levels for crop disease control. *Agric. Systems* 18:207–226.

40. Ostlie, K.R. 1984. Soybean transpiration, vegetative morphology, and yield components following actual and simulated insect defoliation. PhD dissertation, Iowa State University, Ames, Iowa.

41. Ostlie, K.R., & L.P. Pedigo. 1985. Soybean response to simulated green cloverworm (Lepidoptera: Noctuidae) defoliation: progress toward determining comprehensive economic injury levels. *J. Econ. Enotomol.* 78:437–444.

42. Ostlie, K.R., & L.P. Pedigo. 1987. Incorporating pest survivorship into economic thresholds. *Bull. Entomol. Soc. Am.* 33:98–101.

43. Pedigo, L.P. 1994. Time-sequential sampling for taking tactical action. In *Handbook of Sampling Methods for Arthropods in Agriculture.* L.P. Pedigo & G.D. Buntin, eds. Boca Raton, Florida: CRC Press, pp 337–353.

44. Pedigo, L.P., & J.W. van Schaik. 1984. Time-sequential sampling: a new use of the sequential probability ratio test for pest management decisions. *Bull Entomol. Soc. Am.* 38:12–21.

45. Pedigo, L.P., & L.G. Higley. 1992. The economic injury level concept and environmental quality: a new perspective. *Am. Entomologist.* 38:12–21.

46. Pedigo, L.P., L.G. Higley, & P.M. Davis. 1989. Concepts and advances in economic thresholds for soybean entomology. In *Proceedings of the World Soybean*

Research Conference IV. Vol. 3. Buenos Ares: Asociacon Argentina de la Soja, pp 1487-1493.

47. Pedigo, L.P., R.B. Hammond, & F.L. Poston. 1977. Effects of green cloverworm larval intensity on consumption of soybean leaf tissue. *J. Econ. Entomol.* 70:159-162.

48. Pedigo, L.P., S.H. Hutchins, & L.G. Higley. 1986. Economic injury levels in theory and practice. *Annu. Rev. Entomol.* 31:341-368.

49. Peterson, R.K.D. 1997. The status of economic decision level development. In *Economic Thresholds for Integrated Pest Management.* L.G. Higley & L.P. Pedigo, eds. Lincoln, Nebraska: University of Nebraska Press, pp 151-178.

50. Peterson, R.K.D., & L.G. Higley 1993. Arthropod injury and plant gas exchange: current understandings and approaches for synthesis. *Trends Agric. Sci.* 1:93-100.

51. Peterson, R.K.D., L.G. Higley, & S.D. Danielson. 1995. Alfalfa consumption by the adult clover leaf weevil (Coleoptera: Curculionidae) and development of injury equivalents for stubble defoliators. *J. Econ. Entomol.* 88:1441-1444.

52. Peterson, R.K.D., S.D. Danielson, & L.G. Higley. 1992. Photosynthetic responses of alfalfa to actual and simulated alfalfa weevil (Coleoptera: Curculionidae) injury. *Environ. Entomol.* 21:501-507.

53. Poston, F.L., L.P. Pedigo, & S.M. Welch. 1983. Economic injury levels: reality and practicality. *Bull. Entomol. Soc. Am.* 29:49-53.

54. Raupp, M.J., J.A. Davidson, C.S. Koehler, C.S. Sadof, & K. Reichelderfer. 1988. Decision-making considerations for aesthetic damage caused by pests. *Bull. Entomol. Soc. Am.* 34:27-32.

55. Sadof, C.S. & M.J. Raupp. 1997. Aesthetic thresholds and their development. In *Economic Thresholds for Integrated Pest Management.* L.G. Higley & L.P. Pedigo, eds. Lincoln, Nebraska: University of Nebraska Press, pp 203-226.

56. Shoemaker, C.A. 1976. Optimal management of an alfalfa ecosystem. In *Pest Management.* G.A. Norton & C.S. Holling, eds. Oxford, UK: Pergamon Press, pp 301-316.

57. Shelton, A.M., J.T. Andaloro, & J. Barnard. 1982. Effects of cabbage looper, imported cabbageworm and diamondback moth on fresh market and processing cabbage. *J. Econ. Entomol.* 75:742-745.

58. Stern, V.M. 1973. Economic thresholds. *Annu. Rev. Entomol.* 18:259-280.

59. Stern, V.M., R.F. Smith, R. van den Bosch, & K.S. Hagen. 1959. The integrated control concept. *Hilgardia* 29:81-101.

60. Stone, J.D., & L.P. Pedigo. 1972. Development and economic-injury level of the green cloverworm on soybean in Iowa. *J. Econ. Entomol.* 65:197-201.

61. Tammes, P.M.L. 1961. Studies of yield losses. II. Injury as a limiting factor of yield. *Tijdschr. Plantenziekten.* 67:257-263.

62. Trumble, J.T., D.M. Kolodny-Hirsch, & I.P. Ting. 1993. Plant compensation for arthropod herbivory. *Annu. Rev. Entomol.* 38:93-119.

63. Wald, A. 1947. *Sequential Analysis.* New York: Wiley.

64. Welter, S.C. 1989. Arthropod impact on plant gas exchange. In *Insect-Plant Interactions.* Vol. 1. E.A. Bernays, ed. Boca Raton, Florida: CRC Press, pp 135-150.

65. Welter, S.C. 1991. Responses of tomato to simulated and real herbivory by to-bacco hornworm (Lepidoptera: Sphingidae). *Environ. Entomol.* 20:1537–1541.
66. Welter, S.C. 1993. Responses of plants to insects: eco-physiological insights. In *International Crop Science*. D.R. Buxton, R. Shibles, R.A. Forsberg, B.L. Blad, K.H. Asay, G.M. Paulson, & R.F. Wilson, eds. Madison, Wisconsin: Crop Science Society of America, pp 773–778.
67. Welter, S.C. 1997. Thresholds for interseasonal management. In *Economic Thresholds for Integrated Pest Management*. L.G. Higley & L.P. Pedigo, eds. Lincoln, Nebraska: University of Nebraska Press, pp 227–248.
68. Whitman, T.G., J. Maschinski, K.C. Larson, & K.N. Paige, 1991. Plant responses to herbivory; the continuum from negative to positive and underlying physiological mechanisms. In *Plant-Animal Interactions: Evolutionary Ecology in Tropical and Temperate Regions*. P.W. Price, T.M. Lewinsohn, G. Wilson Fernandes, & W.W. Bennson, eds. New York: Wiley, pp 227–256.
69. Wilkerson, G.G., S.A. Modena, & H.D. Coble. 1991. HERB: Decision model for postemergence weed control in soybean. *Agron. J.* 83:413–417.

28
Professional Training and Technology Transfer

Frank G. Zalom
University of California, Davis, California

I. AGRICULTURAL RESEARCH AND EXTENSION

Research and education in agriculture are probably as old as agriculture itself, and they are intimately tied to incremental innovation. Progressive farmers who learned a better way to farm were undoubtedly the subject of intense observation by those who hoped to increase their production or decrease labor and input costs. Farmers' conversations centuries ago might well have been similar to those around rural coffee shops today, addressing relevant issues like the weather, problems with their crop, and such.

Agricultural innovations were relatively slow to develop. It was not until the eighteenth century that large landholders of the day became aware of the possibility of significant improvements in agriculture and sought to utilize experimentation to address problems. Special organizations were formed to discuss their observations and to promote agricultural progress. In the United States, the first was the Philadelphia Society for Promoting Agriculture, which was formed in 1785 to disseminate information through lectures, publications, and newspaper articles. Scheuring [32] presents an interesting summary of how these agricultural societies, fostered by the elite, expanded to new regions becoming broader based and addressing more local issues. By the mid nineteenth century, hundreds of these societies were operating, some supported by state agricultural societies or boards of agriculture. The agricultural societies often sponsored speakers and developed traveling "Farmers' Institutes," providing a structure to deliver information to farmers. By 1899, 45 states sponsored these institutes more or less regularly.

A. Land Grant College System

The Congress of the United States, seeing the need for adult education in agriculture, established the Land Grant Colleges through the Morrill Act of 1862. Each state was given land which was to be used in some way to create a state agricultural college where students could learn about advanced production practices and food science. The Hatch Act established the State Agricultural Experiment Stations in 1887, where agricultural research could be conducted to address the needs of farmers and rural inhabitants. Land grant colleges successfully addressed many issues and created specialized courses and publications to deliver new information, but somehow many farmers did not learn of or adopt the research-based practices that were becoming available.

Some Land Grant Colleges began sponsoring the Farmers' Institutes and developed traveling schools for farmers. Some conducted on-farm demonstrations, and encouraged local county governments to work with agencies and businesses to hire resident county agents who could help with the demonstrations. In 1909, President Theodore Roosevelt's Commission on Country Life recommended federal support of an extension system to direct the resources of the Land Grant Colleges to farmers. The Smith-Lever Act of 1914 established an Agricultural Extension Service within the system of Land Grant Colleges and began to place county agents in communities throughout the country.

Agricultural research and technology transfer in U.S. agriculture has depended upon the cooperation of the Land Grant College system with the U.S. Department of Agriculture (USDA) and various other agricultural organizations and industries for this past century. The cooperation has resulted in the implementation of many incremental advances in agricultural production, leading to the present technically advanced state of U.S. production systems. In doing this, however, indirect effects were largely ignored leading to the social and environmental issues which these groups are being asked to address today [22].

B. Government Advisory Services

Agricultural extension programs in much of the rest of the world have a parallel, although different structure, with the difference being that research is generally carried out under one organizational structure while technology transfer is conducted by another. In Great Britain, for example, Parliament provided for technical instruction in agriculture in 1890, initiating a formal system of delivering technical advice to farmers [9]. In 1946, the National Agricultural Advisory Service was formed to cover all aspects of agriculture including crop protection. Interestingly, the Service underwent a major reorganization in 1971, and it was combined within a new organization, the Agricultural Development and Advisory Service, which emphasized development rather than traditional outreach activities, effectively combining development and educational programs. Crop advisory services in Germany are organized at the state level, with some offices being centralized and some regionalized at the state's discretion [14]. In many instances, advice is given by local crop-oriented advisers working with both agricultural industry interests and individual farmers. The British model and those of other European countries became the model for colonial government extension services, and most of those systems remain in existence in developing countries today.

In the end, both the Land Grant College and the government advisory service models of education are similar in that they support the transfer of technology from institutionalized research and development organizations to farmers in a hierarchical system. These systems work well when the adviser at the local level is aware of farmers' needs as well as possibilities of technological advancement from current research and can effectively communicate this to both growers and researchers. These systems fail when the advisers cannot effectively reach all farmers with the detail and timeliness of information required, and when the technologies being transferred are not relevant to individual farmer's needs.

II. PEST MANAGEMENT IMPLEMENTATION

A comprehensive Integrated Pest Management (IPM) program for an agricultural system requires willing and cooperative growers and dedicated research and extension personnel who can produce a body of relevant information concerning farming practices, pest biology, and pest management tactics. IPM is a challenging educational subject because of its complexity and intensive emphasis on management, which requires that individuals who wish to apply it understand broader ecological relationships in addition to crop production practices. Rather than simply treating pests when they occur or preventatively, IPM emphasizes monitoring of pests, natural enemy population levels, and crop status throughout the season, and applies control measures only when pest abundance is sufficient to cause unacceptable economic loss relative to the cost of control. In addition, IPM tactics include a variety of agents and materials for pest population suppression which may not be promoted through commercial enterprises.

Many studies, such as those of Rogers [29], have considered the adoption of agricultural innovations, but until recently, there were relatively few studies focusing on the implementation process specifically for IPM, including incentives for and constraints to adoption. Most analyses have involved case studies, in part because IPM encompasses such a broad range of strategies and tactics and is, therefore, intimately linked to the production of a specific crop, often in a specific region. Further, the degree to which IPM is implemented on a specific crop can vary as growers may modify innovations to fit their needs, a process termed "adaptive implementation" by Berman [1] or "reinvention" by Rogers and Shoemaker [30], or they may use only selected IPM management techniques from the universe available.

An example of adoption versus adaption was presented by Grieshop et al. [13] in a study of implementing IPM monitoring techniques in California processing tomatoes. They found that over the 5-year period after the techniques were first available, 26% of all growers had adopted the IPM techniques, whereas 31% had adapted (or reinvented) the techniques. In this case, the degree to which the program was modified had little effect on results but better fit the growers' practices. Mixed adoption levels of selected practices among several available in an IPM cropping system was noted by Klonsky et al. [18] in California almonds where some cultural controls (e.g., timely pickup of nuts from the ground, winter sanitation, and harvesting prior to pest flight) for the key pest, navel orangeworm (*Paramyelois transitella*), were broadly adopted, whereas other techniques were not.

A. U.S. Federal Extension IPM Programs

In 1984, a national evaluation of Cooperative Extension IPM programs was undertaken and coordinated by the Virginia Cooperative Extension Service [28]. Fourteen states undertook case history studies of specific IPM programs. Of these, 10 of the studies identified the level of grower IPM adoption. Levels of adoption in excess of 50% were noted in nine of the studies, but "high"-level IPM use by over 50% of the growers was only reported in two studies (Table 1).

In Texas, 82% of the cotton growers studied [28] were identified as IPM users. The Texas IPM program in cotton has a long history dating back to the original extension IPM pilot project for cotton and tobacco which was initiated in 1971 [12]. In the pilot project, state extension services hired and trained scouts who provided IPM services to growers subscribing to the programs. Today, private pest control advisors who are part of the Texas Pest Management Association provide IPM monitoring services to a majority of Texas cotton growers. California also participated in this pilot project, and by 1973, over 200,000 acres of cotton (about 22% of the state total) were serviced by private pest management consultants [24]. This number has increased to an estimated 80% of the California cotton acreage at present. The primary role of "scouts" hired as part of these early extension IPM programs was primarily to monitor pest abundance or incidence and then to recommend pesticide use based on having reached control action thresholds. As these pest management consultants became private, their services often expanded to include consulting on all aspects of pest,

Table 1 Percentage of U.S. Farmers in the 1984 National Evaluation of Extension IPM Programs Identified as IPM Users

State(s)	Crop	Level of IPM use		
		None	Low	High
California	Almond	34.5	16.4	49.1
Georgia	Peanut	20.7	19.7	59.6
Indiana	Corn	31.0	27.2	41.8
Kentucky	Stored Grain	68.0	11.0	21.0
Massachussetts	Apple	48.9	—[a]	—
New York	Apple	19.3	73.4	7.3
Texas	Cotton	18.0	—[b]	—
Virginia	Soybean	25.1	25.1	49.8
Idaho	Alfalfa seed	26.4	34.8	38.8
Oregon				
Washington				

[a]51.1% of farmers were identified as IPM users.
[b]82.0% of farmers were identified as IPM users.
Source: Data from ref. 28.

nutrient, and water management, including the use of biological controls when appropriate.

Insecticide use on cotton has decreased nationally since pilot extension IPM implementation projects were initiated in 1971. In 1971, an average of 6.5 kg of insecticides was applied to each hectare of cotton. According to Frisbie and Adkisson [10], this decreased to 6.2 kg/ha in 1976, and 1.7 kg/ha in 1982. Cotton absorbed 49% of total kilograms of insecticides applied to major field crops in 1976 but received only 24% by 1982. Not all of this reduction can be attributed to IPM, however, as new pesticides such as the synthetic pyrethroids (which utilize substantially lower application rates) were introduced during this time.

It is interesting to note that the case history reporting the greatest percentage of "high"-level IPM users in the 1987 study was tobacco in North Carolina (see Table 1), another target of the original extension IPM pilot projects [4].

Federally sponsored extension IPM programs continue at some level in all states through special project funding from the USDA Extension Service, and an IPM Coordinator is designated at each Land Grant University. These programs target a wide range of pest problems in agricultural as well as urban systems. Some state legislatures, notably those of New York and California, have allocated substantial state funding to expand extension implementation efforts. Most state programs no longer support scouting services operated through extension but rather support training programs and develop materials for farmers and the emerging crop or pest management consulting industry.

B. Crop Consultants

The extension service of each state traditionally provided pest control advisory services to farmers. Following World War II, when synthetic pesticides became widely used for pest control, farm supply dealers who marketed the wide array of pesticides also became important sources of information, often because they could provide more personalized service on a more frequent basis to farmers than could county agents of the extension service who had many other duties. These individuals also had good knowledge of pest control technology and a profit incentive to encourage farmer adoption. The importance of the farm supply dealer in providing pest management information cannot be overstated, and these individuals are significant factors in the pest management decision making of most farmers today. Many of these individuals are trained in IPM, and they incorporate IPM into services they provide growers. However, the role of dealers as an unbiased source of information is often questioned.

As early as the 1950s, a few pioneers began selling knowledge and information rather than products to their farmer clientele, initiating private pest control advising (or consulting) companies. Cox [8], one of these early private consultants, described the role and requirements of a private pest management consultant in the first detailed treatise on the subject. Increased IPM research activities in the 1970s accelerated the development of the private pest management consultant industry [26]. Requirements for continual monitoring of insects and diseases greatly increased the need and opportunity for the consultant.

As mentioned earlier, some extension services began to train field scouts to monitor growers' fields on a fee-for-service basis and later encouraged the services to continue in the private sector. Many of the scouts trained as part of the federal extension IPM pilot projects of the 1970s became private consultants themselves.

Encouraging the development of a crop and pest management consulting industry can be viewed as one of the positive effects of extension IPM implementation efforts, and private consultants are becoming a major force in the delivery of pest management information to growers [2,11].

A survey of extension personnel by Post [26] revealed that over 70% felt that private pest management consultants benefited agriculture and were experienced, educated, and capable. Frisbie and McWhorter [11] attributed the expansion of the private pest management consulting industry in Texas to the expansion of extension IPM programs that demonstrated the need for pest management advice to growers. In contrast, the large number of private consultants in California, perhaps the most of any state, is attributed by Kendrick and Stimmann [16] to the state's regulatory and enforcement system. In the early 1970s, a program for state licensing of "pest control advisers" was initiated by the State of California. The law requires anyone who recommends pesticides or any other pest control method or device for agricultural use to be licensed. To be licensed as pest control advisers, individuals must have a degree in agriculture or biological sciences, a specified amount of experience, and pass a written examination in laws and regulations as well as in various pest control disciplines. They must also receive 20 h of continuing education credits each year. Although this structure is the most comprehensive currently in place, some individuals (e.g., see refs. 39 and 40) have suggested that the registration of pest control advisers in California has not had a material effect on IPM in practice, because the licensing program does not distinguish private consultants from the majority who work for farm supply dealers or other chemical retailers. Post [26], who owns a private crop consulting business in California and was an extension county pomologist, believes that farmers employ private pest management consultants because of:

> The rapid discovery of new technology and the farmer's inability to keep abreast of these findings
> Increasing government controls and the need for help in complying with pesticide and other regulations
> Low profit margins and the necessity to farm efficiently without unexpected loss from insects or diseases
> The need for a "direct line" through the consultant to researchers for accurate and timely agricultural information
> Continual access to an adviser that is totally loyal
> The increasing demands of farm management which do not leave the farmer adequate time for both business and crop production
> Transferring liability associated with pesticide use

1. Sources of Information

Lambur et al. [19] surveyed 136 private pest management consultants from 45 states to determine their sources of pest management information. Extension bulletins, manuals, and handbooks (81.3%, the consultants own research (79.7%, extension specialists (usually Land Grant University campus based) (79.3%), extension newsletters (72.2%), and the consultant's own scouts (70.0%) were reported as the top five information sources. Surprisingly, only 44% of the consultants considered county-based extension agents as useful sources of pest management information. Extension electronic information methods such as radio (11.8%), computer networks (11.4%), telephone recordings (9.7%), videotape (3.0%) videotext (2.3%), and television (2.2%) were considered to be useful by the fewest private consultants surveyed.

The total number of pest management consultants in the United States is not known, but in California over 3900 individuals are licensed. The use of IPM practices by California pest control advisers is a good indicator of overall adoption or potential adoption because of their key role in grower's pest control decisions. However, the number of pest control advisers alone does not guarantee that IPM will be encouraged or adopted but only that IPM exposure is increased.

California pest control advisers are the target audience for research and extension efforts of the University of California's Statewide IPM Project which was funded by the state legislature in 1979 to foster more rapid implementation of IPM practices. A 1987 survey of California pest control advisers showed levels of use of 20 selected IPM monitoring techniques by the advisors ranged from 10.7 to 93.6%. The most widely adopted of the IPM techniques tended to be those that had been available longest (e.g., monitoring cotton squares for *Lygus* damage, 81.5%; codling moth traps and phenology models in pome fruit, 93.6%), or those which had been actively demonstrated by the extension IPM staff (e.g., presence-absence spider mite sampling in cotton, almond, and citrus, 84.6, 70.6, and 52.4%, respectively; peach twig borer traps and phenology models in stone fruit, 66.1%). Over 86% of the active pest control advisers had at least one of the University of California's IPM manuals which are written for specific crops, and the average adviser had three of four manuals.

The key role of crop consultants in pest management decisions has been reinforced many times, yet it is interesting to note that California farmers participating in a series of recent U.S. Environmental Protection Agency (EPA) producer workshops [36] indicated that their involvement in pest management had actually increased over the past 5-10 years because the complexity of pesticide-related issues has made use decisions too important to delegate.

2. Degree Programs

IPM is best achieved by individuals well trained in both pest management and the plant sciences, but most graduate degree programs do not emphasize such interdisciplinary studies. The Land Grant Universities have evolved strong disciplinary departments (e.g., plant sciences, entomology, plant pathology, economics, systems science). These departmental units are typically not well linked and are not encouraged to interact with one another. Successful development and application of IPM programs require interdisciplinary interaction because of their systematic treatment of all components of agricultural production. Degree programs which would lead to professional accreditation in plant health and pest management should be supported at the Land Grant Universities in much the same way as are veterinary schools, and they could be structured as graduate groups or professional schools at those institutions [15]. No comprehensive doctoral programs currently exist, but a few professional programs have been developed at both the undergraduate and master's degree levels. The University of Florida has proposed a Doctor of Plant Health program, and it hopes to receive permission formally to offer the degree. It is envisioned that plant health practitioners who would graduate from such professional programs, like their veterinary or medical counterparts, would be permitted to prescribe the use of restricted agricultural chemicals in response to careful monitoring of crop health and in accordance with label requirements.

3. Professional Associations

There is a need for professional organizations to represent the crop-consulting profession in legislative issues affecting their industry, in providing professional training, and in helping to provide standards and a code of ethics for members. The health professions have developed strong associations which have become powerful lobbying forces for their industry, making certain that the interests of healthcare professionals are considered when legislation is proposed which may influence their industry. These associations also sponsor professional conferences and are involved in professional training and accreditation. In agriculture, lobbying on issues of interest to crop-care professionals (e.g., pesticide use and safety, licensing) has largely been the purview of farming (e.g., American Farm Bureau, National Cotton Council), industry (e.g., National Agricultural Chemicals Association), or public interest (e.g., environmental, consumer) groups. These organizations may not ultimately have the best interests of the pest management advisory profession in mind.

Some statewide organizations of crop advisers function in this way. For example, members of the California Agricultural Crop Production Association (CAPCA) and the Texas Pest Management Association (TPMA) are represented by an Executive Director and elected President. These organizations are involved in responding to state and to some extent national, regulatory issues affecting their membership, sponsoring meetings for their members, and providing other services. Both organizations have a broad-based membership. The National Alliance of Independent Crop Consultants (NAICC) is a professional organization that is envisioned to represent independent crop consultants and contract researchers nationally, and it has an interest in national certification and standards for the industry. Formed in 1978, membership has grown steadily. The Association of Applied Insect Ecologists (AAIE) is a largely California organization of private crop consultants and others whose members are particularly interested in learning about and implementing integrated pest management and biological controls.

4. Certification and Licensing

Voluntary certification of professionals involved in crop advising has been available for specialists through the professional societies of agronomy, soil science, and certain pest management disciplines. For example, within the Entomological Society of America there exists the American Registry of Professional Entomologists (ARPE), whose members must pass detailed qualifying examinations in subdisciplines within entomology to become board certified. Membership in this case implies a proven level of entomological knowledge. Nationally, much recent focus has been on the Certified Crop Adviser Program (CCA) coordinated by the American Society of Agronomy [33]. The CCA program sets base standards for those who provide crop-production advice to farmers in soils and soil fertility, pest management, soil and water management, and crop growth and development, and these are the areas in which performance objectives are established. Those eligible for certification must have a minimum of 4 years of crop advising experience and a high school diploma (or 2 years of crop advising experience and a bachelor of science degree in agriculture or life sciences), and they must pass a national CCA examination and a state or regional CCA examination. The certified crop adviser must participate in 60 h of continuing education every 2 years to remain certified, and these credits must be audited by the state or regional board. Competency areas within pest management are basic pest management practices, weed management, plant disease management, insect management, calibration of pesticide application equipment, pesticide resistance, using pesticides in an environmentally sound way, protecting humans against pesticide exposure, and integrated management.

It is disappointing to note the relatively simplistic approach to pest management at this point in the national certification process. Pesticides and pesticide application seem to be the focus of the competency areas. Control of vertebrate pests and nematodes are not mentioned at all, nor are biological control approaches. Hopefully, these oversights will be recognized and corrected by state and regional boards.

Both Congress and the EPA have suggested on occasion that pesticides be applied only on the recommendation of plant health practitioners, much as are prescription medications for human health. The concept of prescriptive use of pesticides has both ardent supporters and vehement critics, and past proposed amendments to FIFRA which would require such action have been opposed by agricultural interests. Some see tying pesticide use to certification of crop consultants as a vehicle to build confidence with the general public, elected officials, and regulatory agencies, perhaps permitting some agrochemicals to remain available, especially for minor-use crops. Others see it as an unnecessary bureaucracy that would not meet the intended objectives, while increasing costs to producers. Certainly, if the public were to have confidence in the prescriptive use of pesticides, those individuals making use recommendations would have to possess more than a basic knowledge of plant protection practices. State or federal licensing in addition to professional certification may also be needed.

California pesticide laws and regulations have created a system which approaches prescriptive use. As mentioned previously, state law requires anyone who recommends pesticides or any other pest control method or device for agricultural use to be licensed as a pest control adviser. In order for farmers to use a pesticide, they must have a written recommendation from a licensed pest control adviser and file a notice of intent to make a pesticide application with the county agricultural commissioner (the local regulatory official representing the state), who must determine within 24 h if the application should be permitted.

III. EVALUATION OF EXTENSION METHODS

Lasserre [20] stated that education of the user is a key element in successful implementation of most technologies. This is especially true of complex concepts such as IPM [21, 25, 34]. Studies such as those of Wearing [39], who reported on IPM delivery methods in the United States, Europe, and Australia by interviewing IPM research and extension specialists, have generally shown that verbal communication is the most effective method. Wearing's study indicated that among verbal communication methods, one-on-one contact with consultants (including both private consultants and farm supply dealers) and with extension agents was rated highest (approximately 90% favorable), with meetings and demonstrations also highly rated (76 and 73% favorable, respectively). Wearing's study rated written communication sources as moderately to highly effective, with newsletters (83% favorable), articles (58%), and bulletins (56%) considered most valuable. Electronic sources such as film or video, computers, television, and radio were all rated lowest in utility with 82–89% seldom or never using the methods. However, this might have changed since the study was published in 1988.

It is useful to distinguish between the types and complexity of information being discussed. Pest management information might include simple pesticide use recommendations in addition to IPM concepts, whereas IPM information typically also includes a wider variety of control options and management strategies. Table 2 shows the degree of use of various sources of general pest management information. Although there is variability between studies, some generalizations can be made. Extension agents (or products of extension services) and farm supply dealers are often the leading sources of general pest manage-

Table 2 Percentage of Farmers in the United States Utilizing Various Pest Management Informational Sources in Selected Studies

Informational source	State and crop									
	WA, OR (alfalfa seed)	CA (almond)	MA (apple)	NY (apple)	IN (corn)	TX (cotton)	GA (peanut)	VA (soybean)	KY (stored grain)	NC (tobacco)
Extension agent	55.5	69.3	74.4	65.9	52.9	79.5	96.8	74.1	37.5	79.0
Farm supply dealer	66.3	66.8[a]	60.7	78.3	37.7	45.0	75.8	60.9	50.6	61.0
Extension publications	67.9	65.9	83.3	88.0	64.0	59.0	91.5	82.9	51.9	85.8
Extension meetings	74.1	45.8	—[b]	66.8	58.3	—	94.7	45.7	62.2	58.2
Other farmers	73.9	62.6	35.7	37.3	54.1	34.7	74.2	71.1	44.4	62.7
Trade publications	46.9	45.2	29.4	40.7	59.9	25.4	81.9	63.3	31.3	49.7
Farmers/industry	36.1	56.4	—	—	—	—	77.1	—	35.4	—
Electronic sources	18.1	<10.0	—	74.2	23.6	40.2	61.4	25.5	33.3	35.0
Private consultants	—	66.8[a]	—	23.5	26.7	77.2	25.7	10.2	—	14.4

[a]Private consultants and farm supply dealers are not differentiated in this study.
[b]Source not available as a survey choice.
Source: From refs. 18 and 28.

ment information, usually being utilized by over 50% of growers surveyed. Other farmers and trade publications, such as magazines and newspapers, are also widely used sources of pest management information, but their utility was relatively more variable between the studies cited. Farmer group and other agricultural industry associations were usually used less frequently than the other sources mentioned.

When farmers who considered themselves high users of IPM were asked to identify their first source of contact with specific IPM techniques, extension agents and other extension sources were most often identified (Table 3). Trade publications on average were the next most identified sources of first contact with specific IPM techniques. Farm supply dealers were never as highly rated as extension agents or other extension sources for first contact with IPM techniques.

Studies such as these show the potential of extension agents, farm supply dealers, and one might assume private pest management consultants, as change agents who could accelerate movement through the stages (awareness, interest, evaluation, trial, and adoption) through which potential IPM adopters pass in the decision-making process.

IV. CONSTRAINTS TO ADOPTION

Even when IPM systems and practices are available and demonstrated to be effective, they are typically not universally adopted by farmers in spite of factors such as rising pesticide costs [31] and increasing resistance to pesticides [23].

Many studies have addressed the issue of constraints to IPM and have identified significant factors influencing its use (e.g., see refs. 7,13,40, and 41). Wearing [39] categorized these obstacles as technical, financial, educational, organizational (institutional), and social in his study of IPM in the United States, Europe, and Australia/New Zealand. The 600 participants at the 1992 National IPM Forum identified 61 constraints which are listed in its proceedings which were assembled by Sorenson [37]. Four of the top six constraints involved IPM education. These included (a) insufficient funding and support for IPM implementation, demonstration and fundamental infrastructure; (b) insufficient funding and support for long-term interdisciplinary research and extension/education; (c) lack of funding for applied research, regulatory personnel to expedite product registration and education/promotion for growers; and (d) shortage of independent and trained IPM practitioners. Because the subject has been so thoroughly addressed recently, it serves little purpose to survey the literature here. However, it is useful to remember that support for and structure of organizations (i.e., infrastructure) involved in implementation activities are almost universally

Table 3 Percentage of High IPM Use Farmers in the United States Surveyed in Relation to the Sources of Their Initial Contact with IPM

Source of initial contact	State and crop							
	W, OR (alfalfa seed)	CA (almond)	MA (apple)	NY (apple)	GA (peanut)	VA (soybean)	KY (stored grain)	CA (tomato)
University specialist	36.3	—[a]	—	—	—	55.4[b]	10.5	17.8
Extension agent	19.8	27.2	—	—	85.1	55.4[b]	63.2	24.8
Extension newsletter	—	—	34.2	30.8	—	—	—	—
Meetings	—	—	31.7	15.4	—	—	—	5.7
Trade publications	18.0	25.0	—	—	7.6	22.8	15.8	16.6
Private consultant	6.1	7.4	—	—	—	—	—	22.3
Farm supply dealer	5.2	16.9	—	15.4	—	1.0	14.9	5.3
Other farmers	7.5	4.7	0.0	38.4	4.2	3.0	0.0	7.0
Other	7.5	18.8	34.1		2.1	3.9	5.2	5.8

[a]Source not available as a survey choice.
[b]Extension specialists and agents were not distinguished in this study.
Source: From refs. 13, 18, and 28.

identified as factors influencing adoption. For example, significantly more funding is devoted to marketing by manufacturers than for public sector extension activities, resulting in many more chemical dealers than extension IPM staff [35, 38]. It would also be useful to direct public sector research efforts by concentrating more funding in the area of transitional research which represents the interface of more fundamental studies and field level IPM use.

V. PARTICIPATORY TRAINING

Although the primary focus of this chapter is on pest management education and technology transfer in developed countries (primarily the United States), innovative educational systems based on a farmer participation model are being successfully implemented in several developing countries. It is useful briefly to review the concept and to assess its applicability elsewhere.

In traditional extension education programs, the individual training farmers is an expert who determines what is necessary for the farmer to know and often chooses to transfer knowledge and skills in a classroom setting. Farmers attending the meetings may not be accustomed to the classroom setting, where a teacher tends to dominate and lecture. Even if the individual doing the training has a grasp of their needs and interests, the farmers may feel a lack of involvement and become unmotivated.

Chambers et al. [5] believe that the best results are obtained when farmers are involved in every step of the process from identifying research needs though developing technologies to addressing those needs and ultimately implementing them. Farmer participation permits application of their intimate understanding of local conditions and their prior experiences, knowledge which might well be neglected by a system where research and training are directed centrally.

Perhaps the best example of this type of program is the United Nations Food and Agriculture Organization's Inter-Country Program for IPM in Rice, which has made major strides in implementing IPM practices in Southeast Asia [17]. Through this program, farmers learned principles and obtained the knowledge and skills to apply them. Research innovations were the result of farmer-identified needs and incorporated their knowledge.

A series of steps to be followed in implementing participatory training might include:

Basic curriculum which is relevant, simple, and repetitive

Field-oriented training to permit farmers to participate in practices and to practice learned skills

Classroom or laboratory experiments and demonstrations which require farmer involvement

Group meetings and discussion to permit farmers to learn from each other
    and reinforce decisions
Periodic updates to reinforce skills and introduce new concepts

The use of participatory training model as an extension tool could prove to
be an effective structure in developed countries as well. Farmers, as do most
people, like to receive more personal attention, and would undoubtedly be more
receptive and feel more involved in educational situations where they were active participants rather than students. Research and development programs would
also benefit from farmers' experiences and understanding of constraints. The
use of focus groups, already widespread in industry to evaluate products and
marketing approaches, could also be beneficial in identifying research and extension needs.

VI. EDUCATION DIRECTIONS

Public agency budgets, especially for agriculture, are generally in decline, yet
the need for specialized training in pest management has never been greater.
Pest management has become much more complex for farmers and their advisers as options for pesticide use become more restricted and limitations on
use increase. The challenge for those involved in education becomes how to
provide the individualized training desired and proven to be effective while
reducing costs.

A. Hands-on Training

Meetings for farmers and their advisers are typically held during the winter,
presumably when those attending the meetings are less busy than they might
be during the production season. While convenient, this approach offers little
opportunity for hands-on field training. Hands-on training provides an alternative to classroom type instruction. It encourages those attending actually to
observe and practice what is being demonstrated. In order to achieve this, the
group of participants is divided into small groups and assigned a knowledgeable facilitator. The group is given some background information and is charged
with making observations and reaching conclusions. The facilitator asks questions and adds technical information if necessary. All individuals are encouraged to participate and offer comments. At the end of the meeting, the facilita-
tors ask the group to summarize what has been learned as a result of the activities
completed. The entire group attending the meeting might consist of several
hundred individuals, and the curriculum might emphasize four to eight concepts.

If the total number of individuals attending the meeting is this large, it is necessary to train others to serve as trainers.

Hands-on training and train-the-trainer activities have been successfully used for pesticide safety training and pest management training (particularly in landscape and ornamental situations) on a large scale by the Statewide IPM Project in California (P. Marer and M. L. Flint personal communications). Such instruction is also possible in the classroom by developing experiments which the participants can do individually or in small groups. Group dynamics become important, and hopefully the informal atmosphere which is developed through interpersonal interactions will translate into receptivity and enthusiasm. Hands-on training can be very labor intensive, a constraint in organizations such as extension which faces serious budget limitations. However, the approach is justified in terms of training quality. Training other public-sector or private-sector individuals to become trainers at these workshops is one approach of achieving the goal with reduced organizational expense. These trainers could be certified and provided with materials that they could then use to provide education to others in cases where certification or licensing is needed. The development of a Master Gardener program within Cooperative Extension permitted increased public service to urban clientele by expanding the base of knowledgeable individuals at county offices. The program uses volunteers, who are intensively trained by university research and extension staff in horticulture and pest management. Once certified, these volunteers provide reliable advice to homeowners and other gardeners. To maintain certification, the volunteers receive updates on new problems and approaches.

B. Information Systems

Agricultural applications for computers have not been as widely adopted by farmers to date as has been the case in other industries. There are probably multiple reasons for this. Farmers are by nature conservative, and for many of them, computers are a relatively new technology which has yet to be proven. Market penetration in the agricultural community is relatively low [3, 27] and the agricultural market (only 2% of the U. S. population) is relatively small; this has served to limit the interest of larger companies in addressing the industry. With a few noteable exceptions, development of computer applications in agricultural production has largely occurred at Land Grant Universities, with few resources being devoted to the task and limited marketing. These agricultural applications are typically specialized, because each crop and region has its own unique set of production information needs.

In spite of these limitations to use, several computer delivery systems have been developed which are routinely accessed by farmers and their advisors who have adopted computers in their operations. Among the most successful were

FACTS in Indiana, FAIRS in Florida, CENET in New York, and IMPACT in California [42]. In addition, several crop-specific software programs have been produced and are becoming widely used. An excellent example is the Potato Crop Management (PCM) program from Wisconsin [6], which is now being used on almost 60% of the state's potato acreage. Expert systems, which link quantifiable relationships with field data and expert knowledge present additional opportunities for agricultural software. Expert systems for apples and grapes in Pennsylvania and for rice and cotton in California are available for public distribution and are being used to some extent by farmers in those states. Combined with internet and CD-ROM technology, the opportunity exists for photographs and illustrations to be linked to the software making even more applications possible.

Opportunities for enhanced communications exist in satellite conferencing, which is widely used by a few midwestern states to bring experts from different locations to local meetings, in the production of instructional videos, which can be used at meetings or loaned for individual viewing, and with the internet, which presents opportunities for access to bulletin boards and databases at relatively low cost.

The utilization of modern communications and information systems presents an opportunity for reaching audiences with lower labor input. Although printed trade journals, newsletters, leaflets, manuals, and books will likely continue to be the primary methods of mass communication in the near future, emphasis on electronic information dissemination and other types of information systems will undoubtedly increase. This emphasis will strengthen as more powerful technologies develop and more individuals have access to computers and become comfortable with their use.

C. Collaborative Efforts

One way to meet the challenges of providing pest management education at reduced cost is by increasing collaborative public/private sector efforts and perhaps changing organizational roles. The classic approach is for extension to coordinate educational efforts that may include farmers and industry personnel as educators. Other approaches might include farm groups coordinating educational efforts using technical expertise from extension and industry or agricultural industry groups coordinating educational efforts using extension personnel and others as educators. Several models exist which illustrate these alternative approaches.

1. Farmer-Coordinated Education

By directing their own resources to a coordinated effort, it is possible for groups of farmers to help focus research and education efforts on their own local situations. A good example is the Lodi-Woodbridge Winegrape Commission which represents more than 600 winegrape growers who produce dozens of varieties of grapes on 45,000 acres in and around California's Sacramento River Delta. Local growers voted to establish the Commission in 1991, and launched a district-wide IPM program the following year with the objective of promoting effective and rapid adoption of production practices which could lead to the reduced use of pesticides and herbicides. The Commission employs a private crop consultant with a doctor of philosophy degree who assists with the research and educational efforts. Funding from the Commission (which is obtained via an annual assessment from all district farmers) is used to help support research proposals which are solicited from University of California scientists. The Commission works closely with Cooperative Extension to coordinate educational programs and to seek additional funding from organizations such as the Kellogg Foundation and the California Energy Commission.

Farmer participation in the IPM effort is voluntary, but results to date seem to be good. Attendance at meetings and field days is high. Field demonstrations of recommended practices are occurring on about 20 vineyards in the district. More importantly, farmers elsewhere in California are becoming interested in the Lodi-Woodbridge model.

2. Industry-Coordinated Education

As mentioned earlier, CAPCA is a professional organization representing licensed pest control advisors in California. One objective of CAPCA is to provide for their members continuing education credits which are required for licensing. To do this, local chapters of CAPCA work with the statewide office to conduct periodic continuing education meetings for members. These meetings typically involve university and industry researchers in a lecture-type setting.

The food industry has a significant role in farmers' pest management decisions in that it specifies the level of quality that must be met for contracted products. In the past, farmers were often concerned with not meeting quality standards which would result in future contract problems, so they were more likely to make "insurance" treatments to reduce their risk of damage. In the last several years, progressive companies in the food industry have become proactive in support of IPM use by farmers with which they contract. The California Processed Tomato Foundation is one such organization that was initiated by processors specifically to help promote the use of IPM. Individual companies who are members of the group, including Campbell's, Del Monte, Muir Glen, and Tri Valley, have hired IPM coordinators who work with Cooperative Extension to provide on-site training and consultation for farmers and their pest control advisors. In some cases, the processors have actually assumed responsibility for any additional damage that might occur in fields that are farmed using IPM.

D. Impediments and Opportunities

Changing approaches to pest management education is not a simple task, and one might expect resistance to such change. Farmers have received information on technical advances in a fairly structured way for years, and why should either the farmers or their trainers be expected to change the way information is received or delivered?

A major concern is the lack of incentive to change. Recognition and rewards for taking new approaches or for participating in collaborative arrangements are not great. In addition, there is a tendency of individuals and groups to feel threatened by new approaches and to protect their self-interests.

Lack of funding is a significant impediment, but this could also be taken as justification for pursuing new educational paradigms. For example, decreased public funding for agricultural research and education, which is occurring concomitantly with increasing pressure to develop and implement information-intensive pest control alternatives, makes it essential for the agricultural industry to shoulder more of the costs involved in the process and to convince legislators that it is a good investment to support a healthy public program.

In the United States, the Land Grant University system must continue to produce new innovations and to provide professional training. However, these efforts must be linked more closely than ever to the needs of farmers, their advisors, the agricultural industry, and public agencies. Providing quality curriculum rather than wholesaling information should be the primary educational objective. Cooperative Extension should train farmers to make informed decisions and encourage the widespread availability of professional advice through crop consultants, farmer cooperatives, or other elements of the agricultural industry that can act as agents of change. These change agents must conduct their business ethically, recognizing their critical role in the spread of innovations and their responsibility to farmers and the general public. They must also acknowledge the critical role of the Land Grant University in continued professional training and technology transfer linked closely to the research process.

REFERENCES

1.   Berman, P. 1980. Thinking about programmed and adaptive implementation: matching strategies to situations. In *Why Policies Succeed or Fail*. H. Ingram & D. Mann, eds. Beverly Hills, California: Sage, 205–230.

2. Blair, B.D. 1986. Dissemination of pest management information in the midwest, USA. In *Advisory Work in Crop, Pest and Disease Management*. J. Palti & R. Ausher, eds. Berlin: Springer-Verlag, pp 231–233.

3. Burhoe, S.A. 1989. News and views: computer use on the farm. *Agric. Comput.* 10:1.

4. Carlson, G.A., & J. Cooper. 1975. Evaluation of the pest management program for tobacco in North Carolina. In *Evaluation of Pest Management Programs for Cotton, Peanuts and Tobacco in the United States*. R. Von Rumker, G.A. Carlson, R.D. Lacewell, R.B. Norgaard, & D.W. Parvin, eds. Washington, DC: Council on Environmental Quality Final Report Contract No. EQ4ACQ36.

5. Chambers, R.A., A. Pacey, & L.A. Thrupp. 1989. *Farmer First: Farmer Innovation and Agricultural Research*. New York: Bootstrap Press.

6. Connell, T.R., J.P. Koenig, W.R. Stevenson, K.A. Kelling, D. Curwen, J.A. Wyman, & L.K. Binning. 1991. An integrated systems approach to potato crop management. I *Prod. Agric.* 4:453–460.

7. Corbet P.S. 1981. Non-entomological impediments to the adoption of integrated pest management. *Prot. Ecol.* 3:183–202.

8. Cox, R.S. 1971. *The Private Practitioner in Agriculture*. Lake Worth, Florida: Solo Publications.

9. Fletcher, J.T. 1986. Crop protection: the role of the agricultural development and advisory service in England and Wales. In *Advisory Work in Crop Pest and Disease Management*. J. Palti & R. Ausher, eds. Berlin: Springer-Verlag, pp 177–193.

10. Frisbie, R.E., & P.L. Adkisson. 1985. IPM: definitions and current status in U.S. Agriculture. In *Biological Control in Agricultural IPM Systems*. M.A. Hoy & C. Herzog, eds. New York: Academic Press, pp 41–50.

11. Frisbie, R.E., & G.M. McWhorter. 1986. Implementing a statewide pest management program for Texas, USA. In *Advisory Work in Crop, Pest and Disease Management*. J. Palti & R. Ausher, eds. pp. Berlin: Springer-Verlag, pp 234–262.

12. Good, J.M. 1973. Pilot programs for integrated pest management in the United States. In *US-USSR Pest Management Conference Kiev, USSR*. Washington, DC: U.S. Department of Agriculture Extension Service ANR-5-15 (10-73).

13. Grieshop, J.I., F.G. Zalom, & G. Miyao. 1988. Adoption and diffusion of integrated pest management innovations in agriculture. *Bull. Entomol. Soc. Am.* 34:72–78.

14. Hanuss, K., & W. Beicht. 1986. Advisory work in crop protection in the Federal Republic of Germany (FRG). In *Advisory Work in Crop Pest and Disease Management*. J. Palti & R. Ausher, eds. Berlin: Springer-Verlag, pp 196–207.

15. Kendrick, J.B. 1988. A viewpoint on integrated pest management. *Plant Dis.* 82:647.

16. Kendrick, J.B. & M.W. Stimmann. 1984. Impacts of chemicals on agricultural production and the environment. *Beltsville Symposium on Agricultural Research*. Vol. 8. *Agricultural Chemicals of the Future*. J.L. Hilton, ed. Totowa: Rowman and Allanheld, pp 37–42.

17. Kenmore, P.E., J.A. Litsinger, J.P. Bandong, A.C. Santiago, & M.M. Salac. 1987. Phillipine rice farmers and insecticides: thirty years of growing dependency

and new options for change. In *Management of Pests and Pesticides: Farmers' Perceptions and Practices*. J. Tait & B. Napompeth, eds. Boulder, Colorado: Westview Press, pp 98–108.

18. Klonsky, K., F.G. Zalom, & W.W. Barnett. 1990. Evaluation of California's almond IPM program. *Calif. Agric.* 44:21–24.

19. Lambur, M.T., R.F. Kazmierczak & E.G. Rajotte.1989. Analysis of private consulting firms in integrated pest management. *Bull. Entomol. Soc. Am.* 35:5–11.

20. Lasserre, P. 1992. Training: key to technological transfer. *Long Range Planning*. 15:51–60.

21. Lincoln, C., & B.D. Blair. 1977. Extension entomology: a critique. *Annu. Rev. Entomol.* 22:139–155.

22. Madden, J.P. 1987. A new covenant for agricultural academe. In *Public Policy and Agricultural Technology*. D.F. Hadwiger & W.P. Brown, eds. New York: St. Martin's Press, pp 102–103.

23. National Academy of Sciences. 1986. *Pesticide Resistance: Strategies and Tactics for Management*. Washington, DC: National Academy of Sciences Press.

24. Norgaard, R.B. 1975. Evaluation of the pest management program for cotton in California and Arizona. In *Evaluation of Pest Management Programs for Cotton, Peanuts and Tobacco in the United States*. R. Von Rumker, G.A. Carlson, R.D. Lacewell, R.B. Norgaard, & D.W. Parvin, eds. Washington, DC: Council on Environmental Quality Final Report Contract No. EQ4ACQ36.

25. Poe, S.L. 1981. An overview of integrated pest management. *HortScience* 16:501–506.

26. Post, G.R. 1988. The private consultant: benefit or burden? *HortScience* 23:490–492.

27. Putler, D.S., & D. Zilberman. 1988. Computer use in Tulare County agriculture. *Calif. Agric.* 42:16–18.

28. Rajotte, E.G., R.F. Kazmierczak, M.T. Lambur, G.W. Norton, & W.A. Allen. 1987. *the National Evaluation of Extension's Integrated Pest Management (IPM) Programs*. Blacksburg, Virginia: Virginia Cooperative Extension Service Publication 491-010 through 491-024.

29. Rogers, E.M. 1983. *The Diffusion of Innovations*. New York: Free Press.

30. Rogers, E.M., & F.F. Shoemaker. 1971. *Communication Innovations: A Cross-Cultural Approach*. New York: Fress Press.

31. Samuel, A.C.I., B.W. Cox, & H.H. Cramer. 1983. Implications of rising costs of registering agrochemicals. *Crop Protect.* 2:131–141.

32. Scheuring, A.F. 1988. *A Sustaining Comradeship*. Oakland, California: University of California Division of Agriculture and Natural Resources.

33. Seibert, A., & J. Vorst. 1993. *National Certified Crop Advisor Performance Objectives*. West Lafayette, Indiana: Purdue University.

34. Smith, E.H. 1978. Integrating pest management needs- teaching, research, and extension. In *Pest Control Strategies*. E.H. Smith & D. Pimentel, eds. New York: Academic Press, pp 309–328.

35. Smith, E.H., & D. Pimentel. 1978. *Pest Control Strategies*. New York: Academic Press.

36. Sorenson, A.A. 1993. *Regional Producer Workshops: Constraints to the Adoption of Integrated Pest Management*. Austin, Texas: National Foundation for Integrated Pest Management Education.
37. Sorenson, A.A. 1994. *Proceedings of the National Integrated Pest Management Forum*. DeKalb, Illinois: American Farmland Trust.
38. van den Bosch, R. 1978. *The Pesticide Conspiracy*. Garden City, New York: Doubleday.
39. Wearing, C.H. 1988. Evaluating the IPM implementation process. *Annu. Rev. Entomol.* 33:17–38.
40. Willey, W.R.Z. 1978. Barriers to the diffusion of IPM programs in commercial agriculture. In *Pest Control Strategies*. E.H. Smith & D. Pimentel, eds. New York: Academic Press, pp 285–308.
41. Zalom, F.G. 1993. Reorganizing to facilitate the development and use of integrated pest management. *Agric. Ecosystems Environ.* 46:245–256.
42. Zalom, F.G., & J.F. Strand. 1990. Expectations for computer decision aids in IPM. *AI Appl. Natural Resource Manage.* 4:53–58.

Organismal Index

Potato leaf roll virus, 296
Potato leafhopper, 61, 88, 699
Potato stem mottle, 296
Potato tuberworm, 223, 402
Potato X virus (potexvirus), 186
Potato Y virus (potyvirus), 186
Potentilla recta, 90
Potherb butterfly, 389
Powdery mildews, 178, 219, 264, 270, 273,
   310, 312, 313, 341, 347, 348, 353–354,
   357, 360–361, 362, 363, 367, 368, 370
Prairie dog, 666, 672, 679
Prairie threeawn, 570
Pratylenchus spp., 703
Predatory mites, 135
Prickly lettuce, 571, 572
Prickly sida, 87, 554–555, 571
Pristiphora erichsonii, 133
Procyon lotor, 666
Prostigmata, 704
Prostrate knotweed, 571
Prostrate spurge, 571
Prunus spp., 244, 296
Prunus serotina, 616, 618
Pseudaletia separata, 123
Pseudaletia unipuncta, 88, 92, 384, 696
Pseudocercosporella herpotrichoides, 40
Pseudomonads, 313, 318, 320, 322
Pseudomonas spp., 219, 317–318, 320–321,
   322
Pseudomonas aureofaciens, 320–321
Pseudomonas cepacia, 314
Pseudomonas fluorescens, 310, 315, 319
Pseudomonas pyrocinia, 366
Pseudomonas solanacearum, 188, 317, 322,
   599
Pseudomonas syringae, 310, 314, 315
Pseudomonas syringae pv. *lachrymans*, 316,
   323
Pseudomonas syringae pv. *phaseolicola*, 187,
   295, 316, 323
Pseudomonas syringae pv. *syringae*, 185, 295
Pseudomonas syringae pv. *tabaci*, 185, 188
Pseudomonas syringae pv. *tagetis*, 600
Pseudomonas syringae pv. *tomato*, 226, 316
Pseudoperonospora cubensis, 341

Pseudoplusia includens, 61, 167, 385, 405
Psila rosae, 91, 382
Psychidae, 378
Psyllidae, 604
Pteridium aquilinum, 599
Puccinia spp., 341
Puccinia canaliculata, 598, 616, 617
Puccinia carduorum, 614–615
Puccinia chondrillina, 609, 611, 613–614
Puccinia coronata, 64
Puccinia graminis f. sp. *tritici*, 64, 296
Puccinia lagenophorae, 599
Puccinia melanocephala, 64
Puccinia menthae, 295
Puccinia polyspora, 64
Puccinia punctiformis, 598
Puccinia recondita, 64
Puccinia romagnoliana, 598
Puccinia rottboelliae, 599
Puccinia rufipes, 599
Puccinia striiformis, 64, 300
Puccinia xanthii, 600
Pulasan, 245
Puncturevine, 571
Purple loosestrife, 556, 558, 610
Purple nutsedge, 551, 554–555
Purple scale, 429
Pyralidae, 113, 378
Pyrenophora spp., 47, 341
Pyrenophora chaetomioides, 598
Pyrenophora semeniperda, 598, 600
Pyrenophora tritici-repentis, 298, 701
Pyricularia oryzae, 340, 341, 358, 369–370
Pyrus spp., 244
Pythium spp., 47, 48, 310, 317, 319–320, 341,
   702
Pythium aphanidermatum, 41
Pythium oligandrum, 320
Pythium pod rot, 48
Pythium splendens, 320
Pythium ultimum, 39, 320

Quackgrass, 553, 554–555, 573, 576
Quadraspidiotus perniciosus, 382, 386
Queensland fruit fly, 506
Quiscalus quiscula, 666

Subject Index

Abscisic acid, 274
Acarodomatia (*see* Domatia)
Acetohydroxyacid synthase, 636, 642
Acetolactate synthase, 636
Acetylcholine, 459
Acetylcholinesterase, 459
Acetyl-CoA carboxylase, 636, 650–651
Acid soils, 49
Action sites, 89
Adenosine deaminase, 353
Adjuvants, 633–634
Adsorption coefficient (K), 725–726
Adventive pests (*see* Pests)
Aeration (*see* Soil aeration)
Africa:
  ascendancy of humans in, 377
  cassava mealybug in, 422
  entomophagy in, 17
  eradication of screwworm fly in north, 505
  management of African armyworm moth in, 72
  management of desert locust in, 71–72
  management of river blindness in, 71, 72
  management of tsetse fly in, 532
  movement of armyworm moths in, 69, 72
  percent of population involved in agriculture, 6

[Africa]
  source of pyrethrum, 457
  source of sorghum, 380
Agrochemicals (*see* Pesticides)
Agrocin, 321–322
Agroecosystems:
  climate and, 17–18
  components of natural systems, 79–80, 263
  disturbance and diversity in, 80–81, 390–391, 560
  experimental design critical to understand, 26–27
  human role in, 16–19
  natural ecosystems and, 2, 19–21
  need for appropriate experimental design in study of, 26–27
  nutrient accumulations in, 22
  productivity of, 16–17
  sustainability and, 2–3
Air pollution:
  field burning and, 295
  influences resistance of plants to pathogens, 182
  soilborne pests and, 50
Aircraft in agriculture, 451
Alarm pheromone, 528